AF684235

Basic Principles and Calculations in Chemical Engineering

Eighth Edition

BASIC PRINCIPLES AND CALCULATIONS IN CHEMICAL ENGINEERING

EIGHTH EDITION

David M. Himmelblau
James B. Riggs

Pearson

Original edition, entitled *Basic Principles and Calculations in Chemical Engineering*, 8th Edition, by David M. Himmelblau, James B. Riggs., published by Pearson Education, Limited, Inc, Publishing as Prentice Hall, Copyright © 2012by Pearson Education, Inc.

Indian edition

Copyright © 2015 Pearson India Education Services Pvt. Ltd.

All rights reserved. This book is sold subject to the condition that it shall not, by way of trade or otherwise, be lent, resold, hired out, or otherwise circulated without the publisher's prior written consent in any form of binding or cover other than that in which it is published and without a similar condition including this condition being imposed on the subsequent purchaser and without limiting the rights under copyright reserved above, no part of this publication may be reproduced, stored in or introduced into a retrieval system, or transmitted in any form or by any means (electronic, mechanical, photocopying, recording or otherwise), without the prior written permission of both the copyright owner and the above-mentioned publisher of this book.

ISBN 978-93-325-4962-3

First Impression, 2015
Fifth Impression, 2019
Sixth Impression, 2020

This edition is manufactured in India and is authorized for sale only in India, Bangladesh, Bhutan, Pakistan, Nepal, Sri Lanka and the Maldives. Circulation of this edition outside of these territories is UNAUTHORIZED.

Published by Pearson India Education Services Pvt. Ltd, CIN: U72200TN2005PTC057128.

Head Office: 15th Floor, Tower-B, World Trade Tower, Plot No. 1, Sector 16, Noida 201301, Uttar Pradesh, India.
Registered Office: The HIVE, 3rd Floor, Metro Zone, No 44, Pilliayarkoil Street, Jawaharlal Nehru Road, Anna Nagar, Chennai, Tamil Nadu 600040, Tamil Nadu, India. Phone: 044-66540100
Website: in.pearson.com, Email: companysecretary.india@pearson.com

Printed in India by Pushp Print Services.

This book is dedicated to the memory of David M. Himmelblau (1923–2011) and his contribution to the field of chemical engineering.

CONTENTS

PREFACE

This book is intended to serve as an introduction to the principles and techniques used in the field of chemical engineering as well as biological, petroleum, and environmental engineering. Although the range of subjects deemed to be in the province of chemical engineering has broadened over the last twenty years, the basic principles of this field of study remain the same. This book presents the foundation of specific skills and information that are required for the successful undergraduate and postgraduate study of chemical engineering as well as the professional practice of chemical engineering. Moreover, your remaining chemical engineering classes will rely heavily on the skills that you will develop in this course: your ability to solve abstract problems as well as the application of material and energy balances. One can view the study of the field of chemical engineering as a tree with material and energy balances being the trunk and the subjects of thermodynamics, fluid flow, heat transfer, mass transfer, reactor kinetics, process control, and process design being the branches off the trunk. From this perspective, it is easy to see the importance of mastering the material that follows.

The primary objective of this book is to teach you how to systematically formulate and solve material and energy balance problems. More important, you should learn to systematically formulate and solve all types of problems using the methods presented in this text. In addition, this text serves to introduce you to the breadth of processes that chemical engineers work with, from the types of processes found in the refining and chemical industries to those found in bioengineering, nanoengineering, and the microelectronics industries. While the analysis used in this book will be based largely on a macroscopic scale (i.e., representing a complex system as a uniform system), your later engineering courses will teach you how to formulate microscopic material and energy balances that can be used to more completely describe these systems. In fact, you will learn in these classes that to formulate a microscopic balance you only have to apply the balances presented in this textbook to a very small volume inside the process of interest.

This text is organized as follows:

- Part I Introduction: background information (Chapters 1–2)
- Part II Material Balances: how to formulate and solve material balances (Chapters 3–6)
- Part III Gases, Vapors, and Liquids: how to describe gases and liquids (Chapter 7–8)
- Part IV Energy: how to formulate and solve energy balances (Chapters 9–11)

Expecting to "absorb" the information and skills in this text by reading and listening to lectures is a bit naïve. It is well established that one learns by doing, that is, applying what you have been exposed to. In this regard, our text offers a number of resources to assist you in this endeavor. Probably the most important resources for your study of this material are the Self-Assessment Tests at the end of each section in the book. In particular, the Self-Assessment questions and problems are particularly valuable because by answering them and comparing your answers to the answers posted in Appendix A, you can determine what it is that you do not fully understand, which is quite an important piece of information. A number of valuable resources are provided to you on the CD that accompanies this book, which includes the physical property software, which provides timesaving access to physical properties for over 700 compounds and elements; Polymath for solving sets of equations, which comes with a 15-day free trial; and the Supplemental Problems Workbook with over 100 solved problems and process equipment descriptions. For more specific information on the resources available with this textbook and the accompanying CD, refer to the "Read Me" section that follows.

It is our sincere hope that this textbook and materials not only inspire you to continue to pursue your goal to become a chemical engineer, but also make your journey toward that goal easier.

Jim Riggs
Austin, Texas

READ ME

Welcome to *Basic Principles and Calculations in Chemical Engineering.* Several tools exist in the book in addition to the basic text to aid you in learning its subject matter. We hope you will take full advantage of these resources.

Learning Aids

1. Numerous examples worked out in detail to illustrate the basic principles
2. A consistent strategy for problem solving that can be applied to any problem
3. Figures, sketches, and diagrams to provide a detailed description and reinforcement of what you read
4. A list of the specific objectives to be reached at the beginning of each chapter
5. Self-Assessment Tests at the end of each section, with answers so that you can evaluate your progress in learning
6. A large number of problems at the end of each chapter with answers for about a third of them provided in Appendix E
7. Thought and discussion problems that involve more reflection and consideration than the problem sets cited in item 6
8. Appendixes containing data pertinent to the examples and problems
9. Supplementary references for each chapter
10. A glossary following each section

11. A CD that includes some valuable accessories:
    a. Polymath—an equation-solving program that requires minimal experience to use. Polymath is provided with a 15-day free trial. Details on the use of Polymath are provided. A special web site gives significant discounts on educational versions of Polymath for various time periods: 4 months, 12 months, and unlimited use: www.polymath-software.com/himmelblau
    b. Software that contains a physical properties database of over 700 compounds.
    c. A Supplementary Problems Workbook with over 100 completely solved problems and another 100 problems with answers.
    d. The workbook contains indexed descriptions of process equipment and animations that illustrate the functions of the equipment. You can instantly access these pages if you want to look something up by clicking on the page number.
    e. Problem-solving suggestions including checklists to diagnose and overcome problem-solving difficulties that you experience.
    f. Additional chapters and appendixes
12. A set of steam tables (properties of water) in both SI and American Engineering units in the pocket in the back of the book

Scan through the book now to locate these features.

Good Learning Practices (Learning How to Learn)

You cannot put the same shoe on every foot.

Publilius Syrus

Those who study learning characteristics and educational psychologists say that almost all people learn by practicing and reflecting, and not by watching and listening to someone else telling them what they are supposed to learn. "Lecturing is not teaching and listening is not learning." You learn by doing.

Learning involves more than memorizing

Do not equate memorizing with learning. Recording, copying, and outlining notes or the text to memorize problem solutions will be of little help in really understanding how to solve material and energy balance problems. Practice will help you to be able to apply your knowledge to problems that you have not seen before.

Adopt good learning practices

You will find that skipping the text and jumping to equations or examples to solve problems may work sometimes but in the long run will lead to frustration. Such a strategy is called "formula-centered" and is a very poor way to approach a problem-solving subject. By adopting it, you will not be able to generalize, each problem will be a new challenge, and the interconnections among essentially similar problems will be missed.

Various appropriate learning styles (information processing) do exist; hence you should reflect on what you do to learn and adopt techniques best suited to you. Some students learn through thinking things out in solitary study. Others prefer to talk things through with peers or tutors. Some focus best on practical examples; others prefer abstract ideas. Sketches and graphs used in explanation usually appeal to most people. Do you get bored by going over the same ground? You might want to take a battery of tests to assess your learning style. Students often find such inventories interesting and helpful. Look in the CD that accompanies this book to read about learning styles.

Whatever your learning style, what follows are some suggestions to enhance learning that we feel are appropriate to pass on to you.

Suggestions to Enhance Learning

1. Each chapter in this book will require three or more hours to read, assimilate, and practice your skills in solving pertinent problems. Make allowance in your schedule so that you will have read the pertinent material **before** coming to class. Instead of sitting in class and not fully understanding what your professor is discussing, you will be able to raise your understanding to a much higher level. It is not always possible, but it is one of the most efficient ways to spend your study time.
2. If you are enrolled in a class, work with one or more classmates, if permitted, to exchange ideas and discuss the material. But do not rely on someone to do your work for you.
3. Learn every day. Keep up with the scheduled assignments—don't get behind, because one topic builds on a previous one.
4. Seek answers to unanswered questions right away.
5. Employ active reading; that is, every five or ten minutes stop for one or two minutes and summarize what you have learned. Look for connecting ideas. Write a summary on paper if it helps.

Suggestions for How to Use This Book Effectively

How can you make the best use of this book? Read the objectives before and after studying each section. Read the text, and when you get to an example, first cover up the solution and try to solve the stated problem. Some people, those who learn by reading concrete examples, might look at the examples first and then read the text. After reading a section, solve the self-assessment problems at the end of the section. The answers are in Appendix A. After completing a chapter, solve a few of the problems listed at the end of the chapter. R. P. Feynman, the Nobel laureate in physics, made the point: "You do not know anything until you have practiced." Whether you solve the problems using hand calculators or computer programs is up to you, but use a systematic approach to formulating the information leading to a proper solution. Use the supplement on the CD in the back of the book (print it out if you need to) as a source of examples of additional solved problems with which to practice solving problems.

This book functions as a savings account—what you put in, you get out, with interest.

ACKNOWLEDGMENTS

We are indebted to many former teachers, colleagues, and students who directly or indirectly helped in preparing this book, and in particular the present edition of it. We want to thank Professor C. L. Yaws for his kindness in making available the physical properties software database that is the basis of the physical properties package in the CD that accompanies our book, and also thanks to Professors M. B. Cutlip and M. Shacham who graciously made the Polymath software available. Far too many instructors using the text have contributed their corrections and suggestions to list them by name. Any further comments and suggestions for improvement of this textbook would be appreciated.

Jim Riggs
Jim.Riggs@ttu.edu

ABOUT THE AUTHORS

David M. Himmelblau was the Paul D. and Betty Robertson Meek and American Petrofina Foundation Centennial Professor Emeritus in Chemical Engineering at the University of Texas, where he taught for 42 years. He received his B.S. from MIT in 1947 and his Ph.D. from the University of Washington in 1957. He was the author of 11 books and over 200 articles on the topics of process analysis, fault detection, and optimization, and served as President of the CACHE Corporation (Computer Aids for Chemical Engineering Education) as well a Director of the AIChE. His book, *Basic Principles and Calculations in Chemical Engineering,* has been recognized by the American Institute of Chemical Engineers as one of the most important books in chemical engineering.

James B. Riggs earned his B.S. in 1969 and his M.S. in 1972, both from the University of Texas at Austin. In 1977, he earned his Ph.D. from the University of California at Berkeley. Dr. Riggs was a university professor for 30 years, the first five years being spent at West Virginia University and the remainder at Texas Tech University. He was appointed Professor Emeritus of Chemical Engineering at Texas Tech University after he retired in 2008. In addition, he has a total of over five years of industrial experience in a variety of capacities. His research interests centered on advanced process control and online process optimization. During his academic career he served as an industrial consultant and founded the Texas Tech Process Control and Optimization Consortium, which he directed for 15 years. Dr. Riggs is the author of two other popular undergraduate chemical engineering textbooks: *An Introduction to Numerical Methods for Chemical Engineers,* Second Edition, and *Chemical and Bio-Process Control,* Third Edition. He currently resides near Austin in the Texas Hill Country.

PART I

INTRODUCTION

CHAPTER 1

What Are Chemical Engineering and Bioengineering?

Your objectives in studying this chapter are to be able to

1. Appreciate the history of chemical engineering and bioengineering
2. Understand the types of industries that hire chemical and bioengineers
3. Appreciate the diversity of the types of jobs in which chemical and bioengineers engage
4. Understand some of the ways in which chemical and bioengineers can contribute in the future to the resolution of certain of society's problems

Looking Ahead

In this chapter we will present some features of the professions of chemical and bioengineering. First, we will present an overview of the history of these fields. Next, we will consider where graduates of these programs go to work. Finally, we will present types of projects in which chemical and bioengineers might participate now and in the future.

1.1 Introduction

Why did you choose to work toward becoming a chemical or bioengineer? Was it the starting salary? Did you have a role model who was a chemical or bioengineer, or did you live in a community in which engineers were prominent? Or were you advised that you would do well as a chemical or bioengineer because you were adept at math and chemistry and/or biology? In fact, most prospective engineers choose this field without fully understanding the profession (i.e., what chemical and bioengineers actually do and what they are capable of doing). This brief chapter will attempt to shed some light on this issue.

Chemical and bioengineers today hold a unique position at the interface between molecular sciences and macroscopic (large-scale) engineering. They participate in a broad range of technologies in science and engineering projects, involving nanomaterials, semiconductors, and biotechnology. Note that we say "participate" because engineers most often work in multidisciplinary groups, each member contributing his or her own expertise.

1.2 A Brief History of Chemical Engineering

The chemical engineering profession evolved from the industrial applications of chemistry and separation science (the study of separating components from mixtures), primarily in the refining and chemical industry, which we will refer to here as the **chemical process industries (CPI)**. The first high-volume chemical process was implemented in 1823 in England for the production of soda ash, which was used for the production of glass and soap. During the same time, advances in organic chemistry led to the development of chemical processes for producing synthetic dyes from coal for textiles, starting in the 1850s. In the latter half of the 1800s a number of chemical processes were implemented industrially, primarily in Britain.

And in 1887 a series of lectures on chemical engineering which summarized industrial practice in the chemical industry was presented in Britain. These lectures stimulated interest in the United States and to some degree led to the formation of the first chemical engineering curriculum at MIT in 1888. Over the next 10 to 15 years a number of U.S. universities embraced the field of chemical engineering by offering fields of study in this area. In 1908, the American Institute of Chemical Engineers was formed and since then has served to promote and represent the interests of the chemical engineering community.

Mechanical engineers understood the mechanical aspects of process operations, including fluid flow and heat transfer, but they did not have a background in chemistry. On the other hand, chemists understood chemistry and its ramifications but lacked the process skills. In addition, neither mechanical engineers nor chemists had backgrounds in separation science, which is critically important to the CPI. In the United States, a few chemistry departments were training process engineers by offering degrees in industrial chemistry, and these served as models for other departments as the demand for process engineers in the CPI began to increase. As industrial chemistry programs grew, they eventually formed separate degree-granting programs as the chemical engineering departments of today.

The acceptance of the "horseless carriage," which began commercial production in the 1890s, created a demand for gasoline, which ultimately fueled exploration for oil. In 1901, a Texas geologist and a mining engineer led a drilling operation (the drillers were later to be known as "wildcatters") that brought in the Spindletop Well just south of Beaumont, Texas. At the time, Spindletop produced more oil than all of the other oil wells in the United States. Moreover, a whole generation of wildcatters was born, resulting in a dramatic increase in the domestic production of crude oil, which created a need for larger-scale, more modern approaches to crude refining. As a result, a market developed for engineers who could assist in the design and operation of processing plants for the CPI. The success of oil exploration was to some degree driven by the demand for gasoline for the automobile industry, but ultimately the success of the oil exploration and refining industries led to the widespread availability of automobiles to the general population because of the resulting lower cost of gasoline.

These early industrial chemists/chemical engineers had few analytical tools available to them and largely depended upon their physical intuition to perform their jobs as process engineers. Slide rules were used to perform calculations, and by the 1930s and 1940s a number of nomographs were developed to assist them in the design and operation analysis of processes for the CPI. Nomographs are charts that provide a concise and convenient means to represent physical property data (e.g., boiling point temperatures or heat of vaporization) and can also be used to provide simplified solutions of complex equations (e.g., pressure drop for flow in a pipe). The computing resources that became available in the 1960s were the beginnings of the computer-based technology that is commonplace today. For example, since the 1970s **computer-aided design (CAD)** packages have allowed engineers to design complete processes by specifying only a minimum amount of information; all the tedious and repetitive calculations are done by the computer in an extremely short period of time, allowing the design engineer to focus on the task of developing the best possible process design.

During the period 1960 to1980, the CPI also made the transition from an industry based on innovation, in which the profitability of a company depended to a large degree on developing new products and new processing approaches, to a more mature commodity industry, in which the financial success of a company depended on making products using established technology more efficiently, resulting in less expensive products.

Globalization of the CPI markets began in the mid-1980s and led to increased competition. At the same time, developments in computer hardware made it possible to apply process automation (advanced process control, or APC, and optimization) more easily and reliably than ever before. These automation projects provided improved product quality while increasing production rates and overall production efficiency with relatively little capital investment. Because of these economic advantages, APC became widely accepted by industry over the next 15 years and remains an important factor for most companies in the CPI.

Beginning in the mid-1990s, new areas came on the scene that took advantage of the fundamental skills of chemical engineers, including the microelectronics industry, the pharmaceutical industry, the biotechnology industry, and, more recently, nanotechnology. Clearly, the analytical skills and the process training made chemical engineers ideal contributors to the development of the production operations for these industries. In the 1970s, over 80% of graduating chemical engineers took jobs with the CPI industry and government. By 2000, that number had dropped to 50% because of increases in the number taking jobs with biotechnology companies, pharmaceutical/health care companies, and microelectronics and materials companies. The next section addresses the current distribution of jobs for chemical engineers.

1.3 Where Do Chemical and Bioengineers Work?

Table 1.1, which lists the percentages of all chemical engineers by employment sector between 1996 and 2007, shows that the percentage of chemical engineers in these developing industries (pharmaceutical, biomedical, and microelectronics industries) increased from 7.1% in 1997 to 19.9% in 2005.

Chemical engineers are first and foremost process engineers. That is, chemical engineers are responsible for the design and operation of processes that produce a wide range of products from gasoline to plastics to composite materials to synthetic fabrics to computer chips to corn chips. In addition, chemical engineers work for environmental companies, government agencies including the military, law firms, and banking companies.

The trend of chemical engineering graduates taking employment in industries that can be designated as bioengineering is a new feature of the twenty-first century. Not only have separate bioengineering or biomedical departments been established, but some long-standing chemical engineering departments have modified their names to "chemical and bioengineering" to reflect the research and fresh interests of students and faculty.

Table 1.1 Chemical Engineering Employment by Sector (from AIChE Surveys)

	1996	2000	2002	2005	2007
Chemical, industrial gases, rubber, soaps, fibers, glass, metals, paper	33.3	32.5	25.2	28.1	25.5
Food, ag products, ag chemical	4.5	5.1	5.6	5.7	5.0
Energy, petroleum, utilities	14.1	1.9	5.1	4.5	3.7
Electronics, materials, computers	1.4	1.9	5.1	4.5	3.7
Equipment design and construction	13.8	12.6	10.6	12.6	14.3
Environmental, health, and safety	6.4	4.7	4.4	4.2	3.4
Aerospace, automobile	1.1	0.9	1.8	2.0	2.1
Research and development	3.9	3.8	4.4	4.2	3.4
Government	3.6	3.6	3.5	3.7	4.4
Biotechnology	1.5	2.2	2.4	4.4	3.7
Pharmaceutical, health care	4.2	6.5	6.1	8.4	7.6
Professional (including education)	4.7	4.5	8.6	7.0	8.4
Other	7.4	8.6	9.6	-	1.5

A bioengineer uses engineering expertise to analyze and solve problems in chemistry, biology, and medicine. The bioengineer works with other engineers as well as physicians, nurses, therapists, and technicians. Biomedical engineers may be called upon in a wide range of capacities to bring together knowledge from many technical sources to develop new procedures, or to conduct research needed to solve problems in areas such as drug delivery, body imaging, biochemical processing, innovative fermentation, bioinstrumentation, biomaterials, biomechanics, cellular tissue and genetics, system physiology, and so on. They work in industry, hospitals, universities, and government regulatory agencies. It is difficult to find valid surveys of specific companies or topics to classify bioengineering graduates' ultimate locations, but roughly speaking, one-third of graduates go to medical school, one-third continue on to graduate school, and one-third go to work in industry with a bachelor's degree.

1.4 Future Contributions of Chemical and Bioengineering

The solution of many of the pressing problems of society for the future (e.g., global warming, clean energy, manned missions to Mars) will depend significantly on chemical and bioengineers. In order to more fully explain the role of chemical and bioengineers and to illustrate the role of chemical and bioengineers in solving society's technical problems, we will now consider some of the issues associated with carbon dioxide capture and sequestration, which is directly related to global warming.

Because fossil fuels are less expensive and readily available, we would like to reduce the impact of burning fossil fuels for energy, but without significantly increasing the costs. Therefore, it is imperative that we develop low-cost CO_2 capture and sequestration technologies that will allow us to do that.

An examination of Figure 1.1 shows the sources of CO_2 emissions in the United States. What category would you attack first? Electric power generation is the number-one source. Transportation sources are widely distributed. No doubt power generation would be the most fruitful.

Carbon capture and storage (CCS) is viewed as having promise for a few decades as an interim measure for reducing atmospheric carbon emissions relatively quickly and sharply while allowing conventional coal-fired power plants to last their full life cycles. But the energy costs, the disposal challenges, and the fact that adding CCS to an existing plant actually boosts the overall consumption of fossil fuels (because of the increased consumption of energy to collect and sequester CO_2, more power plants have to be built so that the final production of net energy is the same) all suggest that CCS is not an ultimate solution.

One interim measure under serious consideration for CCS that might allow existing conventional coal-fired power plants to keep producing until they can be phased out at the end of their full lives involves various known technologies. An existing plant could be retrofitted with an amine scrubber

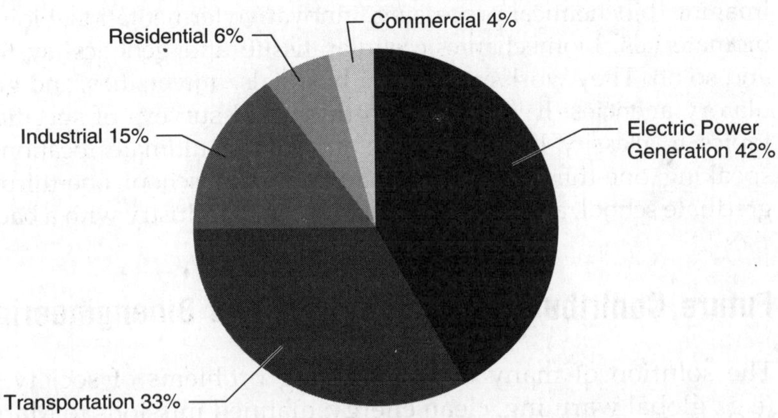

Figure 1.1 Major sources of carbon dioxide emissions in the United States excluding agriculture

to capture 80% to 95% of CO_2 from combustion gases; the CO_2 would then be condensed into a liquid that would be transported and stored somewhere indefinitely where it could not leak into the atmosphere. If several hundreds or thousands of CCS systems were deployed globally this century, each capturing 1 to 5 metric tons of CO_2 per year collectively, they could contribute between 15% and 55% of the worldwide cumulative mitigation effort.

However, the engineering challenges are significant. First, CCS is an energy-intensive process, so power plants require significantly more fuel to generate each kilowatt-hour of electricity produced for consumption. Depending on the type of plant, additional fuel consumption ranges from 11% to 40% more—meaning not only in dollars, but also in additional fossil fuel that would have to be removed from the ground to provide the power for the capture and sequestration, as well as additional CO_2 needing sequestration by doing so. Current carbon-separation technology can increase the price tag of producing electricity by as much as 70%. Put another way, it costs about \$40 to \$55 per ton of carbon dioxide. The annual U.S. output of carbon dioxide is nearly 2 billion tons, which indicates the economic scale of the problem. The U.S. Department of Energy is working on ways to reduce the expenses of separation and capture.

By far, the most cost-effective option is partnering CCS not with older plants, but with advanced coal technologies such as integrated-gasification combined-cycle (IGCC) or oxygenated-fuel (oxyfuel) technology. There is also a clear need to maximize overall energy efficiency if CCS itself is not merely going to have the effect of nearly doubling both demand for fossil fuels and the resultant CO_2 emitted.

Once the CO_2 has been captured as a fairly pure stream, the question is what to do with it that is economical. In view of the large quantity of CO_2 that must be disposed of, disposal, to be considered a practical strategy, has to be permanent.

Any release of gas back into the atmosphere not only would negate the environmental benefits, but it could also be deadly. In large, concentrated quantities, carbon dioxide can cause asphyxiation. Researchers are fairly confident that underground storage will be safe and effective.

This technology, known as carbon sequestration, is used by energy firms as an oil-recovery tool. But in recent years, the Department of Energy has broadened its research into sequestration as a way to reduce emissions. And the energy industry has taken early steps toward using sequestration to capture emissions from power plants.

Three sequestration technologies are actively being developed: storage in saline aquifers in sandstone formations [refer to S. M. Benson and T. Surles, "Carbon Dioxide Capture and Storage," *Proceed. IEEE*, **94**, 1795 (2006)], where the CO_2 is expected to mineralize into carbonates over time;

injection into deep, uneconomic coal seams; and injection into depleted or low-producing oil and natural-gas reservoirs.

Preliminary tests show that contrary to expectations, only 20% maximum of CO_2 precipitates form carbonate minerals, but the majority of the CO_2 dissolves in water. Trapping CO_2 in minerals would be more secure, but CO_2 dissolved in brine is an alternate disposal outcome.

Other suggestions for the reduction of CO_2 emissions include permanent reduction in demand, chemical reaction, various solvents, use of pure O_2 as the oxidant, and so on. See J. Ciferno et al., *Chemical Engineering Progress*, 33–41 (April, 2009), and F. Princiotta, "Mitigating Global Climate Change through Power-Generation Technology," *Chemical Engineering Progress*, 24–32 (November, 2007), who have a large list of possible avenues of approach. The bottom line is that a solution for CO_2 emissions reduction is not just a matter of solving technical problems but a matter of cost and environmental acceptance. Based on the nature of these challenges, it is easy to see that chemical and bioengineers will be intimately involved in these efforts to find effective solutions.

1.5 Conclusion

The chemical engineering profession evolved from society's need for products and energy. Today and into the future, chemical and bioengineers will continue to meet society's needs using their process knowledge, their knowledge of fundamental science, and their problem-solving skills.

Looking Back

In this chapter we reviewed the history of chemical engineering and presented information on the current and projected future status of the profession.

Glossary

Chemical process industries (CPI) The chemical and refining industries.

Computer-aided design (CAD) packages Software programs that are used to design and/or analyze systems including chemical processes.

Web Site

www.pafko.com/history/h_whatis.html

CHAPTER 2

Introductory Concepts

Your objectives in studying this chapter are to be able to

1. Understand and explain the difference between the SI and the AE systems of units
2. Convert a value in one set of units into an equivalent value in another set of units
3. Apply the concepts of dimensional consistency to determine the validity of an equation or function
4. Employ an appropriate number of significant figures in your calculations
5. Convert from mass to moles and vice versa
6. Validate your answer to a problem
7. Choose an appropriate basis to solve a problem
8. Effectively employ the various units associated with density, concentration, temperature, and pressure
9. Calculate the average molecular weight of a mixture
10. Apply the manometer equation

Take care of your units and they will take care of you.

Anonymous

Looking Ahead

In this chapter we review the SI and American Engineering systems of units, show how conversions between units can be accomplished efficiently, and discuss the concept of dimensional homogeneity (consistency). We also provide some comments with respect to the number of significant figures to use in your calculations and how to validate your solutions. Next, we introduce the concept of the selection of a basis to aid in the solution of certain problems. These concepts are applied to relations involving moles, density, specific gravity, and measures of concentration. Finally, we discuss unit conversions for temperatures and pressures.

2.1 Systems of Units

As you will see in this chapter, the proper handling of units is an essential part of being an engineer. Moreover, checking the consistency of units in your equations will prove to be a valuable tool that will reduce the number of errors you commit when performing engineering calculations. Engineers and scientists have to be able to communicate not only with words but also by carefully defined numerical descriptions. Read the following news report that appeared in the *Wall Street Journal* (June 6, 2001):

> SEOUL, South Korea—A mix up in the cockpit over whether altitude guidance was measured in feet or meters led to the crash of a Korean Air Lines McDonnell Douglas MD-11 freighter soon after takeoff in Shanghai in April 1999, investigators said. The crash killed all three crew-members. Five people on the ground were killed and 40 more were injured when the plane went down in light rain onto a construction site near Shanghai's Hongqiao Airport.
>
> According to a summary of the crash report released by South Korean authorities, a Chinese air-traffic controller directed the pilots to an altitude of 1,500 meters (4,950 feet). The plane was climbing rapidly to that level when the co-pilot told the pilot he thought the instructed height was 1,500 feet, equivalent to 455 meters. The international aviation industry commonly measures altitude in feet, and the confusion led the pilot to conclude the jet was almost 1,000 meters too high, so he quickly moved the controls to lower the plane. As the plane descended, the pilot realized the error but couldn't correct the mistake in time.
>
> South Korea's Ministry of Construction and Transportation said Korean Air Lines would lose the right to serve the Seoul-Shanghai cargo route for at least two years because of errors by the pilots. Korean Air Lines said it would appeal the decision . . .

Now you can understand the point of defining your quantities carefully so that your communications are understood.

What are units and dimensions and how do they differ? **Dimensions** are the general expression of a characteristic of measurement such as *length, time, mass, temperature,* and so on; **units** are the means of explicitly expressing the dimensions, such as *feet* or *centimeters* for length, or *hours* or *seconds* for time. Primarily two types of units are used in this text:

1. **SI**, formally called Le Système Internationale d'Unités and informally called SI or, more often (redundantly), the SI system of units
2. **AE**, or American Engineering system of units, not to be confused with what is called the U.S. Conventional System (USCS) or the English system of units

The SI system has certain advantages over the AE system in that fewer names are associated with the dimensions, and conversion of one set of units to another is easier, but in the United States the AE system has deep roots. Most modern process simulators and other software packages (e.g., Mathcad) allow the use of either or mixed sets of units.

Dimensions and their respective units are classified as fundamental or derived:

- **Fundamental** (or basic) **dimensions/units** are those that can be measured independently and are sufficient to describe most physical quantities such as length, mass, time, and temperature.
- **Derived dimensions/units** are those that can be developed in terms of the fundamental dimensions/units.

Tables 2.1 and 2.2 list both the basic and derived units in the SI and AE systems, respectively. These symbols for units should be followed precisely to avoid confusion. Note the use of upper- and lowercase letters presented in Tables 2.1 and 2.2. Unit abbreviations have the same form for both the singular and plural, and they are not followed by a period (except in the case of inches). One of the best features of the SI system is that the prefixes of the units (except for time) are related by multiples of 10 as indicated in Table 2.3.

When quantities with units are added or subtracted, they need to be expressed in the same units. For example,

$$7 \text{ in.} + 2 \text{ in.} = 9 \text{ in.}$$

$$6 \text{ cm} - 1 \text{ cm} = 5 \text{ cm}$$

Table 2.1 SI Units Encountered in This Book

Physical Quantity	Name of Unit	Symbol for Unit*	Definition of Unit
	Basic SI units		
Length	metre, meter	m	
Mass	kilogramme, kilogram	kg	
Time	second	s	
Temperature	kelvin	K	
Amount of substance	mole	mol	
	Derived SI units		
Force	newton	N	$(kg)(m)(s^{-2}) \rightarrow (J)(m^{-1})$
Energy	joule	J	$(kg)(m^2)(s^{-2})$
Power	watt	W	$(kg)(m^2)(s^{-3}) \rightarrow (J)(s^{-1})$
Density	kilogram per cubic meter		$(kg)(m^{-3})$
Velocity	meter per second		$(m)(s^{-1})$
Acceleration	meter per second squared		$(m)(s^{-2})$
Pressure	newton per square meter, pascal		$(N)(m^{-2})$, Pa
Heat capacity	joule per (kilogram × kelvin)		$(J)(kg^{-1})(K^{-1})$
	Alternative units		
Time	minute, hour, day, year	min, h, d, y	
Temperature	degree Celsius	°C	
Volume	litre, liter (dm^3)	L	
Mass	tonne, ton (Mg), gram	t, g	

*Symbols for units do not take a plural form, but plural forms are used for the unabbreviated names. Non-SI units such as day (d), liter or litre (L), and ton or tonne (t) are legally recognized for use with SI.

On the other hand,

$$7 \text{ in.} + 2 \text{ cm}$$

cannot be evaluated until both of the measurements of length are expressed in the same units.

When quantities with units are multiplied or divided, the numerical result is equal to the product or division of the numerical values, and the net units are equal to the product or division of the units themselves. For example,

$$3 \text{ N} \times 2 \text{ m} = 6(\text{N})(\text{m})$$

$$8 \text{ lb}_m \div 2 \text{ ft}^3 = 4 \text{ lb}_m/\text{ft}^3$$

Table 2.2 American Engineering (AE) System Units Encountered in This Book

Physical Quantity	Name of Unit	Symbol
	Some basic units	
Length	foot or inch	ft or in.
Mass	pound (mass)	lb_m
Time	second, hour	s, hr
Temperature	degree Rankine or degree Fahrenheit	°R or °F
Amount of substance	pound mole	lb mol
	Derived units	
Force	pound (force)	lb_f
Energy	British thermal unit, foot pound (force)	Btu, $(ft)(lb_f)$
Power	horsepower	hp
Density	pound (mass) per cubic foot	lb_m/ft^3
Velocity	feet per second	ft/s
Acceleration	feet per second squared	ft/s^2
Pressure	pound (force) per square inch	$lb_f/in.^2$
Heat capacity	Btu per pound (mass) per degree F	$Btu/[(lb_m)(°F)]$
Volume	cubic feet	ft^3

Table 2.3 SI Prefixes

Factor	Prefix	Symbol	Factor	Prefix	Symbol
10^9	giga	G	10^{-1}	deci	d
10^6	mega	M	10^{-2}	centi	c
10^3	kilo	k	10^{-3}	milli	m
10^2	hecto	h	10^{-6}	micro	μ
10^1	deka	da	10^{-9}	nano	n

Did you note that when a compound unit is formed by multiplication of two or more other units, its symbol consists of the symbols for the separate units expressed as a product using parentheses, for example (N)(m)? The parentheses may be omitted in the case of familiar units such as watt-hour (symbol Wh) if no confusion will result, or if the symbols are separated by exponents, as in $N(m^2kg^{-2})$. Hyphens should not be used in symbols for compound units. Also, the SI convention of leaving a space between groups of numbers such as 12,650 instead of inserting a comma, as in 12,650, will be ignored to avoid confusion in handwritten numbers.

Frequently Asked Questions

1. Is the SI system of units the same as the metric system? The answer is no. SI differs from the versions of the metric system (such as CGS) in the number of basic units and in the way the basic units are defined.
2. What does ms mean: millisecond or meter seconds? Mind your use of meters! The letters ms mean millisecond; the combination (m)(s) would mean meter seconds. Similarly, 1 Mm is not 1 mm! Notation such as cm^2 meaning square centimeters frequently has to be written as $(cm)^2$ to avoid confusion.

Self-Assessment Test

(Answers to the self-assessment tests are listed in Appendix A.)

Questions

1. Which of the following best represents the force needed to lift a heavy suitcase?
   a. 25 N
   b. 25 kN
   c. 250 N
   d. 250 kN
2. Pick the correct answer(s); a watt is
   a. 1 joule per second
   b. Equal to 1 $(kg)(m^2)/s^2$
   c. The unit for all types of power
   d. All of the above
   e. None of the above
3. Is kg/s a basic or derived unit in SI?

Problem

Indicate whether each of the following units is correct or incorrect in the SI system:

a. nm
b. °K
c. sec
d. N/mm
e. $kJ/(s)(m^3)$

Thought Problem

What volume of material will a barrel hold?

Discussion Problem

In a letter to the editor, the letter writer says:

> I believe SI notation might be improved so as to make it mathematically more useful by setting SI-sanctioned prefixes in **boldface** type. Then one would write, **1 c** = **10 m** without any ambiguity [**c** = $\mathbf{10^{-2}}$, **m** = $\mathbf{10^{-3}}$] and the meaning of "mm" would be at once clear to any mathematically literate, if scientifically illiterate, citizen, namely, either $\mathbf{10^{-3}}$ m [**m**m], $\mathbf{10^{-6}}$ [**mm**], or (after Gauss and early algebraists) m^2 [mm].
>
> With respect to the "mm" problem and remarks regarding the difference between "one square millimeter" [$(\mathbf{m}\mathrm{m})^2$] and "one mili square-meter" [$\mathbf{m}(\mathrm{m}^2)$], these difficulties are analogous to the confusion between a "camel's-hair" brush and a camel's "hair-brush."

What do you think of the author's proposal?

2.2 Conversion of Units

Mistakes are the usual bridge between inexperience and wisdom.

Phyllis Theroux, *Night Lights* (Viking Penguin)

As an example of a serious unit conversion error, in 1999 the Mars Climate Orbiter was lost because engineers failed to make a simple conversion from English units to SI, an embarrassing lapse that sent the $125 million craft fatally close to the Martian surface, destroying it and causing the mission to be a failure.

As an engineer you must be able to handle all kinds of units and be able to convert a given set of units to another with ease. As you probably already know, the procedure for converting one set of units to another is simply to multiply any number and its associated units by ratios termed **conversion factors** to arrive at the desired answer with its associated units. Conversion factors are statements of equivalent values of different units in the form of ratios. Note that because conversion factors are composed of equivalents between units, multiplying a quantity by one or more conversion factors does not actually change the basic quantity, only its numerical value and its

Table 2.4 Examples of Conversion Factors

Relationship	Conversion Factor
1 ft = 12 in.	$\dfrac{1\text{ ft}}{12\text{ in.}}$
1 in. = 2.54 cm	$\dfrac{1\text{ in.}}{2.54\text{ cm}}$
1 m = 100 cm	$\dfrac{1\text{ m}}{100\text{ cm}}$

units. Table 2.4 demonstrates how several equivalents between units can be converted into ratios that are deemed conversion factors.

On the inside of the front cover of this book you will find tables listing commonly used conversion factors. **To obtain unit conversion factors from these tables, use the following procedure: (1) Locate the name of the current unit in the row on the left-hand side of the table. (2) Locate the name of the desired unit in the column at the top of the table. (3) The number in the box that is at the intersection of the row and column of the table is the value of the conversion factor, that is, is equal to the value of the current unit divided by the desired unit.** For example, look at the first table on the inside cover: "Volume Equivalents." To find the conversion factor to convert U.S. gallons to liters (L), locate "liters" on the top row and "U.S. gal" in the left column. At the intersection of these two units is a box that contains the number 3.785, which means that the conversion factor is U.S. gal/L = 3.785. What is the conversion factor to convert liters to U.S. gallons? Is the numerical value of 0.2642 correct? You will find engineering handbooks as well as certain Web sites to be good sources of tables of conversion factors. We recommend that you memorize a few of the most common conversion factors to avoid looking them up in tables.

In this book, to help you follow the calculations and emphasize the use of units, we frequently make use of a special format in the calculations, as shown below, as a substitute for parentheses, times signs, and so on.

Consider the following problem:

If a plane travels at the speed of sound (assume that the speed of sound is 1100 ft/s), how fast is it going in miles per hour? First, let's convert feet (ft) to miles (mi). From the table "Linear Measure Equivalents," 1 mile is equal to 5280 ft. Therefore, using this conversion factor to convert feet to miles yields

$$\frac{1100\ \cancel{\text{ft}}}{\text{s}}\left|\frac{\text{mi}}{5280\ \cancel{\text{ft}}}\right. = \frac{1100}{5280}\ \text{mi/s} = 0.2083\ \text{mi/s}$$

Note that when the conversion factor is applied correctly, the units of feet are eliminated from the problem, which is indicated by the slashes through "ft." Note the format of these calculations. We have set up the calculations with vertical lines separating each ratio. We will use this formulation frequently in this text to enable you to see clearly how units are handled in each case.

Next, let's convert from seconds to hours to obtain an answer with the desired units. You know that there are 60 s in a minute and 60 min in an hour; therefore, 1 hr is equal to 3600 s. Thus

$$\frac{0.2083\ \text{mi}}{\cancel{\text{s}}}\left|\frac{3600\ \cancel{\text{s}}}{\text{hr}}\right. = (0.2083)(3600)\ \text{mi/hr} = 750\ \text{mi/hr}$$

This unit conversion problem can be more efficiently implemented by applying both unit conversions in one step:

$$\frac{1100\ \cancel{\text{ft}}}{\cancel{\text{s}}}\left|\frac{\text{mi}}{5280\ \cancel{\text{ft}}}\right|\frac{3600\ \cancel{\text{s}}}{\text{hr}} = \frac{(110)(3600)}{5280}\ \text{mi/hr} = 750\ \text{mi/hr}$$

We recommend that you always write down the units next to the associated numerical value to ensure accurate unit conversions. By striking out units that cancel at any stage in the conversion, you can determine the consolidated net units and see what conversions are still required. In this manner, you can reliably perform very complex unit conversions by checking to ensure that all the units have been correctly converted.

Consistent use of units in your calculations throughout your professional career will assist you in avoiding silly mistakes such as converting 10 cm to inches by multiplying by 2.54:

$$(10)\ (2.54) = 25.4\ \text{in. !! instead of}\ \frac{10\ \text{cm}}{}\left|\frac{1\ \text{in.}}{2.54\ \text{cm}}\right. = 3.94\ \text{in.}$$

By three methods we may learn wisdom: First, by reflection, which is noblest; second, by imitation, which is easiest; and third, by experience, which is the bitterest.

Confucius

Some Web sites and computer programs do the conversions for you! In the physical property software on the CD in the back of this book you can insert almost any units you want in order to retrieve property values. Nevertheless, being able to make conversions yourself is important.

Now let's look at some examples of using conversion factors.

Example 2.1 Use of Conversion Factors

Change 400 in^3/day to cm^3/min.

Solution

$$\frac{400\ \text{in}^3}{\text{day}}\left|\left(\frac{2.54\ \text{cm}}{1\ \text{in}}\right)^3\right|\frac{1\ \text{day}}{24\ \text{hr}}\left|\frac{1\ \text{hr}}{60\ \text{min}}\right. = 4.56\ \frac{\text{cm}^3}{\text{min}}$$

In this example note that not only are the numbers raised to a power, but the units also are raised to the same power.

Example 2.2 Nanotechnology

Nanosize materials have become the subject of intensive investigation in the last decade because of their potential use in semiconductors, drugs, protein detectors, and electron transport. **Nanotechnology** is the generic term that refers to the synthesis and application of such small particles. An example of a semiconductor is ZnS with a particle diameter of 1.8 nm. Convert this value to (a) decimeters (dm) and (b) inches (in.).

Solution

(a) $$\frac{1.8\ \text{nm}}{}\left|\frac{10^{-9}\ \text{m}}{1\ \text{nm}}\right|\frac{10\ \text{dm}}{1\ \text{m}} = 1.8 \times 10^{-8}\ \text{dm}$$

(b) $$\frac{1.8\ \text{nm}}{}\left|\frac{10^{-9}\ \text{m}}{1\ \text{nm}}\right|\frac{39.37\ \text{in.}}{1\ \text{m}} = 7.1 \times 10^{-8}\ \text{in.}$$

Example 2.3 Conversion of Units Associated with Biological Materials

In biological systems, enzymes are used to accelerate the rates of certain biological reactions. Glucoamylase is an enzyme that aids in the conversion of starch to glucose (a sugar that cells use for energy). Experiments show that 1 μg mol of glucoamylase in a 4% starch solution results in a production rate of glucose of 0.6 μg mol/(mL)(min). Determine the production rate of glucose for this system in units of lb mol/(ft^3)(day).

Solution

The production rate of glucose is stated in the problem as 0.6 μg mol/(mL)(min). Therefore, to solve this problem, you just have to convert this quantity into the specified units:

$$\frac{0.6\ \mu\text{g mol glucose}}{(\text{mL})(\text{min})}\left|\frac{1\ \text{g mol}}{10^6\ \mu\text{g mol}}\right|\frac{1\ \text{lb mol}}{454\ \text{g mol}}\left|\frac{1\ \text{L}}{3.531\times10^{-2}\ \text{ft}^3}\right|$$

$$\frac{60\ \text{min}}{1\ \text{hr}}\left|\frac{24\ \text{hr}}{1\ \text{day}}\right. = 0.0539\,\frac{\text{lb mol}}{(\text{ft}^3)(\text{day})}$$

In the AE system the conversion of terms involving pound **mass** and pound **force** deserve special attention. Let us start the discussion with Newton's law, which states that force (F) is proportional to the product of mass (m) and acceleration (a), that is,

$$F = Cma \tag{2.1}$$

where C is a constant whose numerical values and units depend on the units selected for F, m, and a. In the SI system, the unit of force is defined to be the newton (N), which corresponds to 1 kg accelerated at 1 m/s². Therefore, the conversion factor $C = 1\ \text{N}/(\text{kg})(\text{m})/\text{s}^2$ results so that the force is expressed in newtons (N):

$$F = \underbrace{\frac{1\ \text{N}}{\frac{(\text{kg})(\text{m})}{\text{s}^2}}}_{C}\left|\underbrace{1\ \text{kg}}_{m}\right|\underbrace{\frac{1\ \text{m}}{\text{s}^2}}_{a} = 1\ \text{N} \tag{2.2}$$

Note that in this case C has the numerical value of 1; hence the conversion factor seems simple, even nonexistent, and the units are usually ignored.

In the American Engineering system an analogous conversion factor is required. In the AE system, **one pound force** ($1\ \text{lb}_f$) corresponds to the action of the Earth's gravitational field on **one pound mass** ($1\ \text{lb}_m$):

$$F = \underbrace{\left(\frac{1(\text{lb}_f)(\text{s}^2)}{32.174(\text{lb}_m)(\text{ft})}\right)}_{C_{AE}}\left(\underbrace{1\ \text{lb}_m}_{m}\left|\underbrace{\frac{g\ \text{ft}}{\text{s}^2}}_{g}\right.\right) = 1\ \text{lb}_f \tag{2.3}$$

where g is the acceleration of gravity, which at 45° latitude and sea level has the following value:

$$g = 32.174\,\frac{\text{ft}}{\text{s}^2} = 9.80665\,\frac{\text{m}}{\text{s}^2}$$

A numerical value of 1/32.174 has been chosen for the value of C_{AE} to maintain the numerical correspondence between lb_f and lb_m on the surface of the Earth. The acceleration caused by gravity, you may recall, varies by a few tenths of 1% from place to place on the surface of the Earth but, of course, is quite different on the surface of the moon.

The inverse of this conversion factor C_{AE} is given the special symbol g_c:

$$g_c = 32.174\,\frac{(\text{ft})(\text{lb}_\text{m})}{(\text{s}^2)(\text{lb}_\text{f})}$$

a conversion factor that you will see included in some texts to remind you that the numerical value of the conversion factor in the AE system is not unity. To avoid confusion, we will not place the symbol g_c in the equations in this book because we will be using both SI and AE units.

To sum up, you can see that the American Engineering system has the convenience that the numerical value of a pound mass is also that of a pound force if the numerical value of the ratio g/g_c is equal to 1, as it is approximately on the surface of the Earth. No one should get confused by the fact that a person 6 feet tall has only two feet. In this book, **we will not subscript the symbol lb with m (for mass)** unless it becomes essential to do so to avoid confusion. We will always mean by the unit lb without a subscript the quantity pound mass. **But never forget that the pound (mass) and pound (force) are not the same units in the American Engineering system** even though we speak of *pounds* to express force, weight, or mass.

What is the difference between mass and weight? Is weight ever expressed in grams or kilograms? The **weight** of a mass is the value of an external force required to maintain the mass at rest in its frame of reference, the Earth usually. People often refer to astronauts in a space station as weightless. They are with respect to the frame of reference of the space station because the centrifugal force present essentially is equivalent to the gravitational force. With respect to the Earth as the frame of reference, they are far from weightless (if they were, how would they get home?). Thus, when an engineer says a drum (on the Earth's surface) weighs 100 lb or 39.4 kg, you can interpret this statement to mean the drum has a mass of 100 lb_m or 39.4 kg. That is, a weight of 39.4 kg would exert a downward force of 915 N (i.e., 39.4 kg $\times$ 9.8 m/s) at the surface of the Earth.

Example 2.4 A Conversion Involving Both lb_m and lb_f

What is the potential energy in $(ft)(lb_f)$ of a 100 lb drum hanging 10 ft above the surface of the Earth with reference to the surface of the Earth?

Solution

The first thing to do is read the problem carefully. What are the unknown quantities? The potential energy (*PE*) is unknown. What are the known quantities? The mass and the height of the drum are known. How are they related? You have to look up the relation unless you recall it from physics:

$$\text{Potential energy} = PE = mgh$$

The 100 lb means 100 lb mass; let g = acceleration of gravity = 32.2 ft/s^2. Figure E2.4 is a sketch of the system.

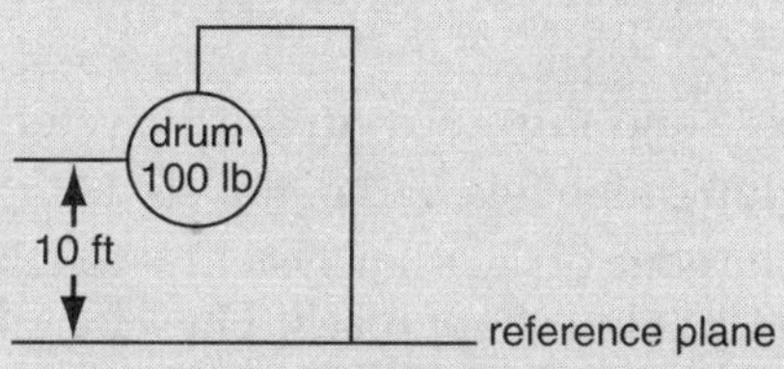

Figure E2.4

Now substitute the numerical values of the variables into the equation and perform the necessary unit conversions.

$$PE = mgh = \frac{100\ lb_m}{} \left| \frac{32.2\ ft}{s^2} \right| \frac{10\ ft}{} \left| \frac{(s^2)(lb_f)}{32.174\ (ft)\ (lb_m)} \right. = 1000\ (ft)(lb_f)$$

Notice that in the ratio of 32.2 ft/s^2 divided by $32.174[(ft)(lb_m)]/[(s^2)(lb_f)]$, the numerical values are essentially equal. A good many engineers would solve the problem by saying that 100 lb × 10 ft = 1000 (ft)(lb) without realizing that in effect they are canceling out the numbers in the g/g_c ratio, and that the lb in the solution means lb_f.

Some Trivia Concerning Unit Conversions

A U.S. frequent flier mile is not the same as a U.S. mile—the former is a nautical mile (1.85 km) whereas the latter is 1.61 km. In the AE system 1 m = 39.37 in., whereas for U.S. land survey applications it is 2×10^{-6} inches shorter.

Self-Assessment Test

Questions

1. What is g_c?
2. Is the ratio of the numerator and denominator of a conversion factor equal to unity?
3. What is the difference between pound force and pound mass in the AE system, if any?
4. Contrast what is involved in converting units within the SI system with what is involved in the AE system.
5. What is the weight of a 1 pound mass at sea level? Would the mass be the same at the center of the Earth? Would the weight be the same at the center of the Earth?
6. What is the mass of an object that weighs 9.80 kN at sea level?

Problems

1. What are the value and units of g_c in the SI system?
2. Electronic communication via radio travels at approximately the speed of light (186,000 mi/s). The edge of the solar system is roughly at Pluto, which is 3.6×10^9 mi from the Earth at its closest approach. How long in hours does it take for a radio signal from Earth to reach Pluto?
3. Convert the following from AE to SI units:
   a. 4 lb_m/ft to kg/m
   b. 1.00 $lb_m/(ft^3)(s)$ to $kg/(m^3)(s)$
4. Convert the following to AE units: 1.57×10^{-2} $g/(cm)(s)$ to $lb_m/(ft)(s)$.
5. Convert 1.1 gal to m^3.

Thought Problems

1. Comment on what is wrong with the following statements from a textbook:
   a. Weight is the product of mass times the force of gravity.
   b. A 67 kg person on Earth will weigh only 11 kg on the moon.
   c. If you have 1 g of water at 4 °C that has a volume of 1.00 mL, you can use 1.00 g water/4 °C as a conversion factor.
2. In *Perry's Handbook* (5th ed.) in the conversion tables is a row showing that the factor 0.10197 converts newtons to kilograms. Can this be correct?

Discussion Problem

In spite of the official adoption of the SI system of units in most countries, people still buy 10 kg of potatoes and inflate automobile tires to a value in kilograms (or kilograms per square centimeter). Why does this usage occur?

2.3 Dimensional Consistency

Now that we have addressed units and dimensions and unit conversions, we can immediately make use of this information in a very practical and important application. **A basic principle exists that equations must be dimensionally consistent**. What the principle means is that each term in an equation must have the same net dimensions and units as every other term to which it is added or subtracted or equated. Consequently, dimensional considerations can be used to help identify the dimensions and units of terms or quantities in an equation.

The concept of dimensional consistency can be illustrated by an equation that represents the pressure/volume/temperature behavior of a gas and is known as van der Waals' equation, an equation to be discussed in more detail in Chapter 7:

$$\left(p + \frac{a}{V^2}\right)(V - b) = \mathrm{R}T$$

Inspection of the equation shows that the constant a must have the units of [(pressure)(volume)^2] because each term inside the first bracket must have units of pressure. If the units of pressure are atmospheres and those of volume are cubic centimeters, a will have the units specifically of [(atm)(cm)^6]. Similarly, b must have the same units as V, or in this particular case the units of cubic centimeters. If T is in kelvin, what must be the units of R? Check your answer by looking up R inside the front cover of the book. All equations must exhibit dimensional consistency.

Example 2.5 Dimensional Consistency

Your handbook shows that microchip etching roughly follows the relation

$$d = 16.2 - 16.2e^{-0.021t} \quad t < 200$$

where d is the depth of the etch in microns [micrometers (μm)] and t is the time of the etch in seconds. What are the units associated with the numbers 16.2 and 0.021? Convert this relation so that d becomes expressed in inches and t can be used in minutes.

Solution

After you inspect the equation that relates d as a function of t, you should be able to reach a decision about the units associated with each term on the

(Continues)

Example 2.5 Dimensional Consistency (*Continued*)

right-hand side of the equation. Based on the concept of dimensional consistency, both values of 16.2 must have the associated units of microns (μm). The exponential term must be dimensionless so that 0.021 must have units of s^{-1}. To carry out the specified unit conversion for this equation, look up suitable conversion factors inside the front cover of this book (i.e., convert 16.2 μm to inches and 0.021 s^{-1} to min^{-1}).

$$d(\text{in.}) = \frac{16.2\ \mu\text{m}}{}\left|\frac{1\text{ m}}{10^6\ \mu\text{m}}\right|\frac{39.37\text{ in.}}{1\text{ m}}\left[1 - \exp\frac{-0.021}{\text{s}}\left|\frac{60\text{ s}}{1\text{ min}}\right|\frac{t(\text{min})}{}\right]$$

$$d(\text{in.}) = 6.38 \times 10^{-4}(1 - e^{-1.26t(\text{min})})$$

As you proceed with the study of chemical engineering, you will find that groups of symbols may be put together, either by theory or based on experiment, that have no net units. Such collections of variables or parameters are called **dimensionless** or **nondimensional groups**. One example is the Reynolds number that arises in fluid mechanics:

$$\text{Reynolds number} = \frac{Dv\rho}{\mu} = N_{RE}$$

where D is the pipe diameter (e.g., cm), v is the fluid velocity (e.g., cm/s), ρ is the fluid density (e.g., g/cm^3), and μ is the viscosity [usually given in the units of centipoise, which itself has the units of g/(cm)(s)]. Introducing this consistent set of units for D, v, ρ, and μ into $Dv\rho/\mu$, you will find that all the units cancel, resulting in a dimensionless number for the Reynolds number:

$$N_{RE} = \frac{Dv\rho}{\mu} = \frac{\cancel{\text{cm}}}{}\left|\frac{\cancel{\text{cm}}}{\cancel{\text{s}}}\right|\frac{\cancel{\text{g}}}{\cancel{\text{cm}^3}}\left|\frac{(\cancel{\text{cm}})(\cancel{\text{s}})}{\cancel{\text{g}}}\right. = \text{a dimensionless quantity}$$

Example 2.6 Dimensional Consistency of an Equation

The following equation is proposed to calculate the pressure drop (Δp) across a length of pipe (L) due to flow through the pipe. Determine the dimensional consistency of this equation:

$$\Delta p = \frac{1}{2}v^2\left(\frac{L}{D}\right)f$$

where v is the average velocity of the fluid flowing through the pipe, D is the diameter of the pipe, and f is a dimensionless coefficient called the friction factor, which is a function of the Reynolds number.

Solution

Let's substitute SI units appropriate for each term into the proposed equation, recognizing that pressure is force per unit area (see Table 2.1). What are the units of Δp? They are

$$\frac{\text{N}}{\text{m}^2} = \frac{(\text{kg})(\text{m})}{\text{s}^2}\bigg|\frac{}{\text{m}^2} \rightarrow \frac{\text{kg}}{(\text{s}^2)(\text{m})}$$

What are the net units of the right-hand side of the proposed equation?

$$\left(\frac{\text{m}}{\text{s}}\right)^2\bigg|\frac{\text{m}}{\text{m}} \rightarrow \frac{\text{m}^2}{\text{s}^2}$$

Therefore, because the units of the left-hand side of the equation do not match the units of the right-hand side, the proposed equation is not dimensionally consistent. By some research or inspection, it was determined that the proposed equation was missing a density term on the right-hand side of the equation; that is, the equation should be

$$\Delta p = {}^1\!/\!_2 v^2 \rho\left(\frac{L}{D}\right)f$$

With this modification, the units on the right-hand side of the equation become

$$\left(\frac{\text{m}}{\text{s}}\right)^2\bigg|\frac{\text{kg}}{\text{m}^3}\bigg|\frac{\text{m}}{\text{m}} \rightarrow \frac{\text{kg}}{(\text{s}^2)(\text{m})}$$

Therefore, if the density is included, this equation is shown to be dimensionally consistent.

Self-Assessment Test

Questions

1. Explain what dimensional consistency means in an equation.
2. Explain why the so-called dimensionless group has no net dimensions.
3. If you divide all of a series of terms in an equation by one of the terms, will the resulting series of terms be dimensionless?

4. How might you make the following variables dimensionless?
   a. Length (of a pipe)
   b. Time (to empty a tank full of water)

Problems

1. An orifice meter is used to measure the rate of flow of a fluid in pipes. The flow rate is related to the pressure drop by the following equation:

$$u = c\sqrt{\frac{\Delta p}{\rho}}$$

where u = fluid velocity
Δp = pressure drop (force per unit area)
ρ = density of the flowing fluid
c = constant

What are the units of c in the SI system of units?

2. The thermal conductivity k of a liquid metal is predicted via the empirical equation

$$k = \mathrm{A} \exp(\mathrm{B}/T)$$

where k is in J/(s)(m)(K)
T is in Kelvin
A and B are constants.

What are the units of A and B?

Thought Problems

1. Can you prove the accuracy of an equation by checking it for dimensional consistency?
2. Suppose that some short time after the "Big Bang" the laws of nature turned out to be different from the laws currently used. In particular, instead of $pV = n\mathrm{R}T$, a different gas law arose, namely, $pVT = n\mathrm{R}$. What comments do you have about such an equation? Hint: See the text above for the units of p, V, n, R, and T.

Discussion Problem

In a letter criticizing an author's equation, the writer said:

> The equation for kinetic energy of the fluid is not dimensionally consistent. I suggest the modification
>
> $$KE = mv^2/2g_c$$

in which g_c is introduced. Then the units in the equation will not be $(\text{ft/s})^2$ which are the wrong units for energy.

What do you think of the comment in the letter?

2.4 Significant Figures

Decimals have a point.

Unknown

You have probably heard the story about the Egyptian tour guide who told visitors that the pyramid they beheld in awe was 5013 years old. "Five thousand and thirteen!" said a visitor. "How do you know?" "Well," said the guide, "when I first began working here 13 years ago, I was told the pyramid was 5000 years old."

What do you think about the accuracy of a travel brochure in which you read that a mountain on the trip is 8000 m (26,246 ft) high?

Responsible physical scientists and engineers agree that a measurement should include three pieces of information:

1. The magnitude of the variable being measured
2. Its units
3. An estimate of its uncertainty

In this book we will interpret a number such as 31.4 in terms of engineering (scientific) notation as 3.14×10^1. How many significant figures occur? Three. Adding zeros in front of the 3 and after the 4 have no effect on the number of significant figures. Table 2.5 contains several examples demonstrating how to determine the number of significant figures for a number. Note that putting the number into scientific notation directly indicates the number of significant figures. **The number of significant figures for a number is a direct indication of its accuracy.**

What should you do about maintaining the proper degree of certainty when you add, subtract, multiply, and divide numbers that implicitly have associated uncertainty? The accuracy you need for the results of a calculation depends on the proposed application of the results. The question is: How close is close enough? For example, on income tax forms you do not need to include cents, whereas in a bank statement cents (two decimals) must be included. In engineering calculations, if the cost of inaccuracy is great (failure, fire, downtime, etc.), knowledge of the uncertainty in the calculated variables is vital. On the other hand, in determining how much fertilizer to put on your lawn in the summer, being off by 10 to 20 lb out of 100 lb is not important.

Table 2.5 Significant Figure Examples

Number	Scientific Notation	Number of Significant Figures
12.44	1.244×10^1	4
53000	5.3×10^4	2
53000.	5.3000×10^4	5*
53000.0	5.30000×10^4	6
0.00034	3.4×10^{-4}	2
0.000340	3.40×10^{-4}	3**

* Did you note the presence of the decimal point in contrast with the row above?

** How does the last row differ from the row above? The right-hand zero was present initially—not added.

The general rule we use in this book is to retain no more significant figures in an answer to a problem than exist in the least accurate number used in the calculations (i.e., the number that has the smallest number of significant figures). For example, multiplication of 25.3 by 2.1 would yield 53.13, which should be replaced with 53 because 2.1 has only two significant figures. Apply the same concept to the result of a series of calculations. However, within each step of a series of calculations you should carry as many significant figures in your calculations as the most accurate number (e.g., 53.13 for the preceding case) before rounding the final answer to the proper number of significant figures.

Keep in mind that some numbers are exact, such as the $\frac{1}{2}$ in the equation for kinetic energy, $KE = \frac{1}{2}mv^2$, and the 2 in the superscript for the operation of square. You will also encounter integers such as 1, 2, 3, and so on, which in some cases are exact (2 reactors, 3 input streams) but in other cases are shortcut substitutes for presumed very accurate measurements in problem solving (3 moles, 10 kg). You should assume that a given mass such as 10 kg, in which the number does not have a decimal point, involves quite a few significant figures, particularly in relation to the other values of the parameters stated in an example or problem. You will also occasionally encounter fractions such as 2/3, which can be treated by a highly accurate decimal approximation. In this text, for convenience we will often use 273 K for the temperature equivalent to 0°C instead of 273.15 K. Keep in mind, however, that in addition, subtraction, multiplication, and division, the errors you introduce propagate into the final answer.

In summary, **be sure to round off your answers to problems to a reasonable number of significant figures** even though in the intermediate calculations numbers are carried out to ten or more digits in your computer or calculator.

Example 2.7 Retention of Significant Figures

If 20,100 kg is subtracted from 22,400 kg, is the answer of 2300 kg good to four significant figures?

Solution

If you note that 22,400, 20,100, and 2300 have no decimal points after the right-hand zero, how many significant figures can you input to 22,400 and 20,100? From the scientific representation, you can conclude that three is the number of significant figures.

$$\begin{array}{r} 2.24 \times 10^4 \text{ kg} \\ -2.01 \times 10^4 \text{ kg} \\ \hline 0.23 \times 10^4 \text{ kg} \end{array}$$

Even though the first two numbers have three significant figures, the result has only two significant figures.

On the other hand, if a decimal point were placed in each number thus—22,400. and 20,100.—indicating that the last zero was significant, then the answer of 2300. would be valid to four significant figures.

Example 2.8 Micro-dissection of DNA

A stretch-and-positioning technique on a carrier layer can be used for micro-dissection of an electrostatically positioned DNA strand. The procedure employs a glass substrate on the top of which a sacrificial layer, a DNA carrier layer, and a pair of electrodes are deposited. The DNA is electrostatically stretched and immobilized onto the carrier layer with one of its molecular ends aligned on the electrode edge. A cut is made through the two layers with a stylus as a knife at an aimed portion of the DNA. By dissolving the sacrificial layer, the DNA fragment on the piece of carrier can be recovered on a membrane filter. The carrier piece can then be melted to obtain the DNA fragment in solution.

If the DNA is stretched out to a length of 48 kb, and a cut made with a width of 3 μm, how many base pairs (bp) should be reported in the fragment? Note: 1 kb is 1000 base pairs (bp), and 3 kb = 1 μm.

(*Continues*)

Example 2.8 Micro-dissection of DNA (*Continued*)

Solution

The conversion is

$$\frac{3\ \mu\text{m}}{}\left|\frac{3\ \text{kb}}{1\ \mu\text{m}}\right|\frac{1000\ \text{bp}}{1\ \text{kb}} = 9000\ \text{bp}$$

However, because the measurement of the number of molecules in a DNA fragment can be determined to three or four significant figures and the 3 μm reported for the cut may well have more than the reported one significant figure if measured properly, the precision in the 9000 value may actually be better than indicated by the calculation.

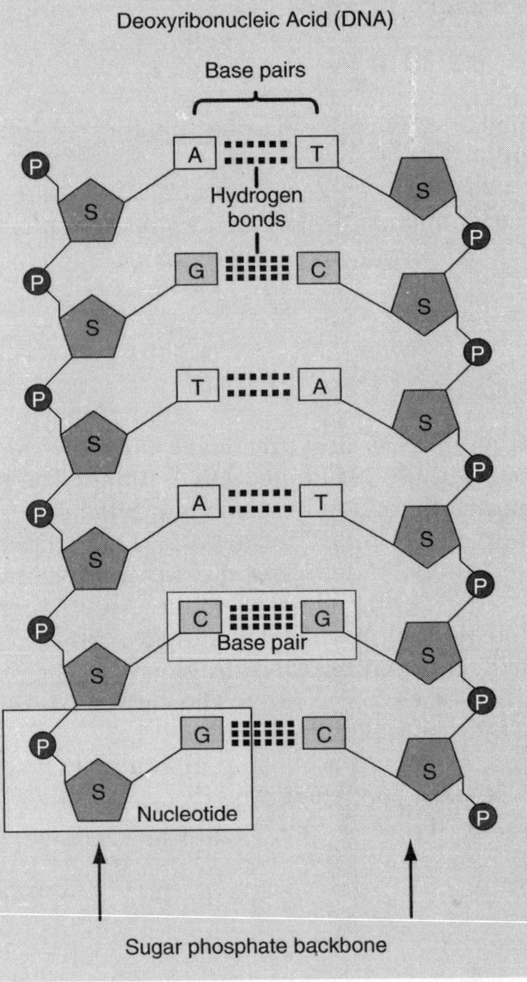

Figure E2.8a A two-dimensional representation of a section of DNA that is shown in Figure E2.8c in three dimensions

Nearly all living things contain DNA (which stands for deoxyribonucleic acid), namely, the molecule that stores **genetic information**. DNA consists of a series of nucleotides (look at Figure E2.8a). Each nucleotide is denoted by the base it contains, abbreviated as follows: A (for adenine), C (for cytosine), G (for guanine), or T (for thymine). Figure E2.8b shows in two dimensions the composition of each base. The most famous form of DNA in a cell is composed of two very long backbones of sugar (S) phosphate (P) molecules forming two intertwined chains called a double helix that are tied together by base pairs as shown in Figure E2.8c. DNA can also take other forms not shown here.

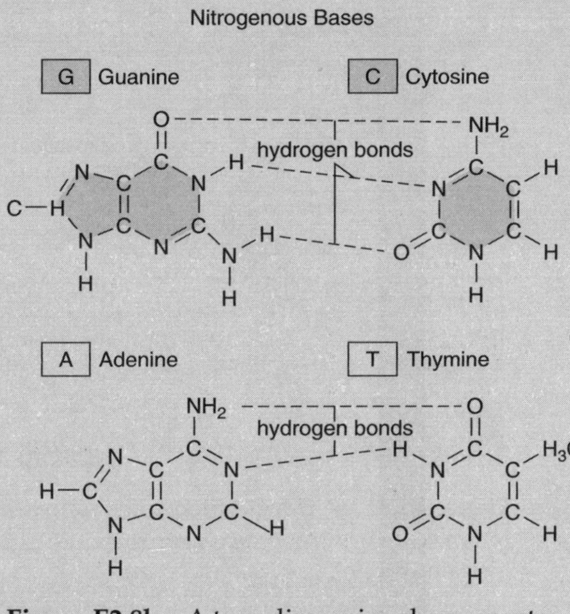

Figure E2.8b A two-dimensional representation of the chemical compounds of the respective bases G, C, A, and T

The length of a segment of DNA is measured in the number of base pairs (as indicated in Figures E2.8b and E2.8c); 1 kb is 1000 base pairs (bp), and 3 kb = 1 μm. The sugar phosphate backbone is connected by successive combinations of A, T, G, and C.

A genome is one section of DNA in an enormously long sequence of A, C, G, and T. Certain parts of the genome correspond to genes that carry the information needed to direct protein synthesis and replication. For a section of the genome to be a gene, the sequence of bases must begin with

(Continues)

Example 2.8 Micro-dissection of DNA (*Continued*)

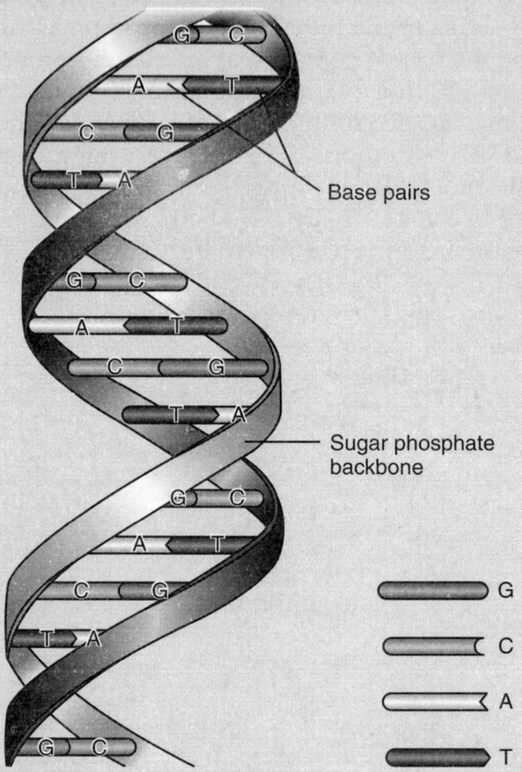

Figure E2.8c A three-dimensional representation of an arbitrary slice of an enormously long strand of DNA

ATG or GTG and must end with TAA, TAG, or TGA. The length of the sequence must be an exact multiple of three. Protein synthesis is the production of the proteins needed by the cell for its activities and development. Replication is the process by which DNA copies itself for each descendant cell, passing on the information needed for protein synthesis. In most cellular organisms, DNA is organized on chromosomes located in the nucleus of a cell.

Self-Assessment Test

Questions

1. How can you avoid a significant loss of accuracy in carrying out calculations involving many repetitive operations (such as addition, multiplication, and so on)?
2. Will adding a decimal point to a reported number that does not have a decimal point, such as replacing 12,600 with 12,600., improve the accuracy of the number?

Problems

1. Identify the number of significant figures in each of the following numbers:
   a. 1.0
   b. 0.353
   c. 1000.
   d. 23
   e. 1000
   f. 1000.0
2. What is the correct sum and the number of significant digits when you add (a) 5750 and 10.3? (b) 2.000 and 0.22?
3. Convert the flow rate of 87.0 kg/min to the units of gal/hr, giving the answer in the proper number of significant figures.
4. A computer chip made in Japan was calculated to cost $78. The calculation to convert the price from yen to dollars was made as follows:

$$\left(\frac{10{,}000 \text{ yen}}{1 \text{ computer chip}}\right)\left(\frac{\$1.00}{128 \text{ yen}}\right) = \$78/\text{computer chip}$$

   Is the number of significant digits shown in the answer correct?
5. What is the answer to 78.3 − 3.14 − 0.388?

Thought Problems

1. Is 6 5/8 inches equivalent to (a) 53/8 inches or (b) 6.625 inches?
2. When you want to calculate the weight of six silicon chips, each weighing 2.35 g, is the answer good only to one significant figure, for example, 6?
3. A textbook mentions the quantity of reactant as being 100 mL. How would you decide on the number of significant figures to associate with the quantity of reactant?

Discussion Problem

In a report of a crew laying fiber-optic cable, the results for the month were listed as follows:

3000 ft
4120 ft
1300 ft
2100 ft
10,520 ft

How many significant figures would you attribute to the sum?

2.5 Validation of Results

When solving either homework problems in school or problems encountered as a professional engineer, it is critically important that you ensure that your answers are accurate enough. By this we mean correct or close enough considering the problem requirements. Unfortunately, there is an almost unlimited number of ways to make errors when solving problems. Being able to eliminate errors when problem solving is an important attribute of a good engineer. Here are some suggestions to help you "catch" errors when problem solving:

- **Make sure your answers are reasonable.** Check to ensure that your answer seems reasonable according to your understanding of the physical system. For example, for a process plant, if you calculated a pipe diameter of 20 feet to transport 100 gal/min, you should know immediately that this is not a reasonable answer. As you gain engineering experience, you will be able to enhance your knowledge of what is a reasonable answer.
- **Check the details of your calculations.** Errors such as keying in the wrong number, confusing the proper decimal point of a number, reading an intermediate number incorrectly, transposing two numbers, and not being careful about the units are among the most common errors. Carefully repeat your calculations using your calculator or computer, perhaps in a different sequence. You can also simplify the numbers in your calculation so that you can calculate a rough approximate answer in your head or in the calculator. For example, the following exact calculation can be compared with an approximate calculation to confirm the validity of the more exact answer:

$$\frac{468}{0.0181}\left|\frac{6250}{2.95}\right. = 5.48 \times 10^7 \qquad \frac{5 \times 10^2}{2 \times 10^{-2}}\left|\frac{6 \times 10^3}{3 \times 10^0}\right. = 5 \times 10^7$$

- **Review the solution procedure.** You should always review your problem solution immediately after it has been completed. Review the problem statement and problem specifications to ensure that you solved the correct problem. Also, make sure that the data used in the problem solution were correctly transferred or selected from a source of data. You should also review the assumptions that you used to make sure that they are reasonable and proper.

It may seem that carrying out these steps to check your solution will cause extra work for you; it will. But if you realize how important it is to reliably develop accurate solutions now and in the future, you will appreciate the point of establishing good habits now.

Self-Assessment Test

Questions

1. What do you have to do to make sure that an answer to a problem is reasonable?
2. What is the most common type of error in a problem solution and how can you correct it?
3. What benefits does reviewing a solution procedure provide?

Problem

Develop an approximate solution for the following calculation of the volume in cubic feet:

$$\text{Volume} = \frac{\pi}{4}\left|\frac{(4\text{ in.})^2}{}\right|\frac{10\text{ ft}}{}$$

2.6 The Mole and Molecular Weight

What is a mole? For our purposes we will say that a **mole** is a certain amount of material corresponding to a specified number of molecules, atoms, electrons, or other specified types of particles.

In the SI system a mole (which we will call a **gram mole** to avoid confusing units) is composed of 6.022×10^{23} (Avogadro's number) molecules. However, for convenience in calculations and for clarity, we will make use of other specifications for moles such as the **pound mole** (lb mol, composed of $6.022 \times 10^{23} \times 453.6$) molecules, the **kg mol** (kilomole, kmol, composed of 1000 moles), and so on. You will find that such nonconforming (to SI) definitions of the amount of material will help avoid excess details in many calculations. What would a metric ton mole of molecules consist of?

One important calculation at which you should become skilled is to convert the number of moles to mass and the mass to moles. To do this you make use of the **molecular weight**—*the mass per mole:*

$$\text{molecular weight (MW)} = \frac{\text{mass}}{\text{mole}}$$

Based on this definition of molecular weight:

$$\text{g mol} = \frac{\text{mass in g}}{\text{molecular weight}}$$

$$\text{lb mol} = \frac{\text{mass in lb}}{\text{molecular weight}}$$

Therefore, from the definition of the molecular weight, you can calculate the mass knowing the number of moles or the number of moles knowing the mass. For historical reasons, the terms *atomic weight* and *molecular weight* are usually used instead of the more accurate terms *atomic mass* and *molecular mass*. Does using *weight* for *mass* make any difference?

Example 2.9 Use of Molecular Weights to Convert Mass to Moles

If a bucket holds 2.00 lb of NaOH:

a. How many pound moles of NaOH does it contain?

b. How many gram moles of NaOH does it contain?

Solution

You can convert pounds to pound moles, and then convert the values to the SI system of units. Look up the molecular weight of NaOH, or calculate it from the atomic weights. (It is 40.0.) Note that the molecular weight is used as a conversion factor in this calculation:

a. $$\frac{2.00 \text{ lb NaOH}}{} \left| \frac{1 \text{ lb mol NaOH}}{40.0 \text{ lb NaOH}} \right. = 0.050 \text{ lb mol NaOH}$$

b1.
$$\frac{2.00 \text{ lb NaOH}}{} \left| \frac{1 \text{ lb mol NaOH}}{40.0 \text{ lb NaOH}} \right| \frac{454 \text{ g mol}}{1 \text{ lb mol}} = 22.7 \text{ g mol}$$

Check your answer by converting the 2.00 lb of NaOH to the SI system first and completing the conversion to gram moles:

b2.
$$\frac{2.00 \text{ lb NaOH}}{} \left| \frac{454 \text{ g}}{1 \text{ lb}} \right| \frac{1 \text{ g mol NaOH}}{40.0 \text{ g NaOH}} = 22.7 \text{ g mol}$$

Example 2.10 Use of Molecular Weights to Convert Moles to Mass

How many pounds of NaOH are in 7.50 g mol of NaOH?

Solution

This problem involves converting gram moles to pounds. From Example 2.9, the MW of NaOH is 40.0:

$$\frac{7.50 \text{ g mol NaOH}}{} \left| \frac{1 \text{ lb mol}}{454 \text{ g mol}} \right| \frac{40.0 \text{ lb NaOH}}{1 \text{ lb mol NaOH}} = 0.661 \text{ lb NaOH}$$

Note the conversion between pound moles and gram moles was to proceed from SI to the AE system of units. Could you first convert 7.50 g mol of NaOH to grams of NaOH, and then use the conversion of 454g = 1 lb to get pounds of NaOH? Of course.

Values of the **molecular weights** (relative molar masses) are built up from the values of atomic weights based on a scale of the *relative* masses of the elements. The **atomic weight** of an element is the mass of an atom based on the scale that assigns a mass of exactly 12 to the carbon isotope ^{12}C. The value 12 is selected in this case because an atom of carbon 12 contains 6 protons and 6 neutrons for a total molecular weight of 12.

Appendix B lists the atomic weights of the elements. On this scale of atomic weights, hydrogen is 1.008, carbon is 12.01, and so on. (In most of our calculations we shall round these off to 1 and 12, respectively, for convenience). The atomic weights of these and other elements are not whole numbers because elements can appear in nature as a mixture of different isotopes. As an example, approximately 0.8% of hydrogen is deuterium (a hydrogen atom with one proton and one neutron); thus the atomic weight is 1.008 instead of 1.000.

A **compound** is composed of more than one atom, and the molecular weight of the compound is nothing more than the sum of the weights of atoms of which it is composed. Thus H_2O consists of 2 hydrogen atoms and 1 oxygen atom, and the molecular weight of water is $(2)(1.008) + 16.000 = 18.016$, or approximately 18.02.

You can compute **average molecular weights** for mixtures of constant composition even though they are not chemically bonded if their compositions are known accurately. Example 2.11 shows how to calculate the fictitious quantity called the average molecular weight of air. Of course, for a material such as fuel oil or coal whose composition may not be exactly known, you cannot determine an exact molecular weight, although you might estimate an approximate average molecular weight, which is good enough for most engineering calculations.

Example 2.11 Average Molecular Weight of Air

Calculate the average molecular weight of air, assuming that air is 21% O_2 and 79% N_2.

Solution

Because the composition of air is given in mole percent, a basis of 1 g mol is chosen. The MW of the N_2 is not actually 28.0 but 28.2 because the value of the MW of the pseudo 79% N_2 is actually a combination of 78.084% N_2 and 0.934% Ar. The masses of the O_2 and pseudo N_2 are

Basis: 1 g mol of air

$$\text{Mass of } O_2 = \frac{1 \text{ g mol air}}{} \left| \frac{0.21 \text{ g mol } O_2}{\text{g mol air}} \right| \frac{32.00 \text{ g } O_2}{\text{g mol } O_2} = 6.72 \text{ g } O_2$$

$$\text{Mass of } N_2 = \frac{1 \text{ g mol air}}{} \left| \frac{0.79 \text{ g mol } N_2}{\text{g mol air}} \right| \frac{28.2 \text{ g } N_2}{\text{g mol } N_2} = 22.28 \text{ g } N_2$$

$$\text{Total} \qquad = 29.0 \text{ g air}$$

Therefore, the total mass of 1 g mol of air is equal to 29.0 g, which is called the average molecular weight of air. (Because we chose 1 g mol of air as the basis, the total mass calculated directly provides the average molecular weight of 29.0.)

Example 2.12 Calculation of Average Molecular Weight

Since the discovery of superconductivity almost 100 years ago, scientists and engineers have speculated about how it can be used to improve the use of energy. Until recently most applications were not economically viable because the niobium alloys used had to be cooled below 23 K by liquid He. However, in 1987 superconductivity in Y-Ba-Cu-O material was achieved at 90 K, a situation that permits the use of inexpensive liquid N_2 cooling.

What is the molecular weight of the cell of a superconductor material shown in Figure E2.12? (The figure represents one cell of a larger structure.)

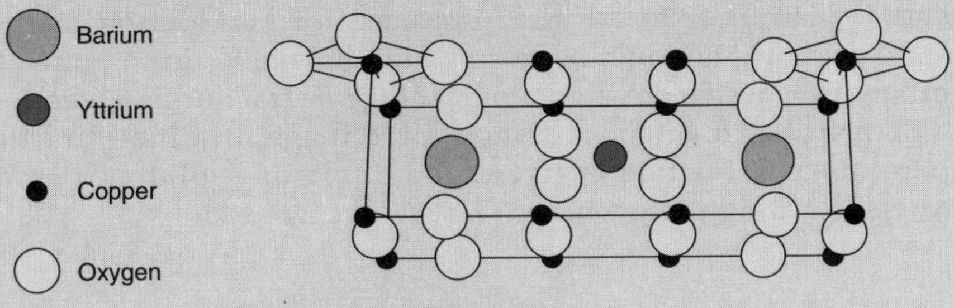

Figure E2.12

Solution

First look up the atomic weights of the elements from the table in Appendix B. Assume that one cell is a molecule. By counting the atoms you can find:

Element	Number of Atoms	Atomic Weight (g)	Mass (g)
Ba	2	137.34	2(137.34)
Cu	16	63.546	16(63.546)
O	37	16.00	37(16.00)
Y	1	88.905	1(88.905)
		Total	1972.3

The molecular weight of the cell is 1972.3 atomic masses/1 molecule or 1972.3 g/g mol. Check your calculations and check your answer to ensure that it is reasonable.

Mole fraction is simply the number of moles of a particular substance in a mixture or solution divided by the total number of moles present in the mixture or solution. This definition holds for gases, liquids, and solids. Similarly, the **mass (weight) fraction** is nothing more than the **mass (weight)** of the substance divided by the total mass (weight) of all of the substances present in the mixture or solution. Although *mass fraction* is the correct term, by custom ordinary engineering usage frequently employs the term **weight fraction.** These concepts can be expressed as

$$\text{mole fraction of A} = \frac{\text{moles of A}}{\text{total moles}}$$

$$\text{mass (weight) fraction of A} = \frac{\text{mass of A}}{\text{total mass}}$$

Mole percent and weight percent are the respective fractions times 100. Be sure to learn how to convert from mass fraction to mole fraction and vice versa without thinking, because you will have to do so quite often. **Unless otherwise specified, when a percentage or fraction is given for a gas, it is assumed that it refers to a mole percentage or a mole fraction. When a percentage or fraction is given for a liquid or a solid, it is assumed that it refers to a weight percentage or a weight fraction.**

Example 2.13 Conversion between Mass (Weight) Fraction and Mole Fraction

An industrial-strength drain cleaner contains 5.00 kg of water and 5.00 kg of NaOH. What are the mass (weight) fraction and mole fraction of each component in the drain cleaner?

Solution

You are given the masses so it is easy to calculate the mass fractions. From these values you can then calculate the desired mole fractions.

A convenient way to carry out the calculations in such conversion problems is to form a table as shown below. Become skilled at doing so, because this type of problem and its inverse—that is, conversion of mole fraction to mass (weight) fraction—will occur more frequently than you would like. List the components, their masses, and their molecular weights in columns.

Basis: 10.0 kg of total solution

Component	kg	Weight Fraction	Mol. Wt.	kg mol	Mole Fraction
H_2O	5.00	$\frac{5.00}{10.0} = 0.500$	18.0	0.278	$\frac{0.278}{0.403} = 0.69$
NaOH	5.00	$\frac{5.00}{10.00} = 0.500$	40.0	0.125	$\frac{0.125}{0.403} = 0.31$
Total	10.00	1.000		0.403	1.00

Self-Assessment Test

Questions

1. Indicate whether the following statements are true or false:
   a. The pound mole is composed of 2.73×10^{26} molecules.
   b. The kilogram mole is composed of 6.023×10^{26} molecules.
   c. Molecular weight is the mass of a compound or element per mole.
2. What is the molecular weight of acetic acid (CH_3COOH)?

Problems

1. Convert the following:
   a. 120 g mol of NaCl to grams
   b. 120 g of NaCl to gram moles
   c. 120 lb mol of NaCl to pounds
   d. 120 lb of NaCl to pound moles
2. Convert 39.8 kg of NaCl per 100 kg of water to kilogram moles of NaCl per kilogram mole of water.
3. How many pound moles of $NaNO_3$ are there in 100 lb?
4. Commercial sulfuric acid is 98% H_2SO_4 and 2% H_2O. What is the mole ratio of H_2SO_4 to H_2O?
5. A solid compound contains 50% sulfur and 50% oxygen. Is the empirical formula of the compound (a) SO, (b) SO_2, (c) SO_3, or (d) SO_4?
6. A gas mixture contains 40 lb of O_2, 25 lb of SO_2, and 30 lb of SO_3. What is the composition of the mixture?
7. Saccharin, an artificial sweetener that is 3000 times sweeter than sucrose, is composed of 45.90% carbon, 2.73% hydrogen, 26.23% oxygen, 7.65% nitrogen, and 17.49% sulfur. Is the molecular formula of saccharin (a) $C_{14}H_{10}O_6N_2S_2$, (b) $C_5H_7O_3NS$, (c) $C_8H_9O_2NS$, or (d) $C_7H_5O_3NS$?

Thought Problem

There is twice as much copper in 480 g of copper as there is in 240 g of copper, but is there twice as much copper in 480 g of copper as there is silver in 240 g of silver?

Discussion Question

In the journal *Physics Education* (July, 1977, p. 276) McGlashan suggested that the physical quantity we call the mole is not necessary. Instead, it would be quite feasible to use molecular quantities, that is, the number of molecules or atoms, directly. Instead of $pV = nRT$ where n denotes the number of moles of a substance, we should write $pV = NkT$ where N denotes the number of molecules and k is the

Boltzmann constant [1.380×10^{-23} J/(molecule)(K)]. Thus, for example, would be the molecular volume for water, a value used instead of 18 cm^3/g mol. Is the proposal a reasonable idea?

2.7 Choosing a Basis

A **basis** is a reference chosen by you for the calculations you plan to make in a particular problem, and a proper choice of basis often can make a problem much easier to solve than a poor choice. The basis may be a period of time such as hours, or a given mass of material, or some other convenient quantity. To select a sound basis (which in many problems is predetermined for you but in some problems is not so clear), ask yourself the following three questions:

1. What do I have to start with (e.g., I have 100 lb of oil; I have 46 kg of fertilizer)?
2. What answer is called for (e.g., the amount of product produced per hour)?
3. What is the most convenient basis to use? (For example, suppose that the composition of a given material is known in mole percent. Then selecting 100 kg moles of the material as basis would make sense. On the other hand, if the composition of the material in terms of mass is known, then 100 kg of the material would be an appropriate basis.)

These questions and their answers will suggest suitable bases. Sometimes when several bases seem appropriate, you may find it is best to use a unit basis of 1 or 100 of something, for example, kilograms, hours, moles, or cubic feet. For liquids and solids in which a mass (weight) analysis applies, a convenient basis is often 1 or 100 lb or kg; similarly, because gas compositions are usually provided in terms of moles, 1 or 100 moles is often a good choice for a gas.

Always state the basis you have chosen for your calculations by writing it prominently on your calculation sheets or in the computer program used to solve the problem.

Example 2.14 Choosing a Basis

The dehydration of the lower-molecular-weight alkanes can be carried out using a ceric oxide (CeO) catalyst. What are the mass fraction and mole fraction of Ce and O in the catalyst?

Solution

Start the solution by selecting a basis. Because no specific amount of material is specified, the question "What do I have to start with?" does not help determine a basis. Neither does the question about the desired answer. Thus, selecting a convenient basis becomes the best choice. What do you know about CeO? You know from the formula that 1 mole of Ce is combined with 1 mole of O. Consequently, a basis of 1 kg mol (or 1 g mol, or 1 lb mol, etc.) would make sense. You can get the atomic weights for Ce and O from Appendix B, and then you are prepared to calculate the respective masses of Ce and O in CeO. The calculations for the mole and mass fractions for Ce and O in CeO are presented in the following table:

Basis: 1 kg mole of CeO

Component	kg mol	Mole Fraction	Mol. Wt.	kg	Mass Fraction
Ce	1	0.50	140.12	140.12	0.8975
O	1	0.50	16.0	16.0	0.1025
Total	2	1.00	156.12	156.12	1.0000

Example 2.15 Choosing a Basis

Most processes for producing high-energy-content gas or gasoline from coal include some type of gasification step to make hydrogen or synthesis gas. Pressure gasification is preferred because of its greater yield of methane and higher rate of gasification.

Given that a 50.0 kg test run of gas averages 10.0% H_2, 40.0% CH_4, 30.0% CO, and 20.0% CO_2, what is the average molecular weight of the gas?

Solution

Let's choose a basis. The answer to question 1 is to select a basis of 50.0 kg of gas ("What do I have to start with?"), but is this choice a good basis? A little reflection will show that such a basis is of no use. You cannot multiply the given *mole percent* of this gas (remember that the composition of gases is given in mole percent unless otherwise stated) times kilograms and expect the result to mean anything. Try it, being sure to include the respective units. Thus, the next step is to choose a "convenient basis," which is, say, 100 kg mol of gas, and proceed as follows:

(Continues)

Example 2.15 Choosing a Basis (*Continued*)

Basis: 100 kg mol or lb mol of gas

Set up a table such as the following to make a compact presentation of the calculations. You do not have to, but making individual computations for each component is inefficient and more prone to errors.

Component	Percent = kg mol or lb mol	Mol. Wt.	kg or lb
CO_2	20.0	44.0	880
CO	30.0	28.0	840
CH_4	40.0	16.04	642
H_2	10.0	2.02	20
Total	100.0		2382

$$\text{Average molecular weight} = \frac{2382\text{ kg}}{100\text{ kg mol}} = 23.8\text{ kg/kg mol}$$

Check the solution by noting that an average molecular weight of 23.8 is reasonable because the molecular weights of the components range only from 2 to 44 and the answer is intermediate to these values.

To sum up, be sure to state the basis of your calculations so that you will keep clearly in mind their real nature, and so that anyone checking your problem solution will be able to understand on what basis your calculations were performed.

You no doubt have heard the story of Ali Baba and the 40 thieves. Have you heard about Ali Baba and the 39 camels? Ali Baba gave his four sons 39 camels to be divided among them so that the oldest son got one-half of the camels, the second son a quarter, the third an eighth, and the youngest a tenth. The four brothers were at a loss as to how they should divide the inheritance without killing camels until a stranger came riding along on his camel. He added his own camel to Ali Baba's 39 and then divided the 40 among the sons. The oldest son received 20; the second, 10; the third, 5; and the youngest, 4. One camel was left. The stranger mounted it—for it was his own—and rode off. Amazed, the four brothers watched him ride away. The oldest brother was the first to start calculating. Had his father not willed half of the camels to him? Twenty camels are obviously more than half of 39. One of the four sons must have received less than his due. But figure as they would, each found that he had more than his share. Cover the next few lines of text. What is the answer to the paradox?

After thinking over this problem, you will realize that the sum of $^1/_2$, $^1/_4$, $^1/_8$, and $^1/_{10}$ is not 1 but is 0.975. By adjusting (normalizing) the camel fractions (!) so that they total 1, the division of camels is validated:

Camel Fractions	Normalizing		Correct Fractions				Distributed Camels (Integer)
0.500	$\left(\frac{0.500}{0.975}\right)$	=	0.5128	×	39	=	20
0.250	$\left(\frac{0.250}{0.975}\right)$	=	0.2564	×	39	=	10
0.125	$\left(\frac{0.125}{0.975}\right)$	=	0.1282	×	39	=	5
0.100	$\left(\frac{0.100}{0.975}\right)$	=	0.1026	×	39	=	4
0.975	$\left(\frac{0.975}{0.975}\right)$	=	1.000				39

What we have done is to change the calculations from a basis total of 0.975 to a new basis of 1.000.

More frequently than you probably would like, you will have to change from your original selection of a basis in solving a problem to one or more different bases in order to put together the information needed to solve the entire problem. Consider the following example.

Example 2.16 Changing Bases

Considering a gas containing O_2 (20%), N_2 (78%), and SO_2 (2%), find the composition of the gas on *an SO_2-free basis,* meaning gas without the SO_2 in it.

Solution

First choose a basis of 1 mol of gas (or 100 mol). Why? The composition for the gas is in mole percent. Next you should calculate the moles of each component,

(Continues)

Example 2.16 Changing Bases (*Continued*)

remove the SO_2, and adjust the basis for the calculations so that the gas becomes composed of only O_2 and N_2 with a percent composition totaling 100%:

Basis: 1.0 mol of gas

Components	Mol Fraction	Mol	Mol SO_2-Free	Mol Fraction SO_2-Free
O_2	0.20	0.20	0.20	0.20
N_2	0.78	0.78	0.78	0.80
SO_2	0.02	0.02	—	—
	1.00	1.00	0.98	1.00

The round-off in the last column is appropriate given the original values for the mole fractions.

Self-Assessment Test

Questions

1. What are the three questions you should ask yourself when selecting a basis?
2. Why do you sometimes have to change bases during the solution of a problem?

Problem

What would be good initial bases to select for solving problems 2.6.3, 2.6.7, and 2.6.8 at the end of the chapter?

Thought Problem

Water-based dust-suppression systems are an effective and viable means of controlling dust and virtually eliminate the historic risk of fires and explosions in grain elevators. Water-based safety systems have resulted in cleaner elevators, improved respiratory atmospheres for employees, and reduced dust emissions into the environments surrounding storage facilities. However, some customers have complained that adding water to the grain causes the buyer to pay too much for the grain. The grain elevators argue that all grain shipments unavoidably contain a weight component in the form of moisture. Moisture is introduced to grain and to grain products in a broad variety of practices.

What would you recommend to elevator operators and grain dealers to alleviate this problem?

2.8 Density and Specific Gravity

In ancient times, counterfeit gold objects were identified by comparing the ratio of the weight to the volume of water displaced by the object, which is a way to measure the density of the material of the object, to that of an object known to be made of gold.

A striking example of quick thinking by an engineer who made use of the concept of density was reported by P. K. N. Paniker in the June 15, 1970, issue of *Chemical Engineering*:

> The bottom outlet nozzle of a full lube-oil storage tank kept at a temperature of about 80°C suddenly sprang a gushing leak as the nozzle flange became loose. Because of the high temperature of the oil, it was impossible for anyone to go near the tank and repair the leak to prevent further loss.
>
> After a moment of anxiety, we noticed that the engineer in charge rushed to his office to summon fire department personnel and instruct them to run a hose from the nearest fire hydrant to the top of the storage tank. Within minutes, what gushed out from the leak was hot water instead of valuable oil. Some time later, as the entering cold water lowered the oil temperature, it was possible to make repairs.

Density (we use the Greek symbol ρ) is the ratio of mass per unit volume such as kg/m^3 or lb/ft^3:

$$\rho = \text{density} = \frac{\text{mass}}{\text{volume}} = \frac{m}{V}$$

Density has both a numerical value and units. Densities for liquids and solids do not change significantly at ordinary conditions with pressure, but they can change significantly with temperature for certain compounds if the temperature change is large enough, as shown in Figure 2.1. Note that between 0°C and 70°C, the density of water is relatively constant at $1.0\ g/cm^3$. On the other hand, for the same temperature range, the density of NH_3 changes by approximately 30%. Usually we will ignore the effect of temperature on liquid density unless the density of the material is especially sensitive to temperature or the change in the temperature is particularly large.

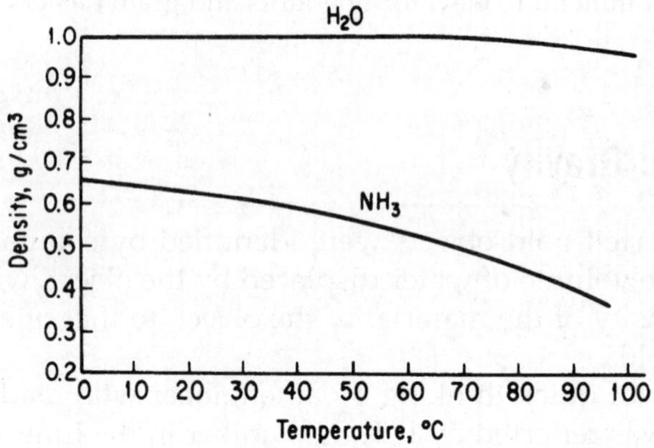

Figure 2.1 Densities of liquid H_2O and NH_3 as a function of temperature

Specific volume (we use the symbol $\hat{V}$) is the inverse of density, such as cm^3/g or ft^3/lb:

$$\hat{V} = \text{specific volume} = \frac{\text{volume}}{\text{mass}} = \frac{V}{m}$$

Because density is the ratio of mass to volume, it can be used to calculate the mass given the volume or the volume knowing the mass. For example, given that the density of *n*-propyl alcohol is 0.804 g/cm^3, what would be the volume of 90.0 g of the alcohol? The calculation is

$$\frac{90.0g}{} \left| \frac{1\ cm^3}{0.804\ g} \right. = 112\ cm^3$$

Some quantities related to density are molar density (ρ/MW) and molar volume (MW/ρ). By analogy, in a packed bed of solid particles containing void spaces, the bulk density is

$$\rho_B = \text{bulk density} = \frac{\text{total mass of solids}}{\text{total empty bed volume}}$$

Now let's turn to specific gravity. **Specific gravity** is the ratio of the density of a substance to the density of a reference material. In symbols for compound A:

$$\text{specific gravity of A} = \text{sp.gr. of A} = \frac{(g/cm^3)_A}{(g/cm^3)_{ref}} = \frac{(kg/m^3)_A}{(kg/m^3)_{ref}} = \frac{(lb/ft^3)_A}{(lb/ft^3)_{ref}}$$

The reference substance for liquids and solids normally is water. Thus, the specific gravity is the ratio of the density of the substance of interest to the density of water, namely, 1.000 g/cm^3, 1000 kg/m^3, or 62.43 lb/ft^3 at 4°C. The specific gravity of gases frequently is referred to air but may be referred to other gases.

To be precise when referring to specific gravity, the data should be accompanied by both the temperature of the substance of interest and the temperature at which the reference density is measured. Thus, for solids and liquids the notation

$$\text{sp.gr.} = 0.73\frac{20^\circ}{4^\circ} = 0.73$$

can be interpreted as follows: The specific gravity when the solution is at 20°C and the reference substance (implicitly water) is at 4°C is 0.73. In case the temperatures for which the specific gravity is stated are unknown, assume ambient temperature for the substance and 4°C for the water. Because the density of water at 4°C is very close to 1.0000 g/cm^3, in units of grams per cubic centimeter, the numerical values of the specific gravity and the density are essentially equal.

Note that the units of specific gravity as used here clarify the calculations, and that the calculation of density from the specific gravity can often be done in your head. Because densities in the American Engineering system are expressed in pounds per cubic foot, and the density of water is about 62.4 lb/ft^3, you can see that the specific gravity and density values are not numerically equal in that system. Yaws et al. (1991)[1] is a good source for values of liquid densities, and the software on the CD in the back of this book contains portions of Yaws's database.

Example 2.17 Calculation of Density Given the Specific Gravity

If penicillin has a specific gravity of 1.41, what is the density in (a) g/cm^3, (b) lb_m/ft^3, and (c) kg/m^3?

Solution

Start with the specific gravity to get the density via a reference substance. No temperatures are cited for the penicillin (P) or the reference compound (presumed to be water); hence for simplicity we assume that the penicillin is at

(Continues)

[1]C. L. Yaws, H. C. Yang, J. R. Hooper, and W. A. Cawley, "Equation for Liquid Density," *Hydrocarbon Processing*, **106**, 103–6 (January, 1991).

Example 2.17 Calculation of Density Given the Specific Gravity (*Continued*)

room temperature (22°C) and that the reference material is water at 4°C. Therefore, the reference density is 62.4 lb/ft³ or 1.00×10^3 kg/m³ (1.00 g/cm³).

$$\text{a.}\quad \frac{1.41\dfrac{\text{g P}}{\text{cm}^3}}{1.00\dfrac{\text{g H}_2\text{O}}{\text{cm}^3}}\left|\frac{1.00\dfrac{\text{g H}_2\text{O}}{\text{cm}^3}}{}\right. = 1.41\frac{\text{g P}}{\text{cm}^3}$$

$$\text{b.}\quad \frac{1.41\dfrac{\text{lb}_\text{m}\text{ P}}{\text{ft}^3}}{1.00\dfrac{\text{lb}_\text{m}\text{ H}_2\text{O}}{\text{ft}^3}}\left|\frac{62.4\dfrac{\text{lb}_\text{m}\text{H}_2\text{O}}{\text{ft}^3}}{}\right. = 88.0\frac{\text{lb}_\text{m}\text{ P}}{\text{ft}^3}$$

$$\text{c.}\quad \frac{1.41\text{ g P}}{\text{cm}^3}\left|\left(\frac{100\text{ cm}}{1\text{ m}}\right)^3\right|\frac{1\text{ kg}}{1000\text{ g}} = 1.41\times10^3\frac{\text{kg P}}{\text{m}^3}$$

You should become acquainted with the fact that in the petroleum industry the specific gravity of petroleum products is often reported in terms of a hydrometer scale called **°API**. The equations that relate the API scale to density and vice versa are

$$°\text{API} = \frac{141.5}{\text{sp. gr.}\dfrac{60°\text{F}}{60°\text{F}}} - 131.5 \qquad \text{(API gravity)} \qquad (2.4)$$

or

$$\text{sp.gr.}\frac{60°}{60°} = \frac{141.5}{°\text{API} + 131.5} \qquad (2.5)$$

The volume and therefore the density of petroleum products vary with temperature, and the petroleum industry has established 60°F as the standard temperature for specific gravity and API gravity. The CD in the back of the book contains data for petroleum products.

Example 2.18 Application of Specific Gravity to Calculate Mass and Moles

In the production of a drug having a molecular weight of 192, the exit stream from the reactor containing water and the drug flows at the rate of 10.5 L/min. The drug concentration is 41.2% (in water), and the specific gravity of the solution is 1.024. Calculate the concentration of the drug (in kilograms per liter) in the exit stream, and the flow rate of the drug in kilogram moles per minute.

Solution

Read the problem carefully because this example is more complicated than the previous examples. You have a problem with some known properties including specific gravity. The strategy for the solution is to use the specific gravity to get the density, from which you can calculate the moles per unit volume.

For the first part of the problem, you want to transform the mass fraction of 0.412 into mass per liter of the drug. Take 1.000 kg of the exit solution as a basis because the mass fraction of the drug in the product is specified in the problem statement. Figure E2.18 shows the output.

Basis: 1.000 kg solution

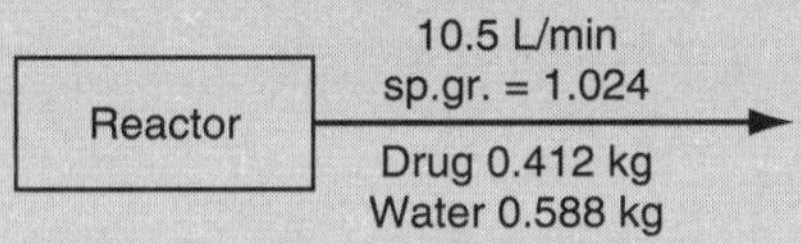

Figure E2.18

How do you get mass of drug per volume of solution (the density) from the given data, which are in terms of the fraction of the drug (0.412)? Use the given specific gravity of the solution. Calculate the density of the solution as follows:

$$\text{density of solution} = (\text{sp.gr.})(\text{density of reference})$$

$$\text{density of solution} = \frac{1.024\ \dfrac{\text{g soln}}{\text{cm}^3\ \text{soln}}}{1.000\ \dfrac{\text{g H}_2\text{O}}{\text{cm}^3\ \text{H}_2\text{O}}} \left| \frac{1.000\ \dfrac{\text{g H}_2\text{O}}{\text{cm}^3\ \text{H}_2\text{O}}}{} \right. = 1.024\ \frac{\text{g soln}}{\text{cm}^3\ \text{soln}}$$

The detail of the calculation of the density of the solution showing the units may seem excessive but is presented to make the calculation clear.

Next, convert the amount of drug in 1.000 kg of solution to mass of drug per volume of solution using the density previously calculated, recognizing that there is 0.412 kg of the drug for the basis of 1.000 kg of solution.

$$\frac{0.412\ \text{kg drug}}{1.000\ \text{kg soln}} \left| \frac{1.0254\ \text{g soln}}{1\ \text{cm}^3\ \text{soln}} \right| \frac{1\ \text{kg soln}}{10^3\ \text{g soln}} \left| \frac{1000\ \text{cm}^3\ \text{soln}}{1\ \text{L soln}} \right. = 0.422\ \text{kg drug/L soln}$$

(Continues)

Example 2.18 Application of Specific Gravity to Calculate Mass and Moles (*Continued*)

Note that a distinction is drawn between properties of the solution (e.g., g soln, L soln) and the mass of the drug to prevent confusion in the cancellation of units.

To get the flow rate, take a different basis, namely, 1 min.

Basis: 1 min = 10.5 L of solution

Convert the selected volume to mass and then to moles using the information previously calculated:

$$\frac{10.5\text{ L soln}}{1\text{ min}}\left|\frac{0.422\text{ kg drug}}{1\text{ L soln}}\right|\frac{1\text{ kg mol drug}}{192\text{ kg drug}} = 0.0231\text{ kg mol/min}$$

How might you check your answers?

Self-Assessment Test

Questions

1. Indicate whether the following statements are true or false:
   a. The inverse of the density is the specific volume.
   b. The density of a substance has the units of the mass per unit volume.
   c. The density of water is less than the density of mercury.
2. A cubic centimeter of mercury has a mass of 13.6 g. What is the density of mercury?
3. For liquid HCN, a handbook gives sp. gr. 10°C/4°C = 1.2675. What does this statement mean?
4. Indicate whether the following statements are true or false:
   a. The density and specific gravity of mercury are the same.
   b. Specific gravity is the ratio of two densities.
   c. If you are given the value of a reference density, you can determine the density of a substance of interest by multiplying by the specific gravity.
   d. The specific gravity is a dimensionless quantity.

Problems

1. The density of a material is 2 kg/m^3. What is its specific volume?
2. If you add 50 g of sugar to 500 mL of water, how do you calculate the density of the sugar solution?

3. For ethanol, a handbook gives sp. gr. 60°F = 0.79389. What is the density of ethanol at 60°F?
4. The specific gravity of steel is 7.9. What is the volume in cubic feet of a steel ingot weighing 4000 lb?
5. A solution in water contains 1.704 kg of HNO_3/kg H_2O, and the solution has a specific gravity of 1.382 at 20°C. How many kilograms of HNO_3 per cubic meter of solution at 20°C are there?

Thought Problems

1. The representative of Lloyd's Register of Shipping testified in a Houston district court in the fraud trial of a Houston businessman. The indictment alleged that the businessman stole 200,000 tons of oil from the Italian owner by delivering the oil to South Africa, and then scuttling the tanker to cover up the theft. The prosecutor asked the insurance representative whether the tanker could have been sunk with a full load of oil. What do you think?
2. A refinery tank that had contained gasoline was used for storing pentane. The tank overflowed when the level indicator said that it was only 85% full. The level indicator was a DP cell that measured weight of fluid. Can you explain what went wrong?

Discussion Question

From *Chemical and Engineering News* (October 12, 1992, p. 10):

> Two Dutch scientists have won government and industry support to explore the possibility of raising the level of the ground in coastal areas of their low-lying country by converting subsurface limestone to gypsum with waste sulfuric acid. The scheme centers on the fact that gypsum, $CaSO_4 \cdot 2H_2O$, occupies twice the volume of a corresponding amount of calcium carbonate. The project envisions drilling holes as deep as 1 km at selected sites above limestone strata for injecting the acid. The resulting gypsum should raise the surface as much as several meters. Instances of ground swelling have already occurred from sulfuric acid spillage in the Netherlands at Pernis, an industrial region near Rotterdam.

What do you think of the feasibility of this idea?

2.9 Concentration

Concentration designates the amount of a component (solute) in a mixture divided by the total of the mixture. The amount of the component of interest is usually expressed in terms of the mass or moles of the component, whereas the amount of the mixture can be expressed as the corresponding volume or mass of the mixture. Some common examples that you will encounter are

- **Mass per unit volume** such as lb_m of solute/ft^3 of solution, g of solute/L, lb_m of solute/bbl, kg of solute/m^3.
- **Moles per unit volume** such as lb mol of solute/ft^3 of solution, g mol of solute/L, g mol of solute/cm^3.
- **Mass (weight) fraction**—the ratio of the mass of a component to the total mass of the mixture, a fraction (or a percent).
- **Mole fraction**—the ratio of the moles of a component to the total moles of the mixture, a fraction (or a percent).
- **Parts per million (ppm) and parts per billion (ppb)**—a method of expressing the concentration of extremely dilute solutions; ppm is equivalent to a mass (weight) ratio for solids and liquids. It is a mole ratio for gases.
- **Parts per million by volume (ppmv) and parts per billion by volume (ppbv)**—the ratio of the volume of the solute per volume of the mixture (usually used only for gases).

Other expressions for concentration from chemistry with which you should be familiar are molality (g mol solute/kg solvent), molarity (g mol/L), and normality (equivalents/L). Note that concentrations expressed in terms of mass (mass per unit volume, mass fraction, ppm) are referred to as **mass concentrations** and those in terms of moles as **molar concentrations** (e.g., moles per unit volume, mole fraction).

Example 2.19 Nitrogen Requirements for the Growth of Cells

In normal living cells, the nitrogen requirement for the cells is provided from protein metabolism (i.e., consumption of protein in the cells). When cells are grown commercially such as in the pharmaceutical industry, $(NH_4)_2SO_4$ is usually used as the source of nitrogen. Determine the amount of $(NH_4)_2SO_4$ consumed in a fermentation medium in which the final cell concentration is 35 g/L in a 500 L volume of fermentation medium. Assume that the cells contain 9 wt % N, and that $(NH_4)_2SO_4$ is the only nitrogen source.

Solution

Basis: 500 L of solution containing 35 g/L

$$\frac{500 \text{ L}}{} \left| \frac{35 \text{ g cell}}{\text{L}} \right| \frac{0.09 \text{ g N}}{\text{g cell}} \left| \frac{\text{g mol}}{14 \text{ g N}} \right| \frac{1 \text{ g mol } (NH_4)_2SO_4}{2 \text{ g mol N}} \Bigg|$$

$$\frac{132 \text{ g } (NH_4)_2SO_4}{\text{g mol } (NH_4)_2SO_4} = 7425 \text{ g } (NH_4)_2SO_4$$

Here is a list of typical measures of concentration given in the set of guidelines by which the Environmental Protection Agency defines the extreme levels at which the five most common air pollutants could harm people if they are exposed to these levels for the stated periods of exposure:

1. **Sulfur dioxide:** 365 $\mu g/m^3$ averaged over a 24 hr period
2. **Particulate matter** (10 μm or smaller): 150 $\mu g/m^3$ averaged over a 24 hr period
3. **Carbon monoxide:** 10 mg/m^3 (9 ppm) when averaged over an 8 hr period; 40 mg/m^3 (35 ppm) when averaged over 1 hr
4. **Nitrogen dioxide:** 100 $\mu g/m^3$ averaged over 1 yr
5. **Ozone:** 0.12 ppm measured over 1 hr

Note that the gas concentrations are mostly mass/volume except for the ppm.

Example 2.20 Use of ppm

The current OSHA 8 hr limit for HCN in air is 10.0 ppm. A lethal dose of HCN in air is (from the *Merck Index*) 300 mg/kg of air at room temperature. How many milligrams of HCN per kilogram of air is 10.0 ppm? What fraction of the lethal dose is 10.0 ppm?

Solution

In this problem you have to convert ppm in a gas (a mole ratio, remember!) to a mass ratio.

Basis: 1 kg of the air-HCN mixture

We can treat the 10.0 ppm as 10.0 g mol HCN/10^6 g mol air because the amount of HCN is so small when added to the air in the denominator of the ratio.

The 10.0 ppm is

$$\frac{10.0 \text{ g mol HCN}}{10^6(\text{air} + \text{HCN})\text{g mol}} = \frac{10.0 \text{ g mol HCN}}{10^6 \text{ g mol air}}$$

(Continues)

Example 2.20 Use of ppm (*Continued*)

Next, get the MW of HCN so that it can be used to convert moles of HCN to mass of HCN; the MW = 27.03. Then

$$\frac{10.0 \text{ g mol HCN}}{10^6 \text{ g mol air}}\left|\frac{27.03 \text{ g HCN}}{1 \text{ g mol HCN}}\right|\frac{1 \text{ g mol air}}{29 \text{ g air}}\left|\frac{1000 \text{ mg HCN}}{1 \text{ g HCN}}\right|\frac{1000 \text{ g air}}{1 \text{ kg air}}$$

$$= 9.32 \text{ mg HCN/kg air}$$

$$\frac{9.32}{300} = 0.031$$

Does this answer seem reasonable? At least it is not greater than 1!

Self-Assessment Test

Questions

1. Do parts per million denote a concentration that is a mole ratio?
2. Does the concentration of a component in a mixture depend on the amount of the mixture?
3. Pick the correct answer. How many ppm are there in 1 ppb?
   a. 1000
   b. 100
   c. 1
   d. 0.1
   e. 0.01
   f. 0.001
4. Is 50 ppm five times greater than 10 ppm?
5. A mixture is reported as 15% water and 85% ethanol. Should the percentages be deemed to be by mass, mole, or volume?
6. In a recent EPA inventory of 20 greenhouse gases that are emitted in the United States, carbon dioxide constituted 5.1 million gigagrams (Gg), which was about 70% of all of the U.S. greenhouse gas emissions. Fossil fuel combustion in the electric utility sector contributed about 34% of all of the carbon dioxide emissions; and the transportation, industrial, and residential-commercial sectors accounted for 34%, 21%, and 11% of the total, respectively. Are the last four percents mole percent or mass percent?

Problems

1. How many milligrams per liter are equivalent to a 1.2% solution of a substance in water?
2. If a membrane filter yields a count of 69 fecal coliform (FC) colonies from 5 mL of well water, what should be the reported FC concentration?
3. The danger point in breathing sulfur dioxide for humans is 2620 $\mu g/m^3$. How many ppm is this value?

Discussion Questions

1. It has been suggested that an alternative to using pesticides on plants is to increase the level of natural plant toxins by breeding or gene manipulation. How feasible is this approach from the viewpoint of mutagenic and carcinogenic effects on human beings? For example, solamine and chaconine, some of the natural alkaloids in potatoes, are present at a level of 15,000 μg per 200 g potatoes. This amount is about 1/6 of the toxic level for human beings. Neither alkaloid has been tested for carcinogenicity. The man-made pesticide intake by humans is estimated to be about 150 μg/day. Only about one-half have been shown to be carcinogenic in test animals. The intake of known natural carcinogens is estimated to be 1 g per day from fruits and vegetables alone, omitting coffee (500 μg per cup), bread (185 μg per slice), and cola (2000 μg per bottle).

   Prepare a brief report ranking possible carcinogenic hazards from man-made and natural substances. List the possible exposure, source of exposure, carcinogenic dose per person, and relative potency and risk of death.
2. Certain trace elements are known to be toxic to humans but at the same time are essential for health. For example, would you knowingly drink a glass of water containing 50 ppb of arsenic? The human body normally contains 40 to 300 ppb. Wines contain 5 to 116 ppb of arsenic. Marine fish contain 2000 to 8000 ppb. Should you stop eating fish? Another compound essential to humans and animals is selenium. We know that 0.1 to 0.3 ppm of selenium is essential to the diet, but that 5 to 10 ppm is a toxic dose. The Delaney Clause of the food additives section of the Food, Drug, and Cosmetic Act has been interpreted as prohibiting the presence in food of any added carcinogen. Can selenium be added to your diet via vitamin pills? What about arsenic?

2.10 Temperature

You can hardly go through a single day without noticing or hearing what the temperature is. Believe it or not, considerable controversy exists among some scientists as to what the correct definition of temperature is (consult some of the references at the end of this chapter for further information). Some scientists prefer to say that temperature is a measure of the energy (mostly kinetic) of the molecules in a system. Other scientists prefer to say that temperature is a property of the state of thermal equilibrium of the system with respect to other systems because temperature tells us about the capability of a system to transfer energy (as heat).

In this book we use four classes of temperature measures, two based on a relative scale, degrees **Fahrenheit** (°F) and **Celsius** (°C), and two based on an absolute scale, degrees **Rankine** (°R) and **kelvin** (K). Relative scales are the ones you hear the TV or radio announcer give and are based on a specified reference temperature (32°F or 0°C) that occurs in an ice-water mixture (the freezing point of water).

Absolute temperature scales have their zero point at the lowest possible temperature that we believe can exist. As you may know, this lowest temperature is related both to the ideal gas laws and to the laws of thermodynamics. The absolute scale that is based on degree units the size of those in the Celsius scale is called the *kelvin* scale (in honor of Lord Kelvin, 1824–1907); the absolute scale that corresponds to the Fahrenheit degree units is called the *Rankine* scale (in honor of W. J. M. Rankine, 1820–1872, a Scottish engineer). Figure 2.2 illustrates the relationships between relative temperature and absolute temperature. We shall usually round off absolute zero on the Rankine scale of −459.67°F to −460°F; similarly, −273.15°C frequently will be rounded off to −273°C. Remember that 0°C and its equivalents are known as **standard conditions of temperature.**

Now we turn to a topic that causes endless difficulty in temperature conversion because of confusing semantics and notation. To start, you should recognize that the unit degree (i.e., the unit temperature difference or division) on the kelvin-Celsius scale is not the same size as that on the Rankine-Fahrenheit scale. If we let $\Delta°F$ represent the unit temperature difference on the Fahrenheit scale and $\Delta°R$ be the unit temperature difference on the Rankine scale, and $\Delta°C$ and ΔK be the analogous units in the other two scales, you probably are aware that

$$\Delta°F = \Delta°R$$

$$\Delta°C = \Delta K$$

Also, because of the temperature difference between boiling water and ice (Celsius: 100°C − 0°C = 100°C; Fahrenheit: 212°F − 32°F = 180°F), the following relationships hold:

$$\Delta°C = 1.8000\ \Delta°F \text{ and } \Delta K = 1.8000\ \Delta°F$$

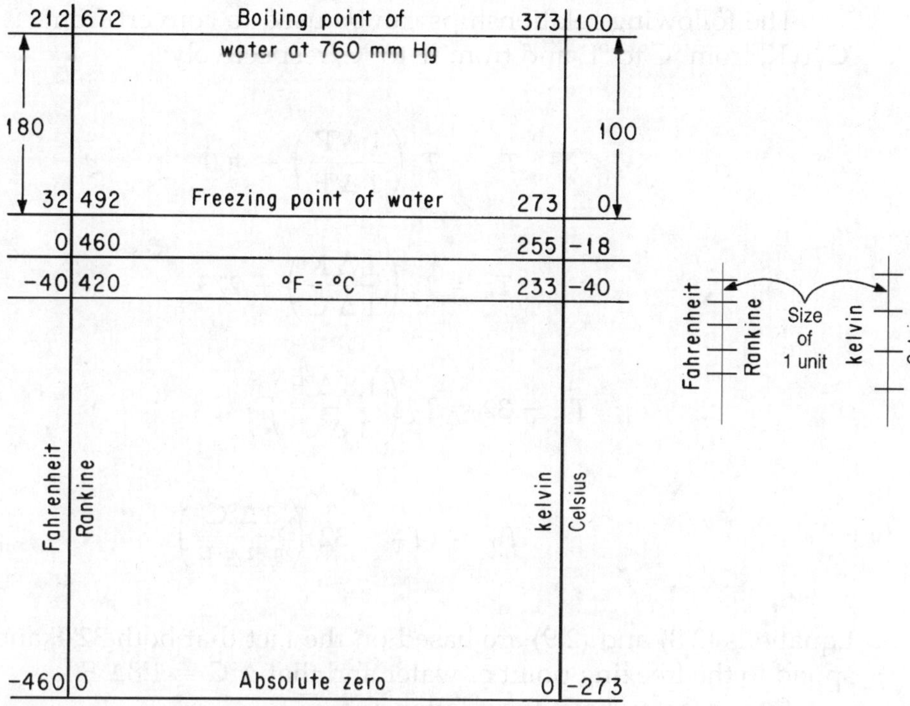

Figure 2.2 Temperature scales

If you keep in mind that the unit degree Δ°C = Δ K is larger than the unit degree Δ°F = Δ°R, you can avoid much confusion.

Now to the final point. When we cite the temperature of a substance, we are **stating the cumulative number of units of the temperature scale that occur** (an enumeration of Δ*T*s) **measured from the reference point** (i.e., absolute zero for K and °R, and the freezing point of water for °C and °F). Reexamine Figure 2.2.

Unfortunately, the symbols Δ°C, Δ°F, Δ K, and Δ°R are not in standard usage; the Δ symbol is normally omitted. A few books try to maintain the difference between degrees of temperature (°C, °F, etc.) and the unit degree by assigning to the unit degree a symbol such as C°, F°, and so on. But most journals and texts use the same symbol for the two different quantities, one the unit temperature difference and the other the temperature itself. Consequently, **the proper meaning of the symbols °C, °F, K, and °R as either the temperature or the unit temperature difference must be interpreted from their usage**. What this statement means is to use some common sense.

The following relationships can be used to convert from °F to °R, from °C to K, from °C to °F, and from °F to °C, respectively:

$$T_{°R} = T_{°F}\left(\frac{1\,\Delta°R}{1\,\Delta°F}\right) + 460 \tag{2.6}$$

$$T_{K} = T_{°C}\left(\frac{1\,\Delta\,K}{1\,\Delta°C}\right) + 273 \tag{2.7}$$

$$T_{°F} - 32 = T_{°C}\left(\frac{1.8\,\Delta°F}{1\,\Delta°C}\right) \tag{2.8}$$

$$T_{°C} = (T_{°F} - 32)\left(\frac{1\Delta°C}{1.8\Delta°F}\right) \tag{2.9}$$

Equations (2.8) and (2.9) are based on the fact that both 32°F and 0°C correspond to the freezing point of water and that $\Delta°C = 1.8\Delta°F$.

Suppose you have the relation

$$T_{°F} = a + bT_{°C}$$

What are the units of *a* and *b*? Certainly from what you have learned so far, the units of *a* must be °F for consistency. Are the units of *b* equal to the units in the ratio $T_{°F}/T_{°C}$? No, because the reference points for °C and °F differ; $T_{°F}/T_{°C}$ is not a valid conversion factor. The correct units for the conversion factor are $\Delta°F/\Delta°C$, a factor that converts the size of a division on each of the respective temperature scales:

$$T_{°F} = a°F + b\left(\frac{1\,\Delta°F}{1\,\Delta°C}\right)T_{°C}$$

Consequently, unfortunately, the units for *b* are usually ignored; just the value of *b* is given.

When you reach Chapter 9, you will note that the heat capacity in the SI system has the units of J/(g mol)(K). Does the K in the heat capacity designate the temperature in degrees K or the unit interval Δ K? Now look at some temperature conversion examples.

Example 2.21 Temperature Conversion

Convert 100°C to (a) K, (b) °F, and (c) °R.

Solution

a. $(100 + 273)°C \frac{1\ \Delta K}{1\ \Delta °C} = 373\ K$

or, with suppression of the Δ symbol,

$$(100 + 273)°C \frac{1\ K}{1°C} = 373\ K$$

b. $(100°C) \frac{1.8\ \Delta °F}{1\ \Delta °C} + 32°F = 212°F$

c. $(212 + 460)°F \frac{1\ \Delta °R}{1\ \Delta °F} = 672°R$

or $(373\ K) \frac{1.8\ \Delta °R}{1\ \Delta K} = 672°R$

Example 2.22 Temperature Conversion

The heat capacity of sulfuric acid in a handbook has the units J/[(g mol) (°C)] and is given by the relation

$$\text{heat capacity} = 139.1 + 1.56 \times 10^{-1} T$$

where T is expressed in degrees Celsius. Modify the formula so that the resulting expression yields the heat capacity with the associated units of Btu/[(lb mol) (°R)] with T in degrees Rankine.

Solution

The symbol °C in the denominator of the heat capacity stands for the unit temperature difference, Δ°C, not the temperature, whereas the units of T in the equation are in °C. First you have to substitute the proper equation in the formula to convert T in °C to T in °R, and then multiply by conversion factors to convert the units in the right-hand side of the equation to Btu/(lb mol) (°R) as requested.

(Continues)

Example 2.22 Temperature Conversion (*Continued*)

heat capacity =

$$\left\{139.1 + 1.56 \times 10^{-1}\left[\overbrace{T_{°R} - 460 - 32)\frac{1}{1.8}}^{T_{°C}}\right]\right\}$$

$$\times \underbrace{\frac{1}{(\text{g mol})(°\text{C})}\left|\frac{1 \text{ Btu}}{1055 \text{ J}}\right|\frac{454 \text{ g mol}}{1 \text{ lb mol}}\left|\frac{1°\text{C}}{1.8°\text{R}}\right.}_{\text{conversion factors}} = 23.06 + 2.07 \times 10^{-2} T_{°R}$$

Note the suppression of the Δ symbol in the original units of the heat capacity and in the conversion between Δ°C and Δ°R.

Self-Assessment Test

Questions

1. What are the reference points of the (a) Celsius and (b) Fahrenheit scales?
2. How do you convert a *temperature difference*, Δ, from Fahrenheit to Celsius?
3. Is the unit temperature difference Δ°C a larger interval than Δ°F? Does 10°C correspond to a higher temperature than 10°F?

Problems

1. Complete the following table with the proper equivalent temperatures:

°C	°F	K	°R
−40.0	___	___	___
___	77.0	___	___
___	___	698	___
___	___	___	69.8

2. In Appendix G, the heat capacity of sulfur is $C_p = 15.2 + 2.68T$, where C_p is in J/(g mol)(K) and T is in kelvin. Convert so that C_p is given in cal/(g mol)(°F) with T in degrees Fahrenheit.

Thought Problems

1. While reading a report on the space shuttle, you find the statement that "the maximum temperature on reentry is 1482.2°C." How many significant figures do you think are represented by this temperature?
2. A vacuum tower used to process residual oil experienced severe coking (carbon formation) on the tower internals when it rained. Coking occurs because the temperature of the fluid gets to be too high. The temperature of the entering residual was controlled by a temperature recorder-controller (TRC) connected to a thermocouple inserted into a thermowell in the pipeline bringing the residual into the column. The TRC was showing 700°F whereas the interior of the column was at 740°F (too hot). What might be the problem?

Discussion Question

In the kelvin or Rankine (absolute temperature) scales, the relation used for temperature is $T = n\Delta T$ where ΔT is the value of the unit temperature and n is the number of units enumerated. When $n = 0, T = 0$. Suppose that temperature is defined by the relation for $\ln(T) = n\Delta T$. Does $T = 0$ occur? What does $n = 0$ mean? Does the equivalent of absolute zero kelvin exist?

2.11 Pressure and Hydrostatic Head

In Florentine Italy in the seventeenth century, well diggers observed that when they used suction pumps, water would not rise more than about 10 m. In 1642 they came to the famous Galileo for help, but he did not want to be bothered. As an alternative, they sought the help of Torricelli. He learned from experiments that water was not being pulled up by the vacuum but rather was being pushed up by the local air pressure. Therefore, the maximum height that water could be raised from a well depended on the atmospheric pressure.

Pressure is defined as "the normal (perpendicular) force per unit area." In the SI system the force is expressed in newtons and the area in square meters; then the pressure is N/m^2 or pascal (Pa). (The value of a pascal is so small that the kilopascal (kPa) is a more convenient unit of pressure.) In the AE system the force is the pounds force and the area used is square inches.

What are other units for pressure? Look at Table 2.6 for some of the most common ones. Note that each pressure unit is expressed as the equivalent of 1 standard atmosphere (atm).

Table 2.6 Convenient Conversion Factors for Pressure

Pressure Units	Conversion Factor
bar	1.013 bar = 1 atm
kPa	101.3 kPa = 1 atm
Torr	760 Torr = 1 atm
mm Hg	760 mm Hg = 1 atm
in. Hg	29.92 in. Hg = 1 atm
ft H_2O	33.94 ft H_2O = 1 atm
in. H_2O	407 in. H_2O = 1 atm
psi	14.69 psi = 1 atm

Keep in mind that the pounds referred to in psi are pounds force, not pounds mass, so the cartoon caption "It says inflate to 12 pounds. How can I throw a 12-pound football?" causes a chuckle.

Examine Figure 2.3. Pressure is exerted on the top of the mercury in the cylinder by the atmosphere. The pressure at the bottom of the column of mercury is equal to the pressure exerted by the mercury plus that of the atmosphere on the mercury.

The pressure at the bottom of the **static** (nonmoving) column of mercury (also known as the hydrostatic pressure) exerted on the sealing plate is

$$p = \frac{F}{A} = \frac{mg}{A} + p_0 = \frac{mgh}{Ah} + p_0 = \frac{mgh}{V} + p_0 = \rho gh + p_0 \tag{2.10}$$

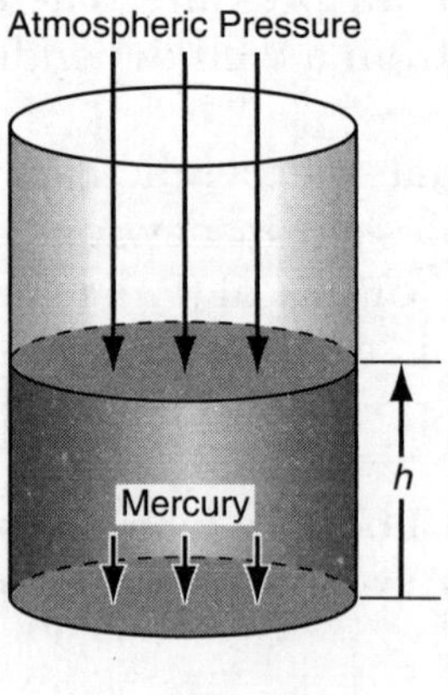

Figure 2.3 Pressure is the normal force per unit area. Arrows show the force exerted on the respective areas.

where the first term after the pressure p is the definition of pressure, the second term is the combination of atmospheric pressure and the pressure change due to the column of liquid, the third shows how h can be added to the numerator and denominator to get the volume in the denominator, in the fourth term volume is substituted for area times height, and in the fifth term the density is substituted for mass divided by volume. The notation used is as follows:

p = pressure at the bottom of the column of fluid
F = force
A = area
ρ = density of fluid
g = acceleration of gravity
h = height of the fluid column
p_0 = pressure at the top of the column of fluid

You can calculate the force exerted at the bottom of a static fluid by applying Equation (2.10). For example, suppose that the cylinder of fluid in Figure 2.3 is a column of mercury with a cross-sectional area of 1 cm^2 and is 50 cm high. From Table D.1 in Appendix D you can find that the specific gravity of mercury at 20°C, and hence the density of the Hg, is 13.55 g/cm^3. Thus, the force exerted by the mercury alone on the 1 cm^2 section of the bottom plate by the column of mercury is

$$F = mg = \rho V g$$

$$= \frac{13.55\ \text{g}}{\text{cm}^3}\left|\frac{980\ \text{cm}}{\text{s}^2}\right|\frac{50\ \text{cm}}{}\left|\frac{1\ \text{cm}^2}{}\right|\frac{1\ \text{kg}}{1000\ \text{g}}\left|\frac{1\ \text{m}}{100\ \text{cm}}\right|\frac{1\ (\text{N})(\text{s}^2)}{1\ (\text{kg})(\text{m})}$$

$$= 6.64\ \text{N}$$

The pressure on the section of the plate covered by the mercury is the force per unit area of the mercury *plus* the pressure (p_0) of the atmosphere that is over the mercury:

$$p = \frac{F}{A} + p_0 = \frac{6.64\ \text{N}}{1\ \text{cm}^2}\left|\left(\frac{100\ \text{cm}}{1\ \text{m}}\right)^2\right|\frac{(1\ \text{m}^2)(1\ \text{Pa})}{(1\ \text{N})}\left|\frac{1\ \text{kPa}}{1000\ \text{Pa}}\right| + p_0$$

$$= 66.4\ \text{kPa} + p_0$$

If we had started with units in the American Engineering system, the pressure would be computed as follows [the density of mercury is $(13.55)(62.4)\text{lb}_m/\text{ft}^3 = 845.5\ \text{lb}_m/\text{ft}^3$]:

$$p = \rho g V + p_0 = \frac{845.5\ \text{lb}_m}{1\ \text{ft}^3}\bigg|\frac{32.2\ \text{ft}}{\text{s}^2}\bigg|\frac{50\ \text{cm}}{}\bigg|\frac{1\ \text{in.}}{2.54\ \text{cm}}\bigg|\frac{1\ \text{ft}}{12\ \text{in.}}\bigg|\frac{1\ (\text{s})^2(\text{lb}_f)}{32.174\ (\text{ft})(\text{lb}_m)} + p_0$$

$$= 1388\ \frac{\text{lb}_f}{\text{ft}^2} + p_0$$

Sometimes in engineering practice, a liquid column is referred to as *head of liquid*, the head being the height of the column of liquid. Thus, the pressure of the column of mercury could be expressed simply as 50 cm Hg, and the pressure on the sealing plate at the bottom of the column would be 50 cm Hg + p_0 (in centimeters of Hg).

Pressure, like temperature, can be expressed in either absolute (**psia**) or relative scales. Rather than using the word *relative*, the relative pressure is usually called **gauge pressure (psig)**. The atmospheric pressure is nothing more than the barometric pressure. The relationship between gauge and absolute pressure is given by the following expression:

$$p_{\text{absolute}} = p_{\text{gauge}} + p_{\text{atmospheric}}$$

Another term with which you should become familiar is **vacuum**. When you measure the pressure in "inches of mercury vacuum," you are reversing the direction of measurement from the reference pressure, the atmospheric pressure, and toward zero absolute pressure, that is,

$$p_{\text{vacuum}} = p_{\text{atmospheric}} - p_{\text{absolute}}$$

As the vacuum value increases, the value of the absolute pressure being measured decreases. What is the maximum value of a vacuum measurement? A pressure that is only slightly below atmospheric pressure may sometimes be expressed as a "draft" in inches of water, as, for example, in the air supply to a furnace or a water cooling tower.

Here are a news article and figure (Figure 2.4) [R. E. Sanders, *Chemical Engineering Progress* (September 1993), p. 54] to remind you about vacuum.

Don't Become Another Victim of Vacuum

Tanks are fragile. An egg can withstand more pressure than a tank. How was a vacuum created inside the vessel? As water was drained from the

Figure 2.4 Close-up of failed stripper column

column, the vent to let in air was plugged up, and the resulting pressure difference between inside and out caused the stripper to fail. (Reproduced through the courtesy of Roy E. Sanders)

Figure 2.5 illustrates the relationships among the pressure concepts. Note that the vertical scale is exaggerated for illustrative purposes. The dashed line illustrates the atmospheric (barometric) pressure, which changes from time to time. The solid horizontal line is the standard atmosphere. Point (1) in the figure denotes a pressure of 19.3 psia, that is, an absolute pressure referred to zero absolute pressure, or 4.6 psig (i.e., 19.3 psia − 14.7 psia) referred to the barometric pressure; (2) is zero pressure; any point on the heavy line such as (3) is a point corresponding to the standard pressure and is below atmospheric pressure; (4) illustrates a negative relative pressure, that is, a pressure less than atmospheric

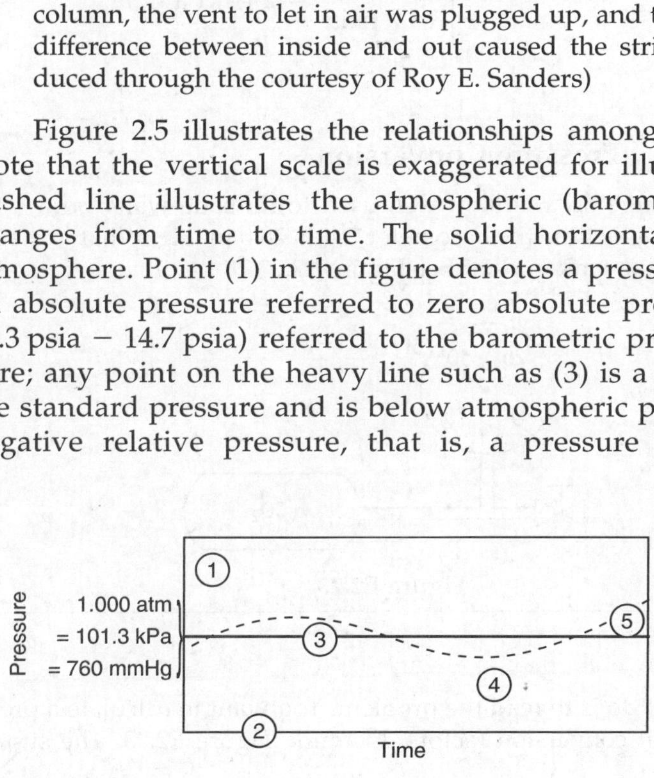

Figure 2.5 Pressure terminology

pressure. This latter type of measurement is described below as a vacuum measurement. Point (5) indicates negative relative pressure that is above the standard atmospheric pressure.

You definitely must not confuse the standard atmosphere with atmospheric pressure. The **standard atmosphere** is defined as the pressure (in a standard gravitational field) equivalent to 1 atm or 760 mm Hg at 0°C or other equivalent value and is fixed, whereas atmospheric pressure is variable and must be obtained from a barometric measurement or equivalent each time you need it.

You can easily convert from one pressure unit to another by using the relation between a pair of different pressure units as they respectively relate to the standard atmospheres so as to form a conversion factor. For example, let us convert 35 psia to inches of mercury and also to kilopascals by using the values in Table 2.6:

$$\frac{35 \text{ psia}}{} \left| \frac{29.92 \text{ in. Hg}}{14.7 \text{ psia}} \right. = 71.24 \text{ in Hg}$$

$$\frac{35 \text{ psia}}{} \left| \frac{101.3 \text{ kPa}}{14.7 \text{ psia}} \right. = 241 \text{ kPa}$$

Example 2.23 Pressure Conversion

The pressure gauge on a tank of CO_2 used to fill soda-water bottles reads 51.0 psi. At the same time the barometer reads 28.0 in. Hg. What is the absolute pressure in the tank in psia? See Figure E2.23.

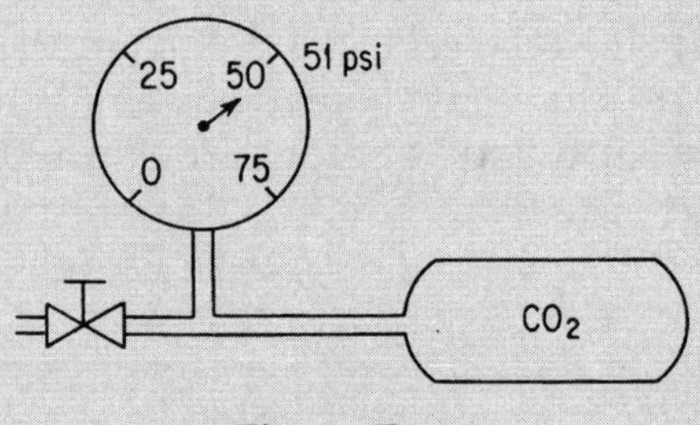

Figure E2.23

Solution

The first thing to do is to read the problem. You want to calculate a pressure using convenient conversion factors. Examine Figure E2.23. The system is the tank plus the line to the gauge. All of the necessary data are known except whether the pressure gauge reads absolute or gauge pressure. What do you think? Is it more probable that the pressure gauge is reading psig, not psia? Yes. Let's at least assume so for this problem. Because the absolute pressure is the sum of the gauge pressure and the atmospheric (barometric) pressure expressed in the same units, you have to make the units the same in each term before adding or subtracting. Let's use psia. Start the calculations by changing the atmospheric pressure to psia:

$$\frac{28.0 \text{ in. Hg}}{} \left| \frac{14.7 \text{ psia}}{29.92 \text{ in. Hg}} \right. = 13.76 \text{ psia}$$

The absolute pressure in the tank is 51.0 + 13.76 = 64.8 psia.

Example 2.23 identifies an important issue. **Whether relative or absolute pressure is measured in a pressure-measuring device depends on the nature of the instrument used to make the measurements.** For example, an open-end **manometer** (Figure 2.6a) would measure a gauge pressure because the reference pressure is the pressure of the atmosphere at the open end of the manometer. On the other hand, closing off the open end of the manometer (Figure 2.6b), and creating a vacuum in that end, results in a

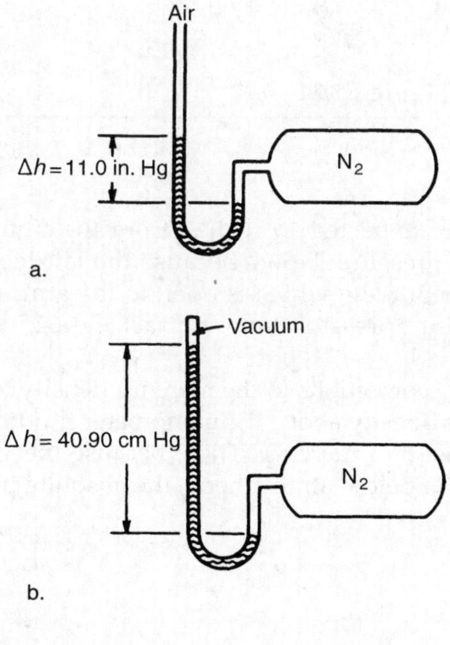

Figure 2.6 (a) Open-end manometer showing a pressure above atmospheric pressure in the tank; (b) manometer measuring absolute pressure in the tank

measurement against a complete vacuum, hence is reported as an **absolute pressure**. In Figure 2.6b, if the pressure of the nitrogen in the tank is atmospheric pressure, what will the manometer read approximately?

Sometimes you have to use common sense as to the units of the pressure being measured if just the abbreviation psi is used without the *a* or the *g* being appended to the abbreviation for pressure. For any pressure unit, be certain to carefully specify whether it is gauge or absolute, although you rarely find people doing so. For example, state "300 kPa absolute" or "12 cm Hg gauge" rather than just 300 kPa or 12 cm Hg, as the latter two without the absolute or gauge notation can on occasion cause confusion.

Example 2.24 Vacuum Pressure Reading

Small animals such as mice can live at reduced air pressures down to 20 kPa absolute (although not comfortably). In a test, a mercury manometer attached to a tank as shown in Figure E2.24 reads 64.5 cm Hg and the barometer reads 100 kPa. Will the mice survive?

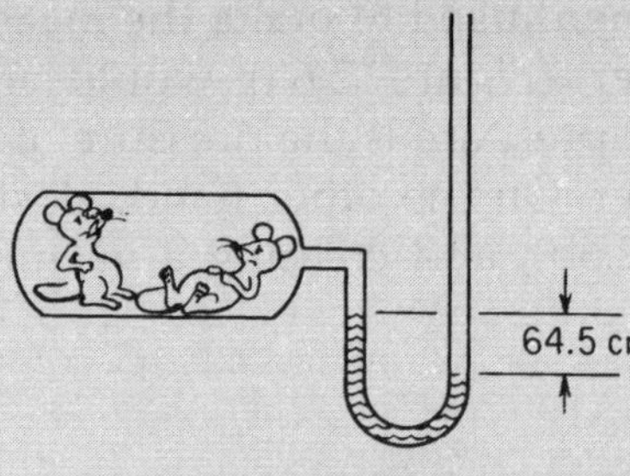

Figure E2.24

Solution

First read the problem. You are expected to realize from the figure that the tank is below atmospheric pressure. How? Because the left leg of the manometer is higher than the right leg, which is open to the atmosphere. Consequently, to get the absolute pressure you subtract the 64.5 cm Hg from the barometer reading.

We ignore any temperature corrections to the mercury density for temperature and also ignore the gas density above the manometer fluid because it is so much less than the density of mercury. Then, because the vacuum reading on the tank is 64.5 cm Hg below atmospheric, the absolute pressure in the tank is

$$p_{\text{absolute}} = p_{\text{atmospheric}} - p_{\text{vacuum}} = 100\text{ kPa} - \frac{64.5\text{ cm Hg}}{} \bigg| \frac{101.3\text{ kPa}}{76.0\text{ cm Hg}}$$

$$= 100 - 86 = 14\text{ kPa absolute}$$

Have you noted in the discussion and examples so far that we have said we can ignore the gas in the manometer tube above the measurement fluid? Is this OK? Let's see. Examine Figure 2.7, which illustrates the section of a U-tube involving three fluids.

When the columns of fluids are static (it may take some time for the dynamic fluctuations to die out!), the relationship among ρ_1, ρ_2, and the heights of the various columns of fluid is as follows. Pick a reference level for measuring pressure, such as the bottom of d_1. (If you pick the very bottom of the U-tube, instead of d_1, the left-hand and right-hand distances up to d_1 are equal, and consequently the pressures exerted by the right-hand and left-hand legs of the U-tube will cancel out in Equation (2.11)—try it and see.)

$$p_1 + \rho_1 d_1 g = p_2 + \rho_2 g d_2 + \rho_3 g d_3 \tag{2.11}$$

If fluids 1 and 3 are gases, and fluid 2 is mercury, because the density of a gas is so much less than that of mercury, you can ignore the terms involving the gases in Equation (2.11) for practical applications.

However, if fluids 1 and 3 are liquids and fluid 2 is a nonmiscible fluid, the density of fluids 1 and 3 cannot be neglected in Equation (2.11). As the densities of fluids 1 and 3 approach that of fluid 2, what happens to

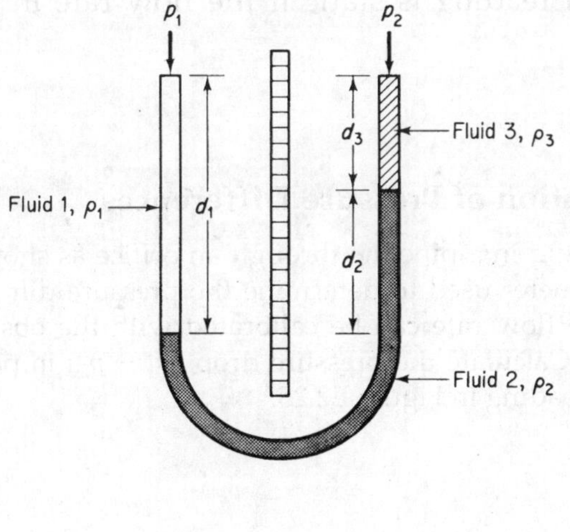

Figure 2.7 Manometer with three fluids

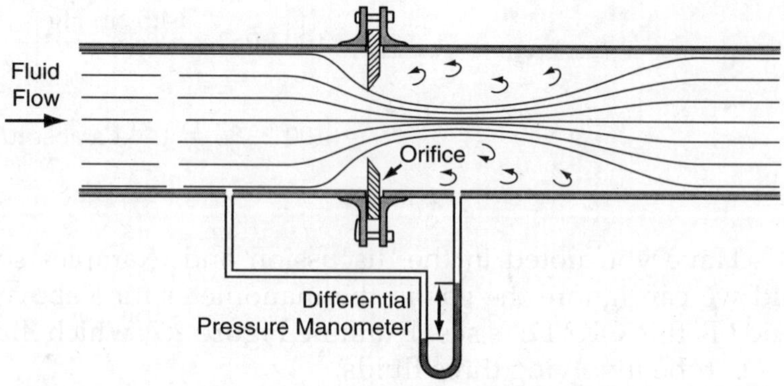

Figure 2.8 Concentric orifice used to restrict flow and measure the fluid flow rate with the aid of a manometer

the value of fluid level 2 (d_2) in Figure 2.7 for a given pressure difference $p_1 - p_2$?

Can you show for the case in which $\rho_1 = \rho_3 = \rho$ that the manometer expression reduces to the well-known differential manometer equation?

$$p_1 - p_2 = (\rho_2 - \rho)gd_2 \tag{2.12}$$

As you may know, a flowing fluid experiences a **pressure drop** when it passes through a restriction such as the orifice in the pipe as shown in Figure 2.8.

The pressure difference can be measured with any instrument connected to the pressure taps, such as a manometer as illustrated in Figure 2.8, which can be used to measure the fluid flow rate in the pipe. Note that the manometer fluid reading is static if the flow rate in the pipe is constant.

Example 2.25 Calculation of Pressure Differences

In measuring the flow of fluid in a pipeline through an orifice as shown in Figure E2.25, with a manometer used to determine the pressure difference across the orifice plate, the flow rate can be calibrated with the observed pressure drop (difference). Calculate the pressure drop ($p_1 - p_2$) in pascals for the steady manometer reading in Figure E2.25.

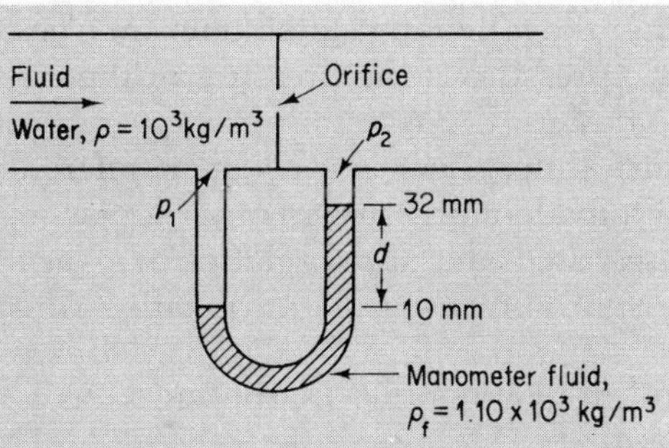

Figure E2.25

Solution

In this problem you cannot ignore the density of the water above the manometer fluid. Thus, we apply Equation (2.11), or simpler (2.12), because the densities of the fluids above the manometer fluid are the same in both legs of the manometer. The basis for solving the problem is the information given in Figure E2.25. Apply Equation (2.12):

$$p_1 - p_2 = (\rho_f - \rho)gd$$

$$= \frac{(1.10 - 1.00)10^3 \text{ kg}}{\text{m}^3}\left|\frac{9.807 \text{ m}}{\text{s}^2}\right|\frac{(22)(10^{-3}) \text{ m}}{}\left|\frac{1 \text{ (N)(s}^2)}{\text{(kg)(m)}}\right|\frac{1\text{(Pa)(m}^2)}{1 \text{ (N)}}$$

$$= 21.6 \text{ Pa}$$

Check your answer. How much error would occur if you ignored the density of the flowing fluid?

Self-Assessment Test

Questions

1. Figure SAT2.11Q1 shows two coffeepots sitting on a level table. Both are cylindrical and have the same cross-sectional area. Which coffeepot will hold more coffee?

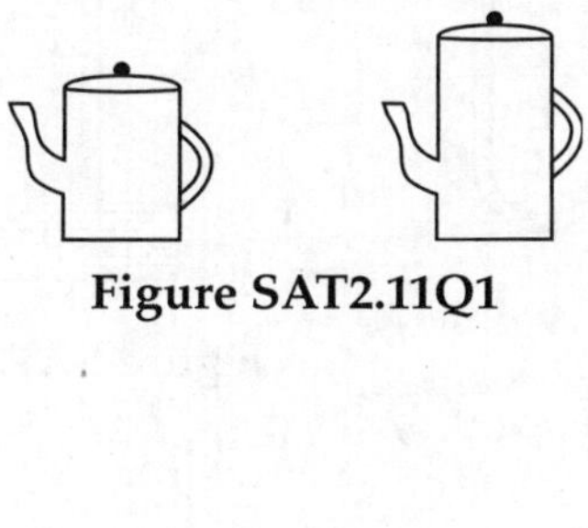

Figure SAT2.11Q1

2. Indicate whether the following statements are true or false:
   a. Atmospheric pressure is the pressure of the air surrounding us and changes from day to day.
   b. The standard atmosphere is a constant reference atmosphere equal to 1.000 atm or the equivalent pressure in other units.
   c. Absolute pressure is measured relative to a vacuum.
   d. Gauge pressure is measured in a positive direction relative to atmospheric pressure.
   e. Vacuum and draft pressures are measured in a positive direction from atmospheric pressure.
   f. You can convert from one type of pressure measurement to another using the standard atmosphere.
   g. A manometer measures the pressure difference in terms of the height of a fluid(s) in the manometer tube.
   h. Air flows in a pipeline, and the manometer containing Hg that is set up as illustrated in Figure 2.7 shows a differential pressure of 14.2 mm Hg. You can ignore the effect of the density of air on the height of the columns of mercury.
3. What is the equation to convert vacuum pressure to absolute pressure?
4. Can a pressure have a value lower than that of a complete vacuum?

Problems

1. Convert a pressure of 800 mm Hg to the following units:
   a. psia
   b. kPa
   c. atm
   d. ft H_2O
2. Your textbook cites five types of pressures: atmospheric, barometric, gauge, absolute, and vacuum pressure.
   a. What kind of pressure is measured in Figure SAT2.11P2A?
   b. In Figure SAT2.11P2B?
   c. What would be the reading in Figure SAT 2.11P2C assuming that the pressure and temperature inside and outside the helium tank are the same as in parts a and b?

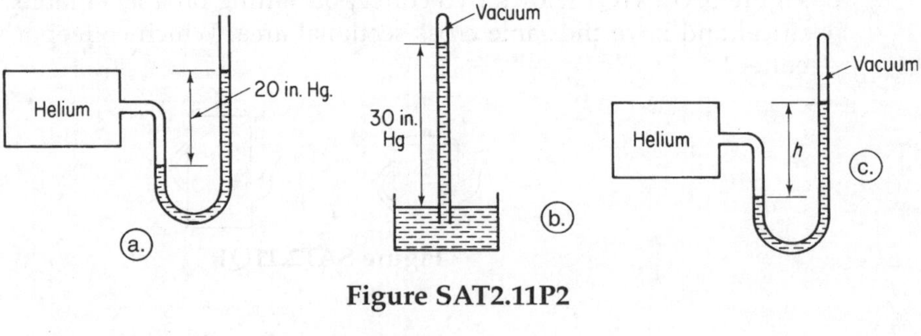

Figure SAT2.11P2

3. An evaporator shows a reading of 40 kPa vacuum. What is the absolute pressure in the evaporator in kilopascals?
4. A U-tube manometer filled with mercury is connected between two points in a pipeline. If the manometer reading is 26 mm Hg, calculate the pressure difference in kilopascals between the points when (a) water is flowing through the pipeline, and (b) air at atmospheric pressure and 20°C with a density of 1.20 kg/m^3 is flowing in the pipeline.

Thought Problems

1. When you lie motionless on a bed, the bed supports you with a force that exactly matches your weight, and when you do the same on the floor, the floor pushes up against you with the same amount of force. Why does the bed feel softer than the floor?
2. If you push the tube down into the water as shown in Figure SAT2.11TP2 so that bend A is below the water level, the tube will become a siphon and drain the water out of the glass. Why does that happen?

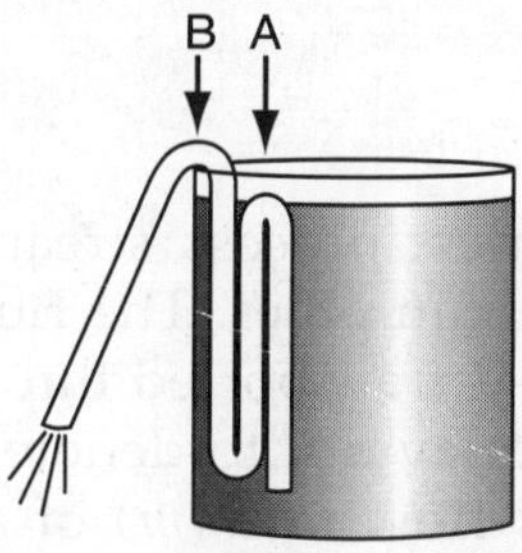

Figure SAT2.11TP2

3. A magic trick is to fill a glass with water, place a piece of paper over the top of the glass to cover the glass completely, and hold the paper in place as the glass is inverted 180°. On the release of your support of the paper, no water runs out! Many books state that the glass should be completely filled with water with no air bubbles present. Then the outside air pressure is said to oppose the weight of the water in the inverted glass. However, the experiment works just as well with a half-filled glass. The trick does not work if a glass plate is substituted for the piece of paper. Can you explain why?
4. A large storage tank was half full of a flammable liquid quite soluble in water. The tank needed maintenance on the roof. Since welding was involved, the foreman attached to the vent pipe on the top of the tank (in which there was a flame arrestor) a flexible hose and inserted the end of the hose into the bottom of a drum of water sitting on the ground to pick up any exhaust vapors. When the tank was emptied, the water rose up in the hose, and the tank walls collapsed inward. What went wrong in this incident?

Discussion Question

Form a study group to investigate the possibility of raising the *Titanic*. The mystique of the sinking and attempts to raise the vessel have crept into literature (*A Night to Remember*), the movies (*Raising the Titanic*), and magazines such as the *National Geographic*. Carry out a literature search to get the basic facts (4000 m deep, 4.86×10^8 N original weight, the density of seawater, and so on).

Prepare a report giving

a. An executive summary, including an estimate of feasibility
b. The proposed method(s) of raising the ship
c. A list of steps to execute to raise the ship
d. A list of the equipment needed (include costs if possible)
e. A time schedule for the entire project (including obtaining the equipment and personnel)
f. A list of (1) all assumption made and (2) problems that might be encountered for which answers are not known
g. As an appendix, all calculations made and references used

2.12 Flow Rate

In the process industries, process streams are normally delivered to or removed from a process in pipes. The **flow rate** of a process stream is the rate at which material is transported through a carrying pipe. In this book we usually use an overlay dot to denote a rate except for the volumetric flow rate F. The **mass flow rate** ($\dot{m}$) of a process stream is the mass (m) transported through a pipe per unit time (t):

$$\dot{m} = \frac{m}{t}$$

The **molar flow rate** ($\dot{n}$) of a process stream is the moles (n) of a substance transported through a pipe per unit time:

$$\dot{n} = \frac{n}{t}$$

The **volumetric flow rate** (F) of a process stream is the volume (V) transported through a line per unit time:

$$F = \frac{V}{t}$$

Self-Assessment Test

Problems

1. Forty gallons per minute of a hydrocarbon fuel having a specific gravity of 0.91 flow into a tank truck with a load limit of 40,000 lb of fuel. How long will it take to fill the tank in the truck to its load limit?
2. Pure chlorine enters a process. By measurement it is found that 2.4 kg of chlorine pass into the process every 3.1 min. Calculate the molar flow rate of the chlorine in kilogram moles per hour.

Looking Back

In this chapter we have reviewed the essential background that you need to become skilled in converting units, applying the concept of dimensional consistency in your work, and reporting numerical values with an appropriate number of significant digits. In addition, we introduced the concept of the selection of a basis to aid in the solution of problems. These concepts are applied to relations involving moles, density, specific gravity, and measures of concentration. Finally, we described certain features of temperature and pressure that will be used in subsequent chapters.

Glossary

Absolute pressure Pressure relative to a complete vacuum.

AE American Engineering system of units.

API Scale used to report specific gravity of petroleum compounds.

Atomic weight Mass of an atom based on ^{12}C being exactly 12.

Average molecular weight A pseudo molecular weight computed by dividing the mass in a mixture or solution by the number of moles in the mixture or solution.

Barometric pressure Pressure measured by a barometer; the same as absolute pressure.

Basis The reference material or time selected to use in making the calculations in a problem.

Celsius (°C) Relative temperature scale based on zero degrees being the freezing point of an air-water mixture.

Changing bases Shifting the basis in a problem from one value to another for convenience in the calculations.

Compound A species composed of more than one atom.

Concentration The quantity of a solute per unit volume, or per a specified amount of a component in a mixture.

Conversion of units Change of units from one set to another.

Density Mass per unit volume of a compound; molar density is the number of moles divided by the total number of moles present.

Derived units Units developed in terms of the fundamental units.

Dimensional consistency Each term in an equation must have the same set of net dimensions.

Dimensionless group A collection of variables or parameters that has no net dimensions (units).

Dimensions The basic concepts of measurement such as length or time.

Fahrenheit (°F) Relative temperature scale with 32 degrees being the freezing point of an air-water mixture.

Flow rate Amount of mass, moles, or volume of a material passing through a pipe or system per unit time.

Force A derived unit for the mass times the acceleration.

Fundamental units Units that can be measured independently.

Gauge pressure Pressure measured above atmospheric pressure.

Gram mole 6.022×10^{23} molecules.

Kelvin (K) Absolute temperature scale based on zero degrees being the lowest possible temperature we believe can exist.

Mass A basic dimension for the amount of material.

Mass concentration A unit of concentration based on the mass of a component.

Molar concentration A unit of concentration based on the moles of a component.

Mole Amount of a substance containing 6.022×10^{23} entities.

Molecular weight Mass of a compound per mole.

Mole fraction Moles of a particular compound in a mixture or solution divided by the total number of moles present.

Nondimensional group See **Dimensionless group**.

Parts per million (ppm) Concentration expressed in terms of parts of the component of interest per million parts of the mixture.

Pound force The unit of force in the AE system.

Pound mass The unit of mass in the AE system.

Pound mole $6.022 \times 10^{23} \times 453.6$ molecules.

Pressure The normal force per unit area that a fluid exerts on a surface.

Rankine (°R) Absolute temperature scale related to degrees Fahrenheit based on zero degrees being the lowest possible temperature we believe can exist.

Relative error Fraction or percent error for a number.

SI Le Système Internationale d'Unités (SI system of units).

Solution Homogeneous mixture of two or more compounds.

Specific gravity Ratio of the density of a compound to the density of a reference compound.

Specific volume Inverse of the density (volume per unit mass).

Standard atmosphere The pressure in a standard gravitational field equivalent to 760 (exactly) mm Hg.

Units Method of expressing a dimension such as feet or hours.

Vacuum A pressure less than atmospheric (but reported as a positive number).

Weight A force opposite to the force required to support a mass (usually in a gravitational field).

Weight fraction The historical term for mass fraction.

Supplementary References

Bhatt, B. I., and S. M. Vora. *Stoichiometry (SI Units)*, Tata McGraw-Hill, New Delhi (1998).

Fogler, H. S. *Elements of Reaction Engineering*, Prentice Hall, Upper Saddle River, NJ (1999).

Gilabert, M. A., and J. Pellicer. "Celsius or Kelvin: Something to Get Steamed Up About," *Phys. Educ.* **31**, 52–55 (1996).

Henley, E. J., and H. Bieber. *Chemical Engineering Calculations*, McGraw-Hill, New York (1959).

Horvath, A. L. *Conversion Tables in Science and Engineering*, Elsevier, New York (1986).

Luyben, W. L., and L. A. Wentzel. *Chemical Process Analysis: Mass and Energy Balances*, Prentice Hall, Englewood Cliffs, NJ (1988).

Michalski, L. (ed.). *Temperature Measurement*, John Wiley & Sons, New York (2001).

National Institute of Standards. *The International System of Units (SI)*, NIST Special Publ. No. 330, U.S. Department of Commerce, Gaithersburg MD, 20899 (1991).

Pellicer, J., M. Ampano Gilabert, and E. Lopez-Baeza. "The Evolution of the Celsius and Kelvin Scales and the State of the Art," *J. Chem. Educ.*, **76**, 911–13 (1999).

Quinn, T. J. *Temperature*, Academic Press, New York (1990).

Reilly, P. M. "A Statistical Look at Significant Figures," *Chem. Eng. Educ.*, 152–55 (Summer, 1992).

Romer, R. H. "Temperature Scales," *The Physics Teacher*, 450 (October, 1982).

Thompson, H. B. "Is 8°C Equal to 50°F?" *J. Chem. Educ.*, **68**, 400 (1991).
Vatavuk, W. M. "How Significant Are Your Figures," *Chem. Eng.*, 97 (August 18, 1986).

Web Sites

www.chemistrycoach.com/tutorials-2.html
http://en.wikibooks.org/wiki/Introduction_to_Chemical_Engineering_Processes/Units
www.lmnoeng.com
http://physics.nist.gov/cuu/units/index.html

Problems

Problems are rate in degree of difficulty from *(easiest) to ****(most difficult).

Section 2.1 Systems of Units

***2.1.1** The following questions will measure your SIQ.

a. Which is (are) a correct SI symbol(s)?
(1) nm (2) °K
(3) sec (4) N/mm

b. Which is (are) consistent with SI usage?
(1) MN/m^2 (2) GHz/s
(3) $kJ/[(s)(m^3)]$ (4) °C/M/s

c. Atmospheric pressure is about
(1) 100 Pa (2) 100 kPa
(3) 10 MPa (4) 1 GPa

d. The temperature 0°C is defined as
(1) 273.15°K (2) Absolute zero
(3) 273.15 K (4) The freezing point of water

e. Which height and mass are those of a petite woman?
(1) 1.50 m, 45 kg (2) 2.00 m, 95 kg
(3) 1.50 m, 75 kg (4) 1.80 m, 60 kg

f. Which is a recommended room temperature in winter?
(1) 15°C (2) 20°C
(3) 28°C (4) 45°C

g. The watt is
(1) 1 joule per second (2) Equal to 1 $(kg)(m^2)/s^3$
(3) The unit for all types of power (4) All of the above

h. What force may be needed to lift a heavy suitcase?
(1) 24 N (2) 250 N
(3) 25 kN (4) 250 kN

Section 2.2 Conversion of Units

*2.2.1 Carry out the following conversions:

a. How many m^3 are there in 1.00 $(mile)^3$?

b. How many gal/min correspond to 1.00 ft^3/s?

2.2.2 Convert

*a. 0.04 $g/[(min)(m^3)]$ to $lb_m/[(hr)(ft^3)]$

*b. 2 L/s to ft^3/day

**c. $\dfrac{6\,(in)(cm^2)}{(yr)(s)(lb_m)(ft^2)}$ to all SI units

*2.2.3 Convert the following:

a. 60.0 mi/h to ft/s

b. 50.0 $lb/in.^2$ to kg/m^2

c. 6.20 cm/hr^2 to nm/s^2

**2.2.4 A technical publication describes a new model 20 hp Stirling (air cycle) engine that drives a 68 kW generator. Is this possible?

**2.2.5 Your boss announced that the speed of the company Boeing 737 is to be cut from 525 mi/hr to 475 mi/hr to "conserve fuel," thus cutting consumption from 2200 gal/hr to 2000 gal/hr. How many gallons are saved in a 1000 mi trip?

**2.2.6 From *Parade* magazine (by Marilyn Vos Savant, August 31, 1997, p. 8): "Can you help with this problem? Suppose it takes one man 5 hours to paint a house, and it takes another man 3 hours to paint the same house. If the two men work together, how many hours would it take them? This is driving me nuts."

**2.2.7 Two scales are shown in Figure P2.2.7, a balance (a) and a spring scale (b). In the balance, calibrated weights are placed in one pan to balance the object to be weighed in the other pan. In the spring scale, the object to be weighed is placed on the pan and a spring is compressed that moves a dial on a scale in kilograms.

State for each device whether it directly measures mass or weight. Underline your answer. State in *one* sentence for each the reason for your answer.

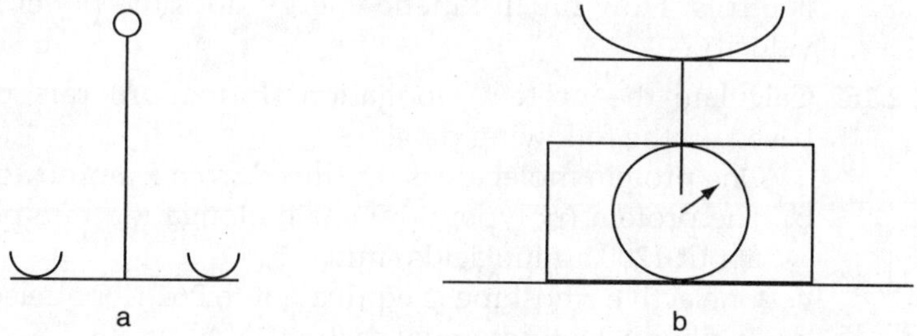

Figure P2.2.7

****2.2.8** In the American Engineering system of units, the viscosity can have the units of $(lb_f)(h)/ft^2$, while in a handbook the units are (g)/[(cm)(s)]. Convert a viscosity of 20.0 (g)/[(m)(s)] to the given American Engineering units.

****2.2.9** Thermal conductivity in the American Engineering system of units is

$$k = \frac{\text{Btu}}{(\text{h})(\text{ft}^2)(°\text{F}/\text{ft})}$$

Determine the conversion factor to convert from AE units to the following units:

$$\frac{\text{kJ}}{(\text{d})(\text{m}^2)(°\text{C}/\text{cm})}$$

*****2.2.10** Water is flowing through a 2-in.-diameter pipe with a velocity of 3 ft/s.

a. What is the kinetic energy of the water in $(ft)(lb_f)/lb_m$?
b. What is the flow rate in gallons per minute?

******2.2.11** Consider water pumped at a rate of 75 gal/min through a pipe which undergoes a 100 ft elevation increase by a 2 hp pump. The rate energy input from the pump that goes into heating the water is approximately equal to the rate of energy input from the pump minus the rate of potential energy generated by pumping the water up the elevation change ($m'gh$, where m' is the mass flow rate of water and h is the elevation change). Estimate the rate of energy input for heating the water for this case in British thermal units per hour.

****2.2.12** What is meant by a scale that shows a weight of "21.3 kg"?

*****2.2.13** A tractor pulls a load with a force equal to 800 lb (4.0 kN) with a velocity of 300 ft/min (1.5 m/s). What is the power required using the given American Engineering system data? The SI data?

*****2.2.14** What is the kinetic energy of a vehicle with a mass of 2300 kg moving at the rate of 10.0 ft/s in British thermal units?

*****2.2.15** A pallet of boxes weighing 10 tons is dropped from a lift truck from a height of 10 ft. The maximum velocity the pallet attains before hitting the ground is 6 ft/s. How much kinetic energy does the pallet have in $(ft)(lb_f)$ at this velocity?

*****2.2.16** Calculate the protein elongation (formation) rate per mRNA per minute based on the following data:

a. One protein molecule is produced from x amino acid molecules.
b. The protein (polypeptide) chain elongation rate per active ribosome uses about 1200 amino acids/min.
c. One active ribosome is equivalent to 264 ribonucleotides.
d. $3x$ ribonucleotides equal each mRNA.

Messenger RNA (mRNA) is a copy of the information carried by a gene in DNA and is involved in protein synthesis.

Section 2.3 Dimensional Consistency

*2.3.1 The density of a certain liquid is given an equation of the following form:

$$\rho = (A + BT)e^{CP}$$

where ρ = density (g/cm^3)
T = temperature (°C)
P = pressure (atm)

For this equation to be dimensionally consistent, what are the units of A, B, and C?

*2.3.2 Explain in detail whether the following equation for flow over a rectangular weir is dimensionally consistent. (This is the modified Francis formula.)

$$q = 0.415\,(L - 0.2h_0)\,h_0^{1.5}\sqrt{2g}$$

where q = volumetric flow rate (ft^3/s)
L = crest height (ft)
h_0 = weir head (ft)
g = acceleration of gravity (32.2 ft/s^2)

**2.3.3 In an article on measuring flows from pipes, the author calculated $q = 80.8\ m^3/s$ using the formula

$$q = CA_1\sqrt{\frac{2gV(p_1 - p_2)}{1 - (A_1/A_2)^2}}$$

where q = volumetric flow rate (m^3/s)
C = dimensionless coefficient (0.6)
A_1 = cross-sectional area 1 (m^2)
A_2 = cross-sectional area 2 (m^2)
V = specific volume ($10^{-3}\ m^3/kg$)
P = pressure
$p_1 - p_2$ = 50 kPa
g = acceleration of gravity (9.80 m/s)

Was the calculation correct? (Answer yes or no and explain briefly the reasoning underlying your answer.)

***2.3.4 Leaking oil tanks have become such an environmental problem that the federal government has implemented a number of rules to reduce the problem. A flow rate from a leak from a small hole in a tank can be predicted from the following relation:

$$Q = 0.61\, S\sqrt{\frac{2\Delta p}{\rho}}$$

where Q is the leakage rate (gal/min)

S is the cross-sectional area of the hole causing the leak (in.^{2})

Δp is the pressure drop between the inside of the tank opposite the leak and the atmospheric pressure (psi)

ρ is the fluid density (lb/ft^3)

To test the tank, the vapor space is pressurized with N_2 to a pressure of 23 psig. If the tank is filled with 73 in. of gasoline (sp. gr. = 0.703) and the hole is $\frac{1}{4}$ in. in diameter, what is the value of Q (in cubic feet per hour)?

***2.3.5 A relation for a dimensionless variable called the compressibility (z), which is used to describe the pressure-volume-temperature behavior for real gases, is

$$z = 1 + B\rho + C\rho^2 + D\rho^3$$

where ρ is the density in g mol/cm^3. What are the units of B, C, and D? Convert the coefficients in the equation for z so that the density can be introduced into the equation in the units of lb$_m$/ft^3 thus:

$$z = 1 + B^*\rho^* + C^*(\rho^*)^2 + D^*(\rho^*)^3$$

ρ^* is in lb$_m$/ft^3. Give the units for B^*, C^*, and D^*, and give the equations that relate B^* to B, C^* to C, and D^* to D.

**2.3.6 The velocity in a pipe in turbulent flow is expressed by the following equation:

$$u = k\left(\frac{\tau}{\rho}\right)^{1/2}$$

where τ is the shear stress in N/m^2 at the pipe wall

ρ is the density of the fluid in kg/m^3

u is the velocity in m/s

k is a coefficient

You are asked to modify the equation so that the shear stress can be introduced in the units of τ which are lb$_f$/ft^2, and the density ρ' for which the units are lb/ft^3 so that the velocity u comes out in the units of ft/s. Show all calculations, and give the final equation in terms of u, τ, and ρ' so a reader will know that American Engineering units are involved in the equation.

**2.3.7 In 1916 Nusselt derived a theoretical relation for predicting the coefficient of heat transfer between a pure saturated vapor and a colder surface:

$$h = 0.943\left(\frac{k^3\rho^2 g\lambda}{L\mu\,\Delta T}\right)^{1/4}$$

where h is the mean heat transfer coefficient, Btu/[(hr)(ft^2)(°F)]
k is the thermal conductivity, Btu/[(hr)(ft)(°F)]
ρ is the density in lb/ft^3
g is the acceleration of gravity, 4.17×10^8 ft/(hr)^2
λ is the enthalpy change of evaporation in Btu/lb
L is the length of tube in ft
m is the viscosity in lb$_m$/[(hr)(ft)]
ΔT is a temperature difference in °F

What are the units of the constant 0.943?

****2.3.8 The efficiency of cell growth in a substrate in a biotechnology process was given in a report as

$$\eta = \frac{Y^c_{x/S}\gamma_b \Delta H^c_b/e^-}{\Delta H_{cat}}$$

In the notation table
η is the energetic efficiency of cell metabolism (energy/energy)
$Y^c_{x/S}$ is the cell yield, carbon basis (cells produced/substrate consumed)
γ_b is the degree of reductance of biomass (available electron equivalents/g mole carbon, such as 4.24 e^- equiv./mol cell carbon)
$\Delta H^c_b/e^-$ is the biomass heat of combustion (energy/available electron equiv.)
ΔH_{cat} is the available energy from catabolism (energy/mole substrate carbon)

Is there a missing conversion factor? If so, what would it be? The author claims that the units in the numerator of the equation are (mol cell carbon/mol substrate carbon) (mol available *e*-/mol cell carbon) (heat of combustion/mol available *e*-). Is this correct?

****2.3.9 The Antoine equation, which is an empirical equation, is used to model the effect of temperature on the vapor pressure of a pure component. The Antoine equation is given by

$$\ln p^* = A - \frac{B}{C + T}$$

where p^* is the vapor pressure, T is the absolute temperature, and A, B, and C are empirical constants specific to the pure component and the units used for the vapor pressure and temperature. Determine under what conditions this equation will be dimensionally consistent.

**2.3.10 A letter to the editor says: "An error in units was made in the article 'Designing Airlift Loop Fermenters.' Equation (4) is not correct."

$$\Delta p = 4f\rho\left[\frac{v^2}{2g}(L/D)\right] \tag{4}$$

Is the author of the letter correct? (f is dimensionless.)

Section 2.4 Significant Figures

*2.4.1 If you subtract 1191 cm from 1201 cm, each number with four significant figures, does the answer of 10 cm have two or four (10.00) significant figures?

*2.4.2 What is the sum of

3.1472
32.05
1234
8.9426
0.0032
9.00

to the correct number of significant figures?

**2.4.3 Suppose you make the following sequence of measurements for the segments in laying out a compressed air line:

4.61 m
210.0 m
0.500 m

What should be the reported total length of the air line?

**2.4.4 Given that the width of a rectangular duct is 27.81 cm, and the height is 20.49 cm, what is the area of the duct with the proper number of significant figures?

**2.4.5 Multiply 762 by 6.3 to get 4800.60 on your calculator. How many significant figures exist in the product, and what should the rounded answer be?

***2.4.6 Suppose you multiply 3.84 times 0.36 to get 1.3824. Evaluate the maximum relative error in (a) each number and (b) the product. If you add the relative errors in the two numbers, is the sum the same as the relative error in their product?

Section 2.6 The Mole and Molecular Weight

***2.6.1** Convert the following:
a. 4g mol of $MgCl_2$ to g
b. 2 lb mol of C_3H_8 to g
c. 16 g of N_2 to lb mol
d. 3 lb of C_2H_6O to g mol

***2.6.2** How many pounds are there in each of the following?
a. 16.1 lb mol of pure HCl
b. 19.4 lb mol of KCl
c. 11.9 g mol of $NaN0_3$
d. 164 g mol of SiO_2

*****2.6.3** A solid compound was found to contain 42.11% C, 51.46% O, and 6.43% H. Its molecular weight was about 341. What is the formula for the compound?

****2.6.4** The structural formulas in Figure P2.6.4 are for vitamins.
a. How many pounds are contained in 2.00 g mol of vitamin A?
b. How many grams are contained in 1.00 lb mol of vitamin C?

Vitamin	Structural formula	Dietary sources	Deficiency symptoms
Vitamin A	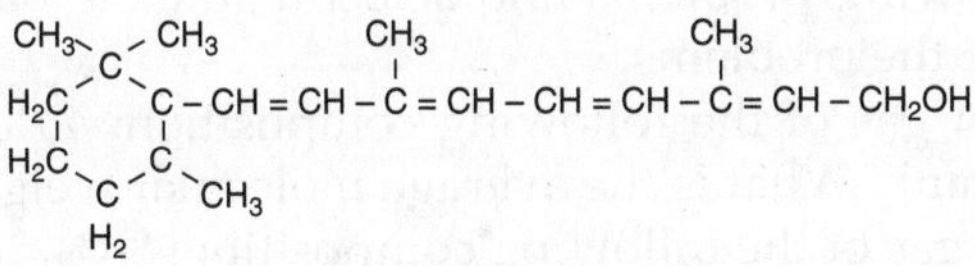 Retinol	Fish liver oils, liver, eggs, fish, butter, cheese, milk; a precursor, β-carotene, is present in green vegetables, carrots, tomatoes, squash	Night blindness, eye inflammation
Ascorbic acid (vitamin C)	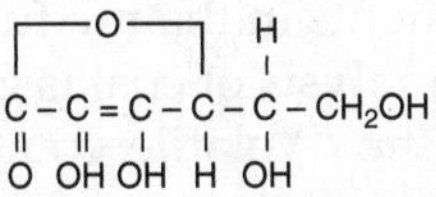	Citrus fruit, tomatoes, green peppers, strawberries, potatoes	Scurvy

Figure P2.6.4

***2.6.5** A sample has a specific volume of 5.2 m^3/kg and a molar volume of 1160 m^3/kg mol. Determine the molecular weight of the material.

****2.6.6** Prepare an expression that converts mass (weight) fraction (ω) to mole fraction (x), and another expression for the conversion of mole fraction to mass fraction, for a binary mixture.

***2.6.7** You have 100 kg of gas of the following composition:

CH_4	30%
H_2	10%
N_2	60%

What is the average molecular weight of this gas?

*2.6.8 You analyze the gas in 100 kg of gas in a tank at atmospheric pressure and find the following:

CO_2	19.3%
O_2	6.5%
H_2O	2.1%
N_2	72.1%

What is the average molecular weight of the gas?

**2.6.9 Suppose you are required to make an analysis of 317 lb of combustion gas and find it has the following composition:

CO_2	60%
CO	10%
N_2	30%

What is the average molecular weight of this gas in the American Engineering system of units?

Section 2.7 Choosing a Basis

2.7.1 Read each of the following problems and select a suitable basis for solving each one. Do not solve the problems.

**a. You have 130 kg of gas of the following composition: 40% N_2, 30% CO_2, and 30% CH_4 in a tank. What is the average molecular weight of the gas?

**b. You have 25 lb of a gas of the following composition: 80%, 10%, 10%. What is the average molecular weight of the mixture? What is the weight (mass) fraction of each of the components in the mixture?

****c. The proximate and ultimate analysis of coal is given in the following table. What is the composition of the "Volatile combustible material" (VCM)? Present your answer in the form of the mass percent of each element in the VCM.

Proximate Analysis (%)		**Ultimate Analysis (%)**	
Moisture	3.2	Carbon	79.90
Volatile combustible material	21.0	Hydrogen	4.85
Fixed carbon	69.3	Sulfur	0.69
Ash	6.5	Nitrogen	1.30
		Ash	6.50
		Oxygen	6.76
Total	100.0	Total	100.00

*d. A fuel gas is reported to analyze, on a mole basis, 20% methane, 5% ethane, and the remainder CO_2. Calculate the analysis of the fuel gas on a mass percentage basis.

***e. A gas mixture consists of three components: argon, B, and C. The following analysis of this mixture is given: 40.0 mol % argon, 18.75 mass % B, 20.0 mol % C. The molecular weight of argon is 40 and the molecular weight of C is 50. Find (1) the molecular weight of B and (2) the average molecular weight of the mixture.

***2.7.2 Two engineers are calculating the average molecular weight of a gas mixture containing oxygen and other gases. One of them uses the correct molecular weight of 32 for oxygen and determines the average molecular weight as 39.2. The other uses an incorrect value of 16 and determines the average molecular weight as 32.8. This is the only error in the calculations. What is the percentage of oxygen in the mixture expressed as mole percent? Choose a basis to solve the problem, but do not solve the problem.

**2.7.3 Choose a basis for the following problem: Chlorine usage at a water treatment plant averages 134.2 lb/day. The average flow rate of water leaving the plant is 10.7 million gal/day. What is the average chlorine concentration in the treatment water leaving the plant (assuming no reaction of the chlorine), expressed in milligrams per liter?

Section 2.8 Density and Specific Gravity

*2.8.1 You are asked to decide what size containers to use to ship 1000 lb of cottonseed oil of specific gravity equal to 0.926. What would be the minimum size drum expressed in gallons?

*2.8.2 The density of a certain solution is 8.80 lb/gal at 80°F. How many cubic feet will be occupied by 10,010 lb of this solution at 80°F?

*2.8.3 Which of the three sets of containers in Figure P2.8.3 represents respectively 1 mol of lead (Pb), 1 mol of zinc (Zn), and 1 mol of carbon (C)?

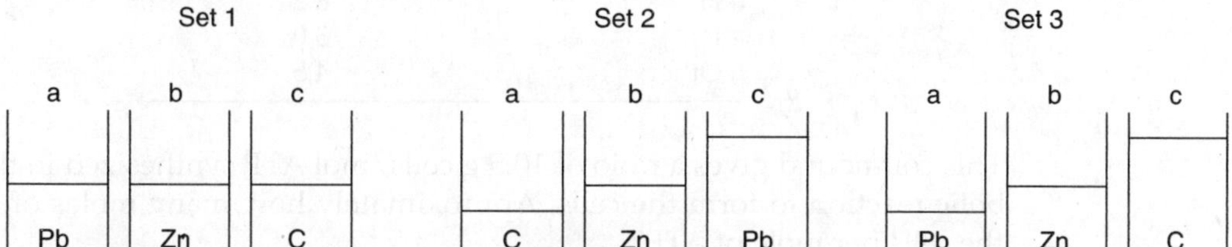

Figure P2.8.3

Section 2.9 Concentration

*2.9.1 Calculate the mass and mole fractions of the respective components in $NaClO_3$.

*2.9.2 The specific gravity of a solution of KOH at 15°C is 1.0824 and contains 0.813 lb KOH per gal of solution. What are the mass fractions of KOH and H_2O in the solution?

*2.9.3 You purchase a tank with a volume of 2.1 ft^3. You pump the tank out and add first 20 lb of CO_2 gas and then 10 lb of N_2 gas. What is the analysis of the gas mixture in the tank?

*2.9.4 How many ppb are there in 1 ppm? Does the system of units affect your answer? Does it make any difference if the material for which the ppb are measured is a gas, liquid, or solid?

**2.9.5 The following table lists values of Fe, Cu, and Pb in Christmas wrapping paper for two different brands. Convert the ppm to mass fractions on a paper-free basis.

	Concentration, ppm		
	Fe	**Cu**	**Pb**
Brand A	1310	2000	2750
Brand B	350	50	5

2.9.6 Harbor sediments in the New Bedford, Massachusetts, area contain PCBs at levels up to 190,000 ppm, according to a report prepared by Grant Weaver of the Massachusetts Coastal Zone Management Office [*Environ. Sci. Technol.*, **16, No. 9, 491A (1982)]. What is the concentration of PCBs in percent? Does this seem reasonable?

**2.9.7 The analysis of a biomass sample gave

Element	% Dry Weight of Cell
C	50.2
O	20.1
N	14.0
H	8.2
P	3.0
Other	4.5

This compound gives a ratio of 10.5 g cells/mol ATP synthesized in the metabolic reaction to form the cells. Approximately how many moles of C are in the cells per mole of ATP?

***2.9.8 A radioactive tracer-labeled microorganism (MMM) decomposes to NN as follows:

$$MMM(s) \rightarrow NN(s) + 3CO_2(g)$$

If the $CO_2(g)$ yields 2×10^7 dpm (disintegrations per minute) in a detection device, how many μCi (microcuries) is this? How many cpm (counts per minute) will be noted if the counting device is 80% efficient in counting disintegrations? Data: 1 curie = 3×10^{10} dps (disintegrations per second).

***2.9.9 Several alternative compounds have been added to gasoline, including methanol, ethanol, and methyl tert-butyl ether (MTBE), to increase the oxygen content of gasoline in order to reduce the formation of CO on combustion. Unfortunately, MTBE has been found in groundwater at concentrations sufficient to cause concern. Persistence of a compound in water can be evaluated from its half-life, $t_{1/2}$, that is, the time for one-half of the compound to leave the system of interest. The half-life depends on the conditions in the system, of course, but for environmental evaluation can be approximated by

$$t_{1/2} = \frac{\ln(2)}{k[\mathrm{OH}^-]}$$

where $[\mathrm{OH}^-]$ is the concentration of hydroxyl radical in the system that for this problem of the contamination of water is equal to 1.5×10^6 molecules/cm^3. The values of k determined from experiment are

	k $\mathrm{cm}^3/[\mathrm{molecule(s)}]$
Methanol	0.15×10^{-12}
Ethanol	1×10^{-12}
MTBE	0.60×10^{-12}

Calculate the half-life of each of the three compounds, and order them according to their persistence.

***2.9.10 The National Institute for Occupational Safety and Health (NIOSH) sets standards for CCl_4 in air at 12.6 mg/m^3 of air (a time-weighted average over 40 hr). The CCl_4 found in a sample is 4800 ppb. Does the sample exceed the NIOSH standard? Be careful!

***2.9.11 The following table shows the annual inputs of phosphorus to Lake Erie:

A	Short Tons/Yr
Source	
Lake Huron	2240
Land drainage	6740
Municipal waste	19,090
Industrial waste	2030
Total of sources	30,100
Outflow	4500
Retained	25,600

a. Convert the retained phosphorus to concentration in micrograms per liter, assuming that Lake Erie contains 1.2×10^{14} gal of water and that the average phosphorus retention time is 2.60 yr.

b. What percentage of the input comes from municipal water?
c. What percentage of the input comes from detergents, assuming they represent 70% of the municipal waste?
d. If 10 ppb of phosphorus triggers nuisance algal blooms, as has been reported in some documents, would removing 30% of the phosphorus in the municipal waste and all the phosphorus in the industrial waste be effective in reducing the eutrophication (i.e., the unwanted algal blooms) in Lake Erie?
e. Would removing all of the phosphate in detergents help?

**2.9.12 A gas contains 350 ppm of H_2S in CO_2. If the gas is liquefied, what is the weight fraction of H_2S?

**2.9.13 Sulfur trioxide (SO_3) can be absorbed in sulfuric acid solution to form a more concentrated sulfuric acid. If the gas to be absorbed contains 55% SO_3, 41% N_2, 3% SO_2, and 1% O_2:
a. How many parts per million of O_2 are there in the gas?
b. What is the composition of the gas on an N_2-free basis?

****2.9.14 Twenty-seven pounds (27 lb) of chlorine gas is used for treating 750,000 gal of water each day. The chlorine used up by the microorganisms in the water is measured to be 2.6 mg/L. What is the residual (excess) chlorine concentration in the treated water?

***2.9.15 A newspaper report says the FDA found 13–20 ppb of acrylonitrile in a soft-drink bottle, and if this is correct, it amounts to only 1 molecule of acrylonitrile per bottle. Is this statement correct?

***2.9.16 Several studies of global warming indicate that the concentration of CO_2 in the atmosphere is increasing by roughly 1% per year. Do we have to worry about the decrease in the oxygen concentration also?

Section 2.10 Temperature

*2.10.1 "Japan, U.S. Aim for Better Methanol-Powered Cars," reads the headline in the *Wall Street Journal*. Japan and the United States plan to join in developing technology to improve cars that run on methanol, a fuel that causes less air pollution than gasoline. An unspecified number of researchers from Japanese companies will work with the EPA to develop a methanol car that will start in temperatures as low as −10°C. What is this temperature in degrees Rankine, kelvin, and Fahrenheit?

*2.10.2 Can negative temperature measurements exist?

**2.10.3 The heat capacity C_p of acetic acid in J/[(g mol)(K)] can be calculated from the equation

$$C_p = 8.41 + 2.4346 \times 10^{-5}\, T$$

where T is in kelvin. Convert the equation so that T can be introduced into the equation in degrees Rankine instead of kelvin. Keep the units of C_p the same.

*2.10.4 Convert the following temperatures to the requested units:

a. 10°C to °F

b. 10°C to °R

c. −25°F to K

d. 150K to °R

***2.10.5 Heat capacities are usually given in terms of polynomial functions of temperature. The equation for carbon dioxide is

$$C_p = 8.4448 + 0.5757 \times 10^{-2}\,T - 0.2159 \times 10^{-5}\,T^2 + 0.3059 \times 10^{-9}\,T^3$$

where T is in °F and C_p is in Btu/[(lb mol)(°F)]. Convert the equation so that T can be in °C and C_p will be in J/[(g mol)(K)].

**2.10.6 In a report on the record low temperatures in Antarctica, *Chemical and Engineering News* said at one point that "the mercury dropped to −76°C." In what sense is that possible? Mercury freezes at −39°C.

Section 2.11 Pressure and Hydrostatic Head

**2.11.1 From the newspaper:

> BROWNSVILLE, TX. Lightning or excessive standing water on the roof of a clothes store are emerging as the leading causes suspected in the building's collapse. Mayor Ignacio Garza said early possibilities include excessive weight caused by standing water on the 19-year-old building's roof. Up to six inches of rain fell here in less than six hours."

Flat-roof buildings are a popular architectural style in dry climates because of the economy of materials of construction. However, during the rainy season water may pool up on the roof decks so that structural considerations for the added weight must be taken into account. If 15 cm of water accumulates on a 10 m by 10 m area during a heavy rainstorm, determine (a) the total added weight from the standing water the building must support and (b) the force of the water on the roof in psi.

***2.11.2 A problem with concrete wastewater treatment tanks set belowground was realized when the water table rose and an empty tank floated out of the ground. This buoyancy problem was overcome by installing a check valve in the wall of the tank so that if the water table rose high enough to float the tank, it would fill with water. If the density of concrete is 2080 kg/m^3, determine the maximum height at which the valve should be installed to prevent a buoyant force from raising a rectangular tank with inside dimensions of 30 m by 27 m that is 5 m deep. The walls and floor have a uniform thickness of 200 mm.

**2.11.3 A centrifugal pump is to be used to pump water from a lake to a storage tank that is 148 ft above the surface of the lake. The pumping rate is to be 25.0 gal/min, and the water temperature is 60°F. The pump on hand can develop a pressure of 50.0 psig when it is pumping at a rate of 25.0 gal/min. (Neglect pipe friction, kinetic energy effects, and factors involving pump efficiency.)

a. How high (in feet) can the pump raise the water at this flow rate and temperature?
b. Is this pump suitable for the intended service?

**2.11.4 A manufacturer of large tanks calculates the mass of fluid in the tank by taking the pressure measurement at the bottom of the tank in psig, and then multiplying that value by the area of the tank in square inches. Can this procedure be correct?

**2.11.5 Suppose that a submarine inadvertently sinks to the bottom of the ocean at a depth of 1000 m. It is proposed to lower a diving bell to the submarine and attempt to enter the conning tower. What must the minimum air pressure be in the diving bell at the level of the submarine to prevent water from entering the bell when the opening valve at the bottom is cracked open slightly? Give your answer in absolute kilopascals. Assume that seawater has a constant density of 1.024 g/cm^3.

**2.11.6 A pressure gauge on a welder's tank gives a reading of 22.4 psig. The barometric pressure is 28.6 in. Hg. Calculate the absolute pressure in the tank in (a) lb/ft^2, (b) in. Hg, (c) N/m^2, and (d) ft water.

**2.11.7 John Long says he calculated from a formula that the pressure at the top of Pikes Peak is 9.75 psia. John Green says that it is 504 mm Hg because he looked it up in a table. Which John is right?

***2.11.8 The floor of a cylindrical water tank was distorted into 7 in. bulges because of the settling of improperly stabilized soil under the tank floor. However, several consulting engineers restored the damaged tank to use by placing plastic skirts around the bottom of the tank wall and devising an air flotation system to move it to an adjacent location. The tank was 30.5 m in diameter and 13.1 m deep. The top, bottom, and sides of the tank were made of 9.35-mm-thick welded steel sheets. The density of the steel is 7.86 g/cm^3.

a. What was the gauge pressure in kilopascals of the water at the bottom of the tank when it was completely full of water?
b. What would the air pressure have to be in kilopascals beneath the empty tank in order to just raise it up for movement?

***2.11.9 Examine Figure P2.11.9. Oil (density = $0.91 g/cm^3$) flows in a pipe, and the flow rate is measured via a mercury (density = $13.546 g/cm^3$) manometer. If the difference in height of the two legs of the manometer is 0.78 in., what is the corresponding pressure difference between points A and B in mm Hg? At which point, A or B, is the pressure higher? The temperature is 60°F.

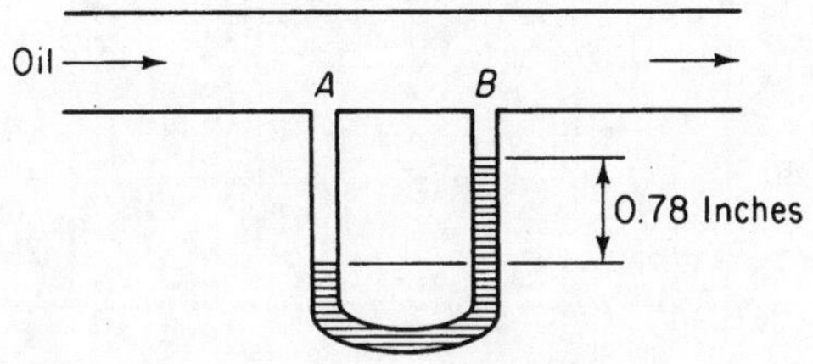

Figure P2.11.9

PART II

MATERIAL BALANCES

CHAPTER 3
Material Balances

Your objectives in studying this chapter are to be able to

1. Develop a conceptual understanding of material balances
2. Understand the features of open, closed, steady-state, and unsteady-state systems
3. Express in words how to form the material balances for processes involving single or multiple components
4. Familiarize yourself with the strategy to assist you in solving material balance problems

If people knew how hard I worked to get my mastery, it wouldn't seem so wonderful after all.

Michelangelo

In Chapter 3 we introduce the concept of a material balance. Material balances are nothing more than the application of the conservation law for mass: "Matter is neither created nor destroyed." Just what this statement means in practice, and how you can use the concept to solve problems of varying degrees of complexity, requires a bit of explanation.

Material balances basically involve accounting—not for money but for material. You will find that solving material balance problems is much easier if you develop a systematic strategy that is applicable to a wide range of problems. Consequently, in this chapter we focus on the strategy of making appropriate decisions, implementing them properly, and assessing whether the implementation has been correct in solving a problem. Our aim is to help you acquire a generalized approach to problem solving so that you may avoid looking upon each new problem, unit operation, or process as entirely new and unrelated to anything you have seen before. In the examples used to illustrate the principles involved in this chapter, explore the method of analysis, but avoid memorizing each example, because, after all, they are only samples of the myriad problems that exist or could be devised on the subject of material balances. Most of the principles we consider are of about the same degree of complexity as the law of compensation, devised by some unknown, self-made philosopher who said: "Things are generally made even somewhere or someplace. Rain always is followed by a dry spell, and dry weather follows rain. I have found it an invariable rule that when a man has one short leg, the other is always longer!"

3.1 Introduction to Material Balances

Once you understand something, it can seem almost trivial. This is true for material balances and problems based on material balances. On the other hand, your first exposure to a new topic can be almost overwhelming. This section is designed to gradually expose you to the concepts and terminology associated with the application of material balances. We will start by drawing parallels between a typical bank statement and a material balance and from this example develop a general material balance equation that you will use throughout the remainder of this text. The general material balance equation will be simplified and applied to a system involving a single component to introduce important terminology that is used to describe systems. Then you will be shown how to apply material balances to systems with more than one component using a variety of examples. In this manner, you should generally become familiar with material balances so that you can more fully understand the systematic method for solving material balance problems presented in Section 3.2.

3.1.1 The Concept of a Material Balance

What are **material balances**? You can get a good idea of what is involved in making a material balance by examining a bank statement such as the following one that illustrates the changes in a checking account.

Bank of the West

Customer Summary Information

Date	Notes	Deposit	Withdrawal	Other	Balance
3/1	Beginning balance				\$1253.89
3/2	Deposit from ABC Co.	\$1500.00			\$2753.89
3/3	Check No. 2133		\$550.00		\$2203.89
3/3	ATM withdrawal 3/2		\$200.00		\$2003.89
3/5	Check No. 2134		\$401.67		\$1602.22
3/15	Check No. 2135		\$321.83		\$1280.39

Date	Notes	Deposit	Withdrawal	Other	Balance
3/18	ATM withdrawal 3/17		$200.00		$1080.39
3/20	Deposit at the Bank	$1250.00			$2330.39
3/23	Check No. 2136		$442.67		$1887.72
3/31	Service charge			−10.00	$1877.72
3/31	Interest earned			1.04	$1878.76
3/31	Closing balance				$1878.76

The checking account initially has a balance of $1253.89, which is the **initial condition** of the account. Deposits (**what goes in, the inputs**) add to the balance, and withdrawals and checks (**what goes out, the outputs**) are subtracted from the balance. The **final condition** is the closing balance ($1878.76). The difference between the initial balance and the final balance is the **accumulation** in the account (which can be negative!). The column headed "Other" represents interest **generated** by the balance in the account and bank charges that **consume** some of the balance in the account. These latter two terms affect the balance but are not withdrawals, checks, or deposits. Can you write an equation in words that expresses the accumulation in the bank account in terms of the initial and final balances, the deposits and withdrawals, and the interest and fees? Finally, the daily balance represents the net effect of the withdrawals, deposits, fees, and interest on the balance since the beginning of the month. Look at the following relation:

Accumulation = Final balance − Initial balance
= Deposits − Withdrawals + Interest − Fees

Industrial processes are quite analogous to the checking account. Material balances pertain to materials rather than money, as you can infer from the name, but the concepts of balancing money and material are exactly the same. The initial conditions for a process involve the amount of material initially present in the process. Deposits into the checking account are analogous to the flow of material into a process, and withdrawals are analogous to the flow of material out of a process. Accumulation of material in a process is the same as accumulation in the bank account. Generation and consumption terms in a process occur because of chemical reaction.

A summary of the nomenclature used for a bank statement and that used in making a material balance is as follows:

Bank Statement	Material Balance
Deposits ($2750.00)	Inputs, flow in
Withdrawals, checks ($2116.17)	Outputs, flow out
Initial balance ($1253.89)	Initial conditions

(Continues)

(Continued)

Bank Statement	Material Balance
Final balance (\$1878.76)	Final conditions
Accumulation (\$624.87)	Accumulation
Interest (\$1.04)	Generation
Fees (\$10.00)	Consumption

The **general material balance** written as an equation that conforms to the bank account statement (using the appropriate nomenclature) is

$$\boxed{\text{Accumulation in the system from } t_1 \text{ to } t_2} = \boxed{\text{Amount in system at } t = t_2} - \boxed{\text{Amount in system at } t = t_1} =$$

$$\boxed{\text{Input to the system from } t_1 \text{ to } t_2} - \boxed{\text{Output from the system from } t_1 \text{ to } t_2} + \boxed{\text{Generation in the system from } t_1 \text{ to } t_2} - \boxed{\text{Consumption in the system from } t_1 \text{ to } t_2} \quad (3.1)$$

Note that the last two terms, which pertain to generation and consumption of material, are generally included in material balances only for chemical components when chemical reaction occurs in the system, a topic that will be deferred until Chapter 5.

Equation (3.1) can be applied to conservation of total mass, mass of a component, moles of a component (but usually not total moles—why?), mass of an atomic species, and moles of an atomic species, but not to volume. Why? Mass is conserved while volume is generally not because different materials have different densities.

Did you notice that Equation (3.1) is written in terms of amounts of material, not rates? You can apply it to any selected time interval you want, whether explicitly specified in a problem or not. By this statement we mean that if a rate is specified in a problem, you can solve the problem on the basis of selecting a convenient time period, or by using 1 or 100 of some quantity in the problem as a basis, and then converting your solution back to the original rate specified in the problem. The form of Equation (3.1) is known as a **difference equation** (for discrete units of time). The sums of the discrete amounts, as you can observe from the daily transactions in the bank statement, become the amounts involved in each term of Equation (3.1) from the beginning to the end of the selected time interval (one month for the bank statement). The times you select for the final and initial conditions can be any arbitrary interval, but for convenience, if given a *flow rate*, you usually select an interval such as 1 min or 1 hr as the basis. Variables that are *flows* will be distinguished from variables that are *flow rates* by placing an overlay dot on the latter. As an example, F represents a flow and $\dot{F}$ the flow rate.

In Equation (3.1) the accumulation term represents the sum of all of a material that has accumulated in the system over the time interval (not the total amount of material in the system at the end of the interval because there may have been some material present at the initial state). The units of the accumulation might be mass or moles, but not mass or moles per unit time. Similarly, the flows in and out represent the sum of all of the flows of mass or moles in and out, respectively. If a constant rate is given in a problem, you can multiply the rate by your selected time interval (such as 1 hr) to get the cumulative entering or exit mass or moles of a material. It does not matter if a flow rate varies with time; each term on the right-hand side of Equation (3.1) represents the cumulative total of a material that flowed in and out, respectively, during the interval, so that the specific flow rates during the interval are not used directly. Look at the bank statement again to relate these comments to the various entries in the statement.

A **differential equation** (a topic that we defer to Chapter 17) would represent a process continuous in time. Because most process measurements are made at discrete time intervals, Equation (3.1) may initially seem a bit strange, but for applications to material (and energy) balances it becomes quite useful. If you have studied calculus, you will note that if both sides of Equation (3.1) are divided by $\Delta t = t_2 - t_1$ when $\Delta t \rightarrow 0$, the result looks like a differential equation with the units of each term being a rate.

3.1.2 Material Balances for a Single Component

To begin our discussion of the material balance equation, let's consider a system composed of a single component for which there are no chemical reactions occurring. Then, the general equation, Equation (3.1), ***with the absence of reaction***, reduces to

$$\boxed{\begin{array}{c}\textbf{Accumulation}\\ \textbf{in the system}\\ \textbf{from } t_1 \textbf{ to } t_2\end{array}} = \boxed{\begin{array}{c}\textbf{Amount in}\\ \textbf{system at}\\ t = t_2\end{array}} - \boxed{\begin{array}{c}\textbf{Amount in}\\ \textbf{system at}\\ t = t_1\end{array}} = \boxed{\begin{array}{c}\textbf{Input to}\\ \textbf{the system}\\ \textbf{from } t_1 \textbf{ to } t_2\end{array}} - \boxed{\begin{array}{c}\textbf{Output from}\\ \textbf{the system}\\ \textbf{from } t_1 \textbf{ to } t_2\end{array}} \qquad (3.2)$$

In other words, the accumulation inside a system is equal to the difference between what enters and what leaves the system. All the comments in Section 3.1.1 with regard to converting flow rates into flows by picking a suitable time interval as the basis apply equally well to Equation (3.2).

Now for a test. Here is a puzzle involving money balances. Three students rent a room the night before the game and pay the desk clerk \$60. A new clerk comes on duty and finds that the discount rate for students should have been \$55. The new clerk gives the bellhop \$5 to return to the students, but the bellhop, not having change and being slightly dishonest, returns only \$1 to each student and keeps the remaining \$2. Now each student paid \$20 − \$1 = \$19, and 3 × \$19 = \$57 paid in total. The bellhop kept \$2 for a total of \$59. What happened to the other \$1?

Apply what you have learned so far to solve the puzzle. It provides a good simple illustration of the confusion that can be cleared up by making a material (here, dollars) balance. (If the puzzle stumps you, look at the end of the next Self-Assessment Test for the answer.)

Example 3.1 Water Balance for a Lake

Water balances on a lake can be used to evaluate the effect of groundwater infiltration, evaporation, or precipitation on the lake. Prepare a water balance, in symbols, for a lake, including the physical processes indicated in Figure E3.1 (all symbols are in mass over the same time interval).

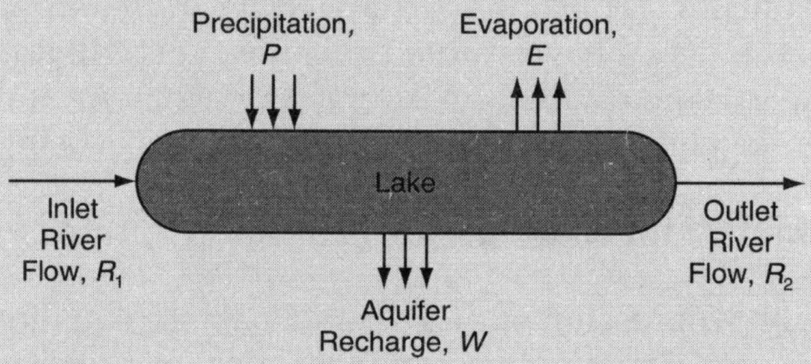

Figure E3.1

Solution

Equation (3.2) applies inasmuch as there is no reaction for the water in the system. Each term in Equation (3.2) can be represented by symbols shown in Figure E3.1. Remember that R_1, for example, is not the *flow rate* but the *total flow* of the river into the system during the interval, and that R_2 is the total flow out of the system during the interval. If there are any creeks flowing into the lake, they could be added as additional *R*s. Any accumulation requires some notation to designate the time. Let us use $S_R(t_1)$ and $S_R(t_2)$ as the initial and final amounts of water in the lake based on the idea that the amount of water in the lake during the time interval may change.

What should the system be? You want an overall balance on the water in the lake, so the system should be the lake. What should the basis be? Select the specified (but unknown) time interval, which is the same as picking one of the labeled flows. Then the water balance is

$$S_R(t_2) - S_R(t_1) = R_1 - R_2 + P - E - W$$

Example 3.2 Mass Balance for Water in a Fructose Storage Tank

Consider the storage tank shown in Figure E3.2. Over a 3 h period, the accumulation of water in the tank was determined to be 6000 kg. Assuming that the feed and removal rates remain constant during the 3 h period of interest, determine the flow rate of the second feed stream, $\dot{F}_2$. $\dot{F}_1$, is 10,000 kg/h and the water removal rate, $\dot{P}$, is 12,000 kg/h.

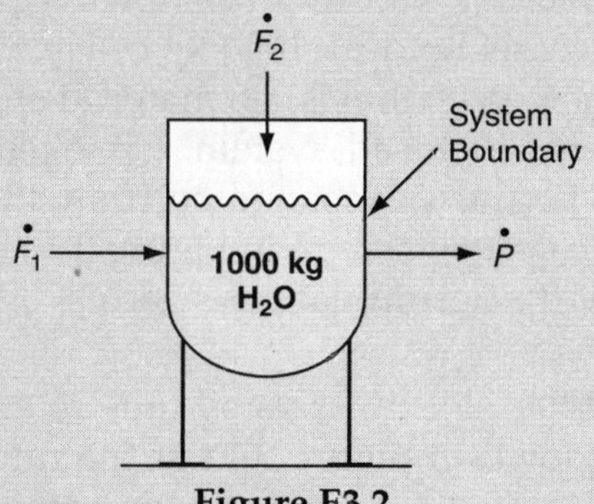

Figure E3.2

Solution

What should be the basis? Pick a basis of Δt equal to 3 h. Apply Equation (3.2) to this problem:

$$S_T(t_2) - S_T(t_1) = \dot{F}_1\,\Delta t + \dot{F}_2\,\Delta t - \dot{P}\Delta t$$

$$6000\text{ kg} = (10{,}000\text{ kg/h})(3\text{ h}) + \dot{F}_2(3\text{ h}) - (12{,}000\text{ kg/h})(3\text{ h})$$

Divide both sides of the equation by 3 to get $\dot{F}_2 = 4000$ kg/h. Note that the amount of water entering the system during the 3 h, F_2, is equal to the flow rate multiplied by the time interval.

3.1.3 Characteristics of Systems

What's in a name? That which we call a rose by
any other name would smell as sweet.
William Shakespeare in *Romeo and Juliet*

Now for the explanation of some terms that you will encounter frequently in the remainder of the text.

System By **system** we mean any arbitrary portion of or a whole process you want to consider for analysis. You can define a system such as a reactor, a section of a pipe, or an entire refinery by stating in words what the system is. Or you can define the limits of the system by drawing the **system boundary**, which is a line that encloses the portion of the process that you want to analyze. The boundary could coincide with the outside of a piece of equipment or some section inside the equipment. Now, let's focus on some important characteristics associated with systems.

Closed System or Process Figure 3.1a shows a two-dimensional view of a three-dimensional vessel holding 1000 kg of H_2O. Note that material neither enters nor leaves the vessel; that is, no material crosses the system boundary. Figure 3.1a represents a **closed system**. Changes can take place inside the closed system, but no mass exchange occurs with the surroundings. Apply Equation (3.2) to the system shown in Figure 3.1a. What is the accumulation? Did you decide that the accumulation is zero?

Open System or Process Next, let us assume in an experiment that you add water to the tank shown in Figure 3.1a at the rate of 100 kg/min and withdraw water at the rate of 100 kg/min as indicated in Figure 3.1b. Figure 3.1b is an example of an **open system** (also called a **flow system**) because the material crosses the system boundary. Apply Equation (3.2) again to Figure 3.1b. What is the value of the accumulation? In this case the accumulation is zero because the total inlet flow is equal to the total outlet flow.

Steady-State System or Process What does **steady-state** mean? A variable, such as an amount of material or a property, is in the steady state if the value of that variable is invariant (does not change) with time. If you look at Figure 3.1b, you will note that the flow rates in and out are just water and presumably constant and equal, respectively; hence the accumulation is zero, and the system is in the steady state.

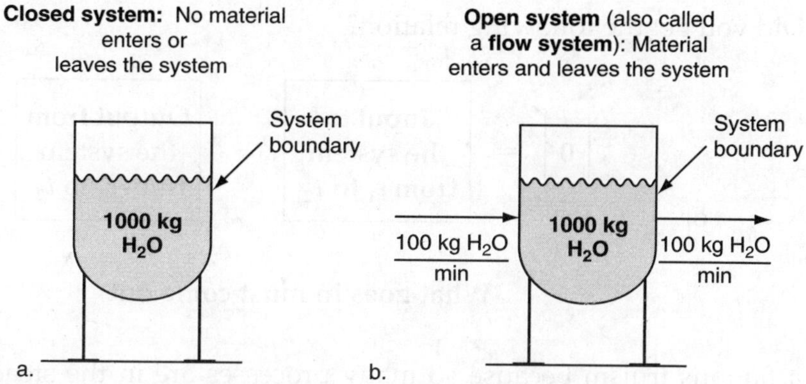

Figure 3.1 Comparison of (a) closed and (b) open systems

Most industrial processes under normal conditions are not really steady-state because the values of the variables involved when measured fluctuate continuously due to low-level disturbances to the process and the action of the control systems to counteract these disturbances. Yet the operation of such processes is frequently referred to as "steady-state behavior." Consequently, we will have to distinguish between an operating process and the model, or mathematical description, of the process. Thus, a looser concept of the steady state, one that should properly be called **pseudo steady-state** (meaning that it is not truly steady-state but for convenience is treated as steady-state) or perhaps **quasi steady-state** (meaning effectively behaving as though it were steady-state), will in this book mean for the selected time interval that (a) the accumulation term in Equation (3.1) will be zero, and (b) the rates of flow and the properties of the flows can be assumed to be constant even if they are not. Also, we will suppress the prefix *pseudo* or *quasi* and just use *steady-state* for convenience. With this assumption, the main criterion of steady-state will be that the accumulation term in Equations (3.1) and (3.2) is zero.

In a steady-state system the accumulation is zero. What will Equation (3.2) reduce to in a steady-state system?

~~Accumulation in the system from t_1 to t_2~~	=	~~Amount in system at $t = t_2$~~	−	~~Amount in system at $t = t_1$~~	=	Input to the system from t_1 to t_2	−	Output from the system from t_1 to t_2

Did you get the following relation?

$$\boxed{0} = \boxed{\begin{array}{c}\textbf{Input to}\\ \textbf{the system}\\ \textbf{from } t_1 \textbf{ to } t_2\end{array}} - \boxed{\begin{array}{c}\textbf{Output from}\\ \textbf{the system}\\ \textbf{from } t_1 \textbf{ to } t_2\end{array}} \quad (3.3)$$

or

What goes in must come out

a famous truism because so many processes are in the steady state (or are approximately steady-state).

Figure 3.1 involves mass for the flows and initial and final conditions. However, mass is not the only quantity that is included in the category of "material balance." The term *material* can apply to moles or any quantity that is conserved. As an example, look at Figure 3.2 in which we have converted all of the mass quantities in Figure 3.1b to their equivalent in moles.

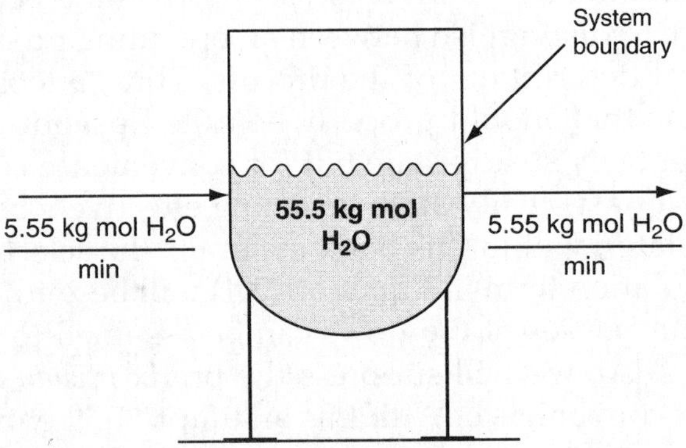

Figure 3.2 Material flows involving moles rather than mass

Unsteady-State System or Process What if you make a change in the process shown previously so that the flow rate out of the system is instantaneously reduced to a constant 90 kg/min from 100 kg/min?

Figure 3.3 shows the initial and final conditions *in* the vessel. Because water accumulates at the rate of 10 kg/min in the system, the amount of water present in the vessel will depend on the period of time during which the rate of accumulation is maintained. Figure 3.3b shows the system after 50 min of accumulation; 50 min of accumulation at 10 kg/min amounts to 500 kg of net accumulation. Because the amount of water in the system at the

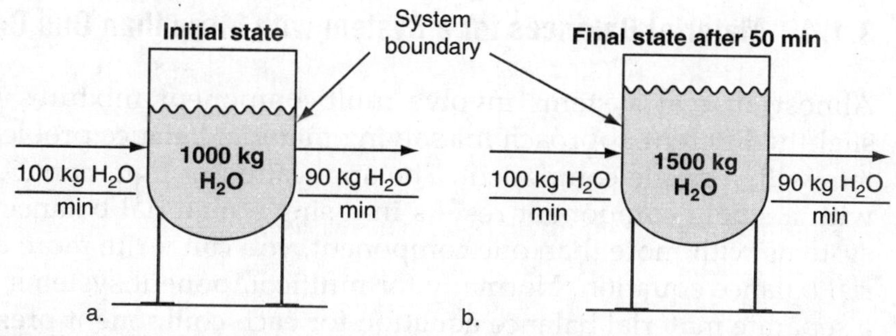

Figure 3.3 (a) Initial and (b) final states of an unsteady-state system

final time differs from that at the initial time, the process and model are deemed to be an **unsteady-state** (or **transient**) process or model.

What else might you change about the process we have been analyzing? Suppose you make the flow rate out of the system 100 kg/min again and reduce the flow rate into the system to 90 kg/min. Note that the amount of water in the system decreases with time at the rate of 10 kg/min. Can you sketch a figure analogous to Figure 3.3b that shows the system after 50 min of operation? Can the accumulation ever be negative? Yes, if some water exists at the start of the interval. Can the amount of water in the tank ever be negative? No. Can your bank balance be negative? Possibly!

Continuous Process A **continuous process** is one in which material enters and leaves the system without interruption. In some problems you know from the problem statement that the flows in and out are not continuous, but because Equation (3.1) is concerned only with the net changes over time, the character of the flows can be assumed to be whatever is convenient.

Batch Process In contrast with a continuous process, a **batch process**, which is a closed process, treats a fixed amount of material each time it operates. The empty system is charged with the batch of material, and at the end of the processing all of the resulting material is removed. In soup preparation, for example, all of the ingredients are placed in an empty pot, heated and stirred, and finally removed from the pot.

Semi-Batch Process In a **semi-batch process**, which is an open process, material enters the process during its operation but none leaves. Similarly, some soup is made by successively placing the necessary ingredients in the pot over time while the cooking occurs.

3.1.4 Material Balances for a System with More than One Component

Almost all real systems involve multicomponent mixtures that require a slightly different approach for solving material balance problems than a system with a single component. The application of Equation (3.2) to a system with a single component results in a single material balance equation. For systems with more than one component, you can write more than one material balance equation. Normally for multicomponent systems, you can write a separate material balance equation for each component present in the system plus one additional material balance equation based on the total mass of the system, but one of the set will be dependent (redundant). In any event, the issue of selecting the right material balance equations to use is addressed in Section 3.2 in the eighth step of the generalized strategy for solving material balance problems.

All the cases considered so far have illustrated the flow of water—a single component. Suppose the input to a vessel contains more than one component, such as 100 kg/min of a 50% water and 50% sugar (sucrose, $C_{12}H_{22}O_{11}$, MW 342.3) mixture. As indicated in Figure 3.4, the vessel initially contains 1000 kg of water, and the exit stream flow is 100 kg/min of water and sugar.

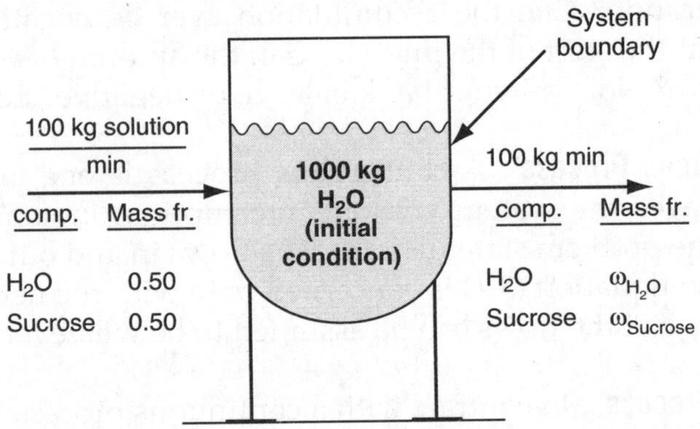

Figure 3.4 A multicomponent (two components) system

How would the material balance for the process in Figure 3.4 differ from the material balances that would be used for the systems displayed in Figures 3.1 through 3.3? Certainly you can write a **total material balance** for the process as before. Will it be a steady-state balance? Yes. In addition, because two components exist, you can write two independent **component material balances**, one for water and one for sucrose. Will they be steady-state balances? No, because sugar starts to accumulate in the system while water is depleted after the addition of the sugar solution.

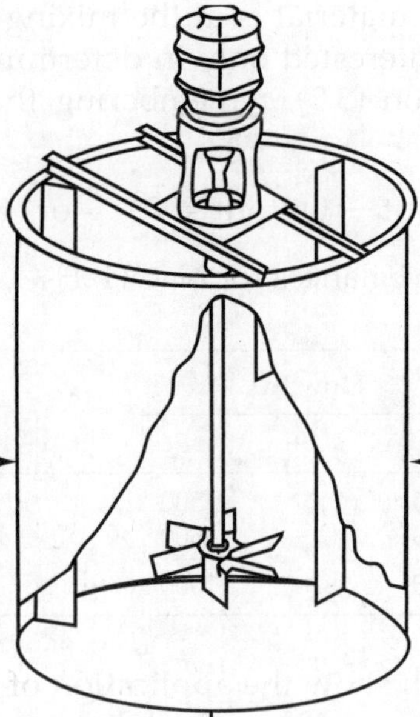

Feed 1 = 9000 kg/hr

Component	Mass fr	kg
NaOH	0.050	450
H_2O	0.950	8550
Total	1.000	9000

Feed 2 = 1000 kg/hr

Component	Mass fr	kg
NaOH	0.50	500
H_2O	0.50	500
Total	1.00	1000

Product = 10,000 kg/hr

Component	Mass fr	kg
NaOH	0.095	950
H_2O	0.905	9050
Total	1.000	10,000

Figure 3.5 Mixing of a dilute stream of NaOH with a concentrated stream of NaOH. Values in the figure are based on 1 hr of operation.

Now look at the mixer shown in Figure 3.5, an apparatus that mixes two streams to increase the concentration of NaOH in a dilute solution. We will use the values of the components listed in Figure 3.5 to show how the total and component balances for mass and moles can be written using Equation (3.2). Note: If the tank is well mixed, the concentration of a component in the output stream will be the same as the concentration of the component inside the tank during mixing, an assumption frequently made that is relatively accurate if adequate mixing is used.

Let's take a convenient basis of 1 hr. Then the values shown in kilograms in the figure next to the mass fractions show the results of 1 hr of mixing. Could you select another basis? Of course, such as 1 min, 1000 kg, and so on, but the chosen basis is the simplest one. Let's also assume that this process is at steady state. What does this assumption mean with respect to the material balances? No *net* accumulation occurs overall. We do not know how much material is in the mixing tank, but does that matter? No, because we are interested only in determining what is leaving the process. Applying Equation (3.3), remembering that the accumulation is equal to zero, yields

$$\text{Input} - \text{Output} = 0 \quad \text{or} \quad \text{Input} = \text{Output}$$

Then the material balances for NaOH, H_2O, and total mass are as follows:

	Flow In				Flow Out		
Balances	F_1	+	F_2	−	**(Product)**	=	**Accum.**
NaOH	450	+	500	−	950	=	0
H_2O	8550	+	500	−	9050	=	0
Total	9000	+	1000	−	10,000	=	0

Next, we will show the application of Equation (3.3) in terms of moles for Figure 3.5. We can convert the kilograms shown in Figure 3.5 to kilogram moles by dividing each compound by its respective molecular weight (NaOH = 40 and H_2O = 18).

	F_1	F_2	Product
NaOH	$\frac{450}{40} = 11.25$	$\frac{500}{40} = 12.50$	$\frac{950}{40} = 23.75$
H_2O	$\frac{8550}{18} = 475$	$\frac{500}{18} = 27.78$	$\frac{9050}{18} = 502.78$

Then the component and total balances in kilogram moles are

	Flow In				Flow Out		
Balances	F_1	+	F_2	−	**(Product)**	=	**Accum.**
NaOH	11.25	+	12.50	−	23.75	=	0
H_2O	475	+	27.78	−	502.78	=	0
Total	486.25	+	40.28	−	536.53	=	0

Following are several examples of material balance problems for systems with two components. We will apply Equation (3.2) to solve them.

Example 3.3 Efficiency of Recovery of DNA

In the development of a procedure to recover DNA from cells and tissue, 20 μg of pure DNA sequences in 500 μg of water was fragmented by sound to 500 bp and smaller sizes. [A bp (base pair) is 0.34 nm (along the helical axis); 10.4 bp is equal to one helical turn in the DNA molecule.] See Figures E3.3a and E3.3b. After cross-linking proteins to the DNA followed by several additional processing and separation steps, the remaining DNA was precipitated from solution, cleaned, and dried, yielding 1.20 μg of DNA. What fraction of the DNA was lost in the processing steps?

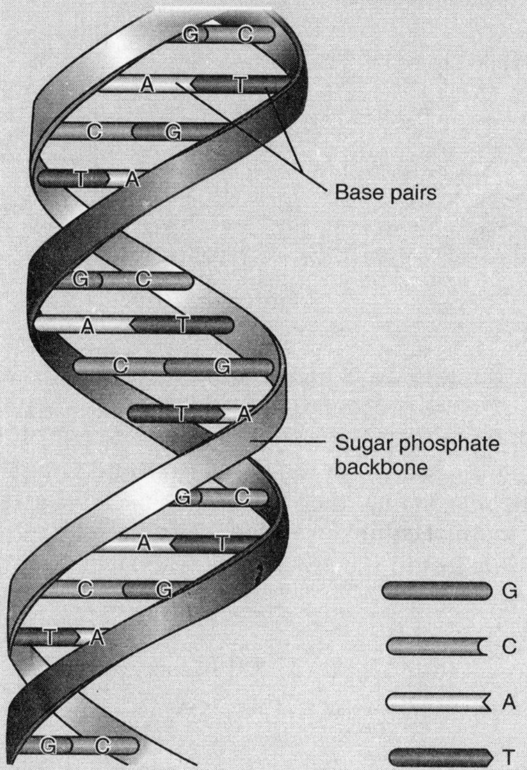

Figure E3.3a Three-dimensional representation of a strand of DNA

In nearly all living organisms, the DNA—which stands for deoxyribonucleic acid—is the molecule that stores genetic information. It forms long linear

(*Continues*)

Example 3.3 Efficiency of Recovery of DNA (*Continued*)

molecules of two intertwined chains called a double helix tied together as shown in Figure E3.3a.

Solution

This is an easy problem, but it illustrates the analysis needed to solve material balance problems. A review of the problem indicates that you can make a DNA balance. Do you have to worry about the water? No, because the processing involves water, and no information is given about that water. Figure E3.3b shows the given information.

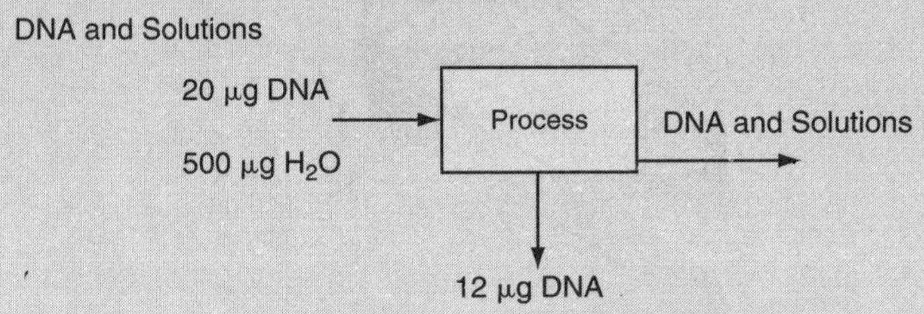

Figure E3.3b

The first thing to do is pick a system, keeping in mind Equation (3.2). Let the system be the process denoted by the box shown in Figure E3.3b. Is the process an open one? Yes. Is it a steady-state one? If you assume that no DNA existed in the process at the initial time and none remained at the final time, as seems reasonable, then the process is a steady-state process. Does "what comes in must come out" apply? Yes. Let x be the output of DNA (the unknown) that is lost in the processing. The DNA balance is on the basis of 20 μg of DNA:

$$\begin{array}{ccccc} \text{Input} & & \text{Output} & & \text{Output} \\ 20\ \mu\text{g DNA} & = & 12\ \mu\text{g DNA} & + & x\ \mu\text{g DNA} \end{array}$$

The solution of the mass balance is $x = 8$ μg DNA, so that the fraction lost in the processing is 8/20 or 0.4.

You, of course, could first calculate the fraction of the DNA recovered, 12/20 or 0.6. How does a material balance enter into the solution then? In effect, 1 μg becomes the basis, and x becomes the desired fraction. The material balance then is $1 = 0.6 + x$ so that x is still equal to 0.4. Could you choose 12 μg as the basis? Yes.

Example 3.4 Concentration of Cells Using a Centrifuge

Centrifuges are used to separate particles in the range of 0.1 to 100 μg in diameter from a liquid using centrifugal force. Yeast cells are recovered from a broth (a liquid mixture containing cells) using a tubular centrifuge (a cylindrical system rotating about a cylindrical axis). Determine the amount of the cell-free discharge per hour if 1000 L/hr are fed to the centrifuge. The feed contains 500 mg cells/L, and the product stream contains 50 wt % cells. Assume that the feed has a density of 1 g/cm^3 and that there are no cells in the broth discharge from the centrifuge.

Solution

Several different types of centrifuges exist. Figure E3.4 implies continuous feed and continuous outputs; hence, you can conclude that the process involves a steady-state, open (flow) system without reaction. Two components are involved: cells and broth. What should you take as a basis? Take a convenient basis of 1 hr. Let P be the desired product and D be the discharge, both in grams.

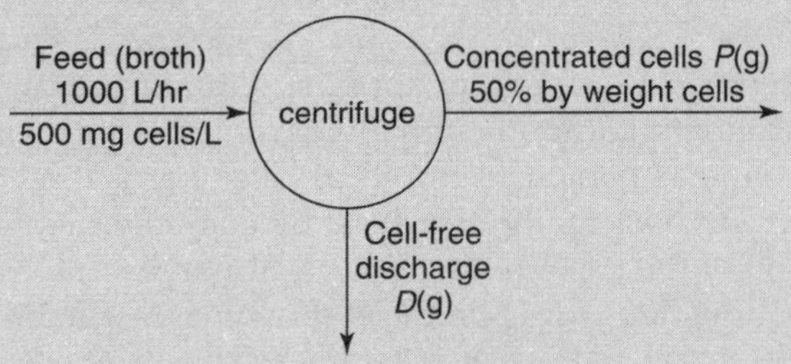

Figure E3.4

Because the accumulation is zero, the material balances (total and both components) are "what goes in must come out." Let us make a cell balance followed by a fluid balance.

Cell balance:

$$\frac{1000 \text{ L feed}}{} \left| \frac{500 \text{ mg cells}}{1 \text{ L feed}} \right| \frac{1 \text{ g}}{1000 \text{ mg}} = \frac{0.50 \text{ g cells}}{1 \text{ g } P} \left| \frac{P \text{ g}}{} \right| \qquad P = 1000 \text{ g}$$

(*Continues*)

Example 3.4 Concentration of Cells Using a Centrifuge (*Continued*)

Fluid balance:

Using the calculated value of P in the fluid balance yields

$$\frac{1000\text{ L}}{}\left|\frac{1000\text{ cm}^3}{1\text{ L}}\right|\frac{1\text{ g fluid}}{1\text{ cm}^3} = \frac{1000\text{ g }P}{}\left|\frac{0.50\text{ g fluid}}{1\text{ g }P}\right. + D\text{ g fluid}$$

$$D = (10^6 - 500)\text{ g fluid}$$

Example 3.5 A Material Balance for the Blending of Gasoline

Will you save money if instead of buying premium 89-octane gasoline at \$2.987 per gallon that has the octane you want, you blend sufficient 93-octane supreme at \$3.137 per gallon with 87-octane regular gasoline at \$2.837 per gallon?

Solution

This problem is an example of applying Equation (3.2) to octane as a component of gasoline, a component that is assumed to be conserved. You can think of the gasoline as being composed of octane and something else, but octane number (grade) is actually measured by engine tests of gasoline. Because the composition of the gasoline is adjusted from season to season, and from location to location, depending on the costs and the availability of its components, making an actual mass balance for each component would involve quite extensive calculations. To get the octane number, we will multiply the units of octane number per gallon by the number of gallons of each grade of gasoline and divide by the number of total gallons to obtain the average octane number, a reasonable step for this example.

What basis should you choose? This can be an example of choosing a basis of what you want, say, 1 gal of 89-octane gasoline, to be added to the tank. Examine Figure E3.5. Could you choose another basis? Of course.

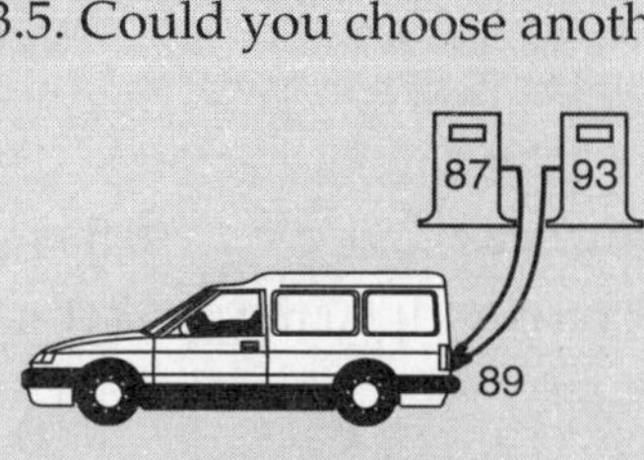

Figure E3.5

First you have to decide whether the system is open or closed, and whether it is steady-state or unsteady-state. In practice, even though gasoline exists in the tank at the start of the blending, we are interested in just the *addition* of gasoline by the choice of the basis. Let us use 1 gal. Clearly the filling is an open system, and it is unsteady-state. The flows into the system will consist of the number of (fractional) gallons of each of the two grades of gasoline and the flow of the octane number as illustrated in Figure E3.5. Thus, the initial octane in the tank for the addition is zero, and the final number is 89. The initial number of gallons of gasoline in the tank is zero, and the final number is 1 gal.

Suppose we let x be the fraction of gallons of 87-octane gasoline added, and y be the fraction of gallons of 93-octane added to the blend.

Octane balance:

Accumulation *Inputs*

$$\left|\frac{89 \text{ octane}}{1 \text{ gal}}\right|\frac{1 \text{ gal}}{} - 0 = \left|\frac{87 \text{ octane}}{1 \text{ gal}}\right|\frac{x \text{ gal}}{} + \left|\frac{93 \text{ octane}}{1 \text{ gal}}\right|\frac{y \text{ gal}}{}$$

Gasoline balance:

Can you make a gasoline balance? You know only the volume of gasoline, not its mass, and each grade will have slightly different components. Because mass is volume times density, will the densities of the input streams and the blend be the same? For all practical purposes, yes. Then the density can be eliminated from each term in the material balance, and you can make a volume balance:

$$x + y = 1$$

The simultaneous solution of the two balances (i.e., two equations containing two unknowns) is $x = 2/3$ gal, and thus $y = 1/3$ gal and the cost of the blended gasoline is

$$2/3\ (\$2.837) + 1/3\ (\$3.137) = \$2.936$$

a value slightly less than the cost of the 89-octane gasoline ($2.987). In practice, refineries take into account the nonlinear blending of gasoline with different octane numbers because one- or two-tenths of an octane number amounts to a significant amount of money for them, considering the volume of gasoline that they sell.

Self-Assessment Test

Questions

1. How does a material balance relate to the concept of the conservation of mass?
2. In an automobile engine, as the valve opens to a cylinder, the piston moves down and air enters the cylinder. Fuel follows and is burned. Thereafter, the combustion gases are discharged as the piston moves up. On a very short timescale, say, a few microseconds, would the cylinder be considered an open or closed system? Repeat for a timescale of several seconds.
3. Without looking at the text, write down the equation that represents a material balance in (a) an open system and (b) a closed system.
4. Can an accumulation be negative? What does a negative accumulation mean?
5. Under what circumstances can the accumulation term in the material balance be zero for a process?
6. What is a transient process? Is it different from an unsteady-state process?
7. Does Equation (3.2) apply to a system involving more than one component?
8. When a chemical plant or refinery uses various feeds and produces various products, does Equation (3.2) apply to each component in the plant?

Problems

1. Draw a sketch of the following processes, and place a dashed line appropriately to designate the system boundary:
   a. A teakettle
   b. A fireplace
   c. A swimming pool
2. Classify the following processes as open, closed, neither, or both:
   a. Oil storage tank at a refinery
   b. Flush tank on a toilet
   c. Catalytic converter on an automobile
   d. Fermentation vessel
3. As an example of a system, consider a bottle of beer. Pick a system.
   a. What is *in* the system?
   b. What is outside the system?
   c. Is the system open or closed?
4. A plant discharges 4000 gal/min of treated wastewater that contains 0.25 mg/L of PCBs (polychloronated biphenyls) into a river that contains no measurable PCBs upstream of the discharge. If the river flow rate is 1500 ft^3/s, after the discharged water has thoroughly mixed with the river water, what is the concentration of PCBs in the river in milligrams per liter?

5. Mr. Ledger deposited $1000 in his bank account and withdrew various amounts of money as listed in the following schedule:

	Withdrawals	Amount Left
	$500	$500
	250	250
	100	150
	80	70
	50	20
	20	0
Total	**$1000**	**$990**

The withdrawals total $1000, but it looks as if Mr. Ledger had only $990 in the bank. Does he owe the bank $10?

Thought Problems

1. Examine Figure SAT3.1TP1. A piece of paper is put into the bell in (1). In picture (2) you set fire to the paper. Ashes are left as shown in picture (3). If everything is weighed (the bell, the dish, and the materials) in each of the three cases, you would find:
   a. Case 1 would have the larger weight.
   b. Case 2 would have the larger weight.
   c. Case 3 would have the larger weight.
   d. None of the above.

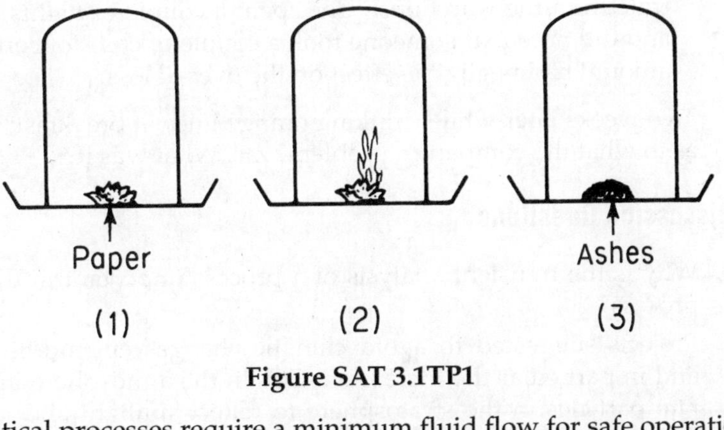

Figure SAT 3.1TP1

2. Certain critical processes require a minimum fluid flow for safe operation during the emergency shutdown of a plant. For example, during normal operation in one process, the chlorine is removed safely from the processing unit along with the flowing liquids. But during an emergency shutdown, the chlorine collects in

the unit and its pipeline headers. Hence a minimum flow rate is needed to remove the chlorine. If the unit and pipelines are considered to be one system, how can a minimum flow rate be obtained for safe operation if the electric power and controller fail?

3. This true story was told to David M. Himmelblau by Professor Gene Woolsey:

   Once upon a time a big manufacturer of canned frozen orange juice concentrate found that 12% to 15% of the concentrate disappeared someplace between Florida and the storage warehouse in the Northeast. A chemical engineer was called on to solve the problem. When she arrived at the Florida plant, the first thing she noticed was a stink caused by a number of dead alligators floating on the surface of an adjacent lake. Ignoring the smell, she carefully followed the processing of the orange juice from squeezing, through concentration, and cooling to slush. She watched the waiting trucks being steam cleaned and sanitized, and then the slush being pumped into the trucks prior to being sent off to the warehouse.

   The trucks were weighed before and after filling. She placed a seal on one or two trucks and flew to the warehouse district to wait for "her" trucks to come in. On arrival she noted that the seals were undisturbed. A pump was attached to the bottom drain of a truck, and the slush was pumped out into a holding tank until the pump started sucking air, at which time the pump was disconnected.

   Then she watched the cans being filled from the holding tank, counted the cans, had some weighed, and noted a minor amount of spillage, but the spillage and filling operation amounted to less than 1% of the overall loss. At this point she concluded that either (a) cheating was going on in Florida, and/or (b) someone was stealing cases or concentrate from the warehouse. She spent a couple of nights in the warehouse and did note that someone took a couple of cases of concentrate, but that amount had negligible effect on the overall loss.

   Two weeks later while drinking orange juice at breakfast, she had a sudden idea as to what the company's problem was. What was it?

Discussion Questions

1. Why is the transient analysis of a process important in the overall analysis of a process?
2. Projects suggested to avoid climatic changes engendered by human activities and in particular the increase in CO_2 in the atmosphere include dispersal of sulfate particles in the stratosphere to reflect sunlight and fertilizing the southern oceans with iron to stimulate phytoplankton growth. It is believed that low levels of iron limit the biological productivity of nutrient-rich southern oceans. Adding iron to these waters would increase the growth of phytoplankton, thus reducing CO_2 levels in the seawater and thereby altering the CO_2 balance between seawater and the atmosphere. What do you think of each suggestion?

Answer to the Puzzle

The additions and subtractions made in the puzzle are invalid. A diagram of the flow of money shows the transfers (amounts on the arrows) and final conditions (the amounts within the boxes):

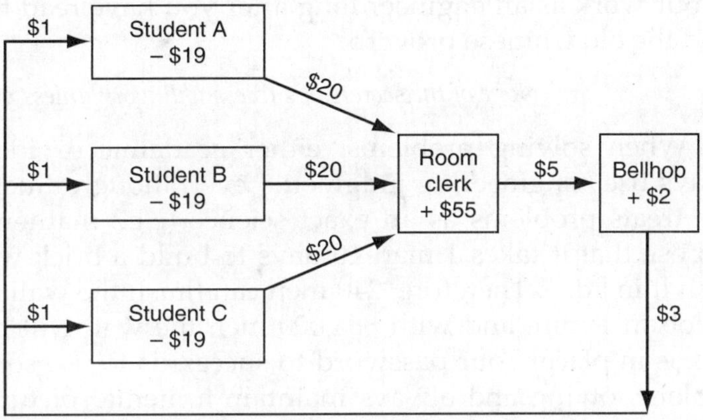

You can see that the money balance should be

$$\$55 + \$2 - (\$19)(3) = 0$$

3.2 A General Strategy for Solving Material Balance Problems

Most of the literature on problem solving views a "problem" as a gap between some initial information (the initial state or problem statement) and the desired information (the final state or the answer to the problem). Problem solving is the activity of finding a path between these two states.

You will find as you go through this book that routine substitution of data into an appropriate equation will not be adequate to solve material (and later energy) balances other than the most trivial ones. You can, of course, formulate your own strategy for solving problems—everyone has a different viewpoint. But adoption of the well-tested general strategy presented in this chapter has been found to significantly ease the difficulty students have when they encounter problems that are not exactly the same as those presented as examples and homework in this book, for example, problems in industrial practice. After all, the problems in this book are only samples, and simple ones at that, of the myriad problems that exist or could be formulated. Even if you pick your individual problem-solving technique, you will find the following steps to be a handy check on your work and a help if you get stuck.

An orderly method of analyzing problems and presenting their solutions represents training in logical thinking that is of considerably greater value than the mere knowledge of how to solve a particular type of problem. Understanding how to approach these problems from a logical viewpoint will help you to develop those fundamentals of thinking that will assist you in your work as an engineer long after you have read this material. Keep in mind the old Chinese proverb:

None of the secrets of success will work unless you do.

When solving problems, either academic or industrial, you should always use "engineering judgment" even though much of your training to date treats problems as an exact science (e.g., mathematics). For instance, suppose that it takes 1 man 10 days to build a brick wall; then 10 men can finish it in l day. Therefore, 240 men can finish the wall in 1 hr, 14,400 can do the job in 1 min, and with 864,000 men the wall will be up before a single brick is in place! Your password to success is to use some common sense in problem solving and always maintain a mental picture of the system that you are analyzing. Do not allow a problem to become abstract and unrelated to physical behavior.

You do not have to follow the steps in the following list in any particular sequence or formally employ every one of them. You can go back several steps and repeat steps at will. You can consolidate steps. As you might expect, when you work on solving a problem, you will experience false starts, encounter extensive preliminary calculations, suspend work for higher-priority tasks, look for missing links, and make foolish mistakes. The strategy outlined here is designed to focus your attention on the main path rather than the detours.

Howe's law: Every person has a scheme which will not work.

Gordon's law: If a project is not worth doing, it is not worth doing well.

1. Read and understand the problem statement.

This means **read the problem carefully** so that you know what is given and what is to be accomplished. Rephrase the problem to make sure you understand it. An anecdote illustrates the point of really understanding the problem.

An English family visiting Khartoum in the Sudan took their young son each day by the statue of General Gordon on a camel. On the last day of their visit to the statue, as the family was leaving, the boy asked, "Who was that man that was sitting on General Gordon?"

Here is a question to answer:

How many months have 30 days?

Now you may remember the mnemonic "30 days hath September . . ." and give the answer as 4, but is that what the question concerns—how many months have exactly 30 days? Or, does the question ask how many months have at least 30 days (the answer being 11)? Individuals reading the same problem frequently have different perspectives. If the streets in your town are numbered consecutively from 1 to 24, and you are asked by a stranger what street comes after 6th Street, you would be likely to respond 7th Street, whereas the stranger, if facing the opposite direction, would more likely be interested in 5th Street.

Be sure to decide if a problem is a simple or complex calculation and involves a steady-state or unsteady-state process, and when your calculations are completed, state your conclusion somewhere on your calculation sheet or computer printout, say, at the end or the beginning.

Example 3.6 Understanding the Problem

A train is approaching a station at 105 cm/s. A man in one car is walking forward at 30 cm/s relative to the seats. He is eating a foot-long hot dog, which is entering his mouth at the rate of 2 cm/s. An ant on the hot dog is running away from the man's mouth at 1 cm/s. How fast is the ant approaching the station? Cover the solution below, and try to determine what the problem requests before peeking.

Solution

As you read the problem, make sure you understand how each piece of information meshes with the others. Would you agree that the following is the correct analysis?

A superficial analysis would ignore the hot dog length but would calculate $105 + 30 - 2 + 1 = 134$ cm/s for the answer. However, on more careful reading it becomes clear that the problem states that the ant is moving away from the man's mouth at the rate of 1 cm/s. Because the man's mouth is moving toward the station at the rate of 135 cm/s, the ant is moving toward the station at the rate of 136 cm/s.

2. Draw a sketch of the process and specify the system boundary.
It is always good practice to begin solving a problem by drawing a sketch of the process or physical system. You do not have to be an artist to make a sketch. A simple box or circle drawn by hand to denote the system boundary with some arrows to designate flows of material will be fine. You can also state what the system is in words or with a label. Figure 3.6 illustrates some examples.

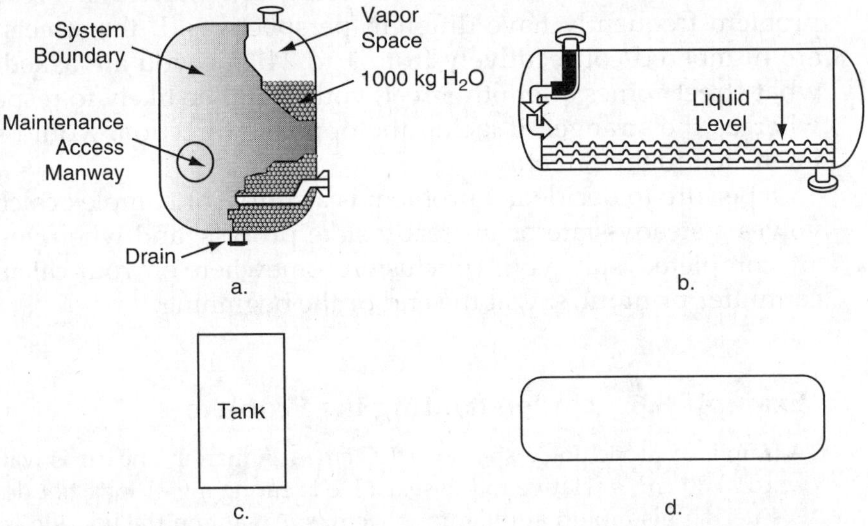

Figure 3.6 Examples of sketches used to represent process equipment

Figure 3.6c adequately represents Figure 3.6a, and Figure 3.6d represents Figure 3.6b because the internal details do not normally affect the application of Equations (3.1) and (3.2). The diagram itself will indicate if the system is open or closed.

3. Place labels (symbols, numbers, and units) on the diagram for all of the known flows, materials, and compositions.
By putting data on the diagram, you will avoid having to look back at the problem statement repeatedly and will also be able to clarify what data are missing. For the unknown flows, materials, and compositions, insert symbols and units. Add any other useful relations or information. What kinds of information might you place on the diagram? Some specific examples are

- Stream flow rates (e.g., $\dot{F} = 100$ kg/min)
- Compositions of each stream (e.g., $x_{H_2O} = 0.40$)
- Given flow ratios (e.g., $F/R = 0.7$)
- Given identities (e.g., $F = P$)
- Yields (e.g., Y kg/X kg = 0.63)
- Efficiency (e.g., 40%)
- Specifications for a variable or a constraint (e.g., $x < 1.00$)
- Conversion (e.g., 78%)
- Equilibrium relationships (e.g., $y/x = 2.7$)
- Molecular weights (e.g., MW = 129.8)

How much data should you place on the diagram? Enough to help solve the problem and be able to interpret the answer. Values of variables that are zero because they are not present in the problem can be ignored. If your diagram becomes too crowded with data, make a separate table and key it to the diagram. Be sure to include the units associated with the flows and other material when you write the numbers on your diagram or in a table. Remember that units make a difference!

Some of the essential data may be missing from the problem statement. If you do not know the value of a variable to put on the figure, you can substitute a symbol such as F_1 for an unknown flow or ω_1 for a mass fraction. The substitution of a symbol for a number will focus your attention on searching for the appropriate information needed to solve the problem.

Example 3.7 Placing the Known Information on the Diagram

A continuous mixer mixes NaOH with H_2O to produce an aqueous solution of NaOH. The problem is to determine the composition and flow rate of the product if the flow rate of NaOH is 1000 kg/hr and the ratio of the flow rate of the H_2O to the product solution is 0.9. Draw a sketch of the process and put the data and unknown variables on the sketch with appropriate labels. We will use this example in subsequent illustrations of the proposed strategy.

Solution

Because no contrary information is provided about the composition of the H_2O and NaOH streams, we will assume that they are 100% H_2O and

(Continues)

Example 3.7 Placing the Known Information on the Diagram (*Continued*)

NaOH, respectively. Look at Figure E3.7 for a typical way the data might be put on a diagram (Figure E3.7 looks nicer than the one that you probably would draw because it was drawn by a draftsman).

Basis: 1 hr ≡ 1000 kg

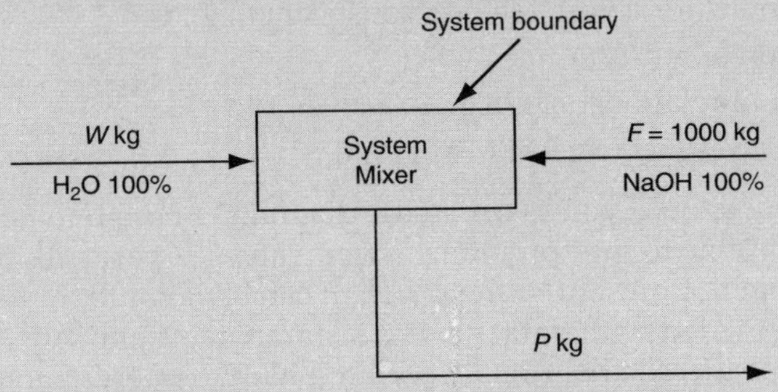

Component	kg	ω (add if useful)
NaOH	P_{NaOH}	$\omega^P_{NaOH} = \frac{P_{NaOH}}{P}$
H_2O	P_{H_2O}	$\omega^P_{H_2O} = \frac{P_{H_2O}}{P}$
Total	P	1.00

Figure E3.7

Note that the composition of the product stream is listed along with the symbols for unknown flows. Could you have listed the mass fractions instead of or in addition to the mass flows? Of course. Because you know the ratio $W/P = 0.9$, why not add that ratio to the diagram at some convenient place?

You will find it convenient to use a consistent set of algebraic symbols to represent the variables whose values are unknown (called the **unknowns**) in a problem. In this book we frequently use mnemonic letters to represent the flow of material, both mass and moles, with the appropriate units attached or inferred, as illustrated in Figure E3.7.

When useful, employ m for the flow of mass and n for the flow of moles with appropriate subscripts and/or superscripts to make the meaning crystal clear. Table 3.1 lists some examples. In specific problems pick obvious or mnemonic letters such as W for water and P for product to avoid confusion. If you run out of suitable letters of the alphabet, you can always insert superscripts to distinguish between streams such as F^1 from F^2 or label streams as $F1$ and $F2$. Letters for flow *rates* should have overlay dots imposed. Make the dot big enough to distinguish it from a speck of something on your paper.

In the beginning there was the symbol
David Hilbert

Table 3.1 Some Examples of the Symbols Used in This Book

Symbol	Designates
F kg	Flow of mass in kilograms
F_{Total} or F_{Tot}	Total flow of material*
F^1	Flow in stream number 1*
F_{A} lb	Flow of component A in stream F in pounds
m_{A}	Mass flow of component A*
m_{Total} or m_{Tot}	Mass flow of the total material*
m_{A}^{F1}	Mass flow of component A in stream $F1$*
n_{A}^{W}	Molar flow of component A in stream W*
ω_{A}^{F}	The mass (weight) fraction of A in stream F (The superscript is not required if the meaning is otherwise clear.)
x_{A}^{F}	The mole fraction of A in stream F, a liquid (The superscript is not required if the meaning is otherwise clear.)
y_{A}^{F}	The mole fraction of A in stream F, a gas

*Units not specified but inferred from the problem statement

4. Obtain any data you know are needed to solve the problem but are missing.

Never assume the obvious is true.
William Safire in The Sleeper Spy *(Random House, 1995)*

An evaporator cost $34,700. How much did it cost per pound? Clearly, something is missing from the problem statement. Table 3.2 is a clever list of the degrees of ignorance. Look at this table and decide what your level of ignorance is for the evaporator problem. Did you pick Level 1? We hope you are not at Level 2! You have to find out what is the weight (mass) of the evaporator.

Table 3.2 Armour's Laws of Ignorance*

Order of Ignorance		State of Mind
0	Lack of ignorance	You know something.
1	Lack of knowledge	You don't know something.
2	Lack of awareness	You don't know that you don't know something.
3	Lack of process	You don't know an efficient way to find out that you don't know that you don't know something.
4	Meta-ignorance	You don't know about the five orders of ignorance.

*P. G. Armour, *Commun. ACM*, **44**, 15 (2001).

Here is another example: How do you pronounce the name of the capital of Kentucky: "Loo-EE-ville" or Loo-ISS-ville"? If you pick either one, you have demonstrated ignorance at Level 2! Hint: Look at a map.

When you review a problem, you may immediately notice that some essential detail is missing in the problem statement such as a physical property (molecular weight, density, etc.). You can look up the values in a physical properties database such as the one on the CD that accompanies this book, or in reference books, on the Web, and in many other places. Or, some value may be missing, but you can calculate the value in your head. For example, you are given a stream flow that contains just two components; one is H_2O and the other, NaOH. You are given the concentration of the NaOH as 22%. There is no point in writing a symbol on the diagram for the unknown concentration of water. Just calculate the value of 78% in your head and put that value on the diagram.

5. Choose a basis.

We discussed the topic of basis in Chapter 2, where we suggested three ways of selecting a basis:

1. What do I have?
2. What do I want to find?
3. What is convenient?

Although picking a basis is listed in Step 5 in the proposed strategy, frequently you know what basis to pick immediately after reading the problem statement and can enter the value on your process diagram at that time. Although the basis we chose for the problem stated in Example 3.7 was 1 hr, you could pick some other basis (but it would not be as convenient).

Be sure to write the word *Basis* on your calculation page, and enter the value and associated units so that you, and anyone who reads the page, can later on (weeks or months later on) know what you did. Choosing a basis should eliminate at least one unknown.

6. Determine the number of variables whose values are unknown (the unknowns).

Plan ahead.

Unknown

Note: Frequently you will find it convenient to combine Steps 6, 7, and 8 as an aggregate to save space, but here we will explain each step separately to focus on the details of the thought process that should occur as you proceed with the solution of a problem.

Determination of the number of unknowns in a problem is somewhat subjective. No unique number exists. Different views of what is known and not known yield different counts. The general objective in solving problems by hand is to reduce the number of simultaneous equations that have to be solved by assigning known values to as many variables as possible at the start of the count. Also, it is sensible to assign values to variables that you can calculate in your head. For example, if an input stream F is assigned a value of 100 kg because that value was selected as the basis, and you know that the input contains 60% NaCl and the other component is KCl, you can easily calculate that 60 kg of NaCl and 40 kg of KCl enter the system. If you plan to fill in a dialog box for a solution by a computer, you may be able to place all of the assigned values first in the proper cells and place any other known facts in the set of simultaneous equations without making any preliminary simple calculations. You can omit from consideration values of variables that are zero. They do not exist.

If you did a thorough job in placing either notation for the variables or values of the variables on a figure as indicated in Figure E3.7, or made a list of them, determining the number of unknowns is easy. Each time you assign a value to a variable, you reduce the number of unknowns by one. In assigning a value to a variable, you are actually making use of a trivial equation; for example, if the input is given as 100 kg, then $F = 100$ kg is the assignment. Proceed to assign values to variables until you run out of specifications to assign. Because each specification is an equation, be careful not to assign specifications so as to generate a redundant equation in carrying out Steps 6, 7, or 8. For example, if you know the mass flow and the mass fraction of one of the two components in the flow, you cannot arbitrarily assign a value to the other component. You can make a calculation in your head and eliminate the second component as an unknown, but in so doing you have used up one of the possible independent equations in making the calculation. Which one? The implicit relation $\sum n_i^F = F$.

The basic idea in Step 6 is to reduce the unknowns to as few variables as possible based on the problem specifications plus calculations that you can carry out in your head so that you have to solve a minimum number of simultaneous equations to complete the solution. In the problem stated in Example 3.7 from which Figure E3.7 was prepared, how many unknowns exist? There are nine variables, but you can assign values to all but four. We do not know the values of the following variables: W, P, P_{NaOH}, and P_{H_2O}, or, alternatively, W, P, ω_{NaOH}, and ω_{H_2O}. In light of the necessary conditions stated in the next step, Step 7, you should be thinking about assembling four independent equations to solve the mixing problem.

7. **Determine the number of independent equations and carry out a degree-of-freedom analysis.**

Traveling through a maze looks easy from above.

Unknown

IMPORTANT COMMENT

Before proceeding with Step 7, we need to call to your attention an important point from mathematics related to solving equations. Steps 6 and 7 focus on determining whether you can solve a set of equations formulated for a material balance problem. For simple problems, if you omit Steps 6 and 7 and proceed directly to Step 8 (writing equations), you probably will not be bothered by skipping the steps. However, for complicated problems, you can easily run into trouble if you neglect them. Computer-based process simulators take great care to make sure that the equations you formulate indeed can be solved.

What does solving a material balance problem mean? For our purposes it means finding a unique answer to a problem. If the material balances you write are linear independent equations (refer to Chapter 12 if you are not clear as to what **linear or independent equation** means), as will be the vast majority of the equations you write, you are guaranteed to get a unique answer if the following necessary condition is fulfilled:

The number of variables whose values are unknown equals the number of independent equations you formulate to solve a problem.

To check the sufficient conditions for this guarantee, refer to Chapter 12.

In Step 7 you want to *preview* the compilation of equations you plan to use to solve a problem, making sure that you have an appropriate number of independent equations. Step 8 pertains to actually formulating the equations. Steps 7 and 8 are frequently merged. What kinds of equations should you be thinking about?

a. The **material balances** themselves
You can write as many independent material balance equations as there are species involved in the system. In the specific case of the problem stated in Example 3.7, you have two species, NaOH and H_2O, and thus can write two independent material balance equations. If for the problem posed in Example 3.7 you write three material balances:

- One for the NaOH
- One for the H_2O
- One total balance (the sum of the two component balances)

only two of the three equations are independent. You can use any combination of two of the three in solving the problem.

b. The **basis** (if not already assigned in Step 6).

c. **Explicit relations** specified (the **specifications**) in the problem statement such as $W/P = 0.9$ stated in Example 3.7, or specified relations among the variables given in the problem statement, if not used in Step 6.

d. **Implicit relations**, particularly the sum of the mass or mole fractions in a stream being unity, or, alternatively, the sum of the amounts of each of the component materials equaling the total material. In Example 3.7 you have

$$\omega_{\text{NaOH}}^{P} + \omega_{\text{H}_2\text{O}}^{P} = 1$$

or, multiplying both sides of the equation by P, you get the equivalent equation

$$P_{\text{NaOH}} + P_{\text{H}_2\text{O}} = P$$

Frequently Asked Questions

1. What does the term *independent equation* mean? You know that if you add two independent equations together to get a third equation, the set of three equations is said to be *not independent*; they are said to be *dependent*.

Only two of the equations are said to be independent because you can add or subtract any two of them to get the third equation. Figures 3.7 and 3.8 illustrate some examples of independent and dependent equations. Chapter 12 explains in considerably more detail what independence means.

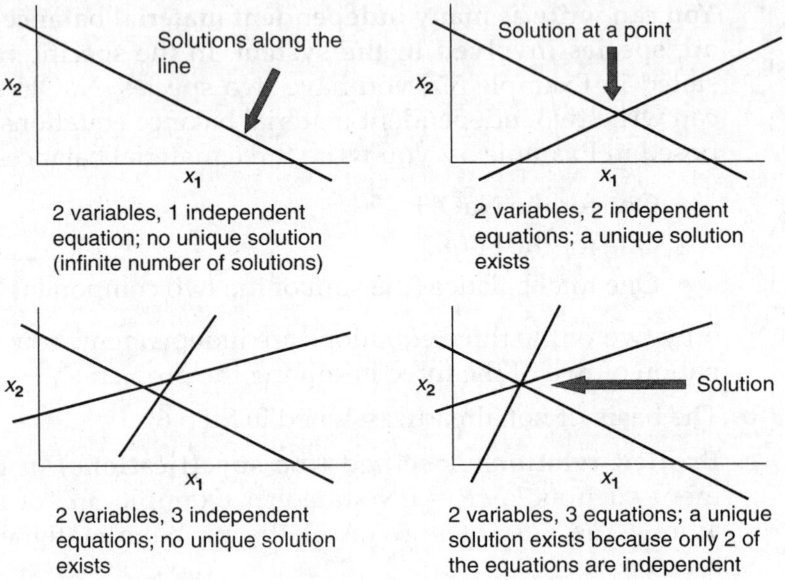

Figure 3.8 Further illustrations of independent and nonindependent equations

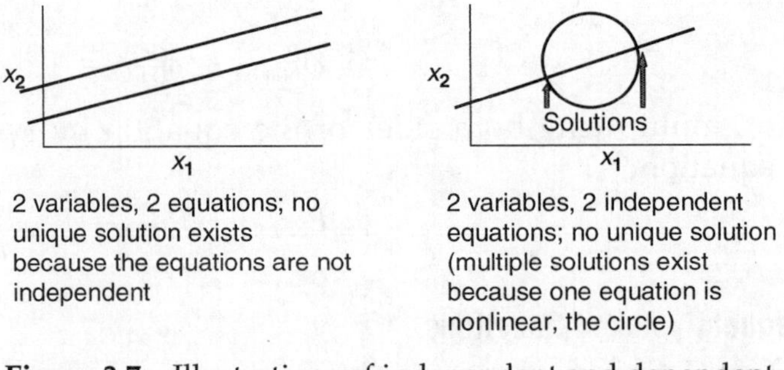

Figure 3.7 Illustrations of independent and dependent equations

2. If I have several equations, how can I tell if they are independent? The best thing to do is use a software program to make the calculations. MATLAB, Mathcad, Mathematica, Excel, Polymath (on the CD in the

back of this book), and many other equation-solving programs will tell you that an error exists, usually that the equations are not independent when you attempt to solve a set of equations that are not independent. For more information on how to use software to determine whether or not equations are independent, refer to Chapter 12.

Once you have determined the number of unknowns and independent equations, it helps to carry out an analysis called a **degree-of-freedom analysis** to determine whether a problem is solvable or not. The difference is called the degrees of freedom available to the designer to specify flow rates, equipment sizes, and so on. You calculate the number of degrees of freedom (N_D) as follows, using the number of unknowns (N_U) and the number of independent equations (N_E):

$$N_D = N_U - N_E$$

When you calculate the number of degrees of freedom (N_D), you can ascertain what the solubility of a problem is. Three outcomes occur:

Case	ND	Classification for Solution
$N_U = N_E$	0	**Exactly specified (determined)**; a solution exists.
$N_U > N_E > 0$	> 0	**Underspecified (determined)**; more independent equations required.
$N_U < N_E < 0$	< 0	**Overspecified (determined)**; in general, no solution exists unless some constraints are eliminated or some additional unknowns are included in the problem.

For the problem in Example 3.7:

From Step 6: $N_U = 4$
From Step 7: $N_E = 4$

so that

$$N_D = N_U - N_D = 4 - 4 = 0$$

and a unique solution exists for the problem.

Example 3.8 Analysis of the Degrees of Freedom

A cylinder containing CH_4, C_2H_6, and N_2 has to be prepared containing a mole ratio of CH_4 to C_2H_6 of 1.5 to 1. Available to prepare the mixture are (1) a cylinder containing a mixture of 80% N_2 and 20% CH_4, (2) a cylinder

(Continues)

Example 3.8 Analysis of the Degrees of Freedom (*Continued*)

containing a mixture of 90% N_2 and 10% C_2H_6, and (3) a cylinder containing pure N_2. What is the number of degrees of freedom, that is, the number of independent specifications that must be made, so that you can determine the relative contributions from each cylinder to get the desired composition in the cylinder with the three components?

Solution

A sketch of the process greatly helps in the analysis of the degrees of freedom. Look at Figure E3.8. No specific amount of gas is required to be prepared; only the relative contribution from each cylinder is needed. Consequently, you can take as a convenient basis any value of the unknowns, although picking one of the *F*s makes the most sense. Pick F_1 = 100 mol as the basis. (Did you contemplate using mass as the basis for a gas stream?)

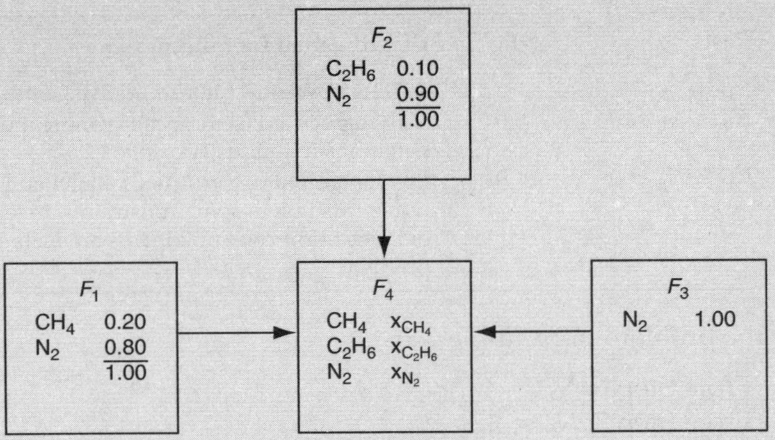

Figure E3.8

First count the number of variables, ignoring the ones whose value is zero. Do you get 12? Look at the following list. The next step is to determine the number of unknowns by assigning all of the known values to their respective variables, values that you can easily find from the problem statement and other sources or can easily calculate in your head. You can assign values to 6 variables. The ?s designate the unknowns.

$$n_{F_1}^{CH_4} = 20 \text{ specified } [(100)(0.20)] \qquad n_{F_2}^{N_2} = 0.90F_2 \text{ specified}$$

$n_{F_1}^{N_2} = 80$ specified $[(100)(0.80)]$ $\quad n_{F_2}^{C_2H_6} = 0.10F_2$ specified

$F_1 = 100$ used sum of n_i in F_1 $\quad F_2 = ?$

$n_{F_3}^{N_2} = (1.00)\ F_3$ $\quad F_3 = ?$

$n_{F_4}^{C_2H_6} = ?$ $\quad n_{F_4}^{CH_4} = ?$

$n_{F_4}^{N_2} = ?$ $\quad F_4 = ?$

Each one of the assignments is equivalent to one equation. The count of the question marks is six. Can you find any other obvious values to assign? If not, what independent equations can you involve to solve the problem? What about

Three (3) species material balances: CH_4, C_2H_6, and N_2
One (1) specified ratio: moles of CH_4 to C_2H_6 equals 1.5
One (1) implicit equation: sum of the mole fractions for product stream

Therefore, a total of five independent equations can be written for this problem. Thus, six minus five equals one degree of freedom. Keep in mind that you must be careful when using equations to formulate a set of independent equations.

Example 3.9 Analysis of the Degrees of Freedom

Examine Figure E3.9, which labels each of the components and streams in a process (say, centrifugation or dielectrophoresis) to separate living cells (superscript *a* for alive) from dead cells (superscript *d*) in water (superscript *W*). If the values of the mass fractions, x_F^W, x_P^W, x_F^a, x_P^d, as well as F, are prespecified (known), how many degrees of freedom remain that can be specified for the process? What values for the unknowns could be specified? All units are in mass.

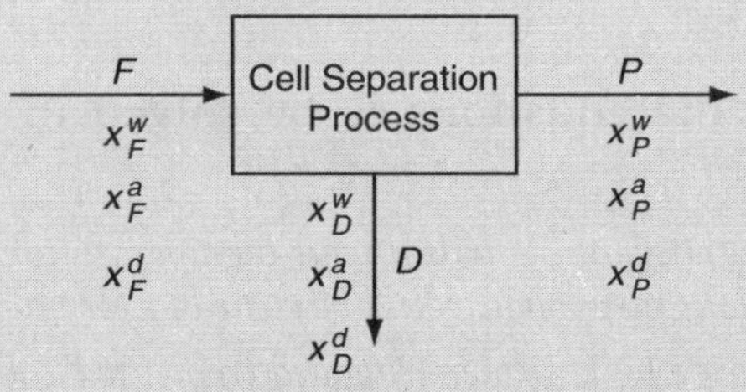

Figure E3.9

(Continues)

Example 3.9 Analysis of the Degrees of Freedom (*Continued*)

Solution

Steps 1–5

See Figure E3.9. The basis is F.

Step 6

Number of unknowns. Each stream has four labels; hence (4)(3) = 12 total variables exist, of which 5 values are prespecified so that 7 unknowns exist:

$$D, P, x^d_F, x^a_P, x^W_D, x^a_D, x^d_D$$

Step 7

Number of independent equations needed: 7

Material balances: You can write four material balances, three component and one total, of which three are independent.

Sum of mass fractions equations: You can write three sum of mass fractions, one for each stream that is independent.

Thus 7 − 6 = 1 degree of freedom exists.

In regard to picking the variable to be specified, you have to be careful. Do not pick a value for an unknown that will be redundant information or render one or more of the equations you have selected in Step 7 to become inadvertently dependent. For example, if you specify the value of x^d_F, because you have already counted the relation $(x^w_F + x^a_F + x^d_F) = 1$ as one of the independent equations, specification of x^d_F will not add any new information to the problem.

A comment: At the start of the analysis in Step 6, if you wanted you could have calculated the values of x^d_F and x^a_P by applying the respective sum of mass fraction equations in your head and reducing the number of unknowns by two and the independent equations by two (because of using two respective sum of mass fraction equations).

8. Write down the equations to be solved in terms of the knowns and unknowns.

Thus [Beatrice] began: "You dull your own perceptions with false imaginings and do not grasp what would be clear but for your preconceptions."

From Beatrice *by H. Rider Haggard (reprinted 2004 by Kessinger Publishing)*

Once you have concluded from the degree-of-freedom analysis that you can solve a problem, you are well prepared to write down the equations to be solved (if you have not already done so as part of Step 7). Bear in mind that some formulations of the equations are easier to solve by hand, and even by using a computer, than others. In particular, **you should attempt to write linear equations rather than nonlinear ones**. Recall that the product of variables, or the ratios of variables, or a logarithm or exponent of a variable, and so on, in an equation causes the equation to be nonlinear.

In many instances you can easily transform a nonlinear equation to a linear one. For instance, in the problem posed in Example 3.7, one constraint given was that $W/P = 0.9$, a nonlinear equation. If you multiply both sides of the equation by P, you obtain a linear equation, $W = 0.9P$.

Another example of the judicious formulation of equations that we mentioned previously occurs in the choice of using a mass or mole flow, such as m or n, versus using the product of $\omega^P_{H_2O}$, the mass fraction of water in P, times P as two variables:

$$m_{H_2O} = \omega^P_{H_2O}P \quad \text{or} \quad n_{H_2O} = y^P_{H_2O}P$$

If you use the product $\omega^P_{H_2O}P$ for m_{H_2O} in the material balance for water, instead of having a linear equation for the water balance

$$F(0) + W(1.000) = m_{H_2O}$$

you would have

$$F(0) + W(1.000) = \omega^P_{H_2O}P$$

a nonlinear equation (which is why we did not use the product).

With these ideas in mind, you can formulate the set of equations to be used to solve the problem in Example 3.7. First, introduce the five specifications into the two material balances and into the summation of moles in P (or its equivalent, the summation of mass fractions).

Then you will obtain a set of four independent equations in four unknowns which were identified in Step 6. The basis is still 1 hr (F = 1000 kg), and the process has been assumed to be steady-state. Recall from Section 3.1 that in such circumstances a material balance simplifies to *in* = *out* or *in* − *out* = 0.

NaOH balance:	$1000 = P_{NaOH}$	or	$1000 - P_{NaOH} = 0$	(1)
H_2O balance:	$W = P_{H_2O}$	or	$W - P_{H_2O} = 0$	(2)
Given ratio:	$W = 0.9P$	or	$W - 0.9P = 0$	(3)
Sum of components in P:	$P_{NaOH} + P_{H_2O} = P$	or	$P_{NaOH} + P_{H_2O} - P = 0$	(4)

Could you substitute the total mass balance $1000 + W = P$ for one of the two component mass balances? Of course. In fact, you could calculate P by solving just two equations:

Total balance:	$1000 + W = P$
Given ratio:	$W = 0.9P$

Substitute the second equation into the first equation and solve for P.

You can conclude that the symbols you select in writing the equations and the particular equations you select to solve a problem do make a difference and require some thought. With practice and experience in solving problems, this issue should resolve itself for you.

9. Solve the equations and calculate the quantities asked for in the problem.

Problems worthy of attack prove their worth by hitting back.

Piet Hein

Industrial-scale problems may involve thousands of equations. Clearly, in such cases efficient numerical procedures for the solution of the set of equations are essential. Process simulators exist to carry out the task on a computer as explained in Chapter 16. Because most of the problems used in this text have been selected for the purpose of communicating ideas, you will find that their solution will involve only a small set of equations and can usually be solved for one unknown at a time using a sequential solution procedure. You can solve two or three equations by successive elimination of unknowns from the equations. For a larger set of equations or for nonlinear equations, use a computer program such as Polymath, Excel, MATLAB, or Mathcad. You will save time and effort by so doing.

Learn to be efficient at problem solving.

For example, when given data in the AE system of units, say, pounds, do not first convert the data to the SI system, say, kilograms, solve the problem, and then convert your results back to the AE system of units. The procedure will work, but it is quite inefficient and introduces unnecessary opportunities for numerical errors to occur.

Select a precedence order for solving the equations you write. One choice of an order can be more effective than another. We showed in Step 8 how the choice of the total balance plus the ratio $W/P = 0.9$ led to two coupled equations that could easily be solved by substitution for P and then W to get

$$P = 10{,}000$$
$$W = 9000$$

From these two values you can calculate the amount of H_2O and NaOH in the product:

$$\text{NaOH balance:} \qquad P_{\text{NaOH}} = 1000 \text{ kg}$$

$$\text{From the} \begin{cases} \text{NaOH balance} \\ H_2O \text{ balance} \end{cases} \text{you get} \begin{cases} P_{\text{NaOH}} = 1000 \text{ kg} \\ P_{H_2O} = 9000 \text{ kg} \end{cases}$$

so that

$$\omega^P_{\text{NaOH}} = \frac{1000 \text{ kg NaOH}}{10{,}000 \text{ kg total}} = 0.1$$

$$\omega^P_{H_2O} = \frac{9000 \text{ kg } H_2O}{10{,}000 \text{ kg total}} = 0.9$$

Examine the set of four equations listed in Step 8. Can you find a shorter or easier sequence of calculations to get a solution for the problem?

10. Check your answer(s).

Error is a hardy plant; it flourishes in every soil.

Martin F. Tupper

Everyone makes mistakes. What distinguishes good engineers is that they are able to find their mistakes before they submit their work. In Chapter 2 we listed several ways to validate your solution. We will not repeat them here. Refer back to Chapter 2. A good engineer uses his or her accumulated knowledge as a primary tool to make sure that the results obtained for a problem (and the data used in the problem) are reasonable. Mass fractions should fall between zero and one. Flow rates normally should be nonnegative.

In any collection of data, the figure that is most obviously correct—beyond all need of checking—is the mistake.

Unknown

To the list of validation techniques that appeared in Chapter 2, we want to add one more very useful one. After solving a problem, use a redundant equation to check your values. In the problem in Example 3.7 that we have been analyzing, one of the three material balances is redundant (not independent), as we have pointed out several times. Suppose you solved the problem using the NaOH and H_2O balances. Then the total balance would have been a redundant balance and could be used to check the answers:

$$P_{NaOH} + P_{H_2O} = P$$

Insert the numbers

$$1000 + 9000 = 10{,}000$$

> Here is a summary of the set of ten steps for solving material balance problems that we have just discussed:
>
> 1. Read and understand the problem statement.
> 2. Draw a sketch of the process and specify the system boundary.
> 3. Place labels for unknown variables and values for known variables on the sketch.
> 4. Obtain any missing needed data.
> 5. Choose a basis.
> 6. Determine the number of unknowns.
> 7. Determine the number of independent equations, and carry out a degree-of-freedom analysis.
> 8. Write down the equations to be solved.
> 9. Solve the equations and calculate the quantities asked for.
> 10. Check your answer(s).

Table 3.3 compares the skills of a novice in problem solving with those of an expert.

Table 3.3 A Comparison of the Problem-Solving Habits of a Novice and an Expert

A Novice	An Expert
Starts solving a problem before fully understanding what is wanted and/or what a good route for solution will be	Reviews the entire plan mentally, explores alternative solution strategies, and clearly understands what result is to be obtained
Focuses only on a known problem set that he or she has seen before and tries to match the problem with one in the set	Concentrates on similarities to and differences from known problems; uses generic principles rather than problem matching
Emphasizes speed of solution; is unaware of blunders	Emphasizes care and accuracy in the solution

A Novice	An Expert
Does not follow an organized plan of attack; jumps about and mixes problem-solving strategies	Goes through the problem-solving process step by step, checking and reevaluating, and backs up from dead ends to another valid path
Is unaware of missing data, concepts, laws	Knows what principles might be involved and where to get missing data
Exhibits bad judgment; makes unsound assumptions	Carefully evaluates the necessary assumptions
Gives up solving the problem because of frustration	Perseveres

Self-Assessment Test

Questions

1. What does the concept "solution of a material balance problem" mean?
2. How many values of unknown variables can you compute (a) from one independent material balance; (b) from three; (c) from four material balances, three of which are independent?
3. What does the concept of independent equations mean?
4. If you want to solve a set of independent equations that contain fewer unknown variables than equations (the overspecified problem), how should you proceed with the solution?
5. What is the major category of implicit constraints (equations) you encounter in material balance problems?
6. If you want to solve a set of independent equations that contain more unknown variables than equations (the underspecified problem), what must you do to proceed with the solution?
7. As I was going to St. Ives,
   I met a man with seven wives,
   Every wife had seven sacks,
   Every sack had seven cats,
   Every cat had seven kits,
   Kits, cats, sacks, and wives
   How many were going to St. Ives?

Problems

1. A water solution containing 10% acetic acid is added to a water solution containing 30% acetic acid flowing at the rate of 20 kg/min. The product P of the

combination leaves at the rate of 100 kg/min. What is the composition of P? For this process:

a. Determine how many independent balances can be written.
b. List the names of the balances.
c. Determine how many unknown variables can be solved for.
d. List their names and symbols.
e. Determine the composition of P.

2. Can you solve these three material balances for F, D, and P?

$$0.1f + 0.3D = 0.2P$$

$$0.9F + 0.7D = 0.8P$$

$$F + D = P$$

3. How many values of the concentrations and flow rates in the process shown in Figure SAT3.2P3 are unknown? List them. The streams contain two components, 1 and 2.

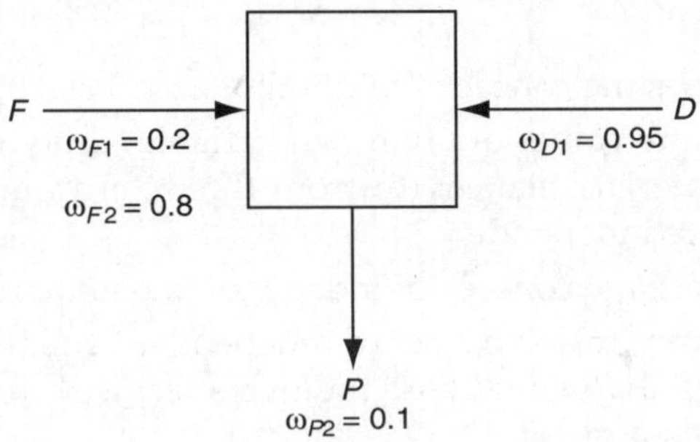

Figure SAT3.2P3

4. How many material balances are needed to solve problem 3? Is the number the same as the number of unknown variables? Explain.

Thought Problem

In the steady-state flow process shown in Figure SAT3.2TP1, a number of values of ω (mass fraction) are not given. Mary says that nevertheless the problem has a unique solution for the unknown values of ω. Kelly says that four values of ω are missing, that you can write three component material balances, and that you can use three relations for $\Sigma_i^n \omega_i = 1$, one for each stream, a total of six equations, so that a unique solution is not possible. Who is right?

Discussion Questions

1. Isotope markers in compounds are used to identify the source of environmental pollutants, investigate leaks in underground tanks and pipelines, and trace the

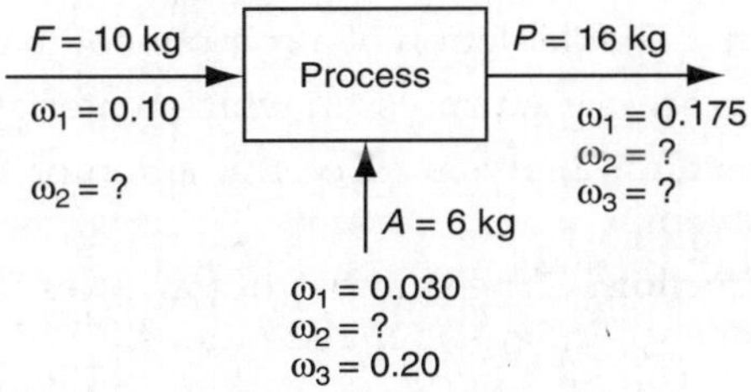

Figure SAT3.2TP1

theft of oil and other liquid products. Both radioactive and isotopic markers are used.

Deuterium is typically used as a marker for organic compounds, by replacing three or more hydrogen atoms on the molecule in a reactor containing heavy water. However, isotopes of carbon and oxygen can also be used. The detection limit of tracers using a combination of gas chromatography and mass spectrometry is about 100 ppb in crude oil and about 20 ppb in refined products.

Explain now such markers might be used in chemical processes.

2. Consider the concept of zero discharge of liquid waste. It would seem to be a good idea for both the environment and the company. What are some of the arguments for and against the zero discharge of wastewater?
3. One proposed method of eliminating waste in solid, liquid, and gas streams is incineration. What are some of the pros and cons regarding disposal of waste by incineration?

Looking Back

In addition to introducing you to the concept of material balances, we covered material balances applied to open and closed systems as well as multicomponent systems. Then a ten-step process was presented for undertaking the solution of material balances to ensure that you solve the right problem correctly.

Glossary

Accumulation An increase or decrease in the material (e.g., mass or moles) in a system.

Batch process A process in which material is neither added to nor removed from a process during its operation.

Closed system A system that does not have material crossing the system boundary.

Component balance A material balance on a single chemical component in a system.

Consumption The depletion of a component in a system due to chemical reaction.

Continuous process A process in which material enters and/or exits continuously.

Degree-of-freedom analysis Determination of the number of degrees of freedom in a problem.

Degrees of freedom The number of variables whose values are unknown minus the number of independent equations.

Dependent equations A set of equations that are not independent.

Exactly specified Describes a problem in which the degrees of freedom are zero.

Final condition The amount of material (e.g., mass or moles) in a process at the end of the processing interval.

Flow system An open system.

Generation The appearance of a component in a system because of chemical reaction.

Implicit equation An equation based on information not explicitly provided in a problem, such as the sum of mass fractions is 1.

Independent equations A set of equations for which the rank of the coefficient matrix formed from the equations is the same as the number of equations.

Initial condition The amount of a material (e.g., mass or moles) in a process at the beginning of the processing interval.

Input Material (e.g., mass, moles) that enters the system.

Knowns Variables whose values are known.

Material balance The balance equation that corresponds to the conservation of mass.

Negative accumulation A depletion of material (usually mass or moles) in a system.

Open system A system in which material crosses the system boundary.

Output Material (e.g., mass, moles) that leaves the system.

Overspecified Describes a set of equations (or a problem) that is composed of more equations than unknowns.

Semi-batch process A process in which material enters the system but product is not removed during operation.

Steady-state system A system in which all the conditions (e.g., temperature, pressure, amount of material) remain constant with time.

System Any arbitrary portion of or whole process that is considered for analysis.

System boundary The closed line that encloses the portion of the process that is to be analyzed.

Transient system A system in which one or more of the conditions (e.g., temperature, pressure, amount of material) of the system vary with time; also known as an **unsteady-state system**.

Underspecified Describes a set of equations (or a problem) that is composed of fewer equations than unknowns.

Unique solution A single solution that exists for a set of equations (or a problem).

Unknowns Variables whose values are unknown.

Unsteady-state system A system in which one or more of the conditions (e.g., temperature, pressure, amount of material) of the system vary with time; also known as a **transient system**.

Supplementary References

CACHE Corp. *Material and Energy Balances 2.0* (a CD), Austin, TX (2000).

Felder, R. M. "The Generic Quiz," *Chem. Eng. Educ.*, 176–81 (Fall, 1985).

Fogler, H. S., and S. M. Montgomery. *Interactive Computer Modules for Chemical Engineering Instruction*, CACHE Corp., Austin, TX (2000).

Veverka, V. V., and F. Madron. *Material and Energy Balances in the Process Industries: From Microscopic Balances to Large Plant*, Elsevier Sciences, Amsterdam (1998).

Woods, D. R., T. Kourti, P. E. Wood, H. Sheardown, C. M. Crowe, and J. M. Dickson. "Assessing Problem Solving Skills," *Chem. Eng. Educ.*, 300–7 (Fall, 2001).

Problems

Section 3.1 Introduction to Material Balances

3.1.1 Examine Figure P3.1.1 [adapted from *Environ. Sci. Technol.*, **27, 1975 (1993)]. What would be a good system to designate for this bioremediation process? Is your system open or closed? Is it steady-state or unsteady-state?

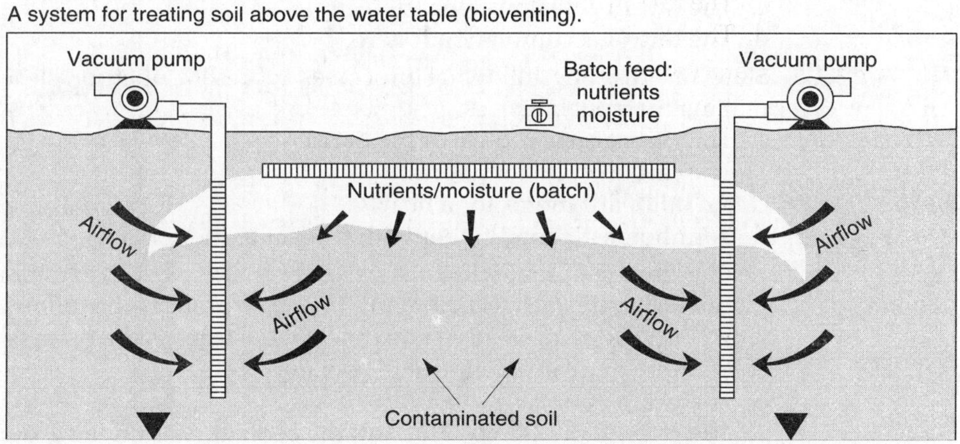

Figure P3.1.1

**3.1.2 Examine Figure P3.1.2, which shows a cylinder that is part of a Ford 2.9-liter V-6 engine. Pick a system and state it. Show by a crude sketch the system boundary. State whether your system is a flow system or a batch system and why (in one sentence).

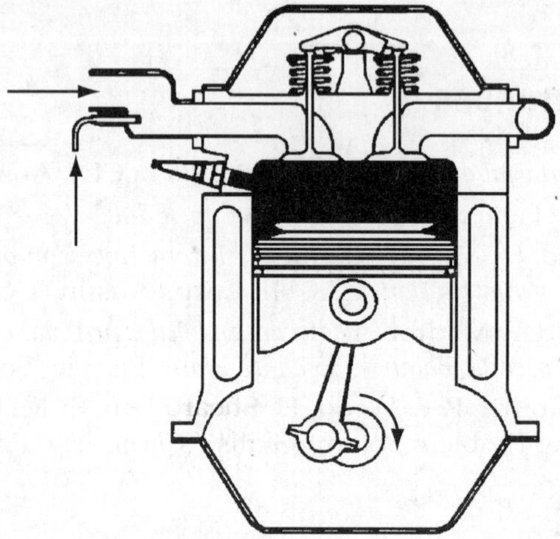

Figure P3.1.2

***3.1.3 State whether the following processes represent open or closed systems, and explain your answer very briefly.
a. Swimming pool (from the viewpoint of the water)
b. Home furnace

*3.1.4 Pick the correct answer(s): For a steady-state system
a. The rate of input is zero.
b. The rate of generation is zero.
c. The rate of consumption is zero.
d. The rate of accumulation is zero.

**3.1.5 State whether the following processes represent open or closed systems in making material balances:
a. The global carbon cycle of the Earth
b. The carbon cycle for a forest
c. An outboard motor for a boat
d. Your home air conditioner with respect to the coolant

**3.1.6 Read each one of the following scenarios. State what the system is. Draw the picture. Classify each as belonging to one or more of the following: open system, closed system, steady-state process, unsteady-state process.
a. You fill your car radiator with coolant.
b. You drain your car radiator.
c. You overfill the car radiator and the coolant runs on the ground.

d. The radiator is full and the water pump circulates water to and from the engine while the engine is running.

***3.1.7 State whether the process of a block of ice being melted by the sun (system: the ice) is an open or closed system, batch or flow, and steady-state or unsteady-state. List the choices vertically, and state beside each entry any assumptions you make.

**3.1.8 Examine the processes in Figure P3.1.8. Each box represents a system. For each, state whether

a. The process is in the
   1. Steady state
   2. Unsteady state
   3. Unknown condition
b. The system is
   1. Closed
   2. Open
   3. Neither
   4. Both

The wavy line represents the initial fluid level when the flows begin. In case (c), the tank stays full.

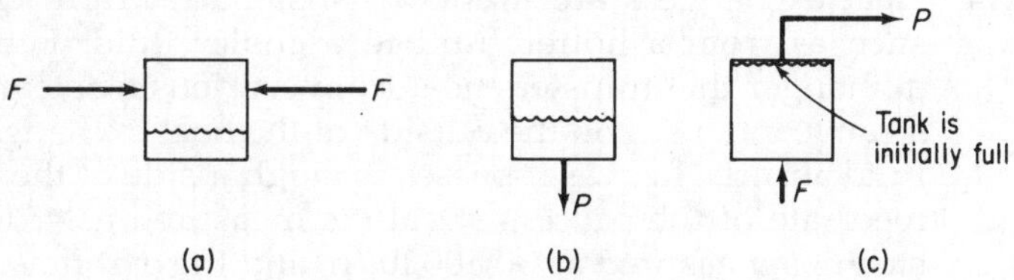

Figure P3.1.8

**3.1.9 In making a material balance, classify the following processes as (1) batch, (2) semi-batch, (3) continuous, (4) open or flow, (5) closed, (6) unsteady-state, or (7) steady-state. More than one classification may apply.

a. A tower used to store water for a city distribution system
b. A can of soda
c. Heating up cold coffee
d. A flush tank on a toilet
e. An electric clothes dryer
f. A waterfall
g. Boiling water in an open pot

***3.1.10 Under what circumstances can a batch process that is carried out repeatedly be considered to be a continuous process?

**3.1.11 A manufacturer blends lubricating oil by mixing 300 kg/min of No. 10 oil with 100 kg/min of No. 40 oil in a tank. The oil is well mixed and is withdrawn at the rate of 380 kg/min. Assume the tank contains no oil at the start of the blending process. How much oil remains in the tank after 1 hr?

****3.1.12** One hundred kilograms of sugar are dissolved in 500 kg of water in a shallow open cylindrical vessel. After standing for 10 days, 300 kg of sugar solution are removed. Would you expect the remaining sugar solution to have a mass of 300 kg?

***3.1.13** A 1.0 g sample of solid iodine is placed in a tube, and the tube is sealed after all of the air is removed (Figure P3.1.13). The tube and the solid iodine together weigh 27.0 g.

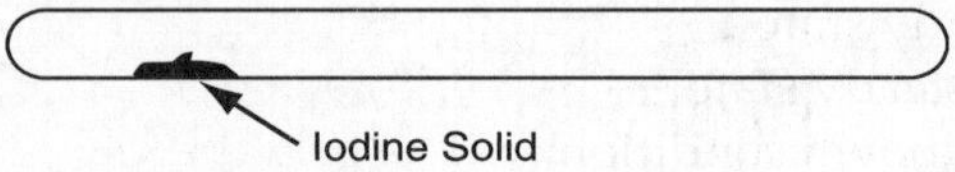

Figure P3.1.13

The tube is then heated until all of the iodine evaporates and the tube is filled with iodine gas. The weight after heating should be

a. Less than 26.0 g
b. 26.0 g
c. 27.0 g
d. 28.0 g
e. More than 28.0 g

*****3.1.14** Heat exchangers are used to transfer heat from one fluid to another fluid, such as from a hotter fluid to a cooler fluid. Figure P3.1.14 shows a heat exchanger that transfers heat from condensing steam to a process stream. The steam condenses on the outside of the heat exchanger tubes while the process fluid absorbs heat as it passes through inside of the heat exchanger tubes. The feed rate of the process stream is measured as 45,000 lb/h. The flow rate of steam is measured as 30,800 lb/h, and the exit flow rate of the process stream is measured as 50,000 lb/h. Perform a mass balance for the process stream. If the balance does not close adequately, what might be a reason for this discrepancy?

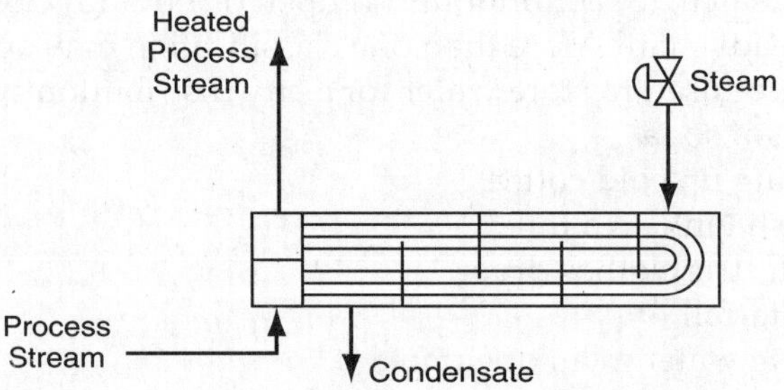

Figure P3.1.14

*****3.1.15** Examine the flowsheet in Figure P3.1.15 [adapted from *Hydrocarbon Processing*, 159 (November, 1974)] for the atmospheric distillation and pyrolysis of all atmospheric distillates for fuels and petrochemicals. Does the mass in equal the mass out? Give one or two reasons why the mass does or does not balance. Note: T/A is metric tons per year.

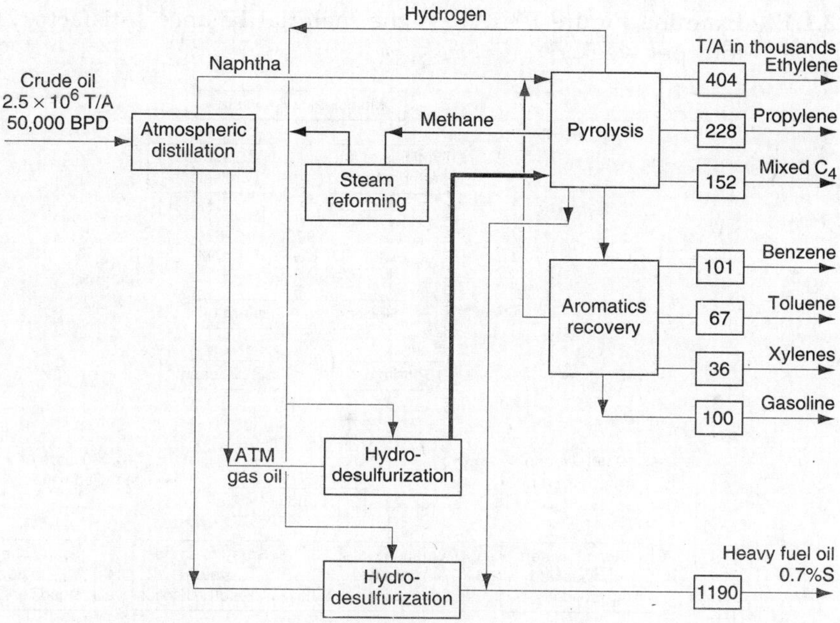

Figure P3.1.15

***3.1.16 Examine the flowsheet in Figure P3.1.16. Does the mass in equal the mass out? Give one or two reasons why the mass does or does not balance.

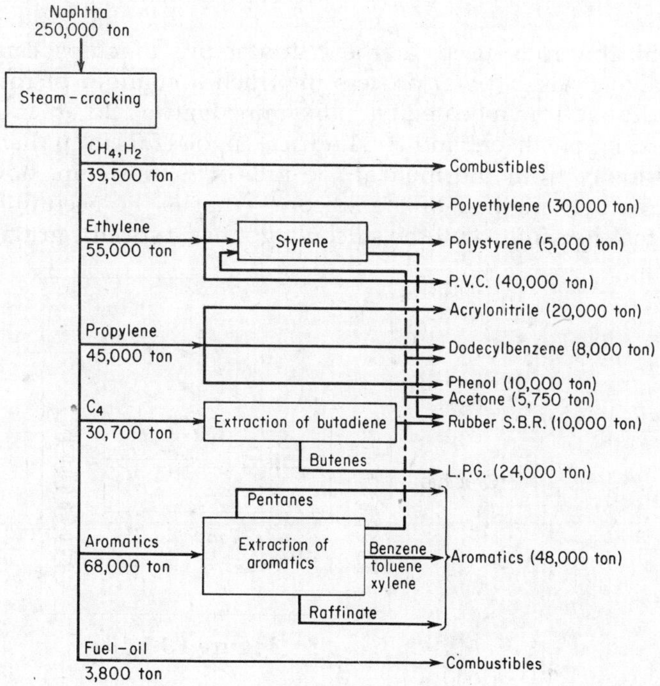

Figure P3.1.16

***3.1.17 Examine Figure P3.1.17. Is the material balance satisfactory? (T/wk means tons per week.)

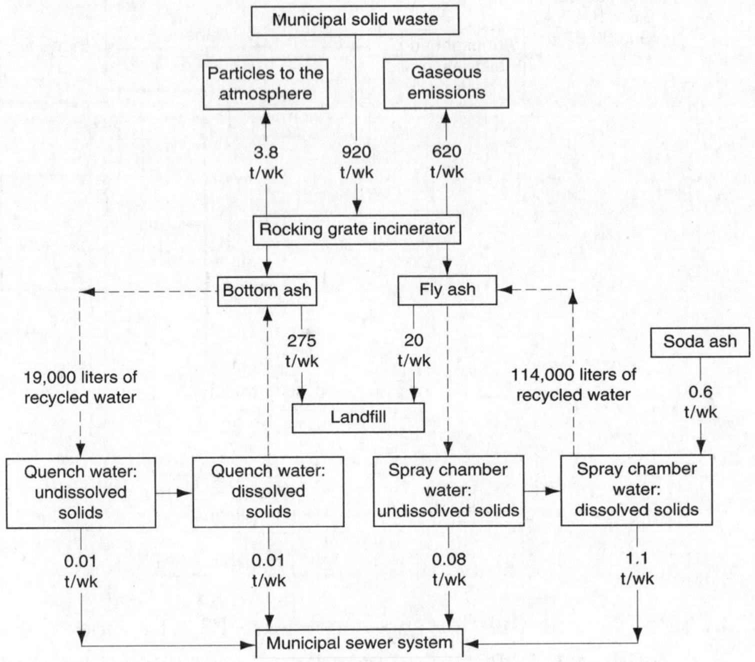

Figure P3.1.17

***3.1.18 Silicon rods used in the manufacture of chips can be prepared by the Czochralski (LEC) process in which a cylinder of rotating silicon is slowly drawn from a heated bath. Examine Figure P3.1.18. If the initial bath contains 62 kg of silicon, and a cylindrical ingot 17.5 cm in diameter is to be removed slowly from the melt at the rate of 3 mm/min, how long will it take to remove one-half of the silicon? What is the accumulation of silicon in the melt? Assume that the silicon ingot has a specific gravity of 2.33.

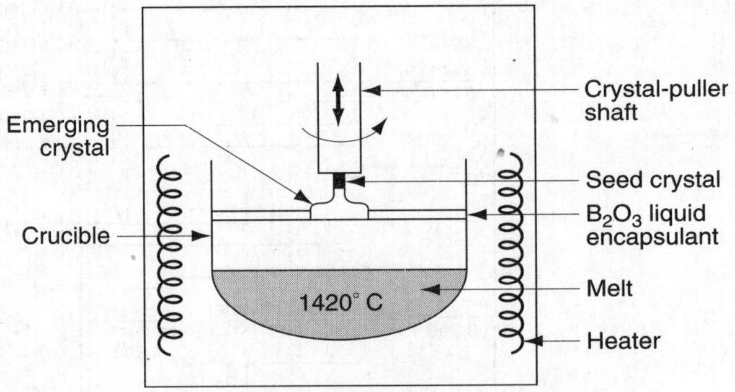

Figure P3.1.18

****3.1.19** Mixers can be used to mix streams with different compositions to produce a product stream with an intermediate composition. Figure P3.1.19 shows a diagram of such a mixing process. Evaluate the closure of the overall material balance and the component material balances for this process. Closure means how closely the inputs agree with the outputs for a steady-state process.

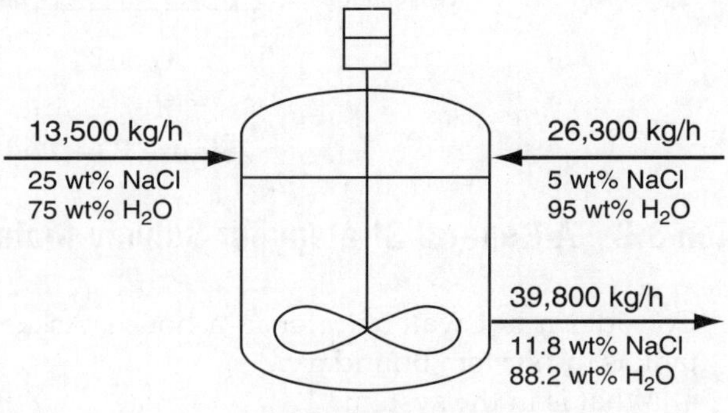

Figure P3.1.19

****3.1.20** Distillation columns are used to separate light boiling components from heavier boiling components and make up over 95% of the separation systems for the chemical process industries. A commonly used distillation column is a propylene-propane splitter. The overhead product from this column is used as a feedstock for the production of polypropylene, which is the largest quantity of plastic produced worldwide. Figure P3.1.20 shows a diagram of a propylene-propane splitter (C_3 refers to propane and $C_3^=$ refers to propylene). The steam is used to provide energy and is not involved in the process material balance. Assume that the composition and flow rates listed on this diagram came from process measurements. Determine if the overall material balance is satisfied for this system. Evaluate the component material balances as well. What can you conclude?

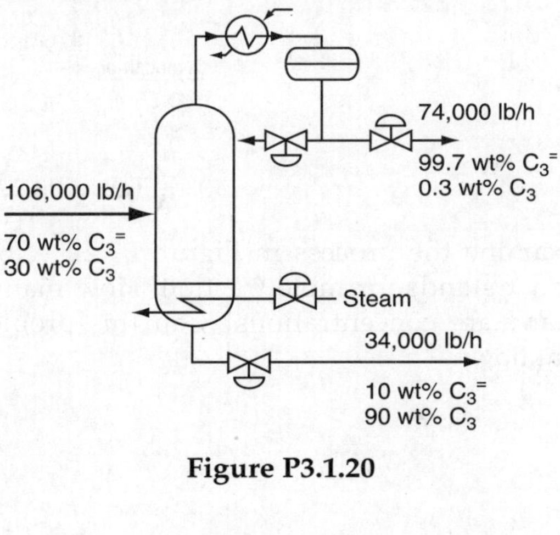

Figure P3.1.20

***3.1.21 A thickener in a waste disposal unit of a plant removes water from wet sewage sludge as shown in Figure P3.1.21. How many kilograms of water leave the thickener per 100 kg of wet sludge that enter the thickener? The process is in the steady state.

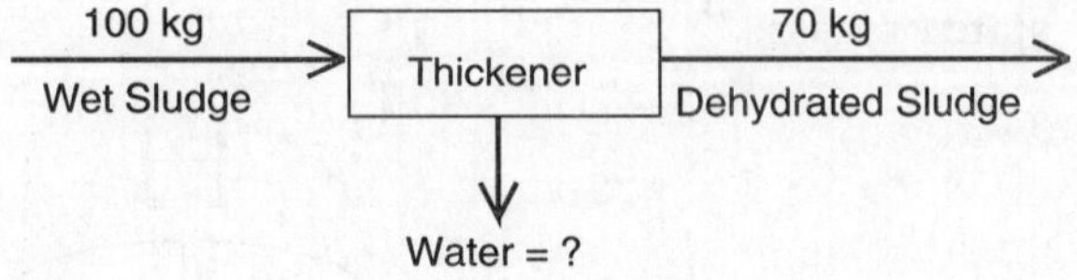

Figure P3.1.21

Section 3.2 A General Strategy for Solving Material Balance Problems

*3.2.1 Consider a hot water heater in a house. Assume that the metal shell of the tank is the system boundary.
a. What is in the system?
b. What is outside the system?
c. Does the system exchange material with the outside of the system?
d. Could you pick another system boundary?

**3.2.2 For the process shown in Figure P3.2.2, how many material balance equations can be written? Write them. How many independent material balance equations are there in the set?

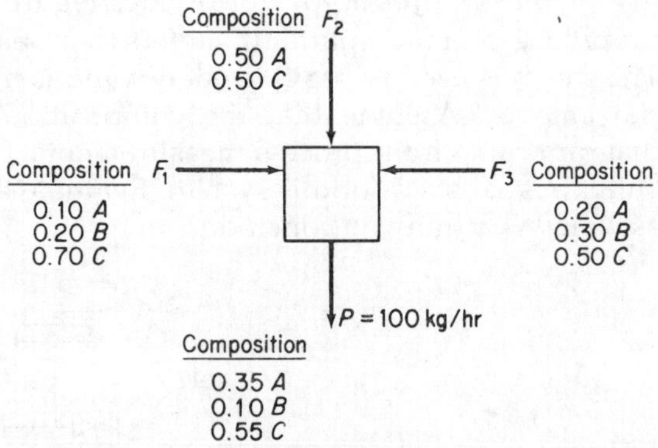

Figure P3.2.2

**3.2.3 Examine the process in Figure P3.2.3. No chemical reaction takes place, and x stands for mole fraction. How many variables are unknown? How many are concentrations? Can this problem be solved uniquely for the unknowns?

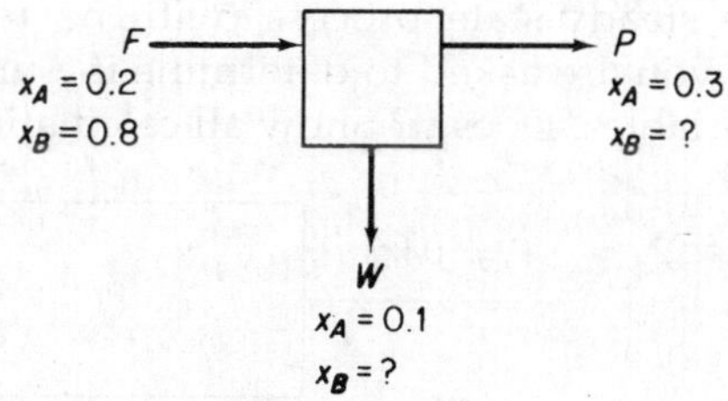

Figure P3.2.3

3.2.4 Are the following equations independent? Do they have a unique solution? Explain your answers.

*a. $x_1 + 2x_2 = 1$

$x_1 + 2x_2 = 3$

****b. $(x_1 - 1)^2 + (x_2 - 1)^2 = 0$

$x_1 + x_2 = 1$

****3.2.5** For one process your assistant has prepared four valid material balances:

$$0.25\, m_{\text{NaCl}} + 0.35\, m_{\text{KCl}} + 0.55\, m_{\text{H}_2\text{O}} = 0.30$$

$$0.35\, m_{\text{NaCl}} + 0.20\, m_{\text{KCl}} + 0.40\, m_{\text{H}_2\text{O}} = 0.30$$

$$0.40\, m_{\text{NaCl}} + 0.45\, m_{\text{KCl}} + 0.05\, m_{\text{H}_2\text{O}} = 0.40$$

$$1.00\, m_{\text{NaCl}} + 1.00\, m_{\text{KCl}} + 1.00\, m_{\text{H}_2\text{O}} = 1.00$$

He says that since the four equations exceed the number of unknowns, three, no solution exists. Is he correct? Explain briefly whether it is possible to achieve a unique solution.

*****3.2.6** Do the following sets of equations have a unique solution?

a. $u + v + w = 0$
$u + 2v + 3w = 0$
$3u + 5v + 7w = 1$

b. $u + w = 0$
$5u + 4v + 9w = 0$
$2u + 4v + 6w = 0$

****3.2.7** Indicate whether the following statements are true or false:

a. When the flow rate of one stream is given in a problem, you must choose it as the basis.

b. If all of the stream compositions are given in a problem, but none of the flow rates are specified, you cannot choose one of the flow rates as the basis.

c. The maximum number of material balance equations that can be written for a problem is equal to the number of species in the problem.

**3.2.8 In the steady-state process (with no reactions occurring) shown in Figure P3.2.8 you are asked to determine if a unique solution exists for the values of the variables. Does it? Show all calculations.

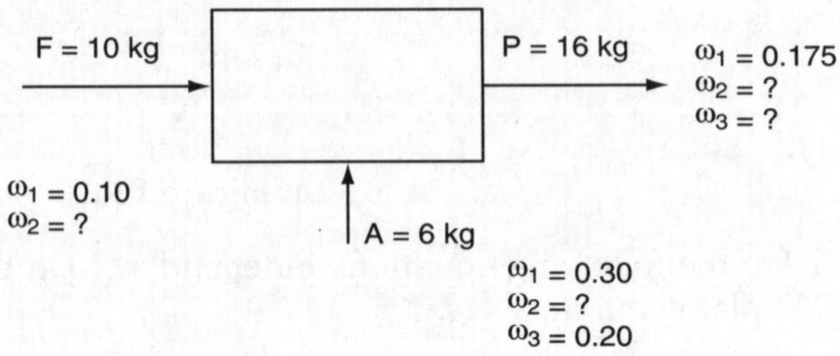

Figure P3.2.8

ω is the mass fraction of component i.

**3.2.9 Three gaseous mixtures, A, B, and C, with the compositions listed in the table are blended into a single mixture.

Gas	A	B	C
CH_4	25	25	60
C_2H_6	35	30	25
C_3H_8	40	45	15
Total	100	100	100

A new analyst reports that the composition of the mixture is 25% CH_4, 25% C_2H_6, and 50% C_3H_8. Without making any detailed calculations, explain how you know the analysis is incorrect.

***3.2.10 A problem is posed as follows: It is desired to mix three LPG (liquefied petroleum gas) streams denoted by A, B, and C in certain proportions so that the final mixture will meet certain vapor-pressure specifications. These specifications will be met by a stream of composition D as indicated in the following table. Calculate the proportions in which streams A, B, and C must be mixed to give a product with a composition of D. The values are liquid volume percent, but the volumes are additive for these compounds.

Component	Stream A	Stream B	Stream C	Stream D
C_2	5.0			1.4
C_3	90.0	10.0		31.2
iso-C_4	5.0	85.0	8.0	53.4
n-C_4		5.0	80.0	12.6
iso-C_5^+			12.0	1.4
Total	100.0	100.0	100.0	100.0

The subscripts on the Cs represent the number of carbons, and the + sign on C_5^+ indicates all compounds of higher molecular weight as well as *iso*-C_5. Does this problem have a unique solution?

***3.2.11 In preparing 2.50 moles of a mixture of three gases, SO_2, H_2S, and CS_2, gases from three tanks are combined into a fourth tank. The tanks have the following compositions (mole fractions):

Combined Tanks Mixture

	Tanks			Combined Mixture
Gas	**1**	**2**	**3**	**4**
SO_2	0.23	0.20	0.54	0.25
H_2S	0.36	0.33	0.27	0.23
CS_2	0.41	0.47	0.19	0.52

In the right-hand column is listed the supposed composition obtained by analysis of the mixture. Does the set of three mole balances for the three compounds have a solution for the number of moles taken from each of the three tanks and used to make up the mixture? If so, what does the solution mean?

**3.2.12 You have been asked to check out the process shown in Figure P3.2.12. What will be the minimum number of measurements to make in order to compute the value of each of the stream flow rates and stream concentrations? Explain your answer.

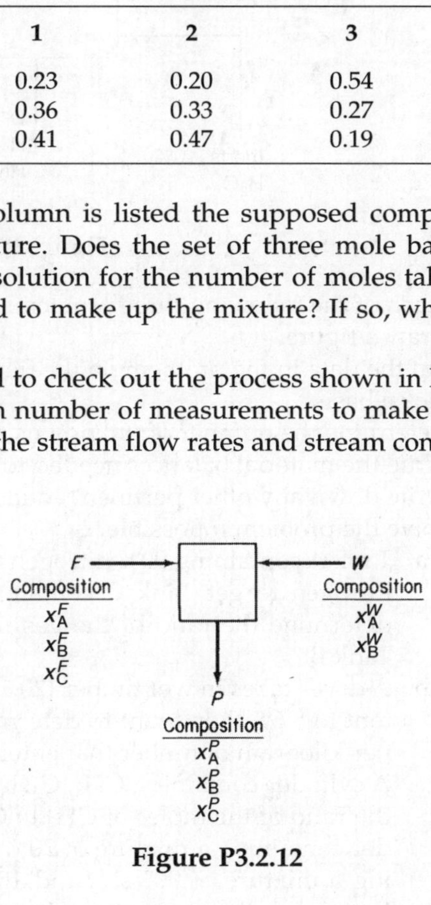

Figure P3.2.12

***3.2.13 Effluent from a fertilizer plant is processed by the system shown in Figure P3.2.13. How many additional concentration and stream flow measurements must be made to completely specify the problem (so that a unique solution exists)? Does only one unique set of specifications exist?

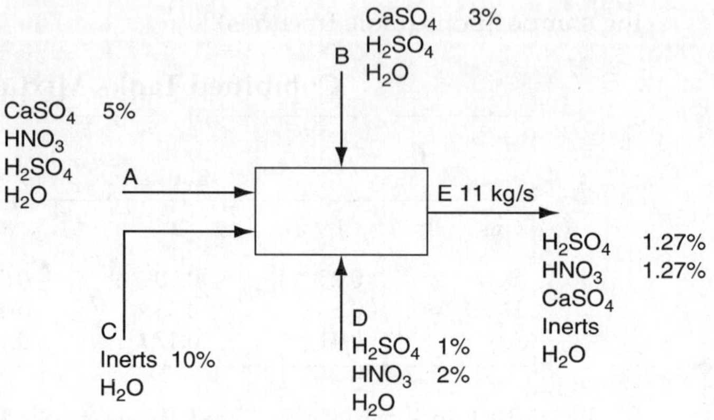

Figure P3.2.13

3.2.14 For each of the following three problems:

1. Draw a figure.
2. Put the data in the problem on the figure.
3. Pick a basis.
4. Determine the number of unknowns and independent equations.
5. Write the material balances needed to solve the problem.
6. Write down any other pertinent equations and specifications.
7. Solve the problem if possible.

**a. Tank A containing 90% nitrogen is mixed with Tank B containing 30% nitrogen to get Tank C containing 65% nitrogen. You are asked to determine the ratio of the gas used from Tank A to that used from Tank B.

**b. A dryer takes in wet timber (20.1% water) and reduces the water content to 8.6%. You want to determine the kilograms of water removed per kilogram of timber that enters the process.

***c. A cylinder containing CH_4, C_2H_6, and N_2 has to be prepared in which the ratio of the moles of CH_4 to C_2H_6 is 1.3 to 1. Available are a cylinder containing a mixture of 70% N_2 and 30% CH_4, a cylinder containing a mixture of 90% N_2 and 10% C_2H_6, and a cylinder of pure N_2. Determine the proportions in which the respective gases from each cylinder should be used.

**3.2.15 After you read a problem statement, what are some of the things you should think about to solve it? List them. This problem does not ask for the ten stages described in the chapter but for brainstorming.

CHAPTER 4

Material Balances without Reaction

Your objectives in studying this chapter are to be able to

1. Analyze a problem statement and organize in your mind the solution strategy
2. Apply the ten-step strategy to solve problems that do not involve chemical reactions

In Chapter 3 you read about solving material balance problems that do not involve chemical reaction. Can you apply these ideas now? You can hone your skills by going through the applications presented in this chapter, first covering up the solution, and then comparing your solution with the one in the text. If you can solve each problem without difficulty, congratulations! If you cannot, analyze places where you had trouble. If you just read the problem and its solution, you will deprive yourself of the learning activity needed to improve your capabilities. You will find additional solved problems on the CD that accompanies this book if you want more extensive practice.

A famous magician stood on a concrete floor and with a flourish pulled a raw egg from his hair. He held the egg in his outstretched hand and said he could drop it 2 meters without breaking its shell and without the aid of any other object. Then he proceeded to do it. What did he do?

Problem solving is what you do when you don't know what to do; otherwise it's not a problem.

G. Bodner, J. Chem. Edu., **63**, 873 (1986)

The use of material balances in a process allows you to calculate the values of the flows of species in the streams that enter and leave the plant equipment, the flow between the pieces of equipment, and also the changes that occur inside the equipment. You want to find out how much of each raw material is used and how much of each product (along with some wastes) is produced by the plant. We will present a wide range of problems in this chapter to demonstrate that no matter what the process is, the problem-solving strategy evolved in Chapter 3 can be effective for all of them. Remember that if the process involves rates of flow, you can pick an interval of time as the basis to avoid carrying along time as a symbol or variable in the analysis and calculations.

Example 4.1 Extraction of Streptomycin from a Fermentation Broth

Streptomycin is used as an antibiotic to fight bacterial diseases in humans as well as to control bacteria, fungi, and algae in crops. First an inoculate is prepared by placing spores of strains of *Streptomyces griseus* in a medium to establish a culture with a high biomass. The culture is then introduced into a fermentation tank which operates at 28°C and a pH of about 7.8 with the nutrients of glucose (the carbon source) and soybean meal (the nitrogen source). High agitation and aeration are needed. After fermentation, the biomass is separated from the liquid, and Streptomycin is recovered by adsorption on activated charcoal followed by extraction with an organic solvent in a continuous extraction process. If we ignore the details of the process and just consider the overall extraction process, Figure E4.1 shows the net result.

Determine the mass fraction of Streptomycin in the exit organic solvent based on the data in Figure E4.1, assuming that no water exits with the solvent and no solvent exits with the aqueous solution. Assume that the density of the aqueous solution is $1\ g/cm^3$ and the density of the organic solvent is $0.6\ g/cm^3$.

Solution

Step 1

Figure E4.1 indicates that the process is an open (flow), steady-state process without reaction. Assume because of the very low concentration of Streptomycin in the aqueous and organic fluids that the volumetric flow rates of the entering fluids equal the volumetric flow rates of the respective exit fluids.

Steps 2–4

All of the data come from Figure E4.1.

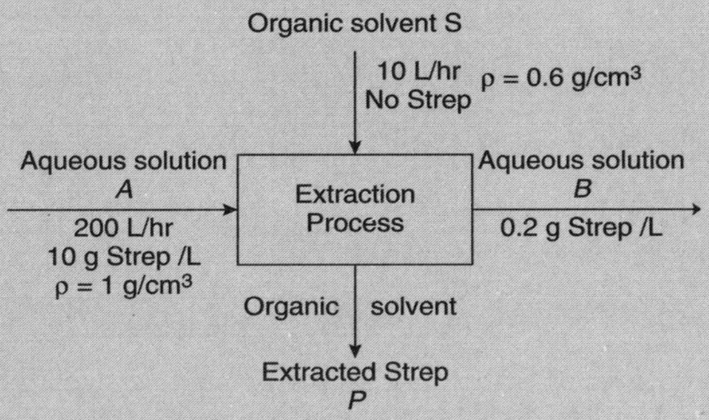

Figure E4.1

Steps 5–7

The basis is 1 hr.

The degree-of-freedom analysis is as follows:

From an analysis of Figure E4.1, you can see that there is a total of eight variables for this problem (i.e., four streams with two components in each stream). You want to reduce the eight variables to as few unknowns as possible. You can assign values to the following variables from the given data (the mass of a component is designated by m with appropriate superscripts and subscripts). Some of the values can be obtained by using the personal computer on top of your neck. Let A and B denote the input and output aqueous streams, respectively, and S and P denote the input and output organic streams, respectively.

$A = 200$ L	$B = 200$ L	$S = 10$ L	$P = 10$ L
$m^{\text{in}}_{\text{Strep.}} = 2000$ g	$m^{\text{out}}_{\text{Strep.}} = 40$ g	$m^{\text{in}}_{\text{Strep.}} = 0$ g	$m^{\text{out}}_{\text{Strep.}} = ?$ g
$m^{\text{in}}_{\text{water}} = 2 \times 10^5$ g	$m^{\text{out}}_{\text{water}} = 2 \times 10^5$ g	$m^{\text{in}}_{\text{solvent}} = 6000$ g	$m^{\text{out}}_{\text{solvent}} = 6000$ g

Number of unknowns: 1
Number of independent equations needed: 1

Note that we used one material balance for the water and one for the solvent in assigning values to variables. What independent balance is left? The material balance for the Streptomycin.

Steps 8 and 9

What equation should you use for the Streptomycin balance? This is an open, steady-state process; therefore, you can use any of Equations (3.1), the most general, (3.2), or (3.3), namely, the truism "What goes in must come out."

(Continues)

Example 4.1 Extraction of Streptomycin from a Fermentation Broth (*Continued*)

$$\underbrace{\frac{200\text{ L of }A}{}\left|\frac{10\text{ g Strep.}}{1\text{ L of }A}\right. + \frac{10\text{ L of }S}{}\left|\frac{0\text{ g Strep.}}{1\text{ L of }S}\right.}_{In} = \underbrace{\frac{200\text{ L of }A}{}\left|\frac{0.2\text{ g Strep.}}{1\text{ L of }A}\right. + \frac{10\text{ L of }S}{}\left|\frac{m^{out}_{Strep.}\text{ g Strep.}}{1\text{ L of }S}\right.}_{Out}$$

$$m^{out}_{Strep.} = 196\text{ g Strep./L of }S$$

To get the grams of Strep. per gram of solvent, you need to convert the volume of S to mass. Use the specified density of the solvent:

$$\frac{196\text{ g Strep.}}{\text{L of }S}\left|\frac{1\text{ L of }S}{1000\text{ cm}^3\text{ of }S}\right|\frac{1\text{ cm}^3\text{ of }S}{0.6\text{ g of }S} = 0.328\text{ g Strep./g of }S$$

$$\text{The mass fraction of Streptomycin} = \frac{0.328}{1 + 0.328} = 0.246$$

Example 4.2 Separation of Gases Using a Membrane

Membranes represent a relatively new technology for the commercial separation of gases. One use that has attracted attention is the separation of nitrogen and oxygen from air. Figure E4.2a illustrates a nanoporous membrane which is made by coating a very thin layer of polymer on a porous graphite supporting layer.

What is the composition of the waste stream if the waste stream amounts to 80% of the input stream?

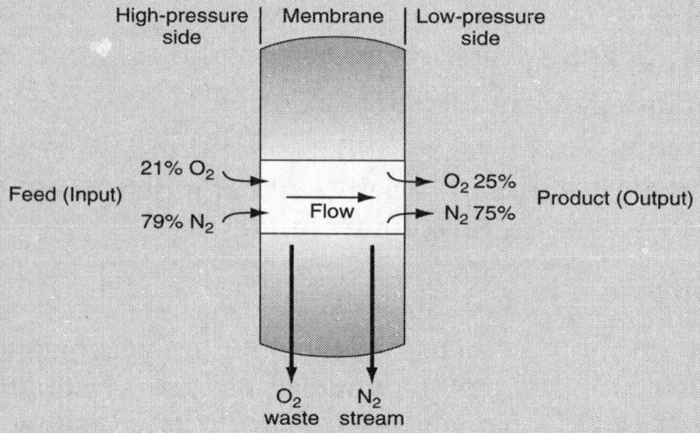

Figure E4.2a

Solution

Step 1

This is an open, steady-state process without chemical reaction. The system is the membrane as depicted in Figure E4.2a. Let y_{O_2} be the mole fraction of oxygen in the waste stream, W, as depicted in Figure E4.2a; let y_{N_2} be the mole fraction of nitrogen in the waste stream; and let n_{O_2} and n_{N_2} be the respective moles in each stream.

Steps 2–4

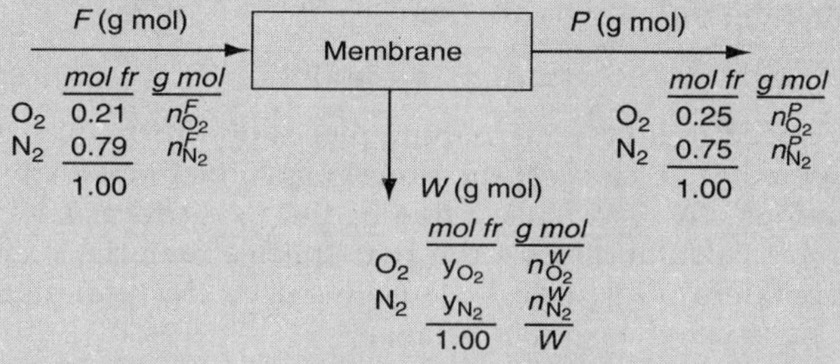

Figure E4.2b

All of the data and symbols have been placed in Figure E4.2b.

Step 5

Pick a convenient basis. The problem does not ask for the actual flow of mass or moles, just for the molar composition of W; hence only relative values have to be calculated. Let

$$\text{Basis: } F = 100 \text{ mol}$$

A degree-of-freedom analysis comes next.

Steps 6–8

To avoid forming nonlinear equations [by including terms such as $(y^W_{O_2})(W)$], let's use as variables the number of moles (n) rather than the mole fractions (y). The total number of variables is nine, but it would be foolish to involve nine unknowns, particularly when you can reduce the number of unknowns at the start by using information in the problem statement to assign values to variables. How many preliminary substitutions and calculations you want to make depends on the information in the problem, of course.

Value	Information Used
$F = 100$ mol	the basis
$W = 0.80\,(100) = 80$ mol	a specification
$P = F - W = 100 - 80 = 20$ mol	the total material balance
$n^F_{O_2} = 0.21(100) = 21$ mol	a specification plus the basis
$n^F_{N_2} = 0.79(100) = 79$ mol	a specification plus the basis

(*Continues*)

Example 4.2 Separation of Gases Using a Membrane (*Continued*)

Note the use of a material balance that involves only one unknown to get P. Now that P has been calculated, we can use another set of specifications to make two more ad hoc calculations:

$$n_{O_2}^{P} = 0.25P = 0.25(20) = 5.0 \text{ mol}$$

$$n_{N_2}^{P} = 0.75P = 0.75\,(20) = 15 \text{ mol}$$

The remaining unknowns are then

$$n_{O_2}^{W} \text{ and } n_{N_2}^{W}$$

Thus, we need to involve two more pieces of information given in the problem statement. Let's look at the process specifications. Are there any unused specifications? We have used five of the five process specifications in the preliminary calculations. Of the two species material balances, how many are independent? Only one, because we used the total material balance previously. Let's use the oxygen balance:

In		Out in P		Out in W
21	=	5	+	$n_{O_2}^{W}$

Solving for the amount of O_2 in the waste stream yields

$$n_{O_2}^{W} = 21 - 5.0 = 16 \text{ mol}$$

What other independent equation might be used? An implicit equation! The sum of the mole fractions in W, or the equivalent, the sum of the moles of the species in W, is an independent equation:

$$W = n_{O_2}^{W} + n_{N_2}^{W} = 80 = 16 + n_{N_2}^{W}$$

Solving for the amount of nitrogen in the waste product yields

$$n_{N_2}^{W} = 80 - 16 = 64 \text{ mol}$$

Are the sums of the mole fractions in F and P independent equations? No, because the information in these two relations is redundant with the specifications that have been previously used.

The result of this analysis is that the values of all of the variables have been determined without solving any simultaneous equations! No residual independent information exists, and the degrees of freedom are zero.

Step 9

The composition of the waste stream is

$$y_{O_2}^{W} = \frac{n_{O_2}^{W}}{W} = \frac{16}{80} = 0.20 \quad y_{N_2}^{W} = \frac{n_{N_2}^{W}}{W} = \frac{64}{80} = 0.80$$

Step 10

Check your results. You can use a redundant equation as a check, that is, one not used previously. For example, let's use the N_2 balance:

$$n^W_{N_2} + n^P_{N_2} = n^F_{N_2}$$

$$64 + 15 = 79 \quad \text{OK}$$

Be careful when formulating and simplifying the equations to be solved to make sure that you use only independent equations.

In the next problem we give an example of distillation. Distillation is the most commonly used process for separating components in the refining and petrochemical industries, and it is based on the separation that results from vaporizing a liquid (see Chapter 8). A liquid mixture is boiled to produce a vapor of a different composition that moves away (is removed) from the liquid. Repeated serial operation of the process leads to purification. Look on the CD accompanying this book for more information about the specialized terminology pertaining to distillation and pictures of distillation equipment.

Example 4.3 Analysis for a Continuous Distillation Column

A new manufacturer of ethyl alcohol—ethanol, denoted as EtOH—for gasohol is having a bit of difficulty with a distillation column. The process is shown in Figure E4.3. They think too much alcohol is lost in the bottoms (waste). Calculate the composition of the bottoms and the mass of the alcohol lost in the bottoms based on the data shown in Figure E4.3 that were collected in 1 hr of operation. Finally, determine the percentage of the EtOH entering the column that is lost in the waste stream.

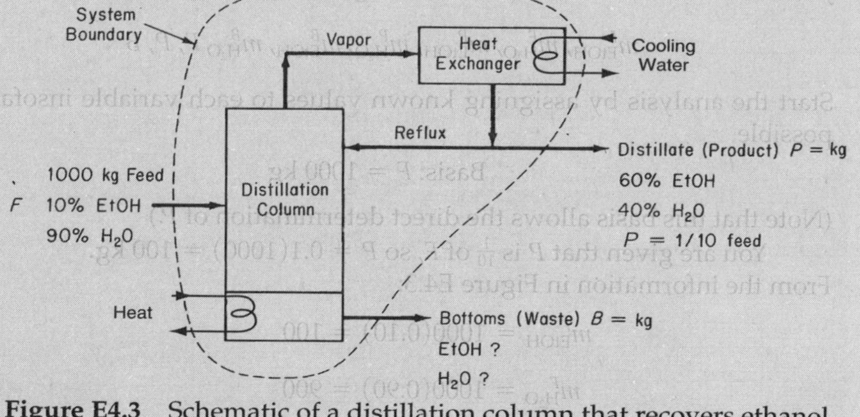

Figure E4.3 Schematic of a distillation column that recovers ethanol

(Continues)

Example 4.3 Analysis for a Continuous Distillation Column (*Continued*)

Solution

Steps 1–4

Although the distillation process shown in Figure E4.3 is composed of more than one unit of equipment, you can select a system composed of all of the equipment included inside the system boundary as one lump. Consequently, you can ignore all of the internal streams for this problem. Let m designate the mass of a component. Clearly the process is an open system, and we assume it is in the steady state. No reaction occurs. The cooling water enters and leaves without mixing with the components being separated and can be ignored for the material balances for the system. In addition, the heat added at the bottom of the column does not involve mass entering or leaving the system and can also be ignored for the material balances.

All of the symbols and known data have been placed on Figure E4.3. This is an open, steady-state process, so you can apply Equation (3.3).

Step 5

Select as the basis the given feed:

$$\text{Basis: } F = 1000 \text{ kg of feed}$$

Steps 6 and 7

The next step is to carry out a degree-of-freedom analysis. Let m with appropriate superscripts and subscripts denote mass in kilograms. From Figure E4.3 you should be able to locate the following variables:

$$m^F_{\text{EtOH}}, m^F_{\text{H}_2\text{O}}, m^P_{\text{EtOH}}, m^P_{\text{H}_2\text{O}}, m^B_{\text{EtOH}}, m^B_{\text{H}_2\text{O}}\; F, P, B$$

Start the analysis by assigning known values to each variable insofar as is possible.

$$\text{Basis: } F = 1000 \text{ kg}$$

(Note that this basis allows the direct determination of P.)

You are given that P is $\frac{1}{10}$ of F, so $P = 0.1(1000) = 100$ kg.

From the information in Figure E4.3:

$$m^F_{\text{EtOH}} = 1000(0.10) = 100$$

$$m^F_{\text{H}_2\text{O}} = 1000(0.90) = 900$$

$$m^P_{\text{EtOH}} = 0.60P = (0.6)(100) = 60 \text{ kg}$$

$$m^P_{H_2O} = 0.40P = (0.40)(100) = 40 \text{ kg}$$

$$P = m^P_{H_2O} + m^P_{EtOH} = 40 \text{ kg} + 60 \text{ kg} = 100 \text{ kg}$$

Thus, values have been assigned to six variables, leaving three unknowns (m^B_{EtOH}, $m^B_{H_2O}$, B). What three independent equations would you suggest using to solve for the remaining unknowns? The usual categories to select from are

Material balances: EtOH, H_2O, and total

$$\text{Implicit equations: } \sum m_i^B = B \text{ or } \sum \omega_i^B = 1$$

Steps 8 and 9

Let's use the EtOH balance to determine m^B_{EtOH}, which yields a value of 40 kg. Then by using the implicit equation for B, $m^B_{H_2O}$ is equal to 860 kg. These results and the results for the mass fractions are shown below.

	kg Feed in		kg Distillate out		kg Bottoms out	Mass Fraction in B
EtOH balance	0.10(1000)	−	0.60(100)	=	40	0.044
H_2O balance	0.90(1000)	−	0.40(100)	=	860	0.956
Total $\sum m_i^B = B$					900	1.000

After all of the unknowns are determined for this problem, the percentage of EtOH can be calculated directly by

$$\text{Percentage of EtOH lost in } B = \frac{\text{EtOH in } B}{\text{EtOH in feed}} = \frac{40}{100} \times 100\% = 40\%$$

Step 10

As a check let's use a redundant equation, the total balance: B = 1000 − 100 = 900 kg.

$$m^B_{EtOH} + m^B_{H_2O} = B \text{ or } \omega^B_{EtOH} + \omega^B_{H_2O} = 1$$

Examine the last two columns of the table above.

The next example represents an open system, but one that can be viewed as either unsteady-state or steady-state depending on how the process is actually carried out. The CD that accompanies this book shows various types of equipment that can be used to mix liquids.

Example 4.4 Mixing of Battery (Sulfuric) Acid

You are asked to prepare a batch of 18.63% battery acid as follows: A mixing vessel of old weak battery acid (H_2SO_4) solution contains 12.43% H_2SO_4 (the remainder is pure water). If 200 kg of 77.7% H_2SO_4 is added (not too fast!) to the vessel, and the final solution is to be 18.63% H_2SO_4, how many kilograms of battery acid have been made? See Figures E4.4a and E4.4b.

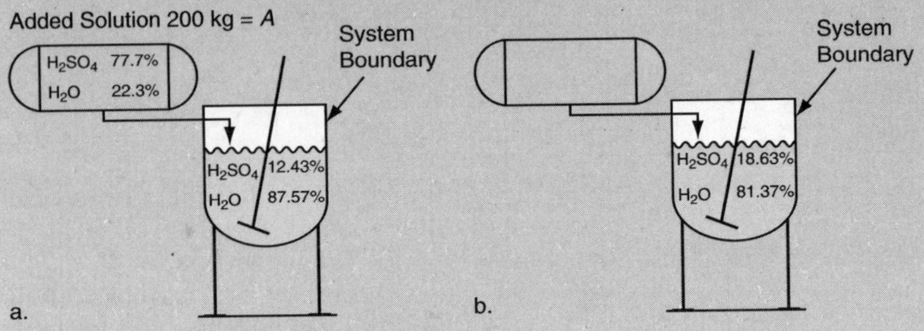

Figure E.4.4 (a) Initial arrangement before mixing the acid solutions; (b) final result after mixing

Solution

Steps 1–4

All of the values of the compositions are known and have been placed on Figure E4.4. No reaction occurs. Should the process be treated as an unsteady-state process or a steady-state process? If the tank with the original weak solution is selected as the system, the system initially contains sulfuric acid solution. Then concentrated solution is added to the system so that accumulation occurs in the system; therefore, you must use an unsteady-state material balance. The total mass increases, and the mass of each component increases so that Equation (3.2) can be applied:

$$\textit{accumulation} = \textit{in} - \textit{out}$$

From another viewpoint, you could select the tank as a system that was initially empty. Then the weak solution is introduced into the system along with the 200 kg of 77.7% solution, the solutions are mixed, and finally the entire contents of the tank are removed, leaving an empty tank. Then, the mass balance reduces to a steady-state flow process so that Equation (3.3) can be applied:

$$\textit{in} = \textit{out}$$

because no accumulation occurs in the tank.

Let us first solve the problem with the mixing treated as an unsteady-state process, and then repeat the solution with the process treated as a steady-state process. Will the answers differ? Let A denote the mass of the strong acid, F denote the mass of the weak acid initially in the mixing vessel, and P denote the mass of the final desired product.

Step 5

Take 200 kg of A as the basis for convenience (the only quantity you know).

Steps 6 and 7

The analysis of the degrees of freedom is analogous to the ones carried out for the previous examples. We will use m with appropriate superscripts and subscripts for mass and not use mass fractions (ω). Why this choice? Look at the comments in some of the previous examples and text.

The number of variables is nine:

$$m^A_{H_2SO_4}, m^A_{H_2O}, m^F_{H_2SO_4}, m^F_{H_2O}, m^P_{H_2SO_4}, m^P_{H_2O}, A, F, P$$

Let's assign known values to their respective variables:

$$m^A_{H_2SO_4} = 155.4 \text{ kg} \qquad m^F_{H_2SO_4} = ? \qquad m^P_{H_2SO_4} = ?$$

$$m^A_{H_2O} = 44.6 \text{ kg} \qquad m^F_{H_2O} = ? \qquad m^P_{H_2O} = ?$$

$$A = 200.0 \text{ kg} \qquad F = ? \qquad P = ?$$

The number of unknowns is six.

Note that we have not yet used all of the easy information available in Figure E4.4, namely, the mass fractions in F and P.

Number of equations needed: 6

What independent equations can you use?

Material balances: 2 H_2SO_4, H_2O, and total (two are independent)

Because the mass fraction specifications in P and F have not been used, four more independent equations exist based on the problem specifications:

$$\frac{m^P_{H_2SO_4}}{P} = 0.1863 \qquad \frac{m^P_{H_2SO_4}}{P} = 0.1243$$

$$\frac{m^F_{H_2O}}{F} = 0.8757 \qquad \frac{m^P_{H_2O}}{P} = 0.8137$$

(*Continues*)

Example 4.4 Mixing of Battery (Sulfuric) Acid (*Continued*)

What about the implicit equations?

$$\sum m_i^F = F \qquad \sum m_i^P = P \qquad \sum m_i^A = A$$

They are all redundant because all of the mass fractions were specified in the problem statement:

$$\sum \omega_i^F = 1 \qquad \sum \omega_i^P = 1 \qquad \sum \omega_i^A = 1$$

The sum of the mass fractions in each stream conveys no new information.

The degrees of freedom are zero because there are six independent equations and six unknowns.

Step 8

Let's insert the assigned values into the mass balances along with the specifications for the four mass fractions. By doing so you will see that we can calculate F and P by solving just two simple simultaneous equations. The balances are in kilograms.

	Final		**Initial**				
H_2SO_4	$P(0.1863)$	−	$F(0.1243)$	=	200(0.777)	−	0
H_2O	$P(0.8137)$	−	$F(0.8757)$	=	200(0.223)	−	0
Total	P	−	F	=	200	−	0

Step 9

Because the equations are linear and only two independent equations exist, you can take the total mass balance, solve it for F, and substitute for F in the H_2SO_4 balance to calculate P and get

$$P = 2110 \text{ kg acid}$$

$$F = 1910 \text{ kg acid}$$

Step 10

You can check the answer using the H_2O balance. Does the H_2O balance?

The problem can also be solved by pretending that during the arbitrary time interval the mixing occurs in a steady-state flow process. The solutions F and A would be inputs to the system composed of an initially empty tank, and the resulting mixture would be an output from the vessel, leaving the tank empty at the end of the process. The mass balance equation would then be "What goes in must come out." The balances are in kilograms.

	A in		*F* in		*P* out
H_2SO_4	200(0.777)	+	$F(0.1243)$	=	$P(0.1863)$
H_2O	200(0.223)	+	$F(0.8757)$	=	$P(0.8137)$
Total	A	+	F	=	P

You can see by inspection that these equations are no different from the first set of mass balances except for the arrangement and labels.

Example 4.5 Separation Using a Chromatographic Column

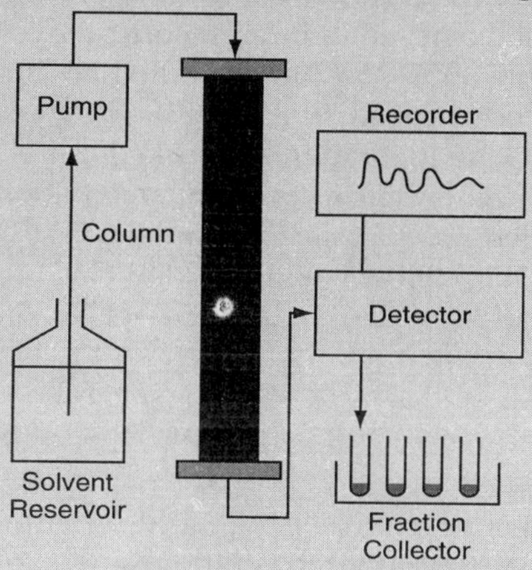

Figure E4.5a

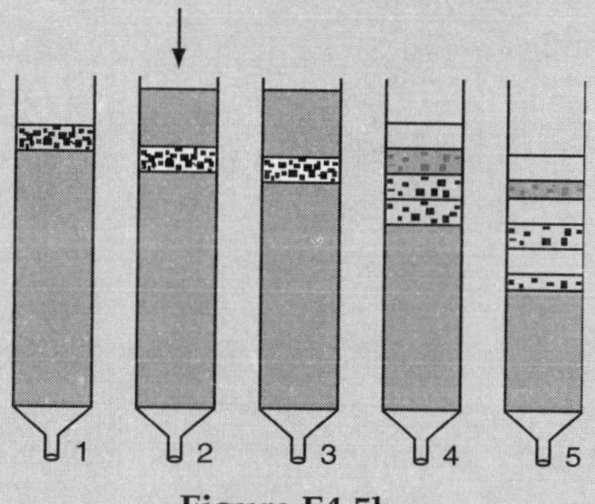

Figure E4.5b

A chromatographic column can be used to separate two or more compounds. Figure E4.5a portrays the main features of such a column, which can be operated in either a horizontal or a vertical position. Figure E4.5b

(Continues)

Example 4.5 Separation Using a Chromatographic Column (*Continued*)

illustrates how the bands of separation occur as the injected mixture of compounds passes through the column. Note that the columns in Figure E4.5b indicate the distribution of the sample at different times, with time increasing from the left to the right. The component that is more strongly absorbed on the packing in the column falls behind the more weakly adsorbed. Some compounds with appropriate packing in a sufficiently long column can be completely separated. A typical small-scale laboratory column might be 5 cm in diameter and 10 cm long with a filled fraction of packing of 0.62 of the column volume. For bio materials in laboratories the columns are much smaller.

In the purification of a protein called bovine serum albumin (BSA) from NaCl (a process called "desalting"), 44.2 g of BSA and 97.7 g of NaCl in 500.0 g of H_2O are injected into a packed initially empty column. If the exit product *P* is collected in one container, after a while you find that you have collected 386.3 g total that contain 302.7 g of H_2O, the remainder being BSA and NaCl. How much BSA and NaCl remained in the column? The ratio of the mass of BSA to the mass of NaCl in *P* is 0.78.

Solution

Steps 1–4

All of the known data have been placed on Figure E4.5c along with the symbols for the masses of each compound. The process is unsteady-state in an open system without reaction. Equation (3.2) can be applied.

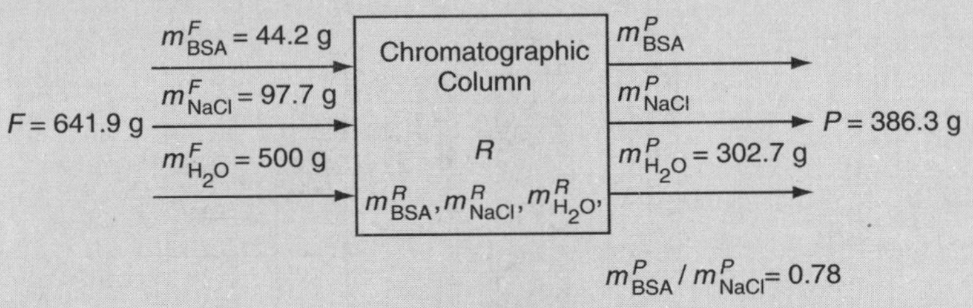

Figure E4.5c

Step 5

Basis: $F = 641.9$ g (feed to the column)

Steps 6 and 7

If you examine Figure E4.5c, you can count the number of variables. Those whose values are not given are m^P_{BSA}, m^P_{NaCl}, m^R_{BSA}, m^R_{NaCl}, $m^R_{H_2O}$, R.

Number of unknowns after all of the known values are assigned their respective variables: 6

Number of independent equations needed: 6

Independent material balances: 3 (BSA, NaCl, H_2O)

Specification: 1 ($m^P_{BSA}/m^P_{NaCl} = 0.78$)

Implicit equations: $2\left(\sum m_i^P = 386.3, \sum m_i^R = R; \sum m_i^F = F \text{ is redundant}\right)$

The degrees of freedom are zero.

Steps 8 and 9

At this point, the set of unknowns can be reduced by some simple calculations using equations involving just one unknown, if you want. For example, you can see from Figure E4.5c that a water balance is such an equation:

H_2O balance: $500 - 302.7 = m^R_{H_2O}$ $\qquad m^R_{H_2O} = 197.3 \text{ g}$

You could use the total balance next to get R, but instead let's solve two simultaneous equations to get the values of the unknowns in P. Remember, if you use the total balance at this stage, one of the two remaining available component balances will be redundant.

$\sum m_i^P = P$ together with the specified ratio ($m^P_{BSA}/m^P_{NaCl} = 0.78$)

$$\left.\begin{aligned} m^P_{BSA} + m^P_{NaCl} + 302.7 &= 386.3 \\ m_{BSA} &= 0.78 m_{NaCl} \end{aligned}\right\} \qquad \begin{aligned} m^P_{BSA} &= 36.7 \text{ g} \\ m^P_{NaCl} &= 47.0 \text{ g} \end{aligned}$$

With the values of these two variables known, you can now solve an equation involving one unknown:

BSA balance: $44.2 - 36.7 = m^R_{BSA}$ $\qquad m^R_{BSA} = 7.5 \text{ g}$

followed by an NaCl balance:

$97.7 - 47.0 = m^R_{NaCl}$ $\qquad m^R_{NaCl} = 50.7 \text{ g}$

All that is left to do is to use the implicit equation $\sum m_i^R = R$ to get R:

$$m^R_{BSA} + m^R_{NaCl} + m^R_{H_2O} = 7.5 + 50.7 + 197.3 = 255.5 \text{ g}$$

(Continues)

Example 4.5 Separation Using a Chromatographic Column (*Continued*)

Step 10

Check using the redundant total balance:

$$R = F - P \qquad R = 641.9 - 386.3 = 255.6$$

The difference is due to round-off.

Example 4.6 Drying

Fish can be turned into fish meal, and the fish meal can be used as animal feed to produce meat or used directly as food for human beings. The direct use of fish meal significantly increases the efficiency of the food chain. However, fish-protein concentrate, primarily for aesthetic reasons, is used mainly as a supplementary protein food. As such, it competes with soy and other oilseed proteins.

In the processing of the fish, after the oil is extracted, the fish cake is dried in rotary drum dryers, finely ground, and packed. The resulting product contains 65% protein. In a given batch of fish cake that contains 80% water (the remainder is dry cake denoted by BDC for "bone dry cake"), 100 kg of water are removed, and it is found that the fish cake is then 40% water. Calculate the weight of the fish cake originally put into the dryer. Figure E4.6 is a diagram of the process.

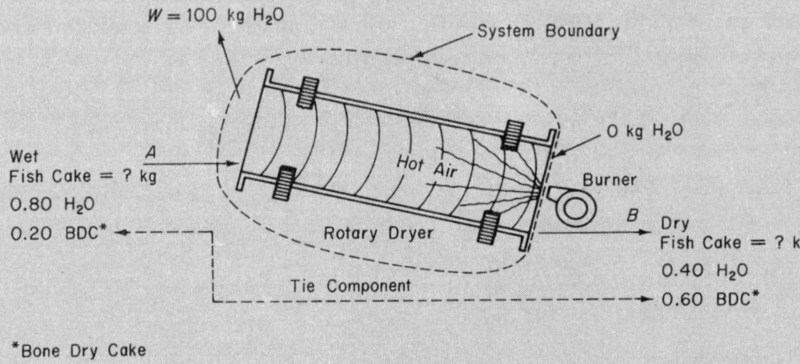

Figure E4.6

Solution

We will abbreviate the solution.

Steps 1–4

The process is a steady-state flow process without reaction. Equation (3.3) can be applied. The system is the dryer. The relation between BDC in the wet and dry fish cake creates a special status for the BDC known as a **tie component** because the BDC enters the process in only one stream and leaves the process unchanged in only one stream. Thus, a tie component allows you to write a material balance for the unique component expressed as a fixed ratio of the streams containing the tie component. An example of the application of a tie component is shown in the following steps.

Step 5

Take a basis of what is given.

$$\text{Basis: 100 kg of water evaporated} = W$$

Steps 6–9

Since the compositions of each stream (A, B, and W) are known, A, B, and W are the only unknown variables, but because W was selected as the basis, only two unknowns remain. We can write two independent material balance equations because there are two components (H_2O and BDC). Therefore, the degree-of-freedom analysis gives zero degrees of freedom. Two streams enter, one containing water and fish cake and the other, just air. Two streams exit, one containing water and fish cake and the other, air and water. The heated air is not used in the material balances for this problem because it is not required. Two independent material balances can be written using the notation in Figure E4.6. We will use the total mass balance plus the BDC balance in kilograms.

You can use the water balance

$$0.80A = 0.40B + 100$$

as a check on the calculations.

	In	Out	
Total balance	A	$= B + W = B + 100$	mass balances
BDC balance	$0.20A$	$= 0.60B$	

The solution is obtained by substituting the second equation into the first equation to eliminate B, and solving for A yields

$$A = 150 \text{ kg initial cake}$$

Applying a material balance for the tie component (BDC) allows you to calculate the ratio of the two streams involving the tie component, that is,

$$\frac{A \text{ kg}}{B \text{ kg}} = \frac{0.60}{0.20} = 3 \quad \text{or} \quad \frac{3\,B \text{ kg}}{A \text{ kg}} = 1$$

(Continues)

Example 4.6 Drying (*Continued*)

Step 10

Check via water balance:

$$0.80(150) \stackrel{?}{=} 0.40\,(150)(1/3) + 100$$

$$120 = 120 \qquad\qquad \text{OK}$$

Example 4.7 Hemodialysis

Hemodialysis is the most common method used to treat advanced and permanent kidney failure. When your kidneys fail, harmful wastes build up in your body, your blood pressure may rise, and your body may retain excess fluid and may not make enough red blood cells. In hemodialysis, your blood flows through a device with a special filter that removes urea and preserves the water balance and the serum proteins in the blood.

The dialyzer itself (refer to Figure E4.7) is a large canister containing thousands of small fibers through which the blood passes.

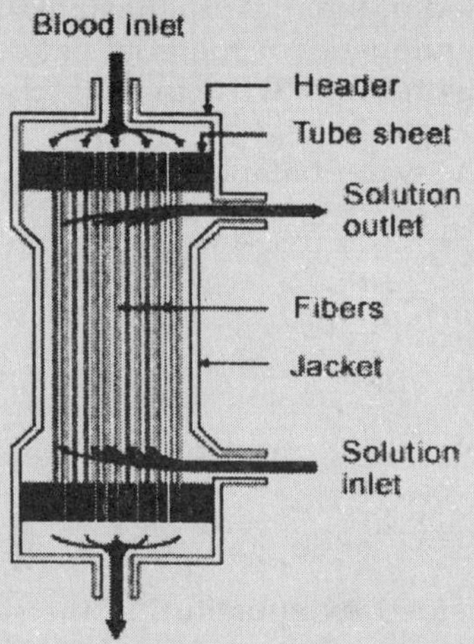

Figure E4.7a

Dialysis solution, the cleansing solution, is pumped around these fibers. The fibers allow wastes and extra fluids to pass from your blood into the solution that carries them away.

This example focuses on the plasma components in streams S (solvent) and B (blood): water, uric acid (UR), creatinine (CR), urea (U), P, K, and Na. You can ignore the initial filling of the dialyzer because the treatment lasts for an interval of two or three hours. Given the measurements obtained from one treatment as shown in Figure E4.7b, calculate the grams per liter of each component of the plasma in the outlet solution.

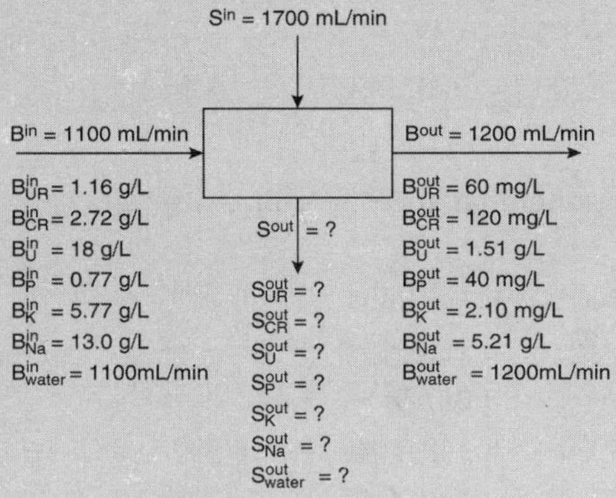

Figure E4.7b

Solution

This is an open, steady-state system.

Step 1

Basis: 1 min

Steps 2–4

The data have been inserted on Figure E4.7b.

You can ignore the effect of the components of the plasma on the density of the solution for this problem. The entering solution can be assumed to be essentially water.

Steps 6 and 7

Number of unknowns: Figure 4.7b shows quite a few variables, but let's count only those variables whose values are not specified in the figure. Only eight unknowns exist, seven values of the components in the exit stream of water and the total exit flow.

Number of independent equations needed: 8

You can make seven component balances and use the implicit equation of the sum of the exit component mass flows to get the total flow out (in grams).

(Continues)

Example 4.7 Hemodialysis (*Continued*)

Steps 8 and 9

The water balance in grams, assuming that 1 mL is equivalent to 1 g (a very convenient assumption), is

$$1100 + 1700 = 1200 + S^{out}_{water} \qquad \text{hence} \qquad S^{out}_{water} = 1600\text{ g}$$

or 1.6 L

The other component balances in grams are

		g/L
UR:	$1.1(1.16) + 0 = 1.2(0.060) + 1.6\, S^{out}_{UR}$	$S^{out}_{UR} = 0.75$
CR:	$1.1(2.72) + 0 = 1.2(0.120) + 1.6\, S^{out}_{CR}$	$S^{out}_{CR} = 1.78$
U:	$1.1(18) + 0 = 1.2(1.51) + 1.6\, S^{out}_{U}$	$S^{out}_{U} = 11.2$
P:	$1.1(0.77) + 0 = 1.2(0.040) + 1.6\, S^{out}_{P}$	$S^{out}_{P} = 0.50$
K:	$1.1(5.77) + 0 = 1.2(0.120) + 16\, S^{out}_{K}$	$S^{out}_{K} = 3.8$
Na:	$1.1(13.0) + 0 = 1.2(3.21) + 1.6\, S^{out}_{Na}$	$S^{out}_{Na} = 6.53$

Step 10

As a consistency check, evaluate the overall mass balance equation.

Frequently Asked Questions

1. All of the examples presented in this chapter have involved only a small set of equations to solve. If you have to solve a large set of equations, some of which may be redundant, how can you tell if the set of equations you select to solve is a set of *independent equations*? You can determine if the set is independent for linear equations by determining the rank of the coefficient matrix of the set of equations, which is equivalent to the number of independent linear equations. Chapter 12 explains how to obtain the rank and shows some examples. Computer programs such as MATLAB, Mathcad, Polymath, and so on provide a convenient way for you to determine the rank of the coefficient matrix without having to carry out the intermediate details of the calculations. Introduce the data into the computer program, and the output from the computer will provide you with some diagnostics if a solution is not obtained. For example, if the equations are not independent, Polymath returns the warning "Error—Singular matrix entered," which means that the rank of the coefficient matrix of the set of linear equations is less than the number of equations.
2. What should you do if the computer solution you obtain by solving a set of equations gives you a negative value for one or more of the unknowns? One

possibility to examine is that you inadvertently reversed the sign of a term in a material balance, say, from + to −. Another possibility is that you forgot to include an essential term(s) so that a zero was entered into the coefficient set for the equations rather than the proper number.

3. In industry, material balances rarely balance when Equations (3.1) to (3.3) are used with process measurements to check process performance. You want to determine what is called *closure*, namely, that the error between "in" and "out" is acceptable. The flow rates and measured compositions for all of the streams entering and exiting a process unit are substituted into the appropriate material balance equations. Ideally, in the steady state in the absence of reaction, the amount (mass) of each component entering a process or group of processes should equal the amount of that component leaving the system. The lack of closure for material balances in industrial process occurs for several reasons:
   a. The process is rarely operating in the steady state. Industrial processes are almost always in a state of flux, rarely approaching steady-state behavior.
   b. The flow and composition measurements have a variety of errors associated with them. Sensor readings include noise (variations in the measurement due to more or less random variations in the readings that do not correspond to changes in the process). Sensor readings can be inaccurate for a wide variety of reasons such as degradation of calibration, poor installment, wrong kind of sensor, dirt and plugging, failure of electronic components, and so on.
   c. A component of interest may be generated or consumed inside the process by reactions that the process engineer has not considered.

As a result, material balance closure to within ±5% for material balances for most industrial processes is considered reasonable. If special care and equipment are employed, discrepancies can be reduced somewhat.

Self-Assessment Test

Questions

1. Indicate whether the following statements are true or false:
   a. The most difficult part of solving material balance problems is the collection and formulation of the data specifying the compositions of the streams into and out of the system, and of the material inside the system.
   b. All open processes involving two components with three streams involve zero degrees of freedom.
   c. An unsteady-state process problem can be analyzed and solved as a steady-state process problem.

d. If a flow rate is given in kilograms per minute, you should convert it to kilogram moles per minute.

2. Under what circumstances do equations or specifications become redundant?

Problems

1. A cellulose solution contains 5.2% cellulose by weight in water. How many kilograms of 1.2% solution are required to dilute 100 kg of the 5.2% solution to 4.2%?
2. A cereal product containing 55% water is made at the rate of 500 kg/hr. You need to dry the product so that it contains only 30% water. How much water has to be evaporated per hour?
3. Salt in crude oil must be removed before the oil undergoes processing in a refinery. The crude oil is fed to a washing unit where freshwater fed to the unit mixes with the oil and dissolves a portion of the salt contained in the oil. The oil (containing some salt but no water), being less dense than the water, can be removed at the top of the washer. If the "spent" wash water contains 15% salt and the crude oil contains 5% salt, determine the concentration of salt in the "washed" oil product if the ratio of crude oil (with salt) to water used is 4 to 1.

Looking Back

In this chapter we explained via examples how to analyze problems involving material balances in the absence of chemical reactions. By using the ten-step procedure outlined in Chapter 3, we explained how the strategy can be used to solve each problem no matter what the process is.

Glossary

Tie component A component in a material balance that enters a process in only one stream and leaves in only one stream and does not react inside the process.

Supplementary References

Felder, R. M., and R. W. Rousseau. *Elementary Principles of Chemical Processes*, 3d ed., John Wiley, New York (2000).

Fogler, H. S., and S. M. Montgomery. *Material and Energy Balances Stoichiometry Software*, CACHE Corp., Austin, TX (1993).

Answer to the Puzzle

The magician held the egg 2.1 m above the floor before dropping it.

Problems

***4.1.1** You buy 100 kg of cucumbers that contain 99% water. A few days later they are found to be 98% water. Is it true that the cucumbers now weigh only 50 kg?

***4.1.2** The fern *Pteris vittata* has been shown [*Nature*, **409**, 579 (2001)] to effectively extract arsenic from soils. The study showed that in normal soil, which contains 6 ppm of arsenic, in two weeks the fern reduced the soil concentration to 5 ppm while accumulating 755 ppm of arsenic. In this experiment, what was the ratio of the soil mass to the plant mass? The initial arsenic in the fern was 5 ppm.

***4.1.3** Sludge is wet solids that result from the processing in municipal sewage systems. The sludge has to be dried before it can be composted or otherwise handled. If a sludge containing 70% water and 30% solids is passed through a dryer, and the resulting product contains 25% water, how much water is evaporated per ton of sludge sent to the dryer?

***4.1.4** Figure P4.1.4 is a sketch of an artificial kidney, a medical device used to remove waste metabolites from the blood in cases of kidney malfunction. The dialyzing fluid passes across a hollow membrane, and the waste products diffuse from the blood into the dialyzing fluid.

If the blood entering the unit flows at the rate of 220 mL/min, and the blood exiting the unit flows at the rate of 215 mL/min, how much water and urea (the main waste product) pass into the dialysate if the entering concentration of urea is 2.30 mg/mL and the exit concentration of urea is 1.70 mg/mL?

If the dialyzing fluid flows into the unit at the rate of 1500 mL/min, what is the concentration of the urea in the dialysate?

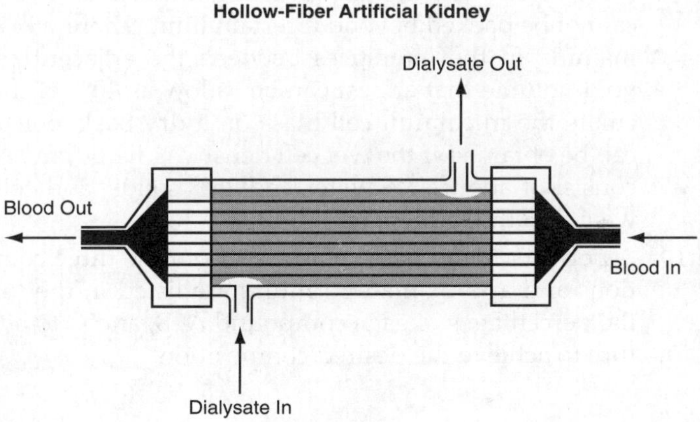

Figure P4.1.4

*4.1.5 A multiple-stage evaporator concentrates a weak NaOH solution from 3% to 18% and processes 2 tons of feed solution per day. How much product is made per day? How much water is evaporated per day?

*4.1.6 A liquid adhesive consists of a polymer dissolved in a solvent. The amount of polymer in the solution is important to the application. An adhesive dealer receives an order for 3000 lb of an adhesive solution containing 13% polymer by weight. On hand are 500 lb of 10% solution and very large quantities of 20% solution and pure solvent. Calculate the weight of each that must be blended together to fill this order. Use all of the 10% solution.

*4.1.7 A lacquer plant must deliver 1000 lb of an 8% nitrocellulose solution. The plant has in stock a 5.5% solution. How much dry nitrocellulose must be dissolved in the solution to fill the order?

**4.1.8 A gas containing 80% CH_4 and 20% He is sent through a quartz diffusion tube (see Figure P4.1.8) to recover the helium. Twenty percent by weight of the original gas is recovered, and its composition is 50% He. Calculate the composition of the waste gas if 100 kg moles of gas are processed per minute.

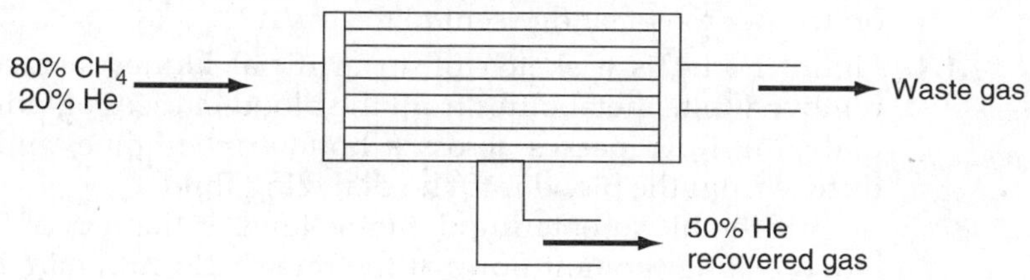

Figure P4.1.8

*4.1.9 In many fermentations, the maximum amount of cell mass must be obtained. However, the amount of mass that can be made is ultimately limited by the cell volume. Cells occupy a finite volume and have a rigid shape so that they cannot be packed beyond a certain limit. There will always be some water remaining in the interstices between the adjacent cells, which represent the void volume that at best can be as low as 40% of the fermenter volume. Calculate the maximum cell mass on a dry basis per liter of the fermenter that can be obtained if the wet cell density is 1.1 g/cm^3. Note that cells themselves consist of about 75% water and 25% solids, and cell mass is reported as dry weight in the fermentation industry.

**4.1.10 A polymer blend is to be formed from the three compounds whose compositions and approximate formulas are listed in the following table. Determine the percentages of each compound A, B, and C to be introduced into the mixture to achieve the desired composition.

	Compound (%)			
Composition	A	B	C	D (Desired Mixture)
$(CH_4)_x$	25	35	55	30
$(C_2H_6)_x$	35	20	40	30
$(C_3H_8)_x$	40	45	5	40
Total	100	100	100	100

How would you decide to blend compounds A, B, and C to achieve the desired mixture D [$(CH_4)_x = 10\%$, $(C_2H_6)_x = 30\%$, $(C_3H_8)_x = 60\%$]?

**4.1.11 Your boss asks you to calculate the flow through a natural-gas pipeline. Since it is 26 in. in diameter, it is impossible to run the gas through any kind of meter or measuring device. You decide to add 100 lb of CO_2 per minute to the gas through a small 1/2 in. piece of pipe, collect samples of the gas downstream, and analyze them for CO_2. Several consecutive samples after 1 hr are given in the following table.

Time	% CO_2
1 hr, 0 min	2.0
10 min	2.2
20 min	1.9
30 min	2.1
40 min	2.0

a. Calculate the flow of gas in pounds per minute at the point of injection.
b. Unfortunately for you, the gas upstream of the point of injection of CO_2 already contained 1.0% CO_2. How much was your original flow estimate in error (in percent)?

Note: In part a the natural gas is all methane, CH_4.

*4.1.12 Ammonia is a gas for which reliable analytical methods are available to determine its concentration in other gases. To measure flow in a natural-gas pipeline, pure ammonia gas is injected into the pipeline at a constant rate of 72.3 kg/min for 12 min. Five miles downstream from the injection point, the steady-state ammonia concentration is found to be 0.382 wt %. The gas upstream from the point of ammonia injection contains no measurable ammonia. How many kilograms of natural gas are flowing through the pipeline per hour?

*4.1.13 Water pollution in the Hudson River has claimed considerable attention, especially pollution from sewage outlets and industrial wastes. To determine accurately how much effluent enters the river is quite difficult because to catch and weigh the material is impossible, weirs are hard to construct, and so on. One suggestion that has been offered is to add a tracer of Br ion to a given sewage stream, let it mix well, and sample the sewage stream after it mixes. On one test of the proposal you add 10 pounds of NaBr per hour for 24 hr to a sewage stream with essentially no Br in it. Somewhat downstream of the introduction point a sampling of the sewage stream shows 0.012% NaBr. The sewage density is 60.3 lb/ft^3 and river water density is 62.4 lb/ft^3. What is the flow rate of the sewage in pounds per minute?

**4.1.14 A process for separating a mixture of incompatible polymers, such as polyethylene terephthalate (PET) and polyvinyl chloride (PVC), promises to expand the recycling and reuse of plastic waste. The first commercial plant, at Celanese's recycling facility in Spartanburg, South Carolina, has been operating at a PET capacity of 15 million lb/yr. Operating cost: 0.5¢/lb.

Targeted to replace the conventional sorting of individual PET bottles from PVC containers upstream of the recycling step, this process first chops the mixed waste with a rotary-blade cutter to 0.5 in. chips. The materials are then suspended in water, and air is forced through to create a bubblelike froth that preferentially entraps the PVC because of its different surface-tension characteristics. A food-grade surfactant is also added to enhance the separation. The froth is skimmed away along with the PVC, leaving behind the PET material. For a feed with 2% PVC, the process has recovered almost pure PET with an acceptable PVC contamination level of 10 ppm.

How many pounds of PVC are recovered per year from this process?

***4.1.15 If 100 g of Na_2SO_4 are dissolved in 200 g of H_2O and the solution is cooled until 100 g of $Na_2SO_4 \cdot 10\,H_2O$ crystallize out, find (a) the composition of the remaining solution (*mother liquor*), and (b) the grams of crystals recovered per 100 g of initial solution.

Hint: Treat the hydrated crystals as a separate stream leaving the process.

****4.1.16 A chemist attempts to prepare some very pure crystals of borax (sodium tetraborate, $Na_2B_4O_7 \cdot 10\,H_2O$) by dissolving 100 g of $Na_2B_4O_7$ in 200 g of boiling water. He then carefully cools the solution slowly until some $Na_2B_4O_7 \cdot 10\,H_2O$ crystallizes out. Calculate the grams of $Na_2B_4O_7 \cdot 10\,H_2O$ recovered in the crystals per 100 g of total initial solution ($Na_2B_4O_7$ plus H_2O), if the residual solution at 55°C after the crystals are removed contains 12.4% $Na_2B_4O_7$.

Hint: Treat the hydrated crystals as a separate stream leaving the process.

***4.1.17 One thousand kilograms of $FeCl_3 \cdot 6\,H_2O$ are added to a mixture of crystals of $FeCl_3 \cdot H_2O$ to produce a mixture of $FeCl_3 \cdot 2.5\,H_2O$ crystals. How much $FeCl_3 \cdot H_2O$ must be added to produce the most $FeCl_3 \cdot 2.5\,H_2O$?

Hint: Treat the hydrated crystals as a separate stream leaving the process.

***4.1.18 The solubility of barium nitrate at 100°C is 34 g/100 g of H_2O and at 0°C is 5.0 g/100 g of H_2O. If you start with 100 g of $Ba(NO_3)_2$ and make a saturated solution in water at 100°C, how much water is required? If the saturated solution is cooled to 0°C, how much $Ba(NO_3)_2$ is precipitated out of solution? The precipitated crystals carry along with them on their surface 4 g of H_2O per 100 g of crystals.

Hint: Treat the hydrated crystals as a separate stream leaving the process.

****4.1.19 A water solution contains 60% $Na_2S_2O_2$ together with 1% soluble impurity. Upon cooling to 10°C, $Na_2S_2O_2 \cdot 5\,H_2O$ crystallizes out. The solubility of this hydrate is 1.4 lb $Na_2S_2O_2 \cdot 5\,H_2O$/lb free water. The crystals removed carry as adhering solution 0.06 lb solution/lb crystals. When dried to remove the remaining water (but not the water of hydration), the final dry $Na_2S_2O_2 \cdot 5\,H_2O$ crystals must not contain more than 0.1% impurity. To meet this specification, the original solution, before cooling, is further diluted with water. On the basis of 100 lb of the original solution, calculate (a) the amount of water added before cooling, and (b) the percentage recovery of the $Na_2S_2O_2$ in the dried hydrated crystals.

Hint: Treat the hydrated crystals as a separate stream leaving the process.

**4.1.20 Paper pulp is sold on the basis that it contains 12% moisture; if the moisture exceeds this value, the purchaser can deduct any charges for the excess moisture and also deduct for the freight costs of the excess moisture. A shipment of pulp became wet and was received with a moisture content of 22%. If the original price for the pulp was $40/ton of air-dry pulp and if the freight is $1.00/100 lb shipped, what price should be paid per ton of pulp delivered?

**4.1.21 A laundry can purchase soap containing 30% water for a price of $0.30/kg FOB the soap manufacturing plant (i.e., at the soap plant before shipping costs, which are owed by the purchaser of the soap). It can also purchase a different grade of soap that contains only 5% water. The freight rate between the soap plant and the laundry is $6.05/100 kg. What is the maximum price the laundry should pay for the 5% soap?

***4.1.22 A manufacturer of briquettes has a contract to make briquettes for barbec ing that are guaranteed to not contain over 10% moisture or 10% ash. T basic material used has this analysis: moisture 12.4%, volatile material 16. carbon 57.5%, and ash 13.5%. To meet the specifications (at their limits) manufacturer plans to mix with the base material a certain amount of p leum coke that has this analysis: volatile material 8.2%, carbon 88.7%, moisture 3.1%. How much petroleum coke must be added per 100 lb of the base material?

****4.1.23 In a gas-separation plant, the feed to the process has the following constitutents:

Component	Mol %
C_3	1.9
i-C_4	51.5
n-C_4	46.0
C^{5+}	0.6
Total	100.0

The flow rate is 5804 kg mol/day. If the overhead and bottoms streams leaving the process have the following compositions, what are the flow rates of the overhead and bottoms streams in kilogram moles per day?

	Mol %	
Component	**Overhead**	**Bottoms**
C_3	3.4	—
i-C_4	95.7	1.1
n-C_4	0.9	97.6
C^{5+}	—	1.3
Total	100.0	100.0

****4.1.24 The organic fraction in the wastewater is measured in terms of the biological oxygen demand (BOD) material, namely, the amount of dissolved oxygen required to biodegrade the organic contents. If the dissolved oxygen (DO) concentration in a body of water drops too low, the fish in the stream or lake may die. The Environmental Protection Agency has set the minimum summer levels for lakes at 5 mg/L of DO.

a. If a stream is flowing at 0.3 m^3/s and has an initial BOD of 5 mg/L before reaching the discharge point of a sewage treatment plant, and the plant discharges 3.785 ML/day of wastewater, with a concentration of 0.15 g/L of BOD, what will be the BOD concentration immediately below the discharge point of the plant?

b. The plant reports a discharge of 15.8 ML/day having a BOD of 72.09 mg/L. If the EPA measures the flow of the stream before the discharge point at 530 ML/day with 3 mg/L of BOD, and measures the downstream concentration of 5 mg/L of BOD, is the report correct?

****4.1.25 Suppose that 100 L/min are drawn from a fermentation tank and passed through an extraction tank in which the fermentation product (in the aqueous phase) is mixed with an organic solvent, and then the aqueous phase is separated from the organic phase. The concentration of the desired enzyme (3-hydroxybutyrate dehydrogenase) in the aqueous feed to the extraction tank is 10.2 g/L. The pure organic extraction solvent runs into the extraction tank at the rate of 9.5 L/min. If the ratio of the enzyme in the exit product stream (the organic phase) from the extraction tank to the concentration of the enzyme in the exit waste stream (the aqueous phase) from the tank is D = 18.5 (g/L organic)/(g/L aqueous), what is the fraction recovery of the enzyme and the amount recovered per minute? Assume negligible miscibility between the aqueous and organic liquids in each other, and ignore any change in density on removal or addition of the enzyme to either stream.

****4.1.26 Consider the following process for recovering NH_3 from a gas stream composed of N_2 and NH_3 (see Figure P4.1.26).

Flowing upward through the process is the gas stream, which can contain NH_3 and N_2 but *not* solvent S, and flowing downward through the device is a liquid stream which can contain NH_3 and liquid S but *not* N_2.

The weight fraction of NH_3 in the gas stream A leaving the process is related to the weight fraction of NH_3 in the liquid stream B leaving the process by the following empirical relationship:

$$\omega_{NH_3}^{A} = 2\omega_{NH_3}^{B}$$

Given the data shown in Figure P4.1.26, calculate the flow rates and compositions of streams A and B.

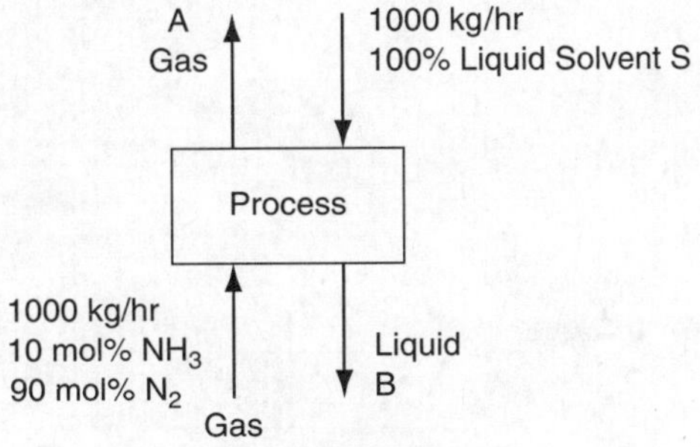

Figure P4.1.26

***4.1.27 MTBE (methyl tert-butyl ether) is added to gasoline to increase the oxygen content of the gasoline. MTBE is soluble in water to some extent and becomes a contaminant when the gasoline gets into surface or underground water. The gasoline used by boats has an MTBE content of 10%. The boats operate in a well-mixed flood control pond having the dimensions 3 km long, 1 km wide, and 3 m deep on the average. Suppose that each of the 25 boats on the pond spills 0.5 L of gasoline during 12 hr of daylight. The flow of water (that contains no MTBE) into the pond is 10 m^3/hr, but no water leaves because the water level is well below the spillway of the pond. By how much will the concentration of MTBE increase in the pond after the end of 12 hr of boating? Data: The specific gravity of gasoline is 0.72.

CHAPTER 5

Material Balances Involving Reactions

Your objectives in studying this chapter are to be able to

1. Write and balance chemical reaction equations
2. Determine the stoichiometric quantities of reactants and products in moles or mass given the chemical reaction
3. Identify the limiting and excess reactants in a reaction, and calculate the fraction or percent excess reactant(s); the percent conversion, or completion; the yield; and the extent of reaction for a chemical reaction with the reactants given in nonstoichiometric proportion
4. Carry out a degree-of-freedom analysis for processes involving chemical reaction(s)
5. Formulate and solve material balances using (a) species balances and (b) element balances
6. Decide when element balances can be used as material balances
7. Understand how the extent of reaction is determined for a process, and how to apply it in material balance problems
8. Understand the meaning of stack gas, flue gas, Orsat analysis, dry basis, wet basis, theoretical air (oxygen), and excess air (oxygen), and employ these concepts in combustion problems

Why are we devoting a whole chapter to discussing material balances for systems with reaction? The heart of many plants is the reactor in which products and by-products are produced. Material balances considering reactions are used to design reactors. Moreover, these material balances can also be used to identify the most efficient operation of the reactors (i.e., process optimization). Of course, computer programs can make the calculations for you, but you have to put the right information into them, and

. . . to calculate is not in itself to analyze.
Edgar Allen Poe, "The Murders in the Rue Morgue"

5.1 Stoichiometry

You are probably aware that chemical engineers differ from most other engineers because of their application of chemistry. When chemical reactions occur in contrast with physical changes of material such as evaporation or dissolution, you want to be able to predict the mass or moles required for the reaction(s), and the mass or moles of each species remaining after the reaction has occurred. Reaction stoichiometry allows you to accomplish this. The word *stoichiometry* (stoi-ki-OM-e-tri) derives from two Greek words: *stoicheion* (meaning "element") and *metron* (meaning "measure"). Stoichiometry provides a quantitative means of relating the amount of products produced by a chemical reaction(s) to the amount of reactants or vice versa.

As you already know, the chemical reaction equation provides both qualitative and quantitative information concerning chemical reactions that occur in a chemical process. Specifically, the chemical reaction equation provides information of two types:

1. It tells you what substances are reactants (those being used up) and what substances are products (those being made).
2. The coefficients of a *balanced* chemical reaction equation tell you what the mole ratios are among the substances that react or are produced. (In 1803, John Dalton, an English chemist, was able to explain much of the experimental results on chemical reactions of the day by assuming that reactions occurred with fixed ratios of elements. This discovery led to the *law of constant proportionality*, which states that chemical reactions proceed with fixed ratios of the number of reactants and products involved in the reaction.)

You should take the following steps when solving problems involving stoichiometry:

1. Make sure the chemical equation is correctly balanced. How can you tell if the reaction equation is balanced? Make sure the total quantities of each of the elements on the left-hand side equal those on the right-hand side. For example,

$$CH_4 + O_2 \rightarrow CO_2 + H_2O$$

is not a balanced stoichiometric equation because there are four atoms of H on the reactant side (left-hand side) of the equation, but only two atoms of H on the product side (right-hand side). In addition, the oxygen atoms do not balance. The balanced equation is given by

$$CH_4 + 2O_2 \rightarrow CO_2 + 2H_2O$$

Note that the sum of each of the elements present in the chemical reaction equation (C, H, and O) is the same for the reactants (left-hand side of the chemical reaction equation) as for the products (right-hand side of the chemical reaction equation). The coefficients in the balanced reaction equation have the units of moles of a species for the particular reaction equation. For example, for the previous chemical reaction equation, for every mole of CH_4 that reacts, 2 moles of O_2 are consumed and 1 mole of CO_2 and 2 moles of H_2O are produced. If you multiply each term in a chemical reaction equation by the same constant, say, 2, the absolute stoichiometric coefficient in each term doubles, but the coefficients still occur in the same relative proportions. For example, for the previous chemical reaction equation, if 2 moles of CH_4 react, 4 moles of O_2 are consumed and 2 moles of CO_2 and 4 moles of H_2O are produced.
2. Use the proper degree of completion for the reaction. If you do not know how much of the reaction has occurred, you may assume a reactant reacts completely in this book.
3. Use molecular weights to convert mass to moles for the reactants and moles to mass for the products.
4. Use the coefficients in the chemical equation to obtain the relative molar amounts of products produced and reactants consumed in the reaction.

Steps 3 and 4 can be applied in a fashion similar to that used in carrying out the conversion of units as explained in Chapter 1. As an example, consider the combustion of heptane:

$$C_7H_{16}(l) + 11\,O_2(g) \rightarrow 7\,CO_2(g) + 8\,H_2O(g)$$

(Note that we have put the states of the compounds in parentheses after the species formula, information not needed for this chapter but that will be vital in subsequent sections of this book.)

What can you learn from this equation? The **stoichiometric coefficients** in the chemical reaction equation (1 for C_7H_{16}, 11 for O_2, and so on) tell you the relative amounts of moles of chemical species that react and are produced by the reaction. The units of a stoichiometric coefficient for species *i* are the change in the moles of species *i* divided by the moles reacting according to a specific chemical equation. In taking ratios of coefficients, the denominators cancel, and you are left with the ratio of the moles of one species divided by another. For example, for the combustion of heptanes:

$$\frac{1 \text{ mol } C_7H_{16}}{\text{moles reacting}} \div \frac{11 \text{ mol } O_2}{\text{moles reacting}} = \frac{1 \text{ mol } C_7H_{16}}{11 \text{ mol } O_2}$$

We will abbreviate the units of a coefficient for species *i* simply as mol *i*/moles reacting when appropriate, but frequently in practice the units are ignored. You can conclude that 1 mole (*not* lb_m or kg) of heptane will react with 11 moles of oxygen to give 7 moles of carbon dioxide plus 8 moles of water. These may be pound moles, gram moles, kilogram moles, or any other type of mole. Another way to use the chemical reaction equation is to conclude that 1 mole of CO_2 is formed from each 1/7 mole of C_7H_{16} and 1 mole of H_2O is formed with each 7/8 mole of CO_2. The ratios indicate the **stoichiometric ratios** that can be used to determine the relative proportions of products and reactants.

Suppose you are asked how many kilograms of CO_2 will be produced as product if 10 kg of C_7H_{16} react completely with the **stoichiometric quantity** of O_2? On the basis of 10 kg of C_7H_{16}:

$$\frac{10 \text{ kg } C_7H_{16}}{} \left| \frac{1 \text{ kg mol } C_7H_{16}}{100.1 \text{ kg } C_7H_{16}} \right| \frac{7 \text{ kg mol } CO_2}{1 \text{ kg mol } C_7H_{16}} \left| \frac{44.0 \text{ kg } CO_2}{1 \text{ kg mol } CO_2} \right. = 30.8 \text{ kg } CO_2$$

Conversion from mass to moles — Mole ratio — Conversion from moles to mass

Let's now write a general chemical reaction equation as

$$c\,\text{C} + d\,\text{D} \rightleftarrows a\,\text{A} + b\,\text{B} \tag{5.1}$$

where *a*, *b*, *c*, and *d* are the stoichiometric coefficients for the species A, B, C, and D, respectively. Equation (5.1) can be written in a general form:

$$\nu_A A + \nu_B B + \nu_C C + \nu_D D = \sum \nu_i S_i = 0 \tag{5.2}$$

where ν_i is the stoichiometric coefficient for species S_i. The products are defined to have *positive values* for stoichiometric coefficients and the reactants to have *negative values* for stoichiometric coefficients. The ratios of stoichiometric coefficients are unique for a given reaction. Specifically for Equation (5.1) written in the form of Equation (5.2):

$$\nu_C = -c \qquad \nu_A = a \qquad \nu_D = -d \qquad \nu_B = b$$

If a species is not present in an equation, the value of its stoichiometric coefficient is deemed to be zero. As an example, in the reaction

$$O_2 + 2CO \rightarrow 2CO_2$$

$$\nu_{O_2} = -1 \qquad \nu_{CO} = -2 \qquad \nu_{CO_2} = 2 \qquad \nu_{N_2} = 0$$

Example 5.1 Balancing a Reaction Equation for a Biological Reaction

The primary energy source for cells is the aerobic catabolism (oxidation) of glucose ($C_6H_{12}O_6$, a sugar). The overall oxidation of glucose produces CO_2 and H_2O by the following reaction:

$$C_6H_{12}O_6 + a\,O_2 \rightarrow b\,CO_2 + c\,H_2O$$

Solution

Basis: The given chemical reaction equation

By inspection, the carbon balance gives $b = 6$, the hydrogen balance gives $c = 6$, and an oxygen balance yields the following equation:

$$6 + 2a = 6 \times 2 + 6$$

which gives $a = 6$. Therefore, the balanced reaction equation is

$$C_6H_{12}O_6 + 6\,O_2 \rightarrow 6\,CO_2 + 6\,H_2O$$

As a consistency check, verify that for each element the number of elements in the reactants is equal to the number of elements in the products.

Example 5.2 Use of the Chemical Reaction Equation to Calculate the Mass of Reactants Given the Mass of Products

In the combustion of heptane with oxygen, CO_2 is produced. Assume that you want to produce 500 kg of dry ice per hour, and that 50% of the CO_2 can be converted into dry ice, as shown in Figure E5.2. How many kilograms of heptane must be burned per hour?

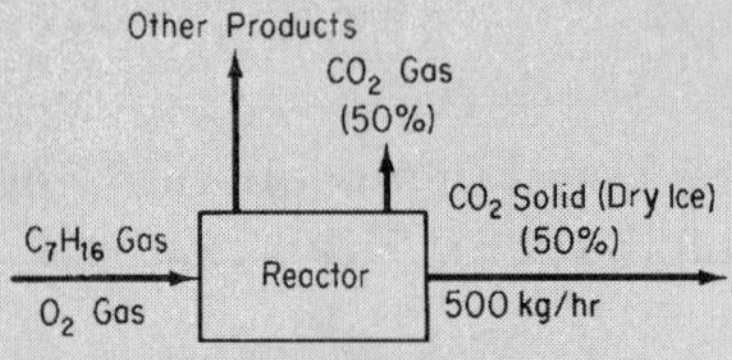

Figure E5.2

Solution

In solving a problem of this sort, the grand thing is to be able to reason backward. This is a very useful accomplishment, and a very easy one, but people do not practice it much.

Sherlock Holmes, in Sir Arthur Conan Doyle's *A Study in Scarlet*

From the problem statement you can conclude that you want to use the product mass of CO_2 to calculate a reactant mass, the C_7H_{16}. The procedure is first to convert kilograms of CO_2 to moles, apply the chemical equation to get moles of C_7H_{16}, and finally calculate the kilograms of C_7H_{16}. We will use Figure E5.2 in the analysis.

Look in Appendix B to get the molecular weight of CO_2 (44.0) and C_7H_{16} (100.1). The chemical equation is

$$C_7H_{16} + 11\,O_2 \rightarrow 7\,CO_2 + 8\,H_2O$$

The next step is to select a basis.

Basis: 500 kg of dry ice (equivalent to 1 hr)

The calculation of the amount of C_7H_{16} can be made in one sequence:

$$\frac{500 \text{ kg dry ice}}{} \left| \frac{1 \text{ kg } CO_2 \text{ formed}}{0.5 \text{ kg dry ice}} \right| \frac{1 \text{ kg mol } CO_2}{44.0 \text{ kg } CO_2} \left| \frac{1 \text{ kg mol } C_7H_{16}}{7 \text{ kg mol } CO_2} \right|$$

$$\left| \frac{100.1 \text{ kg } C_7H_{16}}{1 \text{ kg mol } C_7H_{16}} = 325 \text{ kg } C_7H_{16} \right.$$

Therefore, the answer to this problem is 325 kg C_7H_{16}/hr. Finally, you could check your answer by reversing the sequence of calculations.

Example 5.3 Application of Stoichiometry When More than One Reaction Occurs

A limestone analysis:

$CaCO_3$	92.89%
$MgCO_3$	5.41%
Unreactive	1.70%

By heating the limestone, you recover oxides that together are known as lime.

a. How many pounds of calcium oxide can be made from 1 ton of this limestone?

b. How many pounds of CO_2 can be recovered per pound of limestone?

c. How many pounds of limestone are needed to make 1 ton of lime?

Solution

Steps 1 and 3

Read the problem carefully to fix in mind exactly what is required. The carbonates are decomposed to oxides. You should recognize that lime (oxides of Ca and Mg) will also include all of the impurities present in the limestone that remain after the CO_2 has been driven off.

Step 2

Next, draw a picture of what is going on in this process. See Figure E5.3.

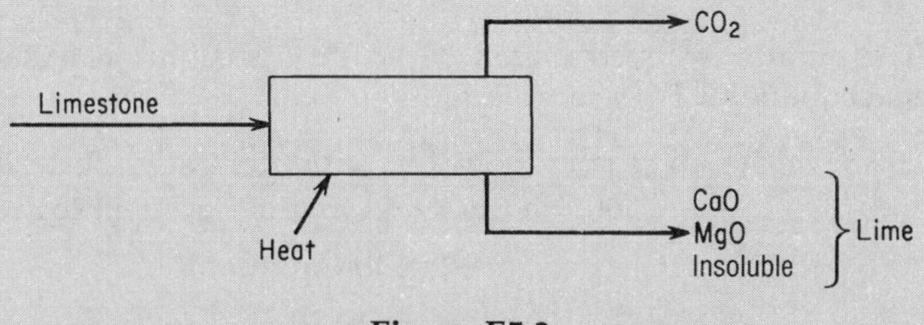

Figure E5.3

Step 4

To complete the preliminary analysis you need the following chemical reaction equations:

$$CaCO_3 \rightarrow CaO + CO_2$$

$$MgCO_3 \rightarrow MgO + CO_2$$

(Continues)

Example 5.3 Application of Stoichiometry When More than One Reaction Occurs (*Continued*)

Additional data that you need to look up (or calculate) are the molecular weights of the species:

	$CaCO_3$	$MgCO_3$	CaO	MgO	CO_2
Mol. wt.:	100.1	84.32	56.08	40.32	44.0

Step 5

The next step is to pick a basis:

Basis: 100 lb of limestone

This basis was selected because pounds of each component will be equal to its weight percent. You could also pick 1 lb of limestone if you wanted, or 1 ton.

Steps 6–9

Calculations of the percent composition and pound moles of the limestone and products in the form of a table will serve as an adjunct to Figure E5.3 and will prove to be most helpful in answering the questions posed.

Limestone			Solid Products		
Component	**lb = percent**	**lb mol**	**Compound**	**lb mol**	**lb**
$CaCO_3$	92.89	0.9280	CaO	0.9280	52.04
$MgCO_3$	5.41	0.0642	MgO	0.0642	2.59
Unreactive	1.70		Unreactive		1.70
Total	100.00	0.9920	Total	0.9920	56.33

The quantities listed under "Solid Products" are calculated from the chemical equations. For example:

$$\frac{92.89 \text{ lb } CaCO_3}{} \left| \frac{1 \text{ lb mol } CaCO_3}{100.1 \text{ lb } CaCO_3} \right| \frac{1 \text{ lb mol CaO}}{1 \text{ lb mol } CaCO_3} \left| \frac{56.08 \text{ lb CaO}}{1 \text{ lb mol CaO}} \right.$$

$$= 52.04 \text{ lb CaO}$$

$$\frac{5.41 \text{ lb } MgCO_3}{} \left| \frac{1 \text{ lb mol } MgCO_3}{84.32 \text{ lb } MgCO_3} \right| \frac{1 \text{ lb mol MgO}}{1 \text{ lb mol } MgCO_3} \left| \frac{40.32 \text{ lb MgO}}{1 \text{ lb mol MgO}} \right.$$

$$= 2.59 \text{ lb MgO}$$

The production of CO_2:

0.9280 lb mol CaO is equivalent to 0.9280 lb mol CO_2
0.0642 lb mol MgO is equivalent to 0.0642 lb mol CO_2
Total 0.9920 lb mol CO_2

$$\frac{0.9920 \text{ lb mol } CO_2}{} \left| \frac{44.0 \text{ lb } CO_2}{1 \text{ lb mol } CO_2} \right. = 44.67 \text{ lb } CO_2$$

Alternatively, you could have calculated the pounds of CO_2 from a total balance: $100 - 56.33 = 44.67$. Note that the total pounds of products equal the 100 lb of entering limestone. If they were not equal, what would you do? Check your molecular weight values and your calculations.

Now let's calculate the quantities originally asked for by converting the units of the previously calculated quantities:

a. $$\text{CaO produced} = \frac{52.04 \text{ lb CaO}}{100 \text{ lb stone}} \left| \frac{2000 \text{ lb}}{1 \text{ ton}} \right. = 1041 \text{ lb CaO/ton limestone}$$

b. $$CO_2 \text{ recovered} = \frac{43.67 \text{ lb } CO_2}{100 \text{ lb limestone}} = 0.437 \text{ lb } CO_2\text{/lb limestone}$$

c. $$\text{Limestone required} = \frac{100 \text{ lb stone}}{56.33 \text{ lb time}} \left| \frac{2000 \text{ lb}}{1 \text{ ton}} \right. = 3550 \text{ lb stone/ton lime}$$

Self-Assessment Test

Question

For the following reaction

$$MnO_4^- + 5\,Fe^{2+} + 8\,H^+ \rightarrow Mn^{2+} + 4\,H_2O + 5\,Fe^{3+}$$

a student wrote the following to determine how many moles of MnO_4^- would react with 3 moles of Fe^{2+}:

1 mol MnO_4^+ = 5 mol Fe^{2+}. Divide both sides by 5 mol Fe^{2+} to get 1 mol MnO_4^-/5 mol Fe^{2+} = 1. The number of mol of MnO_4^- that reacts with 3 mol of Fe^{2+} is

$$\frac{1 \text{ mol } MnO_4^-}{5 \text{ mol } Fe^{2+}} \left| 3 \text{ mol } Fe^{2+} \right. = 0.6 \text{ mol } MnO_4^-.$$

Is the calculation correct?

Problems

1. Write balanced reaction equations for the following reactions:
   a. C_9H_{18} and oxygen to form carbon dioxide and water
   b. FeS_2 and oxygen to form Fe_2O_3 and sulfur dioxide

2. If 1 kg of benzene (C_6H_6) is oxidized with oxygen, how many kilograms of O_2 are needed to convert all of the benzene to CO_2 and H_2O?
3. Can you balance the following chemical reaction equation?

$$a_1\, NO_3 + a_2\, HClO \rightarrow a_3\, HNO_3 + a_4\, HCl$$

5.2 Terminology for Reaction Systems

Nothing is like it seems but everything is like it is.
Yogi Berra

So far we have discussed the stoichiometry of reactions in which the proper stoichiometric ratio of reactants is fed into a reactor, and the reaction goes to completion; no reactants remain in the reactor. What if (a) some other ratio of reactants is fed or (b) the reaction is incomplete? In such circumstances you need to be familiar with a number of terms used to describe these types of problems.

5.2.1 Extent of Reaction, ξ

You will find the **extent of reaction** useful in solving material balances involving chemical reaction if you can ascertain the reaction equations. The extent of reaction applies to each species in the reaction. The extent of reaction, ξ, is **based on a specified stoichiometric equation** and denotes how much reaction occurs. Its units are "moles reacting." The extent of reaction is calculated by dividing the change in the number of moles of a species that occurs in a reaction, for either a reactant or a product, by the associated stoichiometric coefficient (which has the units of the change in the moles of species *i* divided by the moles reacting). For example, consider the chemical reaction equation for the combustion of carbon monoxide:

$$2\, CO + O_2 \rightarrow 2\, CO_2$$

If 20 moles of CO are combined with 10 moles of O_2 to form 15 moles of CO_2, the extent of reaction can be calculated from the amount of CO_2 that is produced.

The value of the change in the moles of CO_2 is $15 - 0 = 15$ mol.

The value of the stoichiometric coefficient for the CO_2 is 2 mol CO_2/moles reacting.
Then the extent of reaction is

$$\frac{(15 - 0) \text{ mol } CO_2}{2 \text{ mol } CO_2/\text{moles reacting}} = 7.5 \text{ moles reacting}$$

Let's next consider a more formal definition of the extent of reaction, one that takes into account incomplete reaction and involves the initial concentrations of reactants and products. The extent of reaction for a reaction is defined as follows for a single reaction involving component i:

$$\xi = \frac{n_i - n_{io}}{\nu_i} \tag{5.3}$$

where n_i = moles of species i present in the system after the reaction occurs
n_{io} = moles of species i present in the system when the reaction starts
ν_i = stoichiometric coefficient for species i in the specified chemical reaction equation (moles of species per moles reacting)
ξ = extent of reaction (moles reacting according to the assumed reaction stoichiometry)

The stoichiometric coefficients of the products in a chemical reaction are assigned positive values and the reactants are assigned negative values. Note that $(n_i - n_{i0})$ is equal to the generation by reaction of component i when the quantity is positive, and the consumption of component i by reaction when it is negative.

Equation (5.3) can be rearranged to calculate the final number of moles of component i from the value of the extent of reaction if known plus the value of the initial amount of component i:

$$n_i = n_{io} + \xi\, \nu_i \tag{5.4}$$

As shown in the next example, the production or consumption of one species can be used to calculate the production or consumption of any of the other species involved in a reaction once you calculate, or are given, the value of the extent of reaction. Remember that Equations (5.3) and (5.4) assume that component i is involved in only one reaction.

Example 5.4 Calculation of the Extent of Reaction

NADH (nicotinamid adenine dinucleotide) supplies hydrogen in living cells for biosynthesis reactions such as

$$CO_2 + 4H \rightarrow CH_2O + H_2O$$

If you saturate 1 L of deaerated water with CO_2 gas at 20°C (the solubility is 1.81 g CO_2/L) and add enough NADH to provide 0.057 g of H into a bioreactor used to imitate the reactions in cells, and obtain 0.7 g of CH_2O, what is the extent of reaction for this reaction? Use the extent of reaction to determine the number of grams of CO_2 left in solution.

Solution

Basis: 1 L water saturated with CO_2

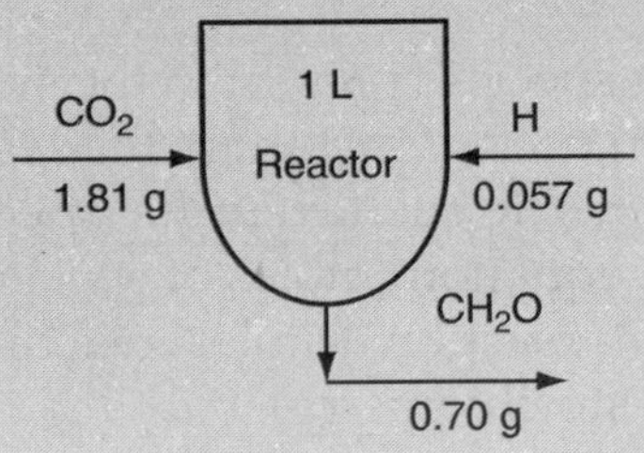

Figure E5.4

The extent of reaction can be calculated by applying Equation (5.3) based on the value given for CH_2O:

$$n_{CH_2O\text{ final}} = \frac{0.70 \text{ g } CH_2O}{} \left| \frac{1 \text{ g mol } CH_2O}{30.02 \text{ g } CH_2O} \right. = 0.0233 \text{ g mol } CH_2O$$

$$n_{CH_2O\text{ initial}} = \frac{0 \text{ g } CH_2O}{} \left| \frac{1 \text{ g mol } CH_2O}{30.02 \text{ g } CH_2O} \right. = 0 \text{ g mol } CH_2O$$

$$\xi = \frac{n_i - n_{i0}}{v_i} = \frac{(0.0233 - 0) \text{ g mol } CH_2O}{1 \text{ g mol } CH_2O/\text{moles reacting}} = 0.0233 \text{ moles reacting}$$

The number of moles of CO_2 left in solution can be obtained by using Equation (5.4) or Equation (5.3) for the CO_2:

$$n_{CO_2\text{ initial}} = \frac{1.81 \text{ g } CO_2}{} \left| \frac{1 \text{ g mol } CO_2}{44.00 \text{ g } CO_2} \right. = 0.041 \text{ g mol } CO_2$$

$$n_{CO_2\,\text{final}} = 0.041 + (-1)(0.0233) = 0.0177 \text{ g mol } CO_2$$

$$m_{CO_2\,\text{final}} = \frac{0.0177 \text{ g mol } CO_2}{} \left| \frac{44.00 \text{ g } CO_2}{1 \text{ g mol } CO_2} \right. = 0.78 \text{ g } CO_2$$

To sum up, the important characteristic of the extent of reaction, ξ, defined in Equation (5.3) is that it has the same value for each molecular species involved in a reaction. Thus, given the initial mole numbers of all species and a value for ξ (or the change in the number of moles of one species from which the value of ξ can be calculated as is done in Example 5.4), you can easily compute the number of all other moles in the system using Equation (5.4).

5.2.2 Limiting and Excess Reactants

In industrial reactors you will rarely find exact stoichiometric amounts of materials used. To make a desired reaction take place or to use up a costly reactant, excess reactants are nearly always used. The excess material comes out together with, or perhaps separately from, the product and sometimes can be used again. The **limiting reactant** is defined as the species in a chemical reaction that theoretically would be the first to be completely consumed if the reaction were to proceed to completion according to the chemical equation—**even if the reaction does not proceed to completion**! All of the other reactants are called **excess reactants**. For example, using the chemical reaction equation in Example 5.2,

$$C_7H_{16} + 11O_2 \rightarrow 7CO_2 + 8H_2O$$

if 1 g mol of C_7H_{16} and 12 g mol of O_2 are mixed so as to react, C_7H_{16} would be the limiting reactant even if the reaction does not take place. The amount of the **excess reactant** O_2 would be calculated as 12 g mol of initial reactant less the 11 g mole needed to react with 1 g mol of C_7H_{16}, or 1 g mol of O_2. Therefore, if the reaction were to go to completion, the amount of product that would be produced is controlled by the amount of the limiting reactant, namely, C_7H_{16} in this example.

As a straightforward way of determining which species is the limiting reactant, you can calculate the **maximum extent of reaction**, a quantity that is based on **assuming the complete reaction** of each reactant. **The reactant with the smallest maximum extent of reaction is the limiting reactant**. For Example 5.2, for 1 g mol of C_7H_{16} and 12 g mol of O_2, you can calculate

$$\xi^{max}(\text{based on } O_2) = \frac{0 \text{ g mol } O_2 - 12 \text{ g mol } O_2}{-11 \text{ g mol } O_2/\text{moles reacting}} = 1.09 \text{ moles reacting}$$

$$\xi^{max}(\text{based on } C_7H_{16}) = \frac{0 \text{ g mol } C_7H_{16} - 1 \text{ g mol } C_7H_{16}}{-1 \text{ g mol } C_7H_{16}/\text{moles reacting}} = 1.00 \text{ moles reacting}$$

Therefore, heptane (C_7H_{16}) is the limiting reactant and oxygen is the excess reactant.

Example 5.5 Calculation of the Limiting and Excess Reactants Given the Mass of Reactants

In this example let's use the same data as in Example 5.4. The basis is the same and the figure is the same.

a. What is the maximum number of grams of CH_2O that can be produced?
b. What is the limiting reactant?
c. What is the excess reactant?

Solution

The first step is to determine the limiting reactant by calculating the maximum extent of reaction based on the complete reaction of both CO_2 and H.

$$\xi^{max}(\text{based on } CO_2) = \frac{0 - 0.041 \text{ g mol } CO_2}{-1 \text{ g mol } CO_2/\text{moles reacting}} = 0.041 \text{ moles reacting}$$

$$\xi^{max}(\text{based on H}) = \frac{0 - 0.057 \text{ g mol H}}{-4 \text{ g mol H}/\text{moles reacting}} = 0.014 \text{ moles reacting}$$

You can conclude that (b) H is the limiting reactant, and that (c) CO_2 is the excess reactant. The excess CO_2 is $(0.041 - 0.014) = 0.027$ g mol. To answer question a, the maximum amount of CH_2O that can be produced is based on assuming complete reaction of the limiting reactant:

$$\frac{0.014 \text{ g mol H}}{} \left| \frac{1 \text{ g mol } CH_2O}{4 \text{ g mol H}} \right| \frac{30.02 \text{ g } CH_2O}{1 \text{ g mol } CH_2O} = 0.017 \text{ g } CH_2O$$

Finally, you should check your answer by working from the answer to the given reactant, or, alternatively, by summing up the mass of the C and the mass of excess H. What should the sums be?

5.2.3 Conversion and Degree of Completion

Conversion and **degree of completion** are terms not as precisely defined as are the extent of reaction and limiting and excess reactant. Rather than cite all of the possible usages of these terms, many of which conflict, we shall define them as follows: **Conversion (or the degree of completion) is the fraction of the limiting reactant in the feed that is converted into products.** Conversion is related to the **degree of completion** of a reaction. The numerator and denominator of the fraction contain the same units, so the fraction conversion is dimensionless. Thus, percent conversion is

$$\% \text{ conversion} = 100\frac{\text{moles (or mass) of the limiting reactant in the feed that reacts}}{\text{moles (or mass) of the limiting reactant introduced in the feed}} \tag{5.5}$$

For example, for the reaction equation used in Example 5.2, if 14.4 kg of CO_2 are formed in the reaction of 10 kg of C_7H_{16}, you can calculate the percent of the C_7H_{16} that is converted to CO_2 (reacts) as follows:

$$\left.\begin{array}{c}C_7H_{16} \text{ in feed equivalent to}\\ \text{the } CO_2 \text{ in the product}\end{array}\right\} \frac{14.4 \text{ kg } CO_2}{}\left|\frac{1 \text{ kg mol } CO_2}{44.0 \text{ kg } CO_2}\right|\frac{1 \text{ kg mol } C_7H_{16}}{7 \text{ kg mol } CO_2}$$

$$= 0.0468 \text{ kg mol } C_7H_{16}$$

$$\left.\begin{array}{c}\text{initial } C_7H_{16}\\ \text{in the reactants}\end{array}\right\} \quad \frac{10 \text{ kg } C_7H_{16}}{}\left|\frac{1 \text{ kg mol } C_7H_{16}}{100.1 \text{ kg } C_7H_{16}}\right. = 0.0999 \text{ kg mol } C_7H_{16}$$

$$\% \text{ conversion} = \frac{0.0468 \text{ kg mol reacted}}{0.0999 \text{ kg mol fed}} 100 = 46.8\% \text{ of the } C_7H_{16}$$

The conversion can also be calculated by using the extent of reaction as follows: Conversion is equal to the extent of reaction based on the formation of CO_2 (i.e., the actual extent of reaction) divided by the extent of reaction, assuming complete reaction of C_7H_{16} (i.e., the maximum possible extent of reaction):

$$\text{conversion} = \frac{\text{extent of reaction that actually occurs}}{\text{extent of reaction that would occur if complete reaction took place}} = \frac{\xi}{\xi^{\max}} \tag{5.6}$$

5.2.4 Selectivity

Selectivity is the ratio of the moles of a particular (usually the desired) product produced to the moles of another (usually undesired or by-product) product produced in a single reaction or group of reactions. For example, methanol (CH_3OH) can be converted into ethylene (C_2H_4) or propylene (C_3H_6) by the reactions

$$2\,CH_3OH \rightarrow C_2H_4 + 2H_2O$$
$$3\,CH_3OH \rightarrow C_3H_6 + 3H_2O$$

Of course, for the process to be economical, the value of the products has to be greater than the value of the reactants. Examine the data in Figure 5.1 for the concentrations of the products of the reactions. What is the selectivity of C_2H_4 relative to the C_3H_6 at 80% conversion of the CH_3OH? Proceed upward at 80% conversion to get for $C_2H_4 \cong 19$ mol % and for $C_3H_6 \cong 8$ mol %. Because the basis for both values is the same, you can compute the selectivity $19/8 \cong 2.4$ mol C_2H_4 per mol C_3H_6.

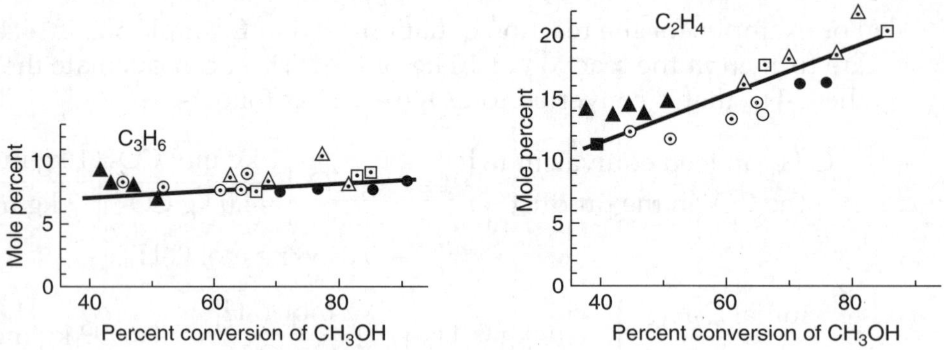

Figure 5.1 Products from the conversion of ethanol

5.2.5 Yield

No universally agreed-upon definitions exist for yield—in fact, quite the contrary. Here are three common ones:

- **Yield** (based on feed): The amount (mass or moles) of desired product obtained divided by the amount of the key (frequently the limiting) reactant fed.
- **Yield** (based on reactant consumed): The amount (mass or moles) of desired product obtained divided by the amount of the key (frequently the limiting) reactant consumed.
- **Yield** (based on 100% conversion): The amount (mass or moles) of a product obtained divided by the theoretical (expected) amount of the product that would be obtained based on the limiting reactant in the chemical reaction equation(s) if it were completely consumed. Note that this is a fractional (dimensionless) yield because the numerator and denominator have

the same units, whereas the previous two definitions of yield are not dimensionless.

Why doesn't the actual yield in a reaction equal the theoretical yield predicted from the chemical reaction equation? Several reasons exist:

- Impurities among the reactants
- Leaks to the environment
- Side reactions
- Reversible reactions

As an illustration, suppose you have a reaction sequence as follows:

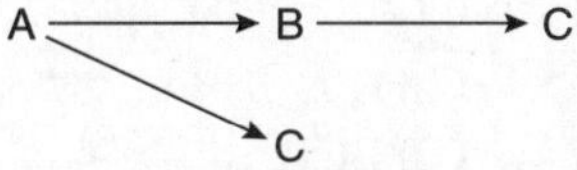

with B being the desired product and C the undesired one. The yield of B according to the first two definitions is the moles (or mass) of B produced divided by the moles (or mass) of A fed or consumed. The yield according to the third definition is the moles (or mass) of B actually produced divided by the maximum amount of B that could be produced in the reaction sequence (i.e., complete conversion of A to B). The selectivity of B is the moles of B divided by the moles of C produced.

The terms *yield* and *selectivity* are terms that measure the degree to which a desired reaction proceeds relative to competing alternative (undesirable) reactions. As a designer of equipment, you want to maximize production of the desired product and minimize production of the unwanted products. Do you want high or low selectivity? Yield?

The next example shows you how to calculate all of the terms discussed in this section.

Example 5.6 Calculation of Various Terms Pertaining to Reactions

Semenov [N. N. Semenov, *Some Problems in Chemical Kinetics and Reactivity*, Vol. II, Princeton University Press, Princeton (1959), pp. 39–42] described some of the chemistry of alkyl chlorides. The two reactions of interest for this example are

$$Cl_2(g) + C_3H_6(g) \rightarrow C_3H_5Cl(g) + HCl(g) \quad (1)$$

$$Cl_2(g) + C_3H_6(g) \rightarrow C_3H_6Cl_2(g) \quad (2)$$

(*Continues*)

Example 5.6 Calculation of Various Terms Pertaining to Reactions (*Continued*)

C_3H_6 is propene (MW = 42.08).
C_3H_5Cl is allyl chloride (3-chloropropene) (MW = 76.53).
$C_3H_6Cl_2$ is propylene chloride (1,2-dichloropropane) (MW = 112.99).

The species recovered after the reaction takes place for some time are listed in Table E5.6.

Table E5.6

Species	g mol
Cl_2	141.0
C_3H_6	651.0
C_3H_5Cl	4.6
$C_3H_6Cl_2$	24.5
HCl	4.6

Based on the product distribution in Table E5.6, assuming that the feed consisted only of Cl_2 and C_3H_6, calculate the following:

a. How much Cl_2 and C_3H_6 were fed to the reactor in gram moles?
b. What was the limiting reactant?
c. What was the excess reactant?
d. What was the fraction conversion of C_3H_6 to C_3H_5Cl?
e. What was the selectivity of C_3H_5Cl relative to $C_3H_6Cl_2$?
f. What was the yield of C_3H_5Cl expressed in grams of C_3H_5Cl to the grams of C_3H_6 fed to the reactor?
g. What was the extent of reaction of Reactions (1) and (2)?

Solution

Steps 1–4

Examination of the problem statement reveals that the amount of feed is not given, and consequently you have to calculate the gram moles fed to the reactor even if the amounts were not asked for. The molecular weights were given. Figure E5.6 illustrates the process as an open-flow system.

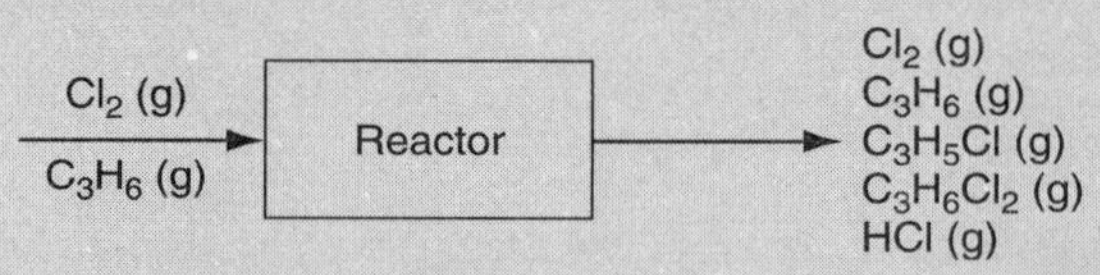

Figure E5.6

Step 5

A convenient basis is what is given in the product list in Table E5.6.

Steps 7–9

Use the chemical equations to calculate the moles of species in the feed. Start with the Cl_2.

Reaction (1):

$$\frac{4.6 \text{ g mol } C_3H_5Cl}{} \left| \frac{1 \text{ g mol } Cl_2}{1 \text{ g mol } C_3H_5Cl} \right. = 4.6 \text{ g mol } Cl_2 \text{ reacts}$$

Reaction (2):

$$\frac{24.5 \text{ g mol } C_3H_6Cl_2}{} \left| \frac{1 \text{ g mol } Cl_2}{1 \text{ g mol } C_3H_6Cl_2} \right. = 24.5 \text{ g mol } Cl_2 \text{ reacts}$$

Total	29.1 g mol Cl_2 reacts
Unreacted Cl_2 in product	141.0
a. Total Cl_2 fed	170.1

What about the amount of C_3H_6 in the feed? From the chemical equations you can see that if 29.1 g mol of Cl_2 reacts in total by Reactions (1) and (2), 29.1 g mol of C_3H_6 must react. Since 651.0 g mol of C_3H_6 exist unreacted in the product, $651.0 + 29.1 = 680.1$ g mol of C_3H_6 were fed to the reactor.

You can check those answers by adding up the gram moles of Cl, C, and H in the product and comparing the value with that calculated in the feed:

In product:

Cl	$2(141.0) + 1(4.6) + 2(24.5) + 1(4.6) = 340.2$
C	$3(651) + 3(4.6) + 3(24.5) = 2040.3$
H	$6(651) + 5(4.6) + 6(24.5) + 1(4.6) = 4080.6$

In feed:

Cl	$2(170.1) = 340.2$	OK
C	$3(680.1) = 2040.3$	OK
H	$6(680.1) = 4080.6$	OK

We will not go through detailed analysis for the remaining calculations but simply determine the desired quantities based on the data prepared for parts a, b, and c. In this particular problem, since both reactions involve the same reaction stoichiometric coefficients, both reactions will have the same limiting and excess reactants:

$$\xi^{max}(\text{based on } C_3H_6) = \frac{-680.1 \text{ g mol } C_3H_6}{-1 \text{ g mol } C_3H_6/\text{mol reacting}} = 680.1 \text{ mol reacting}$$

(Continues)

Example 5.6 Calculation of Various Terms Pertaining to Reactions (*Continued*)

$$\xi^{max}(\text{based on } Cl_2) = \frac{-170.1 \text{ g mol } Cl_2}{-1 \text{ g mol } Cl_2/\text{mol reacting}} = 170.1 \text{ mol reacting}$$

Thus, C_3H_6 was the excess reactant and Cl_2 the limiting reactant.

d. The fraction conversion of C_3H_6 to C_3H_5Cl was

$$\frac{29.1 \text{ g mol } C_3H_6 \text{ that reacted}}{680.1 \text{ g mol } C_3H_6 \text{ fed}} = 0.043$$

e. The selectivity was

$$\frac{4.6 \text{ g mol } C_3H_5Cl}{24.5 \text{ g mol } C_3H_6Cl_2} = 0.19 \frac{\text{g mol } C_3H_5Cl}{\text{g mol } C_3H_6Cl_2}$$

f. The yield was

$$\frac{(76.53)\,(4.6) \text{ g } C_3H_5Cl}{(42.08)\,(680.1) \text{ g } C_3H_6} = 0.012 \frac{\text{g } C_3H_5Cl}{\text{g } C_3H_6}$$

g. Because C_3H_5Cl is produced only by the first reaction, the extent of reaction of the first reaction is

$$\xi_1 = \frac{n_i - n_{io}}{\nu_i} = \frac{4.6 - 0}{1} = 4.6$$

Because $C_3H_6Cl_2$ is produced only by the second reaction, the extent of reaction of the second reaction is

$$\xi_2 = \frac{n_i - n_{io}}{\nu_i} = \frac{24.5 - 0}{1} = 24.5$$

Self-Assessment Test

Questions

1. What is the extent of reaction based upon?
2. How is the extent of reaction used to identify the limiting reactant?

Problem

Two well-known gas phase reactions take place in the dehydration of ethane:

$$C_2H_6 \rightarrow C_2H_4 + H_2 \quad \text{(a)}$$

$$C_2H_6 + H_2 \rightarrow 2CH_4 \quad \text{(b)}$$

Given the product distribution measured in the gas phase reaction of C_2H_6 as follows:

C_2H_6	27%
C_2H_4	33%
H_2	13%
CH_4	27%

a. What species was the limiting reactant?
b. What species was the excess reactant?
c. What was the conversion of C_2H_6 to CH_4?
d. What was the degree of completion of the reaction?
e. What was the selectivity of C_2H_4 relative to CH_4?
f. What was the yield of C_2H_4 expressed in kilogram moles of C_2H_4 produced per kilogram mole of C_2H_6?
g. What was the extent of reaction of C_2H_6?

Thought Problems

1. An accident occurred in which one worker lost his life. A large steel evaporator in magnesium chloride service, containing internal heating tubes, was to be cleaned. It was shut down, drained, and washed. The next day two employees who were involved in the maintenance of the evaporator entered the vessel to repair the tubes. They were overcome, apparently from lack of oxygen. Subsequently, one employee recovered and escaped, but the other never regained consciousness and died several days later.

   What in your opinion might have caused the accident (the lack of oxygen)?

2. OSHA requires the use of breathing apparatus when working in or around tanks containing traces of solvents. While demolishing an old tank, a contractor purchased several cylinders of compressed air, painted gray. After two days he found that he needed more cylinders and sent a truck for another cylinder. The driver returned with a black cylinder. None of the workers, including the man in

charge of the breathing apparatus, noticed the change or, if they did, attached any importance to it. When the new cylinder was brought into use, a welder's facepiece caught fire. Fortunately, he pulled it off at once and was not injured. What happened?

Discussion Question

Diesel pollutants pose a threat to the respiratory tract and are a potential cause of cancer, according to the Environmental Protection Agency. The Clean Air Act of 1990 required that the sulfur content of diesel used on freeways must be lowered from 0.30% by weight to 0.05%—a substantial reduction.

How might this be accomplished economically? In the oil well, at the refinery, at the service station, in the car, or what?

5.3 Species Mole Balances

5.3.1 Processes Involving a Single Reaction

Do you recall from Section 3.1 that the material balance for a species must be augmented to include generation and consumption terms when chemical reactions occur in a system? In terms of moles of species i, a material balance for a general system equivalent to Equation (3.1) is

$$\begin{Bmatrix} \textbf{moles of } i \\ \textbf{at } t_2 \\ \textbf{in the system} \end{Bmatrix} - \begin{Bmatrix} \textbf{moles of } i \\ \textbf{at } t_1 \\ \textbf{in the system} \end{Bmatrix} = \begin{Bmatrix} \textbf{moles of } i \\ \textbf{entering} \\ \textbf{the system} \end{Bmatrix} - \begin{Bmatrix} \textbf{moles of } i \\ \textbf{leaving} \\ \textbf{the system} \end{Bmatrix}$$
$$+ \begin{Bmatrix} \textbf{moles of } i \\ \textbf{generated} \\ \textbf{by reaction} \end{Bmatrix} - \begin{Bmatrix} \textbf{moles of } i \\ \textbf{consumed} \\ \textbf{by reaction} \end{Bmatrix} \quad (5.7)$$

Note that we have written Equation (5.7) in terms of moles rather than mass. The generation and consumption terms are more conveniently represented in terms of moles because reactions are usually written in terms of molar ratios.

Fortunately, you only have to add one additional variable to account for the generation or consumption of each species i present in the system *if you make use of the extent of reaction* that was discussed in Section 5.2. To make the idea clear, let's examine the reaction of N_2 and H_2 to form NH_3 in the gas phase. Figure 5.2 presents the process as a steady-state, open system operating for a fixed interval so that the terms will be zero on the left-hand side of the equal sign in Equation (5.7). Figure 5.2 shows the measured values for the flows in gram moles.

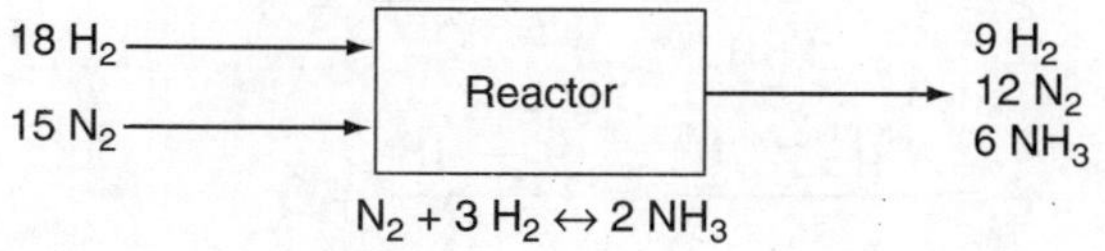

Figure 5.2 A reactor to produce NH_3

For this simple example you can calculate by inspection or by Equation (5.7) a value in gram moles for the generation and/or consumption for each of the three species in the reaction:

NH_3: $6 - 0 = 6$ g mol (product generated)
H_2: $9 - 18 = -9$ g mol (reactant consumed)
N_2: $12 - 15 = -3$ g mol (reactant consumed)

Because of the stoichiometry of the chemical reaction equation

$$N_2 + 3\,H_2 \rightarrow 2\,NH_3$$

the three respective generation and consumption terms are related. For example, given the value for the generation of NH_3, you can calculate the values for the consumption of H_2 and N_2 using the reaction equation. The ratio of hydrogen to nitrogen consumed in the reactants and maintained in the product ammonia is always 3 to 1. Thus, you cannot specify more than one value of the N_2 and H_2 pair left over from the reaction without introducing a redundant or possibly an inconsistent specification. In general, if you specify the value for the generation or consumption of one species in a reaction, you are able to calculate the values of the other species from a solo chemical reaction equation.

Here is where the extent of reaction ξ becomes useful. Recall that Equation (5.3) relates the extent of reaction to the change in moles of a species i divided by the stoichiometric coefficient ν_i of the species in the reaction equation

$$\xi = \frac{n_i^{\text{out}} - n_i^{\text{in}}}{\nu_i} \qquad i = 1, \ldots, N \tag{5.3a}$$

for the NH_3 reaction

$$\nu_{NH_3} = 2$$
$$\nu_{H_2} = -3$$
$$\nu_{N_2} = -1$$

and the extent of reaction calculated via any of the species is

$$\xi = \frac{n_{NH_3}^{out} - n_{NH_3}^{in}}{\nu_{NH_3}} = \frac{6-0}{2} = 3$$

$$\xi = \frac{n_{H_2}^{out} - n_{H_2}^{in}}{\nu_{H_2}} = \frac{9-18}{-3} = 3$$

$$\xi = \frac{n_{N_2}^{out} - n_{N_2}^{in}}{\nu_{N_2}} = \frac{15-12}{1} = 3$$

You can conclude for the case of a single chemical reaction that the specification of the extent of reaction provides one independent quantity that will determine all of the values of the generation and consumption terms for the various species in the respective implementations of Equation (5.7) because the molar ratios of the reactants and products are fixed by the solo independent chemical reaction equation. The three species balances corresponding to the process in Figure 5.2 are listed below based on Equation (5.8), an equation directly derived from Equation (5.1) for an open, steady-state process:

$$n_i^{out} - n_i^{in} = v_i\,\xi \tag{5.8}$$

Component	Out	In	=	Generation or Consumption
I	n_i^{out}	n_i^{in}	=	$\nu_i\,\xi$
NH_3	6	0	=	$2(3) = 6$
H_2	9	18	=	$-3(3) = -9$
N_2	12	15	=	$-1(3) = -3$

The term $\nu_i\xi$ corresponds to the moles of species *i* generated or consumed in Equation 5.7. Can you determine by inspection that the three material balances are independent? Remember, in general you can write *one material balance* equation for each species present in the system. If Equation (5.8) is applied to each species that reacts in a steady-state system, the resulting set of material balances will contain an additional variable, namely, the extent of reaction, ξ. For a species that does not react, $\nu_i = 0$.

Frequently Asked Questions

1. Does it make any difference how the chemical reaction equation is written as long as the equation is balanced? No. For example, write the decomposition of ammonia as

$$NH_3 \rightarrow \frac{1}{2}N_2 + \frac{3}{2}H_2$$

Let's calculate ξ given that zero moles of NH_3 are introduced into a reactor and that 6 moles exit as in Figure 5.2. Then

$$\xi = \frac{6-0}{-1} = -6$$

The material balance for NH_3 is just

$$6 - 0 = (-1)(-6) = 6$$

If you calculate ξ for H_2, what result do you get? The key point here is that product $\nu_i \xi$, which is used in each species material balance, remains unchanged regardless of the size of the stoichiometric coefficients in a chemical reaction equation. Thus, if the chemical reaction equation is multiplied by a factor of 2, so that the stoichiometric coefficients are each twice their previous value, ξ will decrease by a factor of 2. As a result, the product $\nu_i \xi$ remains constant.

2. What does the negative sign in front of ξ mean? The negative sign signifies that the chemical reaction equation was written to represent a direction that is the reverse of the one in which the reaction actually proceeds as in Figure 5.2.
3. Are species balances always independent? Almost always. An example of a nonindependent set of species balances occurs for the decomposition of NH_3 as shown in FAQ No. 1. Given one piece of information, namely, that 1 mole of NH_3 decomposes completely, the products of the reaction would be 3/2 H_2 and 1/2 N_2, and the H_2 and N_2 mole balances will not be independent because their ratio would always be 3 to 1. Would partial decomposition of the 1 mole of NH_3 lead to the same conclusion? Note that if H_2 and N_2 were present in the feed in nonstoichiometric amounts, the species balances for H_2 and N_2 would form independent equations.

You can calculate the total molar flow in, F^{in}, and the total molar flow out, F^{out}, by adding all of the species flows in and out respectively:

$$F^{\text{in}} = \sum_{i=1}^{S} n_i^{\text{in}} \tag{5.9a}$$

$$F^{\text{out}} = \sum_{i=1}^{S} n_i^{\text{out}} \tag{5.9b}$$

where S is the total number of species in the system (n_i may be zero for some species).

Baldy's law: Some of it plus the rest of it is all of it.
Paul Dickson

Neither of the Equations (5.9) is an independent equation since both simply represent the sum of all of the species flows but can be substituted for one of the species balances if convenient. Only S independent equations can be written for the system. Do Equations (5.9) apply to a closed system? No. Equations (5.9) apply only to open, steady-state systems, not to closed systems. Equation (5.7) is used for closed systems.

If the unknown, ξ, occurs in a set of S otherwise independent species equations, you will, of course, have to augment the existing information by one more independent bit of information in order to be able to solve a problem. For example, you might be told that complete conversion of the limiting reactant occurs, or be given the value of the fraction conversion f of the limiting reactant; ξ is related to f by

$$\xi = \frac{(-f)n^{\text{in}}_{\text{limiting reactant}}}{\nu_{\text{limiting reactant}}} \tag{5.10}$$

You can calculate the value of ξ from the value for the fraction conversion (or vice versa) plus information identifying the limiting reactant. In other cases you are given sufficient information about the moles of a species entering and leaving the process so that ξ can be calculated directly from Equation (5.6).

Now let's look at some examples using the concepts discussed so far.

Example 5.7 Reaction in Which the Fraction Conversion is Specified

The chlorination of methane occurs by the following reaction:

$$CH_4 + Cl_2 \rightarrow CH_3Cl + HCl$$

You are asked to determine the product composition if the conversion of the limiting reactant is 67%, and the feed composition in mole percent is 40% CH_4, 50% Cl_2, and 10% N_2.

Solution

Steps 1–4

Assume the reactor is an open, steady-state process. Figure E5.7 is a sketch of the process with the known information placed on it.

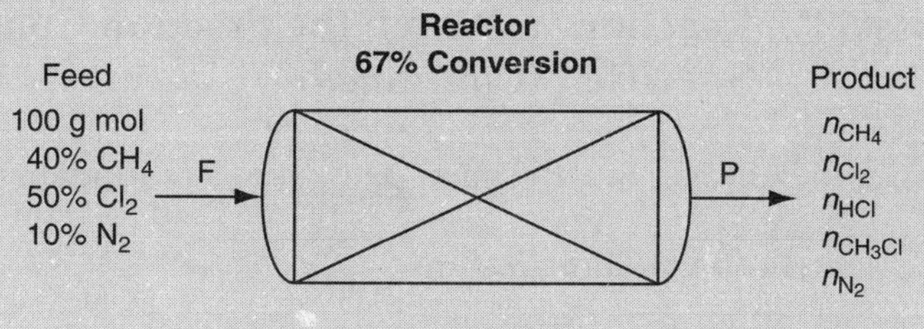

Figure E5.7

Step 5

Select as a basis:

Basis: 100 g mol feed

Step 4

You have to determine the limiting reactant if you are to make use of the information about the 67% conversion. By comparing the maximum extent of reaction for each reactant, you can identify the limiting reactant as being the species that has the smallest ξ^{max}, as explained previously in Section 5.1.

$$\xi^{max}(CH_4) = \frac{n_{i0}}{v_i} = \frac{40}{-(-1)} = 40$$

$$\xi^{max}(Cl_2) = \frac{n_{i0}}{-v_i} = \frac{50}{-(-1)} = 50$$

Therefore, CH_4 is the limiting reactant. You can now calculate the extent of reaction using the specified conversion and Equation (5.10):

$$\xi = \frac{-f\,n^{in}_{\text{lim reactant}}}{v_{\text{lim reactant}}} = \frac{(-0.67)(40)}{-1} = 26.8 \text{ g mol reacting}$$

One unknown can now be assigned a value, namely, ξ.

Steps 6 and 7

The next step is to carry out a degree-of-freedom analysis:

Number of variables: 11

$$n^{in}_{CH_4}, n^{in}_{Cl_2}, n^{in}_{N_2}, n^{out}_{CH_4}, n^{out}_{Cl_2}, n^{out}_{HCl}, n^{out}_{CH_3Cl}, n^{out}_{N_2}, F, P, \xi$$

but you can assign values for the first three variables plus F, and have calculated ξ, hence the *number of unknowns* has been reduced to just 6.

Number of independent equations needed: 6

Species material balances: 5 CH_4, Cl_2, HCl, CH_3Cl, N_2

Specifications: 0 (f was used to calculate ξ in Step 4)

Implicit equations: 1 (Why only 1 instead of 2?)

(Continues)

Example 5.7 Reaction in Which the Fraction Conversion is Specified (*Continued*)

$$\sum n_i^{\text{out}} = P \quad (\text{the sum } \sum n_i^{\text{in}} = F \text{ is redundant})$$

The degrees of freedom are zero.

Steps 8 and 9

The species material balances (in moles) using Equation (5.8) give a direct solution for each species in the product:

$$n_{\text{CH}_4}^{\text{out}} = 40 - 1(26.8) = 13.2$$

$$n_{\text{Cl}_2}^{\text{out}} = 50 - 1(26.8) = 23.2$$

$$n_{\text{CH}_3\text{Cl}}^{\text{out}} = 0 + 1(26.8) = 26.8$$

$$n_{\text{HCl}}^{\text{out}} = 0 + 1(26.8) = 26.8$$

$$n_{\text{N}_2}^{\text{out}} = 10 - 0(26.8) = \underline{10.0}$$

$$P = 100.0$$

Therefore, the composition of the product stream is 13.2% CH_4, 23.2% Cl_2, 26.8% CH_3Cl, 26.8% HCl, and 10% N_2 because the total number of product moles is conveniently 100 g mol. There are 100 g mol of products because there are 100 g mol of feed, and the chemical reaction equation results in the same number of moles for reactants as products. What would you have to do if the total moles in P did not amount to 100 g mol? Remember the camels!

Step 10

The fact that the redundant overall mole balance equation is satisfied can serve as a consistency check for this problem.

5.3.2 Processes Involving Multiple Reactions

In practice, reaction systems rarely involve just a single reaction. There may be a primary reaction (e.g., the desired reaction), but often there are additional or side reactions. To extend the concept of the extent of reaction to processes involving multiple reactions, the question is: Do you just include a ξ_i for every reaction? Usually the answer is yes, but more precisely the answer is no! You should **include in the species material balances only the ξs associated with a (nonunique) set of independent chemical reactions** called the **minimal set**[1] of reaction equations. What this latter term means is the smallest set of chemical reaction equations that can be assembled so as to include *all* of the species involved in the process. It is analogous to a set of independent linear algebraic equations, and you can form any other reaction equation by a linear combination of the reaction equations contained in the minimal set. Usually, the minimal set is equal to the full collection of reaction equations, but you should make sure that each set of reaction equations represents an independent set of reactions.

For example, look at the following set of reaction equations:

$$C + O_2 \rightarrow CO_2$$
$$C + \tfrac{1}{2}O_2 \rightarrow CO$$
$$CO + \tfrac{1}{2}O_2 \rightarrow CO_2$$

By inspection you can see that if you subtract the second equation from the first one, you obtain the third equation. Only two of the three equations are independent; hence the minimal set will be composed of any two of the three equations. Refer to Chapter 12 for information about using a computer to determine whether or not a set of chemical reaction equations are independent.

With these ideas in mind, we can state that for steady-state, open processes with multiple reactions, Equation (5.8) in moles becomes for component i

$$n_i^{\text{out}} = n_i^{\text{in}} + \sum_{j=1}^{R} \nu_{ij}\,\xi_j \tag{5.11}$$

where ν_{ij} is the stoichiometric coefficient of species i in reaction j in the minimal reaction set

ξ_j is the extent of reaction for the jth reaction in which component i is present in the minimal set

R is the number of independent chemical reaction equations (the size of the minimal set)

An equation analogous to Equation (5.11) can be written for a closed, unsteady-state system. Try it, starting with Equation (5.7). Hint: The flow terms are omitted.

The total moles N exiting the reactor are

$$N = \sum_{i=1}^{S} n_i^{\text{out}} = \sum_{i=1}^{S} n_i^{\text{in}} + \sum_{i=1}^{S}\sum_{j=1}^{R} \nu_{ij}\,\xi_j \tag{5.12}$$

where S is the number of species in the system.

[1]Sometimes called the maximal set.

Example 5.8 Material Balances Involving Two Ongoing Reactions

Formaldehyde (CH_2O) is produced industrially by the catalytic oxidation of methanol (CH_3OH) by the following reaction:

$$CH_3OH + \frac{1}{2}O_2 \rightarrow CH_2O + H_2O \qquad (1)$$

Unfortunately, under the conditions used to produce formaldehyde at a profitable rate, a significant portion of the formaldehyde can react with oxygen to produce CO and H_2O:

$$CH_2O + \frac{1}{2}O_2 \rightarrow CO + H_2O \qquad (2)$$

Assume that methanol and twice the stoichiometric amount of air needed for complete oxidation of the CH_3OH are fed to the reactor, that 90% conversion of the methanol results, and that a 75% yield of formaldehyde occurs (based on the theoretical production of CH_2O by Reaction (1)). Determine the composition of the product gas leaving the reactor.

Solution

Steps 1–4

Figure E5.8 is a sketch of the process with y_i indicating the mole fraction of the respective components in P (a gas).

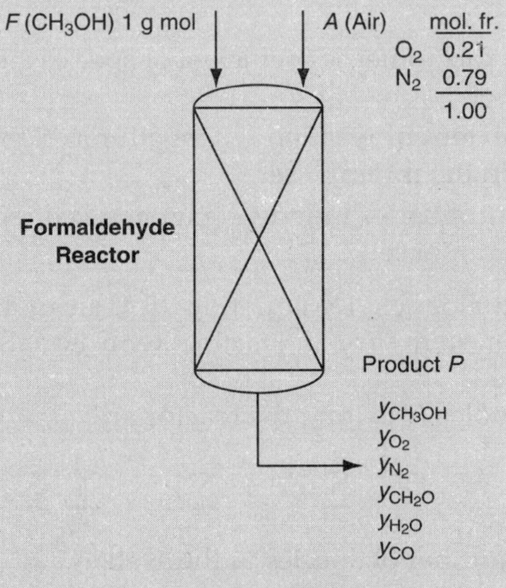

Figure E5.8

Step 5

Basis: 1 g mol F

Step 6

In this step the idea is to assign the values of variables that are specified directly or indirectly (using a related specification). Not all of the calculated values below will be accompanied by their units (to reduce complexity), but the respective units can easily be inferred. The first calculation to make in this problem is to use the specified conversion of methanol and the yield of formaldehyde to determine the extents of reaction for the two reactions. Let ξ_1 represent the extent of reaction for Reaction (1) and ξ_2 represent the extent of reaction for Reaction (2). The limiting reactant is CH_3OH.

$$\text{note } n_{CH_2O}^{out,1} = n_{CH_2O}^{in,2}$$

Based on the specified conversion, the extent of reaction for Reaction (1):

$$\xi_1 = \frac{-0.90}{-1}(1) = 0.9 \text{ g mol reacting}$$

Yield is related to ξ_i as follows:

$$\text{yield of } CH_2O = \frac{n_{CH_2O}^{out,2}}{F}$$

By Reaction (1): $n_{CH_2O}^{out,1} = n_{CH_2O}^{in,1} + 1(\xi_1) = 0 + \xi_1 = \xi_1$

By reaction (2): $n_{CH_2O}^{out,2} = n_{CH_2O}^{in,2} - 1(\xi_2) = n_{CH_2O}^{out,1} - \xi_2 = \xi_1 - \xi_2$

The specified yield: $\dfrac{n_{CH_2O}^{out,2}}{F} = \dfrac{\xi_1 - \xi_2}{1} = 0.75$

$$\xi_2 = (0.90 - 0.75) = 0.15 \text{ g mol reacting}$$

You should next calculate the amount of air (A) that enters the process. The entering oxygen is twice the required oxygen based on Reaction (1), namely,

$$n_{O_2}^{A} = 2\left(\frac{1}{2}F\right) = 2\left(\frac{1}{2}\right)(1.00) = 1.00 \text{ g mol}$$

$$A = \frac{n_{O_2}^{A}}{0.21} = \frac{100}{0.21} = 4.76 \text{ g mol}$$

$$n_{N_2}^{A} = 4.76 - 1.00 = 3.76 \text{ g mol}$$

(Continues)

Example 5.8 Material Balances Involving Two Ongoing Reactions (*Continued*)

Steps 6 and 7

The degree-of-freedom analysis is as follows:

Number of variables: 11

$$F, A, P, y^P_{CH_3OH}, y^P_{O_2}, y^P_{N_2}, y^P_{CH_2O}, y^P_{H_2O}, y^P_{CO}, \xi_1, \xi_2$$

You can assign the following values to certain of the variables:

Calculated values in Step 6 that can be assigned for F, A, ξ_a, ξ_b: total is 4

Number of unknowns: 11 − 4 = 7

Number of independent equations needed: 11 − 4 = 7; here are 7 to select:

Species material balances: 6

CH_3OH, O_2, N_2, CH_2O, H_2O, CO

Implicit equation: 1

$$\sum y^P_i = 1$$

Step 8

Because the variables in Figure E5.8 are y^P_i and not n^P_i, direct use of y^P_i in the material balances will involve the nonlinear terms $y^P_i P$. We could use the variable n^P_i analogous to the material balances in previous examples, but for the purposes of illustration, let us write the equations in terms of y^P_i. Then we will calculate P using Equation (5.12).

$$\begin{aligned} P &= \sum_{i=1}^{S} n^{\text{in}}_i + \sum_{i=1}^{S}\sum_{j=1}^{R} \nu_{ij}\,\xi_j \\ &= 1 + 4.76 + \sum_{i=1}^{6}\sum_{j=1}^{2} \nu_{ij}\,\xi_j \\ &= 5.76 + [(-1) + (-1/2) + (1) + 0 + (1) + 0]\,0.9 \\ &\quad + [0 + (-1/2) + (-1) + 0 + (1) + (1)]\,0.15 = 6.28 \text{ g mol} \end{aligned}$$

The species material balances after entering their assigned values and $P = 6.28$ are

$$n^{\text{out}}_{CH_3OH} = y_{CH_3OH}(6.28) = 1 - (0.9) + 0 = 0.10$$

$$n_{O_2}^{out} = y_{O_2}(6.28) = 1.0 - (1/2)(0.9) - (1/2)(0.15) = 0.475$$

$$n_{CH_2O}^{out} = y_{CH_2O}(6.28) = 0 + 1(0.9) - 1(0.15) = 0.75$$

$$n_{H_2O}^{out} = y_{H_2O}(6.28) = 0 + 1(0.9) + 1(0.15) = 1.05$$

$$n_{CO}^{out} = y_{CO}(6.28) = 0 + 0 + 1(.15) = 0.15$$

$$n_{N_2}^{out} = y_{N_2}(6.28) = 3.76 - 0 - 0 = 3.76$$

Step 9

You can check the value of P by adding all of the n_i^{out} above.

Step 10

The six equations can be solved for y_i. Did you get the following answer (in mole percent)?

$$y_{CH_3OH} = 1.6\%,\ y_{O_2} = 7.6\%,\ y_{N_2} = 59.8\%,\ y_{CH_2O} = 11.9\%,$$

$$y_{H_2O} = 16.7\%,\ \text{and } y_{CO} = 2.4\%$$

Example 5.9 Analysis of a Bioreactor

A bioreactor is a vessel in which biological reactions are carried out involving enzymes, microorganisms, and/or animal and plant cells. In the anaerobic (in the absence of oxygen) fermentation of grain, the yeast *Saccharomyces cerevisiae* digests glucose ($C_6H_{12}O_6$) from plants to form the products ethanol (C_2H_5OH) and propenoic acid ($C_2H_3CO_2H$) by the following overall reactions:

Reaction 1: $C_6H_{12}O_6 \rightarrow 2C_2H_5OH + 2CO_2$

Reaction 2: $C_6H_{12}O_6 \rightarrow 2C_2H_3CO_2H + 2H_2O$

In a process, a tank is initially charged with 4000 kg of a 12% solution of glucose in water. After fermentation, 120 kg of CO_2 have been produced and 90 kg of unreacted glucose remain in the broth. What are the weight (mass) percents of ethanol and propenoic acid in the broth at the end of the fermentation process? Assume that none of the glucose is retained by the microorganisms.

(*Continues*)

Example 5.9 Analysis of a Bioreactor (*Continued*)

Solution

You can treat this process as an unsteady-state process in a closed system. For component i, Equation (5.7) becomes

$$n_i^{\text{final}} = n_i^{\text{initial}} + \sum_{j=1}^{R} \nu_{ij}\,\xi_j \tag{5.13}$$

analogous to Equation (5.11). The bioorganisms do not have to be included in the solution of the problem because they presumably exist in a small amount and are catalysts for the reaction, not reactants.

Steps 1–4

Figure E5.9 is a sketch of the process.

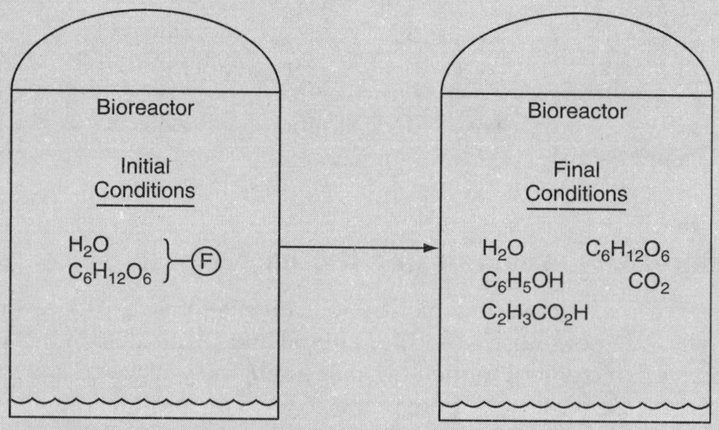

Figure E5.9

Step 5

Basis: 4000 kg F

Step 4

You should first convert the 4000 kg into moles of H_2O and $C_6H_{12}O_6$ because the reaction equations are based on moles:

$$n_{H_2O}^{\text{Initial}} = \frac{4000(0.88)}{18.02} = 195.3 \text{ g mol}$$

$$n^{\text{Initial}}_{C_6H_{10}O_6} = \frac{4000(0.12)}{180.1} = 2.665 \text{ g mol}$$

so that $F = 197.965$ g mol or rounded to 198 g mol.

Steps 6 and 7

The degree-of-freedom analysis is as follows (note that units of gram moles have been suppressed):

Number of variables: 9

$$n^{\text{Initial}}_{H_2O}, n^{\text{Initial}}_{C_6H_{12}O_6}, n^{\text{Final}}_{H_2O}, n^{\text{Final}}_{C_6H_{12}O_6}, n^{\text{Final}}_{C_2H_5OH}, n^{\text{Final}}_{C_2H_3CO_2H}, n^{\text{Final}}_{CO_2}, \xi_1, \xi_2$$

Assign values to their respective variables:

From the specifications you can assign the following values:

$$\left.\begin{aligned} F &= 198 \\ n^{\text{Initial}}_{H_2O} &= 195.3 \\ n^{\text{Initial}}_{C_6H_{10}O_6} &= 2.665 \end{aligned}\right\} \text{ only 2 are independent (Why?)}$$

$$n^{\text{Final}}_{C_6H_{10}O_6} = \frac{90}{180.1} = 0.500$$

$$n^{\text{Final}}_{CO_2} = \frac{120}{44.0} = 2.727$$

Thus, a net of five unknowns exist for this problem.

You can make five species balances:

$$H_2O, C_2H_{12}O_6, C_2H_5OH, C_2H_3CO_2H, CO_2$$

Therefore, the degrees of freedom are zero.

Step 8

The set of material balance equations, after introducing the known values for the variables, is

$$H_2O: \quad n^{\text{Final}}_{H_2O} = 195.3 + (0)\xi_1 + (2)\xi_2 \quad (1)$$

$$C_2H_{12}O_6: \quad 0.500 = 2.665 + (-1)\xi_1 + (-1)\xi_2 \quad (2)$$

$$C_2H_5OH: \quad n^{\text{Final}}_{C_2H_5OH} = 0 + 2\xi_1 + (0)\xi_2 \quad (3)$$

(Continues)

Example 5.9 Analysis of a Bioreactor (*Continued*)

$$C_2H_3CO_2H: \quad n^{Final}_{C_2H_3OH} = 0 + (0)\xi_1 + (2)\xi_2 \tag{4}$$

$$CO_2: \quad 2.727 = 0 + (2)\xi_1 + (0)\xi_2 \tag{5}$$

Step 9

Do any of these five equations involve only one unknown? If so, you could solve it in your head or with a calculator and reduce the number of simultaneous equations to be solved by one. In fact, Equation (5) contains only ξ_1 as an unknown. Therefore, you can determine ξ_1 using Equation (5). Then Equation (2) can be used to calculate the value for ξ_2. Finally, Equations (1), (3), and (4) can be applied to determine the remaining unknowns.
With this solution sequence, you get

$\xi_1 = 1.364$ kg mol reacting $\qquad \xi_2 = 0.801$ kg mol reacting

	Results	Conversion to Mass Percent		
Species	**kg mol %**	**MW**	**kg**	**mass**
H_2O	196.9	18.01	3546.1	89.2
C_2H_5OH	2.728	46.05	125.6	3.2
$C_2H_3CO_2H$	1.602	72.03	115.4	2.9
CO_2	2.272	44.0	100.0	2.5
$C_6H_{12}O_6$	0.500	180.1	90.1	2.3
			3977	1.00

Step 10

The total mass of 3977 kg is close enough to 4000 kg of feed to validate the results of the calculations.

Self-Assessment Test

Questions

1. Indicate whether the following statements are true or false:
   a. If a chemical reaction occurs, the total masses entering and leaving the system for an open, steady-state process are equal.
   b. In the combustion of carbon, all of the moles of carbon that enter a steady-state, open process exit from the process.
   c. The number of moles of a chemical compound entering a steady-state process in which a reaction occurs with that compound can never equal the number of moles of the same compound leaving the process.

2. List the circumstances for a steady-state process in which the number of moles entering the system equals the number of moles leaving the system.
3. If Equation (5.3) is to be applied to a compound, what information must be given about the stoichiometry involved and/or the extent of reaction?
4. Equation (5.3) can be applied to processes in which a reaction and also no reaction occurs. For what types of balances does the simple relation "the input equals the output" hold for *steady-state, open processes* (no accumulation)? Fill in the blanks with yes or no.

Type of Balance	Without Chemical Reaction	With Chemical Reaction
Total balances		
Total mass	[ ]	[ ]
Total moles	[ ]	[ ]
Component balances		
Mass of a pure compound	[ ]	[ ]
Moles of a pure compound	[ ]	[ ]
Mass of an atomic species	[ ]	[ ]
Moles of an atomic species	[ ]	[ ]

5. Explain how the extent of reaction is related to the fraction conversion of the limiting reactant.
6. Explain why for a process you want to determine the rank of the component matrix.

Problems

1. Corrosion of pipes in boilers by oxygen can be alleviated through the use of sodium sulfite. Sodium sulfite removes oxygen from boiler feedwater by the following reaction:

$$2\,Na_2SO_3 + O_2 \rightarrow 2\,NaSO_4$$

How many pounds of sodium sulfite are theoretically required (for complete reaction) to remove the oxygen from 8,330,000 lb of water (10^6 gal) containing 10.0 ppm of dissolved oxygen and at the same time maintain a 35% excess of sodium sulfite?

2. Consider a continuous, steady-state process in which the following two reactions take place:

$$C_6H_{12} + 6H_2O \rightarrow 6CO + 12H_2$$

$$C_6H_{12} + H_2 \rightarrow C_6H_{14}$$

In the process 250 moles of C_6H_{12} and 800 moles of H_2O are fed into the reactor each hour. The yield of H_2 is 40.0% and the selectivity of the first reaction compared to the second reaction is 12.0. Calculate the molar flow rates of all five components in the output stream.

3. Hydrofluoric acid (HF) can be manufactured by treating calcium fluoride (CaF_2) with sulfuric acid (H_2SO_4). A sample of fluorospar (the raw material) contains 75% by weight CaF_2 and 25% inert (nonreacting) materials. The pure sulfuric acid used in the process is in 30% excess of that theoretically required. Most of the manufactured HF leaves the reaction chamber as a gas, but a solid cake is also removed from the reaction chamber that contains 5% of all the HF formed, plus $CaSO_4$, inerts, and unreacted sulfuric acid. How many kilograms of *cake* are produced per 100 kg of fluorospar charged to the process?

5.4 Element Material Balances

In the previous section, you learned how to use species mole balances for reacting systems. Equation (5.7), which included terms for the generation and consumption of each reacting species, was used for these problems to include the effects of species generation and consumption due to reaction. As you probably know, the elements in a process are conserved regardless of whether reactions are occurring, and consequently you can apply a simplified version of Equation (3.1) to the elements. Because elements are neither generated nor consumed, the generation and consumption terms on the right-hand side of Equation (3.1) can be ignored, a result that is often convenient. Thus, for a steady-state process, Equation (3.1) simplifies to Equation (3.3):

$$In = Out$$

Why not use element balances to solve material balance problems rather than species balances? For most problems it is easier to apply mole balances, but for some problems, such as problems with complex or unknown reaction equations, element balances are preferred.

How wonderful that we have met with a paradox.
Now we have some hope of making progress.

Niels Bohr

You can use element balances, but just make sure that the element balances are independent. Here is an illustration of the issue. Let's use the decomposition of ammonia again as the illustration. In the gas phase NH_3 can decompose as follows:

$$NH_3 \rightarrow N_2 + 3H_2$$

Figure 5.3 shows some data for the partial decomposition of NH_3.

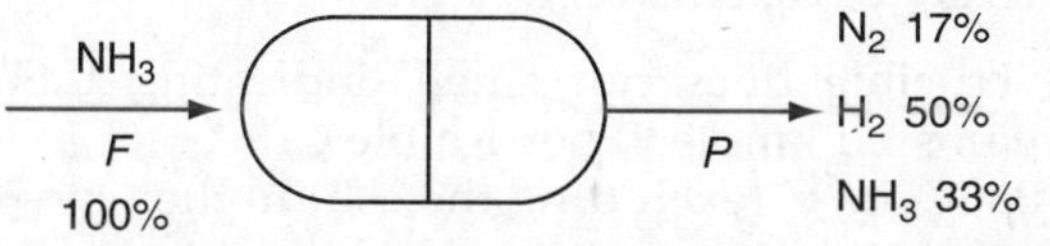

Figure 5.3 Schematic showing a possible case of the decomposition of ammonia. Note: numbers are rounded.

Two unknowns exist, F and P, and the process involves two elements, N and H. It might appear that in Step 7 of the proposed solution strategy you could use the two element balances to solve for the values of the unknowns, F and P, but you can't! Try it. The reason is that the two element balances are not independent. As explained in Section 5.3, only one of the element balances is independent. Look at the element material balances just below (in moles or mass?) for the decomposition of ammonia, and you will observe that the hydrogen balance is three times the nitrogen balance (if you ignore the rounding of the numbers):

$$\text{N balance: } (1)\,F = (2)[P(0.17)] + P(0.33) = 0.67\,P$$

$$\text{H balance: } (3)\,F = (2)[P(0.50)] + (3)[P(0.33)] = 2.0\,P$$

If you add a piece of information, say, by picking a basis of $P = 100$ mol, the degrees of freedom become zero, and then you can solve for F. What happens if you apply species balances to solve the same problem of ammonia decomposition? Refer back to Section 5.3 for a discussion of the same problem. Chapter 12 discusses how to determine whether or not a set of material balance equations is independent.

Example 5.10 Fusion of $BaSO_4$

Barite (entirely composed of $BaSO_4$) is fused (reacted in the solid state) in a crucible with coke (composed of 94.0% C and 6.0% ash). The ash does not react. The final fusion mass of 193.7 g is removed from the crucible, analyzed, and found to have the following composition:

	%
$BaSO_4$	11.1
BaS	72.8
C	13.9
Ash	2.2
	100.0

(Continues)

Example 5.10 Fusion of $BaSO_4$ (*Continued*)

The gas evolved from the crucible does not smell, indicating the absence of sulfur dioxide, and contains 1.13 mole O per 1 mole C.

What was the mass ratio of $BaSO_4$ to C (excluding the ash) in the reactants put into the crucible?

Solution

Steps 1–4

The process is a batch process in which the solid components are loaded together, the reaction is carried out, and the fusion mass is removed. Figure E5.10 is a diagram of the process with the known data entered. You want to calculate the mass ratio of $m^{in}_{BaSO_4}$ to m^{in}_{C}.

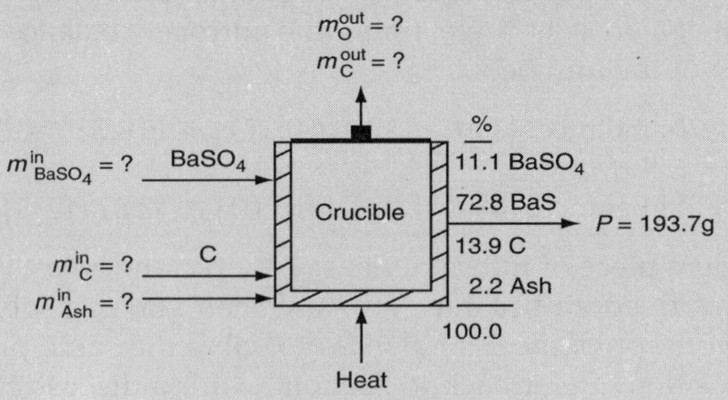

Figure E5.10

Step 5

$$\text{Basis} = 100\text{g } P$$

You could take 193.7 g of P as the basis, but it is slightly easier to take 100 g P as the basis because the analysis of P is given in percent, and you only have to calculate the relative amounts of the inputs of C and BaSO, not the absolute amounts.

Step 4 Again

Some preliminary calculations of the composition of P will be helpful in the solution.

				Composition in P in g mol				
Compound	g	MW	g mol	Ba	S	O	C	Ash
$BaSO_4$	11.1	233.3	0.0477	0.0477	0.0477	0.19	-	-
BaS	72.8	169.3	0.431	0.430	0.430	-	-	-
C	13.9	12.0	1.16	-	-	-	1.16	-

				Composition in P in g mol				
Compound	g	MW	g mol	Ba	S	O	C	Ash
Ash	2.2	-	-	-	-	-	-	*
	100.0			0.479	0.479	0.191	1.16	*
			MW:	137.34	32.064	16.0	12.0	*
			g:	65.8	15.36	3.06	13.9	2.2

Steps 6–8

From Figure E5.10, after assigning known values to variables, it appears that five unknowns exist, ignoring those variables whose value are zero (mass values are in grams below).

Assigned values:

$$P = 100 \qquad m_{O}^{P} = 3.06$$

$$m_{Ba}^{P} = 65.8 \qquad m_{C}^{P} = 13.9$$

$$m_{S}^{P} = 15.36 \qquad m_{Ash}^{P} = 2.2$$

Unknowns:

$$m_{BaSO_4}^{in} \quad m_{C}^{in} \quad m_{Ash}^{in} \quad m_{C}^{out} \quad m_{O}^{out}$$

You can make Ba, S, O, C, and ash balances; hence the problem seems to have zero degrees of freedom. The process is a batch process; hence if the crucible is empty at the beginning and end of the fusion, the material balance equation to use is

"What goes in must come out."

Here are the element mass balances:

$$\text{Ash:} \quad m_{Ash}^{in} = 2.2\text{g}$$

In Out

$$\text{Ba:} \frac{m_{BaSO_4g}^{in}}{} \left| \frac{1 \text{ g mol BaSO}_4}{233.3 \text{ g BaSO}_4} \right| \frac{1 \text{ g mol Ba}}{1 \text{ g mol BaSO}_4} \left| \frac{137.34 \text{ g Ba}}{1 \text{ g mol Ba}} \right. = 65.8 \text{ g}$$

$$m_{BaSO_4}^{in} = 111.7 \text{ g}$$

In Out

$$\text{S:} \frac{m_{BaSO_4g}^{in}}{} \left| \frac{1 \text{ g mol BaSO}_4}{233.3 \text{ g BaSO}_4} \right| \frac{1 \text{ g mol S}}{1 \text{ g mol BaSO}_4} \left| \frac{32.064 \text{ g S}}{1 \text{ g mol S}} \right. = 15.36$$

$$m_{BaSO_4}^{in} = 111.7 \text{ g, a redundant result}$$

(Continues)

Example 5.10 Fusion of $BaSO_4$ (*Continued*)

	In			Out
O:	$\frac{m^{\text{in}}_{BaSO_4}}{}$	$\frac{1 \text{ g mol } BaSO_4}{233.3 \text{ g } BaSO_4}$	$\frac{4 \text{ g mol O}}{1 \text{ g mol } BaSO_4}$	$\frac{16.0 \text{ g 0}}{1 \text{ g mol 0}} = m^{\text{out}}_{O} + 3.06$

Introduce $m^{\text{in}}_{BaSO_4} = 111.7$ into the O balance to get $m^{\text{out}}_{O} = 27.56$ g. The only balance left is the carbon balance.

Step 9

Because the Ba and S balances are not independent, we need one more piece of information, namely, that $m^{\text{out}}_{C} = m^{\text{out}}_{O}/1.13 = 27.56/1.13 = 24.4$, which is needed in the carbon balance.

$$m^{\text{in}}_{C} = 24.4 + 13.92 = 38.3 \text{ g}$$

$$\frac{m^{\text{in}}_{BaSO_4}}{m^{\text{in}}_{C}} = \frac{111.7}{38.3} = 2.92$$

Step 10

Check using the redundant total material balance:

$$m^{\text{in}}_{BaSO_4} + m^{\text{in}}_{C} = m^{\text{out}}_{O} + m^{\text{out}}_{C} + P$$

$$111.7 + 38.3 \stackrel{?}{=} 27.56 + 24.4 + 100.0$$

$$150 \cong 152 \quad \text{close enough}$$

Element balances are especially useful when you do not know what reactions occur in a process. You only know information about the input and output stream components as illustrated by the next example.

Example 5.11 Use of Element Balances to Solve a Hydrocracking Problem

Hydrocracking is an important refinery process for converting low-valued heavy hydrocarbons to more valuable lower-molecular-weight hydrocarbons by exposing the feed to a zeolite catalyst at high temperature and pressure in the presence of hydrogen. Researchers study the hydrocracking of pure components, such as octane (C_8H_{18}), to understand the behavior of cracking reactions. In one such experiment, the cracked products had the following composition in mole percent: 19.5% C_3H_8, 59.4% C_4H_{10}, and 21.1%

C_5H_{12}. You are asked to determine the molar ratio of hydrogen consumed to octane reacted for this experiment.

Solution

We will use element balances to solve this problem because the reactions involved in the process are not specified.

Steps 1–4

Figure E5.11 is a sketch of the laboratory hydrocracker reactor together with the data for the streams. The process is assumed to be open, steady-state.

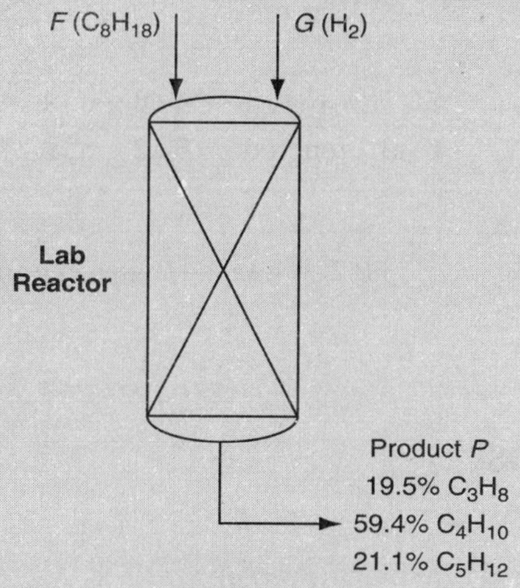

Figure E5.11

Step 5

Basis: $P = 100$ g mol

Steps 6 and 7

The degree-of-freedom analysis is as follows:

Variables: 3 $\quad F, G, P$

Because P is selected as the basis, you can assign $P = 100$ g mol; hence there are two unknowns: *G and P*.

Equations needed: 2

Element balances: 2 $\quad$ H, C

Therefore, this problem has zero degrees of freedom.

(*Continues*)

Example 5.11 Use of Element Balances to Solve a Hydrocracking Problem (*Continued*)

Steps 8 and 9

The element balances after introducing the specifications and the basis are (units are in gram moles):

$$\text{C: } F(8) + G(0) = 100[(0.195)(3) + (0.594)(4) + (0.211)(5)]$$

$$\text{H: } F(18) + G(2) = 100[(0.195)(8) + (0.594)(10) + (0.211)(12)]$$

and solving this set of linear equations simultaneously yields

$$F = 50.2 \text{ g mol} \qquad G = 49.8 \text{ g mol}$$

The ratio

$$\frac{H_2 \text{ consumed}}{C_8H_{18} \text{ reacted}} = \frac{49.8 \text{ g mol}}{50.2 \text{ g mol}} = 0.992$$

Step 10

Check: $50.2 + 49.8 = P = 100 \text{ g mol}$

Self-Assessment Test

Questions

1. Do you have to write element material balances with the units of each term being moles rather than mass? Explain your answer.
2. Will the degrees of freedom be smaller or larger using element balances in place of species balances?
3. How can you determine whether a set of element balances are independent?
4. Can the number of independent element balances ever be larger than the number of species balances in a problem?

Problems

1. Consider a system used in the manufacture of electronic materials (all gases except Si):

   $$SiH_4, Si_2, H_6, SiH_2, H_2, Si$$

   How many independent element balances can you make for this system?
2. Methane burns with O_2 to produce a gaseous product that contains CH_4, O_2, CO_2, CO, H_2O, and H_2. How many independent element balances can you write for this system?

3. What is the size of the minimum reaction set that corresponds to the products listed in problem 2?
4. In the reaction of $KClO_3$ with HCl the following products were measured: KCl, ClO_2, Cl_2, and H_2O. How many element material balances can you make for this system?
5. Solve the self-assessment test problems in Section 5.2 using element balances.

5.5 Material Balances for Combustion Systems

In this section we consider combustion as a special topic involving material balances that include chemical reactions. Combustion is in general the reaction of oxygen with materials containing hydrogen, carbon, and sulfur, and the associated release of heat and the generation of product gases such as H_2O, CO_2, CO, and SO_2. Typical examples of combustion are the combustion of coal, heating oil, and natural gas used to generate electricity in utility power stations, and engines that operate using the combustion of gasoline or diesel fuel. More complicated oxidation processes take place in the human body but are not called combustion. Most combustion processes use air as the source of oxygen. For our purposes you can assume that air contains 79% N_2 and 21% O_2 (see Chapter 2 for a more detailed analysis), neglecting the other components that amount to a total of less than 1.0%, and that air has an average molecular weight of 29. Although a small amount of N_2 oxidizes to NO and NO_2, gases called NO_x, a pollutant, the amount is so small that we treat N_2 as a nonreacting component of air and fuel. Figure 5.4 shows how the CO, unburned hydrocarbons, and NO_x vary with the air-fuel ratio in combustion.

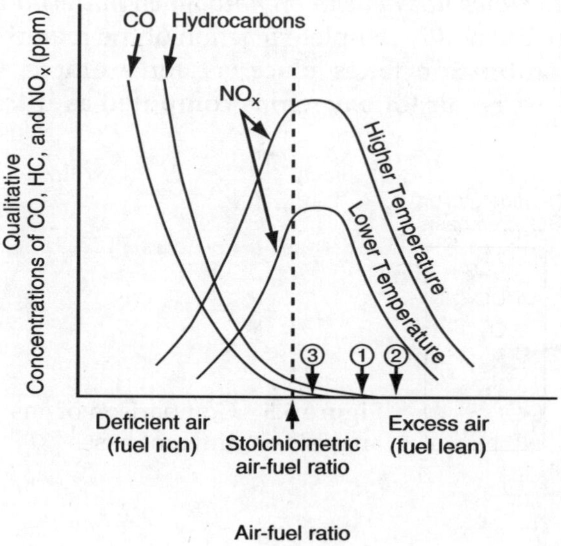

Figure 5.4 Pollutants resulting from combustion of natural gas vary with the air-fuel ratio and the temperature of combustion. The amount of hydrocarbons and CO increases with deficient air. Efficiency goes down with too much excess air, but so does the NO_x. (1) represents the optimal air-fuel ratio (no CO or unburned hydrocarbons). (2) represents inefficient combustion (excess air has to be heated). (3) represents unsatisfactory combustion (because not all of the fuel is burned and the unburned products are released to the atmosphere).

You should become acquainted with some of the special terms associated with combustion:

- **Flue or stack gas:** All of the gases resulting from a combustion process including the water vapor, sometimes known as a **wet basis.**
- **Orsat analysis, or dry basis:** All of the gases resulting from a combustion process *not including the water vapor.* (Orsat analysis refers to a type of gas analysis apparatus in which the volumes of the respective gases are measured over and in equilibrium with water; hence each component is saturated with water vapor. The net result of the analysis is to eliminate water as a component that is measured.) See Figure 5.5. To convert from one analysis to another, you have to adjust the percentages of the components to the desired basis, as explained in Chapter 2.
- **Complete combustion:** The complete reaction of the fuel producing CO_2 and H_2O.
- **Partial combustion:** The combustion of the fuel producing at least some CO from the carbon source. Because CO itself can react with oxygen, the production of CO in a combustion process does not produce as much energy as would be the case if only CO_2 were produced.
- **Theoretical air** (or **theoretical oxygen**): The amount of air (or oxygen) required to be brought into the process **for complete combustion**. Sometimes this quantity is called the **required** air (or **oxygen**).
- **Excess air** (or **excess oxygen**): In line with the definition of excess reactant given in Section 5.1, excess air (or oxygen) is the amount of air (or oxygen) **in excess of the theoretical air required for complete combustion.**

The calculated amount of *excess air does not depend on how much material is actually burned* but what could be burned if complete reaction of the material occurred. **Even if only partial combustion takes place**, as, for example, C burning to both CO and CO_2, **the excess air (or oxygen) is computed as if the**

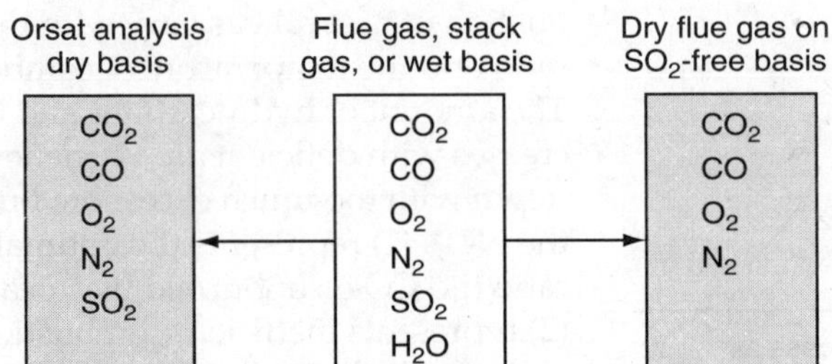

Figure 5.5 Comparison of gas analysis on different bases

process of combustion produced only CO_2. A wooden chair sitting in a room will not burn at normal temperatures, but nevertheless if the volume or mass of the chair is known, the excess air in the room for the combustion of the chair can be calculated.

The percent excess air is identical to the percent excess O_2 (a quantity often more convenient to use in calculations):

$$\% \text{ excess air} = 100\,\frac{\text{excess air}}{\text{required air}} = 100\,\frac{\text{excess } O_2/0.21}{\text{required } O_2/0.21} = 100\,\frac{\text{excess } O_2}{\text{required } O_2} \tag{5.14}$$

Note that the ratio 1/0.21 of air to O_2 cancels out in Equation (5.14). Percent excess air may also be computed as

$$\% \text{ excess air} = 100\,\frac{O_2 \text{ entering process} - O_2 \text{ required}}{O_2 \text{ required}} \tag{5.15}$$

or

$$\% \text{ excess air} = 100\,\frac{\text{excess } O_2}{O_2 \text{ entering} - \text{excess } O_2} \tag{5.16}$$

In calculating the degrees of freedom in a problem, if the percent excess air is specified (and the chemical reaction equation for the process is known), you can calculate how much air enters with the fuel; hence the specification of the amount of excess air can be used to assign a value to one of the variables like other specifications.

Now, let us apply these concepts with some examples.

Example 5.12 Calculation of Excess Air

Fuels for motor vehicles other than gasoline are being evaluated because they generate lower levels of pollutants than does gasoline. Compressed propane has been suggested as a source of power for vehicles. Suppose that in a test 20 kg of C_3H_8 is burned with 400 kg of air to produce 44 kg of CO_2 and 12 kg of CO. What was the percent excess air?

Solution

This is a problem involving the following reaction (is the reaction equation correctly balanced?):

$$C_3H_8 + 5O_2 \rightarrow 3CO_2 + 4H_2O$$

Basis: 20 kg of C_3H_8

(Continues)

Example 5.12 Calculation of Excess Air (*Continued*)

Since the percentage of excess air is based on the *complete combustion* of C_3H_8 to CO_2 and H_2O, the fact that combustion is not complete has no influence on the calculation of excess air. The required O_2 on the basis of 20 kg of C_3H_8 is

$$\frac{20 \text{ kg } C_3H_8}{} \left| \frac{1 \text{ kg mol } C_3H_8}{44.09 \text{ kg } C_3H_8} \right| \frac{5 \text{ kg mol } O_2}{1 \text{ kg mol } C_3H_8} = 2.27 \text{ kg mol } O_2$$

The entering O_2 is

$$\frac{400 \text{ kg air}}{} \left| \frac{1 \text{ kg mol air}}{29 \text{ kg air}} \right| \frac{21 \text{ kg mol } O_2}{100 \text{ kg mol air}} = 2.90 \text{ kg mol } O_2$$

The percentage excess air is

$$100\frac{\text{excess } O_2}{\text{required } O_2} = 100\frac{\text{entering } O_2 - \text{required } O_2}{\text{required } O_2}$$

$$\% \text{ excess air} = \frac{2.90 \text{ lb mol } O_2 - 2.27 \text{ lb mol } O_2}{2.27 \text{ lb mol } O_2} \left| \frac{100}{} \right. = 28\%$$

In calculating the amount of excess air, remember that the excess is the amount of air that enters the combustion process over and above that required for complete combustion. Suppose there is some oxygen in the material being burned. For example, suppose that a gas containing 80% C_2H_6 and 20% O_2 is burned in an engine with 200% excess air. Eighty percent of the ethane goes to CO_2, 10% goes to CO, and 10% remains unburned. What is the amount of the excess air per 100 moles of the C_2H_6? First, you can ignore the information about the CO and the unburned ethane because the basis of the calculation of excess air is *complete combustion* of the C_2H_6. Specifically, the products of reaction are assumed to be the highest oxidation state. For example, C goes to CO_2, S to SO_2, H to H_2O, CO to CO_2, and so on.

Second, the oxygen in the fuel cannot be ignored. Based on the reaction

$$C_2H_6 + \frac{7}{2}O_2 \rightarrow 2CO_2 + 3H_2O$$

80 moles of C_2H_6 require 3.5(80) = 280 moles of O_2 for complete combustion. However, the gas contains 20 moles of O_2, so only 280 − 20 = 260 moles of O_2 are needed in the entering air for complete combustion. Thus, 260 moles of O_2 is the required O_2, and the calculation of the 200% excess O_2 (air) is based on 260, not 280, moles of O_2:

Entering with air	Moles O_2
Required O_2:	260
Excess O_2 (2.00)(260):	520
Total O_2:	780

Example 5.13 A Fuel Cell to Generate Electricity from Methane

"A Fuel Cell in Every Car" is the headline of an article in *Chemical and Engineering News* (March 5, 2001, p. 19). In essence, a fuel cell is an open system into which fuel and air are fed, and out of which come electricity and waste products. Figure E5.13 is a sketch of a fuel cell in which a continuous flow of methane (CH_4) and air (O_2 plus N_2) produces electricity plus CO_2 and H_2O. Special membranes and catalysts are needed to promote the oxidation of the CH_4.

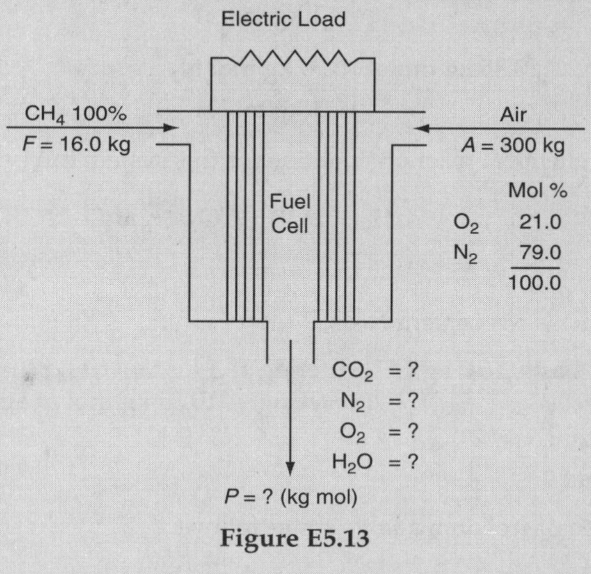

Figure E5.13

Based on the data given in Figure E5.13, you are asked to calculate the composition of the products in *P*.

Solution

Steps 1–4

This is a steady-state process with reaction. Can you assume that a complete reaction occurs? Yes. How? No CH_4 or CO appears in *P*. The system is the

(Continues)

Example 5.13 A Fuel Cell to Generate Electricity from Methane (*Continued*)

fuel cell (open, steady-state). Because the process output is a gas, the composition will be in mole fractions (or moles); hence it is more convenient to use kilogram moles rather than mass in this problem even though the quantities of CH_4 and air are stated in kilograms. You can carry out the necessary conversions from kilograms to kilogram moles as follows:

$$\frac{300 \text{ kg } A}{} \left| \frac{1 \text{ kg mol } A}{29.0 \text{ kg } A} \right. = 10.35 \text{ kg mol } A \text{ in}$$

$$\frac{16.0 \text{ kg } CH_4}{} \left| \frac{1 \text{ kg mol } CH_4}{16.0 \text{ kg } CH_4} \right. = 1.00 \text{ kg mol } CH_4 \text{ in}$$

$$\frac{10.35 \text{ kg mol } A}{} \left| \frac{0.21 \text{ kg mol } O_2}{1 \text{ kg mol } A} \right. = 2.17 \text{ kg mol } O_2 \text{ in } A$$

$$\frac{10.35 \text{ kg mol } A}{} \left| \frac{0.79 \text{ kg mol } N_2}{1 \text{ kg mol } A} \right. = 8.18 \text{ kg mol } N_2 \text{ in } A$$

The chemical reaction equation for this system can be assumed to be

$$CH_4 + 2O_2 \rightarrow CO_2 + 2H_2O$$

Step 5

We will pick a convenient basis.

Basis: 16.0 kg CH_4 entering = 1 kg mol CH_4 plus 300 kg
A entering = 10.35 kg mol of air

Steps 6 and 7

The degree-of-freedom analysis is as follows:

Variables: 10 $F, P, A, n^P_{CO_2}, n^P_{N_2}, n^P_{O_2}, n^P_{H_2O}, n^A_{O_2}, n^A_{N_2}$ plus ξ

Given the basis and the quantities calculated above, four of these variables $(F, A, n^A_{O_2}, n^A_{N_2})$ can be assigned values. Therefore, there are six unknowns remaining.

Equations: 6

Five (5) independent species balances: CH_4, O_2, N_2, CO_2, H_2O

One (1) independent implicit equation for P: $\sum n_i^P = P$

Therefore, the degrees of freedom are zero.

Step 8

The species mole balances are as follows:

Compound	Out		In		$\nu_i \xi$		g mol
CH_4:	$n^P_{CH_4}$	=	1.0	−	ξ	=	0
O_2:	$n^P_{O_2}$	=	2.17	−	2ξ	=	0.17
N_2:	$n^P_{N_2}$	=	8.18	−	$0(\xi)$	=	8.18
CO_2:	$n^P_{CO_2}$	=	0	+	ξ	=	1.0
H_2O:	$n^P_{H_2O}$	=	0	+	2ξ	=	2.0

Step 9

The solution of this set of equations gives

$$n^P_{CH_4} = 0, \; n^P_{O_2} = 0.17, \; n^P_{N_2} = 8.18, \; n^P_{CO_2} = 1.0,$$
$$n^P_{CO_2} = 1.0, \; n^P_{H_2O} = 2.0, \; P = 11.35$$

and the mole percentage composition of P is

$$y_{O_2} = 1.5\%, \; y_{N_2} = 72.1\%, \; y_{CO_2} = 8.8\%, \text{ and } y_{H_2O} = 17.6\%$$

You could also use element balances without knowing the reaction to get the same solution using four element balances and one implicit equation for P (ξ would no longer be a variable).

Step 10

You can check the answer by determining the total mass of the exit gas and comparing it to total mass in (316 kg), but we will omit this step here to save space.

Example 5.14 Combustion of Coal

A local utility burns coal having the following composition on a dry basis. (Note that the coal analysis below is a convenient one for our calculations but is not necessarily the only type of analysis that is reported for coal. Some analyses contain much less information about each element.)

Component	Percent
C	83.05
H	4.45
O	3.36
N	1.08
S	0.70
Ash	7.36
Total	100.0

The average Orsat analysis of the gas from the stack during a 24 hr test was

Component	Percent
$CO_2 + SO_2$	15.4
CO	0.0
O_2	4.0
N_2	80.6
Total	100.0

Moisture (H_2O) in the fuel was 3.90%, and the air on the average contained 0.0048 lb H_2O/lb dry air. The refuse showed 14.0% unburned coal, with the remainder being ash. The unburned coal in the refuse can be assumed to be of the same composition as the coal that serves as fuel.

What is the percent excess air used for this process as shown in Figure E5.14?

Solution

Note that for this problem you are asked only for the percent excess air used. Therefore, Figure E5.14 contains more information than is required to solve this problem.

Steps 1–4

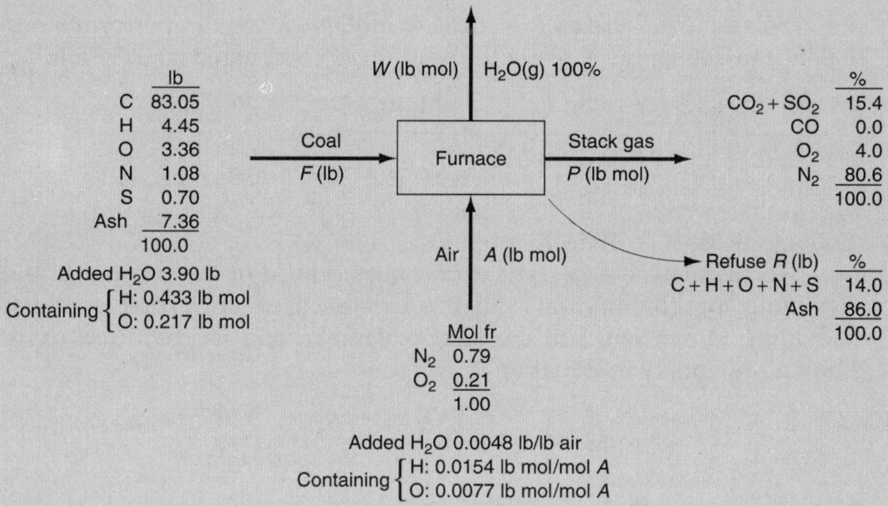

Figure E5.14

Step 5

Pick a basis of $F = 100$ lb as convenient.

Steps 6–9

We need to solve for the feed rate of air (A), but to do this we will first need to solve for the unknown flow rates: R and P. Note that ash is a tie component. Applying a material balance for ash yields

$$7.36 = 0.86R \quad \Rightarrow \quad R = 8.56 \text{ lb}$$

Note that a portion of the coal does not react and leaves the process in R. We will assume here that the portion of R that is not ash has the same composition as F. Therefore, we will simply subtract this from F to determine the amount of coal that is combusted:

$$\text{Amount of coal combusted} = 100. - 0.14\,R = 98.8 \text{ lb}$$

Because all of the C and S in the coal that combusts ends up in the stack gas (P), we can form the following combined mole balance for C and S:

moles of C combusted moles of S combusted moles of C + S in flue gas

$$\frac{(0.8305)(98.8)}{12} + \frac{(0.007)(98.8)}{32} = 0.154P$$

(Continues)

Example 5.14 Combustion of Coal (*Continued*)

Solving for P yields $P = 44.54$ lb mol. Now we can perform a nitrogen balance to determine A using the amount of combusted coal (98.8 lb):

$$\underset{N_2 \text{ in } F}{\frac{(0.0108)(98.8)}{28}} + \underset{N_2 \text{ in } A}{0.79A} = \underset{N_2 \text{ in } P}{(0.806)(44.54)}$$

Solving yields $A = 45.39$ lb mol.

To calculate the percent excess air, because of the oxygen in the coal available for combustion and the existence of unburned combustibles including O, we will use the total oxygen in and the required oxygen as shown previously in Equation (5.15):

$$\% \text{ excess air} = 100\left(\frac{O_2 \text{ entering} - O_2 \text{ required}}{O_2 \text{ required}}\right)$$

The required O_2 is equal to the stoichiometric requirements for complete combustion of C, H, and S minus the O_2 present in the coal (not based on what actually combusted):

Component	Reaction	lb	lb mol	Required O_2 (lb mol)
C	$C + O_2 \rightarrow CO_2$	83.05	6.921	6.921
H	$H_2 + \frac{1}{2}O_2 \rightarrow H_2O$	4.45	4.415	1.104
O	—	3.36	0.210	(0.105)
N	—	—	—	—
S	$S + O_2 \rightarrow SO_2$	0.70	0.022	0.022
Total				7.942

The oxygen in the entering air is $(45.39)(0.21) = 9.532$ lb mol. Therefore,

$$\% \text{ excess air} = 100\left(\frac{9.532 - 7.942}{7.942}\right) = 20.0\%$$

If you (incorrectly) calculated the percent excess air from the wet stack gas analysis alone, you would have ignored the oxygen in the coal.

From the viewpoint of the concern about the increase of the CO_2 concentration in the atmosphere, would the CH_4 in Example 5.13 or the coal in Example 5.14 contribute more CO_2 per kilogram of fuel? The H/C ratio in moles in CH_4 was 4/1 whereas in the coal it was $0.0537/0.0897 = 0.60$. Thus,

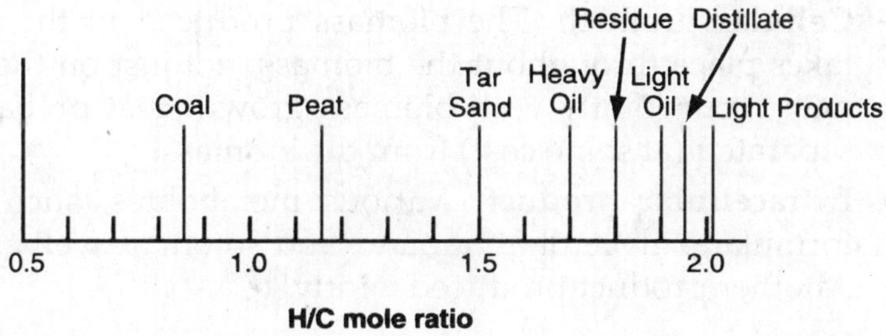

Figure 5.6 Variation of the H/C ratio in selected fuels

the coal results in the larger emission of CO_2. Figure 5.6 shows how the H/C ratio varies with various types of fuel.

Oxygen use in the growth of biomass does not usually focus on the assumption of complete conversion of the reactants to CO_2 and H_2O. The chemical reaction equation includes biomass (in solution or suspension) as a reactant and a product along with the CO_2 and H_2O, and rarely is all of the reactant(s) used up by the dissolved O_2 (acronym DO). In fact, O_2 may be a limiting rather than an excess reactant because its solubility in water, for example, at 25°C is only 8.3 mg/L, that is, 2.3 ppm.

If you write the chemical reaction equation for the growth of biomass and possible by-products as follows, on the basis of 1 mole of C in the substrate, you would get Equation (5.17):

$$\underbrace{CH_xO_yN_z}_{\substack{\text{Substrate}\\ \text{(biomass)}}} + a\,O_2 + b\,\underbrace{H_gO_hN_i}_{\substack{\text{Nitrogen}\\ \text{source}}} \rightarrow c\,\underbrace{CH_jO_kN_\ell}_{\substack{\text{Biomass cells}\\ \text{(from growth),}\\ \text{cellular product}}} + d\,CO_2 \tag{5.17}$$

$$+\, e\,H_2O + f\,\underbrace{C_mH_nO_oN_p}_{\text{Extracellular product}}$$

- **Substrate** The biomass (bacteria, fungi, and so on) that is expected to grow a compound such as glucose in solution (or on a surface) acted upon in a reaction by enzymes or growth promoters, usually to produce a desired product.
- **Nitrogen source** The source of nitrogen, such as ammonia, ammonium, urea, nitrate, secondary or treatment wastes, for biomass growth.

- **Cellular product** The biomass produced in the reaction(s). Growth takes place throughout the biomass, not just on the surface. A product may occur jointly with biomass growth that probably will have to be separated (at some cost) from the biomass.
- **Extracellular product** Various metabolites, such as acetate, citrate, formulate, glycerol, pyramate, and so on, as well as the CO_2 and H_2O. Another product produced jointly.

For the case of a hydrocarbon fuel with components of C and H reacting with O_2 to produce only CO_2 and H_2O, the coefficients b, c, and f are zero in Equation (5.17), leaving only a, d, and e, given the assumption of complete combustion. You can make three element balances, and the degrees of freedom are zero. But if all six terms and associated coefficients are involved in the reaction, you can make only four element balances (C, H, O, N), leaving two degrees of freedom to be specified. Presumably you know the chemical composition of each species in the reaction or additional specifications or assumptions have to be made. Thus the molecular weights (MW) of the substrate and biomass product are usually known, and if the composition of the extracellular product is not known, you can use as an estimate $CH_{1.8}O_{0.5}N_{0.2}$ [J. A. Roels, *Biotechnology Bioengineering*, **22**, 2457(1980)]. A cell contains elements other than C, H, O, and N, but the minor components such as P, S, K, Ca, Mg, Cl, and Fe require so little O_2 that they are often treated as ash, a nonreacting component.

With these assumptions, the values of c and f still have to be obtained by measurements:

$$c = \frac{\dfrac{\text{g biomass produced}}{\text{MW biomass}}}{\dfrac{\text{g substrate consumed}}{\text{MW substrate}}}$$

$$f = \frac{\dfrac{\text{g extracellular product produced}}{\text{MW of extracellular product}}}{\dfrac{\text{g substrate consumed}}{\text{MW substrate}}}$$

It is also possible to use the respiratory quotient (RQ):

$$RQ = \frac{\text{mol } CO_2 \text{ produced}}{\text{mol } O_2 \text{ consumed}} = \frac{d}{a}$$

but the values of RQ are not particularly accurate, and the other stoichiometric coefficients are quite sensitive to errors in RQ.

As a result of the positive degrees of freedom, a number of *empirical* measures of O_2 usage in biosystems are employed to measure what is called **oxygen demand**. The definitions of these measures vary with what you read in books or on the Internet, but they are precisely defined by standard tests that are carried out, or with instruments that serve as substitutes for the standard tests.

- **Total oxygen demand (TOD):** The quantity of oxygen required to completely oxidize all of the organic and inorganic compounds present in a sample (in water) as determined by a COD test (using a strong oxidant such as sulfuric acid or potassium dichromate) or combustion. The amount of O_2 is reported in milligrams O_2 per liter containing the sample, or as grams O_2 per gram sample.
- **Chemical oxygen demand (COD):** The same as TOD except only the organic components are considered. Sometimes the COD is defined to be the same as TOD. The COD value is higher than the BOD value (defined next) because the COD includes slowly biodegradable and recalcitrant organic compounds not degraded by microorganisms as in a BOD test. The value of the COD is reported in units of grams O_2 per gram sample.
- **Biochemical oxygen demand (BOD):** The quantity of oxygen required by microorganisms to oxidize the *organic compounds* in a sample (in water) as determined by a BOD test. The test is carried out for a number of days at 20°C. Five days is the most common duration. The quantity of O_2 used is determined by the difference in the dissolved O_2 (DO) in the water at the beginning of the test and at the end of five days. BOD is used mainly in evaluating water and wastewater quality. The value of the BOD5 (BOD_5) is reported in grams O_2 per gram sample.
- **Theoretical oxygen demand (ThOD):** The quantity of O_2 required according to a valid balanced chemical reaction equation (Equation 5.17) to oxidize the reactant(s) to CO_2, H_2O, and the highest oxidation state of other products, ignoring any extracellular products from a sample ($f = 0$). The ThOD is the same as the theoretical O_2 discussed previously. The highest stage of nitrogen compounds is nitrate. For example, in the oxidation of glycine, $CH_2(NH_3)COOH$, the overall reaction equation is

$$CH_2\,(NH_3)CO\,OH + 15/2\,O_2 = 2\,CO_2 + 5/2\,H_2O + HNO_3$$

and the theoretical O_2 is 15/2 mol.

Self-Assessment Test

Questions

1. Explain the difference for a gas between a flue gas analysis and an Orsat analysis; wet basis and dry basis.
2. What does an SO_2-free basis mean?
3. Write down the equation relating percent excess air to the required air and entering air.
4. Will the percent excess air always be the same as the percent excess oxygen in combustion (by oxygen)?
5. In a combustion process in which a specified percentage of excess air is used, and in which CO is one of the products of combustion, will the analysis of the resulting exit gases contain more or less oxygen than if all the carbon had burned to CO_2?
6. Indicate whether the following statements are true or false:
   a. Excess air for combustion is calculated using the assumption of complete reaction whether or not a reaction takes place.
   b. For the typical combustion process the products are CO_2 gas and H_2O vapor.
   c. In combustion processes, since any oxygen in the coal or fuel oil is inert, it can be ignored in the combustion calculations.
   d. The concentration of N_2 in a flue gas is usually obtained by direct measurement.

Problems

1. Pure carbon is burned in oxygen. The flue gas analysis is

CO_2	75 mol %
CO	14 mol %
O_2	11 mol %

   What was the percent excess oxygen used?
2. Toluene, C_7H_8, is burned with 30% excess air. A bad burner causes 15% of the carbon to form soot (pure C) deposited on the walls of the furnace. What is the Orsat analysis of the gases leaving the furnace?
3. A synthesis gas analyzing CO_2 6.4%, O_2 0.2%, CO 40.0%, and H_2 50.8% (the balance is N_2) is burned with excess dry air. The problem is to determine the composition of the flue gas. How many degrees of freedom exist in this problem; that is, how many additional variables have to have their values specified?
4. A coal analyzing 65.4% C, 5.3% H, 0.6% S, 1.1% N, 18.5% O, and 9.1% ash is burned so that all combustible is burned out of the ash. The flue gas analyzes 13.00% CO_2, 0.76% CO, 6.17% O_2, 0.87% H_2, and 79.20% N_2. All of the sulfur burns to SO_2, which is included in the CO_2 figure in the gas analysis (i.e., $CO_2 + SO_2 = 13.00\%$). Calculate

a. Pounds of coal fired per 100 lb mol of dry flue gas as analyzed
b. Ratio of moles of total combustion gases to moles of dry air supplied
c. Total moles of water vapor in the stack gas per 100 lb of coal if the entering air is dry
d. Percent excess air

5. A hydrocarbon fuel is burned with excess air. The Orsat analysis of the flue gas shows 10.2% CO_2, 1.0% CO, 8.4% O_2, and 80.4% N_2. What is the atomic ratio of H to C in the fuel?

Looking Back

In this chapter we explained how the chemical reaction equation can be used to calculate quantitative relations among reactants and products. We also defined a number of terms used by engineers in making calculations involving chemical reactions. In this chapter we applied Equation (5.6) and its analogs to processes involving reaction. If you make element balances, the generation and consumption terms in Equation (5.6) are zero. If you make species balances, the accumulation and consumption terms are not zero, and you have to use the extent of reaction. You simply apply the general material balance with reaction to these systems, recognizing their characteristics

Glossary

Conversion The fraction of the feed or some *key* material in the feed that is converted into products.

Degree of completion The percent or fraction of the limiting reactant converted into products.

Excess reactant All reactants other than limiting reactants.

Extent of reaction The mole of reactions that occur according to the chemical reaction equation.

Limiting reactant The species in a chemical reaction that would theoretically run out first (would be completely consumed) if the reaction were to proceed to completion according to the chemical equation, even if the reaction did not take place.

Selectivity The ratio of the moles of a particular (usually the desired) product produced to the moles of another (usually undesired or by-product) product produced in a set of reactions.

Stoichiometric coefficient Indicates the relative amounts of moles of chemical species that react and are produced in a chemical reaction.

Stoichiometric ratio Mole ratio obtained by using the coefficients of the species in the chemical equation, including both reactants and products.

Stoichiometry Concerns calculations about the moles and masses of reactants and products involved in a chemical reaction(s).

Yield Based on feed: The amount (mass or moles) of desired product obtained divided by amount of the key (frequently the limiting) reactant fed.

Based on reactant consumed: The amount (mass or moles) of desired product obtained divided by the amount of the key (frequently the limiting) reactant consumed.

Based on theory: The amount (mass or moles) of a product obtained divided by the theoretical (expected) amount of the product that would be obtained based on the limiting reactant in the chemical reaction equation(s) being completely consumed.

Supplementary References

Atkins, P. W. *Physical Chemistry,* 6th ed., Oxford University Press, Oxford, UK (1998).

Atkins, P. W., and L. L. Jones. *Chemistry, Molecules, Matter, and Change,* Freeman, New York (1997).

Brady, J. E. *Liftoff! Chemistry,* Ehrlich Multimedia, John Wiley, New York (1996).

Croce, A. E. "The Application of the Concept of Extent of Reaction," *J. Chem. Educ.,* **79**, 506–9 (2002).

de Nevers, N. *Physical and Chemical Equilibrium for Processes,* Wiley-Interscience, New York (2002).

Kotz, J. C., and P. Treichel. *Chemistry and Chemical Reactivity,* Saunders, Fort Worth, TX (1996).

Moulijin, J. A., M. Markkee, and A. Van Diepen. *Chemical Process Technology,* Wiley, New York (2001).

Peckham, G. D. "The Extent of Reaction—Some Nuts and Bolts," *J. Chem. Educ.,* **78**, 508–10 (2001).

Problems

Section 5.1 Stoichiometry

***5.1.1** $BaCl_2 + Na_2SO_4 \rightarrow BaSO_4 + 2NaCl$

a. How many grams of barium chloride will be required to react with 5.00 g of sodium sulfate?

b. How many grams of barium chloride are required for the precipitation of 5.00 g of barium sulfate?

c. How many grams of barium chloride are needed to produce 5.00 g of sodium chloride?
d. How many grams of sodium sulfate are necessary for the precipitation of 5.00 g of barium chloride?
e. How many grams of sodium sulfate have been added to barium chloride if 5.00 g of barium sulfate is precipitated?
f. How many pounds of sodium sulfate are equivalent to 5.00 lb of sodium chloride?
g. How many pounds of barium sulfate are precipitated by 5.00 lb of barium chloride?
h. How many pounds of barium sulfate are precipitated by 5.00 lb of sodium sulfate?
i. How many pounds of barium sulfate are equivalent to 5.00 lb of sodium chloride?

***5.1.2** $AgNO_3 + NaCl \rightarrow AgCl + NaNO_3$

a. How many grams of silver nitrate are required to react with 5.00 g of sodium chloride?
b. How many grams of silver nitrate are required for the precipitation of 5.00 g of silver chloride?
c. How many grams of silver nitrate are equivalent to 5.00 g of sodium nitrate?
d. How many grams of sodium chloride are necessary for the precipitation of the silver of 5.00 g of silver nitrate?
e. How many grams of sodium chloride have been added to silver nitrate if 5.00 g of silver chloride are precipitated?
f. How many pounds of sodium chloride are equivalent to 5.00 lb of sodium nitrate?
g. How many pounds of silver chloride are precipitated by 5.00 lb of silver nitrate?
h. How many pounds of silver chloride are precipitated by 5.00 lb of sodium chloride?
i. How many pounds of silver chloride are equivalent to 5.00 lb of silver nitrate?

***5.1.3** Balance the following reactions (find the values of a_i):

a. $a_1As_2S_3 + a_2H_2O + a_3HNO_3 \rightarrow a_4NO + a_5H_3ASO_4 + a_6H_2SO_4$
b. $a_1KClO_3 + a_2HCl \rightarrow a_3KCl + a_4ClO_2 + a_5Cl_2 + a_6H_2O$

***5.1.4** The formula for vitamin C is as follows:

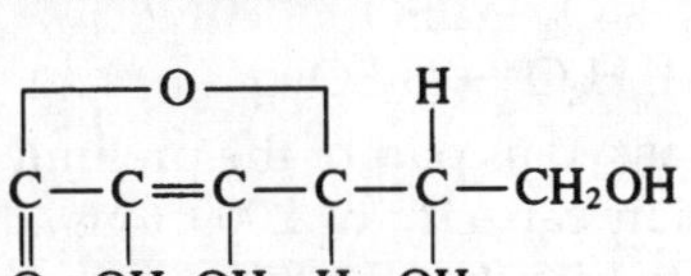

Figure P5.1.4

How many pounds of this compound are contained in 2 g mol?

***5.1.5** Acidic residue in paper from the manufacturing process causes paper based on wood pulp to age and deteriorate. To neutralize the paper, a vapor-phase treatment must employ a compound that would be volatile enough to permeate the fibrous structure of paper within a mass of books but that would have a chemistry that could be manipulated to yield a mildly basic and essentially nonvolatile compound. George Kelly and John Williams successfully attained this objective in 1976 by designing a mass deacidification process employing gaseous diethyl zinc (DEZ).

At room temperature, DEZ is a colorless liquid. It boils at 117°C. When it is combined with oxygen, a highly exothermic reaction takes place:

$$(C_2H_5)_2Zn + 7O_2 \rightarrow ZnO + 4CO_2 + 5H_2O$$

Because liquid DEZ ignites spontaneously when exposed to air, a primary consideration in its use is the exclusion of air. In one case a fire caused by DEZ ruined the neutralization center.

Is the equation shown balanced? If not, balance it. How many kilograms of DEZ must react to form 1.5 kg of ZnO? If 20 cm^3 of water are formed on reaction, and the reaction was complete, how many grams of DEZ reacted?

****5.1.6** The following reaction was carried out:

$$Fe_2O_3 + 2X \rightarrow 2Fe + X_2O_3$$

It was found that 79.847 g of Fe_2O_3 reacted with X to form 55.847 g of Fe and 50.982 g of X_2O_3. Identify the element X.

****5.1.7** A combustion device was used to determine the empirical formula of a compound containing only carbon, hydrogen, and oxygen. A 0.6349 g sample of the unknown produced 1.603 g of CO_2 and 0.2810 g of H_2O. Determine the empirical formula of the compound.

****5.1.8** A hydrate is a crystalline compound in which the ions are attached to one or more water molecules. We can dry these compounds by heating them to get rid of the water. You have a 10.407 g sample of hydrated barium iodide. The sample is heated to drive off the water. The dry sample has a mass of 9.520 g. What is the mole ratio between barium iodide, BaI_2, and water, H_2O? What is the formula of the hydrate?

****5.1.9** Sulfuric acid can be manufactured by the contact process according to the following reactions:

1. $S + O_2 \rightarrow SO_2$
2. $2SO_2 + O_2 \rightarrow 2SO_3$
3. $SO_3 + H_2O \rightarrow H_2SO_4$

You are asked as part of the preliminary design of a sulfuric acid plant with a production capacity of 2000 tons/day of 66° Be (Baumé) (93.2% H_2SO_4 by weight) to calculate the following:

a. How many tons of pure sulfur are required per day to run this plant?
b. How many tons of oxygen are required per day?
c. How many tons of water are required per day for reaction 3?

**5.1.10 Seawater contains 65 ppm of bromine in the form of bromides. In the Ethyl-Dow recovery process, 0.27 lb of 98% sulfuric acid is added per ton of water, together with the theoretical Cl_2 for oxidation; finally, ethylene (C_2H_4) is united with the bromine to form $C_2H_4Br_2$. Assuming complete recovery and using a basis of 1 lb of bromine, find the weights of the 98% sulfuric acid, chlorine, seawater, and ethane dibromide involved.

$$2Br^- + Cl_2 \rightarrow 2Cl^- + Br_2$$

$$Br_2 + C_2H_4 \rightarrow C_2H_4Br_2$$

**5.1.11

BID EVALUATION
TO: *J. Coadwell* DEPT: *Water Waste Water* DATE: 9–29
BID INVITATION: 0374-AV
REQUISITION: 135949 COMMODITY: *Ferrous Sulfate*
DEPARTMENT EVALUATION COMMENTS
It is recommended that the bid from VWR of $83,766.25 for 475 tons of Ferrous Sulfate Heptahydrate be accepted as they were the low bidder for this product as delivered. It is further recommended that we maintain the option of having this product delivered either by rail in a standard carload of 50 tons or by the alternate method by rail in piggy-back truck trailers.

What would another company have to bid to match the VWR bid if the bid they submitted was for ferrous sulfate ($FeSO_4 \cdot H_2O$)? For ($FeSO_4 \cdot 4H_2O$)?

**5.1.12 Three criteria must be met if a fire is to occur: (1) There must be fuel present; (2) there must be an oxidizer present; and (3) there must be an ignition source. For most fuels, combustion takes place only in the gas phase. For example, gasoline does not burn as a liquid. However, when gasoline is vaporized, it burns readily.

A minimum concentration of fuel in air exists that can be ignited. If the fuel concentration is less than this lower flammable limit (LFL) concentration, ignition will not occur. The LFL can be expressed as a volume percent, which is equal to the mole percent under conditions at which the LFL is measured (atmospheric pressure and 25°C). There is also a minimum oxygen concentration required for ignition of any fuel. It is closely related to the LFL and can be calculated from the LFL. The minimum oxygen concentration required for ignition can be estimated by multiplying the LFL concentration by the ratio of the number of moles of oxygen required for complete combustion to the number of moles of fuel being burned.

Above the LFL, the amount of energy required for ignition is quite small. For example, a spark can easily ignite most flammable mixtures. There is also a fuel concentration called the upper flammable limit (UFL) above which the fuel-air mixture cannot be ignited. Fuel-air mixtures in the flammable concentration region between the LFL and the UFL can be ignited. Both the LFL and the UFL have been measured for most of the common flammable gases and volatile liquids. The LFL is usually the more important of the flammability concentrations because if a fuel is present in the atmosphere in concentrations above the UFL, it will certainly be present within the flammable concentration region at some location. LFL concentrations for many materials can be found in the NFPA Standard 325M, *Properties of Flammable Liquids*, published by the National Fire Protection Association.

Estimate the minimum permissible oxygen concentration for *n*-butane. The LFL concentration for *n*-butane is 1.9 mol %. This problem was originally based on a problem in the text *Chemical Process Safety: Fundamentals with Applications*, by D. A. Crowl and J. F. Louvar, published by Prentice Hall, Englewood Cliffs, NJ, and has been adapted from problem 10 of the AIChE publication *Safety, Health, and Loss Prevention in Chemical Processes* by J. R. Welker and C. Springer, New York (1990).

****5.1.13** In a paper mill, soda ash (Na_2CO_3) can be added directly in the causticizing process to form, on reaction with calcium hydroxide, caustic soda (NaOH) for pulping. The overall reaction is $Na_2CO_3 + Ca(OH)_2 \rightarrow 2NaOH + CaCO_3$. Soda ash also may have potential in the on-site production of precipitated calcium carbonate, which is used as a paper filler. The chloride in soda ash (which causes corrosion of equipment) is 40 times less than in regular-grade caustic soda (NaOH), which can also be used; hence the quality of soda ash is better for pulp mills. However, a major impediment to switching to soda ash is the need for excess causticization capacity, generally not available at older mills.

Severe competition exists between soda ash and caustic soda produced by electrolysis. Average caustic soda prices are about \$265 per metric ton FOB (free on board, i.e., without charges for delivery to or loading on carrier), while soda ash prices are about \$130/metric ton FOB.

To what value would caustic soda prices have to drop in order to meet the price of \$130/metric ton based on an equivalent amount of NaOH?

*****5.1.14** A plant makes liquid CO_2 by treating dolomitic limestone with commercial sulfuric acid. The dolomite analyzes 68.0% $CaCO_3$, 30.0% $MgCO_3$, and 2.0% SiO_2; the acid is 94% H_2SO_4 and 6% H_2O. Calculate (a) pounds of CO_2 produced per ton of dolomite treated (b) pounds of acid used per ton of dolomite treated.

*****5.1.15** A hazardous waste incinerator has been burning a certain mass of dichlorobenzene ($C_6H_4Cl_2$) per hour, and the HCl produced was neutralized with soda ash (Na_2CO_3). If the incinerator switches to burning an equal mass of mixed tetrachlorobiphenyls ($C_{12}H_6Cl_4$), by what factor will the consumption of soda ash be increased?

Section 5.2 Terminology for Reaction Systems

*5.2.1 Odors in wastewater are caused chiefly by the products of the anaerobic reduction of organic nitrogen- and sulfur-containing compounds. Hydrogen sulfide is a major component of wastewater odors; however, this chemical is by no means the only odor producer since serious odors can also result in its absence. Air oxidation can be used to remove odors, but chlorine is the preferred treatment because it not only destroys H_2S and other odorous compounds, but it also retards the growth of bacteria that cause the compounds in the first place. As a specific example, HOCl reacts with H_2S as follows in low-pH solutions:

$$HOCl + H_2S \rightarrow S + HCl + H_2O$$

If the actual plant practice calls for 100% excess HOCl (to make sure of the destruction of the H_2S because of the reaction of HOCl with other substances), how much HOCl (5% solution) must be added to 1 L of a solution containing 50 ppm H_2S?

*5.2.2 Phosgene gas is probably most famous for being the first toxic gas used offensively in World War I, but it is also used extensively in the chemical processing of a wide variety of materials. Phosgene can be made by the catalytic reaction between CO and chlorine gas in the presence of a carbon catalyst. The chemical reaction is

$$CO + Cl_2 \rightarrow COCl_2$$

Suppose that you have measured the reaction products from a given reactor and found that they contained 3.00 lb mol of chlorine, 10.00 lb mol of phosgene, and 7.00 lb mol of CO. Calculate the extent of reaction, and using the value calculated, determine the initial amounts of CO and Cl_2 that were used in the reaction.

*5.2.3 In the reaction in which 135 moles of methane and 45.0 moles of oxygen are fed into a reactor, if the reaction goes to completion, calculate the extent of reaction.

$$6CH_4 + O_2 \rightarrow 2C_2H_2 + 2CO + 10H_2$$

*5.2.4 FeS can be roasted in O_2 to form FeO:

$$2FeS + 3O_2 \rightarrow 2FeO + 2SO_2$$

If the slag (solid product) contains 80% FeO and 20% FeS, and the exit gas is 100% SO_2, determine the extent of reaction and the initial number of moles of FeS. Use 100 g or 100 lb as the basis.

**5.2.5 Aluminum sulfate is used in water treatment and in many chemical processes. It can be made by reacting crushed bauxite (aluminum ore) with 77.7 weight percent sulfuric acid. The bauxite ore contains 55.4 weight percent aluminum oxide, the remainder being impurities. To produce crude aluminum sulfate containing 2000 lb of pure aluminum sulfate, 1080 lb of bauxite and 2510 lb of sulfuric acid solution (77.7% acid) are used.

a. Identify the excess reactant.
b. What percentage of the excess reactant was used?
c. What was the degree of completion of the reaction?

**5.2.6 A barite composed of 100% $BaSO_4$ is fused with carbon in the form of coke containing 6% ash (which is infusible). The composition of the fusion mass is

$BaSO_4$	11.1%
BaS	72.8
C	13.9
Ash	2.2
	100.0%

Reaction:

$$BaSO_4 + 4C \rightarrow BaS + 4CO$$

Find the excess reactant, the percentage of the excess reactant, and the degree of completion of the reaction.

**5.2.7 Read problem 5.2.2 again. Suppose that you have measured the reaction products from a given reactor and found that they contained 3.00 kg of chlorine, 10.00 kg of phosgene, and 7.00 kg of CO. Calculate the following:

a. The percent excess reactant used
b. The percentage conversion of the limiting reactant
c. The kilogram moles of phosgene formed per kilogram mole of total reactants fed to the reactor

**5.2.8 The specific activity of an enzyme is defined in terms of the amount of solution catalyzed under a given set of conditions divided by the product of the time interval for the reaction times the amount of protein in the sample:

$$\text{specific activity} = \frac{\mu \text{ mol of solution converted}}{(\text{time interval in minutes})(\text{mg protein in the sample})}$$

A 0.10 mL sample of pure β-galactosidase (β-g) solution that contains 1.00 mg of protein per liter hydrolyzed 0.10 m mol of o-nitrophenyl galactoside (o-n) in 5 min. Calculate the specific activity of the β-g.

**5.2.9 One method of synthesizing the aspirin substitute acetaminophen involves a three-step procedure as outlined in Figure P5.2.9. First, *p*-nitrophenol is catalytically hydrogenated in the presence of aqueous hydrochloric acid to the acid chloride salt of *p*-aminophenol with an 86.9% degree of completion. Next the salt is neutralized to obtain *p*-aminophenol with a 0.95 fractional conversion.

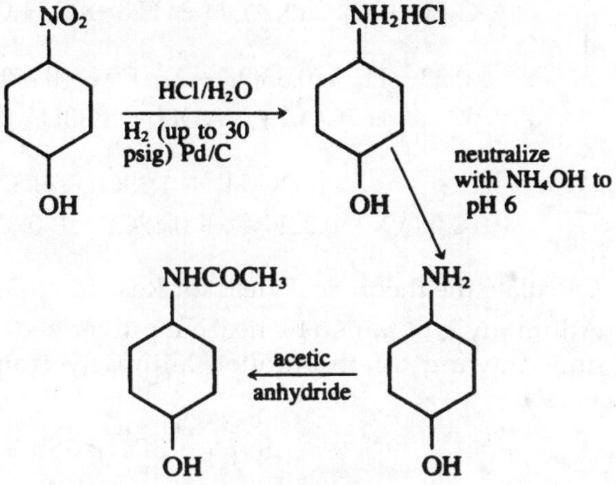

Figure P5.2.9

Finally, the *p*-aminophenol is acetylated by reacting with acetic anhydride, resulting in a yield of 3 kg mol of acetaminophen per 4 kg mol. What is the overall conversion fraction of *p*-nitrophenol to acetaminophen?

****5.2.10** The most economic method of sewage wastewater treatment is bacterial digestion. As an intermediate step in the conversion of organic nitrogen to nitrates, it is reported that the *Nitrosomonas* bacteria cells metabolize ammonium compounds into cell tissue and expel nitrite as a by-product by the following overall reaction:

$$5CO_2 + 55NH_4^+ + 76O_2 \rightarrow C_5H_7O_2N(\text{tissue}) + 54NO_2^- + 52H_2O + 109H^+$$

If 20,000 kg of wastewater containing 5% ammonium ions by weight flows through a septic tank inoculated with the bacteria, how many kilograms of cell tissue are produced, provided that 95% of the NH_4^+ is consumed?

****5.2.11** The overall yield of a product on a substrate in some bioreactions is the absolute value of the production rate divided by the rate of consumption of the feed in the substrate (the liquid containing the cells, nutrients, etc.). The overall chemical reaction for the oxidation of ethylene (C_2H_4) to epoxide (C_2H_4O) is

$$2C_2H_4 + O_2 \rightarrow 2C_2H_4O \qquad \text{(a)}$$

Calculate the theoretical yield (100% conversion C_2H_4) of C_2H_4O in moles per mole for Reaction (a).

The biochemical pathway for the production of epoxide is quite complex. Cofactor regeneration is required, which is assumed to originate by partial further oxidation of the formed epoxide. Thus, the amount of ethylene consumed to produce 1 mole of epoxide is larger than that required by Reaction (a). The following two reactions, (b1) and (b2), when summed approximate the overall pathway:

$$C_2H_4 + O_2 + NADH + H^+ \rightarrow C_2H_4O + H_2O + NAD^+ \qquad \text{(b1)}$$

$$0.33C_2H_4 + 0.33O_2 + NAD^+ + 0.67H_2O + 0.33FAD^+ \rightarrow 0.67CO_2 + NADH + 1.33H^+ + 0.33FADH \qquad \text{(b2)}$$

$$1.33C_2H_4 + 1.33O_2 + 0.33FAD^+ \rightarrow C_2H_4O + 0.33H_2O + 0.67CO_2 + 0.33H^+ + 0.33FADH \qquad \text{(b3)}$$

Calculate the theoretical yield for Reaction (b3) of the epoxide.

***5.2.12 Antimony is obtained by heating pulverized stibnite (Sb_2S_3) with scrap iron and drawing off the molten antimony from the bottom of the reaction vessel:

$$Sb_2S_3 + 3Fe \rightarrow 2Sb + 3FeS$$

Suppose that 0.600 kg of stibnite and 0.250 kg of iron turnings are heated together to give 0.200 kg of Sb metal. Determine

a. The limiting reactant
b. The percentage of excess reactant
c. The degree of completion (fraction)
d. The percent conversion based on Sb_2S_3
e. The yield in kilograms of Sb produced per kilogram of Sb_2S_3 fed to the reactor

***5.2.13 One can view the blast furnace from a simple viewpoint as a process in which the principal reaction is

$$Fe_2O_3 + 3C \rightarrow 2Fe + 3CO$$

but some other undesired side reactions occur, mainly

$$Fe_2O_3 + C \rightarrow 2FeO + CO$$

After 600.0 lb of carbon (coke) are mixed with 1.00 ton of pure iron oxide, Fe_2O_3, the process produces 1200.0 lb of pure iron, 183 lb of FeO, and 85.0 lb of Fe_2O_3. Calculate the following items:

a. The percentage of excess carbon furnished, based on the principal reaction
b. The percentage conversion of Fe_2O_3 to Fe
c. The pounds of carbon used up and the pounds of CO produced per ton of Fe_2O_3 charged
d. The selectivity in this process (of Fe with respect to FeO)

***5.2.14 A common method used in manufacturing sodium hypochlorite bleach is by the reaction

$$Cl_2 + 2NaOH \rightarrow NaCl + NaOCl + H_2O$$

Chlorine gas is bubbled through an aqueous solution of sodium hydroxide, after which the desired product is separated from the sodium chloride (a by-product of the reaction). A water-NaOH solution that contains 1145 lb of pure NaOH is reacted with 851 lb of gaseous chlorine. The NaOCl formed weighs 618 lb.

a. What was the limiting reactant?
b. What was the percentage excess of the excess reactant used?
c. What is the degree of completion of the reaction, expressed as the moles of NaOCl formed to the moles of NaOCl that would have formed if the reaction had gone to completion?
d. What is the yield of NaOCl per amount of chlorine used (on a weight basis)?
e. What was the extent of reaction?

***5.2.15 In a process for the manufacture of chlorine by direct oxidation of HCl with air over a catalyst to form Cl_2 and H_2O (only), the exit product is composed of HCl (4.4%), Cl_2 (19.8%), H_2O (19.8%), O_2 (4.0%), and N_2 (52.0%). What were (a) the limiting reactant; (b) the percent excess reactant; (c) the degree of completion of the reaction; and (d) the extent of reaction?

***5.2.16 A well-known reaction to generate hydrogen from steam is the so-called water gas shift reaction: $CO + H_2O \rightarrow CO_2 + H_2$. If the gaseous feed to a reactor consists of 30 moles of CO per hour, 12 moles of CO_2* per hour, and 35 moles of steam per hour at 800°C, and 18 moles of H_2 are produced per hour, calculate:

a. The limiting reactant
b. The excess reactant
c. The fraction conversion of steam to H_2
d. The degree of completion of the reaction
e. The kilograms of H_2 yielded per kilogram of steam fed
f. The moles of CO_2 produced by the reaction per mole of CO fed
g. The extent of reaction

***5.2.17 In the production of m-xylene (C_8H_{10}) from mesitylene (C_9H_{12}) over a catalyst, some of the xylene reacts to form toluene (C_7H_8):

$$C_9H_{12} + H_2 \rightarrow C_8H_{10} + CH_4$$

$$C_8H_{10} + H_2 \rightarrow C_7H_8 + CH_4$$

The second reaction is undesirable because m-xylene sells for $0.65/lb, whereas toluene sells for $0.22/lb.

The CH_4 is recycled in the plant. One pound of catalyst is degraded per 500 lb of C_7H_8 produced, and the spent catalyst has to be disposed of in a landfill that handles low-level toxic waste at a cost of $25/lb. If the overall selectivity of C_8H_{10} to C_7H_8 is changed from 0.7 mole of xylene produced per mole of toluene produced to 0.8 by changing the residence time in the reactor, what is the gain or loss in dollars per 100 lb of mesitylene reacted?

Section 5.3 Species Mole Balances

*5.3.1 Pure A in gas phase enters a reactor. Fifty percent (50%) of this A is converted to B through the reaction A → 3B. What is the mole fraction of A in the exit stream? What is the extent of reaction?

**5.3.2 A low-grade pyrite containing 32% S is mixed with 10 lb of pure sulfur per 100 lb of pyrites so the mixture will burn readily with air, forming a burner gas that analyzes 13.4% SO_2, 2.7% O_2, and 83.9% N_2. No sulfur is left in the cinder. Calculate the percentage of the sulfur fired that burned to SO_3. (The SO_3 is not detected by the analysis.)

**5.3.3 Examine the reactor in Figure P5.3.3. Your boss says something has gone wrong with the yield of CH_2O, and it is up to you to find out what the problem is. You start by making material balances (naturally!). Show all calculations. Is there some problem?

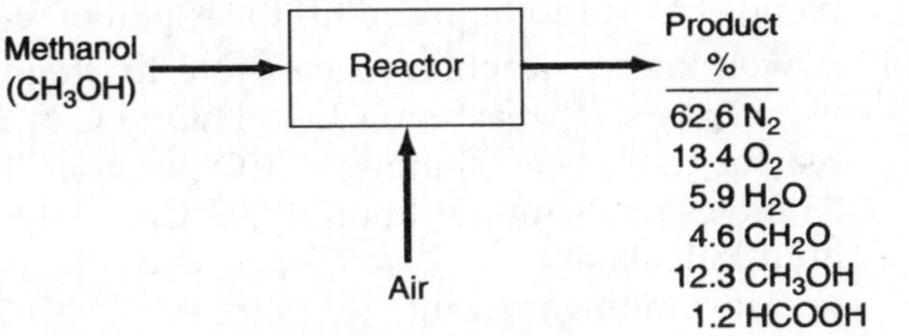

Figure P5.3.3

**5.3.4 A problem statement was the following:

> A dry sample of limestone is completely soluble in HCl and contains no Fe or Al. When a 1.000 g sample is ignited, the loss in weight is found to be 0.450 g. Calculate the percent $CaCO_3$ and $MgCO_3$ in the limestone.

The solution was

$$\frac{x}{84.3} + \frac{(1.000 - x)}{100} = \frac{0.450}{44.}$$

$$100x + 84.3 - 84.3\,x = (0.450)(84.3)(100)/44$$

$$x = 0.121 \quad MgCO_3 = 12.1\%$$

$$CaCO_3 = 87.9\%$$

Answer the following questions:

a. What information in addition to that in the problem statement had to be obtained?

b. What would a diagram for the process look like?
c. What was the basis for the problem solution?
d. What were the known variables in the problem statement, their values, and their units?
e. What were the unknown variables in the problem statement and their units?
f. What are the types of material balances that could be made for this problem?
g. What type(s) of material balance was made for this problem?
h. What was the degree of freedom for this problem?
i. Was the solution correct?

***5.3.5 One of the most common commercial methods for the production of pure silicon that is to be used for the manufacture of semiconductors is the Siemens process (see Figure P5.3.5) of chemical vapor deposition (CVD). A chamber contains a heated silicon rod and a mixture of high-purity trichlorosilane mixed with high-purity hydrogen that is passed over the rod. Pure silicon (EGS—electronic grade silicon) deposits on the rod as a polycrystalline solid. (Single crystals of Si are later made by subsequently melting the EGS and drawing a single crystal from the melt.) The reaction is

$$H_2(g) + SiHCl_3(g) \rightarrow Si(s) + 3HCl(g)$$

The rod initially has a mass of 1460 g, and the mole fraction of H_2 in the exit gas is 0.223. The mole fraction of H_2 in the feed to the reactor is 0.580, and the feed enters at the rate of 6.22 kg mol/hr. What will be the mass of the rod at the end of 20 min?

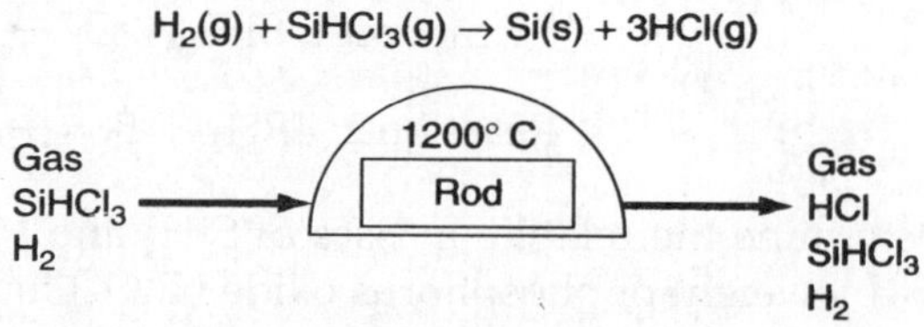

Figure P5.3.5

***5.3.6 Copper as CuO can be obtained from an ore called Covellite which is composed of CuS and gange (inert solids). Only part of the CuS is oxidized with air to CuO. The gases leaving the roasting process analyze SO_2 (7.2%), O_2 (8.1%), and N_2 (84.7%). Unfortunately, the method of gas analysis could not detect SO_3 in the exit gas, but SO_3 is known to exist.

Calculate the percent of the sulfur in the part of the CuS that reacts that forms SO_3. Hint: You can consider the unreacted CuS as a compound that comes in and out of the process untouched and thus is isolated from the process and can be ignored.

***5.3.7 A reactor is used to remove SiO_2 from a wafer in semiconductor manufacturing by contacting the SiO_2 surface with HF. The reactions are

$$6HF(g) + SiO_2(s) \rightarrow H_2SiF_6(l) + 2H_2O(l)$$
$$H_2SiF_6(l) \rightarrow SiF_4(g) + 2HF(g)$$

Assume the reactor is loaded with wafers having a silicon oxide surface, a flow of 50% HF and 50% nitrogen is started, and all the H_2SiF_6 reacts.

In the reaction 10% of the HF is consumed. What is the composition of the exhaust stream?

***5.3.8 In the anaerobic fermentation of grain, the yeast *Saccharomyces cerevisiae* digests glucose from plants to form the products ethanol and propenoic acid by the following overall reactions:

$$\text{Reaction 1: } C_6H_{12}O_6 \rightarrow 2C_2H_5OH + 2CO_2$$
$$\text{Reaction 2: } C_6H_{12}O_6 \rightarrow 2C_2H_3CO_2H + 2H_2O$$

In an open flow reactor 3500 kg of a 12% glucose-water solution flow in. During fermentation, 120 kg of carbon dioxide are produced together with 90 kg of unreacted glucose. What are the weight percents of ethyl alcohol and propenoic acid that exit in the broth? Assume that none of the glucose is assimilated into the bacteria.

***5.3.9 Semiconductor microchip processing often involves chemical vapor deposition (CVD) of thin layers. The material being deposited needs to have certain desirable properties. For instance, to overlay on aluminum or other bases, a phosphorus-pentoxide-doped silicon dioxide coating is deposited as a passivation (protective) coating by the simultaneous reactions

$$\text{Reaction 1: } SiH_4 + O_2 \rightarrow SiO_2 + 2H_2$$
$$\text{Reaction 2: } 4PH_3 + 5O_2 \rightarrow 2P_2O_5 + 6H_2$$

Determine the relative masses of SiH_4 and PH_3 required to deposit a film of 5% by weight of phosphorus oxide (P_2O_5) in the protective coating.

***5.3.10 Printed circuit boards are used in the electronics industry to both connect and hold components in place. In production, 0.03 in. of copper foil is laminated to an insulating plastic board. A circuit pattern made of a chemically resistant polymer is then printed on the board. Next, the unwanted copper is chemically etched away by using selected reagents. If copper is treated with $Cu(NH_3)_4Cl_2$ (cupric ammonium chloride) and NH_4OH (ammonium hydroxide), the products are water and $Cu(NH_3)_4Cl$ (cuprous ammonium chloride). Once the copper is dissolved, the polymer is removed by solvents, leaving the printed circuit ready for further processing. If a single-sided board 4 in. by 8 in. is to have 75% of the copper layer removed using these reagents, how many grams of each reagent will be consumed? Data: The density of copper is 8.96 g/cm^3.

***5.3.11 The thermal destruction of hazardous wastes involves the controlled exposure of waste to high temperatures (usually 900°C or greater) in an oxidizing environment. Types of thermal destruction equipment include high-temperature boilers, cement kilns, and industrial furnaces in which hazardous waste is burned as fuel. In a properly designed system, primary fuel (100% combustible material) is mixed with waste to produce a feed for the boiler.

a. Sand containing 30% by weight of 4,4′-dichlorobiphenyl [an example of a polychlorinated biphenyl (PCB)] is to be cleaned by combustion with excess hexane to produce a feed that is 60% combustible by weight. To decontaminate 8 tons of such contaminated sand, how many pounds of hexane would be required?
b. Write the two reactions that would take place under ideal conditions if the mixture of hexane and the contaminated sand were fed to the thermal oxidation process to produce the most environmentally satisfactory products. How would you suggest treating the exhaust from the burner? Explain.
c. The incinerator is supplied with an oxygen-enriched airstream containing 40% O_2 and 60% N_2 to promote high-temperature operation. The exit gas is found to have a composition of $x_{CO_2} = 0.1654$ and $x_{O_2} = 0.1220$. Use this information and the data about the feed composition to find (a) the complete exit gas concentrations and (b) the percent excess O_2 used in the reaction.

****5.3.12 In order to neutralize the acid in a waste stream (composed of H_2SO_4 and H_2O), dry ground limestone (composition 95% $CaCO_3$ and 5% inerts) is mixed in. The dried sludge collected from the process is only partly analyzed by firing it in a furnace, which results in only CO_2 being driven off. By weight the CO_2 represents 10% of the dry sludge. What percent of the pure $CaCO_3$ in the limestone did not react in the neutralization? Solve this problem using mole balances.

Section 5.4 Element Material Balances

****5.4.1 In order to neutralize the acid in a waste stream (composed of H_2SO_4 and H_2O), dry ground limestone (composition 95% $CaCO_3$ and 5% inerts) is mixed in. The dried sludge collected from the process is only partly analyzed by firing it in a furnace, which results in only CO_2 being driven off. By weight the CO_2 represents 10% of the dry sludge. What percent of the pure $CaCO_3$ in the limestone did not react in the neutralization? Solve this problem using element balances.

Section 5.5 Material Balances Combustion Systems

**5.5.1 A synthesis gas analyzing 6.4% CO_2, 0.2% O_2, 40.0% CO, and 50.8% H_2 (the balance is N_2) is burned with 40% dry excess air. What is the composition of the flue gas?

**5.5.2 Thirty pounds of coal (analysis 80% C and 20% H, ignoring the ash) are burned with 600 lb of air, yielding a gas having an Orsat analysis in which the ratio of CO_2 to CO is 3 to 2. What is the percent excess air?

**5.5.3 A gas containing only CH_4 and N_2 is burned with air yielding a flue gas that has an Orsat analysis of CO_2 8.7%, CO 1.0%, O_2 3.8%, and N_2 86.5%. Calculate the percent excess air used in combustion and the composition of the CH_4N_2 mixture.

**5.5.4 A natural gas consisting entirely of methane (CH_4) is burned with an oxygen-enriched air of composition 40% O_2 and 60% N_2. The Orsat analysis of the product gas as reported by the laboratory is CO_2 20.2%, O_2 4.1%, and N_2 75.7%. Can the reported analysis be correct? Show all calculations.

**5.5.5 Dry coke composed of 4% inert solids (ash), 90% carbon, and 6% hydrogen is burned in a furnace with dry air. The solid refuse left after combustion contains 10% carbon and 90% inert ash (and no hydrogen). The inert ash content does not enter into the reaction.

The Orsat analysis of the flue gas gives 13.9% CO_2, 0.8% CO, 4.3% O_2, and 81.0% N_2. Calculate the percent of excess air based on complete combustion of the coke.

**5.5.6 A gas with the following composition is burned with 50% excess air in a furnace. What is the composition of the flue gas?

$$CH_4\ 60\%,\ C_2H_6\ 20\%,\ CO\ 5\%,\ O_2\ 5\%,\ N_2\ 10\%$$

**5.5.7 In underground coal combustion in the gas phase several reactions take place, including

$$CO + 1/2O_2 \rightarrow CO_2$$

$$H_2 + 1/2O_2 \rightarrow H_2O$$

$$CH_4 + 3/2O_2 \rightarrow CO + 2H_2O$$

where the CO, H, and CH_4 come from coal pyrolysis.

If a gas phase composed of CO 13.54%, CO_2 15.22%, H_2 15.01%, CH_4 3.20%, and the balance N_2 is burned with 40% excess air, (a) how much air is needed per 100 moles of gas, and (b) what will be the analysis of the product gas on a wet basis?

**5.5.8 Solvents emitted from industrial operations can become significant pollutants if not disposed of properly. A chromatographic study of the waste exhaust gas from a synthetic fiber plant has the following analysis in mole percent:

CS	40%
SO	10%
HO	50%

It has been suggested that the gas be disposed of by burning with an excess of air. The gaseous combustion products are then emitted to the air through a smokestack. The local air pollution regulations say that no stack gas is to analyze more than 2% SO_2 by an Orsat analysis averaged over a 24 hr period. Calculate the minimum percent excess air that must be used to stay within this regulation.

****5.5.9** The products and by-products from coal combustion can create environmental problems if the combustion process is not carried out properly. Your boss asks you to carry out an analysis of the combustion in boiler No. 6. You carry out the work assignment using existing instrumentation and obtain the following data:

Fuel analysis (coal): 74% C, 14% H, and 12% ash
Flue gas analysis on a dry basis: 12.4% CO_2, 1.2% CO, 5.7% O_2, and 80.7% N_2

What are you going to report to your boss?

****5.5.10** The Clean Air Act requires automobile manufacturers to warrant their control systems as satisfying the emission standards for 50,000 mi. It requires owners to have their engine control systems serviced exactly according to manufacturers' specifications and to always use the correct gasoline. In testing an engine exhaust having a known Orsat analysis of 16.2% CO_2, 4.8% O_2, and 79% N_2 at the outlet, you find to your surprise that at the end of the muffler the Orsat analysis is 13.1% CO_2. Can this discrepancy be caused by an air leak into the muffler? (Assume that the analyses are satisfactory.) If so, compute the moles of air leaking in per mole of exhaust gas leaving the engine.

****5.5.11** One of the products of sewage treatment is sludge. After microorganisms grow in the activated sludge process to remove nutrients and organic material, a substantial amount of wet sludge is produced. This sludge must be dewatered, one of the most expensive parts of most treatment plant operations.

How to dispose of the dewatered sludge is a major problem. Some organizations sell dried sludge for fertilizer, some spread the sludge on farmland, and in some places it is burned. To burn a dried sludge, fuel oil is mixed with it, and the mixture is burned in a furnace with air. If you collect the following analysis for the sludge and for the product gas:

Sludge (%)		Product Gas (%)	
S	32	SO_2	1.52
C	40	CO_2	10.14
H_2	4	O_2	4.65
O_2	24	N_2	81.67
		CO	2.02

a. Determine the weight percent of carbon and hydrogen in the fuel oil.
b. Determine the ratio of pounds of dry sludge to pounds of fuel oil in the mixture fed to the furnace.

****5.5.12** Many industrial processes use acids to promote chemical reactions or produce acids from the chemical reactions occurring in the process. As a result, these acids many times end up in the wastewater stream from the process and must be neutralized as part of the wastewater treatment process before the water can be discharged from the process. Lime (CaO) is a cost-effective neutralization agent for acid wastewater. Lime is dissolved in water by the following reaction:

$$CaO + 1/2O_2 \rightarrow Ca(OH)_2$$

which reacts directly with acid; for example, for H_2SO_4,

$$H_2SO_4 + Ca(OH)_2 \rightarrow CaSO_4 + 2H_2O$$

Consider an acidic wastewater stream with a flow rate of 1000 gal/min with an acid concentration of 2% H_2SO_4. Determine the flow rate of lime in pounds per minute necessary to neutralize the acid in this stream if 20% excess lime is used. Calculate the production rate of $CaSO_4$ from this process in tons per year. Assume that the specific gravity of the acidic wastewater stream is 1.05.

****5.5.13** Nitric acid (HNO_3) that is used industrially for a variety of reactions can be produced by the reaction of ammonia (NH_3) with air by the following overall reaction:

$$NH_3 + 2O_2 \rightarrow HNO_3 + H_2O$$

The product gas from such a reactor has the following composition (on a water-free basis):

0.8% NH_3

9.5% HNO_3

3.8% O_2

85.9% N_2

Determine the percent conversion of NH_3 and the percent excess air used.

****5.5.14** Ethylene oxide (C_2H_4O) is a high-volume chemical intermediate that is used to produce glycol and polyethylene glycol. Ethylene oxide is produced by the partial oxidation of ethylene (C_2H_4) using a solid catalyst in a fixed-bed reactor:

$$C_2H_4 + {}^1\!/_2O_2 \rightarrow C_2H_4O$$

In addition, a portion of the ethylene reacts completely to form CO_2 and H_2O:

$$C_2H_4 + 3O_2 \rightarrow 2CO_2 + 2H_2O$$

The product gas leaving a fixed-bed ethylene oxide reactor has the following water-free composition: 20.5% C_2H_4O, 72.7 N_2, 2.3 O_2, and 4.5% CO_2. Determine the percent excess air based in the desired reaction, and the pounds per hour of ethylene feed required to produce 100,000 ton/yr of ethylene oxide.

***5.5.15 A flare is used to convert unburned gases to innocuous products such as CO_2 and H_2O. If a gas of the following composition (in percent) is burned in the flare—CH_4 70%, C_3H_8 5%, CO 15%, O_2 5%, N_2 5%—and the flue gas contains 7.73% CO_2, 12.35% H_2O, and the balance is O_2 and N_2, what was the percent excess air used?

***5.5.16 Hydrogen-free carbon in the form of coke is burned (a) with complete combustion using theoretical air, (b) with complete combustion using 50% excess air, or (c) using 50% excess air but with 10% of the carbon burning to CO only. In each case calculate the gas analysis that will be found by testing the flue gases on a dry basis.

***5.5.17 Ethanol (CH_3CH_2OH) is dehydrogenated in the presence of air over a catalyst, and the following reactions take place:

$$CH_3CH_2OH \rightarrow CH_3CHO + H_2$$

$$2CH_3CH_2OH + 3O_2 \rightarrow 4CO_2 + 6H_2$$

$$2CH_3CH_2OH + 2H_2 \rightarrow 4CH_4 + O_2$$

Separation of the product, CH_3CHO (acetaldehyde), as a liquid leaves an output gas with the following Orsat analysis: CO_2 0.7%, O_2 2.1%, CO 2.3%, H_2 7.1%, CH_4 2.6%, and N_2 85.2%. How many kilograms of acetaldehyde are produced per kilogram of ethanol fed into the process?

***5.5.18 Refer to Example 5.14. Suppose that during combustion a very small amount (0.24%) of the entering nitrogen reacts with oxygen to form nitrogen oxides (NO_x). Also, suppose that the CO produced is 0.18% and the SO_2 is 1.4% of the $CO_2 + SO_2$ in the flue gas. The emissions listed by the EPA in the load units (ELU) per kilogram of gas are

NO_x	0.22
CO	0.27
CO_2	0.09
SO_2	0.10

What is the total ELU for the stack gas? Note: The ELU are additive.

****5.5.19 Glucose ($C_6H_{12}O_6$) and ammonia form a sterile solution (no live cells) fed continuously into a vessel. Assume that glucose and ammonia are fed in stoichiometric proportions and that they react completely. One product formed from the reaction contains ethanol, cells ($CH_{1.8}O_{0.5}N_{0.2}$), and water. The gas produced is CO_2. If the reaction occurs anaerobically (without the presence of oxygen), what is the minimum amount in kilograms of feed (ammonia and glucose) required to produce 4.6 kg of ethanol? Only 60% of the moles of glucose are converted to ethanol. The remainder is converted to cell mass, carbon dioxide, and water.

CHAPTER 6

Material Balances for Multi-Unit Systems

Your objectives in studying this chapter are to be able to

1. Write a set of independent material balances for a process involving more than one unit
2. Apply the ten-step strategy to solve multi-unit steady-state problems (with and without chemical reactions) involving sequential, recycle, and/or bypass, and/or purge streams
3. Solve problems by hand in which a modest number of interconnected units are involved by making appropriate balances
4. Use the overall conversion and single-pass (once-through) conversion concepts in solving recycle problems involving reactors
5. Explain the purpose of a recycle stream, a bypass stream, and a purge stream
6. Understand in a general sense how material balances are used in industry

Looking Ahead

In this chapter we consider material balances applied to systems with multiple units, including a sequential arrangement of units, systems with recycle (i.e., instances in which material is returned to a process from downstream), and systems with purge and bypass. In addition, we comment on the application of material balances in industrial systems.

6.1 Primary Concepts

There are nine and sixty ways of constructing tribal lays,
and every single one of them is right.
Rudyard Kipling

A **process flowsheet (flowchart)** is a graphical representation of a process. A flowsheet describes the actual process in sufficient detail that you can use it to formulate material (and energy) balances. Flowsheets are also used for troubleshooting, control of operating conditions, and optimization of process performance. You will find that flowsheets are also prepared for proposed processes that involve new techniques or modifications of existing processes.

Figure 6.1 Section of a large ammonia plant showing the equipment in place

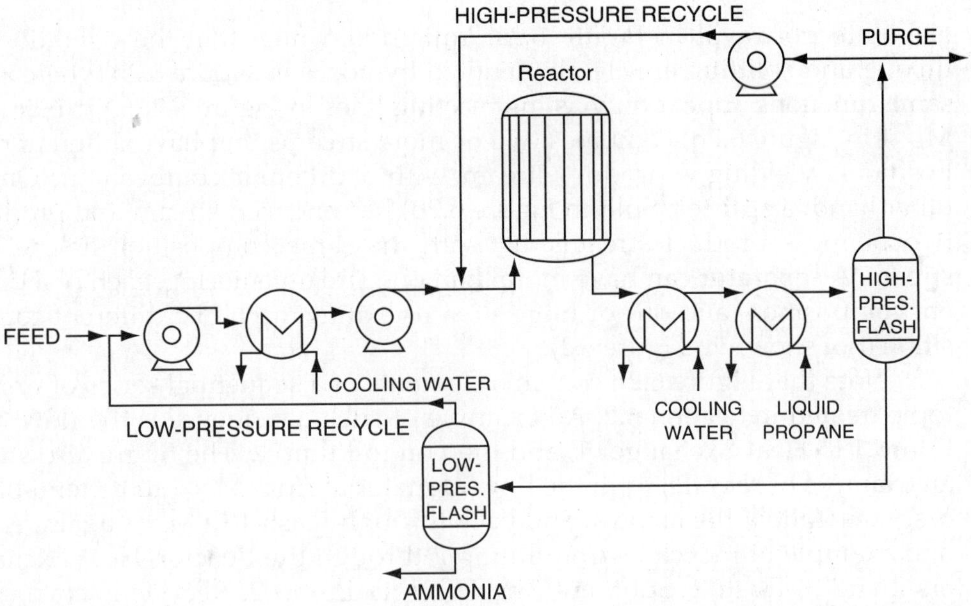

a.

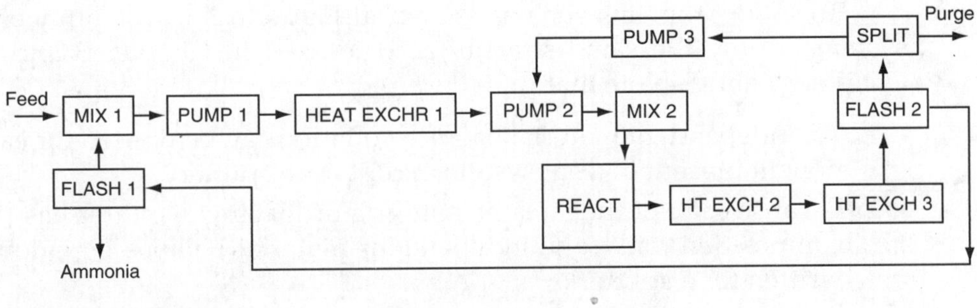

b.

Figure 6.2 (a) Flowsheet of the ammonia plant that includes major pieces of equipment and shows the materials flow; (b) Block diagram corresponding to Figure 6.2a

Figure 6.1 is a picture of a section of an ammonia plant. Figure 6.2a is a flowsheet of the process indicating the equipment sequence and the flow of materials. Figure 6.2b is a **block diagram** corresponding to Figure 6.2a. The units appear as simple boxes called **subsystems** rather than as the more elaborate portrayal in Figure 6.2a. You should note that the operations of mixing and splitting are clearly denoted by boxes in Figure 6.2b whereas the same functions appear only as intersecting lines in Figure 6.2a. A **mixer** (e.g., Mix 1 in Figure 6.2b) combines two or more streams that have different compositions, yielding a product stream with a different composition. On the other hand, a **splitter** (Split in Figure 6.2b) has one feed stream and produces two or more product streams, all with the same composition as the feed stream. A **separator** can have more than one stream entering, each of a different composition, and one or more streams exiting, each of a different composition (not shown in Figure 6.2).

Note that Figure 6.2b contains elements of a sequential series of units, a topic treated in Section 6.2. An example of such a sequence is the flow from Pump 1 to Heat Exchanger 1, and then on to Pump 2. The figure also shows an example of **recycle**, as in the flow of material from Mix 1 to the end of the diagram (follow the arrows) and back through Flash 1 to Mix 1 again. A second example of recycle is from Pump 2 through the Reactor, Heat Exchangers 2 and 3, Flash 2, Split, and then back to Pump 2. Recycle is covered in Section 6.3. In addition, the figure shows an example of **purge** (removing a small portion of a material to prevent its accumulation in the process) at Split, a topic considered in Section 6.4.

To solve problems such as shown in Figure 6.2 with a number of subsystems, apply the same strategy as discussed in Chapter 3 and used in Chapters 4 and 5. Note that

1. An independent material balance equation can be written for each component present in a subsystem except for a splitter
2. Because each of the streams entering or leaving a splitter has the same composition, only one independent material balance equation can be written for a splitter

Warning! For most problems involving multi-units, the number of pertinent *independent* material balance equations that can be written is considerably lower than the total number of material balance equations that could be written. Therefore, when solving multi-unit problems, be careful to choose appropriate equations to use to ensure that the resulting set of equations is independent. Some of the specifications when assigned to variables and/or some of the implicit equations may cause one or more material balance equations to become redundant. Moreover, if you are going to solve multi-unit problems by hand, you have to carefully choose the order in which you solve the various equations to avoid having to solve too many equations simultaneously.

6.2 Sequential Multi-Unit Systems

Let's first examine a multi-unit system composed of a **sequential combination of units**. Figure 6.3a illustrates a sequential combination of mixing and splitting stages. Streams 1 and 2 combine to form the first mixing point, streams 3 and 4 also combine in the box for the second mixing point, and stream 5 splits (at presumably a pipe junction) into streams 6 and 7. Note that the first mixing point occurs at the combination of streams 1 and 2 (presumably a junction of pipes) while the second mixing occurs where streams 3 and 4 enter the box (presumably representing a process). You will encounter both types of mixing in the problems and examples in this book and in flowsheets in professional practice.

Examine Figure 6.3a. Which streams must have the same composition? Do streams 5, 6, and 7 have the same composition? Yes, because streams 6 and 7 flow from a splitter. Do streams 3, 4, and 5 have the same composition? It's quite unlikely. Stream 5 is some type of average of the compositions of streams 3 and 4 (in the absence of reaction). What is the composition inside the system (the box)? It will have the same composition as stream 5 only if streams 3 and 4 are really **well mixed** in the subsystem represented by the box.

How many material balances can you formulate for the system and subsystems shown in Figure 6.3a? First, let's examine how many total material balances you can write. You can write an **overall total material balance,** namely, a total balance on the system that includes all of the three subsystems within the overall system boundary, which is denoted by the dashed line labeled I in Figure 6.3b.

In addition, you can make a total balance on each of the three subsystems that make up the overall system, as denoted by the boundaries

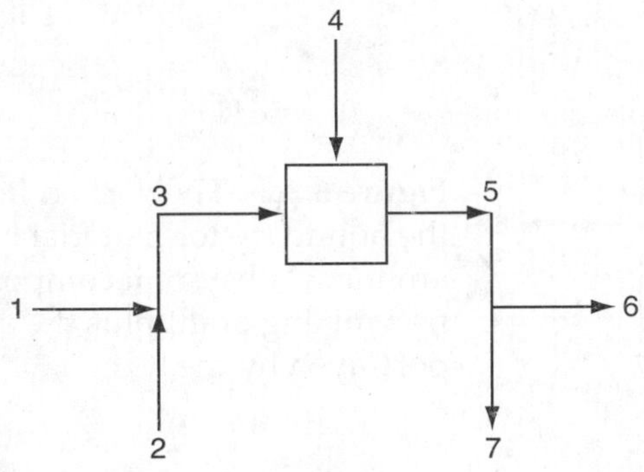

Figure 6.3a Serial mixing and splitting in a process without reaction

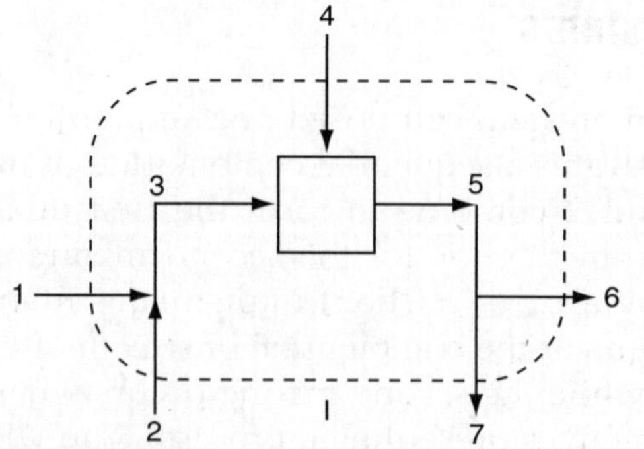

Figure 6.3b The dashed line I designates the boundary for an overall total material balance made on the system in Figure 6.3a.

indicated by the dashed lines II, III, and IV in Figure 6.3c. Finally, you can make a balance about each of the combinations of two subsystems as indicated by the dashed-line boundaries V and VI in Figures 6.3d and 6.3e, respectively.

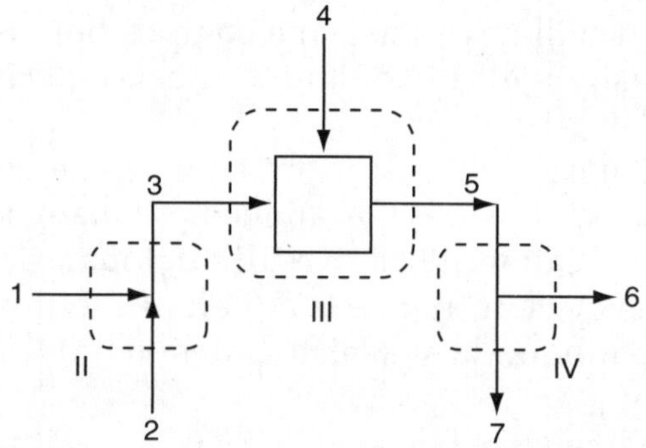

Figure 6.3c Dashed lines II, III, and IV denote the boundaries for material balances around each of the individual units constituting the overall process.

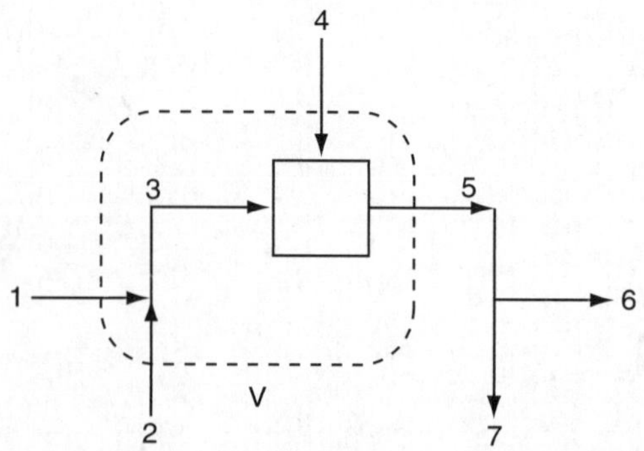

Figure 6.3d The dashed line V denotes the boundary for material balances around a subsystem composed of the first mixing point plus the subsystem portrayed by the box.

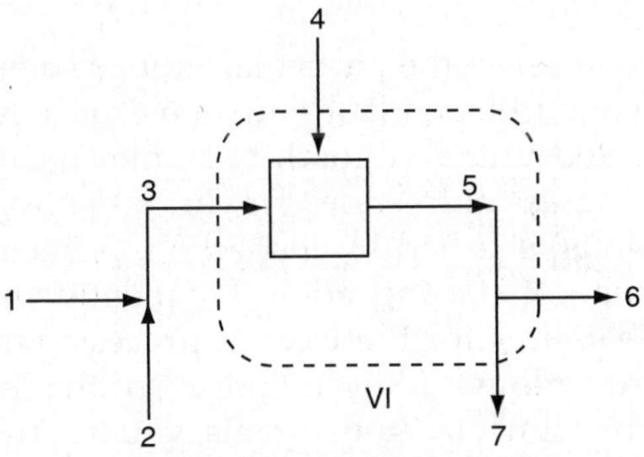

Figure 6.3e The dashed line VI denotes the boundary for material balances about a subsystem composed of the process portrayed by the box plus the splitter.

You can conclude for the system shown in Figure 6.3a that you can make total material balances on six different combinations of subsystems.

The important question is: How many independent material balance equations can be written for the process illustrated in Figure 6.3a if more than one component exists? In Section 6.1 we stated that you can write one independent equation for each component in each subsystem except for the splitter, for which you can write only one independent material balance equation. For Figure 6.3a, assume that three components are present in each of the separate subsystems shown in Figure 6.3c. You can write three independent material balance equations about the first pipe junction, three for the box, plus one independent equation for the splitter, for a total of seven (7) independent material balance equations. How many material balances are possible if you include *all of the redundant equations*, total as well as component balances?

Figure 6.3b:	3 components plus 1 total
Figure 6.3c:	$3 \times 3 = 9$ components plus 3 total
Figure 6.3d:	3 components plus 1 total
Figure 6.3e:	3 components plus 1 total

The total is 24. Which 7 of the 24 equations would be the most appropriate to choose to retain independence and solve easily?

Be careful to **select an independent set of equations**. As an example of what not to do, do not select three component balances for (a) the first pipe junction shown in Figure 6.3c, (b) the box, and (c) one balance for the splitter plus an overall total balance (Figure 6.3b). This set of equations would not be independent because, as you know, the overall balance is just the sum of the respective species balances for the individual unit. However, the total balance could be substituted for one of the component balances.

What strategy should you use to select the particular unit or subsystem with which to start formulating your independent equations for a process composed of a sequence of connected units? A good, but time-consuming, way to decide is to determine the degrees of freedom for various subsystems (single units or combinations of units) selected by inspection. A subsystem with zero degrees of freedom is a good starting point. Frequently, the best way to start is to make material balances for the **overall process**, ignoring information about the internal **connections**. If you ignore all of the internal streams and variables within a set of connected subsystems, you can treat the overall system exactly as you treated a single system in Chapters 3 through 5.

Example 6.1 Determination of the Number of Independent Material Balances in a Process with Multiple Units

Lactic acid ($C_3H_6O_3$) produced by fermentation is used in the food, chemical, and pharmaceutical industries. Figure E6.1 illustrates the mixing of components to form a suitable fermentation broth. The whole system is steady-state and open. The arrows designate the direction of the flows. No reaction occurs in any of the subsystems.

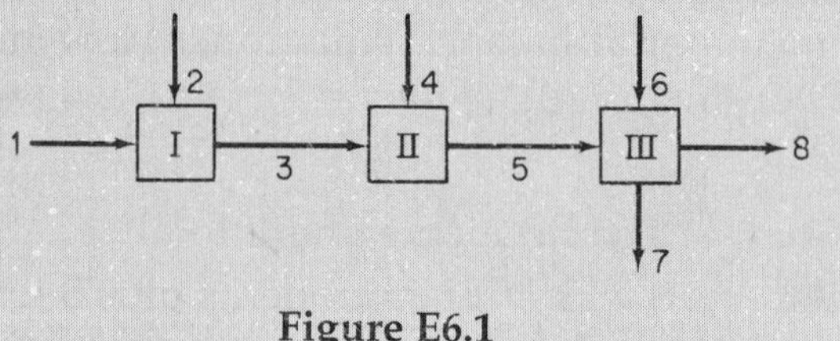

Figure E6.1

The mass compositions of each stream are as follows:

1. Water (W): 100%
2. Glucose (G): 100%
3. W and B, concentrations known: $\omega_W = 0.800$ and $\omega_G = 0.200$
4. *Lactobacillus* (L): 100%
5. W, G, and L, concentrations known: $\omega_W = 0.769$, $\omega_G = 0.192$, $\omega_L = 0.0385$
6. Vitamin G with amino acids and phosphate (V): 100%
7. $\omega_W = 0.962$, $\omega_V = 0.0385$
8. $\omega_G = 0.833$, $\omega_L = 0.167$

What is the maximum number of independent mass balances that can be generated for this system?

Solution

From taking into account each of the three units as subsystems, you certainly can make nine component equations as follows: Let's bypass any total balance for any of the three units as well as any overall component or total balance, or any of the possible balances for combinations of units.

	Total Number of Component Balances
At unit I, two components are involved	2
At unit II, three components are involved	3
At unit III, four components are involved	4
Total	9

However, not all of the balances are independent. In the following list of the component balances, all of the known component concentrations have been inserted. F_i represents the stream flow designated by the subscript.

Subsystem I:

$$\text{Component Balances}\begin{cases}\text{A}: F_1(1.00) + F_2(0) = F_3(0.800) & \text{(a)}\\ \text{B}: F_1(0) + F_2(1.00) = F_3(0.20) & \text{(b)}\end{cases}$$

Subsystem II:

$$\text{Component Balances}\begin{cases}\text{A}: F_3(0.800) + F_4(0) = F_5(0.769) & \text{(c)}\\ \text{B}: F_3(0.200) + F_4(0) = F_5(0.192) & \text{(d)}\\ \text{C}: F_3(0) + F_4(1.00) = F_5(0.0385) & \text{(e)}\end{cases}$$

Subsystem III:

$$\text{Component Balances}\begin{cases}\text{A}: F_5(0.769) + F_6(0) = F_7(0.962) + F_8(0) & \text{(f)}\\ \text{B}: F_5(0.192) + F_6(0) = F_7(0) + F_8(0.833) & \text{(g)}\\ \text{C}: F_5(0.0385) + F_6(0) = F_7(0) + F_8(0.167) & \text{(h)}\\ \text{D}: F_5(0) + F_6(1.00) = F_7(0.086) + F_8(0) & \text{(i)}\end{cases}$$

If you take as an arbitrary basis $F_1 = 100$, seven values of F_i are unknown; hence only seven independent equations need to be written. Can you recognize by inspection that among the entire set of nine equations, two are

(Continues)

Example 6.1 Determination of the Number of Independent Material Balances in a Process with Multiple Units *(Continued)*

indeed redundant, and hence a unique solution can be obtained using the seven independent equations?

If you solved the nine equations by hand *sequentially*, starting with Equation (a) and ending with Equation (i), along the way you would notice that Equation (d) is redundant with Equation (c), and Equation (h) is redundant with Equation (g). The redundancy of Equations (c) and (d) becomes apparent if you recall that the sum of the mass fractions in a stream is unity, hence an implicit relation exists between Equations (c) and (d) so they are not independent. Why are Equations (g) and (h) not independent?

As you inspect the set of Equations (a) through (i) with the viewpoint of solving them sequentially, you will note that each one can be solved for one variable. Look at the following list:

Equation	Determines	Equation	Determines
(a)	F_3	(f)	F_7
(b)	F_2	(g)	F_8
(c)	F_5	(h)	F_8
(d)	F_5	(i)	F_6
(e)	F_4		

If you entered Equations (a) through (i) into a software program that solves equations, you would receive an error notice of some type because the set of equations includes redundant equations.

If the fresh facts which come to our knowledge all fit themselves into the scheme, then our hypothesis may gradually become a solution.

Sherlock Holmes in Sir Arthur Conan Doyle's "The Adventure of Wisteria Lodge," in *The Complete Sherlock Holmes*

If you make one or more component mass balances around the combination of subsystems I plus II, or II plus III, or I plus III in Example 6.1, or around the entire set of three units, no additional *independent* mass balances will be generated. Can you substitute one of the indicated alternative mass balances for an independent species mass balance? In general, yes.

In calculating the degree-of-freedom analysis for problems involving multiple units, you must be careful to involve only independent material balances and not miss any essential unknowns. All the same principles apply to processes with multiple units that were discussed in Chapters 3 through 5. The following is a simplified checklist to help you keep in mind the possible unknowns to include in an analysis, and the possible specifications and equations that should be taken into account.

Make sure that you have accounted for all of the variables involved in the problem and not included irrelevant ones. Check for

- Flow variables entering and leaving each subsystem
- Species or components entering and leaving for each subsystem
- Reaction variables and extents of reaction

Make sure that you have not missed any of the information in the problem that should be included as an equation. Check for

- Selection of a basis for a subsystem or the overall system
- Material balances for each species, element, or component in each subsystem and the overall system as needed
- Specifications for a subsystem and the overall system that can be used to assign values to variables or add equations to the solution set (remember the assignments correspond to possible equations)
- Implicit equations (sum of mole or mass fractions) used explicitly or implicitly

Formally reviewing the checklist will help if you are perplexed or uncertain as to how to identify which variables and equations to employ in solving for the unknowns, and, of course, to ensure that the degrees of freedom are zero before starting to solve the set of equations. Chapter 12 is a detailed discussion of determining the degrees of freedom for a complicated process.

The solution goes on famously; but just as we have got rid of the other unknowns, behold! V disappears as well, and we are left with the indisputable but irritating conclusion—

$$0 = 0$$

This is a favorite device that mathematical equations resort to when we propound stupid questions.

Sir Arthur Eddington

Frequently Asked Questions

1. In carrying out a degree-of-freedom analysis, do you have to include at the start of the analysis every one of the variables and equations that are involved in the entire process? No. What you do is to pick a system for analysis, and you only have to take into account the unknowns and equations pertaining to the streams cut by the system boundary (plus components inside the system if it is an unsteady-state system). For example, note in Example 6.1 that each subsystem could be treated independently. If you picked as the system the overall system (all three of the units) and the system was in the steady state, only the variables and equations pertaining to streams 1, 2, 4, 6, 7, and 8 would be involved in the analysis.
2. Should you use element material balances or species material balances in solving problems that involve multiple units? For processes that do not involve reaction, use species balances; element balances are quite inefficient. For processes that do involve reaction, if you are given the reactions and information that enable you to calculate the extent of reaction, use species balances. Even if you are not specifically given the reactions, you can often formulate them based on your experience, such as C burning with O_2 to yield CO_2. But if such data are missing, element balances are easier to use. Just make sure that they are independent!

We next look at some examples of making and solving material balances for systems composed of multiple units.

Example 6.2 Material Balances for Multiple Units in Which No Reaction Occurs

Acetone is used in the manufacture of many chemicals and also as a solvent. In its latter role, many restrictions are placed on the release of acetone vapor to the environment. You are asked to design an acetone recovery system having the flowsheet illustrated in Figure E6.2. All of the concentrations shown in Figure E6.2 of both the gases and liquids are specified in *weight percent* in this special case to make the calculations simpler. Calculate A, F, W, B, and D in kilograms per hour. Assume that $G = 1400$ kg/hr.

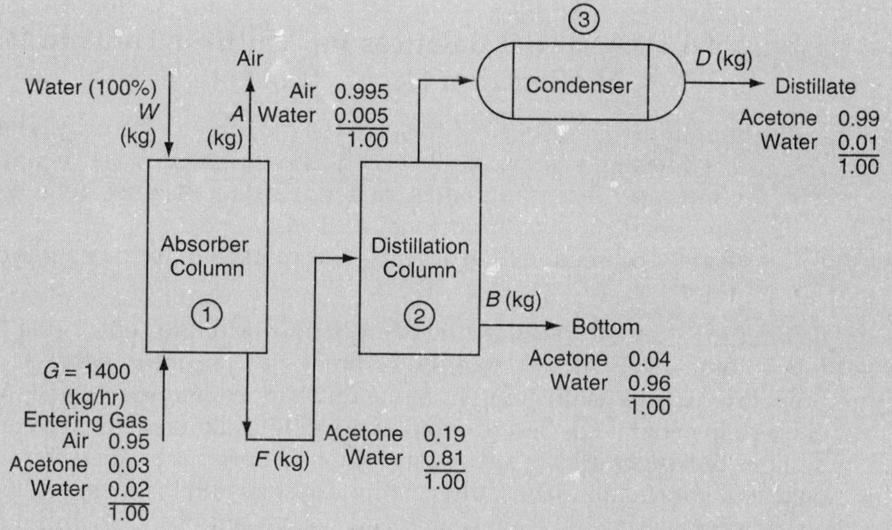

Figure E6.2

Solution

This is an open, steady-state process without reaction. Three subsystems exist as labeled in Figure E6.2.

Steps 1–4

All of the stream compositions are given. All of the unknown stream flows are designated by letter symbols in the figure.

Step 5

Pick 1 hr as a basis so that $G = 1400$ kg.

Steps 6 and 7

We could start the analysis of the degrees of freedom with overall balances, but since the subsystems are connected serially, we will start the analysis with unit 1 (absorber column) and then proceed to unit 2 (distillation column), and then to unit 3 (condenser). In each step of the solution of the problem, as is usual, the variables whose values are not specifically listed in Figure E6.2 will be assumed to be equal to zero and not included in the count of variables.

Unit 1 (absorber)

Variables: Eight components whose values are known and are assigned (see Figure E6.2) plus three variables whose values are not known (the unknowns): W, F, and A (the three unknown flow streams).

(*Continues*)

Example 6.2 Material Balances for Multiple Units in Which No Reaction Occurs (*Continued*)

Equations: Three needed to solve for the three unknowns. The basis, $G = 1400$ kg, has already been assigned a value. What relations are left to use? The component material balances: three (one for each component: air, water, acetone).

Degrees of freedom: Zero. Thus, you can solve for the values of W, F, and A.

Before proceeding to calculate the degrees of freedom for unit 2 (the distillation column), you should note the complete lack of information about the properties of the stream going from the distillation column to unit 3 (the condenser). In general it is best to avoid, if possible, making material balances on systems that include such streams as inputs or outputs, because they contain no useful information. Thus, the substitute system and the degree-of-freedom analysis we select will be for a combined system composed of units 2 and 3.

Units 2 and 3 (distillation column plus condenser)

Variables: D and B (two streams) are the only *unknowns* because F is known from the absorber analysis; the component mass fractions in each stream, $2 \times 3 = 6$, are known values and have already been assigned values.

Equations: Two needed to achieve zero degrees of freedom. You can make two component balances, one for acetone and the other for water.

Degrees of freedom: Zero.

What would happen if a correct analysis of the degrees of freedom for a subsystem gave the result of +1? Then you would hope that the value for one of the unknowns in the subsystem could be determined from another subsystem in the overall system. In fact, for this example, if you started the analysis of the degrees of freedom with the combined units 2 plus 3, you would obtain a value of +1 because the value of F would not be known prior to solving the equations for unit 1.

Step 8 (Unit 1)

The mass balances for unit 1 are as follows:

	In	Out	
Air	400 (0.95)	$= A\,(0.995)$	(a)
Acetone	1400 (0.03)	$= F\,(0.19)$	(b)
Water	$1400\,(0.02) + W\,(1.00)$	$= F\,(0.81) + A\,(0.005)$	(c)

Check to make sure that the equations are independent.

Step 9

Solve Equation (a) for A, solve (b) for F, and then with these results solve (c) for W to get

$$A = 1336.7 \text{ kg/hr}$$
$$F = 221.05 \text{ kg/hr}$$
$$W = 157.7 \text{ kg/hr}$$

Step 10

(Check) Use the total mass balance equation:

$$G + W = A + F$$

$$\begin{array}{rcl} 1400 & = & 1336 \\ 157.7 & = & 221.05 \\ \hline 1557.7 & & 1557.1 \end{array} \quad \text{Close enough}$$

Step 8 (Units 2 and 3)

The mass balances for units 2 plus 3 are

Acetone:	221.05 (0.19)	=	$D(0.99)$ +	$B(0.04)$	(d)
Water:	221.05 (0.81)	=	$D(0.01)$ +	$B(0.96)$	(e)

Step 9

Solve Equations (d) and (e) simultaneously to get

$$D = 34.91 \text{ kg/hr}$$
$$B = 186.1 \text{ kg/hr}$$

Step 10

(Check) Use the total balance:

$F = D + B$ or 221.05 34.91 + 186.1 = 221.01 Close enough

As a matter of interest, what other mass balances could be written for the system and substituted for any one of the Equations (a) through (e)? Typical balances would be the overall balances:

(Continues)

Example 6.2 Material Balances for Multiple Units in Which No Reaction Occurs (*Continued*)

	In	Out	
Air	$G\,(0.95)$	$= A\,(0.995)$	(f)
Acetone	$G\,(0.03)$	$= D\,(0.99) + B\,(0.04)$	(g)
Water	$G\,(0.02)$	$= A\,(0.005) + D\,(0.01) + B\,(0.96)$	(h)
Total	$G + W$	$= A + D + B$	(i)

Equations (f) through (i) do not add any extra information to the problem; the degrees of freedom are still zero. But any of the equations can be substituted for one of Equations (a) through (e) as long as you make sure that the resulting set of equations is composed of independent equations.

Example 6.3 Material Balances for Multiple Units in Which a Reaction Occurs

In the face of higher fuel costs and the uncertainty of the supply of a particular fuel, many companies operate two furnaces, one fired with natural gas and the other with fuel oil. In the RAMAD Corp., each furnace has its own supply of oxygen; the oil furnace uses as a source of oxygen a stream that has the following composition: O_2 20%, N_2 76%, and CO_2 4%. The stack gases from both furnaces exit using a common stack. See Figure E6.3.

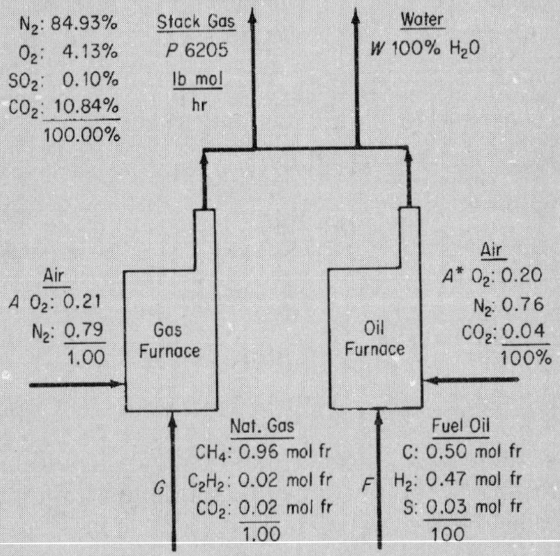

Figure E6.3

Note that two outputs are shown in the common stack to point out that the stack gas analysis is on a dry basis but *water vapor also exists*. The fuel oil composition is given in mole fractions to save you the bother of converting mass fractions to mole fractions.

During one blizzard, all transportation to the RAMAD Corp. was cut off, and officials were worried about the dwindling reserves of fuel oil because the natural-gas supply was being used at its maximum rate possible. The reserve of fuel oil was only 560 bbl. How many hours could the company operate before shutting down if no additional fuel oil was attainable? How many pound moles per hour of natural gas were being consumed? The minimum heating load for the company when translated into the stack gas output was 6205 lb mol/hr of dry stack gas. Analysis of the fuels and stack gas at this time is shown in Figure E6.3. The molecular weight of the fuel oil was 7.91 lb/lb mol, and its density was 7.578 lb/gal.

Solution

This is a steady-state open process with reaction. Two subsystems exist. We want to calculate F and G in pound moles per hour and then F in barrels per hour.

Steps 1–4

Even though species material balances could be used in solving this problem, we will use element material balances because element balances are easier to apply for this problem. The units of all of the variables whose values are unknown will be pound moles. Rather than making balances for each furnace, since we do not have any information about the individual outlet streams of each furnace, we will choose to make overall balances and thus draw the system boundary about both furnaces.

Step 5

$$\text{Basis: 1 hr, so that } P = 6205 \text{ lb mol}$$

Steps 6 and 7

The simplified degree-of-freedom analysis is as follows: You have five elements in the problem and five streams whose values are unknown: A, G, F, A^*, and W; hence, if the elemental mole balances are independent, you can obtain a unique solution for the problem.

Step 8

The overall balances for the elements are (in pound moles):

(Continues)

Example 6.3 Material Balances for Multiple Units in Which a Reaction Occurs (*Continued*)

	In	Out
H	$G(0.96)(4) + F(0.47)(2)$	$= W(2)$
N	$A(0.79)(2) * (0.76)(2)$	$= 6205(0.8493)(2)$
O	$A(0.21)(2) + A^*(0.20 + 0.04)(2)$ $+ G(0.02)(2)$	$= 6205(0.0413 + 0.001 + 0.1084)(2)$ $+ W$
S	$F(0.03)$	$= 6205(0.0010)$
C	$G(0.96 + (2)(0.02) + 0.02)$ $+ F(0.50) + 0.04A^*$	$= 6205(0.1084)$

The balances can be shown to be independent.

Step 9

Solve the S balance for F (inaccuracy in the SO_2 concentrations will cause some error in F, unfortunately); the sulfur is a tie component. Then solve for the other four balances simultaneously for G. The results are

$$F = 207 \text{ lb mol/hr}$$

$$G = 498 \text{ lb mol/hr}$$

Finally, the fuel oil consumption is

$$\frac{207 \text{ lb mol}}{\text{hr}}\left|\frac{7.91 \text{ lb}}{\text{lb mol}}\right|\frac{\text{gal}}{7.578 \text{ lb}}\left|\frac{\text{bbl}}{42 \text{ gal}}\right. = 5.14 \text{ bbl/hr}$$

If the fuel oil reserves were only 560 bbl, they could last at the most

$$\frac{560 \text{ bbl}}{5.14 \dfrac{\text{bbl}}{\text{hr}}} = 109 \text{ hr}$$

Example 6.4 Analysis of a Sugar Recovery Process Involving Multiple Serial Units

Figure E6.4 shows the process and the known data. You are asked to calculate the compositions of every flow stream, and the fraction of the sugar in the cane (F) that is recovered in M.

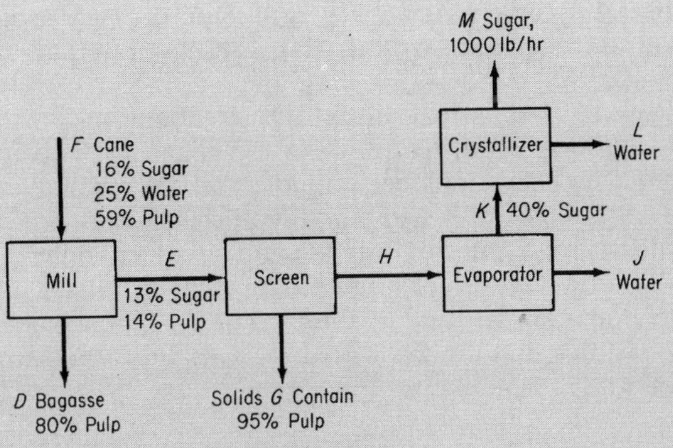

Figure E6.4

Solution

Steps 1–4

All of the known data have been placed on Figure E6.4. The process is an open, steady-state system without reaction. If you examine the figure, two questions naturally arise: What basis should you pick, and what system should you pick to start the analysis? The selection of certain bases and certain systems can lead to more equations that have to be solved simultaneously than others. You could pick as a basis $F = 100$ lb, or $M = 1000$ lb (the same as 1 hr), or the value of any of the intermediate flow streams. You could pick an overall process as the system to start with, or any of the individual units, or any consecutive combinations of units. We will pick M because its value is specified.

Step 5

$$\text{Basis: 1 hr } (M = 1000 \text{ lb})$$

Steps 6 and 7

Another important determination you must make is what the compositions of streams D, E, G, and H are. Stream F has three components, and presumably stream K contains only sugar and water. Does stream H contain pulp? Presumably not, because if you inspect the process flowsheet, you will not find any pulp exiting anywhere downstream of the evaporator. Presumably streams D and G contain sugar and water because the problem implies that not all of the sugar in stream F is recovered. What happens if you assume streams D and G contain neither water nor sugar? Then you would write a

(Continues)

Example 6.4 Analysis of a Sugar Recovery Process Involving Multiple Serial Units (*Continued*)

set of material balances that are not independent and/or are inconsistent (have no solution). Try it. Let S stand for sugar, P stand for pulp, and W stand for water. Their units are pounds. Pick the crystallizer as the initial system to analyze. Why? Because (a) if you check the degrees of freedom for the crystallizer, the values of only a small set of unknowns must be determined to get zero degrees of freedom; (b) the crystallizer involves the feed M whose value is known; and (c) the crystallizer is at one end of the process. Assign the values that are known to the variables (the components in the streams are mass fractions):

$$M = 1000 \qquad L = ? \qquad K = ?$$

$$\omega_{\text{sugar}}^{M} = 1.00 \quad \omega_{\text{water}}^{L} = 1.00 \quad \omega_{\text{sugar}}^{K} = 0.40 \text{ and } \omega_{\text{water}}^{K} = 0.60$$

The unknowns are K and L. You can make two species balance, sugar and water. What about the implicit equations? We have used all three to assign the compositions in M, K, and L. Consequently, the degrees of freedom are zero, and the crystallizer seems to be a good subsystem with which to start.

If you had picked another basis, say, F = 100 lb, and another subsystem, say, the mill, to start with, you would have four unknowns: D, E, $\omega_{\text{sugar}}^{D}$, and $\omega_{\text{water}}^{D}$ (assuming you had assigned $\omega_{\text{water}}^{E} = 0.73$ initially). You could make three species balance and employ one implicit equation, $\sum \omega_i^D = 1$; hence, the degrees of freedom would be zero. But you would have to solve four simultaneous equations. Therefore, **as a general rule, the basis you choose and the unit with which you start the analysis in a multi-unit process affect the degree of complexity of your calculations.**

Steps 8 and 9

For the crystallizer the equations are

$$\text{Sugar: } K\,(0.40) = L\,(0) + 1000$$

$$\text{Water: } K\,(0.60) = L + 0$$

from which you get K = 2500 lb and L = 1500 lb.

Step 10

Check using the total flows:

$$2500 = 1500 + 1000 = 2500$$

The next stage in the solution is to pick the evaporator as the system and repeat the degree-of-freedom analysis. At this point you need to give some thought to the composition of stream H. Is there any pulp in H? If you look at the compositions downstream of H, no pulp emerges; hence, H does not contain pulp, if you use the sum of mass fractions. Then the unknown is H, and you need two independent species balances for the system sugar and water.

The degrees of freedom are zero. You can solve the two equations for H and J, and then can proceed upstream one unit at a time to solve the equations for the screen, and last solve the equations for the mill. The results for all of the variables are as follows:

lb	Mass Fraction
$D = 16{,}755$	$\omega_S^D = 0.174$
$E = 7819$	$\omega_W^D = 0.026$
$F = 24{,}574$	$\omega_W^E = 0.73$
$G = 1152$	$\omega_S^G = 0.014$
$H = 6667$	$\omega_W^G = 0.036$
$J = 4167$	$\omega_W^H = 0.85$
$K = 2500$	$\omega_W^K = 0.60$
$L = 1500$	
$M = 1000$	

The percent sugar recovered is $1000/[(24{,}574)(0.16)] = 0.25$.

Self-Assessment Test

Questions

1. Can a system be composed of more than one unit or piece of equipment?
2. Can one piece of equipment be treated as a set of several subsystems?
3. Does a flowsheet for a process have to show one subsystem for each process unit that is connected to one or more other process units?
4. If you count the degrees of freedom for each individual unit (subsystem) and sum them up, can their total be different from the degrees of freedom for the overall system?

Problems

1. A two-stage separation unit is shown in Figure SAT6.2P1. Given that the input stream $F1$ is 1000 lb/hr, calculate the value of $F2$ and the composition of $F2$.

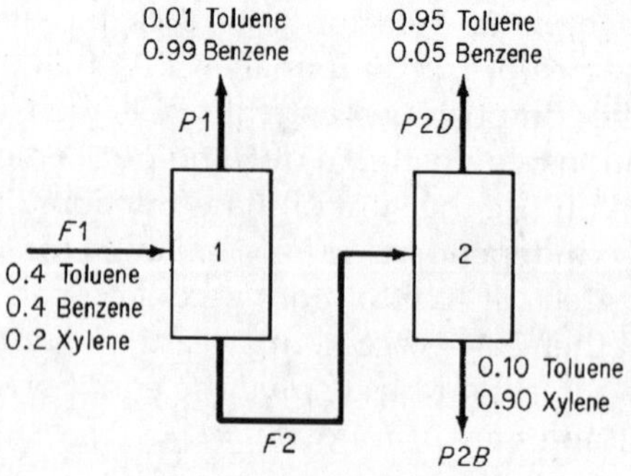

Figure SAT6.2P1

2. A simplified process for the production of SO_3 to be used in the manufacture of sulfuric acid is illustrated in Figure SAT6.2P2. Sulfur is burned with 100% excess air in the burner, but for the reaction $S + O_2 \rightarrow SO_2$, only 90% conversion of the S to SO_2 is achieved in the burner. In the converter, the conversion of SO_2 to SO_3 is 95% complete. Calculate the kilograms of air required per 100 kg of sulfur burned, and the concentrations of the components in the exit gas from the burner and from the converter in mole fractions.

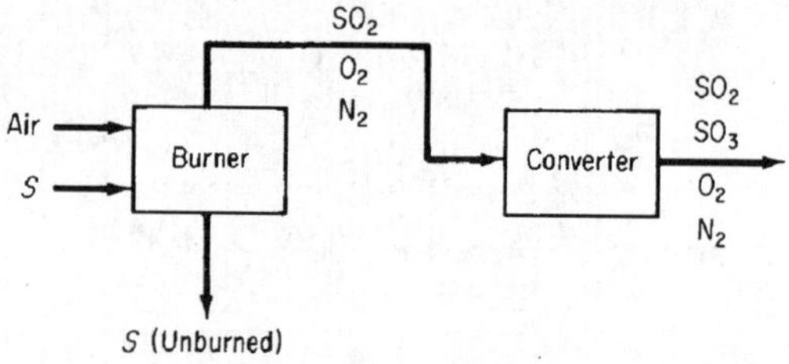

Figure SAT6.2P2

3. In the process for the production of pure acetylene, C_2H_2 (see Figure SAT6.2P3), pure methane (CH_4) and pure oxygen are combined in the burner, where the following reactions occur:

$$CH_4 + 2O_2 \rightarrow 2H_2O + CO_2 \tag{1}$$

$$CH_4 + 1\tfrac{1}{2}O_2 \rightarrow 2H_2O + CO \tag{2}$$

$$2CH_4 \rightarrow C_2H_2 + 3H_2 \tag{3}$$

a. Calculate the ratio of the moles of O_2 to the moles of CH_4 fed to the burner.
b. On the basis of 100 lb mol of gases leaving the condenser, calculate how many pounds of water are removed by the condenser.
c. What is the overall percentage yield of product (pure) C_2H_2, based on the carbon in the natural gas entering the burner?

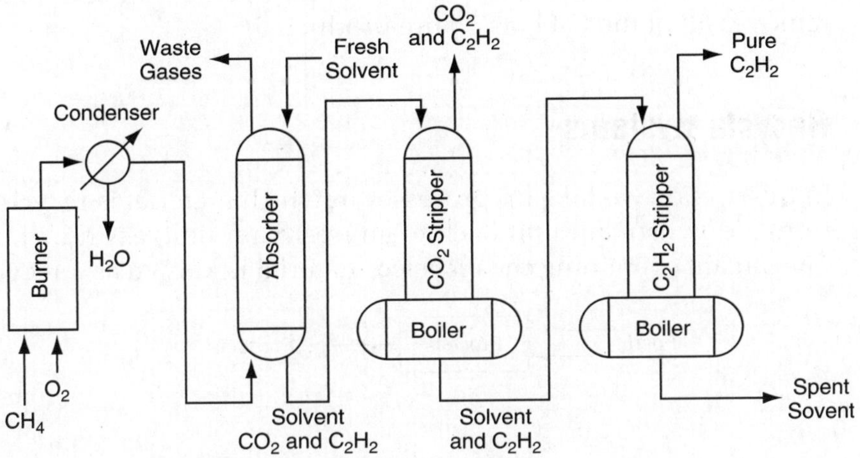

Figure SAT6.2P3

The gases from the burner are cooled in the condenser that removes all of the water. The analysis of the gases leaving the condenser is as follows:

	mol %
C_2H_2	8.5
H_2	25.5
CO	58.3
CO_2	3.7
CH_4	4.0
Total	100.0

These gases are sent to an absorber where 97% of the C_2H_2 and essentially all of the CO_2 are removed with the solvent. The solvent from the absorber is sent to the CO_2 stripper, where all of the CO_2 is removed. The analysis of the gas stream leaving the top of the CO_2 stripper is as follows:

	mol %
C_2H_2	7.5
CO_2	92.5
Total	100.0

The solvent from the CO_2 stripper is pumped to the C_2H_2 stripper, which removes all of the C_2H_2 as a pure product.

6.3 Recycle Systems

In this section we take up processes in which material is **recycled**, that is, fed from a downstream unit back to an upstream unit, as shown in Figure 6.4c. The stream containing the recycled material is known as a **recycle stream**.

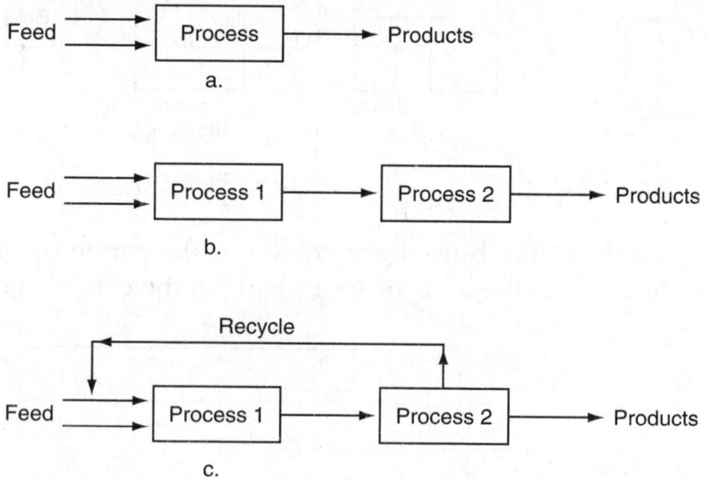

Figure 6.4 (a) A single unit with serial flows
(b) Multiple units but still with serial flows
(c) Multiple units with the addition of recycle

What is a **recycle system**? A recycle system is a system that includes one or more recycle streams.

You can see in Figure 6.4c that the recycle stream is mixed with the feed stream, and the combination is fed to Process 1. In Process 2 (Figure 6.4c), the outputs from Process 1 are separated into (a) the products and (b) the recycle stream. The recycle stream is returned to Process 1 for further processing.

Recycle systems can be found in everyday life. Used newspaper is collected from households, processed to remove the ink, and used to print new newspapers. Clearly, the more newspapers recycled, the fewer trees consumed to produce newspapers. Recycling of glass, aluminum cans, plastics, copper, and iron are also common.

Recycle systems also occur in nature. For example, consider the "water cycle" shown in Figure 6.5. If a region of the Earth is denoted as the system, the recycle stream consists of evaporated water that condenses, falls to Earth as precipitation, and subsequently becomes the flow of water in creeks and rivers into the body of water. Evaporation from a body of water returns (recycles) water into the clouds.

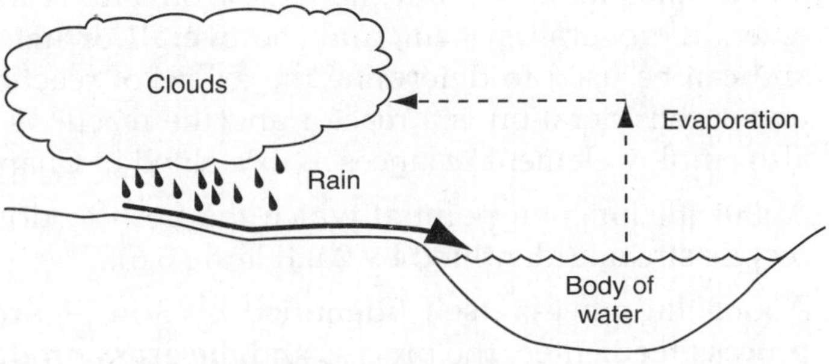

Figure 6.5 A portion of the water cycle

Because of the relatively high cost of industrial feedstocks, when chemical reactions are involved in a process, recycle of unused reactants to the reactor can offer significant economic savings for high-volume processing systems. Heat recovery within a processing unit (comprising energy recycle) reduces the overall energy consumption of the process.

You can formulate material balances for recycle systems without reaction using the same strategy explained in Chapter 3 and applied in subsequent chapters. Examine Figure 6.6.

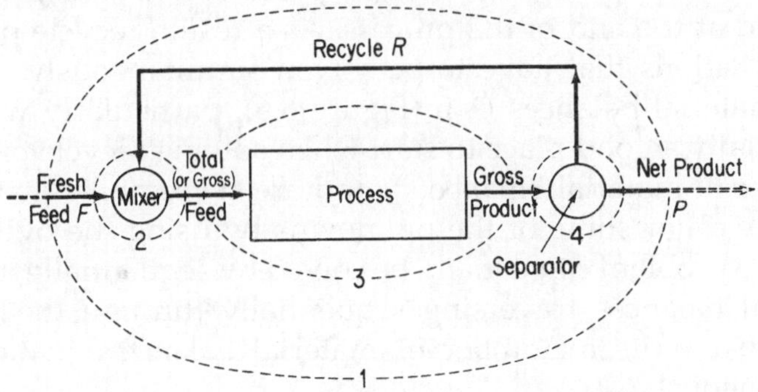

Figure 6.6 Process with recycle. The numbers and dashed lines designate possible system boundaries for the material balances—see the text.

The main new aspect of solving problems involving recycle is picking the proper sequence of systems to analyze. You can write material balances for several different systems, four of which are shown by dashed lines in Figure 6.6, namely:

1. About the entire process including the recycle stream, as indicated by the dashed boundary identified by 1 in Figure 6.6. Such balances contain no information about the recycle stream. Note that the **fresh feed** enters the overall system, and the **overall** or **net product** is removed and can be used to determine the extent of reaction when the process under consideration is a reactor and the reactions are known. You can also employ element balances as explained in Chapter 5.
2. About the junction point at which the fresh feed is combined with the recycle stream (identified by 2 in Figure 6.6).
3. About the process itself (identified by 3 in Figure 6.6). Note that the **process feed** enters the process and the **gross product** is removed.
4. About the separator at which the gross product is separated into recycle and overall (net) product (identified by 4 in Figure 6.6).

In addition, you can make balances about combinations of subsystems, such as the process plus the separator. Only three of the four balances you make for the systems denoted by 1 to 4 are independent, whether made for the total mass or a particular component mass. However, balance 1 will not include the recycle stream, so the balance will not be directly useful in calculating a value for the recycle *R*. Balances 2 and 4 do include *R*. You could write a material balance for the combination of subsystems 2 and 3 or 3 and 4 and include the recycle stream.

Note that in Figure 6.6, the recycle stream is associated both with the mixer, which is located at the beginning of the process, and with the separator, which is located at the end of the process. As a result, recycle problems lead to coupled equations that have to be solved simultaneously. You will find that overall material balances (1 in Figure 6.6), particularly with a tie component, are usually a good place to start when solving recycle problems. If you solve an overall material balance(s), and are successful—that is, you are able to calculate all or some of the unknowns by using the overall balance—the rest of the problem can usually be solved by sequentially applying single-unit material balances traversing sequentially through the process. Otherwise, you must write a number of material balance equations and solve them simultaneously.

Frequently Asked Question

If you feed material to a stream continuously as in Figure 6.4c, why does the value of the material in the recycle stream not increase and continue to build up? What is done in this chapter and subsequent chapters is to assume, often without so stating, that the entire process including all units *is in the steady state*. As the process starts up or shuts down, the flows in many of the streams change, but once the steady state is reached, "what goes in must come out" applies to the recycle stream as well as the other streams in the process.

6.3.1 Recycle without Reaction

Recycle of material occurs in a variety of processes that do not involve chemical reaction, including distillation, crystallization, and heating and refrigeration systems. As an example of a recycle system used to dilute a process stream, look at the process of drying lumber shown in Figure 6.7. If dry air is used to dry the wood, the lumber will warp and crack. By recycling the moist air that exits from the dryer and mixing it with outdoor dry air, the inlet air can be maintained at a safe water content to prevent warping and cracking of the lumber while slowly drying it.

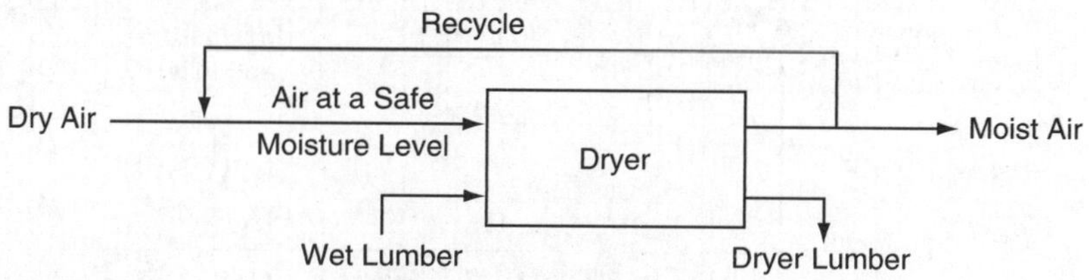

Figure 6.7 Lumber-drying process

Another example, which is shown in Figure 6.8, is a two-product distillation column. Note that a portion of the exit flow from the accumulator is recycled into the column as **reflux** while the reboiler vaporizes part of the liquid in the bottom of the column to create the vapor flow up the column. The recycle of vapor from the reboiler and return of the liquid from the accumulator into the column maintain good vapor and liquid contact on the trays inside the column. This contact makes it possible to concentrate the more volatile components into the overhead vapor stream, leading to the condenser and then to the accumulator, and concentrating the less volatile components into the liquid collected in the bottom of the column.

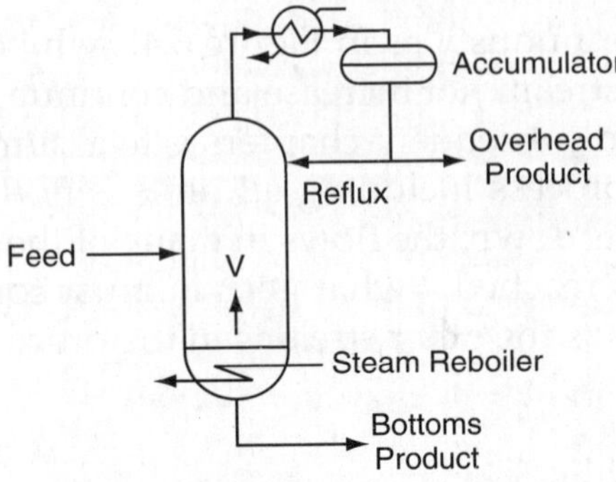

Figure 6.8 Schematic of a two-product distillation column

Example 6.5 Continuous Filtration Involving a Recycle Stream

Figure E6.5 is a schematic of a process for the production of biomass (denoted as Bio) that is to be used in the production of drugs.

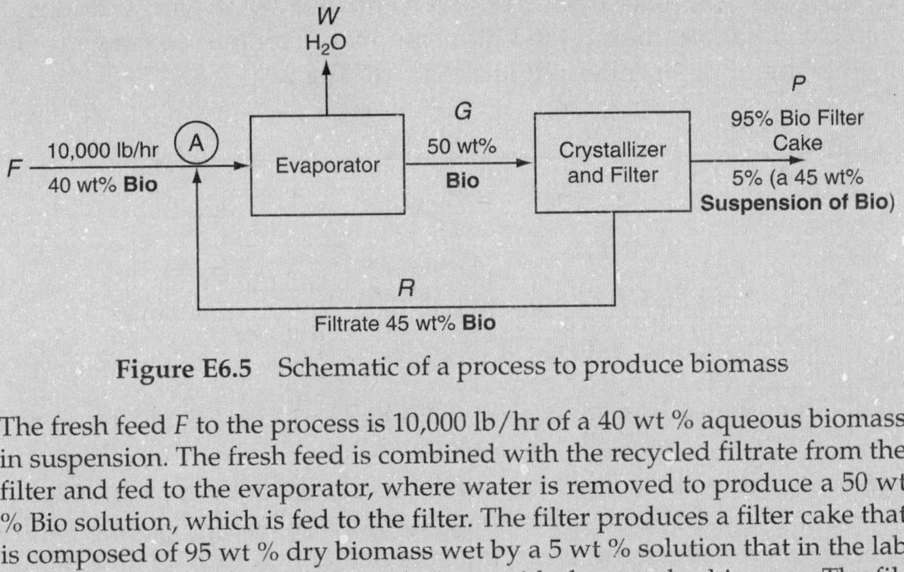

Figure E6.5 Schematic of a process to produce biomass

The fresh feed F to the process is 10,000 lb/hr of a 40 wt % aqueous biomass in suspension. The fresh feed is combined with the recycled filtrate from the filter and fed to the evaporator, where water is removed to produce a 50 wt % Bio solution, which is fed to the filter. The filter produces a filter cake that is composed of 95 wt % dry biomass wet by a 5 wt % solution that in the lab proves to be composed of 55 wt % water with the rest dry biomass. The filtrate contains 45 wt % biomass.

a. Determine the flow rate of water removed by the evaporator and the recycle rate for this process.

b. Assume that the same production rate of filter cake occurs, but that the filtrate is not recycled. What would be the process feed rate of 40 wt % biomass then? Assume that the product solution from the evaporator still contains 50 wt % biomass in water.

Solution a

Steps 1–4

Figure E6.5 contains the information needed to solve the problem.

Step 5

Basis: 10,000 lb fresh feed (equivalent to 1 hr)

Steps 6 and 7

The unknowns are *W*, *G*, *P*, and *R* since you can assign values to *F* and all of the mass fractions in the respective streams using appropriate implicit equations. You can make two component balances about three subsystems: the mixing point A, the evaporator, and the filter (as well as a total balance for each). You can also make similar balances for various combinations of subsystems and the overall system. What balances should you choose to solve the problem? If you put the equations in an equation solver, it does not make any difference as long as the equations are independent. But if you solve the problem by hand, and are going to make two component balances for each of the three subsystems just mentioned and the overall system, you can count the number of unknowns involved for each of the three subsystems and the overall system:

Mixing point: *P* plus feed (and compositions) to evaporator (not labeled in Figure E6.5)

Evaporator: *W*, *G*, and feed to evaporator

Filter: *G*, *P*, and *R*

Overall: *W* and *P*

Can you conclude that by using just two overall component balances you can determine the values of at least *W* and *P*? Consequently, let's start with overall balances.

Steps 8 and 9

Overall Bio balance: $(0.4)(10{,}000) = [0.95 + (0.45)(0.05)]P;$

$$P = 4113 \text{ lb}$$

Overall H_2O balance: $(0.6)(10{,}000) = W + [(0.55)(0.05)](4113);$

$$W = 5887 \text{ lb}$$

The total amount of Bio exiting with *P* is $[(0.95) + (0.45)(0.05)]$ $(4113) = 4000$ lb.

(*Continues*)

Example 6.5 Continuous Filtration Involving a Recycle Stream (*Continued*)

Are you surprised at this result? The amount of water in P is 113 lb. As a check, 113 + 5887 = 6000 lb as expected.

Steps 6 and 7 (repeated)

Now that you know W and P, the next step is to make balances on a subsystem that involves the stream R. Choose either the mixing point A or the filter. Which one would you pick? The filter involves three variables, and you now know the value of P, so only two unknowns would be involved as opposed to more if you chose mixing point A as the system.

$$\text{Bio balance on the filter: } 0.5G = 4000 + 0.45R$$

$$H_2O \text{ balance on the filter: } 0.5G = 113 + 0.55R;\ R = 870 \text{ lb in 1 hr}$$

As a check:

$$10{,}000 + 870 - 5887 = G = 4983 = 4113 + 870 \quad \text{OK}$$

Solution b

Now, suppose that recycle from the filter does not occur, but the production rate of P remains the same. Note that for this case, R would be discharged instead of recycled and mixed with the fresh feed. How should you proceed? Do you recognize that the problem is one of the class that you read about in Section 6.2?

Step 5

The basis is now $P = 4113$ lb (the same as 1 hr).

Steps 6 and 7

The unknowns are now F, W, G, and R. You can make two component balances on the evaporator and two on the filter, plus two overall component balances. Only four are independent. The evaporator balances would involve F, W, and G. The crystallizer balances would involve G and R, while the overall balances would involve F, W, and R. Which balances are best to start with? If you put the equations in an equation solver, it makes no difference which four equations you use as long as they are independent. The filter balances are best because you have to solve just two pertinent equations for G and R.

Steps 8 and 9

$$\text{Bio balance on the filter: } 0.5G = [(0.95) + (0.05)(0.45)](4113) + 0.45R$$

H_2O balance on the filter: $0.5G = [(0.05)(0.55)](4113) + 0.55R$

Solve simultaneously: R = 38,870 lb in 1 hr

Note that without recycle, the feed rate must be 5.37 times larger to produce the same amount of product, not to mention the fact that you would have to dispose of a large volume of filtrate, which is 45 wt % Bio.

6.3.2 Recycle with Chemical Reaction

The most common application of recycle for systems involving chemical reaction is the recycle of reactants, an application that is used to increase the overall conversion in a reactor. Figure 6.9 shows a simple example for the reaction (A → 2B).

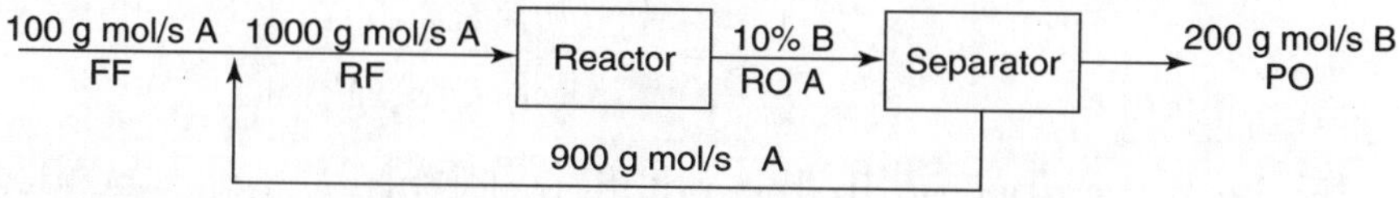

Figure 6.9 A simple recycle system with chemical reaction ($A \rightarrow 2B$); FF (fresh feed); RF (reactor feed); RO (reactor outlet); PO (process outlet)

You will encounter two different types of conversions that can be applied to processes in which reaction occurs:

Overall conversion is based on what enters and leaves the overall process:

Overall conversion = f_{OA} =

$$\frac{\text{moles of limiting reactant in the fresh feed-moles of limiting reactant in the output of the overall process}}{\text{moles of limiting reactant in the fresh feed}} \quad (6.1)$$

The single-pass conversion is based on what enters and leaves the reactor:

Single-pass conversion = f_{SP} =

$$\frac{\text{moles of limiting reactant fed into the reactor-moles of limiting reactant exiting the reactor (gross product)}}{\text{moles of limiting reactant fed into the reactor}} \quad (6.2)$$

For the data in Figure 6.9, what is the overall fraction conversion?

$$f_{OA} = \frac{100\text{ A} - 0}{100\text{ A}} = 1.00$$

$$f_{SP} = \frac{1000\text{ A} - 900}{1000\text{ A}} = 0.10$$

When the fresh feed (FF) consists of more than one reactant, the conversion can be expressed for a defined single component, usually the limiting reactant, or the most important (expensive) reactant. Recall from Chapter 5 that conversion in a reactor (the single-pass conversion) can be limited by chemical equilibrium and/or chemical kinetics. On the other hand, the overall conversion is strongly dependent upon the efficiency of the separator in separating compounds to be recycled from the other compounds.

How are the overall conversion and single-pass conversion related to the extent of reaction, ξ, a term that was discussed in Chapter 5? Do you recall the relation given by Equation (5.10)?

$$\xi = \frac{(-f)n^{\text{in}}_{\text{limiting reactant}}}{\nu_{\text{limiting reactant}}}$$

To be more specific in the notation, let $n^{\text{in}}_{\text{limiting reactant}}$ be denoted by $n^{\text{FF}}_{\text{LR}}$, $\nu_{\text{limiting reactant}}$ be denoted by v_{LR}, and f be denoted by f_{OA} or f_{SP}, respectively. Then

$$\text{Overall conversion for the limiting reactant} = f_{\text{OA}} = \frac{-v_{\text{LR}}\xi}{n^{\text{FF}}_{\text{LR}}} \tag{6.3}$$

and for the single-pass conversion

$$\text{Single-pass conversion} = f_{\text{SP}} = \frac{-v_{\text{LR}}\xi}{n^{\text{RF}}_{\text{LR}}} \tag{6.4}$$

If you solve Equations (6.3) and (6.4) for the extent of reaction, equate the extents of reaction, and use a material balance at the junction of the fresh feed and the recycle stream (a mixing point), $n^{\text{RF}}_{\text{LR}} = n^{\text{FF}}_{\text{LR}} + n^{\text{recycle}}_{\text{LR}}$, you can obtain the following relationship between overall and single-pass conversion:

$$\frac{f_{\text{SP}}}{f_{\text{OA}}} = \frac{n^{\text{FF}}_{\text{LR}}}{n^{\text{FF}}_{\text{LR}} + n^{\text{recycle}}_{\text{LR}}} \tag{6.5}$$

If you now apply Equation (6.3) to the simple recycle example in Figure 6.9, what value do you get for the ratio of the single-pass to overall conversion? Do you get 0.1, which agrees with the result stated previously? The same relationship can be demonstrated mathematically rather than numerically. In general, **the extent of reaction is the same regardless of whether an overall material balance is used or a material balance for the reactor is used.** This important fact can be used in solving material balances for recycle systems with reactions.

Example 6.6 Recycle in a Process in Which a Reaction Occurs

Cyclohexane (C_6H_{12}) can be made by the reaction of benzene (Bz, short for C_6H_{12}) with hydrogen according to the following reaction:

$$C_6H_6 + 3H_2 \rightarrow C_6H_{12}$$

For the process shown in Figure E6.6, determine the ratio of the recycle stream to the fresh feed stream if the overall conversion of benzene is 95% and the single-pass conversion through the reactor is 20%. Assume that 20% excess hydrogen is used in the fresh feed, and that the composition of the recycle stream is 22.74 mol % benzene and 78.26 mol % hydrogen.

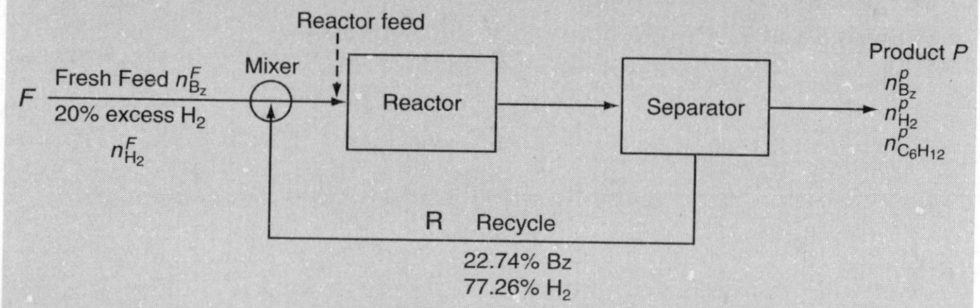

Figure E6.6 Schematic of a recycle reactor

Note that in this example, there is a relatively low conversion per pass (20%) and the overall conversion is relatively high (95%). A low conversion per pass can be desirable in certain cases, for example, a case in which the yield decreases as the conversion increases. By using recycle, you can obtain a high yield and at the same time a high overall conversion (i.e., high utilization of the limiting reactant).

Solution

The process is open and steady-state.

Steps 1–4

Figure E6.6 contains all of the information available about the flow streams except the amount of H_2 feed, which is in 20% excess (for complete reaction, remember),

$$n^F_{H_2} = 100(3)(1 + 0.20) = 360 \text{ mol}$$

and the total fresh feed is 460 mol.

Step 5

From Equation (6.3) for benzene ($\nu_{Bz} = -1$),

$$0.95 = \frac{-(-1)\xi}{100}$$

(Continues)

Example 6.6 Recycle in a Process in Which a Reaction Occurs (*Continued*)

you can calculate that $\xi = 95$ reacting moles.

Steps 6 and 7

The unknowns are R, n^P_{Bz}, $n^P_{H_2}$, and $n^P_{C_6H_{12}}$. You can write three species balances for each of the three systems—the mixing point, the reactor, and the separator—plus overall balances (not all of which are independent, of course). Which systems should you choose to start with? The overall process, because then you can use the value calculated for the extent of reaction.

Steps 8 and 9

The species overall balances are

$$n_i^{out} = n_i^{in} + \nu_i \xi$$

$$\text{Bz:} \quad n^P_{Bz} = 100 + (-1)(95) = 5 \text{ mol}$$

$$H_2\text{:} \quad n^P_{H_2} = 360 + (-3)(95) = 75 \text{ mol}$$

$$C_6H_{12}\text{:} \quad n^P_{C_6H_{12}} = 0 + (1)(95) = 95 \text{ mol}$$

$$\text{Total:} \quad P = 175 \text{ mol}$$

The next step is to use the final piece of information, the information about the single-pass conversion plus Equation (6.4), to get R. The amount of the Bz feed to the reactor is $100 + 0.2274R$, and $\xi = 95$ (the same as was calculated from the overall conversion). Thus

$$0.20 = \frac{95}{100 + 0.2274R}$$

and solving for R yields

$$R = 1649 \text{ mol}$$

Finally, the ratio of recycle to fresh feed is

$$\frac{R}{F} = \frac{1659 \text{ mol}}{360 \text{ mol}} = 3.59$$

Example 6.7 Recycle in a Process with a Reaction Occurring

Immobilized glucose isomerase is used as a catalyst in producing fructose from glucose in a fixed-bed reactor (water is the solvent). For the system shown in Figure E6.7a, what percent conversion of glucose results on one pass through the reactor when the ratio of the exit stream to the recycle stream in mass units is equal to 8.33? The reaction is

$$C_{12}H_{22}O_{11} \rightarrow C_{12}H_{22}O_{11}$$

Glucose Fructose

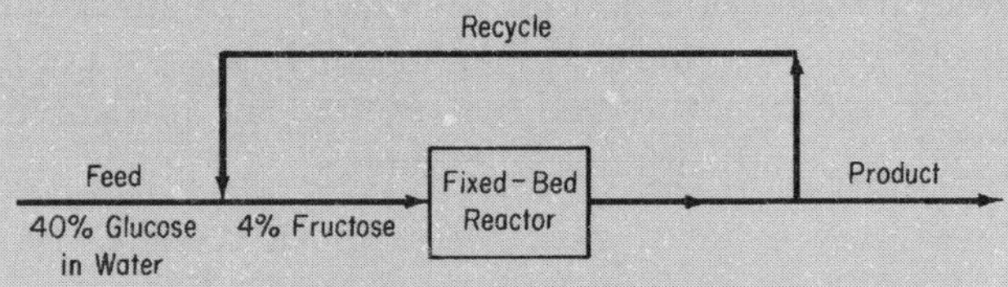

Figure E6.7a

Solution

Steps 1–4

The process is a steady-state process with a reaction occurring and recycle. Figure E6.7b includes all of the known and unknown values of the variables using appropriate notation (*W* stands for water, *G* for glucose, and *F* for fructose). Note that the recycle stream and product stream have the same

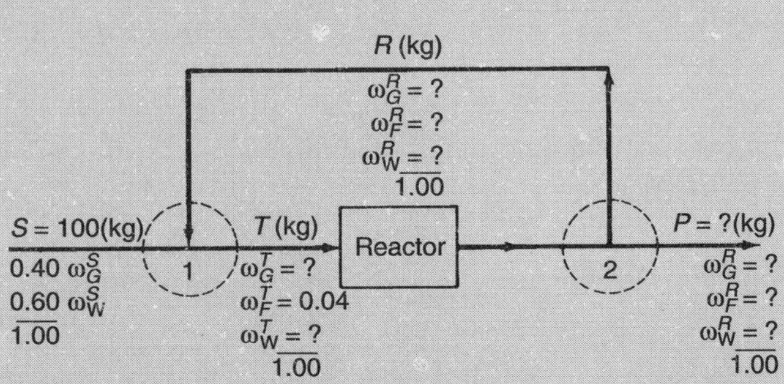

Figure E6.7b

(Continues)

Example 6.7 Recycle in a Process with a Reaction Occurring (*Continued*)

composition, and consequently the same mass symbols are used in the diagram for each stream.

Step 5

Pick as a basis $S = 100$ kg given the data shown in Figure E6.7b.

Step 6

We have not provided notation for the reactor exit stream and composition as we will not be using these values in our balances. Let f be the fraction conversion for one pass through the reactor. The unknowns are R, F, P, T, ω_G^R, ω_F^R, ω_W^R, ω_G^T, ω_W^T, and f, for a total of nine.

Step 7

The balances are $\sum \omega_i^R = 1$, $\sum \omega_i^T = 1$, $R = P/8.33$, plus three species balances each on the mixing point 1, the separator 2, and the reactor, as well as overall balances. We will assume we can find nine independent balances among the lot and proceed. We do not have to solve all of the equations simultaneously. The units are mass (kilograms).

Steps 8 and 9

Overall balances:

Total: $P = 100$ kg (How simple!)

Consequently,

$$R = \frac{100}{8.33} = 12.0 \text{ kg}$$

Overall, no water is generated or consumed (i.e., water is not involved in the reaction); hence

Water
$$100(0.60) = P\omega_W^R = 100\omega_W^R$$
$$\omega_W^R = 0.60$$

We now have six unknowns left for which to solve. We start somewhat arbitrarily with mixing point 1 to calculate some of the unknowns.

Mixing Point 1

No reaction occurs at the mixing point, so species balances can be used without involving the extent of reaction:

$$\text{Total: } 100 + 12 = T = 112$$

$$\text{Glucose: } 100(0.40) + 12\omega_G^P = 112\omega_G^T$$

$$\text{Fructose: } 0 + 12\omega_F^R = 112(0.04)$$

Solving for ω_F^R from the last equation yields $\omega_F^R = 0.373$.

Also, because $\omega_F^R + \omega_G^R + \omega_W^R = 1$,

$$\omega_G^R = 1 - 0.373 - 0.600 = 0.027$$

Next, from the glucose balance, $\omega_G^T = 0.360$.

Next, rather than make separate balances on the reactor and separator, we will combine the two into one system (and thus avoid having to calculate values associated with the reactor exit stream).

Reactor Plus Separator 2

	In	−	Out	=	Consumed
Glucose:	$(0.360)(112)$	−	$(112)(0.027)$	=	$f(0.360)(112)$
					$f = 0.93$

Step 10

Check by using Equation (6.5), including the extent of reaction.

Self-Assessment Test

Questions

1. Explain the purpose of using recycle in a process.
2. Under what circumstances might material be accumulated or depleted in a recycle stream?
3. For what systems can you make material balances in processes that involve recycle?
4. Can you formulate sets of equations that are not independent if recycle occurs in a system?
5. If the components in the feed to a process appear in stoichiometric quantities and the subsequent separation process is complete so that all of the unreacted reactants are recycled, what is the ratio for reactants in the recycle stream?
6. Indicate whether the following statements are true or false:
   a. The general material balance applies for processes that involve recycle with reaction as it does for other processes.

b. The key extra piece of information in material balances on processes with recycle in which a reaction takes place is the specification of the fraction conversion or extent of reaction.
c. The degrees of freedom for a process with recycle that involves chemical reaction are the same as for a process without recycle.

Problems

1. How many recycle streams occur in Figure SAT6.3P1?

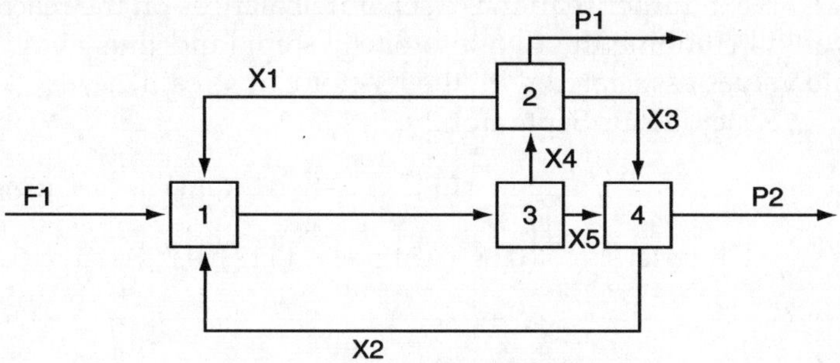

Figure SAT6.3P1

2. A ball mill grinds plastic to make a very fine powder. Look at Figure SAT6.3P2.

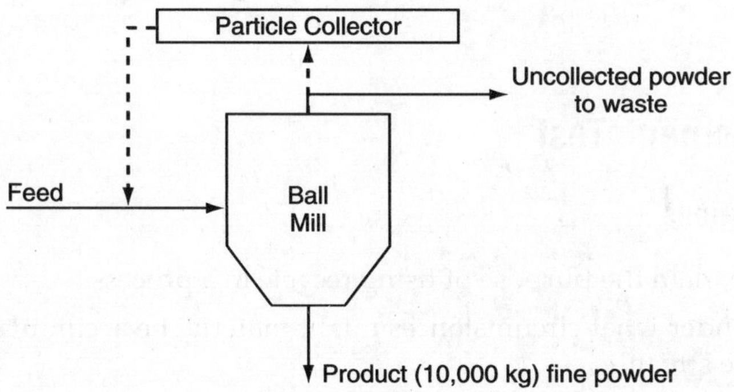

Figure SAT6.3P2

At the present time 10,000 kg of powder are produced per day. You observe that the process (shown by the solid lines) is inefficient because 20% of the feed is not recovered as powder—it goes to waste.

You make a proposal (designated by the dashed lines) to recycle the uncollected material back to the feed so that it can be remilled. You plan to recycle 75% of the 200 kg of uncollected material back to the feed stream. If the feed costs \$1.20/kg, how much money would you save per day while producing 10,000 kg of fine powder?

3. Seawater is to be desalinized by reverse osmosis using the scheme indicated in Figure SAT6.3P3. Use the data given in the figure to determine (a) the rate of waste brine removal (B), (b) the rate of desalinized water (called potable water) production (P), and (c) the fraction of the brine leaving the reverse osmosis cell (which acts in essence as a separator) that is recycled.

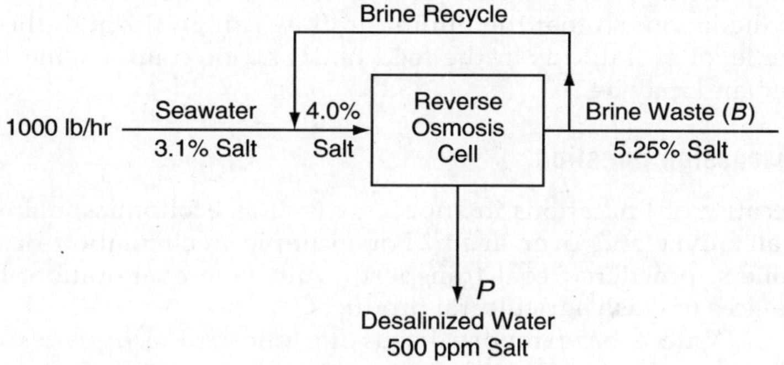

Figure SAT6.3P3

4. A catalytic dehydrogenation process shown in Figure SAT6.3P4 produces 1, 2-butadiene (C_4H_6) from pure normal butane (C_4H_{10}). The product stream contains 75 mol/hr of H_2 and 13 mol/hr of C_4H_{10} as well as C_4H_6. The recycle stream is 30% (mol) C_4H_{10} and 70% (mol) C_4H_6, and the flow is 24 mol/hr.

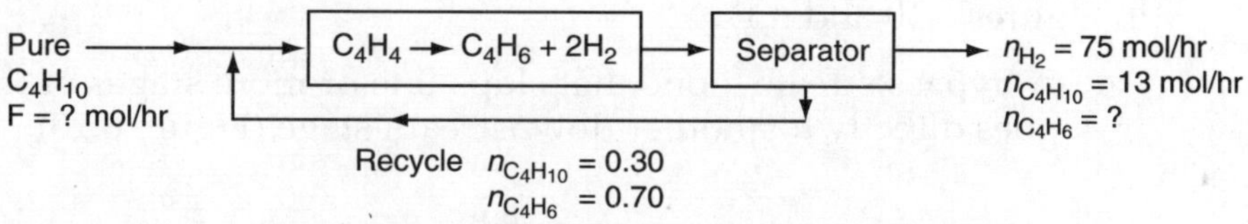

Figure SAT6.3P4

a. What are the feed rate, F, and the product flow rate of C_4H_6 leaving the process?
b. What is the single-pass conversion of butane in the process?

5. Pure propane (C_3H_8) from El Paso is dehydrogenated catalytically in a continuous process to obtain propylene (C_3H_6). All of the hydrogen formed is separated from the reactor exit gas with no loss of hydrocarbon. The hydrocarbon mixture is then fractionated to give a product stream containing 88 mol % propylene and 12 mol % propane. The other stream, which is 70 mol % propane and 30 mol % propylene, is recycled. The one-pass conversion in the reactor is 25%, and 1000 kg of fresh propane are fed per hour. Find (a) the kilograms of product stream per hour, and (b) the kilograms of recycle stream per hour.

Thought Problem

Centrifugal pumps cannot run dry and must have a minimum fluid flow to operate properly—to avoid cavitation and subsequent mechanical damage to the pump. A storage tank is to be set up to provide liquid flow to a process, but sometimes the demand will drop below the minimum flow rate (10% to 15% of the rated capacity of the pump). What equipment setup would you recommend be implemented so that the pump is not damaged by the low flows? Draw a picture of the layout so that the minimum flow can go through the pump no matter what the level of liquid is in the feed tank and no matter what the outlet pressure and demand may be.

Discussion Question

Because of limitations in supply as well as economics, many industries reuse their water over and over again. For example, recirculation occurs in cooling towers, boilers, powdered coal transport, multistage evaporation, humidifiers, and many devices to wash agricultural products.

Write a brief report discussing one of these processes, and include in the report a description of the process, a simplified flowsheet, problems with recycling, the extent of purge, and, if you can find the information, the savings made by recycling.

6.4 Bypass and Purge

Two additional commonly encountered types of process streams are shown in Figures 6.10 and 6.11:

1. A **bypass** stream—one that skips one or more stages of the process and goes directly to another downstream stage (Figure 6.10)

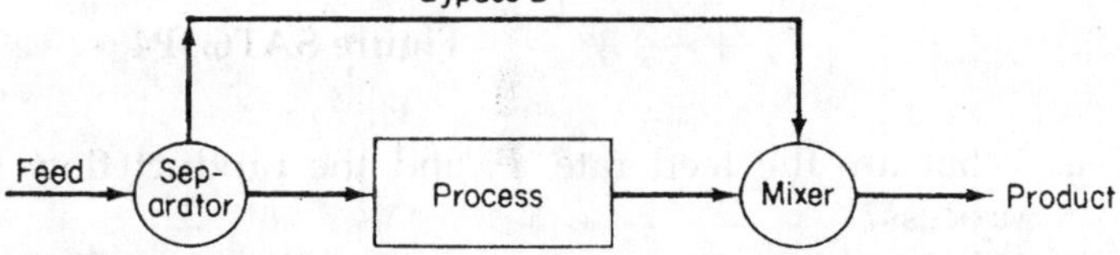

Figure 6.10 A process with a bypass stream

A bypass stream can be used to control the composition of a final exit stream from a unit by mixing the bypass stream and the unit exit stream in suitable proportions to obtain the desired final composition.

2. A **purge** stream—a stream bled off from the process to remove an accumulation of inerts or unwanted material that might otherwise build up in the recycle stream with time of operation (Figure 6.11)

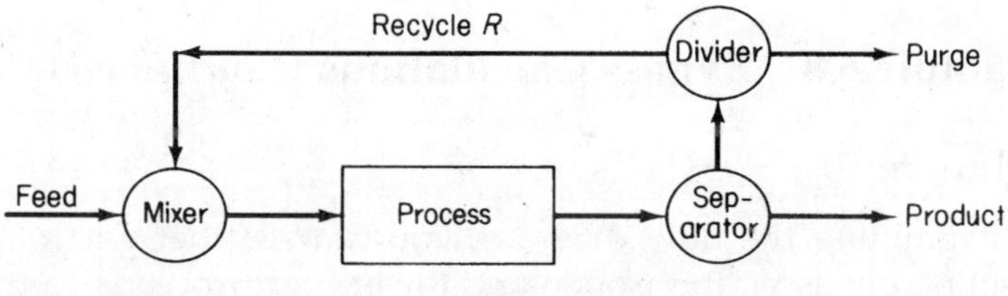

Figure 6.11 A process with a recycle stream with purge

Many companies have had the unfortunate experience that on start-up of a new process, trace components not considered in the material balances used in the design of the process (because the amounts were so small) build up in one or more recycle loops. That is, if a trace species that enters a process is not removed from the process, it will accumulate in the process until the process no longer functions properly. For example, distillation columns without an overhead purge and with light trace impurity in the feed will experience a steady increase in column pressure, undermining the condensation in the overhead condenser, leading to a steady reduction in separation performance of the column. Calculations for processes involving bypass and purge streams introduce no new principles or techniques beyond those presented so far. Two examples will make that clear.

Example 6.8 Bypass Calculations

In the feedstock preparation section of a plant manufacturing natural gasoline, isopentane is removed from butane-free gasoline. Assume for purposes of simplification that the process and components are as shown in Figure E6.8. What fraction of the butane-free gasoline is passed through the isopentane tower? Detailed steps will not be listed in the analysis and solution of this problem. The process is in the steady state and no reaction occurs.

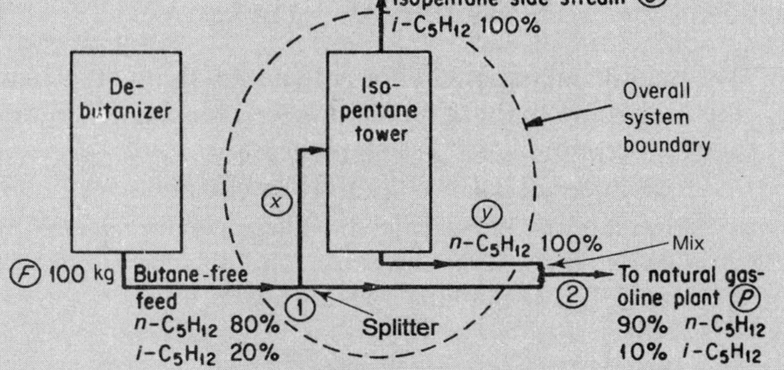

Figure E6.8

(Continues)

Example 6.8 Bypass Calculations (*Continued*)

Solution

By examining the flow diagram, you can see that part of the butane-free gasoline bypasses the isopentane tower and proceeds to the next stage in the natural gasoline plant. All of the compositions (the streams are liquid) are known. The units are all kilograms. Select a basis:

Basis: 100 kg of feed

a. With what system or subsystem should you start the solution? Let's start with the overall system, usually the easiest to work with, especially in view of the information given. What are the unknowns? If you assign $F = 100$ kg, the unknowns are S and P. What balances can you write for the overall system? You can write a total balance and two component balances, but only two of the three are independent.

Total overall material balance:

$$\frac{\text{In}}{100} = \frac{\text{Out}}{S + P} \tag{a}$$

Component material balance for $n - C_5$ (a tie component):

$$\frac{\text{In}}{100(0.80)} = \frac{\text{Out}}{S(0) + P(0.90)} \tag{b}$$

Consequently,

$$P = 100\left(\frac{0.80}{0.90}\right) = 88.9 \text{ kg and}$$

$$S = 100 - 88.9 = 11.1 \text{ kg}$$

b. The overall balances will not tell you anything about the fraction of the feed going to the isopentane tower; for this calculation you need another system. Pick the isopentane tower itself.

Total material balance around isopentane tower:

From Figure E6.8, X is the kilograms of butane-free gasoline going to the isopentane tower, and Y is the kilograms of n-C_5H_{12} leaving the isopentane tower.

$$\text{In} = \text{Out}$$

$$X = 11.1 + Y \tag{c}$$

Component balance (n-C_5, a tie component):

$$X(0.80) = Y \tag{d}$$

Combine (c) and (d) to get $X = 55.5$ kg, so the desired fraction is 0.55.

c. Another approach to this problem would be to make material balances for the subsystem defined by mixing points 1 and 2.

Total material balance around mixing point 2:

$$(100 - X) + Y = 88.9 \tag{e}$$

Component balance (iso-C_5):

$$(100 - X)(0.20) + 0 = 88.9(0.10) \tag{f}$$

Because Equation (f) does not contain Y, it can be solved directly for X: $X = 55.5$ kg as before—would you expect the value to be different?

Example 6.9 Purge

Considerable interest exists in the conversion of coal into more convenient liquid products for subsequent production of chemicals. Two of the main gases that can be generated under suitable conditions from in situ (in the ground) coal combustion in the presence of steam (as occurs naturally in the presence of groundwater) are H_2 and CO. After cleanup, these two gases can be combined to yield methanol according to the following equation:

$$CO + 2H_2 \rightarrow CH_3OH$$

Figure E6.9a illustrates a steady-state, open process for the production of methanol. All of the compositions are in mole fractions or percent. The stream flows will be in moles.

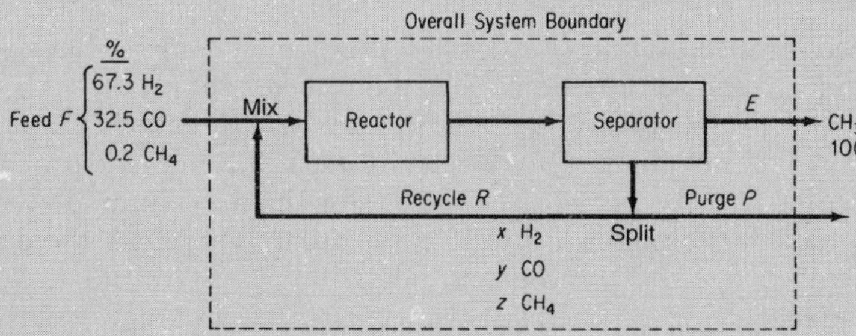

Figure E6.9a

(Continues)

Example 6.9 Purge (*Continued*)

You will note in Figure E6.9a that some CH_4 enters the process. However, the CH_4 does not participate in the reaction. A purge stream is used to maintain the CH_4 concentration in the exit stream from the separator going to R and P at no more than 3.2 mol %, and to prevent H_2 from accumulating in the system. The once-through conversion of the CO in the reactor is 18%.

Compute the moles of recycle, R, the moles of CH_3OH, E, and the moles of purge, P, per 100 moles of feed, and also compute the purge gas composition.

Solution

Steps 1–4

Most of the known information is shown in Figure E6.9a. The process is in the steady state with reaction. The purge and recycle streams have the same composition as implied by the split of one of the outputs of the separator into $P + R$ in the figure. The mole fractions of the components in the purge stream have been designated as x, y, and z for H_2, CO, and CH_4, respectively.

Step 5

Select a convenient basis:

$$\text{Let } F = 100 \text{ mol}$$

Step 6

After assigning known values to variables, the variables whose values are unknown are x, y, z, E, P, R, and the extent of reaction, ξ, the latter being required if you plan to use species material balances rather than element balances. You can ignore the stream between the reactor and separator as no questions are asked about it.

Steps 7–9

One piece of information given in the problem statement that has not been used so far is the information about the upper limit on the CH_4 concentration in the purge stream. This limit can be expressed as $z \leq 0.032$. An inequality will not enable you to get a *unique* solution to the problem; hence, let's assume that the purge stream contains the maximum allowed amount of CH_4 and thus assign a value to z: $z = 0.032$. Therefore, there remain six unknowns, x, y, E, P, R, and ξ, and you will have to formulate six independent equations.

Possible Material Balances to Use	Possible Systems to Use
Component balance for H_2	Overall
Component balance for CO	Junction of F and P
Component balance for CH_4	Split into P and R
Component balance for CH_3OH	Reactor plus separator
Sum of mole fractions for P and F	Reactor alone
Once-through conversion in the reactor	Separator alone

A quick glance at the respective variables should lead you to the conclusion that the overall species material balance for the CH_4 is the simplest to solve because CH_4 is a tie component—all that enters, leaves in one stream without any reaction occurring, and the CH_4 balance involves only one unknown, P:

$$\begin{array}{ccc} \text{In} & & \text{Out} \\ 0.002(100) & = & 0.032P \quad \text{so } P = 6.25 \text{ mol} \end{array}$$

Now that the value of P has been determined (reducing the number of unknowns to five), what should you do next? Are there any other equations that can be solved for a single variable? All of the other overall component balances for an open, steady-state system are of the form given by Equation (5.8) derived from Equation (3.2)—$n_i^{out} = n_i^{in} + v_i\xi$—that will involve at least two unknowns that have to be solved for simultaneously. Also, stream R would not be involved in any of the overall equations. Consequently, the next step would be to consider species material balances for a system that includes F and/or R. A system composed of the reactor plus the separator is usually a good choice (Figure E6.9b), which is formed by moving the left-hand portion of the system boundary to the right, cutting across the recycle stream and the feed to the reactor. By combining the two units, you can avoid involving the direct output of the reactor. The species balances are

$$n_i^{out} \quad = \quad n_i^{in} \quad + \quad v_i\xi$$

$$\text{CO:} \quad (R + 6.25)y = (32.5 + Ry) + (-1)(\xi)$$

$$H_2\text{:} \quad (R + 6.25)x = (657.3 + Rx) + (-2)(\xi)$$

$$CH_3OH\text{:} \quad (E)(1.0)0 + (1)(\xi)$$

(*Continues*)

Example 6.9 Purge (*Continued*)

CH_4: $(R + 6.25)(0.032) = 0.2 + (R)(0.032) + 0$ (redundant equation (used previously))

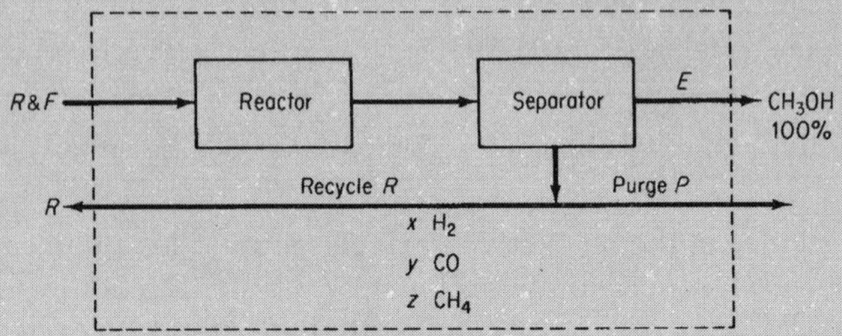

Figure E6.9b

In addition, the fraction conversion f is related to ξ by Equation (6.4):

$$f_{SP} = \frac{-v_{LR}\xi}{n_{LR}^{\text{reactor feed}}} \text{ or } 0.18 = \frac{-(-1)(\xi)}{32.5 + (Ry)}$$

and the implicit equation is

$$x + y + 0.032 = 1$$

You can solve these five equations with an equation editor such as Polymath or by successive substitution to get

E	CH_3OH	31.25
P	purge	6.25
R	recycle	705
x	H_2	0.768
y	CO	0.200
z	CH_4	0.032

Step 10

Check each equation to determine if they all balance.

Self-Assessment Test

Questions

1. Explain what bypassing means in words and also with a diagram.
2. Indicate whether the following statements are true or false:
   a. Purge is used to maintain a concentration of a minor component of a process stream below some set point so that it does not accumulate in the process.
   b. Bypassing means that a process stream enters the process in advance of the feed to the process.
   c. A trace component in a stream or produced in a reactor will have negligible effect on the overall material balance when recycle occurs.
3. Is the waste stream the same as a purge stream in a process?

Problems

1. In the famous Haber process (Figure SAT6.4P1) to manufacture ammonia, the reaction is carried out at pressures of 800 to 1000 atm and at 500°C to 600°C using a suitable catalyst. Only a small fraction of the material entering the reactor reacts on one pass, so recycle is needed. Also, because the nitrogen is obtained from the air, it contains almost 1% rare gases (chiefly argon) that do not react. The rare gases would continue to build up in the recycle until their effect on the reaction equilibrium became adverse, so a small purge stream is used.

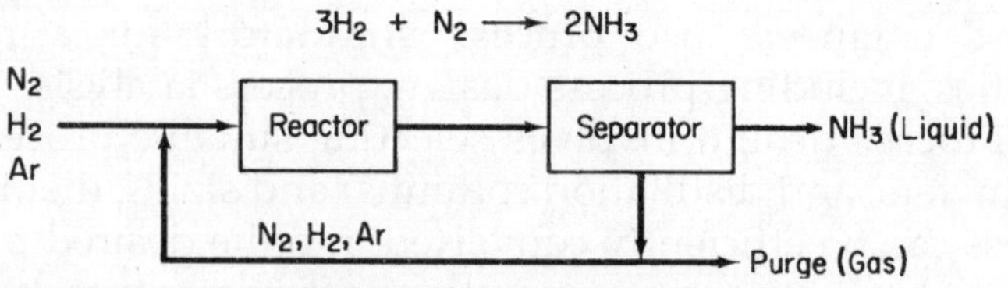

Figure SAT6.4P1

The fresh feed of gas composed of 75.16% H_2, 24.57% N_2, and 0.27% Ar is mixed with the recycled gas and enters the reactor with a composition of 79.52% H_2. The gas stream leaving the ammonia separator contains 80.01% H_2 and no ammonia. The product ammonia contains no dissolved gases. Per 100 moles of fresh feed:
   a. How many moles are recycled and purged?
   b. What is the percent conversion of hydrogen per pass?
2. Figure SAT6.4P2 shows a simplified process to make ethylene dichloride ($C_2H_4Cl_2$). The feed data have been placed on the figure. Ninety percent conversion of the C_2H_4 occurs on each pass through the reactor. The overhead

stream from the separator contains 98% of the Cl_2 entering the separator, 92% of the entering C_2H_4, and 0.1% of the entering $C_2H_4Cl_2$. Five percent of the overhead from the separator is purged.

Calculate (a) the flow rate and (b) the composition of the purge stream.

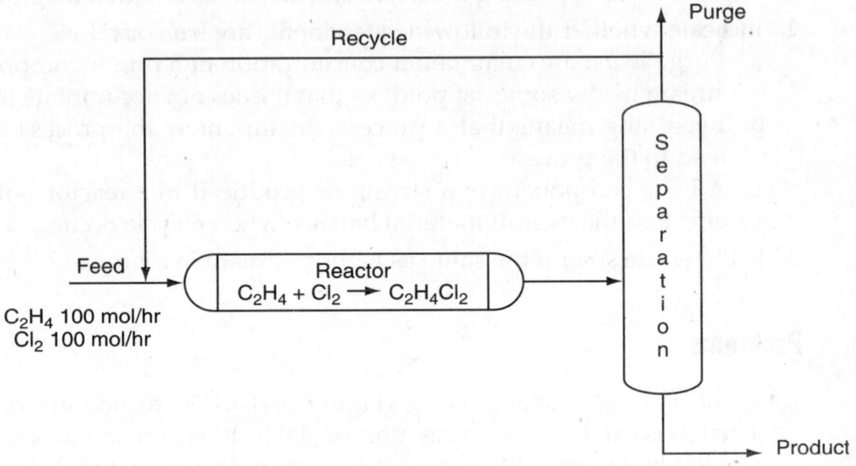

Figure SAT6.4P2

6.5 The Industrial Application of Material Balances

Process engineers use **process simulators** for a number of important activities, including process design, process analysis, and process optimization. **Process design** involves selecting suitable processing units (e.g., reactors, mixers, and distillation columns) and sizing them so that the feed to the process can be efficiently converted into the desired products. **Process analysis** involves comparing predictions of process variables using models of the process units with the measurements made in the operating process. By comparing corresponding values of variables, you can determine if a particular process unit is functioning properly. If discrepancies exist, the predictions from the model can provide insight into the root causes of these discrepancies. In addition, process models can be used to carry out studies that evaluate alternate processing approaches and studies of debottlenecking, that is, methods designed to increase the production rate of the overall process. **Process optimization** is directed at determining the most profitable way to operate the process. For process optimization, models of the major processing units in the process are used to determine the operating conditions, such as product compositions and reactor temperatures, that yield the maximum profit for the process (subject to appropriate constraints).

For each of the three process applications, models of the processing units are based on material balances. For simple equipment, just a few material balances for each component in the system are all that is needed to model the equipment. For more complex equipment such as a distillation column, you will find the model involves a set of material balance equations for each tray in the column, and some industrial columns have over 200 trays. For process design and many applications of process analysis, each processing unit can be analyzed and solved separately. Modern computer codes make it possible to solve extensive sets of simultaneous equations. For example, the optimization model for an ethylene plant usually has over 150,000 equations with material balances constituting over 90% of the equations and requiring many solutions to determine the optimum set of operating conditions for the entire plant.

Now consider material balance closure for industrial processes. One important way in which individual material balances are applied industrially is to check that "in = out"—that is, to determine how well the material balances balance using process measurements in the equations. You look for what is called *closure*, namely, that the error between "in" and "out" is acceptable. The measured flow rates and measured compositions for all of the streams entering and exiting a process unit are substituted into the appropriate material balance equations. Ideally, the amount (mass) of each component entering the system should equal the amount of that component leaving the system. Unfortunately, the amount of a component entering a process rarely equals the amount leaving the process when you make such calculations. The lack of closure for material balances on an industrial process occurs for several reasons:

1. The process is rarely operating in the steady state. Industrial processes are almost always in a state of flux and rarely approach true steady-state behavior.
2. The flow and composition measurements have a variety of errors associated with them. First, sensor readings have noise (variations in the measurement due to more or less random variations in the readings that do not correspond to changes in the process). The sensor readings can also be inaccurate for a wide variety of other reasons. For example, a sensor may require recalibration because it degrades, or it may be used for a measurement for which it was not designed.
3. A component of interest may be generated or consumed inside the process by reactions that the process engineer has not considered.

As a result, material balance closure during a relatively steady-state period of operation to within ±5% for material balances for most industrial processes is considered reasonable. (Here closure is defined as the calculated difference between the amount of a particular material entering and exiting the process divided by the amount entering multiplied by 100). If special attention is paid to calibrating sensors, material balance closure to ±2% to ±3% can be attained. If special high-accuracy sensors are used, smaller closure of the material balances can be attained, but if faulty sensor readings are used, much greater errors in material balances can be observed. In fact, material balances can be used to determine when faulty sensor readings exist, which is known as data reconcilation.

Looking Back

In this chapter you have seen how systems composed of more than one subsystem can be treated by the same principles that you used to treat single systems. Whether you use combinations of material balances from each of several subsystems or lump all of the units into one system, all you have to do is check to see that the number of independent equations you prepared was adequate to solve for the variables whose values are unknown. Flowsheets can help in the preparation of the equation set. The one new factor brought out in this chapter is that recycle for a reactor usually involves information about the fraction conversion of a reactant for the overall process and for the reactor only.

Glossary

Block diagram A sequence of boxes, circles, and other shapes used to represent operational features of a process flowsheet.

Bypass stream A stream that skips one or more units of a process and goes directly to a downstream unit.

Connections Streams flowing between subsystems (units).

Flowchart A graphical representation of a process layout.

Fresh feed The overall feed to a system.

Gross product The product stream that leaves a reactor.

Mixer Apparatus to combine two or more flow streams.

Once-through fraction conversion The conversion of a reactant based on the amount of material that enters and leaves a reactor.

Overall fraction conversion The conversion of a reactant in a process with recycle based on the fresh feed of the reactant and the overall products.

Overall process The entire system composed of subsystems (units).

Overall products The streams that exit a process.

Process feed The feed stream that enters a reactor, usually used in a process with a reactor and recycle.

Process flowsheet A graphical representation of a process. See **flowchart.**

Purge A stream bled off from a process to remove the accumulation of inerts or unwanted material that might otherwise build up in the recycle streams.

Recycle Material (or energy) that leaves a process unit that is downstream and is returned to the same unit or an upstream unit for processing again.

Recycle stream The stream that recycles material (or energy).

Recycle system A system that includes one or more recycle streams.

Separator Apparatus that produces two or more streams of different composition from the fluid(s) entering the apparatus.

Sequential combination of units A set of units arranged in a row.

Single-pass fraction conversion The conversion based on what enters and leaves a reactor. See **once-through fraction conversion.**

Splitter Apparatus that divides a flow into two or more streams.

Subsystem A designated part of a complete system.

Well-mixed system Material within a system (equipment) is of uniform composition, and the exit stream(s) is of the same composition as the material inside the system.

Supplementary References

Cheremisinoff, P. N., and P. Cheremisinoff. *Encyclopedia of Environmental Control Technology: Waste Minimization and Recycling*, Gulf Publishing, Houston, TX (1992).

Lund, H. F. (ed.). *McGraw-Hill Recycling Handbook*, 2nd ed., McGraw-Hill, New York (2000).

Luyben, W. L., and L. A. Wenzel. *Chemical Process Analysis: Mass and Energy Balances* (International Series in Physical and Chemical Engineering Science), Prentice Hall, Englewood Cliffs, NJ (1988).

Myers, A. L., and W. D. Seider. *Introduction to Chemical Engineering and Computer Calculations*, Prentice Hall, Englewood Cliffs, NJ (1976).

Noll, K. E., N. Haas, C. Schmidt, and P. Kodukula. *Recovery, Recycle, and Reuse of Industrial Wastes* (Industrial Waste Management Series), Franklin Book Co. (1985).

Veslind, P. *Unit Operations in Resource Recovery Engineering*, Prentice Hall, Englewood Cliffs, NJ (1981).

Web Site

www.nap.edu/books/0309063779/html/28.html

Problems

6.1 Primary Concepts

*6.1.1 For Figure P6.1.1, how many independent equations are obtained from the overall balance around the entire system plus the overall balances on units *A* and *B*? Assume that only one component exists in each stream.

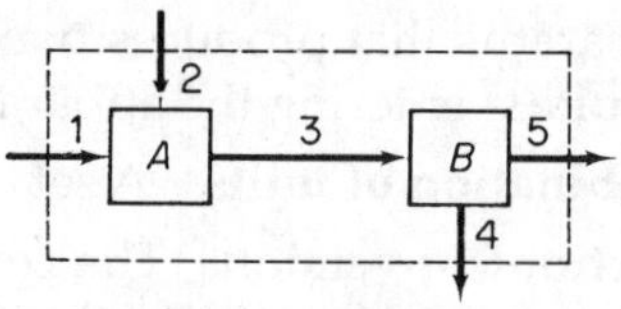

Figure P6.1.1

**6.1.2 What is the maximum number of independent material balances that can be written for the process in Figure P6.1.2?

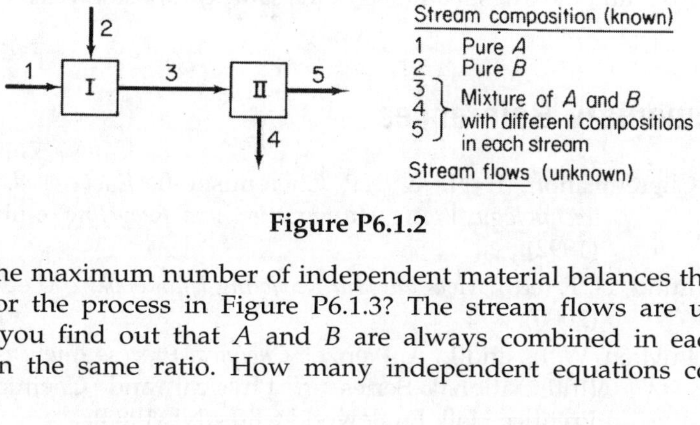

Figure P6.1.2

**6.1.3 What is the maximum number of independent material balances that can be written for the process in Figure P6.1.3? The stream flows are unknown. Suppose you find out that *A* and *B* are always combined in each of the streams in the same ratio. How many independent equations could you write?

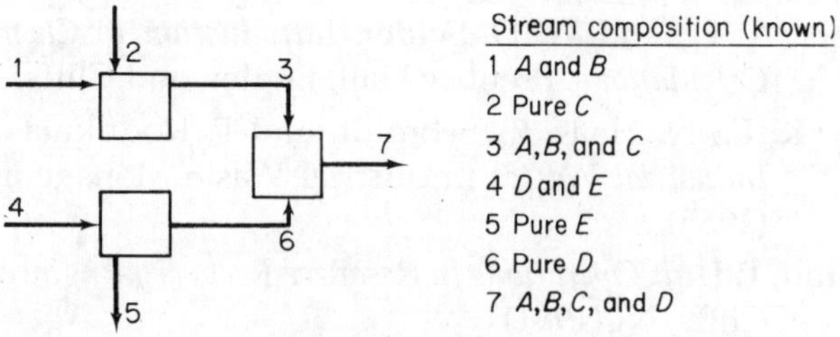

Figure P6.1.3

**6.1.4 The diagram in Figure P6.1.4 represents a typical but simplified distillation column. Streams 3 and 6 consist of steam and water and do not come in contact with the fluids in the column that contains two components. Write the total and component material balances for the three sections of the column. How many independent equations would these balances represent? Assume that stream 1 contains n components.

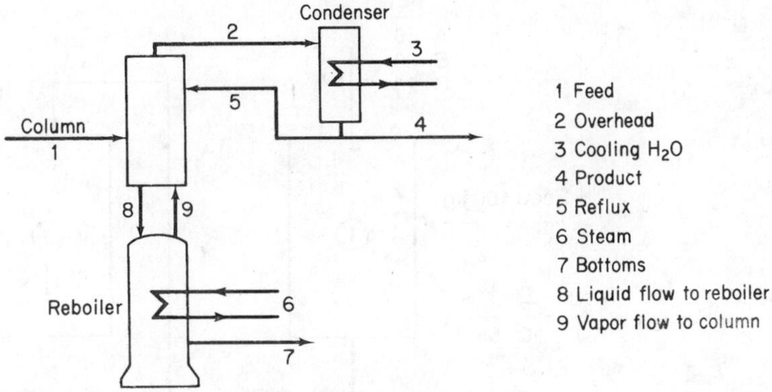

Figure P6.1.4

**6.1.5 A distillation process is shown in Figure P6.1.5. You are asked to solve for all of the values of the stream flows and compositions. How many variables and unknowns are there in the system? How many independent material balance equations can you write? Explain each answer and show all details of how you reached your decision. For each stream (except F), the only components that occur are labeled below the stream

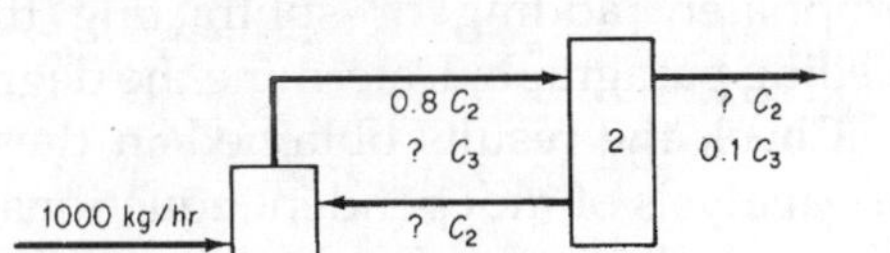

Figure P6.1.5

***6.1.6 In Figure P6.1.6 you see two successive liquid separation columns operating in tandem in the steady state (and with no reaction taking place). The compositions of the feed and products are as shown in the figure. The amount of W_2 is 20% of the feed. Focus only on the material balances. Write down the names of the material balances for each column treated as separate units (one set for each column), along with the balances themselves placed next to the names of the balances. Also, place an asterisk in front of the names of the balances that will constitute a set of independent equations for each column. Write down in symbols the unknowns for each column. Calculate the degrees of freedom for each column separately.

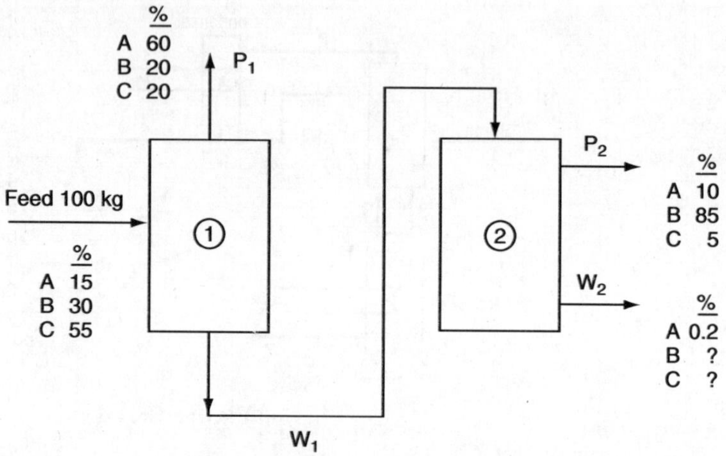

Figure P6.1.6

Then determine the number of independent equations for the overall system composed of the two columns together by listing the duplicate variables (by symbols) and unused specifications as well as the redundant equations, and appropriately adding or subtracting them from the values found in the preceding paragraph. Determine the degrees of freedom for the overall system.

Check the results obtained in the second paragraph by repeating the entire analysis of independent equations, unknowns, and degrees of freedom for the overall system. Do they agree? They should.

Do not solve any of the equations in this problem.

6.2 Sequential Multi-Unit Systems

**6.2.1 Examine Figure P6.2.1, provided through the courtesy of Professor Mike Cutlip.

a. Calculate the molar flow rate of D1, D2, B1, and B2.
b. Reduce the feed flow rate for each one of the compounds by 1% in turn. Calculate the flow rates of D1, D2, B1, and B2 again. Do you notice something unusual?

Explain your results.

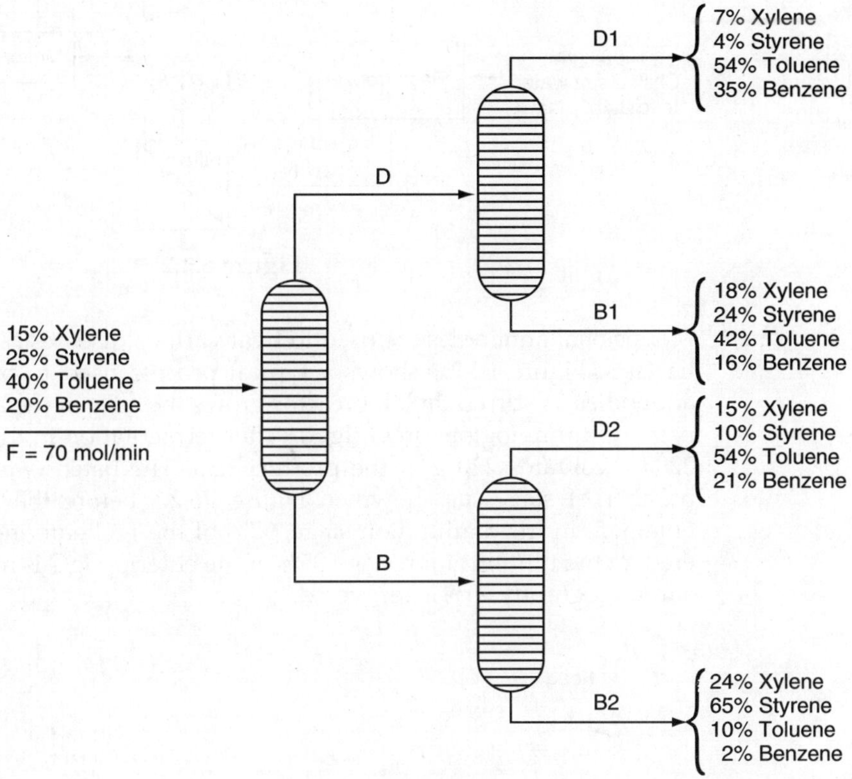

Figure 6.2.1

****6.2.2** Figure P6.2.2 shows a schematic for making freshwater from seawater by freezing. The prechilled seawater is sprayed into a vacuum at a low pressure. The cooling required to freeze some of the feed seawater comes from evaporation of a fraction of the water entering the chamber. The concentration of the brine stream, B, is 4.8% salt. The pure salt-free water vapor is compressed and fed to a melter at a higher pressure, where the heat of condensation of the vapor is removed through the heat of fusion of the ice, which contains no salt. As a result, pure cold water and concentrated brine (6.9%) leave the process as products.

a. Determine the flow rates of streams *W* and *D* if the feed is 1000 kg/hr.
b. Determine the flow rates of streams *C*, *B*, and *A* per hour.

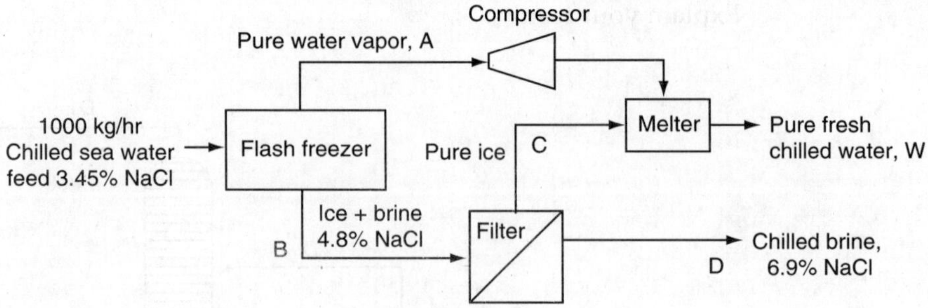

Figure 6.2.2

****6.2.3** Monoclonal antibodies are used to treat various diseases as well as in diagnostic tests. Figure P6.2.3 shows a typical process used to produce monoclonal antibodies. A stirred tank bioreactor grows the cells of the antibody of interest, namely, immunoglobulin G (IgG). After fermentation in the reactor, a batch of 2200 L contains 220 g of the product IgG. The batch is processed through a number of stages as shown in Figure P6.2.3 before the purified product is obtained. In the diafiltration stage, 95% of the IgG entering the filter is recovered; in the ultrafiltration stage 95% of the entering IgG is recovered; and in the chromatography 90% is recovered.

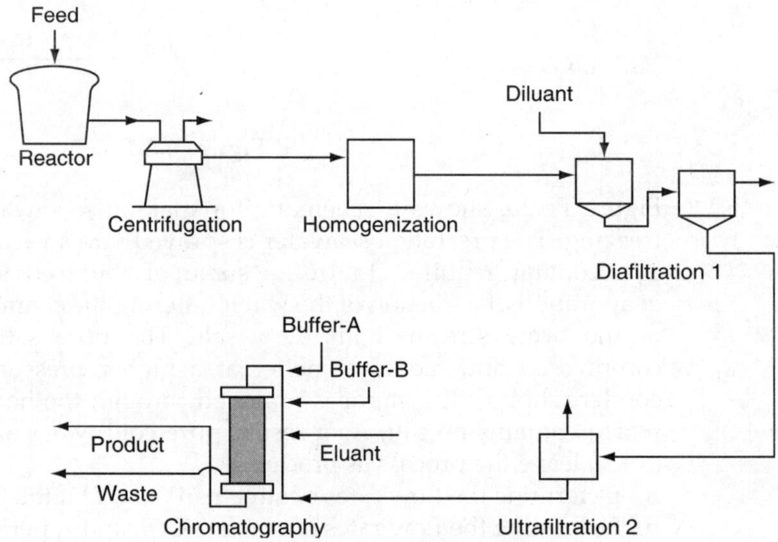

Figure 6.2.3

Table P6.2.3

Component	Total Inlet	Total Outlet	Product
Ammonium sulfate	64.69	64.69	
Biomass	0.00	0.87	
Glycerol	1.85	1.85	
IgG	0.00	0.22	0.14
Growth media	21.76	8.41	
Na_3 citrate	0.80	0.80	
Phosphoric acid	1040.96	1040.96	
Sodium hydrophosphate	6.83	6.81	
Sodium chloride	55.18	55.19	
Tris-HCl	0.69	0.69	
Water	11,459.59	11,458.80	
Injection water	18,269.54	18,269.54	
Total	**30,928.72**	**30,928.72**	**0.14**

Table P6.2.3 lists the essential components entering and leaving the overall process in kilograms per batch. What is the fractional yield of the product IgG of the 220 g produced in the reactor?

****6.2.4** In a tissue paper machine (Figure P6.2.4), stream N contains 85% fiber. Find the unknown fiber values (all values in the figure are in kilograms) in kilograms for each stream.

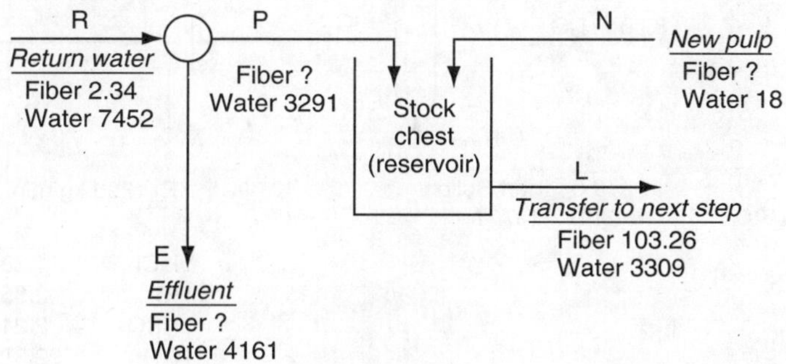

Figure P6.2.4

****6.2.5** An enzyme is a protein that catalyzes a specific reaction, and its activity is reported in a quantity called "units." The specific activity is a measure of the purity of an enzyme. The fractional recovery of an enzyme can be calculated from the ratio of the specific activity (units per milligram) after processing occurs to the initial specific activity. A three-stage process for the purification of an enzyme involves

1. Breakup of the cells in a biomass to release the intercellular products
2. Separation of the enzyme from the intercellular product
3. Further separation of the enzyme from the output of stage 2

Based on the following data for one batch of biomass, calculate the percent recovery of the enzyme after each of the three stages of the process. Also calculate the purification of the enzyme that is defined as the ratio of the specific activity to the initial specific activity. Fill in the blank columns of the following table.

Stage No.	Activity (units)	Protein Present (mg)	Specific Activity (units/mg)	Percent Recovery	Purification
1	6860	76,200			
2	6800	2200			
3	5300	267			

***6.2.6 Several streams are mixed as shown in Figure P6.2.6. Calculate the flows of each stream in kilograms per second.

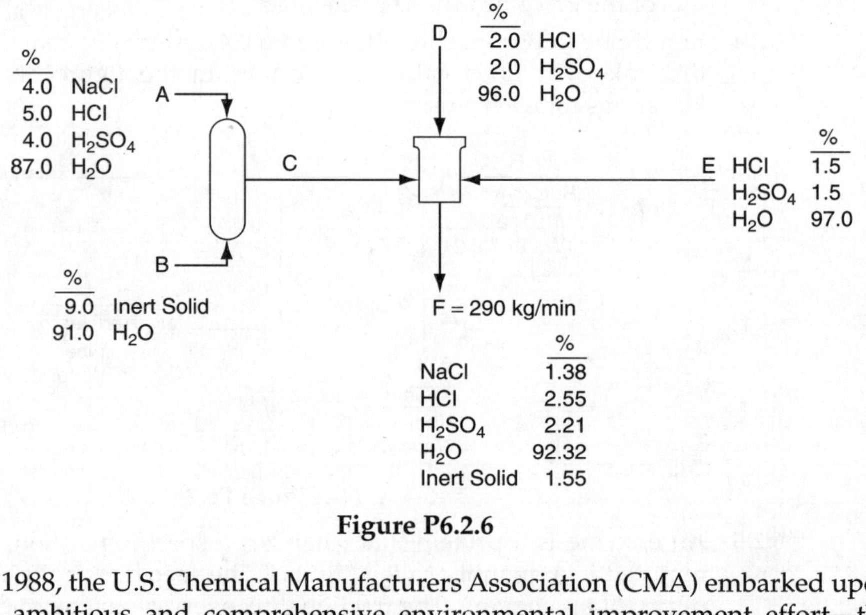

Figure P6.2.6

***6.2.7 In 1988, the U.S. Chemical Manufacturers Association (CMA) embarked upon an ambitious and comprehensive environmental improvement effort—the Responsible Care initiative. Responsible Care committed all of the 185 members of the CMA to ensure continual improvement in the areas of health, safety, and environmental quality, as well as in eliciting and responding to public concerns about their products and operations.

One of the best ways to reduce or eliminate hazardous waste is through source reduction. Generally, this means using different raw materials or redesigning the production process to eliminate the generation of hazardous by-products. As an example, consider the following countercurrent extraction process (Figure P6.2.7) to recover xylene from a stream that contains 10% xylene and 90% solids by weight.

The stream from which xylene is to be extracted enters Unit 2 at a flow rate of 2000 kg/hr. To provide a solvent for the extraction, pure benzene is fed to Unit 1 at a flow rate of 1000 kg/hr. The mass fractions of the xylene in the solids stream (F) and the clear liquid stream (S) have the following relations: $10\,\omega^{F^1}_{Xylene} = \omega^{S^2}_{Xylene}$ and $10\,\omega^{F^2}_{Xylene} = \omega^{S^1}_{Xylene}$.

Determine the benzene and xylene concentrations in all of the streams. What is the percent recovery of the xylene entering the process at Unit 2?

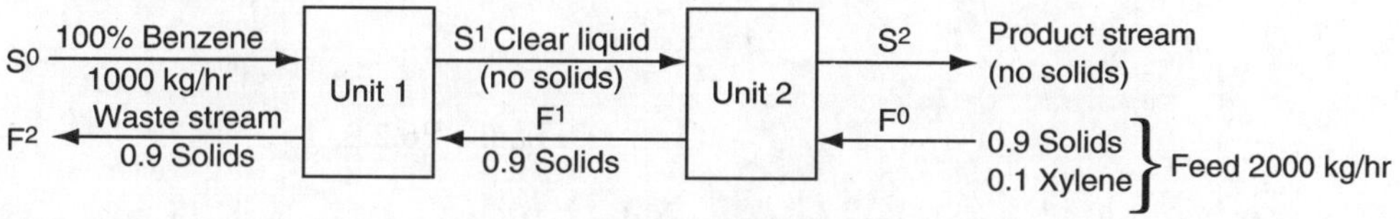

Figure P6.2.7

***6.2.8 Figure P6.2.8 shows a three-stage separation process. The ratio of P_3/D_3 is 3, the ratio of P_2/D_2 is 1, and the ratio of A to B in stream P_2 is 4 to 1. Calculate the composition and percent of each component in stream E.

Hint: Although the problem comprises connected units, application of the standard strategy of problem solving will enable you to solve it without solving an excessive number of equations simultaneously.

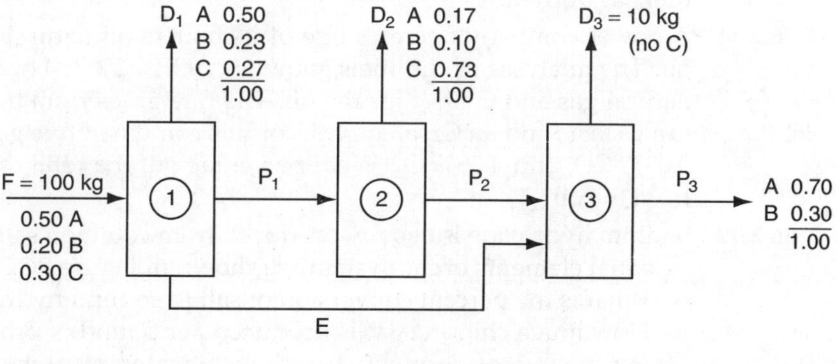

Figure P6.2.8

***6.2.9 Metallurgical-grade silicon is purified to electronic grade for use in the semiconductor industry by chemically separating it from its impurities. The Si metal reacts in varying degrees with hydrogen chloride gas at 300°C to form several polychlorinated silanes. Trichlorosilane is liquid at room temperature and is easily separated by fractional distillation from the other gases. If 100 kg of silicon is reacted as shown in Figure P6.2.9, how much trichlorosilane is produced?

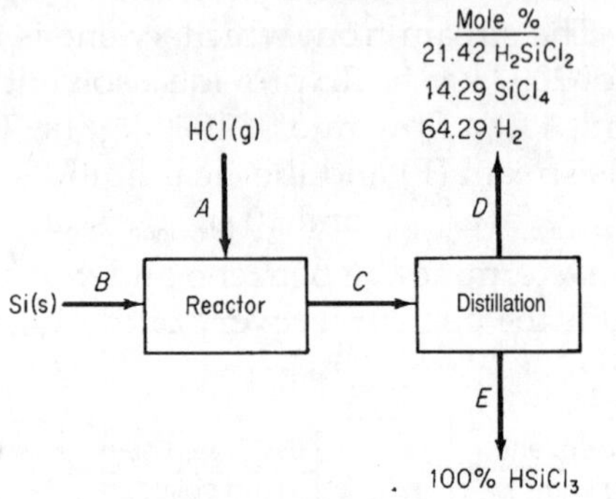

Figure P6.2.9

***6.2.10 A furnace burns fuel gas of the following composition—70% methane (CH_4), 20% hydrogen (H_2), and 10% ethane (C_2H_6)—with excess air. An oxygen probe placed at the exit of the furnace reads 2% oxygen in the exit gases. The gases are then passed through a long duct to a heat exchanger. At the entrance to the heat exchanger the Orsat analysis of the gas reads 6% O_2. Is the discrepancy due to the fact that the first analysis is on a wet basis and the second analysis is on a dry basis (no water condenses in the duct), or due to an air leak in the duct?

If the former, give the Orsat analysis of the exit gas from the furnace. If the latter, calculate the amount of air that leaks into the duct per 100 mol of fuel gas burned.

***6.2.11 A power company operates one of its boilers on natural gas and another on oil. The analyses of the fuels show 96% CH_4, 2% C_2H_2, and 2% CO_2 for the natural gas and $C_nH_{1.8n}$ for the oil. The flue gases from both groups enter the same stack, and an Orsat analysis of this combined flue gas shows 10.0% CO_2, 0.63% CO, and 4.55% O_2. What percentage of the total carbon burned comes from the oil?

***6.2.12 Sodium hydroxide is usually produced from common salt by electrolysis. The essential elements of the system are shown in Figure P6.2.12.

a. What is the percent conversion of salt to sodium hydroxide?
b. How much chlorine gas is produced per pound of product?
c. Per pound of product, how much water must be evaporated in the evaporator?

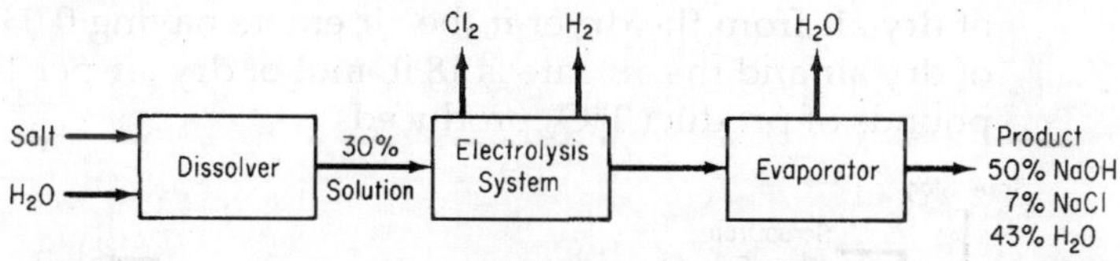

Figure P6.2.12

******6.2.13** The flowsheet shown in Figure P6.2.13 represents the process for the production of titanium dioxide (TiO_2) used by Canadian Titanium Pigments at Varennes, Quebec. Sorel slag of the following analysis:

	wt %
TiO_2	70
Fe	8
Inert silicates	22

is fed to a digester and reacted with H_2SO_4, which enters as 67% by weight H_2SO_4 in a water solution. The reactions in the digester are as follows:

$$TiO_2 + H_2SO_4 \rightarrow TiOSO_4 + H_2O \quad (1)$$

$$Fe + \tfrac{1}{2}O_2 + H2SO_2 \rightarrow FeSO_4 + H_2O \quad (2)$$

Both reactions are complete. The theoretically required amount of H_2SO_4 for the Sorel slag is fed. Pure oxygen is fed in the theoretical amount for all of the Fe in the Sorel slag. Scrap iron (pure Fe) is added to the digester to reduce the formation of ferric sulfate to negligible amounts. Thirty-six pounds of scrap iron are added per pound of Sorel slag.

The products of the digester are sent to the clarifier, where all of the inert silicates and unreacted Fe are removed. The solution of $TiOSO_4$ and $FeSO_4$ from the clarifier is cooled, crystallizing the $FeSO_4$, which is completely removed by a filter. The product $TiOSO_4$ solution from the filter is evaporated down to a slurry that is 82% by weight $TiOSO_4$.

The slurry is sent to a dryer from which a product of pure hydrate, $TiOSO_4 \cdot H_2O$, is obtained. The hydrate crystals are sent to a direct-fired rotary kiln, where the pure TiO_2 is produced according to the following reaction:

$$TiOSO_4 \cdot H_2O \rightarrow TiO_2 + H_2SO_4 \quad (3)$$

Reaction (3) is complete.

On the basis of 100 lb of Sorel slag feed, calculate (a) the pounds of water removed by the evaporator; (b) the exit pounds of H_2O per pound of dry air from the dryer if the air enters having 0.036 moles of H_2O per mole of dry air and the air rate is 18 lb mol of dry air per 100 lb of Sorel slag; (c) the pounds of product TiO_2 produced.

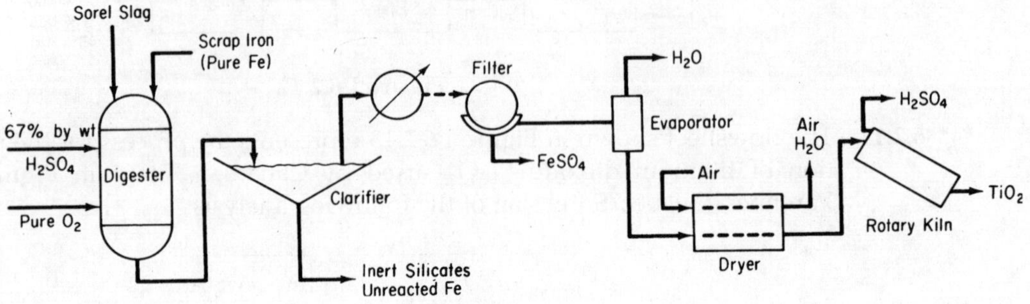

Figure P6.2.13

6.3 Recycle Systems

*6.3.1 How many recycle streams exist in each of the following processes?

a.

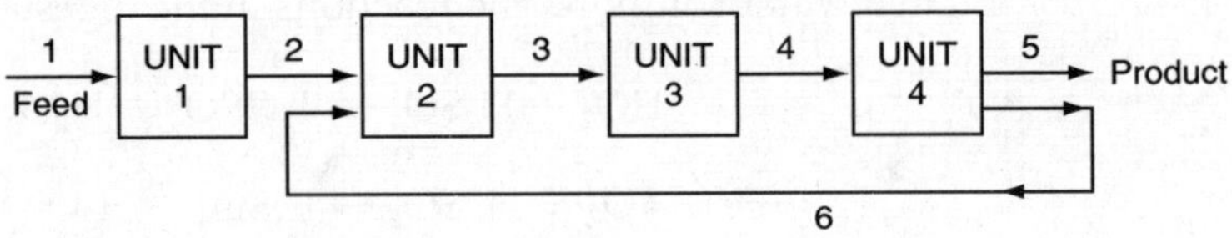

Figure P6.3.1a

b.

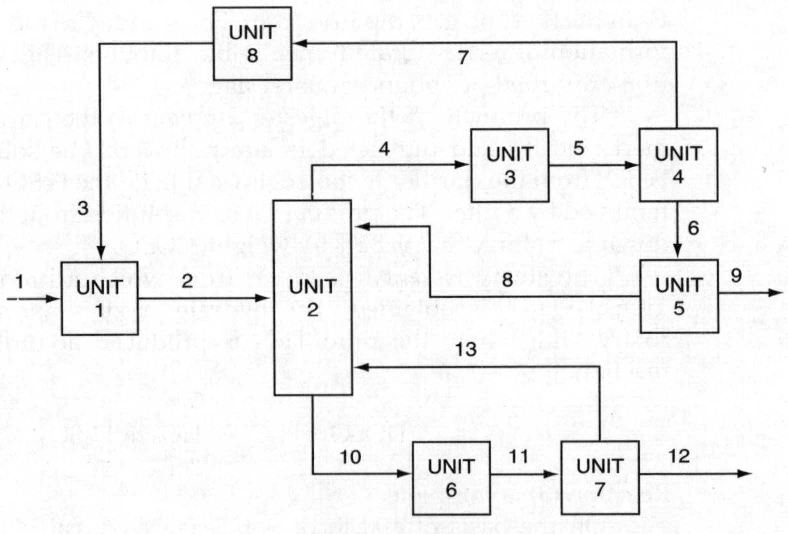

Figure P6.3.1b

c.

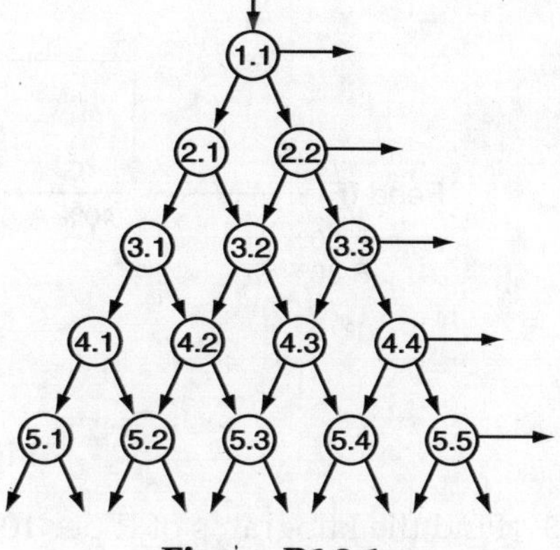

Figure P6.3.1c

d.

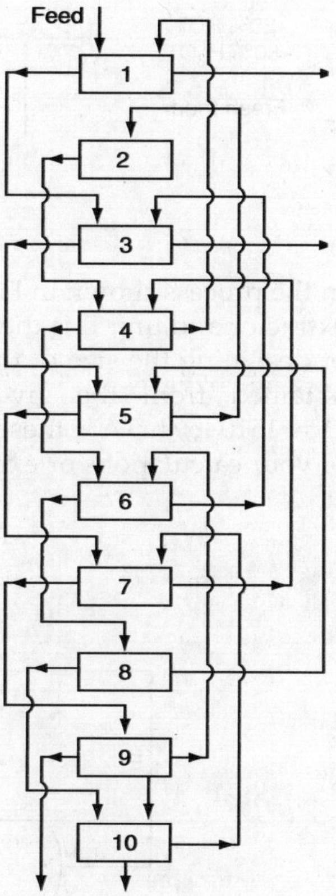

Figure P6.3.1d

*6.3.2 Find the kilograms of recycle per kilogram of feed if the amount of waste (W) is 60 kg of A.

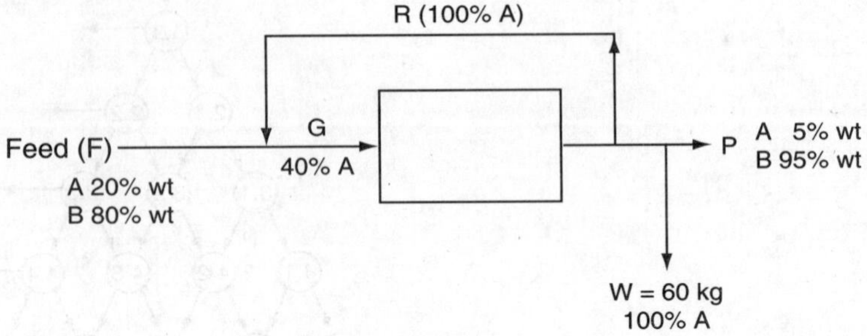

Figure P6.3.2

*6.3.3 Find the kilograms of R per 100 kg of fresh feed.

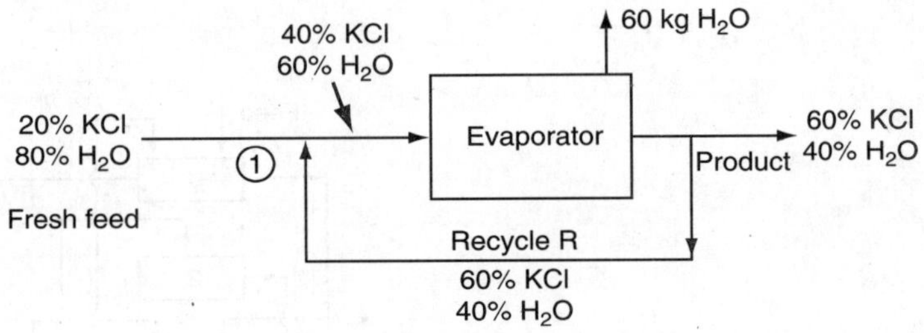

Figure P6.3.3

**6.3.4 In the process shown in Figure P6.3.4, unit I is a liquid-liquid solvent extractor and unit II is the solvent recovery system. For the purposes of designing the size of the pipes for streams C and D, the designer obtained from the given data values of C = 9630 lb/hr and D = 1510 lb/hr. Are these values correct? Be sure to show all details of your calculations or explain if you do not use calculations.

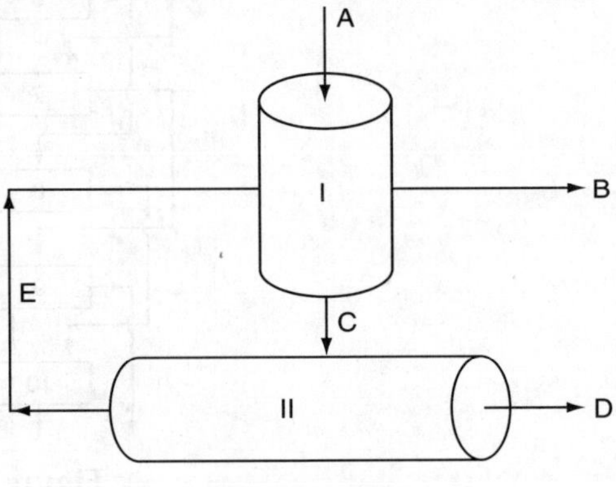

Figure P6.3.4

Known Data

	Flow Rate (lb/hr)	Composition		
		Butene	Butadiene	Solvent
A	5000	0.75	0.25	
B			1.00	
C				
D		0.05	0.95	
E	10,000		0.01	0.99

**6.3.5 The ability to produce proteins through genetic engineering of microbial and mammalian cells and the need for high-purity therapeutic proteins has established a need for efficient large-scale protein purification schemes.

The system of continuous affinity-recycle extraction (CARE) combines the advantages of well-accepted separation methods, such as affinity chromatography, liquid extraction, and membrane filtration, while avoiding the drawbacks inherent in batch and column operations.

The technical feasibility of the CARE system was studied using β-galactosidase affinity purification as a test system. Figure P6.3.5 shows the process. What is the recycle flow rate in milliliters per hour in each stream? Assume that the concentrations of U are equivalent to the concentrations of the β-galactosidase in solution, and that steady state exists.

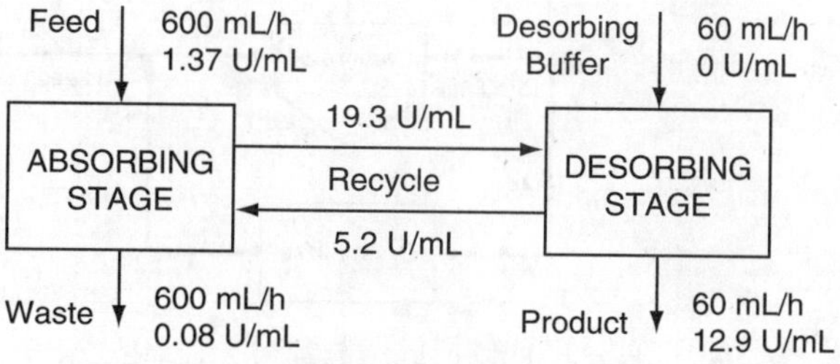

Figure P6.3.5

**6.3.6 Cereal is being dried in a vertical dryer by air flowing countercurrent to the cereal. To prevent breakage of the cereal flakes, exit air from the dryer is recycled. For each 1000 kg/hr of wet cereal fed to the dryer, calculate the input of moist fresh air in kilograms per hour and the recycle rate in kilograms per hour.

Data on stream compositions (note that some are mass and others mole fractions):

	Fresh Air	Wet Cereal	Exit Air	Dried Cereal	Air Entering Dryer
H_2O	0.0132	0.200	0.263	0.050	0.066

	Fresh air	Wet cereal	Exit air	Dried cereal	Air entering drier
H_2O	0.0132	0.200	0.263	0.050	0.066

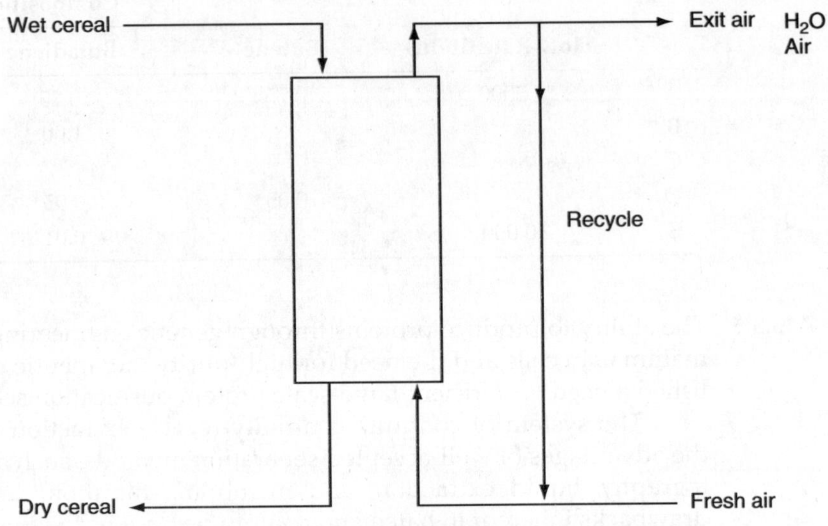

Figure P6.3.6

****6.3.7** Examine Figure P6.3.7. What is the quantity of the recycle stream in kilograms per hour? In stream C the composition is 4% water and 96% KNO_3.

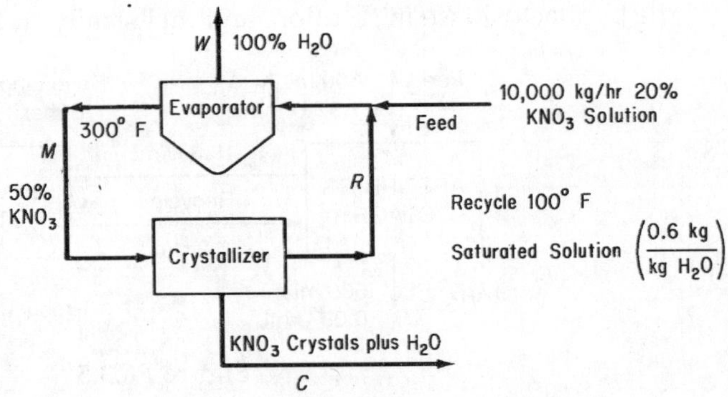

Figure P6.3.7

****6.3.8** Examine Figure P6.3.8 (data for 1 hr).

a. What is the single-pass conversion of H_2 in the reactor?
b. What is the single-pass conversion of CO?
c. What is the overall conversion of H_2?
d. What is the overall conversion of CO?

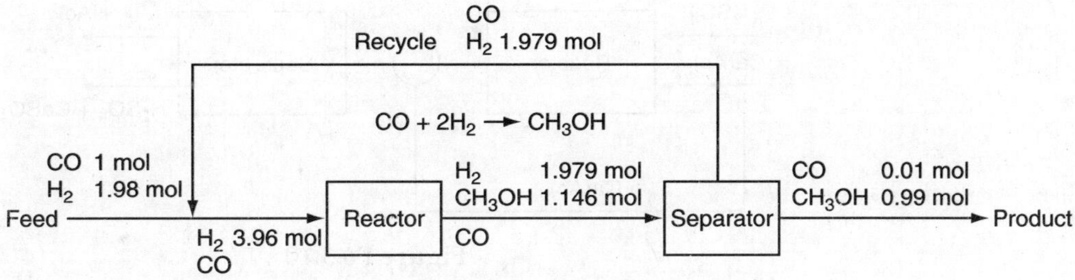

Figure P6.3.8

****6.3.9** Hydrogen, important for numerous processes, can be produced by the shift reaction:

$$CO + H_2O \rightarrow CO_2 + H_2$$

In the reactor system shown in Figure P6.3.9, the conditions of conversion have been adjusted so that the H_2 content of the effluent from the reactor is 3 mol %. Based on the data in Figure P6.3.9:

a. Calculate the composition of the fresh feed.
b. Calculate the moles of recycle per mole of hydrogen produced.

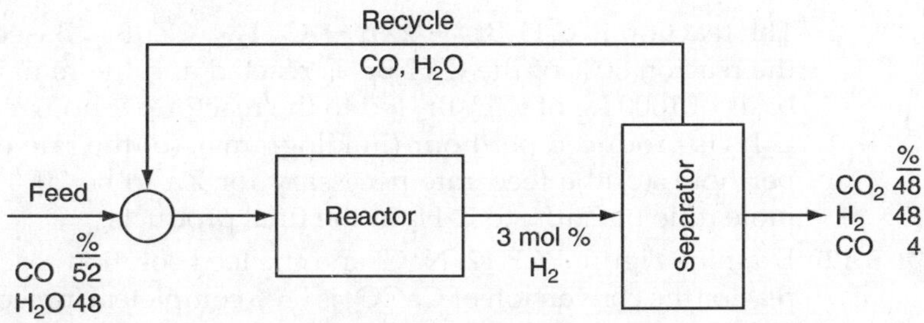

Figure P6.3.9

****6.3.10** Acetic acid (HAc) is to be generated by the addition of 10% excess sulfuric acid to calcium acetate ($Ca(Ac)_2$). The reaction $Ca(Ac)_2 + H_2SO_4 \rightarrow CaSO_4 + 2HAc$ goes to 90% completion based on a single pass through the reactor. The unused $Ca(Ac)_2$ is separated from the products of the reaction and recycled. The HAc is separated from the remaining products. Find the amount of recycle per hour based on 1000 kg of $Ca(Ac)_2$ feed per hour, and also calculate the kilograms of HAc manufactured per hour. See Figure P6.3.10, which illustrates the process.

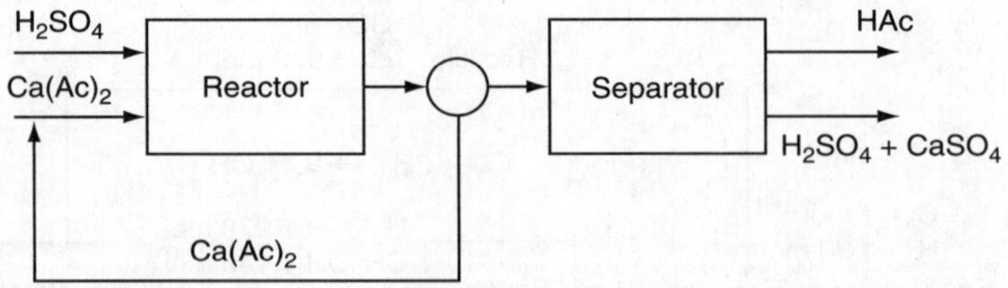

Figure P6.3.10

****6.3.11** The reaction of ethyl-tetrabromide with zinc dust proceeds as shown in Figure P6.3.11.

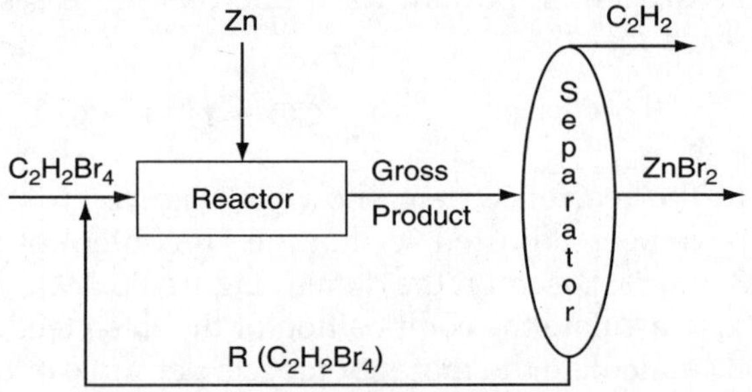

Figure P6.3.11

The reaction is $C_2H_2Br_4 + 2Zn \rightarrow C_2H_2 + 2ZnBr_2$. Based on one pass through the reactor, 80% of the $C_2H_2Br_4$ is reacted and the remainder recycled. On the basis of 1000 kg of $C_2H_2Br_4$ fed to the reactor per hour, calculate (a) how much C_2H_2 is produced per hour (in kilograms), (b) the rate of recycle in kilograms per hour, (c) the feed rate necessary for Zn to be 20% in excess, and (d) the mole ratio of $ZnBr_2$ to C_2H_2 in the final products.

****6.3.12** Examine Figure P6.3.12. NaCl and the feed solution react to form $CaCl_2$. In the reactor the conversion of $CaCO_3$ is 76% complete. Unreacted $CaCO_3$ is recycled. Calculate (a) the kilograms of Na_2CO_3 exiting the separator per 1000 kg of feed, and (b) the kilograms of $CaCO_3$ recycled per 1000 kg of feed.

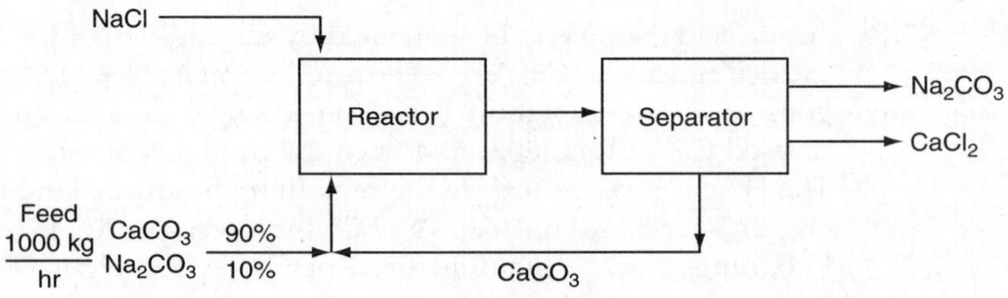

Figure P6.3.12

**6.3.13 Natural gas (CH_4) is burned in a furnace using 15% excess air based on the complete combustion of CH_4. One of the concerns is that the exit concentration of NO (from the combustion of N_2) is about 415 ppm. To lower the NO concentration in the stack gas to 50 ppm it is suggested that the system be redesigned to recycle a portion of the stack gas back through the furnace. You are asked to calculate the amount of recycle required. Will the scheme work? Ignore the effect of temperature on the conversion of N_2 to NO; that is, assume the conversion factor is constant.

***6.3.14 A plating plant has a waste stream containing zinc and nickel in quantities in excess of that allowed to be discharged into the sewer. The proposed process to be used as a first step in reducing the concentration of Zn and Ni is shown in Figure P6.3.14. Each stream contains water. The concentrations of several of the streams are listed in the table. What is the flow (in liters per hour) of the recycle stream R_{32} if the feed is 1 L/hr?

	Concentration (g/L)	
Stream	**Zn**	**Ni**
F	100	10.0
P_0	190.1	17.02
P_2	3.50	2.19
R_{32}	4.35	2.36
W	0	0
D	0.10	1.00

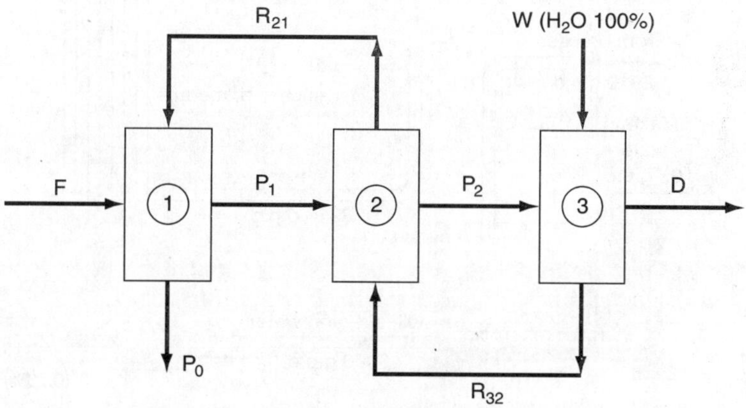

Figure P6.3.14

***6.3.15 Ultrafiltration is a method for cleaning up input and output streams from a number of industrial processes. The lure of the technology is its simplicity, merely putting a membrane across a stream to sieve out physically undesirable oil, dirt, metal particles, polymers, and the like. The trick, of course, is coming up with the right membrane. The screening material has to meet a formidable set of conditions. It has to be very thin (less than 1 μ), highly porous, yet strong enough to hold up month after month under severe stresses of liquid flow, pH, particle abrasion, temperature, and other plant operating characteristics.

A commercial system consists of standard modules made up of bundles of porous carbon tubes coated on the inside with a series of proprietary inorganic compositions. A standard module is 6 in. in diameter and contains 151 tubes, each 4 ft long, with a total working area of 37.5 ft^2 and daily production of 2000 to 5000 gal of filtrate. Optimum tube diameter is about 0.25 in. A system probably will last at least two to three years before the tubes need replacing from too much residue buildup over the membrane. A periodic automatic chemical cleanout of the tube bundles is part of the system's normal operation. On passing through the filter, the exit stream concentration of oil plus dirt is increased by a factor of 20 over the entering stream.

Calculate the recycle rate in gallons per day (g.p.d.) for the setup shown in Figure P6.3.15, and calculate the concentration of oil plus dirt in the stream that enters the filtration module. The circled values in Figure P6.3.15 are the known concentrations of oil plus dirt.

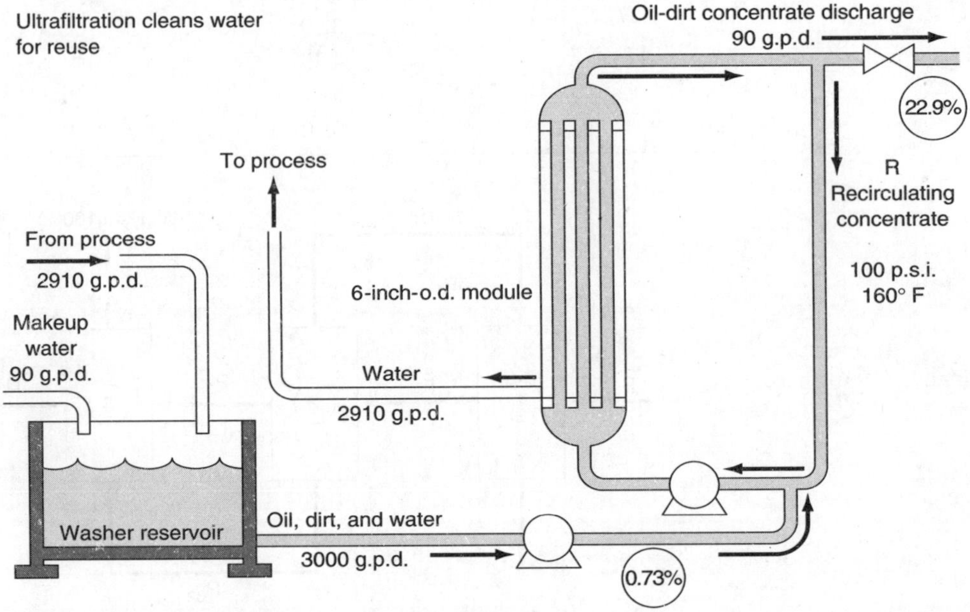

Figure P6.3.15

****6.3.16** To save energy, stack gas from a furnace is used to dry rice. The flowsheet and known data are shown in Figure P6.3.16. What is the amount of recycle gas (in pound moles) per 100 lb of P if the concentration of water in the gas stream entering the dryer is 5.20%?

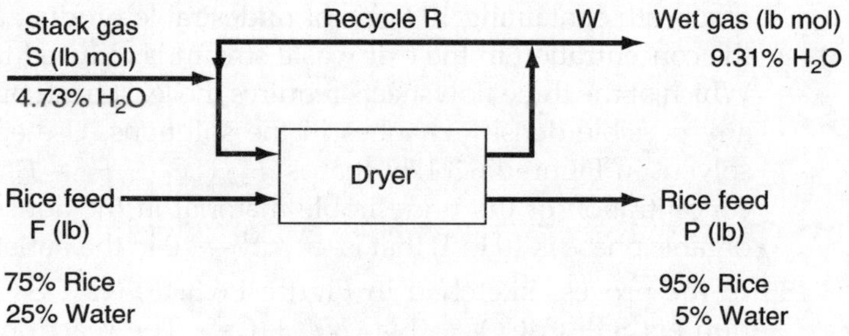

Figure P6.3.16

***6.3.17 This problem is based on the data of G. F. Payne ["Bioseparations of Traditional Fermentation Products," in *Chemical Engineering Problems in Biotechnology*, edited by M. L. Schuler, American Institute of Chemical Engineers, New York (1989)]. Examine Figure P6.3.17. Three separation schemes are proposed to separate the desired fermentation products from the rest of the solution. Ten liters per minute

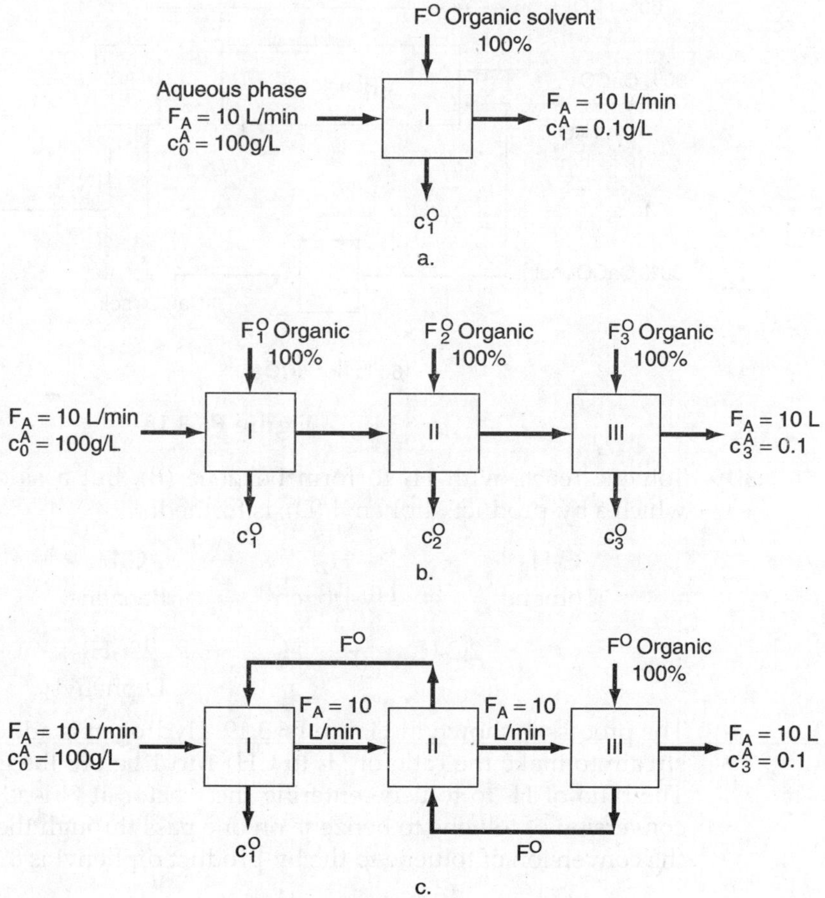

Figure P6.3.17

of a broth containing 100 g/L of undesirable product are to be separated so that the concentration in the exit waste stream is reduced to (not more than) 0.1 g/L. Which of the three flowsheets requires the least fresh pure organic solvent? Ignore any possible density changes in the solutions. Use equal values of the organic solvent in Figure P6.3.17b, that is, $F_1^o + F_2^o + F_3^o = F^o$. The relation between the concentration of the undesirable material in the aqueous phase and that in the organic phase is 10 to 1, that is, $c^A/c^O = 10$ in the outlet streams of each unit.

***6.3.18 In the process sketched in Figure P6.3.18, Na_2CO_3 is produced by the reaction $Na_2S + CaCO_3 \rightarrow Na_2CO_3 + CaS$. The reaction is 90% complete on one pass through the reactor, and the amount of $CaCO_3$ entering the reactor is 50% in excess of that needed. Calculate on the basis of 1000 lb/hr of fresh feed (a) the pounds of Na_2S recycled, and (b) the pounds of Na_2CO_3 solution formed per hour.

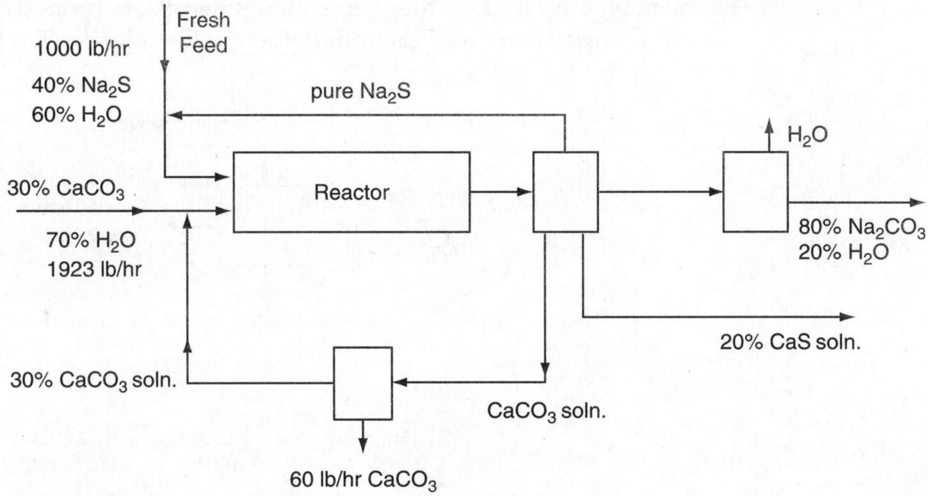

Figure P6.3.18

***6.3.19 Toluene reacts with H_2 to form benzene (B), but a side reaction occurs in which a by-product, diphenyl (D), is formed:

$$\underset{\text{Toluene}}{C_7H_8} + \underset{\text{Hydrogen}}{H_2} \rightarrow \underset{\text{Benzene}}{C_6H_6} + \underset{\text{Methane}}{CH_4} \qquad \text{(a)}$$

$$2C_7H_8 + H_2 \rightarrow \underset{\text{Diphenyl}}{C_{12}H_{10}} + 2CH_4 \qquad \text{(b)}$$

The process is shown in Figure P6.3.19. Hydrogen is added to the gas recycle stream to make the ratio of H_2 to CH_4 1 to 1 before the gas enters the mixer. The ratio of H_2 to toluene entering the reactor at G is $4H_2$ to 1 toluene. The conversion of toluene to benzene on one pass through the reactor is 80%, and the conversion of toluene to the by-product diphenyl is 8% on the same pass.

Calculate the moles of R_G and moles of R_L per hour.

Data:	Compound:	H_2	CH_4	C_6H_6	C_7H_8	$C_{12}H_{10}$
	MW:	2	16	78	92	154

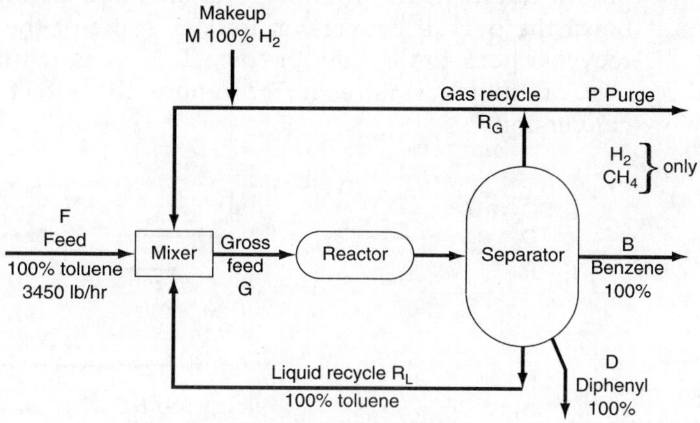

Figure P6.3.19

***6.3.20 The process shown in Figure P6.3.20 is the dehydrogenation of propane (C_3H_8) to propylene (C_3H_6) according to the reaction

$$C_3H_8 \rightarrow C_3H_6 + H_2$$

The conversion of propane to propylene based on the *total* propane feed into the reactor at F_2 is 40%. The product flow rate F_5 is 50 kg mol/hr.

a. Calculate all six flow rates F_1 to F_6 in kilogram moles per hour.
b. What is the percent conversion of propane in the reactor based on the fresh propane fed to the process (F_1)?

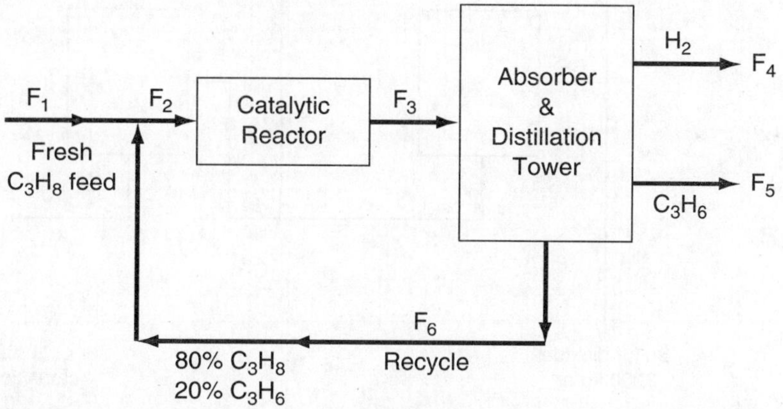

Figure P6.3.20

***6.3.21 Sulfur dioxide may be converted to SO_3, which has many uses, including the production of H_2SO_4 and sulfonation of detergent. A gas stream having the composition shown in Figure P6.3.21 is to be passed through a two-stage converter. The fraction conversion of the SO_2 to SO_3 (on one pass though) in the first stage is 0.75 and in the second stage, 0.65. To boost the overall conversion to 0.95, some of the exit gas from stage 2 is recycled back to the inlet of stage 2. How much must be recycled per 100 mol of inlet gas (stream F)? Ignore the effect of temperature on the conversion.

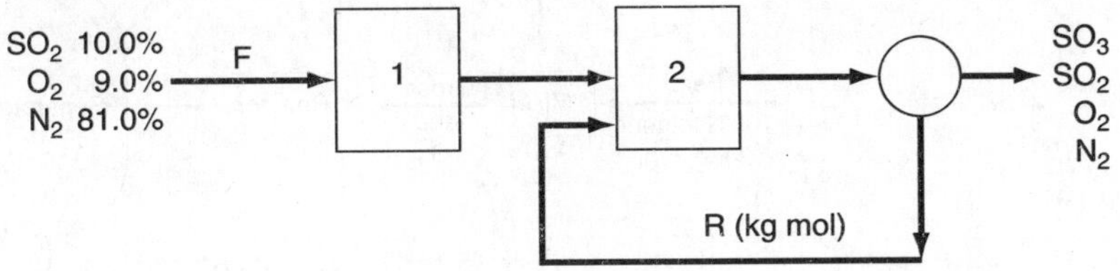

Figure P6.3.21

****6.3.22 Benzene, toluene, and other aromatic compounds can be recovered by solvent extraction with sulfur dioxide. As an example, a catalytic reformate stream containing 70% by weight benzene and 30% non-benzene material is passed through the countercurrent extractive recovery scheme shown in Figure P6.3.22. One thousand kilograms of the reformate stream and 3000 kg

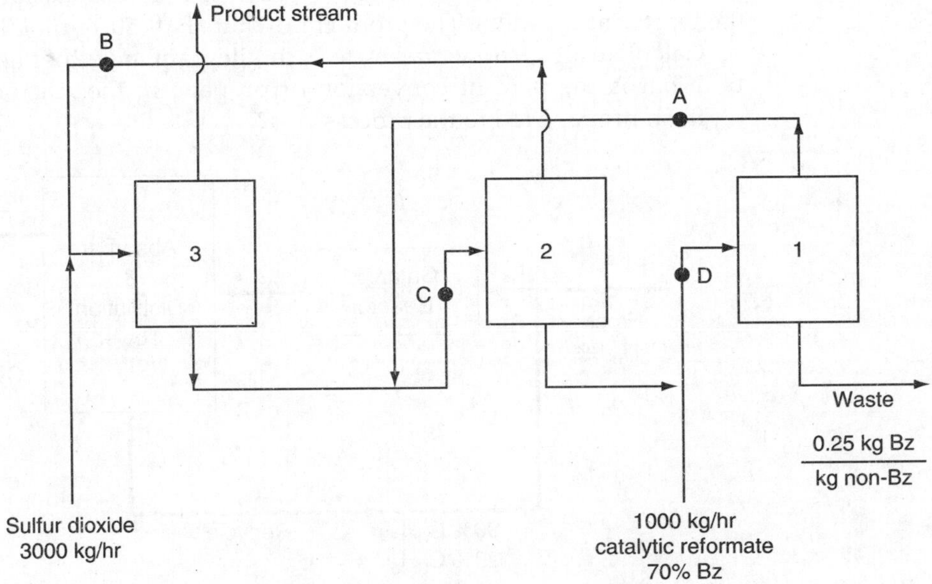

Figure P6.3.22

of sulfur dioxide are fed to the system per hour. The benzene product stream contains 0.15 kg of sulfur dioxide per kilogram of benzene. The waste stream contains all of the initially charged non-benzene material as well as 0.25 kg of benzene per kilogram of the non-benzene material. The remaining component in the waste stream is the sulfur dioxide.

a. How many kilograms of benzene are extracted per hour (are in the product stream)?
b. If 800 kg of benzene containing 0.25 kg of the non-benzene material per kilogram of benzene are flowing per hour at point A and 700 kg of benzene containing 0.07 kg of the non-benzene material per kilogram of benzene are flowing at point B, how many kilograms (exclusive of the sulfur dioxide) are flowing at points C and D?

Product benzene includes

$$\frac{0.15 \text{ kg } SO_2}{\text{kg Bz}}$$

****6.3.23 Nitroglycerine, a widely used high explosive, when mixed with wood flour is called "dynamite." It is made by mixing high-purity glycerine (99.9+% pure) with nitration acid, which contains 50.00% H_2SO_4, 43.00% HNO_3, and 7.00% water by weight. The reaction is

$$C_3H_8O_3 + 3HNO_3 + (H_2SO_4) \rightarrow C_3H_5O_3(NO_2)_3 + 3H_2O + (H_2SO_4)$$

The sulfuric acid does not take part in the reaction but is present to "catch" the water formed. Conversion of the glycerine in the nitrator is complete, and there are no side reactions, so all of the glycerine fed to the nitrator forms nitroglycerine. The mixed acid entering the nitrator (stream G) contains 20.00% excess HNO_3 to assure that all of the glycerine reacts. Figure P6.3.23 is a process flow diagram.

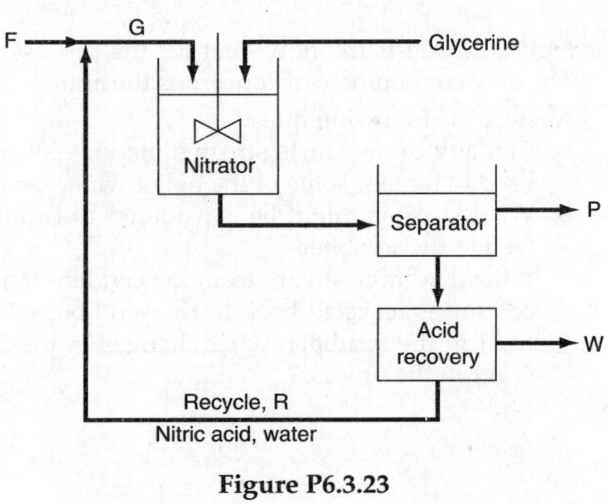

Figure P6.3.23

After nitration, the mixture of nitroglycerine and spent acid (HNO_3, H_2SO_4, and water) goes to a separator (a settling tank). The nitroglycerine is insoluble in the spent acid, and its density is less, so it rises to the top. It is carefully drawn off as product stream P and sent to wash tanks for purification. The spent acid is withdrawn from the bottom of the separator and sent to an acid recovery tank, where the HNO_3 and H_2SO_4 are separated. The $H_2SO_4 - H_2O$ mixture is stream W and is concentrated and sold for industrial purposes. The recycle stream to the nitrator is a 70.00% by weight solution of HNO_3 in water. In the diagram, product stream P is 96.50% nitroglycerine and 3.50% water by weight.

To summarize:

Stream F = 50.00 wt % H_2SO_4, 43.00% HNO_3, 7.00% H_2O

Stream G contains 20.00% excess nitric acid

Stream P = 96.50 wt % nitroglycerine, 3.50 wt % water

Stream R = 70.00 wt % nitric acid, 30.00% water

a. If 1.000×10^3 kg of glycerine per hour are fed to the nitrator, how many kilograms per hour of stream P result?
b. How many kilograms per hour are in the recycle stream?
c. How many kilograms of fresh feed, stream F, are fed per hour?
d. Stream W is how many kilograms per hour? What is its analysis in weight percent? Molecular weights: glycerine = 92.11, nitroglycerine = 227.09, nitric acid = 63.01, sulfuric acid = 98.08, and water = 18.02.

Caution: Do not try this process at home.

****6.3.24 The following problem is condensed from Example 10.3-1 in the book by D. T. Allen and D. R. Shonnard, *Green Engineering* [Prentice Hall, Upper Saddle River, NJ (2002)]. Acrylonitrile (AN) can be produced by the reaction of propylene with ammonia in the gas phase:

$$C_3H_6 + NH_3 + 1.5O_2 \rightarrow C_3H_3N + 3H_2O$$

Figure P6.3.24 is the flowsheet for the process with the data superimposed. The only contaminate of concern is the ammonia.

Answer the following questions:

a. Can any of the waste streams that are collected and sent to treatment be used to replace some of the boiler water feed?
b. What streams might be considered as candidates to replace some of the feed to the scrubber?
c. If the discharge stream from the condenser associated with the distillation column is recycled back to the scrubber to replace 0.7 kg/s of the water used in the scrubber, what changes in the flows and concentrations will occur in the process?

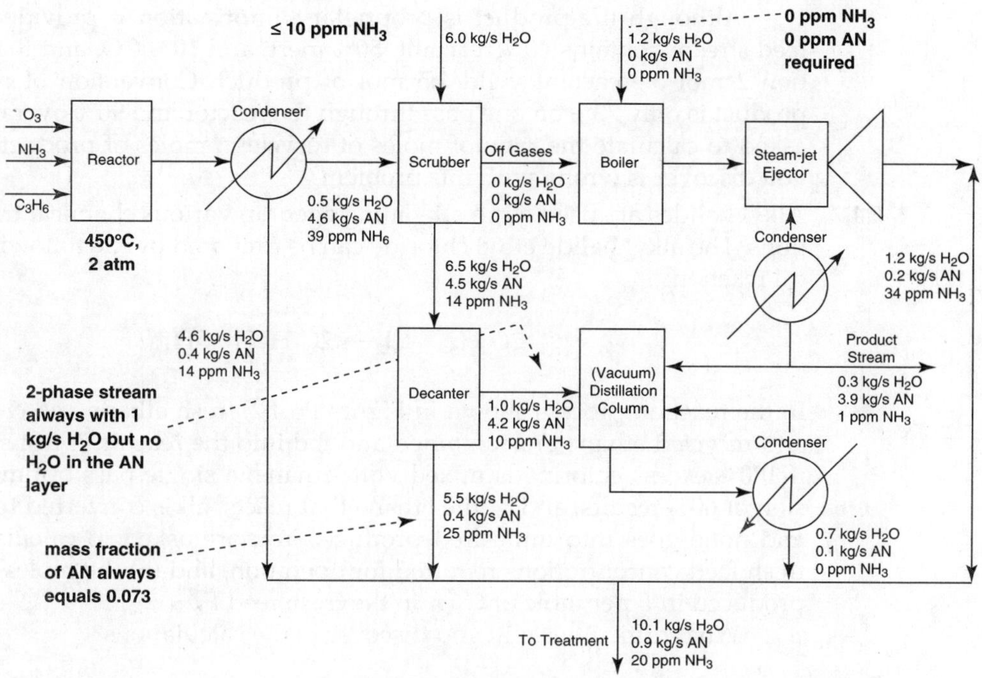

Figure P6.3.24

6.4 Bypass and Purge

****6.4.1** Many chemical processes generate emissions of volatile compounds that need to be controlled. In the process shown in Figure P6.4.1, the exhaust of CO is eliminated by its separation from the reactor effluent and recycling of 100% of the CO generated in the reactor together with some reactant back to the reactor feed.

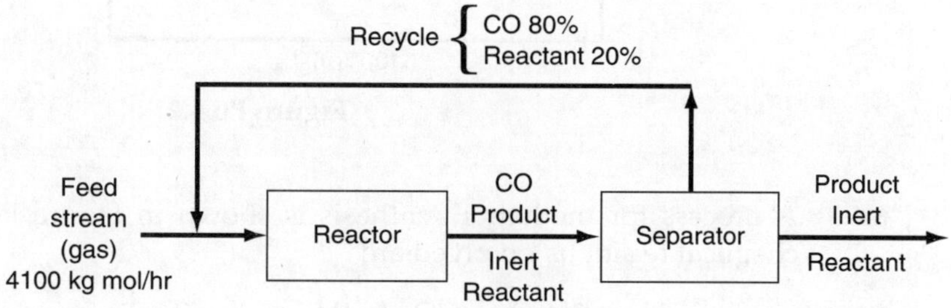

Figure P6.4.1

Although the product is proprietary, information is provided that the feed stream contains 40% reactant, 50% inert, and 10% CO, and that on reaction 2 mol of reactant yield 2.5 mol of product. Conversion of reactant to product is only 73% on one pass through the reactor and 90% overall. You are asked to calculate the ratio of moles of recycle to moles of product. What do you discover is wrong with this problem?

**6.4.2 Alkyl halides are used as an alkylating agent in various chemical transformations. The alkyl halide ethyl chloride can be prepared by the following chemical reaction:

$$2C_2H_6 + Cl_2 \rightarrow 2C_2H_5Cl + H_2$$

In the reaction process shown in Figure P6.4.2, fresh ethane and chlorine gas and recycled ethane are combined and fed into the reactor. A test shows that if 100% excess chlorine is mixed with ethane, a single-pass optimal conversion of 60% results, and of the ethane that reacts, all is converted to products and none goes into undesired products. You are asked to calculate (a) the fresh feed concentrations required for operation, and (b) the moles of C_2H_5Cl produced in P per mole of C_2H_6 in the fresh feed F_1.

What difficulties will you discover in the calculations?

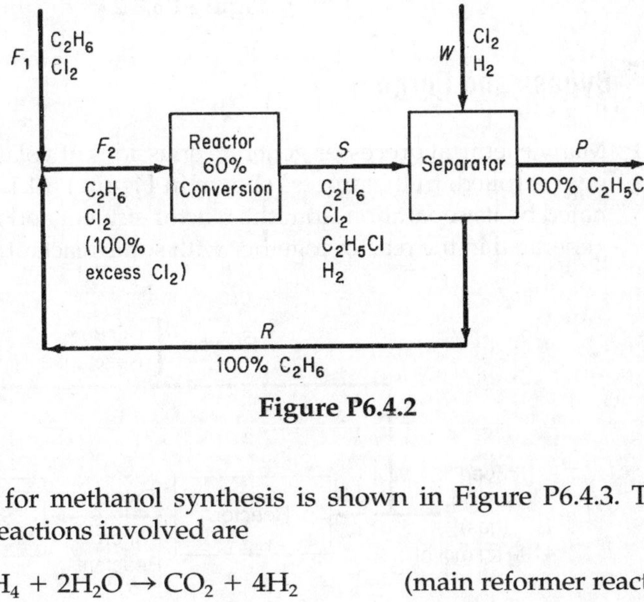

Figure P6.4.2

****6.4.3 A process for methanol synthesis is shown in Figure P6.4.3. The pertinent chemical reactions involved are

$$CH_4 + 2H_2O \rightarrow CO_2 + 4H_2 \quad \text{(main reformer reaction)} \quad \text{(a)}$$

$$CH_4 + H_2O \rightarrow CO + 3H_2 \quad \text{(reformer side reaction)} \quad \text{(b)}$$

$$2CO + O_2 \rightarrow 2CO_2 \qquad \text{(CO converter reaction)} \qquad \text{(c)}$$

$$CO_2 + 3H_2 \rightarrow CH_3OH + H_2O \qquad \text{(methanol synthesis reaction)} \qquad \text{(d)}$$

Ten percent excess steam, based on Reaction (a), is fed to the reformer, and conversion of methane is 100%, with a 90% yield of CO_2. Conversion in the methanol reactor is 55% on one pass through the reactor.

A stoichiometric quantity of oxygen is fed to the CO converter, and the CO is completely converted to CO_2. Additional makeup CO_2 is then introduced to establish a 3-to-1 ratio of H_2 to CO_2 in the feed stream to the methanol reactor.

The methanol reactor effluent is cooled to condense all of the methanol and water, with the noncondensable gases recycled to the methanol reactor feed. The H_2/CO_2 ratio in the recycle stream is also 3 to 1.

Because the methane feed contains 1% nitrogen as an impurity, a portion of the recycle stream must be purged as shown in Figure P6.4.3 to prevent the accumulation of nitrogen in the system. The purge stream analyzes 5% nitrogen. On the basis of 100 mol of methane feed (including the N_2):

a. How many moles of H_2 are lost in the purge?
b. How many moles of makeup CO_2 are required?
c. What is the recycle-to-purge ratio in moles per mole?
d. How much methanol solution (in kilograms) of what strength (weight percent) is produced?

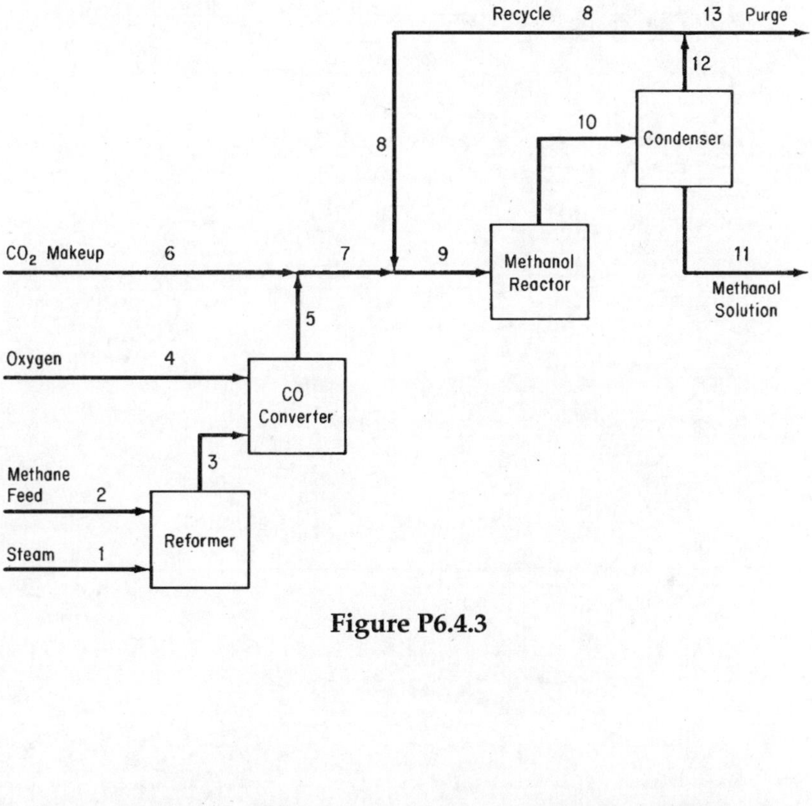

Figure P6.4.3

PART III

GASES, VAPORS, AND LIQUIDS

CHAPTER 7

Ideal and Real Gases

Your objectives in studying this chapter are to be able to

1. Understand the conditions under which the ideal gas law applies, and the conditions for which real gas relations must be used
2. Remember that the values of p and T used in relations to determine gas properties are absolute, not relative, values
3. Use partial pressure in calculations involving multicomponent ideal gases
4. Solve material balances involving ideal or real gases

In this chapter we begin to consider the physical properties of pure components and mixtures. By **property** we mean any measurable characteristic of a substance, such as pressure, volume, or temperature, or a characteristic that can be calculated or deduced, such as internal energy, which is discussed in Chapter 9. The **state** of a system gives the condition of a system as specified by its properties. You can find values for properties of compounds and mixtures in many formats, including

1. Experimental data
2. Tables (developed from experimental data or theory)
3. Graphs
4. Equations

Not everything that can be counted counts, and not everything that counts can be counted.

Albert Einstein

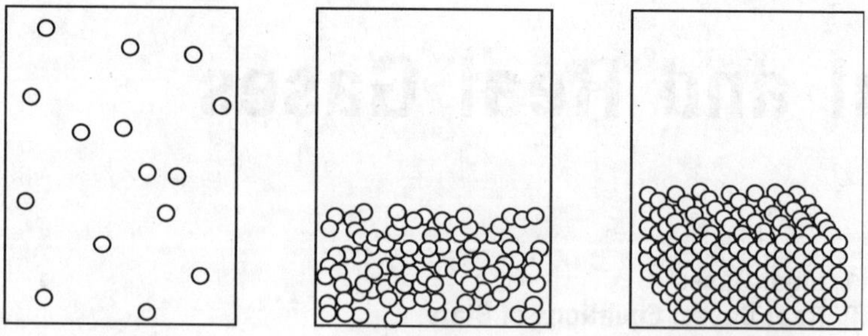

Figure 7.1 Three phases of a compound showing the classification by degree of organization

Clearly you cannot realistically expect to have reliable, detailed experimental data at hand or in a database for the properties of all of the useful pure compounds and mixtures with which you will be involved. Consequently, in the absence of experimental information, you have to estimate (predict) properties based on empirical correlations or graphs so that you can introduce appropriate parameters in material and energy balances.

A compound (or a mixture of compounds) may consist of one or more phases. A **phase** is defined as a completely homogeneous and uniform state of matter. Look at Figure 7.1. Liquid water would be a phase, and ice would be another phase. Two immiscible liquids in the same container, such as mercury and water, would represent two different phases because the liquids, although each is homogeneous, have different properties.

7.1 Ideal Gases

You have no doubt been exposed to the concept of the ideal gas in chemistry and physics. Why go over ideal gases again? At least two reasons exist. First, the experimental and theoretical properties of ideal gases are far simpler than the corresponding properties of liquids and solids. Second, use of the ideal gas concept is of considerable industrial importance.

In this section we explain how the ideal gas law can be used to calculate the pressure, temperature, volume, or number of moles in a quantity of gas, and we define the partial pressure of a gas in a mixture of gases. We also discuss how to calculate the specific gravity and density of a gas. Then we apply the concepts to solving material balances.

7.1.1 Ideal Gas

The most famous and widely used equation that relates p, V, n, and T for a gas is the **ideal gas law:**

$$pV = nRT \tag{7.1}$$

where p is the **absolute pressure** of the gas
V is the total volume occupied by the gas
n is the number of moles of the gas
R is the ideal (universal) gas constant in appropriate units
T is the **absolute temperature** of the gas

You can find values of R in various units inside the front cover of this book. Sometimes the ideal gas law is written as

$$p\hat{V} = RT \tag{7.1a}$$

Note that in Equation (7.1a) $\hat{V}$ is the *specific molar* volume (volume per mole, V/n) of the gas. When gas volumes are involved in a problem, $\hat{V}$ will be the volume per mole and not the volume per mass. The inverse of $\hat{V}$ is the molar density, moles per volume. Figure 7.2 illustrates the surface generated by

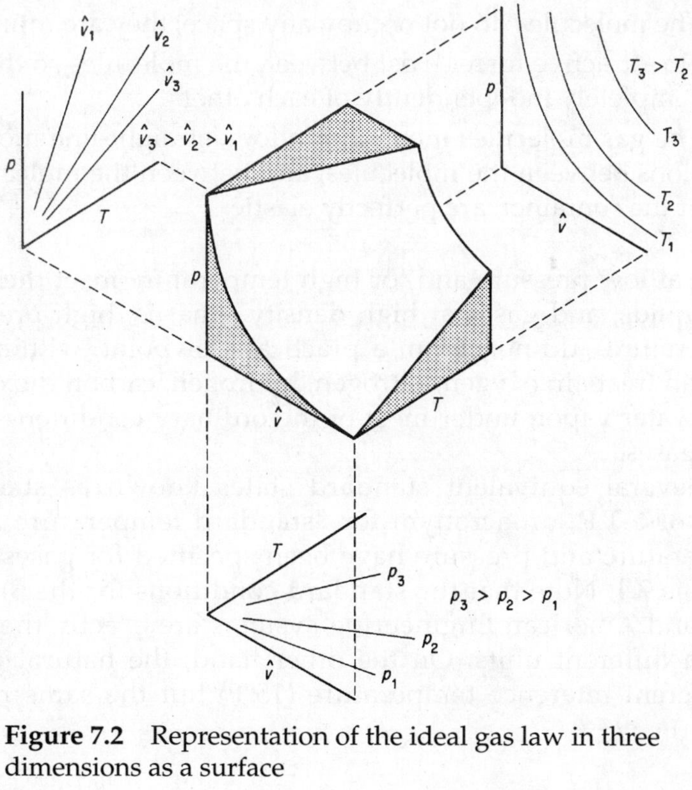

Figure 7.2 Representation of the ideal gas law in three dimensions as a surface

Equation (7.1a) in terms of the three properties p, $\hat{V}$, and T. Look at the projections of the surface in Figure 7.2 onto the two-parameter planes. The interpretation is as follows:

1. The projection to the upper left onto the $p - T$ plane shows straight lines for constant values of $\hat{V}$. Why? Equation (7.1a) for constant specific volume reduces to $p = (\text{constant})(T)$, the equation of a straight line that passes through the origin.
2. The projection to the upper right onto the $p - \hat{V}$ plane shows curves for values of constant T. What kinds of curves are they? For constant T, Equation (7.1a) becomes $p\hat{V} = \text{constant}$, namely, a hyperbola.
3. The projection downward onto the $T - \hat{V}$ plane again shows straight lines. Why? Equation (7.1a) for constant p is $\hat{V} = (\text{constant})(T)$.

For an ideal gas, Equation (7.1) can be applied to a pure component or to a mixture.

What are the conditions for a gas to behave as predicted by the ideal gas law? The major ones for a gas to be ideal are as follows:

1. The molecules do not occupy any space; they are infinitesimally small.
2. No attractive forces exist between the molecules so the molecules move completely independently of each other.
3. The gas molecules move in random, straight-line motion and the collisions between the molecules, and between the molecules and the walls of the container, are perfectly elastic.

Gases at low pressure and/or high temperature meet these conditions. Solids, liquids, and gases at high density—that is, high pressure and/or low temperature—do not. From a practical viewpoint, within reasonable error, you can treat air, oxygen, nitrogen, hydrogen, carbon dioxide, methane, and even water vapor, under most of the ordinary conditions you encounter, as ideal gases.

Several equivalent standard states known as **standard conditions** (S.C., or S.T.P., an acronym for "standard temperature and pressure") of temperature and pressure have been specified for gases by custom. Refer to Table 7.1. Note that the standard conditions for the SI, Universal Scientific, and American Engineering systems are exactly the same conditions, but in different units. On the other hand, the natural gas industry uses a different reference temperature (15°F) but the same reference pressure (1 atm).

Table 7.1 Common Standard Conditions for the Ideal Gas

System	T	p	$\hat{V}$
SI	273.15 K	101.325 kPa	22.415 m^3/kg mol
Universal Scientific	0.0°C	760 mm Hg	22.415 L/g mol
American Engineering	491.67°R (32°F)	1 atm	359.05 ft^3/lb mol
Natural gas industry	59.0°F (15.0°C)	14.696 psia (101.325 kPa)	379.4 ft^3/lb mol

You can insert the values at S.C. into the ideal gas equation to calculate R in any units you want.

For example, what is R for 1 g mol of ideal gas with a volume in cubic centimeters, pressure in atmospheres, and temperature in kelvin?

$$R = \frac{p\hat{V}}{T} = \frac{1 \text{ atm}}{273.15 \text{ K}}\left|\frac{22{,}415 \text{ cm}^3}{1 \text{ g mol}}\right. = 82.06 \frac{(\text{cm}^3)(\text{atm})}{(\text{K})(\text{g mol})}$$

The fact that a substance cannot exist as a gas at 0°C and 1 atm is immaterial. Thus, as we shall see later, water vapor at 0°C cannot exist at a pressure greater than its vapor pressure of 0.61 kPa (0.18 in. Hg) without condensation occurring. However, you can calculate the imaginary volume at standard conditions, and it is just as useful a quantity in the calculation of volume-mole relationships as though it could exist. In what follows, the symbol V will stand for total volume and the symbol $\hat{V}$ for volume per mole.

Because the SI, Universal Scientific, and AE standard conditions all refer to the same point in the p, $\hat{V}$, and T space, you can use the values in Table 7.1 with their units to change from one system of units to another. If you memorize the standard conditions, you will find it easy to work with mixtures of units from different systems.

The following example illustrates how you can use the standard conditions to convert mass or moles to volume. After reading it, see if you can explain to someone else how to convert volume to moles or mass.

Example 7.1 Use of Standard Conditions to Calculate Volume from Mass

Calculate the volume, in cubic meters, occupied by 40 kg of CO_2 at standard conditions, assuming CO_2 acts as an ideal gas.

(Continues)

Example 7.1 Use of Standard Conditions to Calculate Volume from Mass (*Continued*)

Solution

Basis: 40 kg of CO_2

$$\frac{40 \text{ kg } CO_2}{} \left| \frac{1 \text{ kg mol } CO_2}{44 \text{ kg } CO_2} \right| \frac{22.42 \text{ m}^3 \text{ } CO_2}{1 \text{ kg mol } CO_2} = 20.4 \text{ m}^3 \text{ } CO_2 \text{ at S.C.}$$

Notice in this problem that the information that 22.42 m^3 of gas at S.C. = 1 kg mol of gas is applied to transform a known number of moles into an equivalent number of cubic meters. An alternate way to calculate the volume at standard conditions is to use Equation (7.1). Incidentally, whenever you report a volumetric value, **you must establish the conditions of temperature and pressure at which the volumetric measure exists**, since the term m^3 or ft^3, standing alone, is really not any particular *quantity* of material.

In many processes going from an initial state to a final state, you use the ratio of the ideal gas laws in the respective states and thus eliminate R as follows (the subscript 1 designates the initial state, and the subscript 2 designates the final state):

$$\frac{p_1 V_1}{p_2 V_2} = \frac{n_1 R T_1}{n_2 R T_2}$$

or

$$\left(\frac{p_1}{p_2}\right)\left(\frac{V_1}{V_2}\right) = \left(\frac{n_1}{n_2}\right)\left(\frac{T_1}{T_2}\right) \tag{7.2}$$

Note that Equation (7.2) involves ratios of the same variable. This result has the convenient feature that the pressures may be expressed in any system of units you choose, such as kilopascals, inches of Hg, millimeters of Hg, atmospheres, and so on, as long as the same units are used for both conditions of pressure (do not forget that the pressure must be *absolute* pressure in both cases). Similarly, the ratio of the *absolute* temperatures and ratio of the volumes result in dimensionless ratios. Note how the ideal gas constant R is eliminated in taking the ratios.

Let's see how you can apply the ideal gas law to problems in the form of both Equation (7.2) and Equation (7.1).

Example 7.2 Application of the Ideal Gas Law to Calculate a Volume

Calculate the volume occupied by 88 lb of CO_2 at 15°C and a pressure of 32.2 ft of water.

Solution

Examine Figure E7.2. To use Equation (7.2) the initial volume has to first be calculated as shown in Example 7.1.

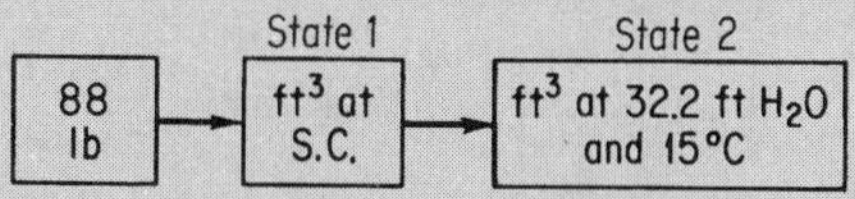

Figure E7.2

Then the final volume can be calculated via Equation (7.2) in which both R and (n_1/n_2) cancel. Table 7.1 does not list the pressure in feet of H_2O at S.C. Where do you get the value? Look in Chapter 2 at the discussion of pressure, or calculate it ($\rho gh = p$).

$$V_2 = V_1\left(\frac{p_1}{p_2}\right)\left(\frac{T_2}{T_1}\right)$$

Assume that the given pressure is absolute pressure.

At S.C. (state 1)	At p and T (state 2)
$p = 33.91$ ft H_2O	$p = 32.2$ ft H_2O
$T = 273$ K	$T = 273 + 15 = 288$ K

Basis: 88 lb of CO_2

$$\frac{88 \text{ lb } CO_2}{} \left| \frac{}{\frac{44 \text{ lb } CO_2}{1 \text{ lb mol } CO_2}} \right| \frac{359 \text{ ft}^3}{1 \text{ lb mol}} \left| \frac{288}{273} \right| \frac{33.91}{32.2} = \frac{798 \text{ ft}^3\ CO_2}{\text{at } 32.2 \text{ ft } H_2O \text{ and } 15°C}$$

You can mentally check your calculations by saying to yourself: The temperature goes up from 0°C at S.C. to 15°C at the final state; hence, the volume must increase from S.C., and the temperature ratio must be greater than unity. Similarly, you can say: The pressure goes down from S.C. to the final state, so the volume must increase from S.C.; hence the pressure ratio must be greater than unity.

(Continues)

Example 7.2 Application of the Ideal Gas Law to Calculate a Volume (*Continued*)

The same result can be obtained by using Equation (7.1). First obtain the value of R in the same units as the variables p, $\hat{V}$, and T. Look it up or calculate the value from p, $\hat{V}$, and T at S.C.:

$$R = \frac{p\hat{V}}{T}$$

At S.C.: $p = 33.91$ ft H_2O $\quad \hat{V} = 359$ ft^3/lb mol $\quad T = 273$ K

$$R = \frac{33.91}{} \left| \frac{359}{273} \right. = 44.59 \frac{(\text{ft } H_2O)(\text{ft}^3)}{(\text{lb mol})(\text{K})}$$

Now, using Equation (7.1), insert the given values, and perform the necessary calculations.

Basis: 88 lb of CO_2

$$V = \frac{nRT}{p} = \frac{88 \text{ lb } CO_2}{\frac{44 \text{ lb } CO_2}{\text{lb mol } CO_2}} \left| \frac{44.59\ (\text{ft } H_2O)(\text{ft}^3)}{(\text{lb mol})(\text{K})} \right| \frac{288 \text{ K}}{32.2 \text{ ft } H_2O}$$

$$= 798 \text{ ft}^3\ CO_2 \text{ at } 32.2 \text{ ft } H_2O \text{ and } 15°C$$

If you inspect the two solutions, you will observe that in both the same numbers appear, and that the results are identical.

To calculate the **volumetric flow rate** of a gas, $\dot{V}$, such as in cubic meters or cubic feet per second, through a pipe, you divide the volume of the gas passing through the pipe in a time interval by the value of the time interval. To get the **velocity**, $\dot{v}$, of the flow, you divide the volumetric flow rate by the area, A, of the pipe

$$\dot{V} = A\dot{v} \quad \text{hence} \quad \dot{v} = \dot{V}/A \tag{7.3}$$

The (mass) **density of a gas** is defined as the mass per unit volume and can be expressed in various units, including kilograms per cubic meter, pounds per cubic foot, grams per liter, and so on. Inasmuch as the mass contained in a unit volume varies with the temperature and pressure, as we have previously mentioned, you should always be careful to specify these two conditions in calculating density. If not otherwise specified, the densities are presumed to be at S.C. As an example, what is the density of N_2 at 27°C and 100 kPa in SI units?

Basis: 1 m^3 of N_2 at 27°C and 100 kPa

$$\frac{1\ \text{m}^3}{} \left| \frac{273\ \text{K}}{300\ \text{K}} \right| \frac{100\ \text{kPa}}{101.3\ \text{kPa}} \left| \frac{1\ \text{kg mol}}{22.4\ \text{m}^3} \right| \frac{28\ \text{kg}}{1\ \text{kg mol}} = 1.123\ \text{kg/m}^3$$

of N_2 at 27°C (300 K) and 100 kPa

In addition to the mass density, sometimes the "density" of a gas refers to the molar density, namely, moles per unit volume. How can you tell the difference if the same symbol is used for the density?

The **specific gravity** of a gas is usually defined as the ratio of the density of the gas at a desired temperature and pressure to that of air (or any specified reference gas) at a certain temperature and pressure. The use of specific gravity occasionally may be confusing because of the sloppy manner in which the values of specific gravity are reported without citing T and p, such as "What is the specific gravity of methane?" The answer to the question is not clear; hence, assume S.C. for both the gas and the reference gas:

$$\text{sp. gr.} = \frac{\text{density of methane at S.C.}}{\text{density of air at S.C.}}$$

7.1.2 Ideal Gas Mixtures

Frequently, as an engineer, you will want to make calculations for *mixtures of gases* instead of individual gases. You can use the ideal gas law, under the proper assumptions, of course, for a mixture of gases by interpreting p as the total absolute pressure of the mixture, V as the volume occupied by the mixture, n as the total number of moles of all components in the mixture, and T as the absolute temperature of the mixture. As the most obvious example, air is composed of N_2, O_2, Ar, CO_2, Ne, He, and other trace gases, but you can treat air as a single compound in applying the ideal gas law.

Engineers use a fictitious but useful quantity called the **partial pressure** in many of their calculations involving gases. The partial pressure of Dalton, p_i, namely, the pressure that would be exerted by a single component in a gaseous mixture if it existed alone in the *same volume* as that occupied by the mixture and at the *same temperature* as the mixture, is defined by

$$p_i V_{\text{total}} = n_i \text{R} T_{\text{total}} \tag{7.4}$$

where p_i is the partial pressure of component i in the mixture. If you divide Equation (7.4) by Equation (7.1), you find that

$$\frac{p_i V_{\text{total}}}{p_{\text{total}} V_{\text{total}}} = \frac{n_i \text{R} T_{\text{total}}}{n_{\text{total}} \text{R} T_{\text{total}}}$$

and

$$p_i = p_{\text{total}} \frac{n_i}{n_{\text{total}}} = p_{\text{total}}\, y_i \tag{7.5}$$

where y_i is the mole fraction of component i. In air the percent oxygen is 20.95; hence, at the standard condition of 1 atm, the partial pressure of oxygen is $p_{O_2} = 0.2095(1) = 0.2095$ atm. Can you show that Dalton's law of the summation of partial pressures is true using Equation (7.5)?

$$p_1 + p_2 + \cdots + p_n = p_{\text{total}} \tag{7.6}$$

Although you cannot easily measure the partial pressure of a gaseous component directly with commercial instruments, you can calculate the value from Equation (7.5) and/or Equation (7.6). To illustrate the significance of Equation (7.5) and the meaning of partial pressure, suppose that you carried out the following experiment with two nonreacting ideal gases. Examine Figure 7.2. Two tanks of 1.50 m^3 volume, one containing gas A at 300 kPa and the other gas B at 400 kPa (both gases being at the same temperature of 20°C), are connected to an empty third tank (C) of the same volume. All of the gas in tanks A and B is forced into tank C isothermally. Now you have a 1.50 m^3 tank of A + B at 700 kPa and 20°C for this mixture. According to Equation (7.5), you could say that gas A exerts a partial pressure of 300 kPa

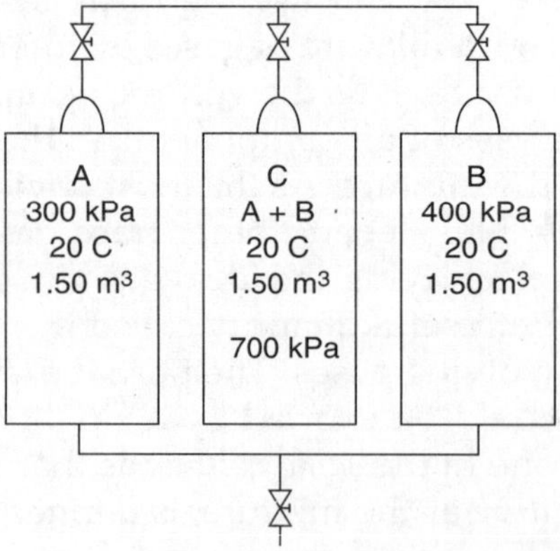

Figure 7.3 Illustration of the meaning of partial pressure of the components of an ideal gas mixture

and gas B exerts a partial pressure of 400 kPa in tank C. Of course, you cannot put a pressure gauge on the tank and check this conclusion because the pressure gauge will read only the total pressure. These partial pressures are hypothetical pressures in tank C that the individual gases would exert if each was put into separate but identical volumes at the same temperature.

When the $150 million Biosphere project in Arizona began in September 1991, it was billed as a sealed utopian planet in a bottle, where everything would be recycled. Its eight inhabitants lived for two years in the first large self-contained habitat for humans. But slowly the oxygen disappeared from the air—four women and four men in the 3.15 acres of glass domes eventually were breathing air with an oxygen content similar to that found at an altitude of about 13,400 ft. The "thin" air left the group members so fatigued and aching that they sometimes gasped for breath. Finally, the leaders of Biosphere 2 had to pump 21,000 lb of oxygen into the domes to raise the oxygen level from 14.5% to 19.0%. Subsequent investigation of the cause of the decrease in oxygen concluded that microorganisms in the soil that took up oxygen, a factor not accounted for in the design of the biosphere, were the cause of the problem.

Example 7.3 Calculation of the Partial Pressures of the Components in a Gas from a Gas Analysis

Few organisms are able to grow in solution using organic compounds that contain just one carbon atom such as methane or methanol. However, the bacterium *Methylococcus capsulates* can grow under aerobic conditions (in the presence of air) on C-1 carbon compounds. The resulting biomass is a good protein source that can be used directly as feed for domestic animals or fish.

In one process the off-gas analyzes 14.0% CO_2, 6.0% O_2, and 80.0% N_2. It is at 300°F and 765.0 mm Hg pressure. Calculate the partial pressure of each component.

Solution

Use Equation (7.5): $p_i = p_{\text{total}} y_i$.

Basis: 1.00 kg (or lb) mol of off-gas

Component	kg (or lb) mol	p_i (mm Hg)
CO_2	0.140	107.1
O_2	0.060	45.9
N_2	0.800	612.0
Total	1.000	765.0

(Continues)

Example 7.3 Calculation of the Partial Pressures of the Components in a Gas from a Gas Analysis (*Continued*)

On the basis of 1.00 mol of off-gas, the mole fraction y_i of each component, when multiplied by the total pressure, gives the partial pressure of that component. If you find that the temperature measurement of the flue gas was actually 337°F but the total pressure measurement was correct, would the partial pressures change? Hint: Is the temperature involved in Equation (7.5)?

7.1.3 Material Balances Involving Ideal Gases

Now that you have had a chance to review the ideal gas law applied to simple problems, let's apply the ideal gas law in material balances. The only difference between the subject matter of Chapters 3 through 6 and this chapter is that here the amount of material can be specified in terms of p, V, and T rather than solely as mass or moles. For example, the basis for a problem, or the quantity to be solved for, might be a volume of gas at a given temperature and pressure rather than a mass of gas. The next two examples illustrate balances for problems similar to those you have encountered before, but now involving gases.

Example 7.4 Material Balances for a Process Involving Combustion

To evaluate the use of renewable resources, an experiment was carried out with rice hulls. After pyrolysis, the product gas analyzed 6.4% CO_2, 0.1% O_2, 39% CO, 51.8% H_2, 0.6% CH_4, and 2.1% N_2. It entered a combustion chamber at 90°F and a pressure of 35.0 in. Hg and was burned with 40% excess air (dry) at 70°F and an atmospheric pressure of 29.4 in. Hg; 10% of the CO remains. How many cubic feet of air were supplied per cubic foot of entering gas? How many cubic feet of product gas were produced per cubic foot of entering gas if the exit gas was at 29.4 in. Hg and 400°F?

Solution

This is an open, steady-state system with reaction. The system is the combustion chamber.

Steps 1–4

Figure E7.4 illustrates the process and notation. With 40% excess air, certainly all of the CO, H_2, and CH_4 should burn to CO_2 and H_2O; apparently, for some unknown reason, not all of the CO burns to CO_2. No CH_4 or H_2 appears in the product gas. The components of the product gas are shown in the figure.

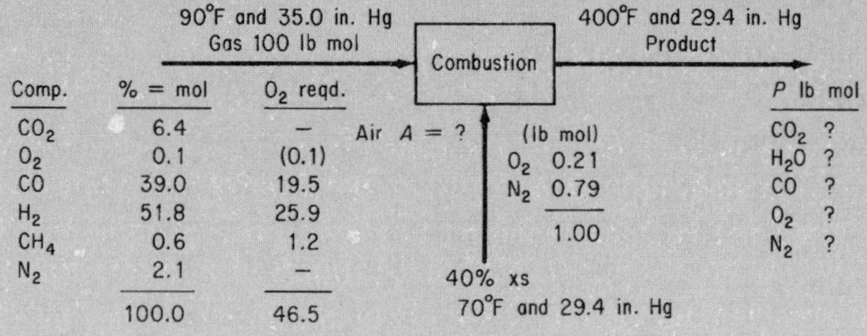

Figure E7.4

Step 5

You could take 1 ft^3 at 90°F and 35.0 in. Hg as the basis and convert the volume to moles, but it is just as easy to take 100 lb (or kg) mol as a basis because then % = lb (or kg) mol. Because only ratios of volumes are asked for, not absolute amounts, at the end of the problem you can convert pound (or kilogram) moles to cubic feet.

Basis: 100 lb mol of pyrolysis gas

Step 4 (continued)

The entering air can be calculated from the specified 40% excess air; the reactions for complete combustion are

$$CO + \frac{1}{2}O_2 \rightarrow CO_2 \quad (1)$$

$$H_2 + \frac{1}{2}O_2 \rightarrow H_2O \quad (2)$$

$$CH_4 + 2O_2 \rightarrow CO_2 + 2H_2O \quad (3)$$

The moles of oxygen required are listed in Figure E7.4. (We will omit the units—pound moles—in what follows.) The excess oxygen is

$$\text{Excess } O_2\text{: } 0.4(46.3) = 18.6$$

$$\text{Total } O_2\text{: } 46.5 + 18.6 = 65.1$$

$$N_2 \text{ in is } 65.1\left(\frac{79}{21}\right) = 244.9$$

(Continues)

Example 7.4 Material Balances for a Process Involving Combustion (*Continued*)

Total moles of air in are 244.9 + 65.1 = 310.0 lb mol.

Finally, use the specification for CO to assign $n_{co} = 0.10(39.0) = 3.9$.

Steps 6 and 7

Degree-of-freedom analysis:

Unknowns (5): $n_{CO_2}, n_{O_2}, n_{N_2}, n_{H_2O}, P$
Equations (5):
Element balances (4): C, H, O, N
Implicit equations (1): $P = \sum n_i$

Steps 8 and 9

Make the element balances in moles to calculate the unknown quantities, and substitute the value of 3.9 for the number of moles of CO exiting.

	In		Out
N	(2)(2.1) + (2)(244.9)	=	$2n_{N_2}$
C	6.4 + 39.0 + 0.6	=	$n_{CO_2} + 3.9$
H	(2)(51.8) + (0.6)(2)	=	$2n_{H_2O}$
O	(2)(6.4) + (2)(0.1) + 39 + (2)(65.1)	=	$2n_{O_2} + 2n_{CO_2} + n_{H_2O} + n_{CO}$

The solutions of these equations are

$$n_{N_2} = 247 \quad n_{CO_2} = 42.1 \quad n_{H_2O} = 53.0 \quad n_{O_2} = 20.55$$

The total moles exiting calculated from the implicit equation sum to 366.6 mol.

Finally, you can convert the pound moles of air and products that were calculated on the basis of 100 lb mol of pyrolysis gas to the volumes of gases at the states requested using the ideal gas law:

$$T_{gas} = 90 + 460 = 550°R \rightarrow 306 \text{ K}$$

$$T_{air} = 70 + 460 = 530°R \rightarrow 294 \text{ K}$$

$$T_{product} = 400 + 460 = 860°R \rightarrow 478 \text{ K}$$

$$\text{ft}^3 \text{ of gas: } \frac{100 \text{ lb mol entering gas}}{} \left| \frac{359 \text{ ft}^3 \text{ at S.C.}}{1 \text{ lb mol}} \right| \frac{550°R}{492°R} \left| \frac{29.92 \text{ in. Hg}}{35.0 \text{ in. Hg}} \right. = 343 \times 10^2$$

$$\text{ft}^3 \text{ of air: } \frac{310 \text{ lb mol air}}{} \left| \frac{359 \text{ ft}^3 \text{ at S.C.}}{1 \text{ lb mol}} \right| \frac{530°\text{R}}{492°\text{R}} \left| \frac{29.92 \text{ in. Hg}}{29.4 \text{ in. Hg}} \right. = 1220 \times 10^2$$

$$\text{ft}^3 \text{ of product: } \frac{366.6 \text{ lb mol } P}{} \left| \frac{359 \text{ ft}^3 \text{ at S.C.}}{1 \text{ lb mol}} \right| \frac{860°\text{R}}{429°\text{R}} \left| \frac{29.92 \text{ in. Hg}}{35.0 \text{ in. Hg}} \right. = 2331 \times 10^2$$

The answers to the questions are

$$\frac{1220 \times 10^2}{343 \times 10^2} = 3.56 \frac{\text{ft}^3 \text{ air at } 530°\text{R and } 29.4 \text{ in. Hg}}{\text{ft}^3 \text{ gas at } 550°\text{R and } 35.0 \text{ in. Hg}}$$

$$\frac{2255 \times 10^2}{343 \times 10^2} = 6.57 \frac{\text{ft}^3 \text{ product at } 860°\text{R and } 29.4 \text{ in. Hg}}{\text{ft}^3 \text{ gas at } 550°\text{R and } 35.0 \text{ in. Hg}}$$

Example 7.5 Material Balance without Reaction

Gas at 15°C and 105 kPa is flowing through an irregular duct. To determine the rate of flow of the gas, CO_2 from a tank is steadily passed into the gas stream. The flowing gas, just before mixing with the CO_2, analyzes 1.2% CO_2 by volume. Downstream, after mixing, the flowing gas analyzes 3.4% CO_2 by volume. As the CO_2 that was injected exited the tank, it was passed through a rotameter and found to flow at the rate of 0.0917 m³/min at 7°C and 131 kPa. What was the rate of flow of the entering gas in the duct in cubic meters per minute?

Solution

This is an open, steady-state system without reaction. The system is the duct. Figure E7.5 is a sketch of the process.

Steps 1–4

The data are presented in Figure E7.5.

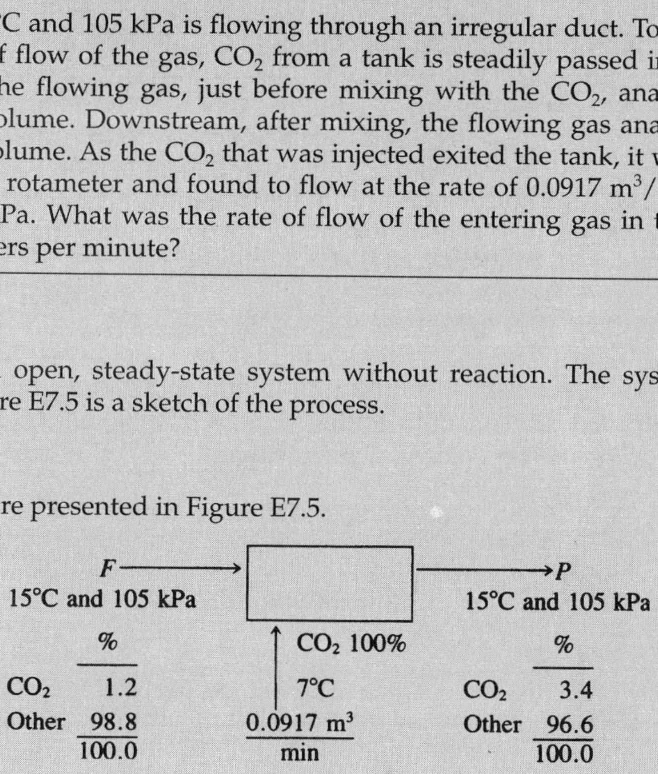

Figure E7.5

(Continues)

Example 7.5 Material Balance without Reaction (*Continued*)

Both F and P are at the same temperature and pressure.

Step 5

Should you take as a basis 1 min → 0.0917 m^3 of CO_2 at 7°C and 131 kPa? The gas analysis is in volume percent, which is the same as mole percent. We could convert all of the gas volumes to moles and solve the problem in terms of moles, but there is no need to do so because we can just as easily convert the known flow rate of the addition of CO_2 to 15°C and 105 kPa and solve the problem using cubic meters for each stream since all of the streams will be at the same conditions.

Steps 6 and 7

The unknowns are F and P, and you can make two independent component balances, CO_2 and "other"; hence, the problem has zero degrees of freedom.

Steps 7–9

The "other" balance (in cubic meters at 15°C and 105 kPa) is

$$F(0.988) = P(0.966) \tag{a}$$

The CO_2 balance (in cubic meters at 15°C and 105 kPa) is

$$F(0.012) + 0.1177 = P(0.034) \tag{b}$$

The total balance (in cubic meters at 15°C and 105 kPa) is

$$F + 0.1177 = P \tag{c}$$

Note that the "other" is a tie component. Select Equations (a) and (c) to solve. The solution of Equations (a) and (c) gives

$$F = 5.17 \text{ m}^3/\text{min at } 15°\text{C and } 105 \text{ kPa}$$

Step 10 (check)

Use the redundant Equation (b):

$$5.17(0.012) + 0.1177 = 0.180 \stackrel{?}{=} \left(5.17\frac{0.988}{0.966}\right)(0.034) = 0.180$$

The equation checks out to a satisfactory degree of precision.

Self-Assessment Test

Questions

1. What are the dimensions of T, P, V, n, and R?
2. List the standard conditions for a gas in the SI and American Engineering systems of units.
3. How do you calculate the density of an ideal gas at S.C.?
4. Can you use the respective specific molar densities (mole/volume) of the gas and the reference gas to calculate the specific gravity of a gas?
5. A partial pressure of oxygen in the lungs of 100 mm Hg is adequate to maintain oxygen saturation of the blood in a human. Is this value higher or lower than the partial pressure of oxygen in the air at sea level?
6. An exposure to a partial pressure of N_2 of 1200 mm Hg in air has been found by experience not to cause the symptoms of N_2 intoxication to appear. Will a diver at 60 m be affected by the N_2 in the air being breathed?

Problems

1. Calculate the volume in cubic feet of 10 lb mol of an ideal gas at 68°F and 30 psia.
2. A steel cylinder of volume 2 m^3 contains methane gas (CH_4) at 50°C and 250 kPa absolute. How many kilograms of methane are in the cylinder?
3. What is the value of the ideal gas constant R to use if the pressure is to be expressed in atmospheres, the temperature in kelvin, the volume in cubic feet, and the quantity of material in pound moles?
4. Twenty-two kilograms per hour of CH_4 are flowing in a gas pipeline at 30°C and 920 mm Hg. What is the volumetric flow rate of the CH_4 in cubic meters per hour?
5. A gas has the following composition at 120°F and 13.8 psia:

Component	mol %
N_2	2
CH_4	79
C_2H_6	19

   a. What is the partial pressure of each component?
   b. What is the volume fraction of each component?

6. A furnace is fired with 1000 ft^3/hr at 60°F and 1 atm of a natural gas having the following volumetric analysis: CH_4 80%, C_2H_6 16%, O_2 2%, CO_2 1%, and N_2 1%. The exit flue gas temperature is 800°F and the pressure is 760 mm Hg absolute; 15% excess air is used and combustion is complete. Calculate (a) the volume of

CO_2 produced per hour; (b) the volume of H_2O vapor produced per hour; (c) the volume of N_2 produced per hour; (d) the total volume of flue gas produced per hour.

Thought Problems

1. In a test of the flow of gases through a pipe, pure hydrogen was found to flow at a volumetric flow 22 times that of carbon dioxide. When the hydrogen was diluted with carbon dioxide entering the pipe midway from the ends, the exit flow rate was less than that of pure hydrogen. Explain the observed differences.
2. A pair of identical balloons are inflated with air to the same pressure and tied to a stick that is held in the center by a string. The balloons are the same distance from the center of the stick so that the stick remains horizontal to the ground. When the left-hand balloon is carefully punctured, will the stick rotate down from the left, rotate up from the left, or remain horizontal?

Discussion Question

In a demonstration, a 30-cm-diameter balloon was filled to two-thirds of its maximum pressure with SF_6, a gas. Students measured the balloon's diameter for 10 days, at which time the balloon burst. No one ever touched it. Explain how this could happen. (Note: The balloon was not defective.)

7.2 Real Gases: Equations of State

Predicting gas properties has appeal
By an old law that appears to be real
But somehow the law
In practice has flaws
Because some gases are seldom ideal

DMH

Gases whose properties cannot be represented by the ideal gas law are called **nonideal** gases or **real gases.** Real gas properties are predicted by equations called **equations of state.**

7.2.1 Equations of State

The simplest example of what is called an equation of state is the ideal gas law itself. Equations of state for nonideal gases can be just empirical relations selected to fit a data set, or they can be based on theory, or a combination of

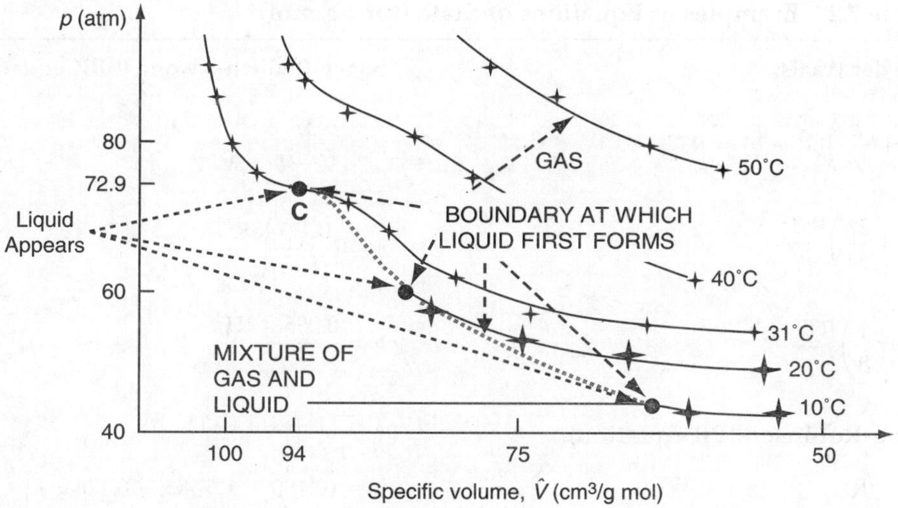

Figure 7.4 Experimental measurements of carbon dioxide by Andrews (+). The solid lines represent smoothed data. C is the highest temperature at which any liquid exists. At the big solid dots liquid and vapor start to coexist. Note the nonlinear scale on the horizontal axis.

the two. Figure 7.4 shows the measurements by Andrews in 1863 of the pressure versus the specific volume for CO_2 at various constant temperatures. Note that point C at 31°C is the highest temperature at which liquid and gaseous CO_2 can coexist in equilibrium. Above 31°C only **critical** fluid exists, so that what is called the **critical temperature** for CO_2 is 31°C (304 K). The corresponding gas (fluid) **pressure** is 72.9 atm (7385 kPa). Also note that at higher temperatures, such as 50°C, the data can be represented by the ideal gas law because pV is constant, a hyperbola. You can find experimental values of the critical temperature (T_c) and the critical pressure (p_c) for various compounds on the CD that accompanies this book. If you cannot find a desired critical value in this text or in a handbook, you can consult Reid et al. (refer to the references at the end of this chapter), who describe and evaluate methods of estimating critical constants for various compounds.

How can you predict the p, $\hat{V}$, and T properties of a gas between the region in which the ideal gas law is valid and the region in which the gas condenses into liquid? One way is to use one or more equations of state. Where substantial changes in curvature occur, perhaps several different equations must be used to cover a region accurately. Table 7.2 lists some of the well-known single equations of state.

Table 7.2 Examples of Equations of State (for 1 g mol)*

van der Waals:

$$\left(p + \frac{a}{\hat{V}^2}\right)(\hat{V} - b) = RT$$

$$a = \left(\frac{27}{64}\right)\frac{R^2T_c^2}{p_c}$$

$$b = \left(\frac{1}{8}\right)\frac{RT_c}{p_c}$$

Peng-Robinson (PR equation):

$$p = \frac{RT}{\hat{V} - b} - \frac{a\alpha}{\hat{V}(\hat{V} + b) + b(\hat{V} - b)}$$

$$a = 0.45724\left(\frac{R^2T_c^2}{p_c}\right)$$

$$b = 0.07780\left(\frac{RT_c}{p_c}\right)$$

$$\alpha = [1 + \kappa(1 - T_r^{1/2})]^2$$

$$\kappa = 0.37464 + 1.54226\omega - 0.26992\omega^2$$

Benedict-Webb-Rubin (BWR equation):

$$p\hat{V} = RT + \frac{\beta}{\hat{V}} + \frac{\sigma}{\hat{V}^2} + \frac{\eta}{\hat{V}^4} + \frac{\omega}{\hat{V}^5}$$

$$\beta = RTB_0 - A_0 - \frac{C^0}{T^2}$$

$$\sigma = bRT - a + \frac{c}{T^2}\exp\left(-\frac{\gamma}{\hat{V}^2}\right)$$

$$\eta = c\gamma \exp\left(-\frac{\gamma}{\hat{V}^2}\right)$$

$$w = a\,\alpha$$

Soave-Redlich-Kwong (SRK equation):

$$p = \frac{RT}{\hat{V} - b} - \frac{a'\lambda}{\hat{V}(\hat{V} + b)}$$

$$a' = \frac{0.42748R^2T_c^2}{p_c}$$

$$b = \frac{0.08664\,RT_c}{p_c}$$

$$\lambda = [1 + \kappa(1 - T_r^{1/2}]^2)$$

$$\kappa = (0.480 + 1.574\omega - 0.176\omega^2)$$

Redlich-Kwong (RK equation):

$$p = \frac{RT}{(\hat{V} - b)} - \frac{a}{T^{1/2}\hat{V}(\hat{V} + b)}$$

$$a = 0.42748\frac{R^2T_c^{2.5}}{p_c}$$

$$b = 0.08664\frac{RT_c}{p_c}$$

Kammerlingh-Onnes (a virial equation):

$$p\hat{V} = RT\left(1 + \frac{B}{\hat{V}} + \frac{C}{\hat{V}^2} + \cdots\right)$$

Holborn (a virial equation):

$$p\hat{V} = RT(1 + B'p + C'p^2 + \cdots)$$

*$\hat{V}$ is the specific volume, T_c and p_c are explained in the text, and ω is the acentric factor, also explained in the text.

The units used in calculating the coefficients in the equations are determined by the units selected for R.

Some of the classical equations of state are formulated as a power series (called the **virial** form) with p being a function of $1/\hat{V}$ or $\hat{V}$ being a function of p with three to six terms. You should note that the coefficients in the van der Waals, Peng-Robinson (PR), Soave-Redlich-Kwong (SRK), and Redlich-Kwong (RK) equations can be calculated from certain physical properties (discussed below), whereas in the virial equations the coefficients are strictly determined from experimental measurement. Because of its accuracy, the databases in many commercial process simulators make extensive use of the SRK equation of state. These equations in general will not predict p-$\hat{V}$-T values across a phase change from gas to liquid very well. Keep in mind that under conditions such that the gas starts to liquefy, the *gas laws apply only to the vapor phase portion* of the system for this book.

How accurate are equations of state? Cubic equations of state such as Redlich-Kwong, Soave-Redlich-Kwong, and Peng-Robinson listed in Table 7.2 can exhibit an accuracy of 1%–2% over a large range of conditions for many compounds. Equations of state in databases may have as many as 30 or 40 coefficients to achieve high accuracy (see, for example, the AIChE DIPPR reports that can be located on the AIChE Web site). Keep in mind that you must know the region of validity of any equation of state and not extrapolate outside that region, particularly not into the liquid region, by ignoring the possibility of condensation for gases such as CO_2, NH_3, and low-molecular-weight organic compounds such as acetone, ethyl alcohol, and so on. If you plan to use a specific equation of state such as one of those listed in Table 7.2, you have numerous choices, no one of which will consistently give the best results.

> *Although this may seem a paradox, all exact science is dominated by the idea of approximation.*
>
> Bertrand Russell

Other than the use of equations of state to make predictions of values of p, $\hat{V}$, and T, what good are they?

1. They permit a concise summary of a large mass of experimental data and also permit accurate interpolation between experimental data points.
2. They provide a continuous function to facilitate calculation of physical properties based on differentiation and integration of p-$\hat{V}$-T relationships.
3. They provide a point of departure for the treatment of the properties of mixtures.

In addition, some of the advantages and disadvantages of using equations of state versus other methods to make predictions are the following:

Advantages:

1. Values of p-$\hat{V}$-T can be predicted with reasonable error in regions where no data exist.
2. Only a few values of coefficients are needed in the equation to be able to predict gas properties, versus collecting large amounts of data by experiment for tables and graphs.
3. The equations can be manipulated on a computer whereas graphics methods of prediction cannot.

Disadvantages:

1. The form of an equation is hard to change to fit new or better data.
2. Inconsistencies may exist between equations for p-$\hat{V}$-T and equations for other physical properties.
3. Usually the equation is quite complicated and may not be easy to solve for p, $\hat{V}$, or T because of its nonlinearity.

> *I consider that I understand an equation when I can predict the properties of its solutions, without actually solving it.*
>
> Paul Dirac

Disadvantage 3 prevented the widespread use of equations of state until computers and computer programs for solving nonlinear algebraic equations came into the picture. You can see that it is easy to solve the SRK equation in Table 7.2 for p given values for T and $\hat{V}$, or for T given values of p and $\hat{V}$, but quite difficult and tedious to solve for $\hat{V}$ given values for T and p without the aid of computers. Similar remarks apply to virial equations.

For example, look at the Redlich-Kwong (RK) equation. Given p and T, is the RK equation cubic in $\hat{V}$? Yes. Given p, T, and $\hat{V}$, is the RK equation cubic in n? Yes. Given p and $\hat{V}$, is it cubic in T? No. We will not focus in this book on how to solve cubic and more complex equations for $\hat{V}$ but instead use an equation solver such as Polymath on the CD in the back of the book.

Example 7.6 Use of the RK Equation to Calculate p or $\hat{V}$

Determine the pressure (in atmospheres) of 1 g mol of C_2H_4 at 300 K with $V = 0.674$ L using the RK equation. From the CD that accompanies the book, for C_2H_4: $T_c = 282.8$ K and $p_c = 50.44$ atm.

Solution

The RK equation is (from Table 7.2)

$$p = \frac{RT}{(\hat{V} - b)} - \frac{a}{T^{1/2}\hat{V}(\hat{V} + b)}$$

What should you do first? Take a basis of the given values of V, n, and T. Then calculate a, b, and $\hat{V}$:

$$a = 0.42748\,\frac{R^2T_c^{2.5}}{p_c} = \frac{0.42748}{} \left| \frac{(0.08206)^2(L)^2(atm)^2}{(g\ mol)^2(K)^2} \right| \frac{(282.8)^{2.5}(K)^{2.5}}{50.44\ atm}$$

$$= 76.75\,\frac{(L)^2(atm)(K)^{0.5}}{(g\ mol)^2}$$

$$b = 0.08664\,\frac{RT_c}{p_c} = \frac{0.08664}{} \left| \frac{(0.08206)(L)(atm)}{(g\ mol)(K)} \right| \frac{282.8\ K}{50.44\ atm} = 0.03986\,\frac{L}{g\ mol}$$

$$\hat{V} = \frac{0.674\ L}{1\ g\ mol} = 0.674\ L/g\ mol$$

Next, insert the known values of the variables and coefficients:

$$p = \frac{(0.08206)(L)^2(atm)}{1\ (g\ mol)(K)} \left| \frac{300\ K}{} \right| \frac{1\ g\ mol}{(0.674 - 0.03986)\ L} -$$

$$\frac{76.75(L)^2(atm)(K)^{0.5}}{1\ (g\ mol)^2} \left| \frac{}{(300\ K)^2} \right| \frac{1\ g\ mol}{0.674\ L} \left| \frac{1\ g\ mol}{(0.674 + 0.03986)\ L} \right. = 29.6\ atm$$

If you know p and V instead of p and T, you can solve explicitly for T by rearranging the equation or by using an equation solver. Figure E7.6 compares the prediction of p by the ideal gas law and the RK equation.

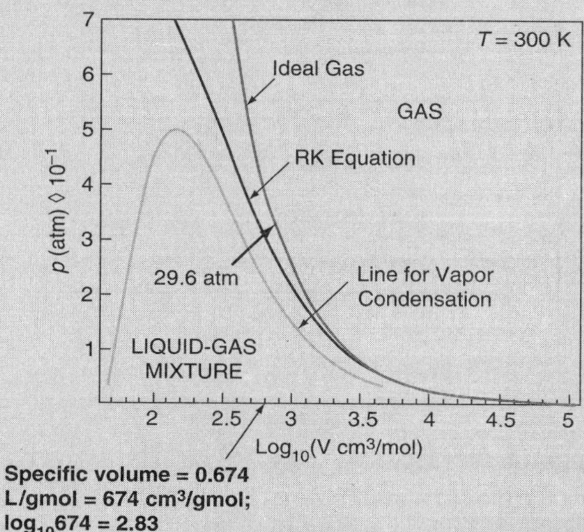

Figure E7.6

(Continues)

Example 7.6 Use of the RK Equation to Calculate p or $\hat{V}$ (*Continued*)

Next, let's determine the specific volume of C_2H_4 at 300 K and 30 atm using the RK equation. Because with the RK equation you cannot explicitly solve for $\hat{V}$, what should you do now? You can solve the RK equation in the usual format after introducing the known values and then use Polymath to determine $\hat{V} = 0.6748$ L/g mol.

7.2.2 The Critical State and Compressibility

We mentioned the critical pressure p_c and critical temperature T_c in connection with Figure 7.4. The **critical state (point)** for the gas-liquid transition is the set of physical conditions at which the density and other properties of the liquid and vapor become identical. In Figure 7.5 the points on the constant temperature lines at which the gas (vapor) starts to condense have been connected by a dashed line (---). On the opposite side of the figure, the dot-dashed curve (___ . ___) shows the locus of points at the respective temperatures at which completion of the condensation occurs; that is, the vapor becomes all liquid. Between the two bounds a mixture of vapor and liquid exists.

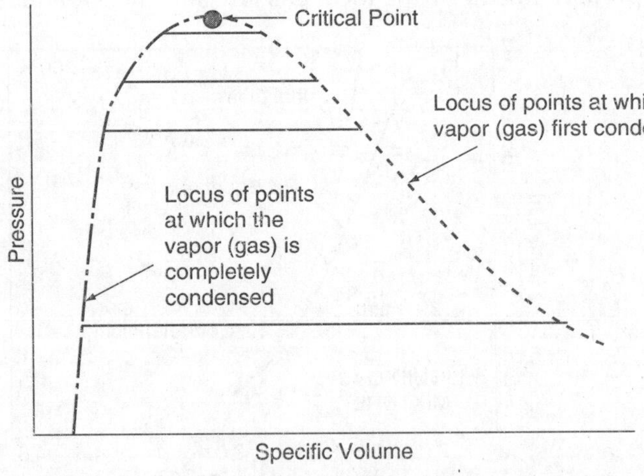

Figure 7.5 The critical point is located where the lengths of the (solid) lines are zero. The solid lines connect the points at which condensation starts at various temperatures to the corresponding points for that temperature at which condensation is complete.

The intersection of the two bounds is denoted as the **critical point**, and it occurs at the highest temperature and pressure possible ($T_r = 1, p_r = 1$) at which gas and liquid can coexist.

A **supercritical fluid** is a compound in a state above its critical point. Supercritical fluids are used to replace solvents such as trichloroethylene and methylene chloride, the emissions from which, and the contact with which, have been severely limited. For example, coffee decaffeination, the removal of cholesterol from egg yolk with CO_2, the production of vanilla extract, and the destruction of undesirable organic compounds all can take place using supercritical water. Supercritical water has been shown to destroy 99.99999% of all of the major types of toxins in these organic compounds.

Other terms with which you should become familiar are the **reduced variables**. These are conditions of temperature, pressure, or specific volume *normalized* (divided) by their respective critical conditions, as follows:

Reduced temperature: $$T_r = \frac{T}{T_c}$$

Reduced pressure: $$p_r = \frac{p}{p_c}$$

Reduced specific volume: $$\widetilde{V}_r = \frac{\hat{V}}{\hat{V}_c}$$

In theory, the **law of corresponding states** indicates that any compound should have the same reduced volume at the same reduced temperature and reduced pressure so that a universal gas law might be

$$P_r \widetilde{V}_r = kT_r \tag{7.7}$$

Unfortunately Equation (7.7) does not universally make accurate predictions. You can check this conclusion by selecting a compound such as water, applying Equation (7.7) at some low temperature and high pressure to calculate $\hat{V}$, and comparing your results with the value obtained for $\hat{V}$ with the corresponding conditions from the tables for water vapor that are in the folder in the back of this book.

The concept of reduced variables nevertheless has been applied to prediction of real gas properties. One common way is to modify the ideal gas law by inserting an adjustable coefficient z, the **compressibility factor**, a factor that compensates for the nonideality of the gas, and can be looked at as a measure of nonideality. Thus, the ideal gas law is turned into a real gas law called a **generalized equation of state**:

$$pV = znRT \tag{7.8}$$

or

$$p\hat{V} = zRT \tag{7.8a}$$

One way to look at z is to consider it to be a factor that makes Equation (7.8) an equality. Note that $z = 1$ is for an ideal gas. Although we treat only gases in this chapter, Equation (7.8) has been applied to liquids.

If you plan on using Equation (7.8), where can you find the values of z to use in it? Equations exist in the literature for specific compounds and classes of compounds, such as those found in petroleum refining. Theoretical calculations based on molecular structure sometimes prove to be useful. Refer to the references at the end of this chapter. Usually you will find graphs or tables of z to be quite convenient sources for engineering purposes. If the compressibility factor derived from experiment is plotted for a given temperature against the pressure for different gases, figures such as Figure 7.6a result. However, if the compressibility factor is plotted against the reduced pressure as a function of the reduced temperature, then for like gases the compressibility values at the same reduced temperature and reduced pressure fall at about the same point, as illustrated in Figure 7.6b.

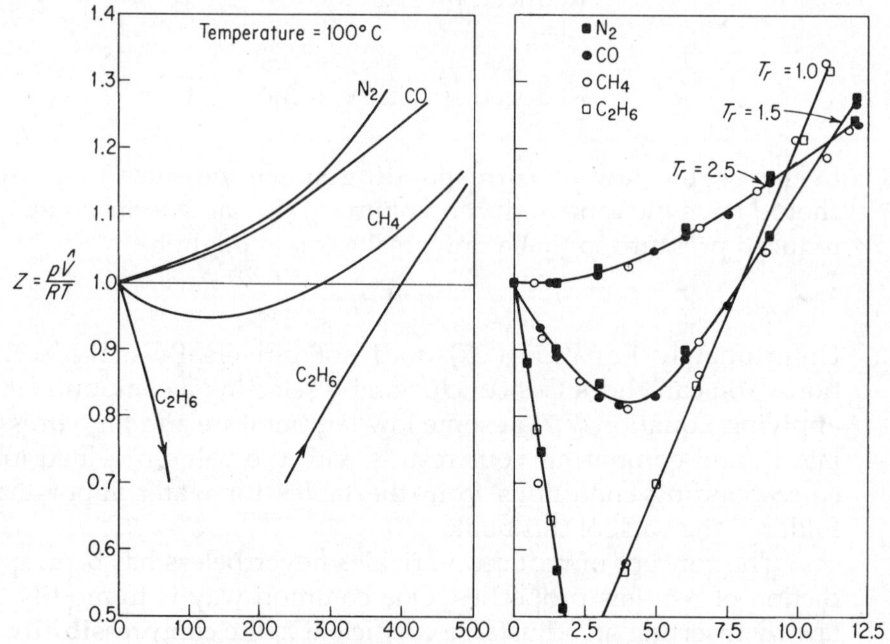

Figure 7.6 (a) Compressibility at 100°C for several gases as a function of pressure; (b) compressibility factor for several gases as a function of reduced temperature and reduced pressure

You can use the charts described in the next section, Section 7.3, to get approximate values of z as a function of the reduced temperature and pressure. Or, you can use one of the numerous methods that have appeared in the literature and in process simulation codes to calculate z via an equation in order to obtain more accurate values of z than can be obtained from charts. Equation (7.9) employs the Pitzer acentric factor, ω:

$$z = z^0 + z^1\omega \tag{7.9}$$

Tables in Appendix C list values of z^0 and z^1 as a function of T_r and p_r; ω is unique for each compound, and you can find values for it on the CD that accompanies this book. Table 7.3 is an abbreviated table of the acentric factors from Pitzer.

The acentric factor ω indicates the degree of acentricity or nonsphericity of a molecule. For helium and argon, ω is equal to zero. For higher-molecular-weight hydrocarbons and for molecules with increased polarity, the value of ω increases.

Table 7.3 Selected Values of the Pitzer* Acentric Factor

Compound	Acentric Factor	Compound	Acentric Factor
Acetone	0.309	Hydrogen sulfide	0.100
Benzene	0.212	Methane	0.008
Ammonia	0.250	Methanol	0.559
Argon	0.000	*n*-Butane	0.193
Carbon dioxide	0.225	*n*-Pentane	0.251
Carbon monoxide	0.049	Nitric oxide	0.607
Chlorine	0.073	Nitrogen	0.040
Ethane	0.098	Oxygen	0.021
Ethanol	0.635	Propane	0.152
Ethylene	0.089	Propylene	0.148
Freon-12	0.176	Sulfur dioxide	0.251
Hydrogen	−0.220	Water vapor	0.344

*K. S. Pitzer, *J. Am. Chem. Soc.*, **77**, 3427 (1955).

As an example of calculating z via the Pitzer correlation, calculate the compressibility factor z for ethylene (C_2H_4) at 300 K and 3000 kPa. First get T_c (283.1 K) and p_c (50.5 atm) from the CD, and then calculate T_r and p_r:

$$T_r = \frac{300\text{ K}}{282.5\text{ K}} = 1.06 \qquad p_r = \frac{3000\text{ kPa}}{50\text{ atm}}\left|\frac{1\text{ atm}}{101.3\text{ kPa}}\right. = 0.586$$

Next, look up (using interpolation) z^0 (0.812) and $z^1(-0.01)$ as well as ω (0.089) from Appendix C or Table 7.3.

$$z = 0.812 + (-0.01)(0.089) = 0.812$$

The calculation of z is easy if you know the values of p and T for the gas and just want to calculate z for those two conditions. But if you know p and $\hat{V}$ or T and $\hat{V}$, you have to employ a trial-and-error solution to get z. What you do is assume a sequence of values of p, calculate the related sequence of values of p_r, and next calculate the associated sequence of values of z. Then you calculate values of $\hat{V}$ from $p\hat{V} = zRT$. When you find the value of the specific volume that matches the value specified in the problem, you have an appropriate z (and $\hat{V}$).

A different way of predicting p, $\hat{V}$, and T properties is the **group contribution method** which has been successful in estimating properties of pure components. This method is based on combining the contribution of each functional group of a compound. The key assumption is that a group such as $-CH_3$ or $-OH$ behaves identically irrespective of the molecule in which it appears. This assumption is not quite true, so any group contribution method yields approximate values for gas properties. Probably the most widely used group contribution method is UNIFAC,[1] which forms a part of many computer databases. UNIQUAC is a variant of UNIFAC and is widely used in the chemical industry in the modeling of nonideal systems (systems with strong interaction between the molecules).

Self-Assessment Test

Questions

1. Explain why the van der Waals and Peng-Robinson equations (of state) are easy to solve for p and hard to solve for V.
2. Under what conditions will an equation of state be the most accurate?
3. What are the units of a and b in the SI system for the Redlich-Kwong equation?

Problems

1. Convert the virial (power series) equations of Kammerlingh-Onnes and Holborn (in Table 7.2) to a form that yields an expression for z.
2. Calculate the temperature of 2 g mol of a gas using van der Waals' equation with $a = 1.35 \times 10^{-6}\ m^6\,(atm)(g\ mol^{-2})$, $b = 0.0322 \times 10^{-3}\ (m^3)(g\ mol^{-1})$ if the pressure is 100 kPa and the volume is 0.0515 m^3.
3. Calculate the pressure of 10 kg mol of ethane in a 4.86 m^3 vessel at 300 K using two equations of state: (a) ideal gas and (b) Soave-Redlich-Kwong. Compare your answer with the experimentally observed value of 34.0 atm.

---

[1]A. Fredenslund, J. Gmehling, and P. Rasmussen, *Vapor-Liquid Equilibria Using UNIFAC*, Elsevier, Amsterdam (1977); D. Tiegs, J. Gmehling, P. Rasmussen, and A. Fredenslund, *Ind. Eng. Chem. Res.*, **26**, 159 (1987).

Thought Problems

1. Data pertaining to the atmosphere on Venus (which has a gravitational field only 0.81 that of the Earth) shows that the temperature of the atmosphere at the surface is 474 ± 20°C and the pressure is 90 ± 15 atm. What do you think is the reason(s) for the difference between these figures and those at the Earth's surface?
2. A method of making protein nanoparticles (0.5 to 5.0 μ in size) has been patented by the Aphio Corp. Anti-cancer drugs that small can be used in novel drug delivery systems. In the process described in the patent, protein is mixed with a gas such as carbon dioxide or nitrogen at ambient temperature and 20,000 kPa pressure. When the pressure is released, the proteins break up into fine particles.

   What are some of the advantages of such a process versus making powders by standard methods?

Discussion Question

Fossil fuels provide most of our power, and the carbon dioxide produced is usually discharged to the atmosphere. The Norwegian company Statoil separates carbon dioxide from its North Sea gas production and, since 1996, has been pumping it at the rate of 1 million tons per year into a layer of sandstone 1 km below the seabed. The sandstone traps the gas in a gigantic bubble that in 2001 contained 4 million tons of carbon dioxide.

Will the bubble of carbon dioxide remain in place? What problems exist with regard to the continuous addition of carbon dioxide in future years? Is the carbon dioxide actually a gas?

7.3 Real Gases: Compressibility Charts

Calculation of any of the variables p, T, $\hat{V}$, and z using the generalized equation $p\hat{V} = zRT$ can be assisted by using graphs called **generalized compressibility charts**, or z factor charts.

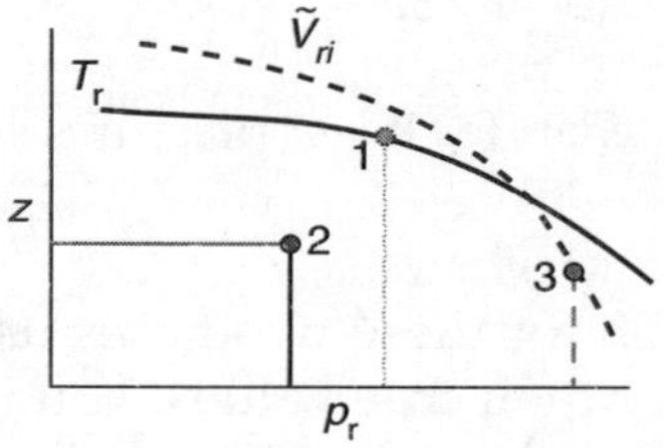

Figure 7.7 A compressibility chart involves four parameters: z, p_r, T_r, and $\tilde{V}_{ri}$.

Four parameters are displayed in Figure 7.7. Any two values will fix a point from which you can determine the other two. For example, if p_r and T_r are known (point 1), the value of $\tilde{V}_{ri}$ can be determined by interpolating between the two closest curves of $\widetilde{V}_{ri}$, and z can be determined by drawing a horizontal line from point 1 to the z axis.

Figures 7.8a and 7.8b show two examples of the **generalized compressibility factor** charts prepared by Nelson and Obert.[2] These charts are based on data for 30 gases. Figure 7.8a represents z for 26 gases (excluding H_2, He, NH_3, and H_2O) with a maximum deviation of 1%, and H_2 and H_2O within a deviation of 1.5%. Figure 7.8b is for 9 gases and errors can be as high as 5%. Note that the vertical axis in Figure 7.8b is not z but zT_r. To use the charts for H_2 and He (only), make corrections to the actual constants to get **pseudocritical** constants as follows:

$$T'_c = T_c + 8\text{ K}$$

$$p'_c = p_c + 8\text{ atm}$$

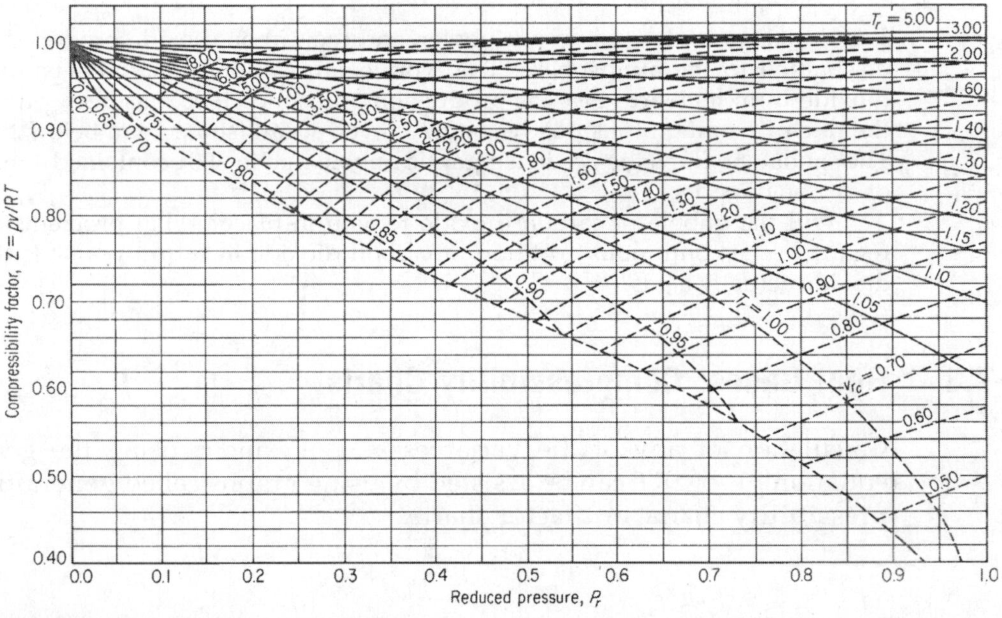

Figure 7.8a Generalized compressibility chart for lower pressures showing z as a function of p_r, T_r, and $\hat{V}_{ri}$

Then you can use Figures 7.8a and 7.8b for these two gases using the pseudocritical constants as replacements for their true values. You will find these two charts and additional charts for other ranges of p_r and T_r on the CD that

[2]L. C. Nelson and E. F. Obert, *Chem. Eng.*, **61**, No. 7, 203–8 (1954). Figures 7.8a and 7.8b include data reported by P. E. Liley, *Chem. Eng.*, 123 (July 20, 1987).

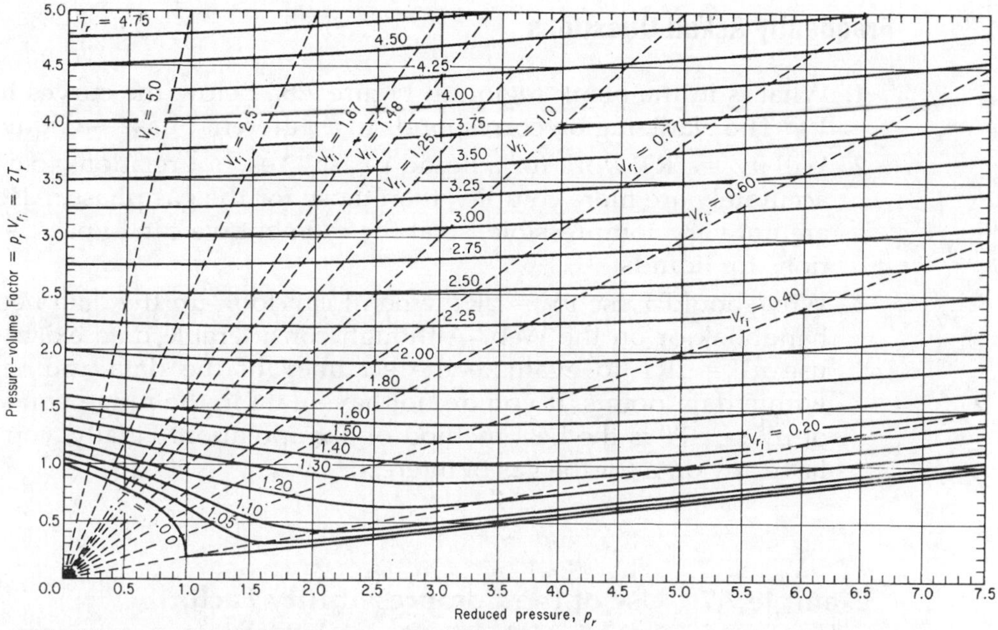

Figure 7.8b Generalized compressibility chart for higher values of p_r

accompanies this book in a format that can be expanded to get better accuracy.

Instead of the reduced specific volume, a third parameter shown on the charts is the dimensionless ideal reduced volume defined by

$$\widetilde{V}_{ri} = \frac{\hat{V}}{\widetilde{V}_{ci}} \quad \text{a dimensionless quantity}$$

where $\widetilde{V}_{ci}$ is the **ideal critical specific volume** (not the experimental value of the critical specific volume which yields poorer predictions) and is calculated from

$$\widetilde{V}_{ci} = \frac{RT_c}{p_c} \tag{7.10}$$

Both $\widetilde{V}_{ri}$ and $\widetilde{V}_{ci}$ are easy to calculate since T_c and p_c are presumed known or can be estimated for a compound. The development of the generalized compressibility charts is of considerable practical as well as pedagogical value because their existence enables you to make engineering calculations with considerable ease, and it also permits the development of thermodynamic functions for gases for which no experimental data are available.

Frequently Asked Questions

1. **What is in the blank region in Figure 7.8a below the curves for T_r and $\hat{V}_{ri}$?** The blank region corresponds to a different phase—a liquid.
2. **Will $p\hat{V} = zRT$ work for a liquid phase?** Yes, but relations to calculate z accurately are more complex than those for the gas phase. Also, liquids are not very compressible, so at the moment we can bypass p-$\hat{V}$-T relations for liquids.
3. **Why should I use $p\hat{V} = zRT$ when I can look up the data needed in a handbook or on the Web?** Although considerable data exists, you can use $p\hat{V} = zRT$ to evaluate the accuracy of the data and interpolate within data points. If you do not have data in the range you want, use of $p\hat{V} = zRT$ is the best method of extrapolation. Finally, you may not have any data for the gas of interest.

Example 7.7 Use of the Compressibility Factor in Calculating a Specific Volume

In spreading liquid ammonia fertilizer, the charges for the amount of NH_3 used are based on the time involved plus the pounds of NH_3 injected into the soil. After the liquid has been spread, there is still some ammonia left in the source tank (volume = 120 ft^3), but in the form of a gas. Suppose that your weight tally, which is obtained by difference, shows a net weight of 125 lb of NH_3 left in the tank at 292 psig. Because the tank is sitting in the sun, the temperature in the tank is 125°F.

Your boss complains that his calculations show that the specific volume of the NH_3 gas is 1.20 ft^3/lb, and hence there are only 100 lb of NH_3 in the tank. Could he be correct? See Figure E7.7.

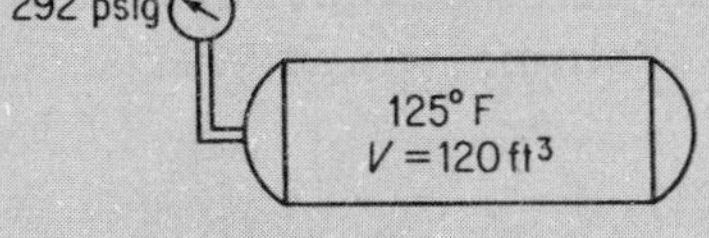

Figure E7.7

Solution

The simplest calculation to make to get the specific volume of the ammonia in the tank is to select a pound or pound mole as a basis:

Basis: 1 lb of NH_3

Apparently, your boss used the ideal gas law ($z = 1$) in getting the figure of 1.20 ft^3/lb of NH_3 gas:

$$R = 10.73\frac{(\text{psia})(\text{ft}^3)}{(\text{lb mol})(°\text{R})} \qquad p = 292 + 14.7 = 306.7 \text{ psia}$$

$$T = 125°\text{F} + 460 = 585°\text{R} \qquad n = \frac{1 \text{ lb}}{17 \text{ lb/lb mol}}$$

$$\hat{V} = \frac{nRT}{p} = \frac{\frac{1}{17}(10.73)\,(585)}{306.7} = 1.20 \text{ ft}^3/\text{lb}$$

What should you do? Ammonia probably does not behave like an ideal gas under the observed conditions of temperature and pressure. You can apply $pV = znRT$ to calculate n and determine the real amount of NH_3 in the tank if you include the correct compressibility factor in the real gas law. Let's compute z; z is a function of T_r and p_r. You can look up all of the values of the necessary parameters in Appendix F or on the CD.

$$T_c = 405.5 \text{ K} \Rightarrow 729.9°\text{R} \qquad p_c = 111.3 \text{ atm} \Rightarrow 1636 \text{ psia}$$

Then since

$$T_r = \frac{T}{T_c} = \frac{585°\text{R}}{729.9°\text{R}} = 0.801 \qquad p_r = \frac{p}{p_c} = \frac{306.7 \text{ psia}}{1636 \text{ psia}} = 0.187$$

From the Nelson and Obert (N&O) chart, Figure 7.8a, you can read $z \cong 0.855$. The value may be somewhat in error because ammonia was not one of the gases included in the preparation of the figure. Rather than calculating the specific volume directly, let's calculate it from the ratio of $pV_{\text{real}} = z_{\text{real}}nRT$ to $pV_{\text{ideal}} = z_{\text{ideal}}nRT$, the net result of which is

$$\frac{\hat{V}_{\text{real}}}{\hat{V}_{\text{ideal}}} = \frac{z_{\text{real}}}{z_{\text{ideal}}}$$

On the basis of 1 lb NH_3,

$$\hat{V}_{\text{real}} = \frac{1.20 \text{ ft}^3 \text{ ideal}}{\text{lb}}\bigg|\frac{0.855}{1} = 1.03 \text{ ft}^3/\text{lb NH}_3$$

On the basis of 120 ft^3 in the tank,

$$\frac{1 \text{ lb NH}_3}{1.03 \text{ ft}^3}\bigg|\frac{120 \text{ ft}^3}{} = 117 \text{ lb NH}_3$$

(Continues)

Example 7.7 Use of the Compressibility Factor in Calculating a Specific Volume (*Continued*)

Certainly 117 lb is a more realistic figure than 100 lb, but it still could be in error, considering that the residual weight of the NH_3 in the tank is determined by difference.

As a matter of interest, as an alternative to making these calculations, you could look up the specific volume of NH_3 at the conditions in the tank in a handbook. You would find that $\hat{V} = 0.973 \text{ ft}^3/\text{lb}$, equivalent to 123 lb of NH_3, the correct value. Would you tell your boss to use the right compressibility factor, or state that you used the handbook value of $\hat{V}$?

If you calculated z from Equation (7.3), you would get

$$z = z^0 + z^1\omega = 0.864 - 0.107(0.250) = 0.837$$

What would the mass of ammonia in the tank be using $z = 0.873$?

Example 7.8 Use of the Compressibility Factor in Calculating a Pressure

Liquid oxygen is used in the steel industry, in the chemical industry, in hospitals, as rocket fuel oxidant, and for wastewater treatment as well as in many other applications. A tank sold to hospitals contains 0.0284 m^3 of volume filled with 3.500 kg of liquid O_2 that will vaporize at −25°C. After all of the O_2 in the tank vaporizes, will the pressure in the tank exceed the safety limit for the tank specified as 104 kPa?

Solution

Basis: 3.500 kg of O_2

You can find from Appendix F on the CD that for oxygen

$$T_c = 154.4 \text{ K} \qquad p_c = 49.7 \text{ atm} \Rightarrow 5035 \text{ kPa}$$

However, you cannot proceed to solve this problem in exactly the same way as the preceding problem because you do not know the pressure of the O_2 in the tank to begin with. But you can use the pseudoparameter, $\widetilde{V}_{ri}$, which is available as a parameter on the Nelson and Obert charts, as a second parameter to fix a point on the compressibility charts.

First calculate

$$\hat{V}(\text{specific molal volume}) = \frac{0.0284 \text{ m}^3}{3.500 \text{ kg}}\left|\frac{32 \text{ kg}}{1 \text{ kg mol}}\right. = 0.260 \text{ m}^3/\text{kg mol}$$

Note that the *specific molar volume* must be used in calculating $\widetilde{V}_{ri}$ since $\hat{V}_{ci}$ is the volume per mole.

$$\widetilde{V}_{ci} = \frac{RT_c}{p_c} = \frac{8.313(\text{m}^3)(\text{kPa})}{(\text{kg mol})(\text{K})}\left|\frac{154.4\text{ K}}{5{,}035\text{ kPa}}\right. = 0.255\frac{\text{m}^3}{\text{kg mol}}$$

Then

$$\widetilde{V}_{ri} = \frac{\hat{V}}{\widetilde{V}_{ci}} = \frac{0.260}{0.255} = 1.02$$

Now you know the values of two parameters, $\widetilde{V}_{ri}$ and

$$T_r = \frac{248\text{ K}}{154.4\text{ K}} = 1.61$$

From the Nelson and Obert chart (Figure 7.8b) you can read

$$p_r = 1.43$$

Then

$$p = p_r p_c$$

$$= 1.43(5035) = 7200\text{ kPa}$$

The pressure of 10^4 kPa will not be exceeded. Even at room temperature the pressure will be less than 10^4 kPa.

To get one snapshot of the difference between estimates of z by three of the methods discussed in this chapter, Table 7.4 compares the experimental values of z for ethylene with predictions by three methods: N&O charts, Pitzer's relation, and the ideal gas laws.

Table 7.4 A Comparison of Values of the Compressibility Factor z for Ethylene* Determined via Three Different Methods with the Associated Experimental Values

	At 350 K and 500 kPa		At 300 K and 3000 kPA		At 274 K and 3600 kPa	
	z	% Deviation	z	% Deviation	z	% Deviation
Experimental value	0.983	-	0.812	-	0.563	-
N&O chart	0.982	0.0	0.815	0.0	0.57	1
Ideal gas law	1	1.8	1	23.1	1	78
Equation (7.9)	0.983	0	0.812	0.0	0.537	−4.6

* $w = 0.089$; $T_c = 282.8$ K; $p_c = 50.5$ atm; z^0 and z^1 are from the tables in Appendix C.

7.4 Real Gas Mixtures

To this point, we have discussed predicting *p-V-T* properties for *pure* components of real gases. How should you treat mixtures of real gases? The actual critical points of binary mixtures are not linear combinations of the properties of the two components as shown in Figure 7.9 for combinations of CO_2 and SO_2. Too many dimensions are involved to draw pictures for three or more components.

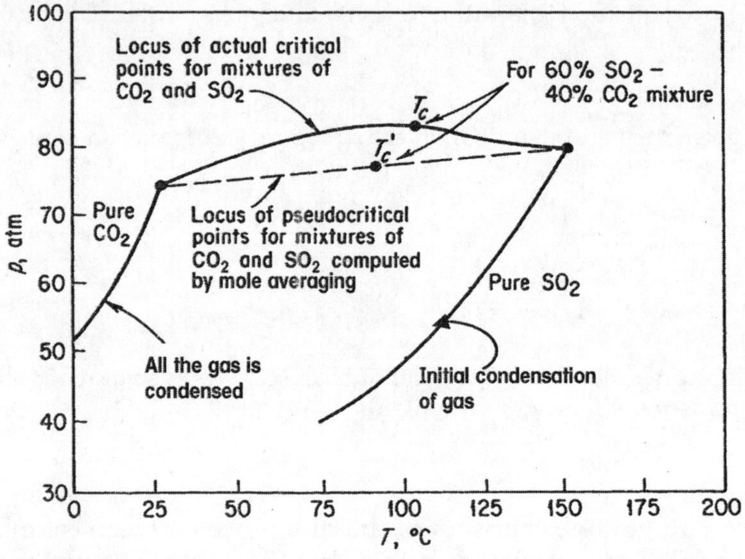

Figure 7.9 Critical and pseudocritical points for mixtures of CO_2 and SO_2

One way you can make reasonable predictions for z and $\widetilde{V}_{ri}$ for engineering purposes is to use Kay's method[3] and the compressibility charts. In Kay's method, **pseudocritical** values for mixtures of gases are calculated on the assumption that each component in the mixture contributes to the pseudocritical value in the same proportion as the mole fraction of that component in the gas. Thus, the pseudocritical values are computed as mole averages as follows:

$$p'_c = p_{c_A} y_A + p_{c_B} y_B + \cdots \quad (7.11)$$

$$T'_c = T_{c_A} y_A + T_{c_B} y_B + \cdots \quad (7.12)$$

[3]W. B. Kay, "Density of Hydrocarbon Gases and Vapors at High Temperature and Pressure," *Ind. Eng. Chem.*, **28**, 1014–19 (1936).

where y_i is the mole fraction, p'_c is the pseudocritical pressure, and T'_c is the pseudocritical temperature. You can see that these are linearly weighted mole average pseudocritical properties. Look at Figure 7.9, which compares the true critical values of a gaseous mixture of CO_2 and SO_2 with the respective pseudocritical values. The respective **pseudoreduced** variables are

$$p'_r = \frac{p}{p'_c}$$

$$T'_r = \frac{T}{T'_c}$$

Kay's method is known as a two-parameter rule since only p_c and T_c for each component are involved in the calculation of z. If a third parameter such as z_c, or the Pitzer acentric factor, or $\hat{V}_{ci}$, is included in the determination of the compressibility factor, you would have a three-parameter rule. Other pseudocritical methods with additional parameters provide better accuracy in predicting p-V-T properties than Kay's method, but Kay's method can suffice for our work, and it is easy to use.

In instances in which the temperature or pressure of a gas mixture is unknown, to avoid a trial-and-error solution using the generalized compressibility charts, you can compute the **pseudocritical ideal volume** and a **pseudoreduced ideal volume** $\widetilde{V}_{ri}$ thus

$$\widetilde{V}'_{ci} = \frac{RT'_c}{p'_c} \quad \text{and} \quad \widetilde{V}'_{ri} = \frac{\hat{V}}{\widetilde{V}_{ci}}$$

$\widetilde{V}_{ri}$ can be used in lieu of p'_r or T'_r in the compressibility charts.

An enormous literature exists describing proposals for *mixing rules* for equations of state, that is, rules to weight the coefficients or the predictions of each pure component so that the weighted values can be used with the same equations of state as are used for a pure component. Refer to the references at the end of this chapter, or look on the Internet for examples.

Even if there is only one possible unified theory it is just a set of rules and equations.

Stephen Hawking

Example 7.9 Calculation of p-V-T Properties for a Real Gas Mixture

A gaseous mixture has the following composition (in mole percent):

Methane, CH_4	20
Ethylene, C_2H_4	30
Nitrogen, N_2	50

(*Continues*)

Example 7.9 Calculation of *p-V-T* Properties for a Real Gas Mixture (*Continued*)

at 90 atm pressure and 100°C. Compare the volume per mole as computed by the methods of (a) the ideal gas law and (b) the pseudoreduced technique (Kay's method). What other types of averaging might you use?

Solution

Basis: 1 g mol of gas mixture

Additional data needed are

Component	T_c (K)	p_c (atm)
CH_4	191	45.8
C_2H_4	283	50.5
N_2	126	33.5

The units used are fixed by the units of R. Let R be $R = 82.06 \frac{(\text{cm}^3)(\text{atm})}{(\text{g mol})(\text{K})}$.

a. Ideal gas law:

$$\hat{V} = \frac{RT}{p} = \frac{(82.06)(373)}{90} = 340 \text{ cm}^3/\text{g mol at 90 atm and 373 K}$$

b. According to Kay's method, you first calculate the pseudocritical values for the mixture:

$$p'_c = p_{c_A} y_A + p_{c_B} y_B + p_{c_C} y_C = (45.8)(0.2) + (50.5)(0.3) + (33.5)(0.5)$$
$$= 41.1 \text{ atm}$$

$$T'_c = T_{c_A} y_A + T_{c_B} y_B + T_{c_C} y_C = (191)(0.2) + (283)(0.3) + (126)(0.5) = 186 \text{ K}$$

Then you calculate the pseudoreduced values for the mixture:

$$p'_r = \frac{p}{p'_c} = \frac{90}{41.2} = 2.19, \qquad T'_r = \frac{T}{T'_c} = \frac{373}{186} = 2.01$$

With the aid of these two parameters you can find from Figure 7.8b that $zT'_r = 1.91$ and thus $z = 0.95$. Then

$$\hat{V} = \frac{zRT}{p} \approx \frac{0.95(1)(82.06)(373)}{90} = 323 \text{ cm}^3/\text{g mol at 90 atm and 373 K}$$

Two of the many possible ways of averaging are to use an equation of state with mole-averaged coefficients, or use the mole-averaged predictions of $\hat{V}$ obtained from the individual equation of state. If you decided to use Equation (7.9) to calculate z for the mixture, how might you average z^0, z^1, and w?

Self-Assessment Test

Questions

1. What is the pseudocritical volume? What is the advantage of using V_{ci}?
2. Indicate whether the following statements are true or false:
   a. Two fluids, which have the same values of reduced temperature and pressure and the same reduced volume, are said to be in corresponding states.
   b. It is expected that all gases will have the same z at a specified T_r and p_r. Thus a correlation of z in terms of T_r and p_r would apply to all gases.
   c. The law of corresponding states states that at the critical state (T_c, p_c) all substances should behave alike.
   d. The critical state of a substance is the set of physical conditions at which the density and other properties of the liquid and vapor become identical.
   e. Any substance (in theory) by the law of corresponding states should have the same reduced volume at the same reduced T_r and P_r.
   f. The equation $pV = znRT$ cannot be used for ideal gases.
   g. By definition a fluid becomes supercritical when its temperature and pressure exceed the critical point.
   h. Phase boundaries do not exist under supercritical conditions.
   i. For some gases under normal conditions, and for most gases under conditions of high pressure, values of the gas properties that might be obtained using the ideal gas law would be at wide variance with the experimental evidence.
3. Explain the meaning of the following equation for the compressibility factor:

$$z = f(T_r, p_r)$$

4. What is the value of z at $p_r = 0$?

Problems

1. Calculate the compressibility factor z, and determine whether or not the following gases can be treated as ideal at the listed temperature and pressure:
   a. Water at 1000°C and 2000 kPa
   b. Oxygen at 35°C and 1500 kPa
   c. Methane at 10°C and 1000 kPa
2. A carbon dioxide fire extinguisher has a volume of 40 L and is to be charged to a pressure of 20 atm at a storage temperature of 20°C. Determine the mass in kilograms of CO_2 in the fire extinguisher.
3. Calculate the pressure of 4.00 g mol of CO_2 contained in a 6.25×10^{-3} m^3 fire extinguisher at 25°C.
4. One pound mole of a mixture containing 0.400 lb mol of N_2 and 0.600 lb mol of C_2H_4 at 50°C occupies a volume of 1.44 ft^3. What is the pressure in the vessel? Compute your answer by Kay's method.

Thought Problems

1. Pressure vessels and rigid piping have to be protected against overpressure by using safety devices. For example, when liquid is trapped within a rigid piping system, it expands, and a small expansion caused by a temperature increase will produce a large pressure rise in the system in the vapor above the liquid. What happens in a space containing 400 L of liquid and a gas bubble of 4.0 L with an initial pressure of 1 atm when the liquid expands 1% so that the gas volume is compressed? Will the piping system fail?
2. The sum of the mass fractions for an ideal gas mixture is equal to 1. Is the sum also equal to 1 for a real gas mixture?

Discussion Question

A letter to the editor was headed "Not Sold on Hydrogen." In part it said:

> Your innovative story "Fuel Cells: A Lot of Hot Air?" concerned me. Hydrogen under pressure is difficult to contain. Leaks are difficult to detect, and you need to obtain virtual zero leakage. Hydrogen is explosive and flammable, and burns with an invisible flame. Finally, your economics do not address the entire process from beginning to end, and come to less than realistic conclusions.

Comment on the points that the author of the letter makes. Do they damage the potential of hydrogen-based fuel cells? In what aspects is he correct and in what aspects wrong?

Looking Back

We reviewed the ideal gas law and showed how to use it in conjunction with material balances. The law of corresponding states was introduced, and it was shown how to correct the ideal gas law by calculating compressibility factors from tables based on reduced conditions. Several commonly used equations of state were presented, and it was shown how to use them to calculate unknown properties of nonideal gases.

Glossary

Acentric factor A parameter that indicates the degree of nonsphericity of a molecule.

Benedict-Webb-Rubin equation of state An eight-parameter equation of state that relates the physical properties p, V, T, and n for a gas.

Compressibility charts Graphs of the compressibility factor as a function of reduced temperature, pressure, and ideal reduced volume.

Compressibility factor A factor that is introduced into the ideal gas law to compensate for the nonideality of a gas.

Corrected Normalized.

Corresponding states Any gas should have the same reduced volume at the same reduced temperature and reduced pressure.

Critical state The set of physical conditions at which the density and other properties of liquid and vapor become identical.

Dalton's law The summation of each of the partial pressures of the components in a system equals the total pressure. The other related law (of partial pressures) is that the total pressure times the mole fraction of a component in a system is the partial pressure of the component.

Density of gas Mass per unit volume expressed in kilograms per cubic meter, pounds per cubic foot, grams per liter, or equivalent units.

Generalized compressibility See **compressibility charts.**

Generalized equation of state The ideal gas law converted to a real gas law by inserting a compressibility factor.

Group contribution method A technique of estimating physical properties of compounds by using properties of molecular groups of elements in the compound.

Holborn A multiple-parameter equation of state expanded in p.

Ideal critical volume $\hat{V}_{ci} = \mathrm{R}T_c/p_c$.

Ideal gas constant The constant in the ideal gas law (and other equations) denoted by the symbol R.

Ideal gas law Equation relating p, V, n, and T that applies to many gases at low density (high temperature and/or low pressure).

Ideal reduced volume $V_{ri} = \hat{V}/\hat{V}_{ci}$.

Kammerlingh-Onnes A multiple-parameter equation of state expanded in V^{-1}.

Kay's method Rule for calculating the compressibility factor for a mixture of gases.

Law of corresponding states See **corresponding states.**

Partial pressure The pressure that would be exerted by a single component in a gaseous mixture if it existed alone in the same volume as occupied by the mixture and at the same temperature as the mixture.

Peng-Robinson A three-parameter equation of state.

Pitzer acentric factor See **acentric factor.**

Pseudocritical Temperatures, pressures, and/or specific volumes adjusted to be used with charts or equations used to calculate the compressibility factor.

Real gases Gases whose behavior does not conform to the assumptions underlying ideality.

Reduced variables Corrected or normalized conditions of temperature, pressure, and volume, normalized by their respective critical conditions.

Soave-Redlich-Kwong (SRK) A three-parameter equation of state.

Specific gravity Ratio of the density of a gas at a temperature and pressure to the density of a reference gas at some temperature and pressure.

Standard conditions (S.C.) Arbitrarily specified standard states of temperature and pressure established for gases by custom.

Supercritical fluid Material in a state above its critical point.

UNIFAC A group contribution method of estimating physical properties.

UNIQUAC An extension of the UNIFAC method of estimating physical properties.

Van der Waals A two-parameter equation of state.

Virial equation of state An equation of state expanded in successive terms of one of the physical properties.

Supplementary References

Ben-Amotz, Dor, Alan Gift, and R. D. Levine. "Updated Principle of Corresponding States," *J. Chem. Edu.*, **81**, No. 1 (2004).

Castillo, C. A. "An Alternative Method for the Estimation of Critical Temperatures of Mixtures," *AIChE J.*, **33**, 1025 (1987).

Chao, K. C., and R. L. Robinson. *Equations of State in Engineering and Research*, American Chemical Society, Washington, DC (1979).

Copeman, T. W., and P. M. Mathias. "Recent Mixing Rules for Equations of State," *ACS Symposium Series*, **300**, 352–69, American Chemical Society, Washington, DC (1986).

Eliezer, S., et al. *An Introduction to Equations of State: Theory and Applications*, Cambridge University Press, Cambridge, UK (1986).

Elliott, J. R., and T. E. Daubert. "Evaluation of an Equation of State Method for Calculating the Critical Properties of Mixtures," *Ind. Eng. Chem. Res.*, **26**, 1689 (1987).

Gibbons, R. M. "Industrial Use of Equations of State," in *Chemical Thermodynamics in Industry*," edited by T. I. Barry, Blackwell Scientific, Oxford, UK (1985).

Lawal, A. S. "A Consistent Rule for Selecting Roots in Cubic Equations of State," *Ind. Eng. Chem. Res.*, **26**, 857–59 (1987).

Manavis, T., M. Volotopoulos, and M. Stamatoudis. "Comparison of Fifteen Generalized Equations of State to Predict Gas Enthalpy," *Chem. Eng. Commun.*, **130**, 1–9 (1994).

Masavetas, K. A. "The Mere Concept of an Ideal Gas," *Math. Comput. Modelling*, **12**, 651–57 (1989).

Mathias, P. M., and M. S. Benson. "Computational Aspects of Equations of State," *AIChE J.*, **32**, 2087 (1986).

Mathias, P. M., and H. C. Klotz. "Take a Closer Look at Thermodynamic Property Models," *Chem. Eng. Progress*, 67–75 (June, 1994).

Orbey, H., S. I. Sander, and D. S. Wong. "Accurate Equation of State Predictions at High Temperatures and Pressures Using the Existing UNIFAC Model," *Fluid Phase Equil.*, **85**, 41–54 (1993).

Reid, R. C., J. M. Prausnitz, and B. E. Poling. *The Properties of Gases and Liquids*, 4th ed., McGraw-Hill, New York (1987).

Sandler, S. I., H. Orbey, and B. I. Lee. "Equations of State," in *Modeling for Thermodynamic and Phase Equilibrium Calculations*, Chapter 2, edited by S. I. Sander, Marcel Dekker, New York (1994).

Span, R. *Multiparameter Equations of State*, Springer, New York (2000).

Sterbacek, Z., B. Biskup, and P. Tausk. *Calculation of Properties Using Corresponding State Methods*, Elsevier Scientific, New York (1979).

Yaws, C. L., D. Chen, H. C. Yang, L. Tan, and D. Nico. "Critical Properties of Chemicals," *Hydrocarbon Processing*, **68**, 61 (July, 1989).

Web Sites

http://en.citizendium.org/wiki/Compressibility_factor_(gases)
www.procdev.com/zcalcs/zclient.aspx
http://en.wikipedia.org/wiki/Equation_of_state

Problems

7.1 Ideal Gases

***7.1.1** How many pounds of H_2O are in 100 ft^3 of vapor at 15.5 mm Hg and 23°C?

***7.1.2** One liter of a gas is under a pressure of 780 mm Hg. What will be its volume at standard pressure, the temperature remaining constant?

***7.1.3** A gas occupying a volume of 1 m^3 under standard pressure is expanded to 1.200 m^3, the temperature remaining constant. What is the new pressure?

***7.1.4** Determine the mass specific volume and molal specific volume for air at 78°F and 14.7 psia.

***7.1.5** Divers work as far as 500 ft below the water surface. Assume that the water temperature is 45°F. What is the molar specific volume (in cubic feet per pound mole) for an ideal gas under these conditions?

***7.1.6** A 25 L glass vessel is to contain 1.1 g mol of nitrogen. The vessel can withstand a pressure of only 20 kPa above atmospheric pressure (taking into account a suitable safety factor). What is the maximum temperature to which the N_2 can be raised in the vessel?

****7.1.7** An oxygen cylinder used as a standby source of oxygen contains O_2 at 70°F. To calibrate the gauge on the O_2 cylinder, which has a volume of 1.01 ft^3, all of the oxygen, initially at 70°F, is released into an evacuated tank of known volume (15.0 ft^3). At equilibrium, the gas pressure was measured as 4 in. H_2O gauge and the gas temperature in both cylinders was 75°F. See Figure P7.1.7. The barometer read 29.99 in. Hg.

What did the pressure gauge on the oxygen tank initially read in psig if it was a Bourdon gauge?

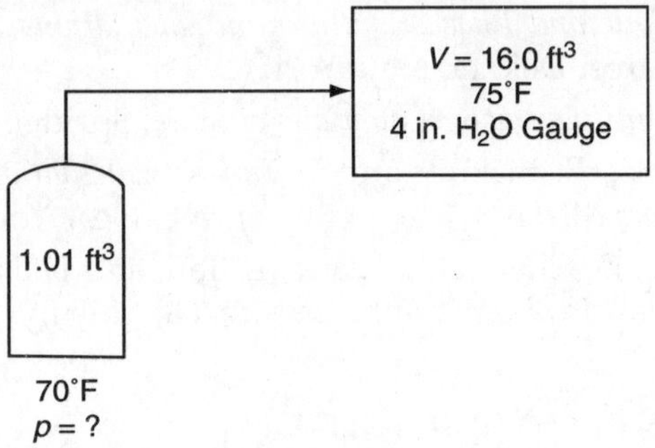

Figure P7.1.7

***7.1.8** An average person's lungs contain about 5 L of gas under normal conditions. If a diver makes a free dive (no breathing apparatus), the volume of the lungs is compressed when the pressure equalizes throughout the body. If compression occurs below 1 L, irreversible lung damage will occur. Calculate the maximum safe depth for a free dive in seawater (assume the density is the same as freshwater).

***7.1.9** An automobile tire when cold (at 75°F) reads 30 psig on a tire gauge. After driving on the freeway, the temperature in the tire becomes 140°F. Will the pressure in the tire exceed the pressure limit of 35 psi the manufacturer stamps on the tire?

***7.1.10** You are making measurements on an air conditioning duct to test its load capacity. The warm air flowing through the circular duct has a density of 0.0796 lb/ft^3. Careful measurements of the velocity of the air in the duct disclose that the average air velocity is 11.3 ft/s. The inside radius of the duct is 18.0 in. What are (a) the volumetric flow rate of the air in cubic feet per hour and (b) the mass flow rate of the air in pounds per day?

***7.1.11** One pound mole of flue gas has the following composition. Treat it as an ideal gas.

$$CO_2(11.2\%), CO(1.2\%), SO_2(1.2\%), O_2(5.3\%), N_2(81.0\%), H_2O(0.1\%)$$

How many cubic feet will the gas occupy at 100°F and 1.54 atm?

*7.1.12 From the known standard conditions, calculate the value of the gas law constant R in the following sets of units:

a. cal/(g mol)(K)
b. Btu/(lb mol)(°R)
c. (psia)(ft^3)/(lb mol)(°R)
d. J/(g mol)(K)
e. (cm^3)(atm)/(g mol)(K)
f. (ft^3)(atm)/(lb mol)(°R)

*7.1.13 What is the density of O_2 at 100°F and 740 mm Hg in (a) pounds per cubic foot and (b) grams per liter?

*7.1.14 What is the density of propane gas (C_3H_8) in kilograms per cubic meter at 200 kPa and 40°C? What is the specific gravity of propane?

*7.1.15 What is the specific gravity of propane gas (C_3H_8) at 100°F and 800 mm Hg relative to air at 60°F and 760 mm Hg?

*7.1.16 What is the mass of 1 m^3 of H_2 at 5°C and 110 kPa? What is the specific gravity of this H_2 compared to air at 5°C and 110 kPa?

*7.1.17 A gas used to extinguish fires is composed of 80% CO_2 and 20% N_2. It is stored in a 2 m^3 tank at 200 kPa and 25°C. What is the partial pressure of the CO_2 in the tank in kilopascals?

*7.1.18 A natural gas has the following composition by volume:

CH_4	94.1%
N_2	3.0
H_2	1.9
O_2	1.0
	100.0%

This gas is piped from the well at a temperature of 20°C and a pressure of 30 psig. It may be assumed that the ideal gas law is applicable. Calculate the partial pressure of the oxygen.

*7.1.19 A liter of oxygen at 760 mm Hg is forced into a vessel containing a liter of nitrogen at 760 mm Hg. What will be the resulting pressure? What assumptions are necessary for your answer?

*7.1.20 Indicate whether the following statements are true or false:

a. The volume of an ideal gas mixture is equal to the sum of the volumes of each individual gas in the mixture.
b. The temperature of an ideal gas mixture is equal to the sum of the temperatures of each individual gas in the mixture.
c. The pressure of an ideal gas mixture is equal to the sum of the partial pressures of each individual gas in the mixture.

****7.1.21** An oxygen cylinder used as a standby source of oxygen contains 1.000 ft^3 of O_2 at 70°F and 200 psig. What will be the volume of this O_2 in a dry-gas holder at 90°F and 4.00 in. H_2O above atmospheric? The barometer reads 29.92 in. Hg.

****7.1.22** You have 10 lb of CO_2 in a 20 ft^3 fire extinguisher tank at 30°C. Assuming that the ideal gas law holds, what will the pressure gauge on the tank read in a test to see if the extinguisher is full?

****7.1.23** The U-tube manometer depicted in Figure P7.1.23 has a left leg 20 in. high and a right leg 40 in. high. The manometer initially contains mercury to a depth of 12 in. in each leg. Then the left leg is closed with a cork, and mercury is poured in the right leg until the mercury in the left (closed) leg reaches a height of 14 in. How deep is the mercury in the right leg from the bottom of the manometer?

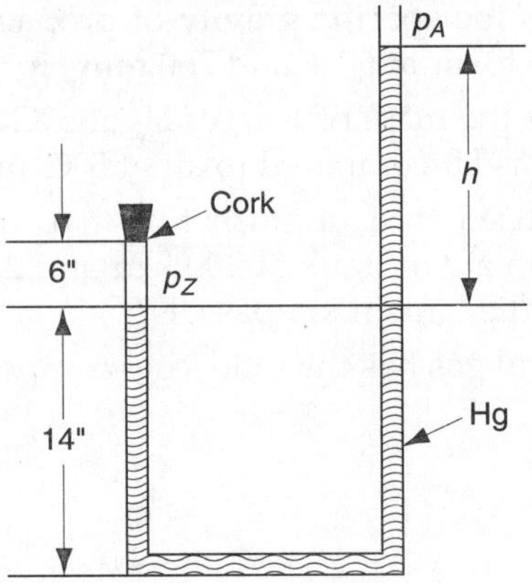

Figure P7.1.23

****7.1.24** One of the experiments in the fuel-testing laboratory has been giving some trouble because a particular barometer gives erroneous readings owing to the presence of a small amount of air above the mercury column. At a true atmospheric pressure of 755 mm Hg the barometer reads 748 mm Hg, and at a true 740 mm Hg the reading is 736 mm Hg. What will the barometer read when the actual pressure is 760 mm Hg?

****7.1.25** Flue gas at a temperature of 1800°F is introduced to a scrubber through a pipe that has an inside diameter of 4.0 ft. The inlet velocity to and the outlet velocity from the scrubber are 25 ft/s and 20 ft/s, respectively. The scrubber cools the flue gas to 550°F. Determine the duct size required at the outlet of the unit.

****7.1.26** Calculate the number of cubic meters of hydrogen sulfide, measured at a temperature of 30°C and a pressure of 15.71 cm Hg, which may be produced from 10 kg of iron sulfide (FeS).

****7.1.27** Monitoring of hexachlorobenzene (HCB) in a flue gas from an incinerator burning 500 lb/hr of hazardous wastes is to be conducted. Assume that all of the HCB is removed from a sample of the flue gas and concentrated in 25 mL of solvent. The analytical detection limit for HCB is 10 μg/ml in the solvent. Determine the minimum volume of flue gas that has to be sampled to detect the existence of HCB in the flue gas. Also, calculate the time needed to collect a gas sample if you can collect 1.0 L/min. The flue gas flow rate is 427,000 ft^3/hr measured at standard conditions.

****7.1.28** Ventilation is an extremely important method of reducing the level of toxic airborne contaminants in the workplace. Since it is impossible to eliminate absolutely all leakage from a process into the workplace, some method is always needed to remove toxic materials from the air in closed rooms when such materials are present in the process streams. The Occupational Safety and Health Administration (OSHA) has set the permissible exposure limit (PEL) of vinyl chloride (VC, MW = 78) at 1.0 ppm as a maximum time-weighted average (TWA) for an 8 hr workday, because VC is believed to be a human carcinogen. If VC escapes into the air, its concentration must be maintained at or below the PEL. If dilution ventilation were to be used, you can estimate the required airflow rate by assuming complete mixing in the workplace air, and then assuming that the volume of airflow through the room will carry VC out with it at the concentration of 1.0 ppm.

If a process loses 10 g/min of VC into the room air, what volumetric flow rate of air will be necessary to maintain the PEL of 1.0 ppm by dilution ventilation? (In practice we must also correct for the fact that complete mixing will not be realized in a room, so you must multiply the calculated airflow rate by a safety factor, say, a factor of 10.)

If the safety analysis or economics of ventilation do not demonstrate that a safe concentration of VC exists, the process might have to be moved into a hood so that no VC enters the room. If the process is carried out in a hood with an opening of 30 in. wide by 25 in. high, and the "face velocity" (average air velocity through the hood opening) is 100 ft/s, what is the volumetric airflow rate at S.C.? Which method of treating the pollution problem seems to be better to you? Explain why dilution ventilation is not recommended for maintaining air quality. What might be a problem with the use of a hood? The problem is adapted with permission from the publication *Safety, Health, and Loss Prevention in Chemical Processes* published by the American Institute of Chemical Engineers, New York (1990).

****7.1.29** Ventilation is an extremely important method of reducing the level of toxic airborne contaminants in the workplace. Trichloroethylene (TCE) is an excellent solvent for a number of applications and is especially useful in degreasing. Unfortunately, TCE can lead to a number of harmful health effects, and ventilation is essential. TCE has been shown to be carcinogenic in animal tests. (Carcinogenic means that exposure to the agent might increase the likelihood of the subject getting cancer at some time in the future.) It is also an irritant to the eyes and respiratory tract. Acute exposure causes depression of the central nervous system, producing symptoms of dizziness, tremors, and irregular heartbeat, plus others.

Since the molecular weight of TCE is approximately 131.5, it is much denser than air. As a first thought, you would not expect to find a high concentration of this material above an open tank because you might assume that the vapor would sink to the floor. If this were so, we would place the inlet of a local exhaust hood for such a tank near the floor. However, toxic concentrations of many materials are not much denser than the air itself, so where there can be mixing with the air we may not assume that all the vapors will go to the floor. For the case of trichloroethylene OSHA has established a time-weighted average 8 hr PEL of 100 ppm. What is the fraction increase in the density of a mixture of TCE in air over that of air if the TCE is at a concentration of 100 ppm and at 25°C? This problem has been adapted from *Safety, Health, and Loss Prevention in Chemical Processes*, Vol. 3, American Institute of Chemical Engineers, New York (1990).

****7.1.30** Benzene can cause chronic adverse blood effects such as anemia and possibly leukemia with chronic exposure. Benzene has a PEL for an 8 hr exposure of 1.0 ppm. If liquid benzene is evaporating into the air at a rate of 2.5 cm^3 of liquid/min, what must the ventilation rate be in volume per minute to keep the concentration below the PEL? The ambient temperature is 68°F and the pressure is 740 mm Hg. This problem has been adapted from *Safety, Health, and Loss Prevention in Chemical Processes*, Vol. 6, American Institute of Chemical Engineers, New York (1990).

****7.1.31** A recent newspaper report states:

> Home meters for fuel gas measure the volume of gas usage based on a standard temperature, usually 60 degrees. But gas contracts when it's cold and expands when warm. East Ohio Gas Co. figures that in chilly Cleveland, the homeowner with an outdoor meter gets more gas than the meter says he does, so that's built into the company's gas rates. The guy who loses is the one with an indoor meter: If his home stays at 60 degrees or over, he'll pay for more gas than he gets. (Several companies make temperature-compensating meters, but they cost more and aren't widely used. Not surprisingly, they are sold mainly to utilities in the North.)

Suppose that the outside temperature drops from 60°F to 10°F. What is the percentage increase in the mass of the gas passed by a noncompensated outdoor meter that operates at constant pressure? Assume that the gas is CH_4.

****7.1.32** Soft ice cream is a commercial ice cream mixture whipped usually with CO_2 (as the O_2 in the air causes deterioration). You are working in the Pig-in-a-Poke Drive-In and need to make a machine full (4 gal) of soft ice cream. The unwhipped mix has a specific gravity of 0.95 and the local ordinance forbids you to make ice cream of less than a specific gravity of 0.85. Your CO_2 tank is a No. 1 cylinder (9 in. diameter by 52 in. high) of commercial-grade

(99.5% min. CO_2) carbon dioxide. In examining the pressure gauge, you note it reads 68 psig. Do you have to order another cylinder of CO_2? Be sure to specifically state all the assumptions you make for this problem.

Additional data:
Atmospheric pressure = 752 mm Hg
Sp. gr. cream = 0.84
Sp. gr. milk = 0.92
Butterfat content of soft ice cream < 14%

****7.1.33** A natural gas has the following composition:

CH_4 (methane)	87%
C_2H_6 (ethane)	12%
C_3H_8 (propane)	1%

a. What is the composition in weight percent?
b. What is the composition in volume percent?
c. How many cubic meters will be occupied by 80.0 kg of the gas at 9°C and 600 kPa?
d. What is the density of the gas in kilograms per cubic meter at S.C.?
e. What is the specific gravity of this gas at 9°C and 600 kPa referred to air at S.C.?

****7.1.34** A mixture of bromine vapor in air contains 1% bromine by volume.
a. What weight percent bromine is present?
b. What is the average molecular weight of the mixture?
c. What is its specific gravity?
d. What is its specific gravity compared to bromine?
e. What is its specific gravity at 100°F and 100 psig compared to air at 60°F and 30 in. Hg?

****7.1.35** The contents of a gas cylinder are found to contain 20% CO_2, 60% O_2, and 20% N_2 at a pressure of 740 mm Hg and at 20°C. What are the partial pressures of each of the components? If the temperature is raised to 40°C, will the partial pressures change? If so, what will they be?

****7.1.36** Methane is completely burned with 20% excess air, with 30% of the carbon going to CO. What is the partial pressure of the CO in the stack gas if the barometer reads 740 mm Hg, the temperature of the stack gas is 300°F, and the gas leaves the stack at 250 ft above the ground level?

****7.1.37** A 0.5 m^3 rigid tank containing hydrogen at 20°C and 600 kPa is connected by a valve to another 0.5 m^3 rigid tank that holds hydrogen at 30°C and 150 kPa. Now the valve is opened and the system is allowed to reach thermal equilibrium with the surroundings, which are at 15°C. Determine the final pressure in the tank.

****7.1.38** A 400 ft^3 tank of compressed H_2 is at a pressure of 55 psig. It is connected to a smaller tank with a valve and short line. The small tank has a volume of 50 ft^3 and contains H_2 at 1 atm absolute and the same temperature. If the interconnecting valve is opened and no temperature change occurs, what is the final pressure in the system?

****7.1.39** A tank of N_2 has a volume of 100 ft^3 and an initial temperature of 80°F. One pound of N_2 is removed from the tank, and the pressure drops to 100 psig while the temperature of the gas in the tank drops to 60°F. Assuming N_2 acts as an ideal gas, calculate the initial pressure reading on the pressure gauge.

****7.1.40** Measurement of flue gas flow rates is difficult by traditional techniques for various reasons. Tracer gas flow measurements using sulfur hexafloride (SF_6) have proved to be more accurate. Figure P7.1.40 shows the stack arrangement and the injection and sampling points for the SF_6. Here are the data for one experiment:

Volume of SF_6 injected (converted to S.C.):	28.8 m^3/min
Concentration of SF_6 at the flue gas sample point:	4.15 ppm
Relative humidity correction:	none

Calculate the volume of the exit flue gas per minute.

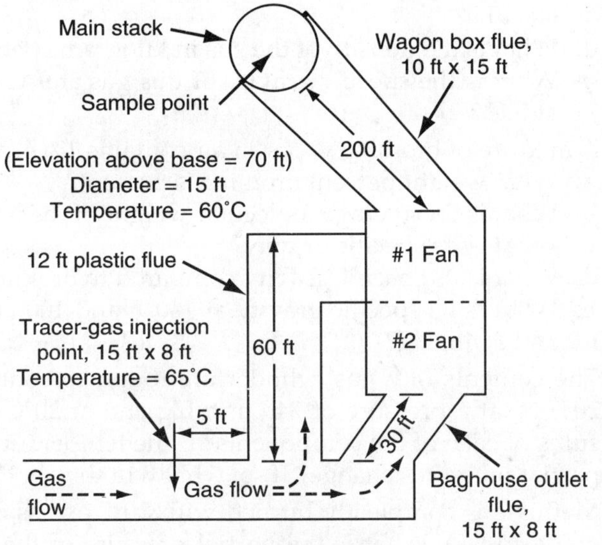

Figure P7.1.40

****7.1.41** An ideal gas at 60°F and 31.2 in. Hg (absolute) is flowing through an irregular duct. To determine the flow rate of the gas, CO_2 is passed into the gas stream. The gas analyzes 1.2 mol % CO_2 before and 3.4 mol % after addition. The CO_2 tank is placed on a scale and found to lose 15 lb in 30 min. What is the flow rate of the entering gas in cubic feet per minute?

****7.1.42** In the manufacture of dry ice, a fuel is burned to a flue gas which contains 16.2% CO_2, 4.8% O_2, and the remainder N_2. This flue gas passes through a heat exchanger and then goes to an absorber. The data show that the analysis of the flue gas entering the absorber is 13.1% CO_2 with the remainder

O_2 and N_2. Apparently something has happened. To check your initial assumption that an air leak has developed in the heat exchanger, you collect the following data on a dry basis on the heat exchanger:

Entering flue gas in a 2 min period: 47,800 ft^3 at 600°F and 740 mm of Hg
Exiting flue gas in a 2 min period: 30,000 ft^3 at 60°F and 720 mm of Hg

Was your assumption about an air leak a good one, or was perhaps the analysis of the gas in error? Or both?

****7.1.43** Three thousand cubic meters per day of a gas mixture containing methane and *n*-butane at 21°C enters an absorber tower. The partial pressures at these conditions are 103 kPa for methane and 586 kPa for *n*-butane. In the absorber, 80% of the butane is removed and the remaining gas leaves the tower at 38°C and a total pressure of 550 kPa. What is the volumetric flow rate of gas at the exit? How many moles per day of butane are removed from the gas in this process? Assume ideal behavior.

****7.1.44** A heater burns normal butane ($n\text{-}C_4H_{10}$) using 40.0% excess air. Combustion is complete. The flue gas leaves the stack at a pressure of 100 kPa and a temperature of 260°C.

a. Calculate the complete flue gas analysis.
b. What is the volume of the flue gas in cubic meters per kilogram mole of *n*-butane?

****7.1.45** The majority of semiconductor chips used in the microelectronics industry are made of silicon doped with trace amounts of materials to enhance conductivity. The silicon initially must contain less than 20 ppm of impurities. Silicon rods are grown by the following chemical deposition reaction of trichlorosilane with hydrogen:

$$HSiCl_3 + H_2 \underset{1000°C}{\rightarrow} 3HCl + Si$$

Assuming that the ideal gas law applies, what volume of hydrogen at 1000°C and 1 atm must be reacted to increase the diameter of a rod 1 m long from 1 cm to 10 cm? The density of solid silicon is 2.33 g/cm^3.

****7.1.46** The oxygen and carbon dioxide concentrations in the gas phase of a 10 L bioreactor operating in the steady state control the dissolved oxygen and pH in the liquid phase where the biomass exists.

a. If the rate of oxygen uptake by the liquid is 2.5 3 1027 g mol/(1000 cells)(hr), and if the culture in the liquid phase contains 2.9×10^6 cells/mL, what is the rate of oxygen uptake in millimoles per hour?
b. If the gas supplied to the gas phase is 45 L/hr containing 40% oxygen at 110 kPa and 25°C, what is the rate of oxygen supplied to the bioreactor in millimoles per hour?
c. Will the oxygen concentration in the gas phase increase or decrease by the end of 1 hr compared to the initial oxygen concentration?

****7.1.47** When natural gas (mainly CH_4) is burned with 10% excess air, in addition to the main gaseous products of CO_2 and H_2O, other gaseous products result in minor quantities. The Environmental Protection Agency (EPA) lists the following data:

Emission Factors (kg/10^6 m^3 at S.C.)

	SO_2	NO_2	CO	CO_2
Large utility boiler, uncontrolled	9.6	3040	1344	1.9×10^6
Large utility boiler, controlled gas recirculation	9.6	1600	1344	1.9×10^6
Residential furnace	9.6	1500	640	1.9×10^6

The data are based on 10^6 m^3 measured at S.C. of methane burned. What is the approximate mole fraction of SO_2, NO_2, and CO (on a dry basis) for each class of combustion equipment?

****7.1.48** Estimate the emissions of each compound produced in cubic meters measured at S.C. per metric ton (1000 kg) of No. 6 fuel oil burned in an oil-fired burner with no emission controls given the following data:

Pollution Emission Factors from the EPA (kg/10^3 L oil)

SO_2	SO_3	NO_2	CO	CO_2	Particulate Matter
19S*	0.69S*	8	0.6	3025	1.5

*S = weight percent sulfur

The No. 6 fuel oil contains 0.84% sulfur and has a specific gravity of 0.86 at 15°C.

****7.1.49** The composition from Perry of No. 6 fuel oil with a specific gravity of 0.86 is in mass percent:

C	87.26
H	10.49
O	0.64
N	0.28
S	0.84
Ash	0.04

Compute the kilograms of each component per 10^3 L of oil, and compare the resulting emissions with those listed in the EPA analysis in problem P7.1.48

*****7.1.50** One important source of emissions from gasoline-powered automobile engines that causes smog is the nitrogen oxides NO and NO_2. They are formed whether combustion is complete or not as follows: At the high temperatures that occur in an internal combustion engine during the burning process, oxygen and nitrogen combine to form nitric oxide (NO). The higher the peak temperatures and the more oxygen available, the more NO is formed. There is insufficient time for the NO to decompose back to O_2 and N_2 because the burned gases cool too rapidly during the expansion and exhaust cycles in the engine. Although both NO and nitrogen dioxide (NO_2) are significant air pollutants (together termed NO_x), the NO_2 is formed in the atmosphere as NO is oxidized.

Suppose that you collect a sample of a NO-NO_2 mixture (after having removed the other combustion gas products including N_2, O_2, and H_2O by various separation procedures) in a 100 cm^3 standard cell at 30°C. Certainly some of the NO will have been oxidized to NO_2

$$2NO + O_2 \rightarrow 2NO_2$$

during the collection, storage, and processing of the combustion gases, so measurement of NO alone will be misleading. If the standard cell contains 0.291 g of NO_2 plus NO and the pressure measured in the cell is 170 kPa, what percent of the NO + NO_2 is in the form of NO?

***7.1.51 Ammonia at 100°C and 150 kPa is burned with 20% excess O_2:

$$4NH_3 + 5O_2 \rightarrow 4NO + 6H_2O$$

The reaction is 80% complete. The NO is separated from the NH_3 and water, and the NH_3 is recycled as shown in Figure P7.1.51.

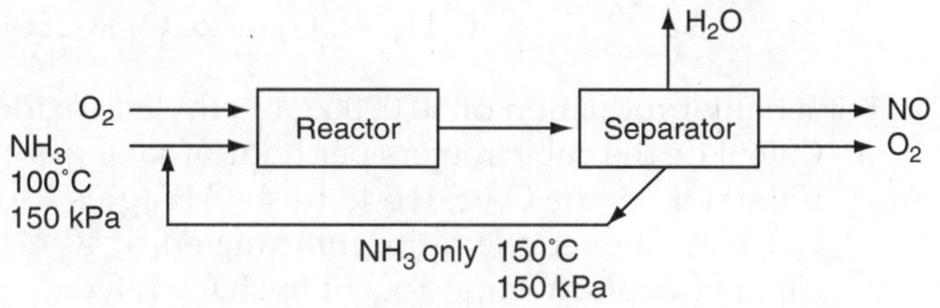

Figure P7.1.51

Calculate the cubic meters of NH_3 recycled at 150°C and 150 kPa per cubic meter of NH_3 fed at 100°C and 150 kPa.

***7.1.52 Benzene (C_6H_6) is converted to cyclohexane (C_6H_{12}) by direct reaction with H_2. The fresh feed to the process is 260 L/min of C_6H_6 plus 950 L/min of H_2 at 100°C and 150 kPa. The single-pass conversion of H_2 in the reactor is 48% while the overall conversion of H_2 in the process is 75%. The recycle stream contains 90% H_2 and the remainder benzene (no cyclohexane). See Figure P7.1.52.

a. Determine the molar flow rates of H_2, C_6H_6, and C_6H_{12} in the exiting product.
b. Determine the volumetric flow rates of the components in the product stream if it exits at 100 kPa and 200°C.
c. Determine the molar flow rate of the recycle stream, and the volumetric flow rate if the recycle stream is at 100°C and 100 kPa.

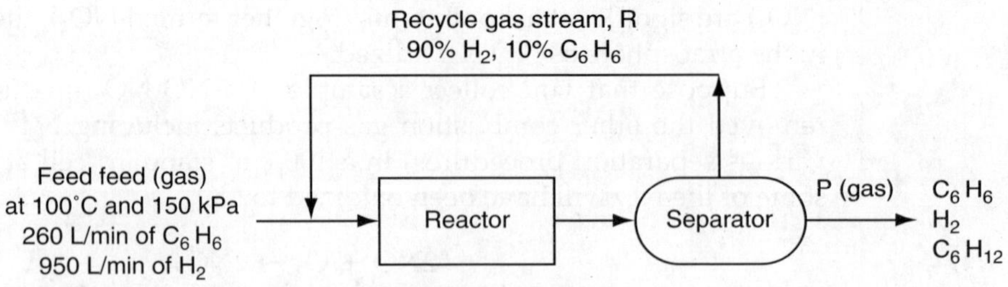

Figure P7.1.52

***7.1.53 Pure ethylene (C_2H_4) and oxygen are fed to a process for the manufacture of ethylene oxide (C_2H_4O):

$$C_2H_4 + \frac{1}{2}O_2 \rightarrow C_2H_4O$$

Figure P7.1.53 is the flow diagram for the process. The catalytic reactor operates at 300°C and 1.2 atm. At these conditions, single-pass measurements on the reactor show that 50% of the ethylene entering the reactor is consumed per pass, and of this, 70% is converted to ethylene oxide. The remainder of the ethylene reacts to form CO_2 and water.

$$C_2H_4 + 3O_2 \rightarrow 2CO_2 + 2H_2O$$

For a daily production of 10,000 kg of ethylene oxide:

a. Calculate the cubic meters per hour of total gas entering the reactor at S.C. if the ratio of the $O_2(g)$ fed to fresh $C_2H_4(g)$ is 3 to 2.
b. Calculate the recycle ratio, cubic meters at 10°C and 100 kPa of C_2H_4 recycled per cubic meter at S.C. of fresh C_2H_4 fed.
c. Calculate the cubic meters of the mixture of O_2, CO_2, and H_2O leaving the separator per day at 80°C and 100 kPa.

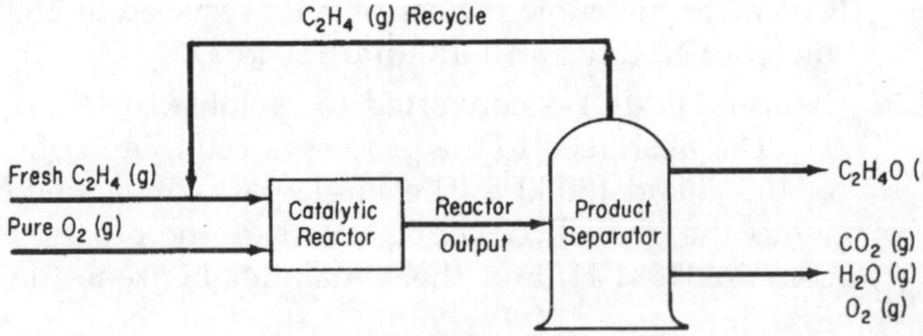

Figure P7.1.53

***7.1.54 An incinerator produces a dry exit gas of the following Orsat composition measured at 60°F and 30 in. Hg absolute: 4.0% CO_2, 26.0% CO, 2.0% CH_4, 16.0% H_2, and 52.0% N_2. A dry natural gas of the following (Orsat) composition—80.5% CH_4, 17.8% C_2H_6, and 1.7% N_2—is used at the rate of 1200 ft^3/min at 60°F and 30 in. Hg absolute to burn the incineration off-gas with air. The final products of combustion analyze on a dry basis 12.2% CO_2, 0.7% CO, 2.4% O_2, and 84.7% N_2.

Calculate (a) the rate of flow in cubic feet per minute of the incinerator exit gas at 60°F and 30 in. Hg absolute on a dry basis, and (b) the rate of airflow in cubic feet per minute, dry, at 80°F and 29.6 in. Hg absolute.

***7.1.55 A gaseous mixture consisting of 50 mol % hydrogen and 50 mol % acetaldehyde (C_2H_4O) is initially contained in a rigid vessel at a total pressure of 760 mm Hg absolute. The formation of ethanol (C_2H_6O) occurs according to

$$C_2H_4O + H_2 \rightarrow C_2H_6O$$

After a time it was noted that the total pressure in the rigid vessel had dropped to 700 mm Hg absolute. Calculate the degree of completion of the reaction at that time using the following assumptions: (a) All reactants and products are in the gaseous state, and (b) the vessel and its contents were at the same temperature when the two pressures were measured.

****7.1.56 Biomass ($CH_{1.8}O_{0.5}N_{0.5}$) can be converted to glycerol by anaerobic (in the absence of air) reaction with ammonia and glucose. In one batch of reactants, 52.4 L of CO_2 measured at 300 K and 95 kPa were obtained per mole of glucose in the reactor. The molar stoichiometric ratio of nitrogen produced to ammonia reacted in the reaction equation is 1 to 1, and the mol CO_2/mol $C_6H_{12}O_6 = 2$.

In gram moles, (a) how much glycerol was produced, and (b) how much biomass reacted to produce the 52.4 L of CO_2?

7.2 Real Gases: Equations of State

**7.2.1 You want to obtain an answer immediately as to the specific volume of ethane at 700 kPa and 25°C. List in descending order the techniques you would use with the most preferable one at the top of the list:

a. Ideal gas law
b. Compressibility charts
c. An equation of state
d. Look up the value on the Web
e. Look up the value in a handbook

Explain your choices.

**7.2.2 Which procedure would you recommend to calculate the density of carbon dioxide at 120°F and 1500 psia? Explain your choice.

a. Ideal gas law
b. Redlich-Kwong equation of state
c. Compressibility charts
d. Look up the value on the Web
e. Look up the value in a handbook

****7.2.3** Finish the following sentence:
Equations of state are preferred in *P-V-T* calculations because __________.

****7.2.4** Use the Kammerlingh-Onnes virial equation with four terms to answer the following questions for CH_4 at 273 K:

a. Up to what pressure is one term (the ideal gas law) a good approximation?
b. Up to what pressure is the equation truncated to two terms a good approximation?
c. What is the error in using a and using b for CH_4?

Data: At 273 K the values of the virial coefficients are

$$B = -53.4\ \text{cm}^3/\text{mol}$$
$$C = 2620\ \text{cm}^6/\text{mol}^{-2}$$
$$D = 5000\ \text{cm}^9/\text{mol}^{-3}$$

****7.2.5** The Peng-Robinson equation is listed in Table 7.2. What are the units of a, b, and α in the equation if p is in atmospheres, $\hat{V}$ is in liters per gram mole, and T is in kelvin?

****7.2.6** The pressure gauge on an O_2 cylinder stored outside at 0°F in the winter reads 1375 psia. By weighing the cylinder (whose volume is 6.70 ft^3) you find that the net weight, that is, the O_2, is 63.9 lb. Is the reading on the pressure gauge correct? Use an equation of state to make your calculations.

****7.2.7** First commercialized in the 1970s as extractants in "natural" decaffeination processes, SCFs (supercritical fluids)—particularly carbon dioxide and ater—are finding new applications, as better, less expensive equipment lowers processing costs, and regulations drive the chemical process industries away from organic solvents.

SCFs' extraction capabilities are now being exploited in a range of new pharmaceutical and environmental applications, while supercritical extraction, oxidation, and precipitation are being applied to waste cleanup challenges.

A compressor for carbon dioxide compresses 2000 m^3/min at 20°C and 500 kPa to 110°C and 4800 kPa. How many cubic meters per minute are produced at the high pressure? Use van der Waals' equation.

*****7.2.8** You are asked to design a steel tank in which CO_2 will be stored at 290 K. The tank is 10.4 m^3 in volume and you want to store 460 kg of CO_2 in it. What pressure will the CO_2 exert? Use the Redlich-Kwong equation to calculate the pressure in the tank. Repeat using the SRK equation. Is there a significant difference in the predictions of pressure between the equations?

*****7.2.9** What pressure would be developed if 100 ft^3 of ammonia at 20 atm and 400°F were compressed into a volume of 5.0 ft^3 at 350°F? Use the Peng-Robinson equation to get your answer.

***7.2.10 An interesting patent (U.S. 3,718,236) explains how to use CO_2 as the driving gas for aerosol sprays in a can. A plastic pouch is filled with small compartments containing sodium bicarbonate tablets. Citric acid solution is placed in the bottom of the pouch, and a small amount of carbon dioxide is charged under pressure into the pouch as a starter propellant. As the CO_2 is charged into the pouch, it ruptures the lowest compartment membrane, thus dropping bicarb tablets into the citric acid. That generates more carbon dioxide, giving more pressure in the pouch, which expands and helps push out more product. (The CO_2 does not escape from the can, just the product.)

How many grams of $NaHCO_3$ are needed to generate a residual pressure of 81.0 psig in the can to deliver the very last cubic centimeter of product if the cylindrical can is 8.10 cm in diameter and 17.0 cm high? Assume the temperature is 25°C. Use the Peng-Robinson equation.

***7.2.11 Find the molar volume (in cubic centimeters per gram mole) of propane at 375 K and 21 atm. Use the Redlich-Kwong and Peng-Robinson equations, and solve for the molar volume using the nonlinear equation solver on the CD in the pocket at the back of this book. The acentric factor for propane to use in the Peng-Robinson equation is 0.1487.

***7.2.12 The tank cited in problem 7.2.8 is constructed and tested, and your boss informs you that you forgot to add a safety factor in the design of the tank. It tests out satisfactorily to 3500 kPa, but you should have added a safety factor of 3 to the design; that is, the tank pressure should not exceed (3500/3) = 1167 kPa, say, 1200 kPa. How many kilograms of CO_2 can be stored in the tank if the safety factor is applied? Use the Redlich-Kwong equation. Hint: Polymath will solve the equation for you.

***7.2.13 A graduate student wants to use van der Waals' equation to express the pressure-volume-temperature relations for a gas. Her project requires a reasonable degree of precision in the *p-V-T* calculations. Therefore, she made the following experimental measurements with her setup to get an idea of how easy the experiment would be:

Temperature, K	Pressure, atm	Volume, ft^3/lb mol
273.1	200	1.860
273.1	1000	0.741

Determine values of constants *a* and *b* to be used in van der Waals' equation that best fit the experimental data.

***7.2.14 An 80 lb block of ice is put into a 10 ft^3 container and heated to 900 K. What is the final pressure in the container? Do this problem two ways: (a) Use the compressibility factor method, and (b) use the Redlich-Kwong equation. Compare your results.

***7.2.15 What weight of ethane is contained in a gas cylinder that is 1.0 ft^3 in volume if the gas is at 100°F and 2000 psig? Do this problem two ways: (a) Use van der Waals' equation, and (b) use the compressibility factor method. The experimental value is 21.4 lb.

***7.2.16 Answer the following questions:

a. Will the constant a in van der Waals' equation be higher or lower for methane than for propane? Repeat for the other van der Waals constant b.
b. Will the constant a' in the SRK equation be higher or lower for methane than for propane? Repeat for the other SRK constant b.

****7.2.17 A 5 L tank of H_2 is left out overnight in Antarctica. You are asked to determine how many gram moles of H_2 are in the tank. The pressure gauge reads 39 atm gauge and the temperature is −50°C. How many gram moles of H_2 are in the tank?

Use the van der Waals and Redlich-Kwong equations of state to solve this problem. (Hint: The nonlinear-equation-solving program on the CD in the pocket at the back of this book will make the execution of the calculations quite easy.)

****7.2.18 A 6250 cm^3 vessel contains 4.00 g mol of CO_2 at 298.15 K and 14.5 atm. Use the nonlinear equation solver on the CD in the back of the book to solve the Redlich-Kwong equation for the molar volume. Compare the calculated molar volume of the CO_2 in the vessel with the experimental value.

7.3 Real Gases: Compressibility Charts

**7.3.1 Seven pounds of N_2 are stored in a cylinder 0.75 ft^3 volume at 120°F. Calculate the pressure in the cylinder in atmospheres (a) assuming N_2 to be an ideal gas and (b) assuming N_2 is a real gas and using compressibility factors.

**7.3.2 Two gram moles of ethylene (C_2H_4) occupy 418 cm^3 at 95°C. Calculate the pressure. (Under these conditions ethylene is a nonideal gas.) Data: T_c = 283.1 K, p_c = 50.5 atm.

**7.3.3 The critical temperature of a real gas is known to be 500 K, but its critical pressure is unknown. Given that 3 lb mol of the gas at 252°C occupy 50 ft^3 at a pressure of 463 psia, estimate the critical pressure.

**7.3.4 The volume occupied by 1 lb of n-octane at 27 atm is 0.20 ft^3. Calculate the temperature of the n-octane.

**7.3.5 A block of dry ice weighing 50 lb was dropped into an empty steel tank, the volume of which was 5.0 ft^3. The tank was heated until the pressure gauge read 1600 psi. What was the temperature of the gas? Assume all of the CO_2 became gas.

**7.3.6 A cylinder containing 10 kg of CH_4 exploded. It had a bursting pressure of 14,000 kPa gauge and a safe operating pressure of 7000 kPa gauge. The cylinder had an internal volume of 0.0250 m^3. Calculate the temperature when the cylinder exploded.

**7.3.7 A cylinder has a volume of 1.0 ft^3 and contains dry methane at 80°F and 200 psig. What weight of methane (CH_4) is in the cylinder? The barometric pressure is 29.0 mm Hg.

**7.3.8 How many kilograms of CO_2 can be put into a 25 L cylinder at room temperature (25°C) and 200 kPa absolute pressure?

**7.3.9 A natural gas composed of 100% methane is to be stored in an underground reservoir at 1000 psia and 120°F. What volume of reservoir is required for 1,000,000 ft^3 of gas measured at 60°F and 14.7 psia?

**7.3.10 Calculate the specific volume of propane at a pressure of 6000 kPa and a temperature of 230°C.

**7.3.11 State whether or not the following gases can be treated as ideal gases in calculations:

a. Nitrogen at 100 kPa and 25°C
b. Nitrogen at 10,000 kPa and 25°C
c. Propane at 200 kPa and 25°C
d. Propane at 2000 kPa and 25°C
e. Water at 100 kPa and 25°C
f. Water at 1000 kPa and 25°C
g. Carbon dioxide at 1000 kPa and 0°C
h. Propane at 400 kPa and 0°C

**7.3.12 One gram mole of chlorobenzene (C_6H_5Cl) just fills a tank at 230 kPa and 380 K. What is the volume of the tank?

**7.3.13 You have been asked to settle an argument. The argument concerns the maximum allowable working pressure (MAWP) permitted in an A1 gas cylinder. One of your coworkers says that calculating the pressure in a tank via the ideal gas law is best because it gives a conservative (higher) value of the pressure than can actually occur in the tank. The other coworker says that everyone knows the ideal gas law should not be used to calculate real gas pressures as it gives a lower value than the true pressure. Which coworker is correct?

***7.3.14 A size A1 cylinder of ethylene ($T_c = 9.7°C$) costs $45.92 FOB New Jersey. The outside cylinder dimensions are 9 in. diameter, 52 in. high. The gas is 99.5% (minimum) C_2H_4, and the cylinder charge is $44.00. Cylinder pressure is 1500 psig, and the invoice says it contains "165 cu.ft." of gas. An identical cylinder of CP-grade methane at a pressure of 2000 psig is 99.0% (minimum) CH_4 and costs $96.00 FOB Illinois. The CH_4 cylinder contains "240 cu.ft." of gas. The ethylene cylinder is supposed to have a gross weight (including cylinder) of 163 lb while the CH_4 cylinder has a gross weight of 145 lb. Answer the following questions:

a. What do the "165 cu.ft." and "240 cu.ft." of gas probably mean? Explain with calculations.
b. Why does the CH_4 cylinder have a gross weight less than the C_2H_4 cylinder when it seems to contain more gas? Assume the cylinders are at 80°F.
c. How many pounds of gas are actually in each cylinder?

***7.3.15 Safe practices in modern laboratories call for placing gas cylinders in hoods or in utility corridors. In case of leaks, a toxic gas can be properly taken care of. A cylinder of CO that has a volume of 175 ft^3 at 1 atm and 25°C is received from the distributor of gases on Friday with a gauge reading of 2000 psig and is placed in the utility corridor. On Monday when you are ready to use the gas, you find the gauge reads 1910 psig. The temperature has remained constant at 76°F as the corridor is air-conditioned, so you conclude that the tank has leaked CO (which does not smell).

a. What has been the leak rate from the tank?
b. If the tank was placed in a utility corridor whose volume is 1600 ft^3, what would be the minimum time that it would take for the CO concentration in the hallway to reach the Ceiling Threshold Limit Value (TLV-C) of 100 ppm set by the state Air Pollution Control Commission if the air conditioning did not operate on the weekend?
c. In the worst case, what would be the concentration of CO in the corridor if the leak continued from Friday, 3 PM, to Monday, 9 AM?
d. Why would either case b or c not occur in practice?

***7.3.16 Levitating solid materials during processing is the best way known to ensure their purity. High-purity materials, which are in great demand in electronics, optics, and other areas, usually are produced by melting a solid. Unfortunately, the containers used to hold the material also tend to contaminate it. And heterogeneous nucleation occurs at the container walls when molten material is cooled. Levitation avoids these problems because the material being processed is not in contact with the container.

Electromagnetic levitation requires that the sample be electrically conductive, but with a levitation method based on buoyancy, the density of the material is the only limiting factor.

Suppose that a gas such as argon is to be compressed at room temperature so that silicon (sp. gr. 2.0) just floats in the gas. What must the pressure of the argon be? If you wanted to use a lower pressure, what different gas might be selected? Is there a limit to the processing temperature for this manufacturing strategy?

***7.3.17 While determining the temperature that occurred in a fire in a warehouse, the arson investigator noticed that the relief valve on a methane storage tank had popped open at 3000 psig, the rated value. Before the fire started, the tank was presumably at ambient conditions, about 80°F, and the gauge read 1950 psig. If the volume of the tank was 240 ft^3, estimate the temperature during the fire. List any assumptions you make.

7.4 Real Gas Mixtures

**7.4.1 A gas has the following composition:

CO_2	10%
CH_4	40%
C_2H_4	50%

It is desired to distribute 33.6 lb of this gas per cylinder. Cylinders are to be designed so that the maximum pressure will not exceed 2400 psig when the temperature is 180°F. Calculate the volume of the cylinder required by Kay's method.

**7.4.2 A gas composed of 20% ethanol and 80% carbon dioxide is at 500 K. What is its pressure if the volume per gram mole is 180 cm^3/g mol?

**7.4.3 A sample of natural gas taken at 3500 kPa absolute and 120°C is separated by chromatography at standard conditions. It was found by calculation that the grams of each component in the gas were as follows:

Component	G
Methane (CH_4)	100
Ethane (C_2H_6)	240
Propane (C_3H_8)	150
Nitrogen (N_2)	50
Total	540

What was the density of the original gas sample?

**7.4.4 A gaseous mixture has the following composition (in mole percent):

C_2H_4	57
Ar	40
He	3

at 120 atm pressure and 25°C. Compare the experimental volume of 0.14 L/g mol with that computed by Kay's method.

***7.4.5 You are in charge of a pilot plant using an inert atmosphere composed of 60% ethylene (C_2H_4) and 40% argon (Ar). How big a cylinder (or how many) must be purchased if you are to use 300 ft^3 of gas measured at the pilot plant conditions of 100 atm and 300°F? Buy the cheapest array.

Cylinder Type	Cost	Pressure (psig)	lb Gas
1A	$52.30	2000	62
2	42.40	1500	47
3	33.20	1500	35

State any additional assumptions. You can buy only one type of cylinder.

***7.4.6 A feed for a reactor has to be prepared composed of 50% ethylene and 50% nitrogen. One source of gas is a cylinder containing a large amount of gas with the composition 20% ethylene and 80% nitrogen. Another cylinder that contains pure ethylene at 1450 psig and 70°F has an internal volume of 2640 in^3. If all the ethylene in the latter cylinder is used up in making the mixture, how much reactor feed was prepared and how much of the 20% ethylene mixture was used?

***7.4.7 A gas is flowing at a rate of 100,000 scfh (standard cubic feet per hour). What is the actual volumetric gas flow rate if the pressure is 50 atm and the temperature is 600°R? The critical temperature is 40.0°F and the critical pressure is 14.3 atm.

***7.4.8 A steel cylinder contains ethylene (C_2H_4) at 200 psig. The cylinder and gas weigh 222 lb. The supplier refills the cylinder with ethylene until the pressure reaches 1000 psig, at which time the cylinder and gas weigh 250 lb. The temperature is constant at 25°C. Calculate the charge to be made for the ethylene if the ethylene is sold at $0.41 per pound, and what the weight of the cylinder is for use in billing the freight charges. Also find the volume of the empty cylinder in cubic feet.

****7.4.9 In a high-pressure separation process, a gas having a *mass* composition of 50% benzene, 30% toluene, and 20% xylene is fed into the process at the rate of 483 m^3/hr at 607 K and 26.8 atm. One exit stream is a vapor containing 91.2% benzene, 7.2% toluene, and 1.6% xylene. A second exit stream is a liquid containing 6.0% benzene, 9.0% toluene, and 85.0% xylene.

What is the composition of the third exit stream if it is liquid flowing at the rate of 9800 kg/hr, and the ratio of the benzene to the xylene in the stream is 3 kg benzene to 2 kg xylene?

CHAPTER 8

Multiphase Equilibrium

Your objectives in studying this chapter are to be able to

1. Recognize the connection between multiphase equilibrium and separation technology
2. Understand phase diagrams and the associated terminology as well as the phase rule
3. Determine the vapor pressure of a pure component and to use it to determine the degree of vaporization into or condensation from a condensable gas
4. Understand vapor-liquid equilibrium for a binary system

Looking Ahead

In this chapter, we will introduce separation technology, which is used extensively in the process industries. Phase diagrams and the phase rule are first introduced, then the characteristics of a variety of systems of single-component two-phase systems, concluding with a discussion of multicomponent two-phase systems.

8.1 Introduction

The most common pieces of equipment in the process industries are separation devices, which remove one or more components from a stream and concentrate them in another stream. Mixing of components occurs regularly in nature (e.g., minerals dissolve in rainwater as the water flows down a creek bed), but to separate components requires separation equipment that uses energy and materials to accomplish the separation. It is well known that the value of products can greatly increase when the key component in a product is taken from a dilute solution by a separation device and transformed into a highly concentrated form. Therefore, separation technologies can provide significant economic advantages for processing companies. This chapter deals with the description of multiphase systems that are used in the development and design of various types of separation systems. Examples of the application of separation systems include

- **Drinking water from seawater.** One way drinking water can be produced from seawater is by boiling the seawater to produce water vapor, which is then condensed, yielding drinking water. This process of boiling and condensing is a simple example of **distillation**.
- **Gasoline from crude oil.** Part of the gasoline produced by an oil refinery comes directly from the distillation of crude oil. The crude is distilled into a number of products, each with a different boiling point range; one of these products is gasoline.
- **Removal of pollutants from effluent streams.** Plants that discharge water and gas streams into the environment are required to reduce the concentration of pollutants to specified levels, a step that usually requires the application of a separation system(s). As an example, when coal is burned in a power plant, SO_2 is produced from the sulfur in the coal. SO_2 in the atmosphere is converted to sulfite, which forms acid rain. Therefore, coal-fired power plants are required to remove SO_2 from their flue gas (i.e., the combustion gases after most of the thermal energy has been removed) before discharging it to the atmosphere. Many power plants pass their flue gas through a scrubbing process, which exposes the flue gas to a lime-water mixture to absorb the SO_2. The column that accomplishes this contacting for SO_2 removal is known as an **absorber** because the lime-water mixture absorbs the SO_2 from the flue gas.
- **Pharmaceuticals.** Certain prescription drugs are produced by concentrating a dilute solution of the desired product, which was produced in a bioreactor, using an **extraction process**. An extraction process uses a liquid that has a much greater affinity for the desired product than the components in the reactor effluent. Thus, an extraction process is able to produce a very nearly pure product.
- **Typical chemical plant.** In a typical chemical plant, a reactor produces a mixture of products and unconverted feed which is fed to a separation train (i.e., a series of separation equipment) that concentrates the

products into salable form and returns the unreacted feed to the reactor. For most chemical plants, the separation train is primarily composed of distillation columns with some absorbers and extractors.

8.2 Phase Diagrams and the Phase Rule

You can conveniently display the properties of compounds via phase diagrams. A pure substance can exist in many phases simultaneously of which, as you know, solid, liquid, and gas are the most common. Phase diagrams enable you to view the properties of two or more phases as functions of temperature, pressure, specific volume, concentration, and other variables.

We are going to discuss phase diagrams in terms of water because presumably you are familiar with the three phases of water, namely, ice, water, and water vapor (steam), but the discussion applies to all other pure substances. The terms *vapor* and *gas* are used very loosely in practice. A gas that exists below its critical temperature is usually called a vapor because it can condense. We will reserve the word **vapor** to describe a gas below its critical point in a process in which the phase change is of primary interest, while the words **gas** or **noncondensable gas** will be used to describe a gas above the critical point or a gas in a process at conditions under which it cannot condense.

Phase diagrams are based on equilibrium conditions. That is, for phase equilibrium, it is assumed that each phase remains invariant (i.e., constant quantity under constant conditions). On a molecular level, when two or more phases are present, there will always be molecules that move from one phase to another, but under phase equilibrium, the net flux is zero. For example, for a liquid and a vapor in phase equilibrium, the flux of molecules from the liquid into the vapor must be equal to the flux from the vapor to the liquid. In fact, when multiple phases exist, continuous exchange between phases occurs, even for equilibrium.

Suppose you carry out some experiments with the apparatus shown in Figure 8.1. Place a lump of ice in the chamber below the piston, and evacuate the chamber to remove all air (you want to retain only pure water in the chamber). Fix the volume of the chamber by fixing the position of the piston, and start slowly (so that the phases of water that result will be in equilibrium) heating the ice. If you plot the measured pressures as a function of temperature, you will get Figure 8.2—a phase diagram—in which all of the measurements made have been fitted by a continuous smooth curve for clarity.

The initial conditions of p and T in the chamber are at 0 in Figure 8.2 with the solid in equilibrium with the vapor.

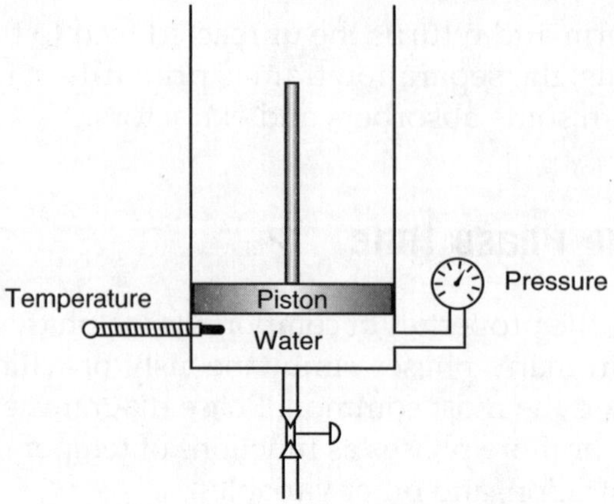

Figure 8.1 Apparatus used to explore the p, $\hat{V}$, and T properties of water

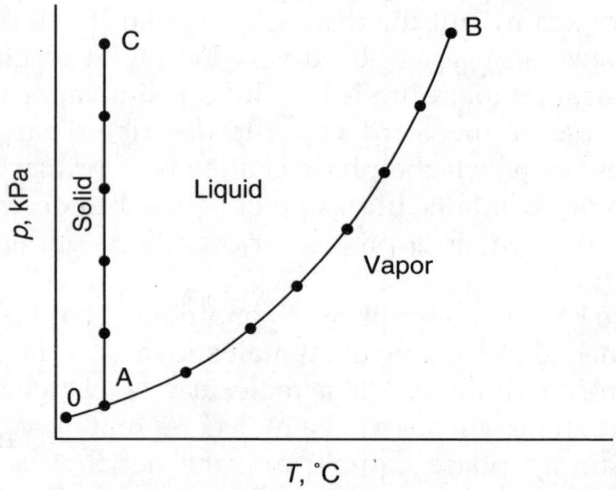

Figure 8.2 Results of the experiment of heating at constant volume shown on a phase diagram (p versus T at constant $\hat{V}$)

As you raise the temperature, the ice would start to melt at point A, the **triple point**, the one p-T-$\hat{V}$ combination at which solid, liquid, and vapor can be in equilibrium. Further increase in the temperature causes the ice to abruptly melt before forming water vapor and the pressure to rise, which is indicated by the curve AB. B is the critical point at which vapor and liquid properties become the same.

If you had kept the temperature almost constant and raised the pressure on the ice, ice would still exist and be in equilibrium with liquid water along the line AC. The line AC is so vertical that *you can use the saturated liquid properties for the properties of the compressed liquid*. Ice skating is possible because the high pressure exerted by the thin blade on ice forms a liquid layer with low friction on the blade.

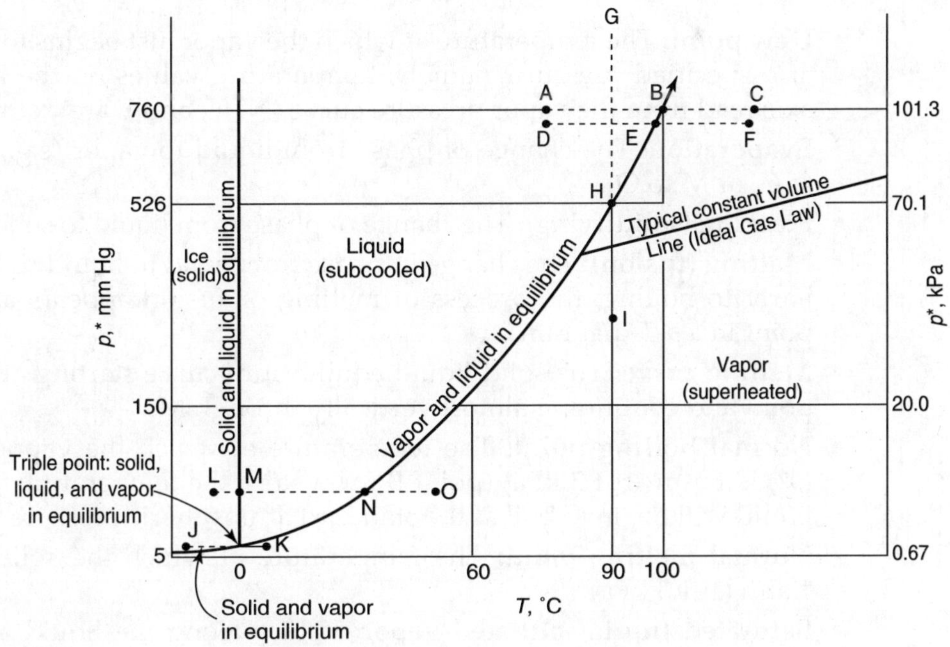

Figure 8.3 Various common processes as represented on a p^*-T diagram

If the vapor and liquid of a pure component are in **equilibrium**, the **equilibrium pressure is called the vapor pressure** which we will denote by p^*. **At a given temperature there is only one pressure at which the liquid and vapor phases of a pure substance may exist in equilibrium**. Either phase alone may exist, of course, over a wide range of conditions.

We next take up some terminology associated with processes that are conveniently represented on a p^*-T phase chart such as Figure 8.3 (in the definitions of terms that follow, the letters in parentheses refer to the corresponding process denoted in Figure 8.3 by the same sequence of letters):

- **Boiling:** The change of phase from liquid to vapor (e.g., B, E, N; note that because boiling occurs at a constant temperature and pressure, the process of boiling appears as a point in a p-T diagram).
- **Bubble point:** The temperature at which a liquid just starts to vaporize (N, H, and E are examples).
- **Condensation:** The change of phase from vapor to liquid (e.g., N, E, B; note that because condensation occurs at a constant temperature and pressure, the process of condensation appears as a point in a p-T diagram).

- **Dew point:** The temperature at which the vapor just begins to condense at a specified pressure, namely, temperature values on the horizontal axis read from the vapor pressure curve (N, H, and E are examples).
- **Evaporation:** The change of phase from liquid to vapor (e.g., D to F, A to C, or M to O).
- **Freezing (solidifying):** The change of phase from liquid to solid (N to L).
- **Melting (fusion):** The change in phase from solid to liquid (L to M; similarly to boiling, the process of melting or fusion appears as a single point in a *p-T* diagram).
- **Melting curve:** The solid-liquid equilibrium curve starting at the triple point and continuing almost vertically through M.
- **Normal boiling point:** The temperature at which the vapor pressure (p^*) is 1 atm (101.3 kPa) (point B for water); the temperature at which a liquid will begin to boil at the standard atmospheric pressure.
- **Normal melting point:** The temperature at which the solid melts at 1 atm (101.3 kPa).
- **Saturated liquid/saturated vapor:** Values along the liquid and vapor equilibrium curve (vapor-pressure curve, e.g., N to B).
- **Subcooled liquid:** T and p values for the liquid between the melting curve and the vapor-pressure curve (liquid D is an example).
- **Sublimation:** Change in phase from solid to vapor (J to K).
- **Sublimation curve:** The solid-vapor equilibrium curve from J (and lower) to the triple point.
- **Sublimation pressure:** The pressure along the melting curve (a function of temperature).
- **Supercritical region:** *p-T* values above the critical point (not shown in Figure 8.3).
- **Superheated vapor:** Values of vapor at temperatures and pressure exceeding those at saturation; I is an example. The **degrees of superheat** are the differences in temperature between the actual T and the saturated T at the given pressure. For example, steam at 500°F and 100 psia (the saturation temperature for 100 psia is 327.8°F) has $(500 - 327.8) = 172.2$°F of superheat.
- **Vaporization:** The change of phase from liquid to vapor (for example, D to F).

In Figure 8.3 the process of evaporation and condensation of water at 1 atm is represented by the line ABC with the phase transformation occurring at 100°C. Suppose that you went to the top of Pikes Peak and repeated the process of evaporation and condensation in the open air. What would happen then? The process would be the same (points DEF) with the exception of the temperature and pressure at which the water would begin to boil, or condense. Since the pressure of the atmosphere at the top of Pikes Peak is lower than 101.3 kPa, the water would start to boil at a lower temperature. Some unfortunate consequences might result if you expected to kill certain types of disease-causing bacteria by boiling the water! In addition, it will take longer to cook rice at that elevation due to the lower boiling point for water at higher elevations.

To conclude, *at equilibrium* you can see that (a) at any given temperature water exerts its unique vapor pressure; (b) as the temperature goes up, the vapor pressure goes up, and vice versa; and (c) it makes no difference whether water vaporizes into air, into a cylinder closed by a piston, into an evacuated cylinder, or into the atmosphere; at any temperature it still exerts the same vapor pressure as long as the liquid water is in equilibrium with its vapor.

A pure compound can change phase at constant volume from a liquid to a vapor, or the reverse, via a constant temperature process as well as a constant pressure process. A process of **vaporization**, or **condensation, at constant temperature** is illustrated by the lines GHI or IHG, respectively, in Figure 8.3. Water would vaporize or condense at constant temperature as the pressure reached point H on the vapor-pressure curve. The change that occurs at H is the increase or decrease in the fraction of vapor, or liquid, respectively, at the fixed temperature. The pressure does not change until all of the vapor, or liquid, has completed the phase transition.

Now let's go back to the experimental apparatus and collect data to prepare a p-$\hat{V}$ phase chart. This time you want to hold the temperature in the chamber constant and adjust the volume while measuring the pressure. Start with compressed liquid water (subcooled water) rather than ice, and raise the piston so that water eventually vaporizes. Figure 8.4 illustrates by dashed lines the measurements for two different temperatures, T_1 and T_2. As the pressure is reduced at constant T_1, $\hat{V}$ increases very slightly (liquids are not very compressible) until the liquid pressure reaches p^*, the vapor pressure, at point A.

Then, as the piston still rises (i.e., as $\hat{V}$ increases), both the pressure and temperature remain constant until all of the liquid is vaporized by point B on the saturated vapor line. Subsequently, starting from point B, as the pressure reduces, the value of $\hat{V}$ can be calculated via an ideal or real gas equation. Compression at constant T_2 is just a reversal of the process at T_1. The dots in Figure 8.4 represent just the measurements made when

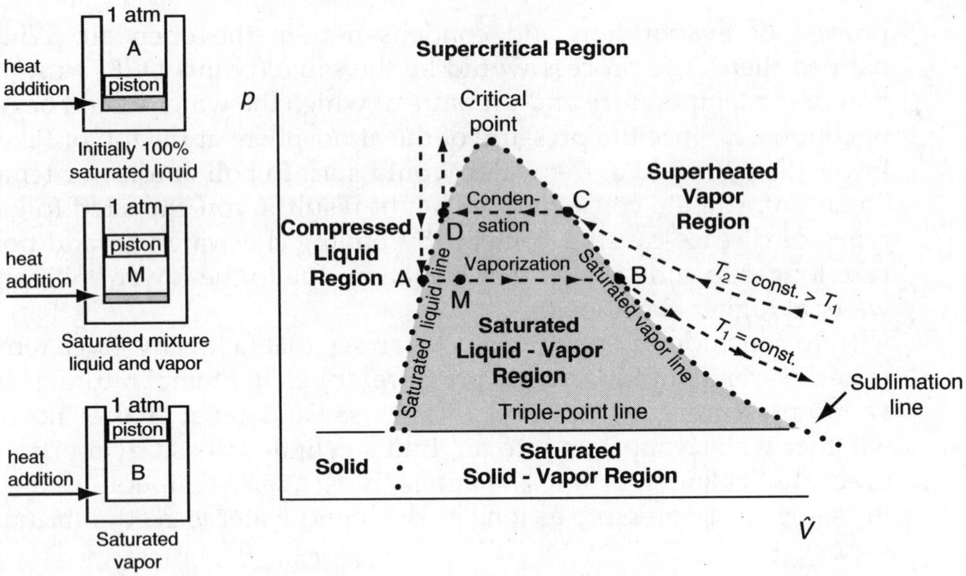

Figure 8.4 Experiments to obtain a p-$\hat{V}$ phase diagram. The dashed lines are measurements made at constant temperatures T_1 and T_2. The dots represent the points at which vaporization, or condensation, respectively, of the saturated liquid, or vapor, occurs; they form an envelope about the two-phase region.

saturation of liquid and vapor coexist and are deemed to form the envelope for the **two-phase** region that from a different angle appears in Figures 8.2 and 8.3 as the vapor-pressure curve. The two-phase region (e.g., A to B or D to C) represents the conditions under which liquid and vapor can exist at equilibrium. Note from Figure 8.4 the discontinuous change in the specific volume in going from a liquid to a solid at the triple point. In other words, water expands when it freezes, and this is why ships trapped in the polar ice can be crushed by the force of the expanding ice. By comparing Figures 8.3 and 8.4, you can see that lines AB and CD in the p-$\hat{V}$ phase diagram (Figure 8.4) correspond to a single point in the p-T diagram (Figure 8.3).

Figure 8.4 involves a new term, **quality**, the fraction or percent of the total vapor and liquid mixture that is vapor (wet vapor). Examine Figure 8.5. You can calculate the volume of the liquid-vapor mixture at B in Figure 8.5 by adding a volume fraction of material that is saturated liquid to the volume fraction that is saturated vapor:

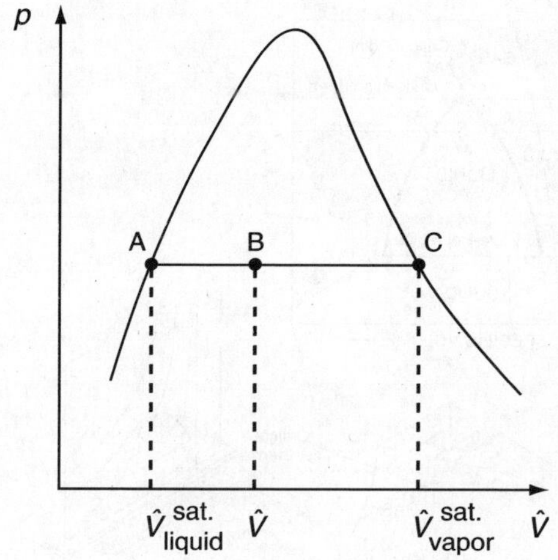

Figure 8.5 Representation of quality on a p-$\hat{V}$ phase diagram. A is saturated liquid and C is saturated vapor. The compound at B is part liquid and part vapor, and the fraction vapor is called the quality.

$$\hat{V} = (1 - x)\hat{V}^{sat.}_{liquid} + x\,\hat{V}^{sat.}_{vapor} \tag{8.1}$$

where x is the fractional quality. Solving for x yields

$$x = \frac{\hat{V} - \hat{V}^{sat.}_{liquid}}{\hat{V}^{sat.}_{vapor} - \hat{V}^{sat.}_{liquid}}$$

That is, by examining the location of $\hat{V}$ in relation to $\hat{V}^{sat.}_{vapor}$ and $\hat{V}^{sat.}_{liquid}$, you can determine the quality.

Figures 8.3 and 8.4 can be reconciled by looking at the three-dimensional surface that illustrates the p-$\hat{V}$-T (see Figure 8.6).

You can see that vapor pressure is the two-dimensional projection, yielding a curve, of a three-dimensional surface into the p-T plane. Note that the vapor-pressure curve in a p-T plane is actually a surface in the three-dimensional representation because a vapor and liquid at equilibrium are at the same temperature (see Figure 8.6). Figure 8.3 thus proves to be a portion of the complete region shown in Figure 8.6.

Let us help one another to see things better.

Claude Monet

Now we will consider the phase rule, which defines key relationships between the phases in a phase diagram. The phase rule pertains only to systems at equilibrium. Equilibrium means

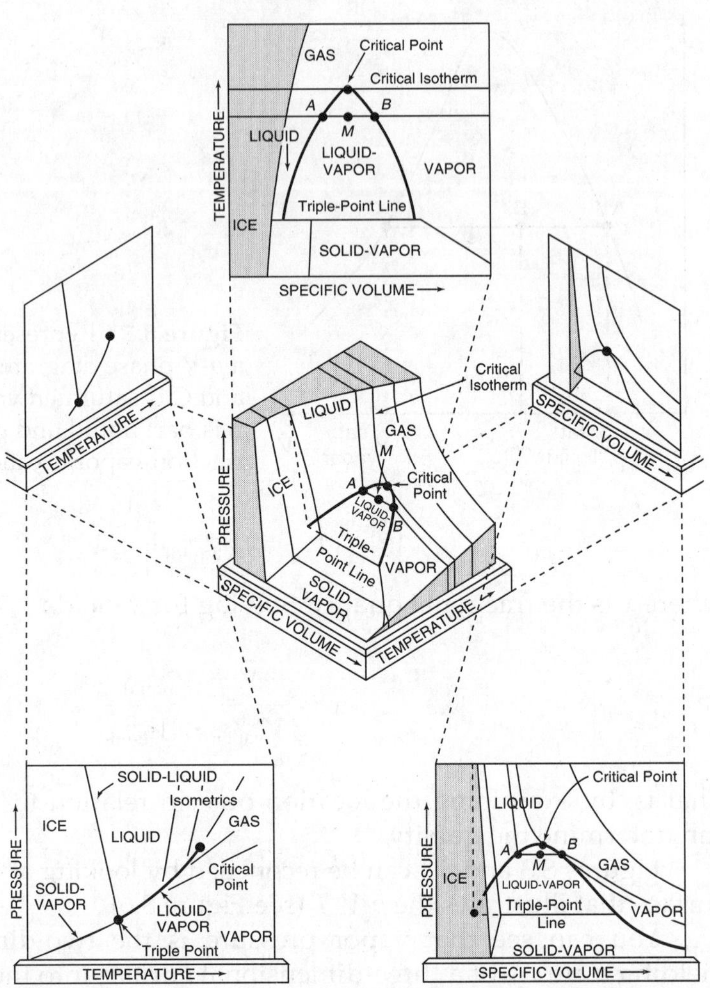

Figure 8.6 The p-$\hat{V}$-T surface for water (a compound that expands on freezing) in three dimensions showing also two-dimensional projections for sequential pairs of the three variables

- A state of absolute rest
- No tendency to change state
- No processes operating (physical equilibrium)
- No fluxes of energy, mass, or momentum
- No temperature, pressure, or concentration gradients
- No reactions occurring (chemical equilibrium)

Thus, **phase equilibrium** means that the phases present in a system are invariant as are the phase properties. By **phase** we mean a part of a system that is *chemically and physically* uniform throughout. This definition does not necessarily imply that a phase is continuous. For example, ice cubes in water represent a system that consists of two phases. The important concept of phase for you to retain is that a gas and liquid at equilibrium can each be treated as having a uniform domain. Each ice cube is chemically and physically the same; hence, all the cubes are considered to make up one phase. The decision about whether a solid is one or more phases is not always clear.

If you mechanically mix table salt and sugar, you have a solid system, but it consists of two distinct solid phases. Small particles of one phase are intermingled with small particles of the other. Particles of sugar are not the same chemically as those of salt, even though they may appear to be the same physically. On the other hand, it should be emphasized here that most gases and liquids at equilibrium can be assumed to be uniform.

The phase rule is concerned only with the **intensive** properties of the system. By intensive we mean **properties that do not depend on the quantity of material present**. If you think about the properties we have employed so far in this book, do you get the feeling that pressure and temperature are independent of the amount of material present? Concentration is an intensive variable, but what about volume? The total volume of a system is called an **extensive variable** because it does depend on how much material you have; the **specific volume** or the **density**, on the other hand—the cubic meters per kilogram, for example—is an **intensive property** because it is independent of the amount of material present. You should remember that the specific (per unit mass) values are intensive properties; the total quantities are extensive properties. Furthermore, the state of a system is specified by the intensive variables, not the extensive ones.

You will find **Gibbs' phase rule** to be a useful guide in establishing how many intensive properties, such as pressure and temperature, have to be specified to definitely fix all of the remaining intensive properties and number of phases that can coexist for any physical system. **The rule can be applied only to systems in equilibrium** and is given by Equation (8.2), assuming that **no chemical reaction occurs**:

$$\mathcal{F} = 2 - \mathcal{P} + \mathcal{C} \tag{8.2}$$

where $\mathcal{F}$ = number of degrees of freedom (i.e., the number of independent properties that have to be specified to determine all of the intensive properties of each phase of the system of interest)—*not to be confused* with the degrees of freedom calculated in solving material balances that can involve both intensive *and* extensive variables

$\mathcal{P}$ = number of phases that can exist in the system; a phase is a homogeneous quantity of material such as a gas, a pure liquid, a solution, or a homogeneous solid

$\mathcal{C}$ = number of independent components (chemical species) in the system

Let's look at Figure 8.7, which shows the surface of part of Figure 8.6. Consider the vapor phase.

You will remember for a pure gas that we had to specify three of the four variables in the ideal gas equation $pV = nRT$ in order to be able to determine the remaining one unknown. You might conclude that $\mathcal{F} = 3$. If we apply the phase rule, for a single phase $\mathcal{P} = 1$, and for a pure gas $\mathcal{C} = 1$.

$$\mathcal{F} = 2 - \mathcal{P} + \mathcal{C} = 2 - 1 + 1 = 2 \text{ variable to be specified}$$

How can we reconcile this apparent paradox with our previous statement? Easily! Since the phase rule is concerned with *intensive properties* only, the following are the phase rule variables to be included in the ideal gas law:

$$\left.\begin{array}{l} p \\ \hat{V} \text{ (specific molar volume)} \\ T \end{array}\right\} \text{3 intensive properties}$$

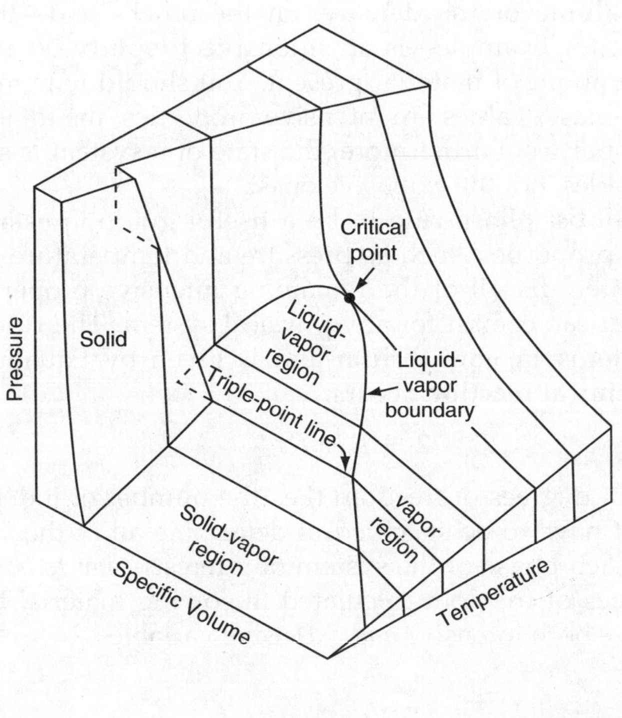

Figure 8.7 The surface of the solid-liquid-vapor phases of water with the coordinates of p, $\hat{V}$, and T

Thus the ideal gas law would be written

$$p\hat{V} = RT \tag{8.3}$$

and in this form you can see that when two intensive variables are specified ($\mathcal{F} = 2$), the third can be calculated. Thus, in the superheated region in the steam tables, you can fix all of the properties of the water vapor by specifying two intensive variables.

An **invariant** system is one in which no variation of conditions is possible without one phase disappearing. In Figure 8.7 a system that is composed of ice, water, and water vapor exists at only one temperature (0.01°C) and pressure (0.611 kPa), namely, along the triple-point line (a point in a p-T diagram), and represents one of the invariant states in the water system:

$$\mathcal{F} = 2 - \mathcal{P} + \mathcal{C} = 2 - 3 + 1 = 0$$

With all three phases present, none of the physical conditions of p, T, or $\hat{V}$ can be varied without one phase disappearing. As a corollary, if the three phases are present, the temperature, the specific volume, and so on must always be fixed at the same values. This phenomenon is useful in calibrating thermometers and other instruments. Now let's look at some examples of the application of the phase rule.

Example 8.1 Application of the Phase Rule

Calculate the number of degrees of freedom (how many additional intensive variables must be specified to fix the system) from the phase rule for the following materials at equilibrium:

a. Pure liquid benzene
b. A mixture of ice and water only
c. A mixture of liquid benzene, benzene vapor, and helium gas
d. A mixture of salt and water designed to achieve a specific vapor pressure

What variables might be specified in each case?

Solution

a. $\mathcal{P} = 1$, and $\mathcal{C} = 1$; hence $\mathcal{F} = 2 - 1 + 1 = 2$. The temperature and pressure might be specified in the range in which benzene remains a liquid.

(*Continues*)

Example 8.1 Application of the Phase Rule (*Continued*)

b. $\mathcal{P} = 2$, and $\mathcal{C} = 1$; hence $\mathcal{F} = 2 - 2 + 1 = 1$. Once either the temperature or the pressure is specified, the other intensive variables are fixed.

c. $\mathcal{P} = 2$, and $\mathcal{C} = 2$; hence $\mathcal{F} = 2 - 2 + 2 = 2$. A pair from temperature, pressure, or mole fraction can be specified.

d. $\mathcal{P} = 2$, and $\mathcal{C} = 2$; hence $\mathcal{F} = 2 - 2 + 2 = 2$. Since a particular pressure is to be achieved, you would adjust the salt concentration and the temperature of the solution.

Note that in a and b it would be likely that a vapor phase would exist in practice, increasing $\mathcal{P}$ by 1 and reducing $\mathcal{F}$ by 1.

Self-Assessment Test

Questions

1. Why does dry ice sublime at room temperature and pressure?
2. List two intensive and two extensive properties.
3. Indicate whether the following statements are true or false:
   a. A phase is an agglomeration of matter having distinctly identifiable properties such as a distinct refractive index, viscosity, density, X-ray pattern, and so on.
   b. A solution containing two or more compounds constitutes a single phase.
   c. A mixture of real gases constitutes a single phase.
4. Fill in the following table for water:

Number of Phases *P*	Example	Degrees of Freedom *F*	Number of Variables That Can Be Adjusted at Equilibrium
1	Steam		
2	Steam and water		
3	Steam, water, and ice		

Problems

1. Determine the number of degrees of freedom from the phase rule for the following systems at equilibrium:
   a. Liquid water, water vapor, and nitrogen
   b. Liquid water with dissolved acetone in equilibrium with their vapors
   c. $O_2(g)$, $CO(g)$, $CO_2(g)$ and $C(s)$ at high temperature

2. A tank contains 1000 kg of acetone (C_3H_6O), half of which is liquid and the other half of which is in the vapor phase. Acetone vapor is withdrawn slowly from the tank, and a heater in each phase maintains the temperature of each of the two phases at 50°C. Determine the pressure in the tank after 100 kg of vapor have been withdrawn.
3. Draw a p-T phase diagram for water. Label the following clearly: vapor-pressure curve, dew point curve, saturated region, superheated region, subcooled region, and triple point. Show the processes of evaporation, condensation, and sublimation by arrows.

8.3 Single-Component Two-Phase Systems (Vapor Pressure)

You can understand the behavior of single-component two-phase systems by examining the phase diagram of the component of interest. For example, consider the p^*-versus-T diagram (at constant $\hat{V}$) for water shown in Figure 8.3. The relationship between temperature and pressure for steam and liquid water phases in equilibrium is represented by the line from the triple point up to point B. In the remainder of this section we will explain how to determine values for the vapor pressure given the temperature, or the temperature given the vapor pressure.

8.3.1 Prediction via Equations

You can see from Figure 8.2 (line AB) that the function of p^* versus T is not a linear function (except as an approximation over a very small temperature range). Many functional forms have been proposed to predict p^* from T, but we will use the **Antoine equation** in this book—it has sufficient accuracy for our needs, and coefficients for the equation exist in the literature for over 5000 compounds:

$$\ln(p^*) = A - \frac{B}{C + T} \tag{8.4}$$

where A, B, C = constants for each substance
T = temperature, kelvin

Refer to Appendix H on the CD that accompanies this text for the values of A, B, and C for various compounds. In addition, the physical property software on the CD is based on data provided by Yaws[1] and will enable you to retrieve vapor pressures for over 700 compounds.

---

[1]C. L. Yaws and H. C. Yang, "To Estimate Vapor Pressure Easily," *Hydrocarbon Processing*, 65 (October, 1989).

You can estimate the values of A, B, and C in Equation (8.4) from experimental data by using a regression program such as Polymath. With just three experimental values for the vapor pressure versus temperature you can fit Equation (8.4). More values are better!

Example 8.2 Vaporization of Metals for Thin Film Deposition

Three methods of providing vaporized metals for thin film deposition are evaporation from a boat, evaporation from a filament, and transfer via an electronic beam. Figure E8.2 illustrates evaporation from a boat placed in a vacuum chamber.

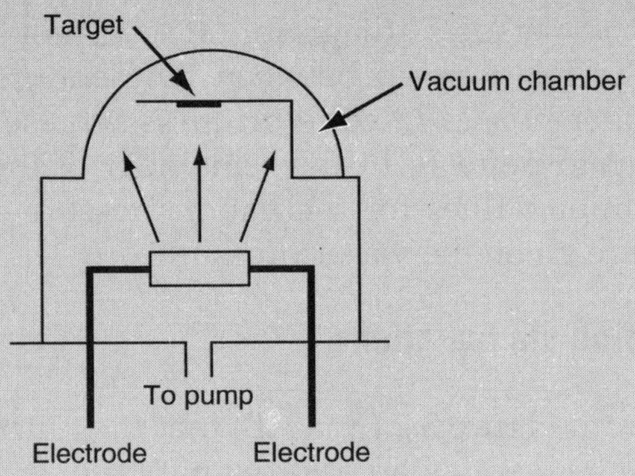

Figure E8.2

The boat made of tungsten has a negligible vapor pressure at 972°C, the operating temperature for the vaporization of aluminum (which melts at 660°C and fills the boat). The approximate rate of evaporation m is given in g/(cm^2)(s) by

$$m = 0.437 \frac{p^*(MW)^{1/2}}{T^{1/2}}$$

where p^* is the vapor pressure in kilopascals and T is the temperature in kelvin. What is the vaporization rate for Al at 972°C in g/(cm^2)(s)?

Solution

You have to calculate p^* for Al at 972°C. The Antoine equation is suitable if data are known for the vapor pressure of Al. Considerable variation exists in the data for Al at high temperatures, but we will use $A = 8.779$, $B = 1.615 \times 10^4$, and $C = 0$ with p^* in millimeters of Hg and T in kelvin.

$$\ln p^*_{972°C} = 8.799 - \frac{1.615 \times 10^4}{972 + 273} = 0.0154 \text{ mm Hg } (0.00201 \text{ kPa})$$

$$m = 0.437 \frac{(0.00201)(26.98)^{1/2}}{(972 + 273)^{1/2}} = 1.3 \times 10^{-4} \text{ g/(cm}^2\text{)(s)}$$

8.3.2 Retrieving Vapor Pressures from the Tables

You can find the vapor pressures of substances listed in tables in handbooks, physical property books, and Web sites. We will use water as an example. Tabulations of the properties of water and steam (water vapor) are commonly called the **steam tables**, although the tables are as much about water as they are about steam. Furthermore, when you retrieve the properties from a CD, such as the American Society of Mechanical Engineers' *Properties of Steam*, they probably were generated by an equation. Furthermore, we will often refer to the tables on the CD in the back of this book as the "steam tables." In this book you will also find a foldout in the back pocket that contains abbreviated steam tables in both AE and SI units. From the CD you can obtain values of the properties of water in mixed units that are continuous over the permitted range of values, thus avoiding single or double interpolation in tables. The properties of water and steam from the CD may not agree precisely with other sources because the values from the CD are generated by simpler equations than those of the other sources.

Three classes of tables exist in the foldout:

1. A table of p^* versus T (saturated water and vapor) listing other properties such as $\hat{V}$

Properties of Saturated Water

Press. kPa	T K	Volume, m³/kg V_t	V_g
0.80	276.92	0.001000	159.7
1.0	280.13	0.001000	129.2
1.2	282.81	0.001000	108.7
1.4	285.13	0.001001	93.92
1.6	287.17	0.001001	82.76
1.8	288.99	0.001001	74.03

(Continues)

Properties of Saturated Water (*Continued*)

Press. kPa	T K	Volume, m^3/kg V_t	V_g
2.0	290.65	0.001002	67.00
2.5	294.23	0.001002	54.25
3.0	297.23	0.001003	45.67
4.0	302.12	0.001004	34.80

2. A table of T versus p^* (saturated water and vapor) containing other properties such as $\hat{V}$

Properties of Saturated Water

T K	Press. kPa	Volume, m^3/kg V_t	V_g
273.16	0.6113	0.001000	206.1
275	0.6980	0.001000	181.7
280	0.9912	0.001000	130.3
285	1.388	0.001001	94.67
290	1.919	0.001001	69.67
295	2.620	0.001002	51.90
300	3.536	0.001004	39.10
305	4.718	0.001005	29.78
310	6.230	0.001007	22.91
315	8.143	0.001009	17.80

3. A table listing superheated vapor (steam) properties as a function of T and p

Superheated Steam

Abs. Press. lb/in.^{2} (Sat. Temp.)		Sat. Water	Sat. Steam	400°	420°	440°
	Sh			29.23	49.23	69.23
175	*v*	0.0182	2.601	2.730	2.814	2.897
(370.77)	*h*	343.61	1196.7	1215.6	1227.6	1239.9

Superheated Steam (*Continued*)

Abs. Press. lb/in.^{2} (Sat. Temp.)		Sat. Water	Sat. Steam	400°	420°	440°
	Sh			26.92	46.92	66.92
180	*v*	0.0183	2.532	2.648	2.731	2.812
(373.08)	*h*	346.07	1197.2	1214.6	1226.8	1239.2
	Sh			24.66	44.66	64.66
185	*v*	0.0183	2.466	2.570	2.651	2.731
(375.34)	*h*	348.47	1197.6	1213.7	1226.0	1238.4

4. A table of subcooled water (liquid) properties as a function of p and T (h and u in this table are the enthalpy and the specific internal energy, respectively, which will be introduced in Chapter 9)

Properties of Liquid Water

p kPa P_{sat}, kPa		400	425	450
P_{sat}, kPa Sat.		0.2456	0.4999	0.9315
	ρ, kg/m^3	937.35	915.08	890.25
	h, kJ/kg	532.69	639.71	748.98
	u, kJ/kg	532.43	639.17	747.93
500	ρ, kg/m^3	937.51	915.08	
	h, kJ/kg	532.82	639.71	
	u, kJ/kg	532.29	639.17	
700	ρ, kg/m^3	937.62	915.22	
	h, kJ/kg	532.94	639.84	
	u, kJ/kg	532.19	639.07	

Locate each table in the foldout, and use the tables to follow the following explanations.

How can you tell which table to use to get the properties you want? One way is to look at one of the phase diagrams for water.

For example, do the conditions of 25°C and 4 atm refer to liquid water, a saturated liquid-vapor mixture, or water vapor? You can use the values in the steam tables plus what you know about phases to reach a decision about the state of the water. In the SI steam tables of T versus p^* for saturated

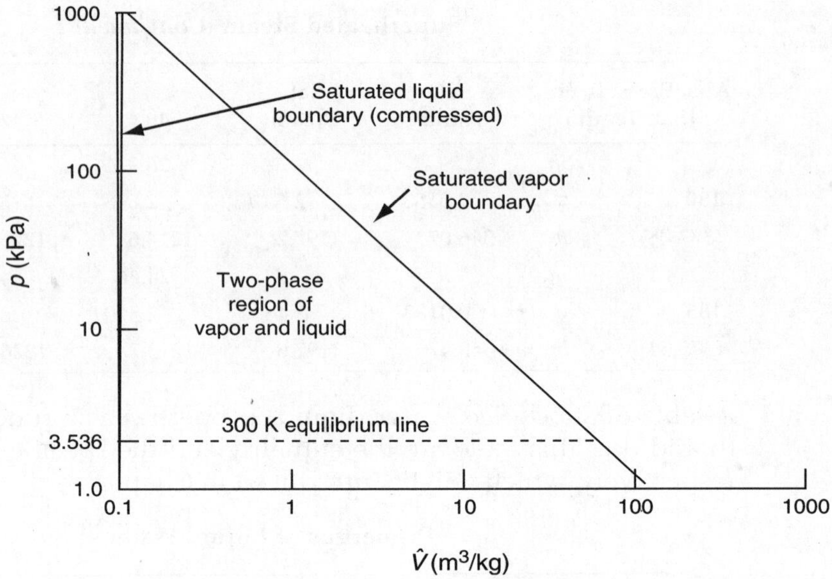

Figure 8.8 Portion of the p-$\hat{V}$ phase diagram for water (note that the axes are logarithmic scales)

water, T is just less than 300 K at which $p^* = 3.536$ kPa. Because the given pressure was about 400 kPa, much higher than the saturation pressure at 298 K, clearly the water is subcooled (compressed liquid).

Can you locate the point $p^* = 250$ kPa and $\hat{V} = 1.00\ \text{m}^3/\text{kg}$ using Figure 8.8? Do you find the water is in the superheated region? The specified volume is larger than the saturated volume of $0.7187\ \text{m}^3/\text{kg}$ at 250 kPa. What about the point $T = 300$ K and $\hat{V} = 0.505\ \text{m}^3/\text{kg}$? Water at that state is a mixture of saturated liquid and vapor. You can calculate the quality of the water–water vapor mixture using Equation (8.1) as follows: From the steam tables the specific volumes of the saturated liquid and vapor are

$$\hat{V}_\ell = 0.001004\ \text{m}^3/\text{kg} \quad \hat{V}_g = 39.10\ \text{m}^3/\text{kg}$$

Basis: 1 kg of wet steam mixture

Let x = mass fraction vapor. Then

$$\frac{0.001004\ \text{m}^3}{1\ \text{kg liquid}}\left|\frac{(1-x)\ \text{kg liquid}}{}\right. + \frac{39.10\ \text{m}^3}{1\ \text{kg vapor}}\left|\frac{x\ \text{kg vapor}}{}\right. = 0.505\ \text{m}^3$$

$$x = 0.0129 \text{ (the fractional quality)}$$

If you are given a specific mass of saturated water plus steam at a specified temperature or pressure so that you know the state of the water is in the two-phase region, you can use the steam tables for various calculations. For example, suppose a 10.0 m^3 vessel contains 2000 kg of water plus steam at 10 atm, and you are asked to calculate the volume of each phase. Let the volume of water be V_ℓ and the volume of steam be V_g; then the masses of each phase are $V_\ell/\hat{V}_\ell$ and $V_g/\hat{V}_g$ respectively. From your knowledge of the total volume and total mass:

$$V_\ell + V_g = 10$$

and

$$V_\ell/\hat{V}_\ell + V_g/\hat{V}_g = 2000$$

From the steam tables $\hat{V}_\ell = 0.0011274\ m^3/kg$ and $\hat{V}_g = 0.19430\ m^3/kg$. Solving these simultaneous equations for V_ℓ and V_g gives the volume of the liquid as 2.21 m^3 and the volume of the steam as 7.79 m^3. The mass of the liquid is 1960 kg, and the mass of the steam is 40 kg.

Because the values in the steam tables are tabulated in discrete increments, for intermediate values you will have to interpolate to retrieve values between the discrete values. (If interpolation does not appeal to you, use the physical property software on the CD in the back of this book.) The next example shows how to carry out interpolations in tables.

Example 8.3 Interpolating in the Steam Tables

What is the saturation pressure of water at 312 K?

Solution

To solve this problem you have to carry out a single interpolation. Look in the steam tables under the properties of saturated water to get p^* so as to bracket 312 K:

T (K)	p^* (kPa)
310	6.230
315	8.143

Figure E8.3 shows the concept of a linear interpolation between 310 K and 315 K. Find the change of p^* per unit change in T.

(*Continues*)

Example 8.3 Interpolating in the Steam Tables (*Continued*)

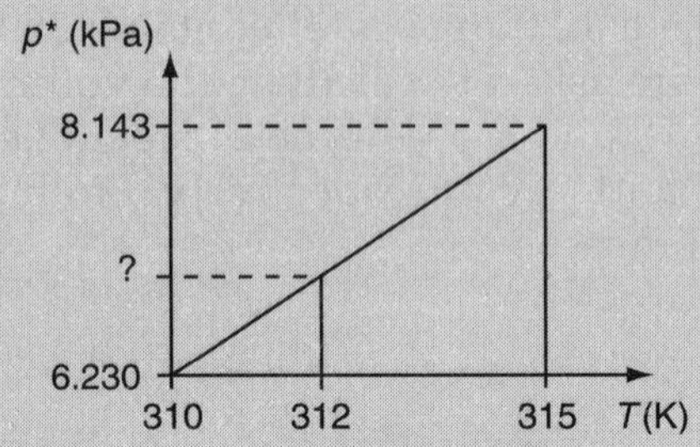

Figure E8.3

$$\frac{\Delta p^*}{\Delta T} = \frac{8.143 - 6.230}{315 - 310} = \frac{1.91}{5} = 0.383$$

Multiply the fractional change times the number of degrees increase from 310 K to get change in p^*, and add the result to the value of p^* at 310 K of 6.230 kPa:

$$p^*_{312\,\text{K}} = p^*_{310\,\text{K}} + \frac{\Delta p^*}{\Delta T}(T_{312} - T_{310}) = 6.230 + 0.383(2) = 7.00 \text{ kPa}$$

8.3.3 Predicting Vapor Pressures from Reference Substance Plots

Because of the curvature of the vapor-pressure data versus temperature (see Figure 8.2), no simple equation with two or three coefficients will fit the data accurately from the triple point to the critical point. Othmer proposed in numerous articles [see, for example, D. F. Othmer, *Ind. Eng. Chem.*, **32**, 841 (1940); and J. H. Perry and E. R. Smith, *Ind. Eng. Chem.*, **25**, 195 (1933)] that **reference substance plots** (the name will become clear in a moment) could convert the vapor-pressure-versus-temperature curve into a straight line. One well-known example is the Cox chart [E. R. Cox, *Ind. Eng. Chem.*, **15**, 592 (1923)]. You can use the Cox chart to retrieve vapor-pressure values as well as to test the reliability of experimental data, to interpolate, and to extrapolate. Figure 8.9 is a Cox chart.

Here is how you can make a Cox chart:

1. Mark on the horizontal scale values of log p^* so as to cover the desired range of p^*.
2. Next, draw a straight line on the plot at a suitable angle, say, 45°, that covers the range of T that is to be marked on the vertical axis.

3. To calibrate the vertical axis in common integers such as 25, 50, 100, 200 degrees, and so on, you use a **reference substance**, usually water. For the first integer, say, T = 100°F, you look up the vapor pressure of water in the steam tables, or calculate it from the Antoine equation, to get 0.9487 psia. Locate this value on the horizontal axis, and proceed vertically until you hit the 45° straight line. Then proceed horizontally left until you hit the vertical axis. Mark the scale there as 100°F.
4. Pick the next temperature, say, 200°F, and get 11.525 psia. Proceed vertically from p^* = 11.525 to the 45° straight line, and then horizontally to the vertical axis. Mark the scale as 200°F.
5. Continue as in steps 3 and 4 until the vertical scale is established over the desired range for the temperature.

Other compounds will give straight lines for p^* versus T as shown in Figure 8.9. What proves useful about the Cox chart is that the vapor pressures of other substances plotted on this specially prepared set of coordinates will yield straight lines over extensive temperature ranges and thus facilitate the extrapolation and interpolation of vapor-pressure data. It has

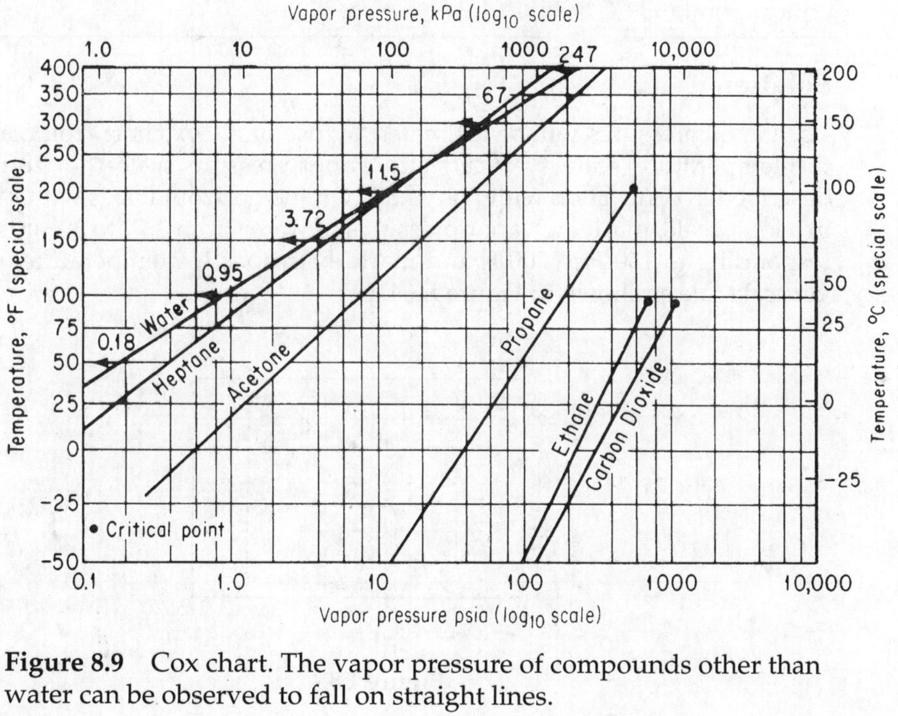

Figure 8.9 Cox chart. The vapor pressure of compounds other than water can be observed to fall on straight lines.

been found that lines so constructed for closely related compounds, such as hydrocarbons, all meet at a common point. Since straight lines can be obtained in a Cox chart, only *two points* of vapor-pressure data are needed to provide adequate information about the vapor pressure of a substance over a considerable temperature range.

Let's look at an example of using a Cox chart.

Example 8.4 Extrapolation of Vapor-Pressure Data

The control of solvents was first described in the *Federal Register* [**36**, No. 158 (August 14, 1971)] under Title 42, Chapter 4, Appendix 4.0, "Control of Organic Compound Emissions." Chlorinated solvents and many other solvents used in industrial finishing and processing, dry-cleaning plants, metal degreasing, printing operations, and so forth can be recycled and reused by the introduction of carbon adsorption equipment. To predict the size of the adsorber, you first need to know the vapor pressure of the compound being adsorbed at the process conditions.

The vapor pressure of chlorobenzene is 400 mm Hg absolute at 110°C and 5 atm at 205°C. Estimate the vapor pressure at 245°C and also at the critical point (359°C).

Solution

The vapor pressures will be estimated by use of a Cox chart. You construct the temperature scale (vertical) and vapor-pressure scale (horizontal) as described in connection with Figure 8.9. On the horizontal axis with p^* given in a $\log_{10}$ scale, mark the vapor pressures of water from 3.72 to 3094 psia corresponding to 150°F to 700°F, and mark the respective temperatures on the vertical scale as shown in Figure E8.4.

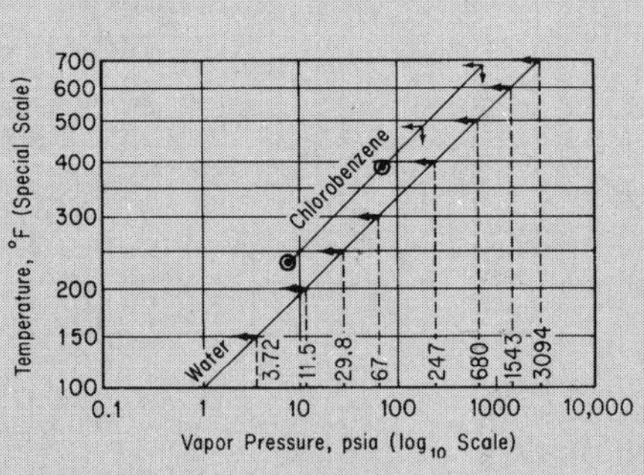

Figure E8.4

Next, convert the two given vapor pressures of chlorobenzene into psia:

$$\frac{400 \text{ mm Hg}}{} \left| \frac{14.7 \text{ psia}}{760 \text{ mm Hg}} \right. = 7.74 \text{ psia} \quad 110°\text{C} = 230°\text{F}$$

$$\frac{5 \text{ atm Hg}}{} \left| \frac{14.7 \text{ psia}}{1 \text{ atm Hg}} \right. = 73.5 \text{ psia} \quad 205°\text{C} = 401°\text{F}$$

and plot these two points on the graph paper. Examine the encircled dots. Finally, draw a straight line between the encircled points and extrapolate to 471°F (245°C) and 678°F (359°C). At these two temperatures, you can read off the estimated vapor pressures.

	471°F (245°C)	678° (359°C)
Estimated:	150 psia	700 psia
Experimental:	147 psia	666 psia

Experimental values are given for comparison.

Although this chapter treats the vapor pressure of a pure component, we should mention that the term *vapor pressure* has been applied to solutions of multiple components as well. For example, to meet emission standards, refiners formulate gasoline and diesel fuel differently in the summer than in the winter. The rules on emissions are related to the vapor pressure of a fuel, which is specified in terms of the *Reid vapor pressure* (RVP), a value that is determined at 100°F in a bomb that permits partial vaporization. For a pure component the RVP is the true vapor pressure, but for a mixture (as are most fuels) the RVP is lower than the true vapor pressure of the mixture (by roughly 10% for gasoline). Refer to J. J. Vazquez-Esparragoza, G. A. Iglesias-Silva, M. W. Hlavinka, and J. Bulin, "How to Estimate RVP of Blends," *Hydrocarbon Processing*, 135 (August, 1992), for specific details about estimating the RVP.

Self-Assessment Test

Questions

1. As the temperature is increased, what happens to the vapor pressure of a compound?
2. Do the steam tables, and similar tables for other compounds, provide more accurate values of the vapor pressure than use of the Antoine equation, or modifications of it?

3. When you need a vapor pressure outside the range of known values, what is the best way to estimate it?
4. Is it possible to prepare a Cox chart for water?

Problems

1. Describe the state and values of the pressure of water initially at 20°F as the temperature is increased to 250°F in a fixed volume.
2. Use the Antoine equation to calculate the vapor pressure of ethanol at 50°C, and compare with the experimental value.
3. Determine the normal boiling point of benzene from the Antoine equation.
4. Prepare a Cox chart from which the vapor pressure of toluene can be predicted over the temperature range −20°C to 140°C.

Thought Problem

In the start-up of a process, Dowtherm, an organic liquid with a very low vapor pressure, was being heated from room temperature to 335°F. The operator suddenly noticed that the gauge pressure was not the expected 15 psig but instead was 125 psig. Fortunately, a relief valve in the exit line ruptured into a vent (expansion) tank so that a serious accident was avoided.

Why was the pressure in the exit line so high?

Discussion Problem

Many distillation columns are designed to withstand a pressure of 25 or 50 psig. The reboiler at the bottom of the column is where the heat used to vaporize the fluid in the column is introduced. What would you recommend for the type of heat source among these three: (a) steam (heat exchanger), (b) fired heater (analogous to a boiler), or (c) hot oil (heat exchanger)?

8.4 Two-Component Gas/Single-Component Liquid Systems

From a single-component two-phase system, let's extend the discussion to a more complicated system, namely, a system with two components in the gas phase together with a single-component liquid system. An example of such a system is water and a noncondensable gas, such as air. The equilibrium relationships for the water and air help explain how rain is formed and lead to a number of meteorological terms, such as the *dew point* and the *humidity* of the air. Moreover, the equilibrium relationship for two-component gas/single-component liquid systems is used industrially to describe and design many systems, including cooling towers, in which water is cooled by evaporation, and stripping systems, in which a volatile component is removed from a liquid by contacting the liquid with a noncondensable gas.

8.4.1 Saturation

When any noncondensable gas (or a gaseous mixture) comes in contact with a liquid, the gas will acquire molecules from the liquid. If contact is maintained for a sufficient period of time, vaporization continues until equilibrium is attained, at which time the *partial pressure of the vapor in the gas will equal the vapor pressure* of the liquid at the temperature of the system. Regardless of the duration of contact between the liquid and gas, after equilibrium is reached no more net liquid will vaporize into the gas phase. The gas is then said to be **saturated** with the particular vapor at the given temperature. We also say that the gas mixture is at its **dew point. The dew point for the mixture of pure vapor and noncondensable gas means the temperature at which the vapor would just start to condense** if the temperature were very slightly reduced. **At the dew point the partial pressure of the vapor is equal to the vapor pressure of the volatile liquid.**

Consider a gas partially saturated with water vapor at p and T. If the partial pressure of the water vapor is increased by increasing the total pressure on the system, eventually the partial pressure of the water vapor will equal p^* at T of the system. Because the partial pressure of water cannot exceed p^* at that temperature, a further attempt to increase the pressure will result in water vapor condensing at constant T and p. Thus, p^* represents the maximum partial pressure that water can attain at that temperature, T.

Do you have to have liquid present for saturation to occur? Really, no; only a minute drop of liquid at equilibrium with its vapor will suffice.

What use can you make of the information or specification that a noncondensable gas is saturated? Once you know that a gas is saturated, you can determine the composition of the vapor-gas mixture from knowledge of the vapor pressure of the vapor (or the temperature of the saturated mixture) to use in material balances. From Chapter 7 you should recall that the ideal gas law applies to both air and water vapor at atmospheric pressure with excellent precision. Thus, we can say that the following relations hold *at saturation*:

$$\frac{p_{H_2O}V}{p_{air}V} = \frac{n_{H_2O}RT}{n_{air}RT} \tag{8.5}$$

or

$$\frac{p_{H_2O}}{p_{air}} = \frac{p^*_{H_2O}}{p_{air}} = \frac{n_{H_2O}}{n_{air}} = \frac{p_{total} - p_{air}}{p_{air}} \tag{8.6}$$

because V and T are the same for the air and water vapor.

Also,

$$y_{H_2O} = \frac{p_{H_2O}}{p_{\text{total}}} = \frac{p_{H_2O}}{p_{\text{air}} + p_{H_2O}} = 1 - y_{\text{air}} \tag{8.7}$$

As a numerical example, suppose you have a saturated gas, say, water in air at 51°C, and the pressure on the system is 750 mm Hg absolute. What is the partial pressure of the air? If the air is saturated, you know that the partial pressure of the water vapor is p^* at 51°C. You can use the physical property software on the CD, or use the steam tables, and find that $p^* = 98$ mm Hg. Then

$$P_{\text{air}} = 750 - 98 = 652 \text{ mm Hg}$$

Furthermore, the vapor-air mixture has the following composition:

$$y_{H_2O} = \frac{p_{H_2O}}{p_{\text{total}}} = \frac{98}{750} = 0.13$$

$$y_{air} = \frac{p_{\text{air}}}{p_{\text{total}}} = \frac{652}{750} = 0.87$$

Example 8.5 Calculation of the Dew Point of the Products of Combustion

Oxalic acid ($H_2C_2O_4$) is burned at atmospheric pressure with 4% excess air so that 65% of the carbon burns to CO. Calculate the dew point of the product gas.

Solution

The solution of the problem involves the following steps:

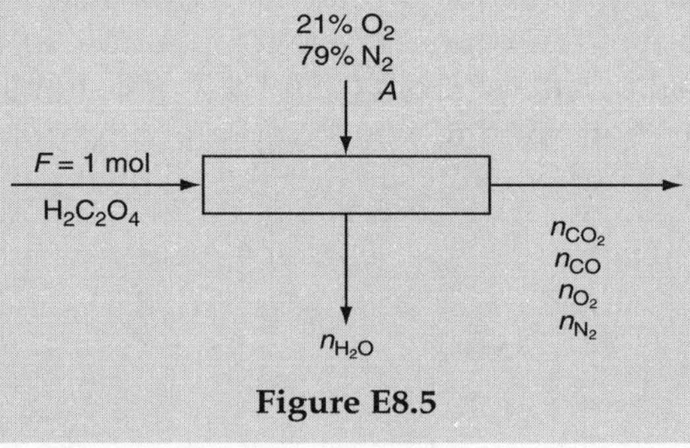

Figure E8.5

1. Calculate the combustion products via material balances.
2. Calculate the mole fraction of the water vapor in the combustion products as indicated just prior to this example.
3. Assume a total pressure, say, 1 atm, and calculate $y_{H_2O}\, p_{total} = p_{H_2O}$ in the combustion products. At equilibrium p_{H_2O} will be the vapor pressure $p^*_{H_2O}$.
4. Condensation (at constant total pressure) would be possible when $p^*_{H_2O}$ equals the calculated p_{H_2O}. This value is the dew point.
5. Look up the temperature corresponding to p_{H_2O} in the saturated steam tables.

Steps 1–5

Basis: 1 mol of $H_2C_2O_4$

The figure and data are given. The chemical reaction equation for the combustion of oxalic acid is given as

$$H_2C_2O_4 + 0.5\,O_2 \rightarrow 2\,CO_2 + H_2O$$

$$H_2C_2O_4 \rightarrow 2\,CO + H_2O + 0.5\,O_2$$

Step 4

O_2 required:

$$\frac{1 \text{ mol } H_2C_2O_4}{1}\;\frac{0.5 \text{ mol } O_2}{1 \text{ mol } H_2C_2O_4} = 0.5 \text{ mol } O_2 \text{ (note oxygen in oxalic acid)}$$

Moles of O_2 in with air including excess:

$$(1 + 0.04)(0.5 \text{ mol } O_2) = 0.52 \text{ mol } O_2$$

Therefore, $0.52/0.21 = 2.48$ mol air enters containing 1.96 mol n_{N_2}.

Specifications: 65% of the carbon burns to CO: $(0.65)(2) = 1.30$.

Basis: 1 mol of $H_2C_2O_4$ enters

Element material balances:

Element	In (mol)	Out (mol)
C	2	$n_{CO_2} + n_{CO}$ or $0.70 + 1.30$
H	2	$2n_{H_2O}$
N	1.96	1.96
O	$0.52 + 4$	$2n_{CO_2} + n_{CO} + n_{O_2}$ or $2(0.70) + 1.30 + n_{O_2}$

(Continues)

Example 8.5 Calculation of the Dew Point of the Products of Combustion (*Continued*)

The results:

$$n_{H_2O} = 1.0; n_{CO_2} = 0.7; n_{CO} = 1.3; n_{O_2} = 1.82; n_{N_2} = 1.96; \text{total mol} = 6.78$$

$$y_{H_2O} = 1 \text{ mol } H_2O/6.78 \text{ mol} = 0.147$$

The partial pressure of the water in the product gas (at an assumed atmospheric pressure) determines the dew point of the stack gas; that is, the temperature of saturated steam that equals the partial pressure of the water is equal to the dew point of the product gas:

$$p^*_{H_2O} = y_{H_2O}(p_{\text{total}}) = 0.147\ (101.3\ \text{kPa}) = 14.9\ \text{kPa}\ (2.16\ \text{psia})$$

From the steam tables, $T = 129°F$.

Self-Assessment Test

Questions

1. What does the term *saturated gas* mean?
2. If a gas is saturated with water vapor, describe the state of the water vapor and the air if it is (a) heated at constant pressure, (b) cooled at constant pressure, (c) expanded at constant temperature, and (d) compressed at constant temperature.
3. How can you lower the dew point of a pollutant gas before analysis?
4. In a gas-vapor mixture, when is the vapor pressure the same as the partial pressure of the vapor in the mixture?

Problems

1. The dew point of water in atmospheric air is 82°F. What is the mole fraction of water vapor in the air if the barometric pressure is 750 mm Hg?
2. Calculate the composition in mole fractions of air that is saturated with water vapor at a total pressure of 100 kPa and 21°C.
3. An 8.00 L cylinder contains a gas saturated with water vapor at 25.0°C and a pressure of 770 mm Hg. What is the volume of the gas when dry at standard conditions?

Thought Problem

Why is it important to know the concentration of water in the air entering a boiler?

8.4.2 Condensation

From Tom and Ray Magliozzi (Click and Clack's *Car Talk* on PBS):

> **Question:** I have a 1994 Buick LeSabre with 19,000 miles. The car runs perfectly, except after it is parked in our carport. Occasionally, in the morning, I find a water puddle under the exhaust pipe about 6 inches in diameter. There seems to be a black carbon substance on top. It is not greasy and does not seem to be oil. What is this stuff? – Sidney.
>
> **Ray:** It's good old H_2O, Sidney. Water is one of the by-products of combustion, so it's produced whenever you run the engine.
>
> **Tom:** And when you use the car for short trips, the exhaust system never really gets hot enough to evaporate the water so some of it condenses and drips out the end of the tailpipe.
>
> **Ray:** And since carbon (or "soot") is also a by-product of (incomplete) combustion, all exhaust systems have some carbon in them. So the water takes a little bit of carbon with it, and that's what you see on the puddle.
>
> **Tom:** It's perfectly normal, Sidney. You might even take advantage of it by parking the car with the tailpipe hanging over your perennial bed. That'll save you from watering it a couple of times a week.

Examine the setup for combustion gas analysis shown in Figure 8.10. What error has been made in the setup? If you do not heat the sample of gas collected by the probe and/or put an intermediate condenser before the pump, the analyzer will fill with liquid as the gas sample cools and will not function.

Condensation is the change of vapor in a noncondensable gas to liquid. Some typical ways of condensing a vapor that is in a gas are to

1. Cool it at constant total system pressure (the volume changes, of course)
2. Cool it at constant total system volume (the pressure changes)
3. Compress it isothermally (the volume changes)

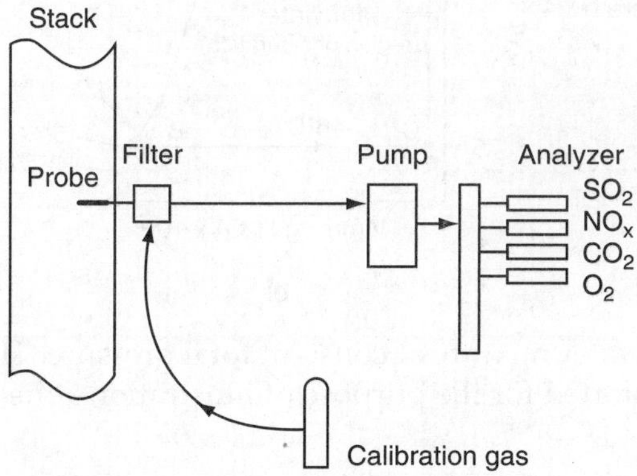

Figure 8.10 Instrumentation for stack gas analysis

Combinations of the three as well as other processes are possible, of course.

As an example of condensation let's look at cooling at constant total system pressure for a mixture of air and 10% water vapor. Pick the air-water vapor mixture as the system. If the mixture is cooled at constant total pressure from 51°C and 750 mm Hg absolute (point A in Figures 8.11a and b for the water), how low can the temperature go before condensation starts (at point B)? You can cool the mixture until the temperature reaches the dew point associated with the partial pressure of water of

$$p^*_{H_2O} \equiv p_{H_2O} = 0.10(750) = 75 \text{ mm Hg}$$

From the steam tables you can find that the corresponding temperature is $T = 46$°C (point B on the vapor curve). After reaching $p^* = 75$ mm Hg at point B, if the condensation process continues, it continues at constant pressure (75 mm Hg) and constant temperature (46°C) until all of the water vapor has been condensed to liquid (point C). Further cooling will reduce the temperature of the liquid water below 46°C.

If the air-water mixture with 10% water vapor starts at 60°C and 750 mm Hg, and is cooled at constant pressure, at what temperature will condensation occur for the same process? Has the dew point changed? It is the same because $p_{H_2O} = 0.10\ (750) = 75$ mm Hg still. The volume of both the air and the water vapor can be calculated from $pV = nRT$ until condensation starts, at which point the ideal gas law applies only to the residual water vapor, not the liquid. The number of moles of H_2O in the gas phase does not change

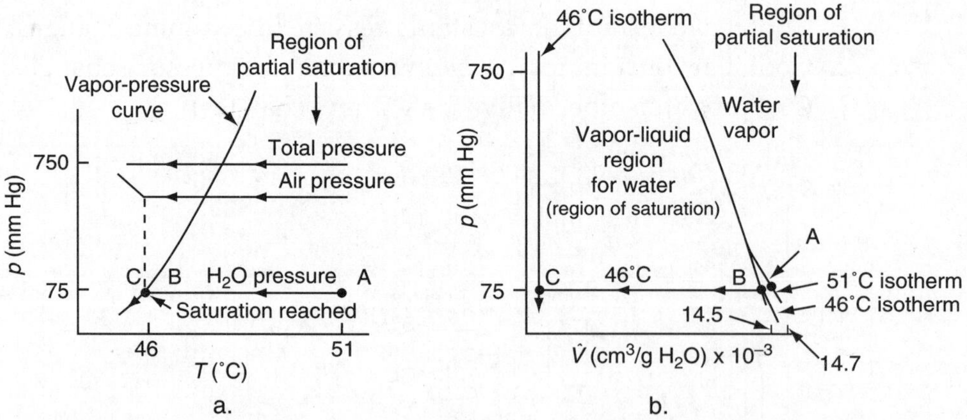

Figure 8.11 Cooling of an air-water mixture at constant total pressure. The lines and curves for the water are distorted for the purpose of illustration—the scales are not arithmetic units.

from the initial number of moles until condensation occurs, at which point the number of moles of water in the gas phase starts to decrease. The number of moles of air in the system remains constant throughout the process.

Condensation can also occur when the pressure on a vapor-gas mixture is increased. If a pound of saturated air at 75°F is isothermally compressed (with a reduction in volume, of course), liquid water will be condensed out of the air (see Figure 8.12).

For example, if a pound of saturated air at 75°F and 1 atm (the vapor pressure of water is 0.43 psia at 75°F) is compressed isothermally to 4 atm (58.8 psia), almost three-fourths of the original content of water vapor now will be in the form of liquid, and the air still has a dew point of 75°F. Remove the liquid water, expand the air isothermally back to 1 atm, and you will find that the dew point has been lowered to about 36°F. Mathematically (1 = state at 1 atm, 4 = state at 4 atm), with $z = 1.00$ for both components:

For saturated air at 75°F and 4 atm:

$$\left(\frac{n_{H_2O}}{n_{air}}\right)_4 = \left(\frac{p^*_{H_2O}}{p_{air}}\right)_4 = \frac{0.43}{58.4}$$

For the same air saturated at 75°F and 1 atm:

$$\left(\frac{n_{H_2O}}{n_{air}}\right)_1 = \left(\frac{p^*_{H_2O}}{p_{air}}\right)_1 = \frac{0.43}{14.3}$$

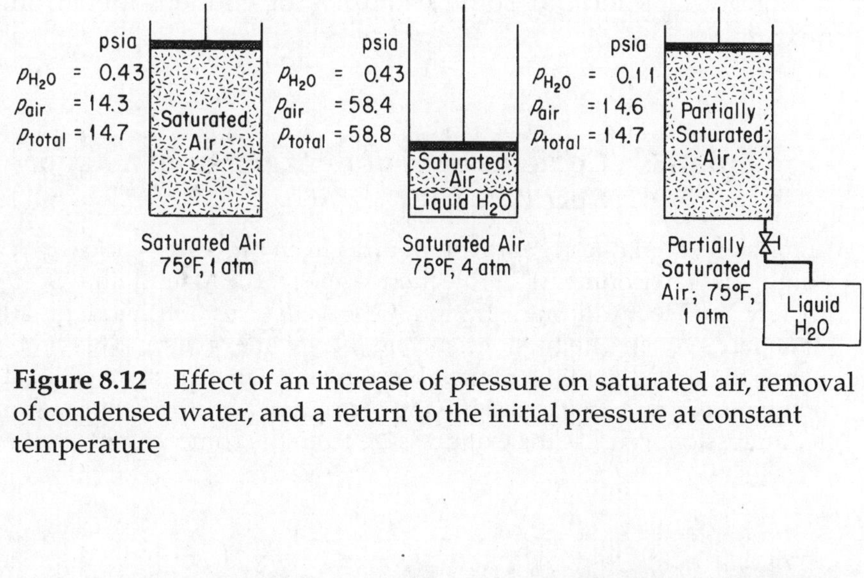

Figure 8.12 Effect of an increase of pressure on saturated air, removal of condensed water, and a return to the initial pressure at constant temperature

The material balance gives for the H_2O:

$$\left(\frac{n_4}{n_1}\right)_{H_2O} = \frac{\dfrac{0.43}{58.4}}{\dfrac{0.43}{14.3}} = \frac{14.3}{58.4} = 0.245$$

In other words, 24.5% of the original water will remain as vapor after compression.

After the air-water vapor mixture is returned to a total pressure of 1 atm, the following two equations now apply at 75°F:

$$p_{H_2O} + p_{air} = 14.7$$

$$\frac{p_{H_2O}}{p_{air}} = \frac{n_{H_2O}}{n_{air}} = \frac{0.43\,(0.245)}{58.4} = 0.00737$$

From these two relations you can find that

$$p_{H_2O} = 0.108 \text{ psia}$$
$$p_{air} = 14.6 \text{ psia}$$
$$p_{total} = 14.7 \text{ psia}$$

The pressure of the water vapor represents a dew point of about 36°F.

Now let's look at some examples of condensation from a gas-vapor mixture.

Example 8.6 Condensation of Benzene from a Vapor Recovery Unit

Emission of volatile organic compounds from processes is closely regulated. Both the Environmental Protection Agency (EPA) and the Occupational Safety and Health Administration (OSHA) have established regulations and standards covering emissions and frequency of exposure. This problem concerns the first step of the removal of benzene vapor from an exhaust stream, shown in Figure E8.6a, designed to recover 95% of the benzene from air by compression. What is the exit pressure from the compressor?

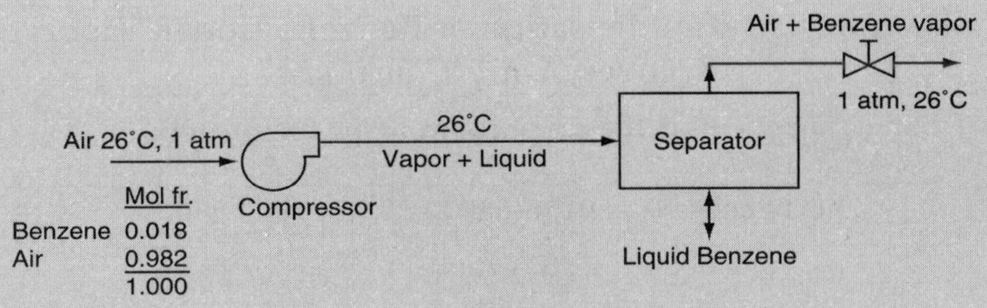

Figure E8.6a

Solution

Figure E8.6b illustrates on a p-versus-$\hat{V}$ chart for benzene what occurs to the benzene vapor during the process. The process is an isothermal compression.

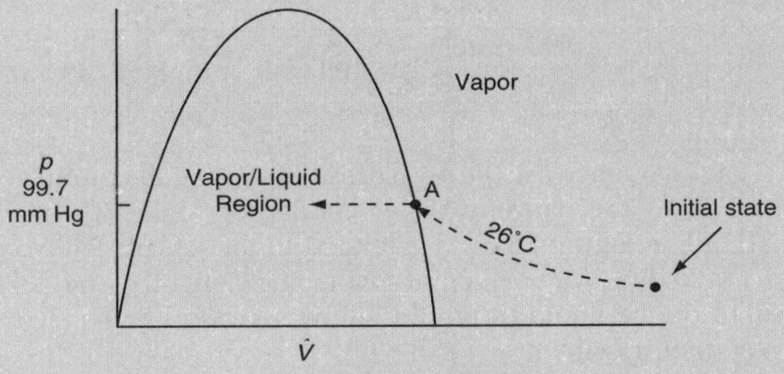

Figure E8.6b

If you pick the compressor as the system, the compression is isothermal and yields a saturated gas. You can look up the vapor pressure of benzene at 26°C in a handbook or get it from the CD in the back of this book. It is $p^* = 99.7$ mm Hg.

Next you have to carry out a short material balance to determine the outlet concentrations from the compressor:

Basis: 1 g mol of entering gas at 26°C and 1 atm

Entering components to the compressor:

$$\text{mol of benzene} = 0.018(1) = 0.018 \text{ g mol}$$

$$\text{mol of air} = 0.982(1) = \underline{0.982} \text{ g mol}$$

$$\text{total gas} = 1.000 \text{ g mol}$$

(*Continues*)

Example 8.6 Condensation of Benzene from a Vapor Recovery Unit (*Continued*)

Exiting components in the gas phase from the compressor:

$$\text{mol of benzene} = 0.018(0.05) = 0.90 \times 10^{-3} \text{ g mol}$$

$$\text{mol of air} = \underline{0.982 \text{ g mol}}$$

$$\text{total gas} = 0.983 \text{ g mol}$$

$$y_{\text{Benzene exiting}} = \frac{0.90 \times 10^{-3}}{0.983} = 0.916 \times 10^{-3} = \frac{p_{\text{Benzene}}}{p_{\text{Total}}}$$

Now the partial prèssure of the benzene is 99.7 mm Hg, so

$$p_{\text{total}} = \frac{99.7 \text{ mm Hg}}{0.916 \times 10^{-3}} = 108 \times 10^{3} \text{ mm Hg (143 atm)}$$

Could you increase the pressure at the exit of the pump above 143 atm? Only if all of the benzene vapor condenses to liquid. Imagine that the dashed line in Figure E8.6b is extended to the left until it reaches the saturated liquid line (bubble point line). Subsequently, the pressure can be increased on the liquid (it would follow a vertical line as liquid benzene is not very compressible).

Example 8.7 Smokestack Emissions and Pollution

A local pollution-solutions group has reported the Simtron Co. boiler plant as being an air polluter and has provided as proof photographs of heavy smokestack emissions on 20 different days. As the chief engineer for the Simtron Co., you know that your plant is not a source of pollution because you burn natural gas (essentially methane) and your boiler plant is operating correctly. Your boss believes the pollution-solutions group has made an error in identifying the stack—it must belong to the company next door that burns coal. Is he correct? Is the pollution-solutions group correct? See Figure E8.7a.

Figure E8.7a

Solution

Methane (CH_4) contains 2 kg mol of H_2/kg mol of C; you can look in Chapter 2 and see that coal contains 71 kg of C/5.6 kg of H_2 in 100 kg of coal. The coal analysis is equivalent to

$$\frac{71 \text{ kg C}}{} \left| \frac{1 \text{ kg mol C}}{12 \text{ kg C}} \right. = 5.92 \text{ kg mol C} \qquad \frac{5.6 \text{ kg } H_2}{} \left| \frac{1 \text{ kg mol } H_2}{2.016 \text{ kg } H_2} \right. = 2.78 \text{ kg mol } H_2$$

or a ratio of $2.78/5.92 = 0.47$ kg mol of H_2/kg mol of C. Suppose that each fuel burns with 40% excess air and that combustion is complete. We can compute the mole fraction of water vapor in each stack gas and thus get the respective partial pressures.

Steps 1–4

The process is shown in Figure E8.7b.

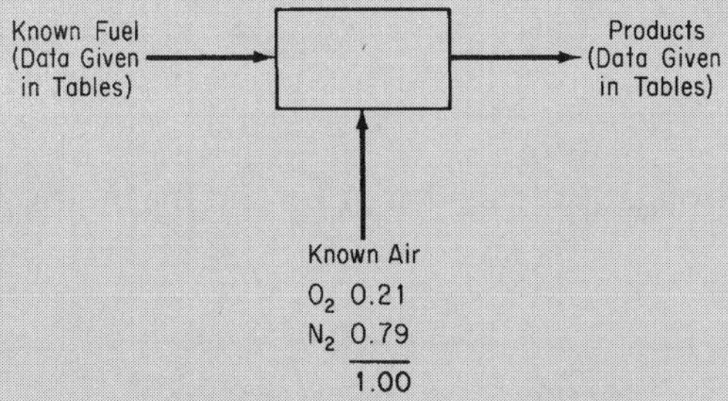

Figure E8.7b

Step 5

Basis: 1 kg mol of C

(*Continues*)

Example 8.7 Smokestack Emissions and Pollution (*Continued*)

Steps 6 and 7

The combustion problem is a standard type of problem having zero degrees of freedom in which both the fuel and airflows are given, and the product flows are calculated directly.

Steps 7–9

Tables will make the analysis and calculations compact.

Natural gas:

$$CH_4 + 2O_2 \rightarrow CO_2 + 2H_2O$$

$$\text{Required } O_2\text{: } 2$$

$$\text{Excess } O_2\text{: } 2(0.40) = 0.80$$

$$N_2\text{: } (2.80)(79/21) = 10.5$$

Composition of combustion gases (kilogram moles):

Combustion Product Gases (kg mol)

Components	kg mol in	CO_2	H_2O	Excess O_2	N_2
C	1.0	1.0			
H	4.0		2.0		
Air				0.80	10.5
Total		1.0	2.0	0.80	10.5

The total kilogram moles of gas produced are 14.3, and the mole fraction H_2O is

$$\frac{2.0}{14.3} = 0.14$$

Coal:

$$C + O_2 \rightarrow CO_2 \qquad H_2 + \frac{1}{2}O_2 \rightarrow H_2O$$

$$\text{Required } O_2\text{: } 1 + 0.47(1/2) = 1.24$$

$$\text{Excess } O_2\text{: } (1.24)(0.40) = 0.49$$

$$N_2\text{: } 1.40(79/21)[1 + 0.47(1/2)] = 6.50$$

Composition of combustion gases (kilogram moles)

Components	kg mol in	CO_2	H_2O	Excess O_2	N_2
C	1	1			
H	0.94		0.47		
Air				0.49	6.5
Total		1	0.47	0.49	6.5

The total kilogram moles of gas produced are 8.46 and the mole fraction H_2O is

$$\frac{0.47}{8.46} = 0.056$$

If the barometric pressure is, say, 100 kPa, and if the stack gas became saturated so that water vapor would start to condense at $p^*_{H_2O}$, condensed vapor could be photographed:

	Natural Gas	Coal
Partial pressure (p^*)	100(0.14) = 14 kPa	100(0.056) = 5.6 kPa
Equivalent temperature	52.5°C	35°C

Thus, the stack will emit condensed water vapor at higher ambient temperatures for a boiler burning natural gas than for one burning coal. The public, unfortunately, sometimes concludes that all the emissions they perceive are pollution. Natural gas could appear to the public to be a greater pollutant than either oil or coal, whereas, in fact, the emissions are just water vapor. The sulfur content of coal and oil can be released as sulfur dioxide to the atmosphere, and the polluting capacities of mercury and heavy metals in coal and oil are much greater than those of natural gas when all three are being burned properly. The sulfur contents as delivered to consumers are as follows: natural gas, 4×10^{-4} mol % (as added mercaptans to provide smell for safety); No. 6 fuel oil, up to 2.6%; and coal, from 0.5% to 5%. In addition, coal may release particulate matter into the stack plume. By mixing the stack gas with fresh air, and by convective mixing above the stack, the mole fraction water vapor can be reduced, and hence the condensation temperature can be reduced. However, for equivalent dilution, the coal-burning plant will always have a lower condensation temperature.

With the information calculated above, how would you resolve the questions that were originally posed?

Self-Assessment Test

Questions

1. Is the dew point of a vapor-gas mixture the same variable as the vapor pressure?
2. What variables can be changed, and how, in a vapor-gas mixture to cause the vapor to condense?
3. Can a gas containing a superheated vapor be made to condense?

Problems

1. A mixture of air and benzene contains 10 mol % benzene at 43°C and 105 kPa pressure. At what temperature can the first liquid form? What is the liquid?
2. Two hundred pounds of water out of 1000 lb is electrolytically decomposed into hydrogen and oxygen at 25°C and 740 mm Hg absolute. The hydrogen and oxygen are separated at 740 mm Hg and stored in two different cylinders, each at 25°C and a pressure of 5.0 atm absolute. How much water condenses from the gas in each cylinder?

Thought Problem

Water was drained from the bottom of a gasoline tank into a sewer, and shortly thereafter a flash fire occurred in the sewer. The operator took special care to make sure that none of the gasoline entered the sewer.

What caused the sewer fire?

Discussion Problem

Expired breath is at body temperature, that is, 37°C, and essentially saturated. Condensation of moisture in respiratory equipment from the expired breath occurs as the breath cools. Such a high humidity level has several implications for developing space suits, diving equipment, oxygen masks in hospitals, and so on. What might they be?

8.4.3 Vaporization

At equilibrium you can vaporize a liquid into a noncondensable gas and raise the partial pressure of the vapor in the gas until the saturation pressure (vapor pressure) is reached. Figure 8.13 shows how the partial pressures of water and air change with time as water evaporates into dry air. On a *p-T* diagram such as Figure 8.2 the liquid would vaporize at the saturation temperature C (the bubble point temperature which is equal to the dew point temperature) until the air was saturated.

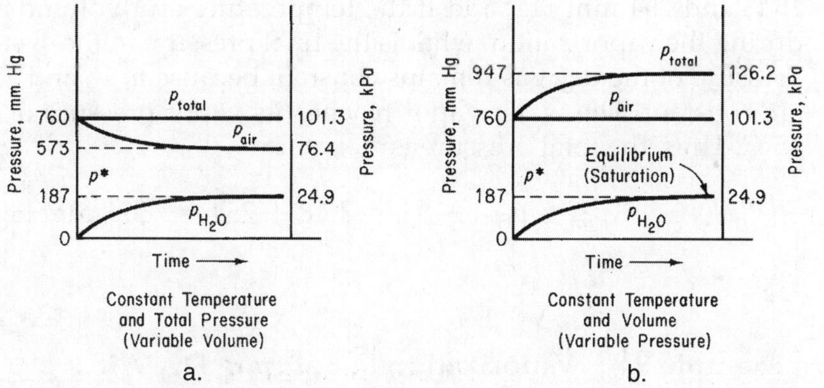

Figure 8.13 Change of partial and total pressure during the vaporization of water into initially dry air (a) at constant temperature and total pressure (variable volume); (b) at constant temperature and volume (variable pressure)

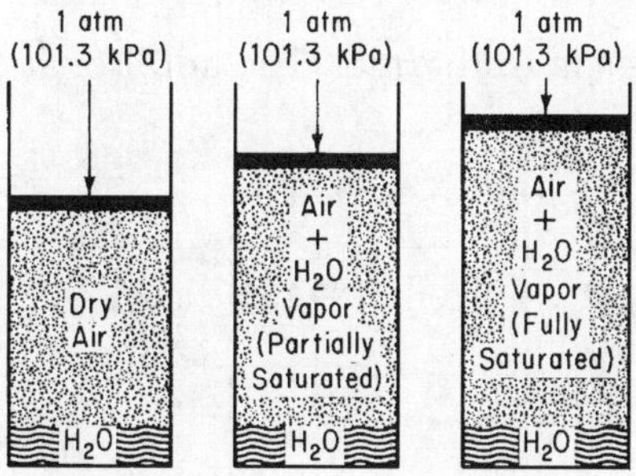

Figure 8.14 Evaporation of water at constant pressure and temperature of 65°C

On the p-$\hat{V}$ diagram (Figure 8.4) evaporation would occur from A to B at constant temperature and pressure until the air was saturated. At constant total pressure, as shown in Figure 8.14, the volume of the air would remain constant, but the volume of water vapor would increase, so the total volume of the mixture would increase.

You might ask: Is it possible to have the water evaporate into air and saturate the air, and yet maintain a constant temperature, pressure, and volume in the cylinder? (Hint: What would happen if you let some of the gas-vapor mixture escape from the cylinder?)

You can use Equations (8.5)–(8.7) to solve vaporization problems. For example, if sufficient liquid water is placed in a volume of dry gas that is at

15°C and 754 mm Hg, and if the temperature and volume remain constant during the vaporization, what is the final pressure in the system? The partial pressure of the dry gas remains constant because *n*, *V*, and *T* for the dry gas are constant. The water vapor reaches its vapor pressure of 12.8 mm Hg at 15°C. Thus, the total pressure is

$$p_{\text{tot}} = p_{H_2O} + p_{\text{air}} = 12.8 + 754 = 766.8 \text{ mm Hg}$$

Example 8.8 Vaporization to Saturate Dry Air

What is the minimum number of cubic meters of dry air at 20°C and 100 kPa that are necessary to evaporate 6.0 kg of ethyl alcohol if the total pressure remains constant at 100 kPa and the temperature remains 20°C? Assume that the air is blown through the alcohol to evaporate it in such a way that the exit pressure of the air-alcohol mixture is at 100 kPa.

Solution

Look at Figure E8.8. The process is isothermal. The additional data needed are

$$p^*_{\text{alcohol}} \text{ at } 20°\text{C} = 5.93 \text{ kPa}$$

$$\text{Mol. wt. ethyl alcohol} = 46.07$$

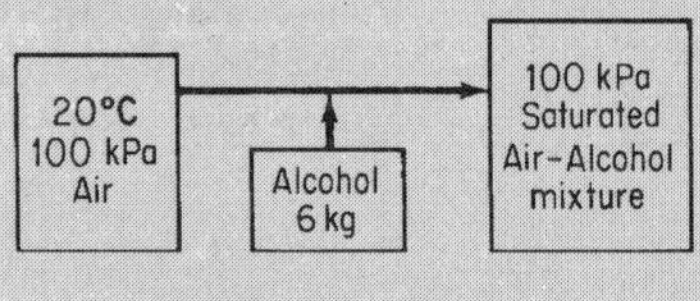

Figure E8.8

The minimum volume of air means that the resulting mixture is saturated; any condition less than saturated would require more air.

Basis: 6.0 kg of alcohol

The ratio of moles of ethyl alcohol to moles of air in the final gaseous mixture is the same as the ratio of the partial pressures of these two substances. Since we know the moles of alcohol, we can find the number of moles of air needed for the vaporization.

$$\frac{p^*_{\text{alcohol}}}{p_{\text{air}}} = \frac{n_{\text{alcohol}}}{n_{\text{air}}}$$

Once you calculate the number of moles of air, you can apply the ideal gas law. Since $p^*_{\text{alcohol}} = 5.93$ kPa,

$$p_{\text{air}} = p_{\text{total}} - p^*_{\text{alcohol}} = (100 - 5.93)\ \text{kPa} = 94.07\ \text{kPa}$$

$$\frac{6.0\ \text{kg alcohol}}{} \left| \frac{1\ \text{kg mol alcohol}}{46.07\ \text{kg alcohol}} \right| \frac{94.07\ \text{kg mol air}}{5.93\ \text{kg mol alcohol}} = 2.07\ \text{kg mol air}$$

$$V_{\text{air}} = \frac{2.07\ \text{kg mol air}}{} \left| \frac{8.314\ (\text{kPa})(\text{m}^3)}{(\text{kg mol})(\text{K})} \right| \frac{293\ \text{K}}{100\ \text{kPa}} = 50.3\ \text{m}^3 \text{ at } 20°\text{C and } 100\ \text{kPa}$$

Sublimation into a noncondensable gas can occur as well as vaporization.

The Chinook: A Wind That Eats Snow

Each year the area around the Bow River Valley in Southwestern Canada experiences temperatures that go down to −40°F. And almost every year, when the wind called the Chinook blows, the temperature climbs as high as 60°F. In just a matter of a few hours, this Canadian area experiences a temperature increase of about 100°F. How does this happen?

Air over the Pacific Ocean is always moist due to the continual evaporation of the ocean. This moist air travels from the Pacific Ocean to the foot of the Rocky Mountains because air masses tend to move from west to east.

As this moist air mass climbs up the western slopes of the Rocky Mountains, it encounters cooler temperatures, and the water vapor condenses out of the air. Rain falls, and the air mass becomes drier as it loses water in the form of rain. As the air becomes drier, it becomes heavier. The heavier dry air falls down the eastern slope of the Rockies, and the atmospheric pressure on the falling gas increases as the gas approaches the ground.

. . . the air gets warmer as you increase the pressure. As the air falls down the side of the mountain, its temperature goes up 5.5°F for every thousand-foot drop. So we have warm, heavy, dry air descending upon the Bow River Valley at the base of the Rockies. The Chinook is this warm air (wind) moving down the Rocky Mountains at 50 mph.

Remember, it is now winter in the Bow River Valley, the ground is covered with snow, and it is quite cold (−40°F). Since the wind is warm, the temperature of the Bow River Valley rises very rapidly. Because the wind is very dry, it absorbs water from the melting snow. The word Chinook is an Indian word meaning "snow eater". The Chinook can eat a foot of snow off the ground overnight. The weather becomes warmer and the snow is cleared away. It's the sort of thing they have fantasies about in Buffalo, New York.

Reprinted with permission. Adapted from an article originally appearing in *Problem Solving in General Chemistry,* 2nd ed., by Ronald DeLorenzo, Wm. C. Brown Publ. (1993), pp. 130–32.

Self-Assessment Test

Questions

1. If a dry gas is isothermally mixed with a liquid in a fixed volume, will the pressure remain constant with time?
2. If dry gas is placed in contact with a liquid phase under conditions of constant pressure and allowed to come to equilibrium:
   a. Will the total pressure increase with time?
   b. Will the volume of the gas plus liquid plus vapor increase with time?
   c. Will the temperature increase with time?

Problems

1. Carbon disulfide (CS_2) at 20°C has a vapor pressure of 352 mm Hg. Dry air is bubbled through the CS_2 at 20°C until 4.45 lb of CS_2 are evaporated. What was the volume of the dry air required to evaporate this CS_2 (assuming that the air becomes saturated) if the air was initially at 20°C and 10 atm and the final pressure on the air—CS_2 vapor mixture is 750 mm Hg?
2. In an acetone recovery system, the acetone is evaporated into dry N_2. The mixture of acetone vapor and nitrogen flows through a 2-ft-diameter duct at 10 ft/s. At a sampling point the pressure is 850 mm Hg and the temperature is 100°F. The dew point is 80°F. Calculate the pounds of acetone per hour passing through the duct.
3. Toluene is used as a diluent in lacquer formulas. Its vapor pressure at 30°C is 36.7 mm Hg absolute. If the barometer falls from 780 mm Hg to 740 mm Hg, will there be any change in the volume of dry air required to evaporate 10 kg of toluene?
4. What is the minimum number of cubic meters of dry air at 21°C and 101 kPa required to evaporate 10 kg of water at 21°C?

Thought Problems

1. To reduce problems of condensation associated with continuous monitoring of stack gases, a special probe and flow controller were developed to dilute the flue gas with outside air in a controlled ratio (such as 10 to 1). Does this seem like a sound idea? What problems might occur with continuous operation of the probe?
2. A large fermentation tank fitted with a 2 in. open vent was sterilized for 30 min by blowing in live steam at 35 psia. After the steam supply was shut off, a cold liquid substrate was quickly added to the tank, at which point the tank collapsed inward. What happened to cause the tank to collapse?

8.5 Two-Component Gas/Two-Component Liquid Systems

In Section 8.3 we discussed vapor-liquid equilibria of a pure component. In Section 8.4 we covered equilibria of a pure component in the presence of a non-condensable gas. In this section we consider certain aspects of a more general set of circumstances, namely, cases in which both the liquid and vapor have two components; that is, the vapor and liquid phases each contain both components. Distilling moonshine from a fermented grain mixture is an example of binary vapor-liquid equilibrium in which water and ethanol are the primary components in the system and are present in both the vapor and the liquid.

The primary result of vapor-liquid equilibrium is that the more volatile component (the component with the larger vapor pressure at a given temperature) tends to accumulate in the vapor phase while the less volatile component tends to accumulate in the liquid phase. Distillation columns, which are used to separate a mixture into its components, are based on this principle. A distillation column is composed of a number of trays that provide contacting between liquid and vapor streams inside the column. At each tray, the concentration of the more volatile component is increased in the vapor stream leaving the tray, and the concentration of the less volatile component is increased in the liquid leaving the tray. In this manner, applying a number of trays in series, the more volatile component is concentrated in the overhead stream from the column while the less volatile component is concentrated in the bottom product. In order to design and analyze distillation, you must be able to quantitatively describe vapor-liquid equilibrium.

8.5.1 Ideal Solution Relations

An **ideal solution** is a mixture whose properties such as vapor pressure, specific volume, and so on can be calculated from the knowledge of only the corresponding properties of the pure components and the composition of the solution. For a solution to behave as an ideal solution:

- All of the molecules of all types should have the same size.
- All of the molecules should have the same intermolecular interactions.

Most solutions are not ideal, but some real solutions are nearly ideal.

Raoult's law The best-known relation for ideal solutions is

$$p_i = x_i p_i^*(T) \tag{8.8}$$

where p_i = partial pressure of component i in the vapor phase
x_i = mole fraction of component i in the liquid phase
$p_i^*(T)$ = vapor pressure of component i at T

Figure 8.15 shows how the vapor pressure of the two components in an ideal binary solution sum to the total pressure at 80°C. Compare Figure 8.15 with Figure 8.16, which displays the pressures for a nonideal solution.

Raoult's law is used primarily for a component whose mole fraction approaches unity (look at methylal in Figure 8.16 as $x_{CS_2} \rightarrow 0.0$) where the

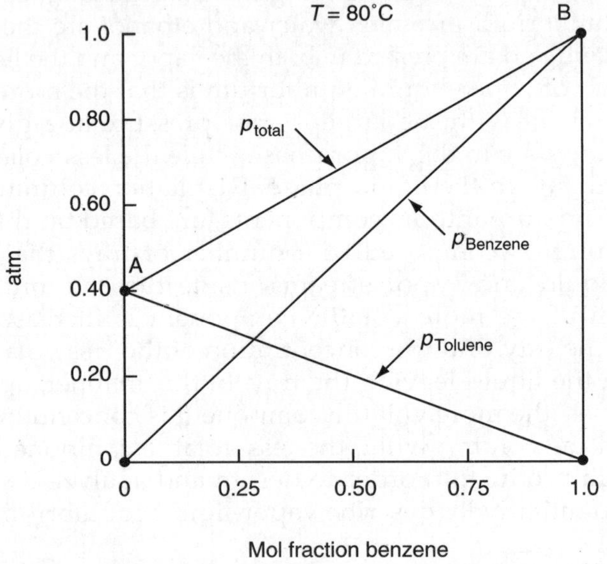

Figure 8.15 Application of Raoult's law to an ideal solution of benzene and toluene (like species) to get the total pressure as a function of composition. The respective vapor pressures are shown by points A and B at 0 and 1.0 mole fraction benzene.

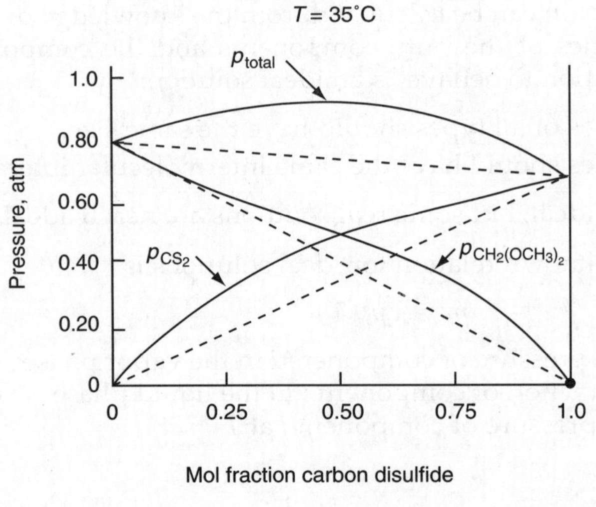

Figure 8.16 Plot of the partial pressures and total pressure (solid lines) exerted by a solution of carbon disulfide (CS_2)–methylal ($CH_2(OCH_3)_2$) as a function of composition. The dashed lines represent the pressures that would exist if the solution were ideal.

data show a linear trend for a short range, or for solutions of components quite similar in chemical nature, such as straight chain hydrocarbons.

Henry's law Henry's law is used primarily for a component whose mole fraction approaches zero, such as a dilute gas dissolved in a liquid:

$$p_i = H_i x_i \tag{8.9}$$

where p_i is the partial pressure in the gas phase of the dilute component at equilibrium at some temperature, and H_i is the *Henry's law constant*. Note that in the limit where $x_i \rightarrow 0$, $p_i \rightarrow 0$. Values of H_i can be found in several handbooks and on the Internet.

Henry's law is quite simple to apply when you want to calculate the partial pressure of a gas that is in equilibrium with the gas dissolved the liquid phase. Take, for example, CO_2 dissolved in water at 40°C for which the value of H is 69,600 atm/mol fraction. (The large value of H shows that $CO_2(g)$ is only sparing soluble in water.) If $x_{CO_2} = 4.2 \times 10^{-6}$, the partial pressure of the CO_2 in the gas phase is

$$p_{CO_2} = 69{,}600(4.2 \times 10^{-6}) = 0.29 \text{ atm}$$

8.5.2 Vapor-Liquid Equilibria Phase Diagrams

Adventurers are easier of entrance than exit; and it is but common prudence to see our way out before we venture in.

Aeson

The phase diagrams discussed in Section 8.2 for a pure component can be extended to cover binary mixtures. Experimental data usually are presented as pressure as a function of composition at a constant temperature, or temperature as a function of composition at a constant pressure. For a pure component, vapor-liquid equilibrium occurs with only one degree of freedom:

$$\mathcal{F} = 2 - \mathcal{P} + \mathcal{C} = 2 - 2 + 1 = 1$$

At 1 atm pressure, vapor-liquid equilibrium will occur at only one temperature—the normal boiling point. However, if you have a binary solution, you have two degrees of freedom:

$$\mathcal{F} = 2 - 2 + 2 = 2$$

For a system at a fixed pressure both the phase compositions and the temperature can be varied over a finite range.

Figures 8.17 and 8.18 show the vapor-liquid envelope for a binary mixture of benzene and toluene, which is essentially ideal.

You can interpret the information on the phase diagrams as follows: Suppose you start in Figure 8.17 at a 50-50 mixture of benzene-toluene at 80ºC and 0.30 atm in the vapor phase. Then you increase the pressure on the system until you reach the dew point at about 0.47 atm, at which point the vapor starts to condense. At 0.62 atm the mole fraction in the vapor phase will be about 0.75 and the mole fraction in the liquid phase will be about 0.38 as indicated by the tie line. As you increase the pressure from 0.70 atm, all of the vapor will have condensed to liquid. What will the composition of the liquid be? 0.50 benzene, of course! Can you carry out an analogous conversion of vapor to liquid on Figure 8.18, the temperature-composition diagram?

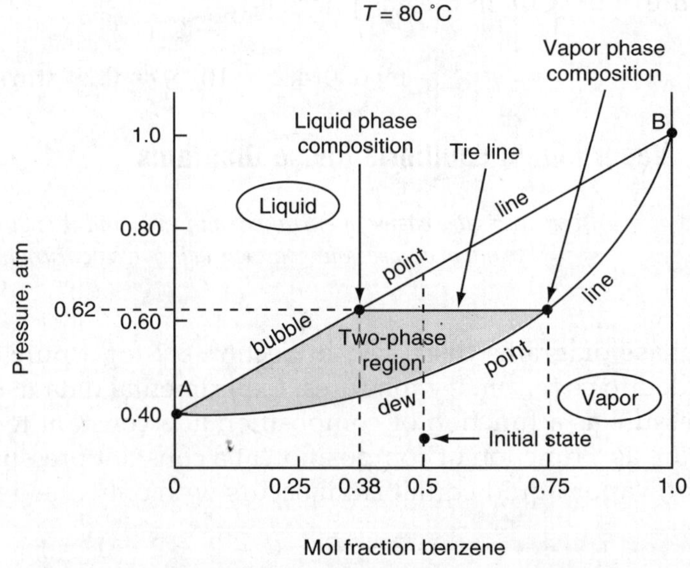

Figure 8.17 Phase diagram for a mixture of benzene and toluene at 80°C. At 0 mol fraction of benzene (point A) the pressure is the vapor pressure of toluene at 80°C. At a mole fraction of benzene of 1 (point B) the pressure is the vapor pressure of benzene at 80°C. The tie line shows the liquid and vapor compositions that are in equilibrium at a pressure of 0.62 atm (and 80°C).

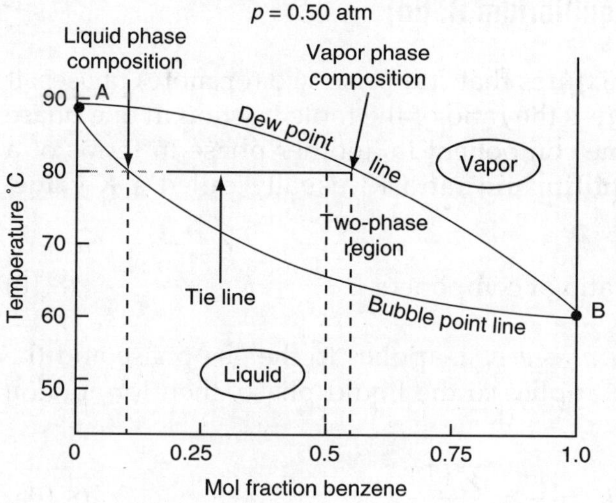

Figure 8.18 Phase diagram for a mixture of benzene and toluene at 0.50 atm. At 0 mole fraction of benzene (point A) the temperature is that when the vapor pressure of toluene is 0.50 atm. At a mole fraction of benzene of 1 (point B) the temperature is that when the vapor pressure of benzene is 0.50 atm. The tie line shows the liquid and vapor compositions that are in equilibrium at a temperature of 80°C (and 0.50 atm).

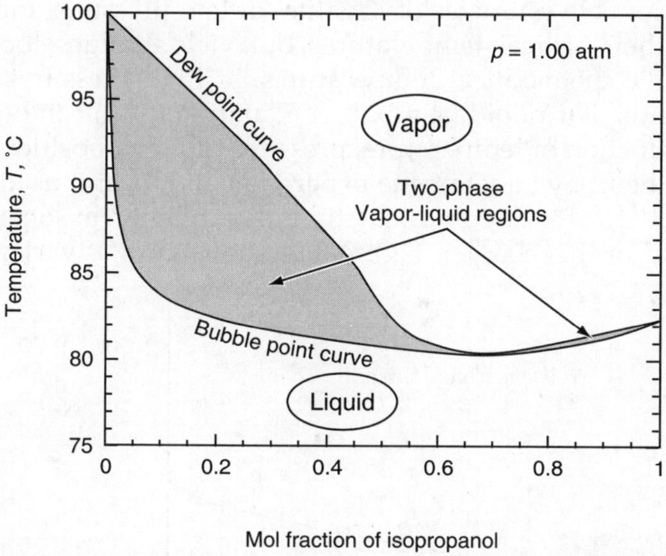

Figure 8.19 Phase diagram for a nonideal mixture of isopropanol and water at 1 atm

Phase diagrams for nonideal solutions abound. Figure 8.19 shows the temperature-composition diagram for isopropanol in water at 1 atm. Note the minimum boiling point at a mole fraction of isopropanol of about 0.68, a point called an **azeotrope** (a point at which on a y_i-versus-x_i plot the function of (y_i/x_i) crosses the function $y_i = x_i$, a straight line). An azeotrope makes separation by distillation difficult.

8.5.3 *K*-value (Vapor-Liquid Equilibrium Ratio)

For nonideal as well as ideal mixtures that comprise two (or more) phases, it proves to be convenient to express the ratio of the mole fraction in one phase to the mole fraction of the same component in another phase in terms of a **distribution coefficient** or **equilibrium ratio** ***K***, usually called a ***K*-value.** For example:

$$\text{Vapor-liquid ratio of component } i\text{:} \quad \frac{y_i}{x_i} = K_i \tag{8.10}$$

and so on. If the ideal gas law $p_i = y_i\, p_{\text{total}}$ applies to the gas phase and the ideal Raoult's law $p_i = x_i p_i^*(T)$ applies to the liquid phase, then for an *ideal* system

$$K_i = \frac{y_i}{x_i} = \frac{p_i^*(T)}{p_{\text{Total}}} \tag{8.10a}$$

Equation (8.10a) gives reasonable estimates of K_i values at low pressures for components well below their critical temperatures but yields values too large for components above their critical temperatures, at high pressures, and/or for polar compounds. For nonideal mixtures Equation (8.10) can be employed if K_i is made a function of temperature, pressure, and composition so that relations for K_i can be fit by equations to experimental data and used directly, or in the form of charts, for design calculations as explained in some of the references at the end of this chapter. Figure 8.20 shows how K varies

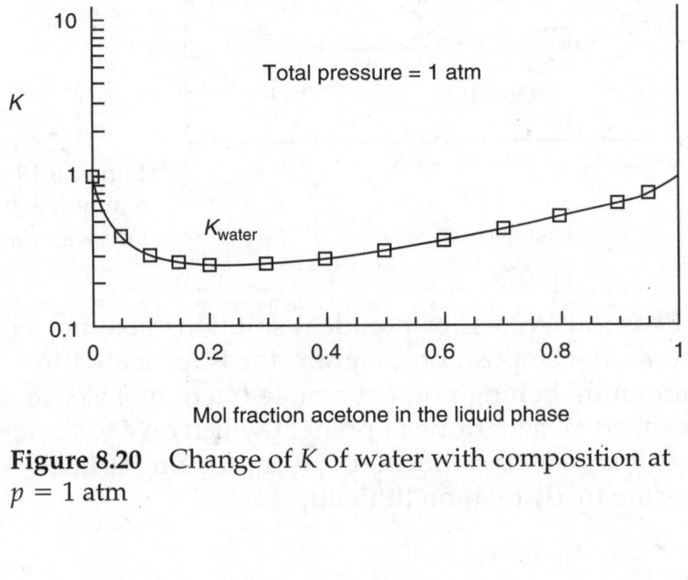

Figure 8.20 Change of K of water with composition at $p = 1$ atm

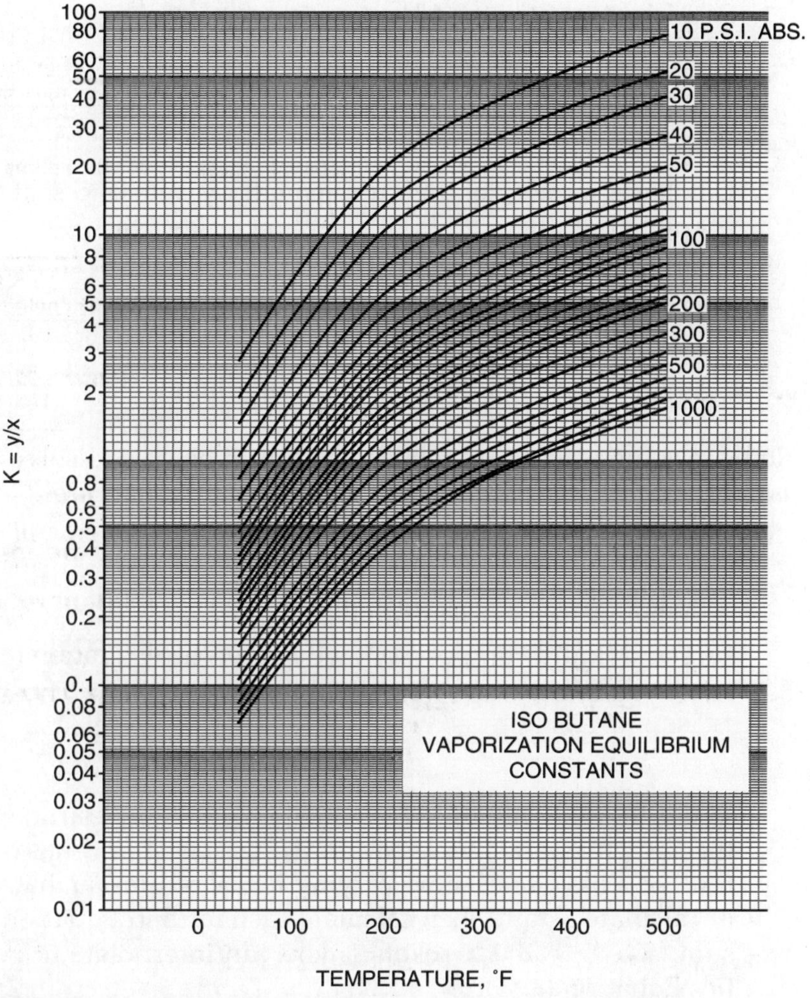

Figure 8.21 *K*-values for isobutane as a function of temperature and pressure. From *Natural Gasoline Association of America Technical Manual*, 4th ed. (1941), with permission (based on data provided by George Granger Brown)

for the nonideal mixture of acetone and water at 1 atm. *K* can be greater or less than 1 but never negative.

For ideal solutions you can calculate values of *K* using Equation (8.10a). For nonideal solutions you can get approximate *K*-values from

1. Empirical equations such as[2]

$$\text{If } T_{c,i}/T > 1.2: \qquad K_i = \frac{(p_{c,i})\exp[7.224 - 7.534/T_{r,i} - 2.598 \ln T_{r,i}]}{p_{\text{total}}}$$

2. Databases—refer to the supplementary references at the end of the chapter
3. Charts such as Figure 8.21
4. Thermodynamic relations—refer to the references at the end of the chapter

8.5.4 Bubble Point and Dew Point Calculations

Here are some typical problems you should be able to solve that involve the use of the equilibrium coefficient K_i and material balances:

1. **Calculate the bubble point temperature of a liquid mixture given the total pressure and liquid composition.**

   To calculate the **bubble point temperature** (given the total pressure and liquid composition), you can write Equation (8.10) as $y_i = K_i x_i$. Also, you know that $\sum y_i = 1$ in the vapor phase. Thus for a binary,

$$1 = K_1 x_1 + K_2 x_2 \tag{8.11}$$

   in which the K_i are functions of solely the temperature. Because each of the K_i increases with temperature, Equation (8.11) has only one positive root. You have to assume varying temperatures so that you can look up or calculate K_i, and then calculate each term in Equation (8.11). After the sum $(K_1x_1 + K_2x_2)$ brackets 1, you can interpolate to get a T that satisfies Equation (8.11).

   For an ideal solution, Equation (8.11) becomes

$$p_{\text{total}} = p_1^* x_1 + p_2^* x_2 \tag{8.12}$$

   and you might use Antoine's equation for p_i^*. Once the bubble point temperature is determined, the vapor composition can be calculated from

$$y_i = \frac{p_i^* x_i}{p_{\text{Total}}}$$

[2]S. I. Sandler, in *Foundations of Computer Aided Design*, Vol. 2, edited by R. H. S. Mah and W. D. Seider, American Institute of Chemical Engineers, New York (1981), p. 83.

A degree-of-freedom analysis for the bubble point temperature for a binary mixture shows that the degrees of freedom are zero:

Total variables $= 2 \times 2 + 2 = 6$ variables: $x_1, x_2; y_1, y_2; p_{\text{Total}}; T$

Prespecified values of variables $= 2 + 1 = 3$ variables: $x_1, x_2; p_{\text{Total}}$

Independent equations $= 2 + 1 = 3$ equations: $y_1 = K_1 x_1, y_2 = K_2 x_2;$ $y_1 + y_2 = 1$

Therefore, there are three unknowns and three equations with which to determine their values.

2. **Calculate the dew point temperature of a vapor mixture given the total pressure and vapor composition.**
3. To calculate the **dew point temperature** (given the total pressure and vapor composition), you can write Equation (8.10) as $x_i = y_i$, and you know $\sum x_i = 1$ in the liquid phase. Consequently, you want to solve the equation

$$1 = \frac{y_1}{K_1} + \frac{y_2}{K_2} \tag{8.13}$$

in which the Ks are functions of temperature as explained for the bubble point temperature calculation. For an ideal solution,

$$1 = p_{\text{total}} \left[\frac{y_1}{p_1^*} + \frac{y_2}{p_2^*} \right] \tag{8.13a}$$

The degree-of-freedom analysis is similar to that for the bubble point temperature calculation.

In selecting a particular form of the equation to be used for your equilibrium calculations, you must select a method of solving the equation that has desirable convergence characteristics. Convergency to the solution should

- Lead to the desired root if the equation has multiple roots
- Be stable, that is, approach the desired root asymptotically rather than by oscillating
- Be rapid, and not become slower as the solution is approached

Polymath, Excel Solver, MATLAB, and Mathcad all have software to solve nonlinear equations.

Table 8.1 summarizes the usual phase equilibrium calculations.

The next example illustrates the details of vapor-liquid equilibrium calculations.

Table 8.1 Summary of the Information Associated with Typical Phase Equilibrium Calculations

Type	Known* Information	Variables to Be Calculated	Equation(s) to Use	Convergence Characteristics
Bubble point temperature	p_{Total}, x_i	T, y_i	8.10	Good
Dew point temperature	p_{Total}, y_i	T, x_i	8.12	Good
Bubble point pressure	T, x_i	p_{Total}, y_i	8.10	Fair
Dew point pressure	T, y_i	p_{Total}, x_i	8.12	Fair

*K_i is assumed to be a known function of T and p_{Total}.

Example 8.9 Bubble Point Calculation

Suppose that a liquid mixture of 4.0% *n*-hexane in *n*-octane is vaporized. What is the composition of the first vapor formed if the total pressure is 1.00 atm?

Solution

Refer back to Figure 8.18 to view a relation of T versus x at constant p such as is involved in this problem. The mixture can be treated as an ideal mixture because the components are quite similar. As an intermediate step, you must calculate the bubble point temperature using Equation (8.12). You have to look up the coefficients of the Antoine equation to obtain the vapor pressures of the two components:

$$\ln(p^*) = A - \frac{B}{C + T}$$

where p^* is in millimeters of Hg and T is in kelvin:

	A	B	C
n-hexane (C_6):	15.8737	2697.55	−48.784
n-octane (C_8):	15.9798	3127.60	−63.633

Basis: 1 kg mol of liquid

You have to solve the following formidable equation to get the bubble point temperature. Use a nonlinear equation solver such as Polymath:

$$760 = \exp\left(15.8737 - \frac{2697.55}{-48.784 + T}\right)0.040 + \exp\left(15.9787 - \frac{3127.60}{-63.633 + T}\right)0.960$$

The solution is $T = 393.3$ K, for which the vapor pressure of a hexane is 3114 mm Hg and the vapor pressure of octane is 661 mm Hg. The mole fractions are

$$y_{C_6} = \frac{p^*_{C_6}}{p_{tot}} x_{C_6} = \frac{3114}{760} = 0.164$$

$$y_{C_8} = 1 - 0.164 = 0.836$$

Self-Assessment Test

Questions

1. When should you use Henry's law and when should you use Raoult's law?
2. As you know, the higher the fuel volatility, the higher the emissions from the fuel. Refiners adjust the butane content of gasoline because it is a high-octane hydrocarbon, which is relatively cheap, that helps cars to start and warm up easily. Similarly, blending ethanol (more expensive than butane) with gasoline raises the volatility of the blend above that of straight gasoline, which makes the emissions problem worse. If you ignore costs, will adding 2 mol % ethanol or butane to octane yield a product that has a higher pressure over the liquid solution?
3. Can you make a plot of the partial pressures and total pressure of a mixture of heptane and octane, given solely that the vapor pressure of heptane is 92 mm Hg and that of octane is 31 mm Hg at a temperature?

Problem

Calculate the boiling point temperature of 1 kg of a solution of 70% ethylene glycol (antifreeze, $C_2H_6O_2$) in water at 1 atm. Assume the solution is ideal.

Thought Problems

1. The fluid in a large tank caught on fire 40 min after the start of a blending operation in which one grade of naphtha was being added to another. The fire was soon put out and the naphtha was moved to another tank. The next day blending was resumed in the second tank; 40 min later another fire started. Can you explain the reason for this sequence of events? What might be done to prevent such accidents?
2. CO_2 can be used to clean optical or semiconductor surfaces and remove particles or organic contaminants. A bottle of CO_2 at 4000 kPa is attached to a jet that sprays onto the optical surface. Two precautions must be taken with this technique. The surface must be heated to about 30°C −35°C to minimize moisture condensation, and you must employ a CO_2 source with no residual heavy

hydrocarbons (lubricants) to minimize recontamination in critical cleaning applications. Describe the physical conditions of the CO_2 as it hits the optical surface. Is it gas, liquid, or solid? How does the decontamination take place?

Discussion Problems

1. Gasoline tanks that have leaked have posed a problem in cleaning up the soil at the leak site. To avoid digging up the soil around the tank, which is located 5–10 m deep, it has been suggested that high-pressure steam be injected underneath the gasoline site via wells to drive the trapped gasoline into a central extraction well which, under vacuum, would extract the gasoline. How might you design an experiment to test the concept of removal? What kinds of soils might be hard to treat? Why do you think steam was proposed for injection rather than water?
2. How to meet increasingly severe federal and state regulations for gasoline, oxygenated fuels, and low-sulfur diesel fuel represents a real challenge. The following table shows some typical values for gasoline components prior to the implementation of the regulations in the state of California, and the limits afterward.

Fuel Parameter	Former (Typical Gasoline)	Current (Limit for Refineries)
Sulfur (ppmw)	150	40
Benzene (vol %)	2	1
Olefins (vol %)	9.9	6
Oxygen (wt %)	0	2.2
Boiling point for 90% of the gasoline (°F)	330	300

Read some of the chemical engineering literature, and prepare a brief report on some of the feasible and economic ways that have been proposed or used to meet the new standards. Will enforcing emission standards on old automobiles (or junking them) be an effective technique of reducing emissions relative to modifying the gasoline? What about control of evaporative emissions from the fuel tank. What about degradation or malfunction of emission controls? And so forth.

8.6 Multicomponent Vapor-Liquid Equilibrium

So-called "white oil" is pressurized and cooled gas well vapors. In Texas awhile back, independent producers operated refrigeration units (to as low as −20°F) at their wells, exploiting a 1977 letter from the legal counsel of the Texas Railroad Commission (the agency that governed oil and gas production) that said "white oil" could be deemed oil rather than gas. There were advantages to having a well classified as an oil well; namely, one oil well could be drilled on 10 acres but a gas well required 640 acres, and at that time "white oil" could be sold for six times the price of gas (which was under price control). In 1984, after years of contention, Judge Clark overruled the Railroad Commission. As a result it issued a new order that said for a liquid to be counted as crude oil it must be liquid in the reservoir, liquid in the well bore, and liquid at the surface. Over $27 billion of gas reserves were involved in this controversy. You can see that determining the true state and composition of petroleum products can be a serious matter.

Multicomponent vapor-liquid equilibrium pertains to systems that contain three or more components in the vapor or liquid phases. Vapor-liquid equilibrium calculations for multicomponent systems are performed in a manner analogous to the ones for binary systems. For bubble point calculations, $\sum_i y_i = 1$; therefore, the bubble point equation can be written in terms of the known liquid compositions (x_i) and the component K-values (K_i):

$$\sum_i^n x_i K_i = 1$$

For dew point calculations, $\sum_i x_i = 1$; therefore, the dew point equation can be written in terms of the known vapor composition (y_i) and the component K-values (K_i):

$$\sum_i^n \frac{y_i}{K_i} = 1$$

Note that both **the bubble point and dew point equations are, in general, nonlinear equations with temperature as the only unknown.**

To calculate the bubble point or the dew point using these equations, you will need the K-values for each component in the system. Bubble point and dew point calculations are used by commercial process simulators to model distillation columns and other separation processes. Equations of state are used by commercial process simulators to calculate component K-values for bubble and dew point calculations. For example, the SRK method is routinely used to calculate the K-values for nonpolar systems (e.g., hydrocarbons) and the UNIQUAC equation of state is used to calculate the K-values for a wide range of polar systems (i.e., systems with strong intermolecular interactions, such as hydrogen bonding).

Looking Back

In this chapter we introduced phase diagrams and the phase rule and showed the connection to pure component properties, such as vapor pressure. Pure component vapor pressures are used to represent the behavior of vapor-liquid systems with a noncondensable gas, allowing one to describe saturation conditions, condensation, and vaporization. Finally, binary vapor-liquid equilibrium relations were considered.

Glossary

Absolute pressure Pressure relative to a complete vacuum.

Antoine equation Equation that relates vapor pressure to absolute temperature.

Azeotrope Minimum boiling point of a liquid composed of two or more components at a constant pressure.

Boiling Change from liquid to vapor.

Bubble point The temperature at which liquid changes to vapor (at some pressure).

Condensation The change of phase from vapor to liquid.

Degrees of superheat The difference in temperature between the actual T and the saturated T at a given pressure.

Dew point The temperature at which the vapor just begins to condense at a specified pressure, that is, the value of the temperature along the vapor-pressure curve.

Equilibrium A state of the system in which there is no tendency to spontaneous change.

Evaporation The change of phase of a substance from liquid to vapor.

Freezing The change of phase of a substance from liquid to solid.

Fusion See **melting**.

Gibbs' phase rule A relation that gives the degrees of freedom for intensive variables in a system in terms of the number of phases and number of components.

Henry's law A relation between the partial pressure of a gas in the gas phase and the mole fraction of the gas in the liquid phase at equilibrium.

Ideal solution A system whose properties, such as vapor pressure, specific volume, and so on, can be calculated from the knowledge only of the corresponding properties of the components and the composition of the solution.

Invariant A system in which no variation of conditions is possible without one phase disappearing.

***K*-value** A parameter (distribution coefficient) used to express the ratio of the mole fraction in one phase to the mole fraction of the same component in another phase.

Melting The change of phase from solid to liquid of a substance.

Noncondensable gas A gas at conditions under which it cannot condense to a liquid or solid.

Normal boiling point The temperature at which the vapor pressure of a substance (p^*) is 1 atm (101.3 kPa).

Normal melting point The temperature at which a solid melts at 1 atm (101.3 kPa).

Phase diagram Representation of the different phases of a compound on a two- (or three-) dimensional graph.

Quality Fraction or percent of the liquid-vapor mixture that is vapor (wet vapor.)

Raoult's law A relation that relates the partial pressure of one component in the vapor phase to the mole fraction of the same component in the liquid phase.

Reference substance The substance used as the reference in a reference substance plot.

Reference substance plot A plot of a property of one substance versus the same property of another (reference) substance that results in an approximate straight line.

Saturated liquid Liquid that is in equilibrium with its vapor.

Saturated vapor Vapor that is in equilibrium with its liquid.

Steam tables Tabulations of the properties of water and steam (water vapor).

Subcooled liquid Liquid at values of temperature and pressure less than those that exist at saturation.

Sublimation Change of phase of a solid directly to a vapor.

Sublimation pressure The pressure given by the melting curve (a function of temperature).

Supercritical region A portion of a physical properties plot in which the substance is at combined p-T values above the critical point.

Superheated vapor Vapor at values of temperature and pressure exceeding those that exist at saturation.

Triple point The one p-T-V combination at which solid, liquid, and vapor are in equilibrium.

Two-phase (region) Region on a plot of physical properties where both the liquid and vapor exist simultaneously.

Vapor A gas below its critical point in a system in which the vapor can condense.

Vaporization The change of a substance from liquid to vapor.

Vapor-liquid equilibria Graphs showing the concentration of a component in a vapor-liquid system as a function of temperature and/or pressure.

Supplementary References

American National Standards, Inc. *ASTM D323-79 Vapor Pressure of Petroleum Products (Reid Method)*, Philadelphia (1979).

Bhatt, B. I., and S. M. Vora. *Stoichiometry (SI Units)*, Tata McGraw-Hill, New Delhi (1998).

Henley, E. J., and J. D. Seader. *Equilibrium—Stage Separation Operations in Chemical Engineering*, Wiley, New York (1981).

Horvath, A. L. *Conversion Tables in Science and Engineering*, Elsevier, New York (1986).

Jensen, W. B. "Generalizing the Phase Rule," *J. Chem. Educ.*, **78**, 1369–70 (2001).

Perry, R. H., et al. *Perry's Chemical Engineers' Handbook*, McGraw-Hill, New York (2000).

Rao, Y. K. "Extended Form of the Gibbs Phase Rule," *Chem. Engr. Educ.*, 40–49 (Winter, 1985).

Yaws, C. L. *Handbook of Vapor Pressure* (4 volumes), Gulf Publishing Co., Houston, TX (1993–1995).

Web Sites

www.youtube.com/watch?v=gbUTffUsXOM
www.et.byu.edu/~rowley/VLEfinal/VLE_home.htm
http://lorien.ncl.ac.uk/ming/webnotes/sp3/bubdew/bub.htm

Problems

8.2 Phase Diagrams and the Phase Rule

*8.2.1 Select the correct answer(s) in the following statements:

a. In a container of 1.00 L of toluene, the vapor pressure of the toluene is 103 mm Hg. The same vapor pressure will be observed in a container of (1) 2.00 L of toluene at the same temperature; (2) 1.00 L of toluene at one-half the absolute temperature; (3) 1.00 L of alcohol at the same temperature; (4) 2.00 L of alcohol at the same temperature.
b. The temperature at which a compound melts is the same temperature at which it (1) sublimes; (2) freezes; (3) condenses; (4) evaporates.
c. At what pressure would a liquid boil first? (1) 1 atm; (2) 2 atm; (3) 200 mm Hg; (4) 101.3 kPa.
d. When the vapor pressure of a liquid reaches the pressure of the atmosphere surrounding it, it will (1) freeze; (2) condense; (3) melt; (4) boil.
e. Sublimation is the phase change from (1) the solid phase to the liquid phase; (2) the liquid phase to the solid phase; (3) the solid phase to the gas phase; (4) the gas phase to the solid phase.
f. A liquid that evaporates rapidly at ambient conditions is more likely than not to have a (1) high vapor pressure; (2) low vapor pressure; (3) high boiling point; (4) strong attraction among the molecules.

*8.2.2 Draw a *p-T* diagram for a pure component. Label the curves and points that are listed in Figure 8.3 on it.

*8.2.3 A vessel contains liquid ethanol, ethanol vapor, and N_2 gas at equilibrium. How many phases, components, and degrees of freedom are there according to the phase rule?

*8.2.4 What is the number of degrees of freedom according to the phase rule for each of the following systems: (a) solid iodine in equilibrium with its vapor; (b) a mixture of liquid water and liquid octane (which is immiscible in water), both in equilibrium with their vapors?

*8.2.5 Liquid water in equilibrium with water vapor is a system with how many degrees of freedom?

*8.2.6 Liquid water in equilibrium with moist air is a system with how many degrees of freedom?

*8.2.7 You have a closed vessel that contains $NH_4Cl(s)$, $NH_3(g)$, and $HCl(g)$ in equilibrium. How many degrees of freedom exist in the system?

*8.2.8 In the decomposition of $CaCO_3$ in a sealed container from which the air was initially pumped out, you generate CO_2 and CaO. If not all of the $CaCO_3$ decomposes at equilibrium, how many degrees of freedom exist for the system according to the Gibbs phase rule?

**8.2.9 Based on the phase diagrams in Figure P8.2.9, answer the questions and explain your answers.

a. What is the approximate normal melting point for compound A?
b. What is the approximate normal boiling point for compound A?
c. What is the approximate triple point temperature for compound B?
d. Which compounds sublime at atmospheric pressure?

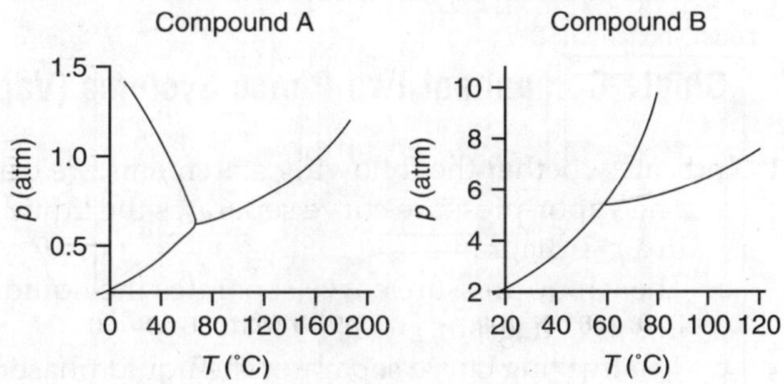

Figure P8.2.9

**8.2.10 One form of cooking is to place the food in a pressure cooker (a sealed pot). Pressure cookers decrease the time require to cook the food.

Some explanations of how a pressure cooker works are as follows. Which of the explanations are correct?

a. We know $p_1T_1 = p_2T_2$. So if the pressure is doubled, the temperature should be doubled and result in quicker cooking.
b. Pressure cookers are based on the principle that $p \propto T$, that is, pressure is directly proportional to temperature. With the volume kept constant, if you increase the pressure, the temperature also increases, and it takes less time to cook.
c. Food cooks faster because the pressure is high. This means that there are more impacts of molecules per surface area, which in turn increases the temperature of the food.
d. If we increase the pressure under which food is cooked, we have more collisions of hot vapor with the food, cooking it faster. On an open stove, vapor escapes into the surroundings, without affecting the food more than once.
e. As the pressure inside the sealed cooker builds, as a result of the vaporization of water, the boiling point of water is increased, thereby increasing the temperature at which the food cooks—hotter temperature, less time.

****8.2.11** A mixture of water, acetic acid, and ethyl alcohol is placed in a sealed container at 40°C at equilibrium. How many degrees of freedom exist according to the phase rule for this system? List a specific variable for each degree of freedom.

****8.2.12** a. A system contains two components at equilibrium. What is the maximum number of phases possible with this system? Give reasons for your answer.
b. A two-phase system is specified by fixing the temperature, the pressure, and the amount of one component. How many components are there in the system at equilibrium? Explain.

8.3 Single-Component Two-Phase Systems (Vapor Pressure)

***8.3.1** Indicate whether the following statements are true or false:
a. The vapor-pressure curve separates the liquid phase from the vapor phase in a p-T diagram.
b. The vapor-pressure curve separates the liquid phase from the vapor phase in a p-V diagram.
c. The freezing curve separates the liquid phase from the solid phase in a p-T diagram.
d. The freezing curve separates the liquid phase from the solid phase in a p-V diagram.
e. At equilibrium at the triple point, liquid and solid coexist.
f. At equilibrium at the triple point, solid and vapor coexist.

***8.3.2** Explain how the pressure for a pure component changes (higher, lower, no change) for the following scenarios:

a. A system containing saturated liquid is compressed at constant temperature.
b. A system containing saturated liquid is expanded at constant temperature.
c. A system containing saturated liquid is heated at constant volume.
d. A system containing saturated liquid is cooled at constant volume.
e. A system containing saturated vapor is compressed at constant temperature.
f. A system containing saturated vapor is expanded at constant temperature.
g. A system containing saturated vapor is heated at constant volume.
h. A system containing saturated vapor is cooled at constant volume.
i. A system containing vapor and liquid in equilibrium is heated at constant volume.
j. A system containing vapor and liquid in equilibrium is cooled at constant volume.
k. A system containing a superheated gas is expanded at constant temperature.
l. A system containing superheated gas is compressed at constant temperature.

***8.3.3** Ice skates function because a lubricating film of liquid forms immediately below the small contact area of the skate blade. Explain by means of diagrams and words why this liquid film appears on ice at 25°F.

***8.3.4** Methanol has been proposed as an alternate fuel for automobile engines. Proponents point out that methanol can be made from many feedstocks such as natural gas, coal, biomass, and garbage, and that it emits 45% less ozone precursor gases than does gasoline. Critics say that methanol combustion emits toxic formaldehyde and that methanol rapidly corrodes automotive parts. Moreover, engines using methanol are hard to start at temperatures below 40°F. Why are engines hard to start? What would you recommend to ameliorate the situation?

***8.3.5** Calculate the vapor pressure of each compound listed below at the designated temperature using the Antoine equation and the coefficients in Appendix H. Compare your results with the corresponding values of the vapor pressures obtained from the Antoine equation found in the physical properties package on the CD accompanying this book.

a. Acetone at 0°C
b. Benzene at 80°F
c. Carbon tetrachloride at 300 K

****8.3.6** Estimate the vapor pressure of ethyl ether at 40°C using the Antoine equation based on the experimental values as follows:

p^* (kPa):	2.53	15.0	58.9
T (°C):	−40.0	−10.0	20.0

****8.3.7** At the triple point, the vapor pressures of liquid and solid ammonia are respectively given by $\ln p^* = 15.16 - 3063/T$ and $\ln p^* = 18.70 - 3754/T$, where p is in atmospheres and T is in kelvin. What is the temperature at the triple point?

****8.3.8** In a handbook the vapor pressure of solid decaborane ($B_{10}H_{14}$) is given as

$$\log_{10} p^* = 8.3647 - \frac{2642}{T}$$

and of liquid $B_{10}H_{14}$ as

$$\log_{10} p^* = 10.3822 - \frac{3392}{T}$$

The handbook also shows that the melting point of $B_{10}H_{14}$ is 89.8°C. Can this be correct?

****8.3.9** Calculate the normal boiling point of benzene and of toluene using the Antoine equation. Compare your results with listed data in a handbook or database.

****8.3.10** Numerous methods are employed to evaporate metals in thin film deposition. The rate of evaporation is

$$W = 5.83 \times 10^{-2} \frac{p_v M^{1/2}}{T^{1/2}} \text{g}/(\text{cm}^2)(\text{s}) \quad (p_v \text{ in torr, } T \text{ in K, } M = \text{molecular weight})$$

Since p_v is also temperature-dependent, it is necessary to define further the vapor pressure–temperature relationship for this rate equation. The vapor-pressure model is

$$\log_{10} p_v = A - \frac{B}{T}$$

where T is in kelvin.

Calculate the temperature needed for an aluminum evaporation rate of 10^{-4} g/(cm^2)(s). Data: $A = 8.79$, $B = 1.594 \times 10^4$.

****8.3.11** Calculate the specific volume for water that exists at the following conditions:

a. $T = 100°\text{C}$, $p = 101.4$ kPa, $x = 0.5$ (in cubic meters per kilogram)
b. $T = 406.70$ K, $p = 300.0$ kPa, $x = 0.5$ (in cubic meters per kilogram)
c. $T = 100.0°\text{F}$, $p = 0.9487$ psia, $x = 0.3$ (in cubic feet per pound)
d. $T = 860.97°\text{R}$, $p = 250$ psia, $x = 0.7$ (in cubic feet per pound)

****8.3.12** Indicate whether the following statements are true or false:

a. A pot full of boiling water is tightly closed by a heavy lid. The water will stop boiling.
b. Steam quality is the same thing as steam purity.
c. Liquid water that is in equilibrium with its vapor is saturated.
d. Water can exist in more than three different phases.
e. Superheated steam at 300°C means steam at 300 degrees above the boiling point.
f. Water can be made to boil without heating it.

**8.3.13 A vessel that has a volume of 0.35 m^3 contains 2 kg of a mixture of liquid water and water vapor at equilibrium with a pressure of 450 kPa. What is the quality of the water vapor?

**8.3.14 A vessel with an unknown volume is filled with 10 kg of water at 90°C. Inspection of the vessel at equilibrium shows that 8 kg of the water is in the liquid state. What is the pressure in the vessel, and what is the volume of the vessel?

**8.3.15 What is the velocity in feet per second when 25,000 lb/hr of superheated steam at 800 psia and 900°F flow through a pipe of inner diameter 2.9 in.?

**8.3.16 Maintenance of a heater was carried out to remove water that had condensed in the bottom of the heater. By accident hot oil at 150°C was released into the heater when the maintenance man opened the wrong valve. The resulting explosion caused serious damage both to the maintenance man and to the equipment he was working on. Explain what happened during the incident when you write up the accident report.

***8.3.17 In a vessel with a volume of 3.00 m^3 you put 0.030 m^3 of liquid water and 2.97 m^3 of water vapor so that the pressure is 101.33 kPa. Then you heat the system until all of the liquid water just evaporates. What are the temperature and pressure in the vessel at that time?

***8.3.18 In a vessel with a volume of 10.0 ft^3 you put a mixture of 2.01 lb of liquid water and water vapor. When equilibrium is reached, the pressure in the vessel is measured as 80 psia. Calculate the quality of the water vapor in the vessel, and the respective masses and volumes of liquid and vapor at 80 psia.

***8.3.19 Take 10 data points from the steam tables for the vapor pressure of water as a function of temperature from the freezing point to 500 K, and fit the following function:

$$p^* = \exp[a + b \ln T + c(\ln T)^2 + d(\ln T)^3]$$

where p is in kilopascals and T is in kelvin.

***8.3.20 For each of the conditions of temperature and pressure listed below for water, state whether the water is a solid phase, liquid phase, superheated, or a saturated mixture, and if the latter, indicate how you would calculate the quality. Use the steam tables (inside the back cover) to assist in the calculations.

State	p (kPa)	T (K)	$\hat{V}$(m^3/kg)
1	2000	475	—
2	1000	500	0.2206
3	101.3	200	—
4	245.6	400	0.7308
5	1000	453.06	0.001127
6	200	393.38	0.8857

***8.3.21 Repeat problem 8.3.20 for the following conditions:

State	p (psia)	T (°F)	$\hat{V}$ (ft^3/lb)
1	0.3388	68	927.0
2	1661.6	610	0.0241
3	308.82	420	0.4012
4	180.0	440	2.812

***8.3.22 Prepare a Cox chart for (a) acetone vapor, (b) heptane, (c) ammonia, (d) ethane from 0°C to the critical point (for each substance). Compare the estimated vapor pressure at the critical point with the critical pressure.

***8.3.23 Estimate the vapor pressure of benzene at 125°C from the following vapor-pressure data:

T (°F):	102.6	212
p^* (psia):	3.36	25.5

by preparing a Cox chart.

***8.3.24 Estimate the vapor pressure of aniline at 350°C based on the following vapor-pressure data:

T (°C):	184.4	212.8	254.8	292.7
p^* (atm):	1.00	2.00	5.00	10.00

***8.3.25 Exposure in the industrial workplace to a chemical can come about by inhalation and skin adsorption. Because skin is a protective barrier for many chemicals, exposure by inhalation is of primary concern. The vapor pressure of a compound is one commonly used measure of exposure in the workplace. Compare the relative vapor pressures of three compounds added to gasoline—methanol, ethanol, and MTBE (methyl tert-butyl ether)—with their respective OSHA permissible exposure limits (PEL) that are specified in parts per million (by volume):

Methanol	200
Ethanol	1000
MTBE	100

8.4 Two-Component Gas/Single-Component Liquid Systems

*8.4.1 Suppose that you place in a volume of dry gas that is in a flexible container a quantity of liquid and allow the system to come to equilibrium at constant temperature and total pressure. Will the volume of the container increase, decrease, or stay the same from the initial conditions? Suppose that the container is of a fixed instead of flexible volume, and the temperature is held constant as the liquid vaporizes. Will the pressure increase, decrease, or remain the same in the container?

**8.4.2 A large chamber contains dry N_2 at 27°C and 101.3 kPa. Water is injected into the chamber. After saturation of the N_2 with water vapor, the temperature in the chamber is 27°C.

a. What is the pressure inside the chamber after saturation?
b. How many moles of H_2O per mole of N_2 are present in the saturated mixture?

**8.4.3 The vapor pressure of hexane (C_6H_{14}) at −20°C is 14.1 mm Hg absolute. Dry air at this temperature is saturated with the vapor under a total pressure of 760 mm Hg. What is the percent excess air for combustion?

**8.4.4 In a search for new fumigants, chloropicrin (CCl_3NO_2) has been proposed. To be effective, the concentration of chloropicrin vapor must be 2.0% in air. The easiest way to get this concentration is to saturate air with chloropicrin from a container of liquid.

Assume that the pressure on the container is 100 kPa. What temperature should be used to achieve the 2.0% concentration? From a handbook, the vapor-pressure data are (T,°C; vapor pressure, mm Hg): 0, 5.7; 10, 10.4; 15, 13.8; 20, 18.3; 25, 23.8; 30, 31.1.

At this temperature and pressure, how many kilograms of chloropicrin are needed to saturate 100 m^3 of air?

**8.4.5 What is the dew point of a mixture of air and water vapor at 60°C and 1 atm in which the mole fraction of the air is 12%? The total pressure on the mixture is constant.

**8.4.6 Hazards can arise if you do not calculate the pressure in a vessel correctly. One gallon of a hazardous liquid that has a vapor pressure of 13 psia at 80°F is transferred to a tank containing 10 ft^3 of air at 10 psig and 80°F. The pressure seal on the tank containing air will rupture at 30 psia. When the transfer takes place, will you have to worry about the seal rupturing?

**8.4.7 A room contains 12,000 ft^3 of air at 75°F and 29.7 in. Hg absolute. The air has a dew point of 60°F. How many pounds of water vapor are in the air?

**8.4.8 One gallon of benzene (C_6H_6) vaporizes in a room that is 20 ft by 20 ft by 9 ft in size at a constant barometric pressure of 750 mm Hg absolute and 70°F. The lower explosive limit for benzene in air is 1.4%. Has this value been exceeded?

**8.4.9 A mixture of acetylene (C_2H_2) with an excess of oxygen measured 350 ft^3 at 25°C and 745 mm Hg absolute pressure. After explosion the volume of the dry gaseous product was 300 ft^3 at 60°C and the same pressure. Calculate the volume of acetylene and of oxygen in the original mixture. The final gas was saturated. Assume that all of the water resulting from the reaction was in the gas phase after the reaction.

**8.4.10 In a science question-and-answer column, the following question was posed: On a trip to see the elephant seals in California, we noticed that when the male elephant seals were bellowing, you could see their breath. But we couldn't see our own breath. How come?

**8.4.11 One way that safety enters into specifications is to specify the composition of a vapor in air that could burn if ignited. If the range of concentration of benzene in air in which ignition could take place is 1.4% to 8.0%, what would be the corresponding temperatures for air saturated with benzene in the vapor space of a storage tank? The total pressure in the vapor space is 100 kPa.

**8.4.12 When you fill your gas tank or any closed vessel, the air in the tank rapidly becomes saturated with the vapor of the liquid entering the tank. Consequently, as air leaves the tank and is replaced by liquid, you can often smell the fumes of the liquid around the filling vent such as with gasoline.

Suppose that you are filling a closed 5 gal can with benzene at 75°F. After the air is saturated, what will be the moles of benzene per mole of air expelled from the can? Will this value exceed the OSHA limit for benzene in air (currently 0.1 mg/cm^3)? Should you fill a can in your garage with the door shut in the winter?

**8.4.13 All of the water is to be removed from moist air (a process called dehydration) by passing it through silica gel. If 50 ft^3/min of air at 29.92 in. Hg absolute with a dew point of 50°F are dehydrated, calculate the pounds of water removed per hour.

***8.4.14 In a dry-cleaning establishment warm dry air is blown through a revolving drum in which clothes are tumbled until all of the Stoddard solvent is removed. The solvent may be assumed to be *n*-octane (C_8H_{18}) and have a vapor pressure of 2.36 in. Hg at 120°F. If the air at 120°F becomes saturated with octane, calculate (a) the pounds of air required to evaporate 1 lb of octane, (b) the percent octane by volume in the gases leaving the drum, and (c) the cubic feet of inlet air required per pound of octane. The barometer reads 29.66 in. Hg.

***8.4.15 When people are exposed to certain chemicals at relatively low but toxic concentrations, the toxic effects are experienced only after prolonged exposures. Mercury is such a chemical. Chronic exposure to low concentrations of mercury can cause permanent mental deterioration, anorexia, instability, insomnia, pain and numbness in the hands and feet, and several other symptoms. The level of mercury that can cause these symptoms can be present in the atmosphere without a worker being aware of it because such low concentrations of mercury in the air cannot be seen or smelled.

Federal standards based on the toxicity of various chemicals have been set by OSHA for PEL, the maximum level of exposure permitted in the workplace based on a time-weighted average (TWA) exposure. The TWA exposure is the average concentration permitted for exposure day after day without causing adverse effects. It is based on exposure for 8 hr/day for the worker's lifetime.

The present federal standard (OSHA/PEL) for exposure to mercury in air is 0.1 mg/m^3 as a ceiling value. Workers must be protected from concentrations greater than 0.1 mg/m^3 if they are working in areas where mercury is being used.

Mercury manometers are filled and calibrated in a small storeroom that has no ventilation. Mercury has been spilled in the storeroom and is not completely cleaned up because the mercury runs into cracks in the floor covering. What is the maximum mercury concentration that can be reached in the storeroom if the temperature is 20°C? You may assume that the room has no ventilation and that the equilibrium concentration will be reached. Is this level acceptable for worker exposure?

Data: $p^*_{Hg} = 1.729 \times 10^{-4}$ kPa; the barometer reads 99.5 kPa. This problem has been adapted from the problems in the publication *Safety, Health, and Loss Prevention in Chemical Processes* published by the American Institute of Chemical Engineers, New York (1990) with permission.

*****8.4.16** Figure P8.4.16 shows a typical *n*-butane loading facility. To prevent explosions either (a) additional butane must be added to the intake lines (a case not shown) to raise the concentration of butane above the upper explosive limit (UEL) of 8.5% butane in air, or (b) air must be added (as shown in the figure) to keep the butane concentration below the lower explosive limit (LEL) of 1.9%. The *n*-butane gas leaving the water seal is at a concentration of 1.5%, and the exit gas is saturated with water (at 20°C). The pressure of the gas leaving the water seal is 120.0 kPa. How many cubic meters of air per minute at 20.0°C and 100.0 kPa must be drawn through the system by the burner if the joint leakage from a single tank car and two trucks is 300 cm^3/min at 20.0°C and 100.0 kPa?

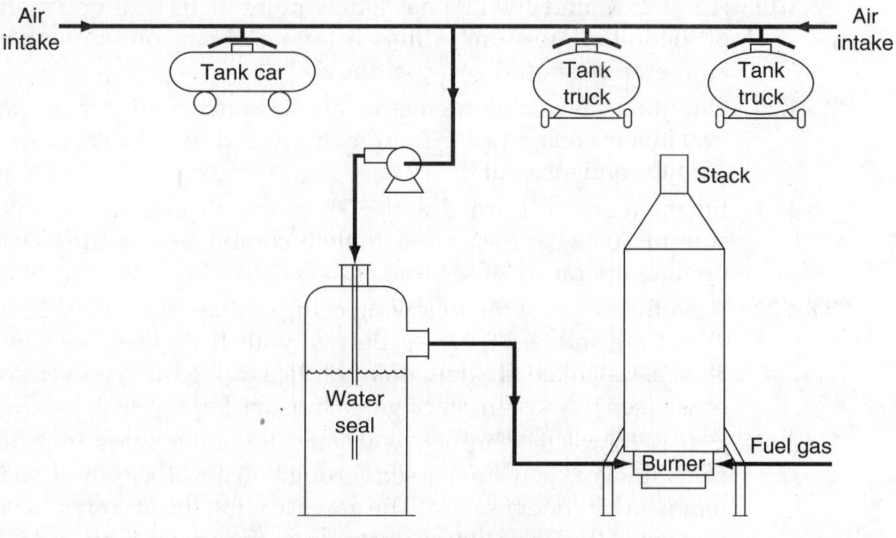

Figure P8.4.16

*****8.4.17** Sludge containing mercury is burned in an incinerator. The mercury concentration in the sludge is 0.023%. The resulting gas (MW = 32) is 40,000 lb/hr, at 500°F, and is quenched with water to bring it to a temperature of 150°F. The resulting stream is filtered to remove all particulates. What happens to the mercury? Assume the process pressure is 14.7 psia. (The vapor pressure of Hg at 150°F is 0.005 psia.)

*****8.4.18** To prevent excessive ice formation on the cooling coils in a refrigerator room, moist air is partially dehydrated and cooled before passing it through the room (see Figure P8.4.18). The moist air from the cooler is passed into the refrigerator room at the rate of 20,000 ft^3/24 hr measured at the entrance temperature and pressure. At the end of 30 days the refrigerator room must be allowed to warm in order to remove the ice from the coils. How many pounds of water are removed from the refrigerator room when the ice on the coils in it melts?

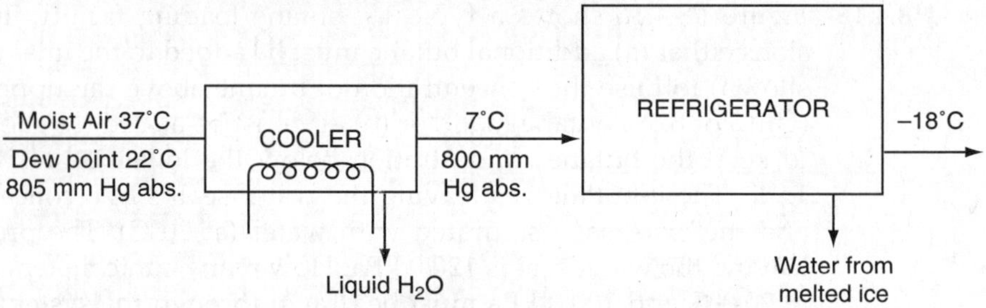

Figure P8.4.18

*****8.4.19** Air at 25°C and 100 kPa has a dew point of 16°C. If you want to remove 50% of the initial moisture in the air (at a constant pressure of 100 kPa), to what temperature should you cool the air?

*****8.4.20** One thousand cubic meters of air saturated with water vapor at 30°C and 99.0 kPa is cooled to 14°C and compressed to 133 kPa. How many kilograms of H_2O condense out?

*****8.4.21** Ethane (C_2H_6) is burned with 20% excess air in a furnace operating at a pressure of 100 kPa. Assume complete combustion occurs. Determine the dew point temperature of the flue gas.

*****8.4.22** A synthesis gas of the following composition—4.5% CO_2, 26.0% CO, 13.0% H_2, 0.5% CH_4, and 56.0% N_2—is burned with 10% excess air. The barometer reads 98 kPa. Calculate the dew point of the stack gas. To prevent condensation and consequent corrosion, stack gases must be kept well above their dew point.

*****8.4.23** CH_4 is completely burned with air. The outlet gases from the burner, which contain no oxygen, are passed through an absorber where some of the water is removed by condensation. The gases leaving the absorber have a nitrogen mole fraction of 0.8335. If the exit gases from the absorber are at 130°F and 20 psia:

a. To what temperature must this gas be cooled at constant pressure in order to start condensing more water?

b. To what pressure must this gas be compressed at constant temperature before more condensation will occur?

***8.4.24 3M removes benzene from synthetic resin base sandpaper by passing it through a dryer where the benzene is evaporated into hot air. The air comes out saturated with benzene at 40°C (104°F). p^* of benzene at 40°C = 181 mm Hg; the barometer = 742 mm Hg. They recover the benzene by cooling to 10°C (p^* = 45.4 mm Hg) and compressing to 25 psig. What fraction of the benzene do they recover? The pressure is then reduced to 2 psig, and the air is recycled in the dryer. What is the partial pressure of the benzene in the recycled air?

***8.4.25 Wet solids containing 40% moisture by weight are dried to 10% moisture content by weight by passing moist air over them at 200°F, 800 mm Hg pressure. The partial pressure of water vapor in the entering air is 10 mm Hg. The exit air has a dew point of 140°F.

How many cubic feet of moist air at 200°F and 800 mm Hg must be used per 100 lb of wet solids entering?

***8.4.26 Aerobic growth (growth in the presence of air) of a biomass involves the uptake of oxygen and the generation of carbon dioxide. The ratio of the moles of carbon dioxide produced per mole of oxygen consumed is called the respiratory quotient (RQ). Calculate the RQ for yeast cells suspended in the liquid in a well-mixed steady-state bioreactor based on the following data:

a. Volume occupied by the liquid: 600 m^3
b. Air (dry) flow rate into the gas (head) space: 600 m^3/hr at 120 kPa and 300 K
c. Composition of the entering air: 21.0% O_2 and 0.055% CO_2
d. Pressure inside the bioreactor: 120 kPa
e. Temperature inside the bioreactor: 300 K
f. Exit gas: saturated with water vapor and contains 8.04% O_2 and 12.5% CO_2
g. Exit gas pressure: 110 kPa
h. Exit gas temperature: 300 K

****8.4.27 Coal as fired contains 2.5% moisture. On a dry basis the coal analysis is C 80%, H 6%, O 8%, ash 6%. The flue gas analyzes CO_2 14.0%, CO 0.4%, O_2 5.6%, N_2 80.0%. The air used has a dew point of 50°F. The barometer is 29.90 in. Hg. Calculate the dew point of the stack gas.

8.5 Two-Component Gas/Two-Component Liquid Systems

*8.5.1 Indicate whether the following statements are true or false:

a. The critical temperature and pressure are the highest temperature and pressure at which a binary mixture of vapor and liquid can exist at equilibrium.
b. Raoult's law is best used for a solute in dilute solutions.
c. Henry's law is best used for a solute in concentrated solutions.
d. A mixture of liquid butane and pentane can be treated as an ideal solution.
e. The liquid phase region is found above the vapor phase region on a p-x-y chart.

f. The liquid phase region is found below the vapor phase region on a T-x-y chart.

*8.5.2 Examine the following statements:

a. "The vapor pressure of gasoline is about 14 psia at 130°F."

b. "The vapor pressure of the system, water–furfural diacetate, is 760 mm Hg at 99.96°C."

Are the statements correct? If not, correct them. Assume the numerical values are correct.

**8.5.3 Determine if Henry's law applies to H_2S in H_2O based on the following measurements at 30°C:

Liquid Mole Fraction × 10^2	Pressure (kPa)
0.0003599	20
0.0004498	30
0.0005397	40
0.0008273	50
0.0008992	60
0.001348	90
0.001528	100
0.003194	200
0.004712	300
0.007858	500
0.01095	700
0.01376	900
0.01507	1000

**8.5.4 Determine the equilibrium concentration in milligrams per liter of chloroform in water at 20°C and 1 atm, assuming that gas and liquid phases are ideal and the mole fraction of the chloroform in the gas phase is 0.024. The Henry constant for chloroform is $H = 170$ atm/mol fraction.

**8.5.5 Water in an enclosed vessel at 17°C contains a concentration of dissolved oxygen of 6.0 mg/L. At equilibrium, determine the concentration of oxygen in the space above the water in mole fraction and the total pressure in kilopascals. Henry's law constant is 4.02×10^6 kPa/mol fraction.

**8.5.6 A tank contains a liquid composed of 60 mol % toluene and 40 mol % benzene in equilibrium with the vapor phase and air at 1 atm and 60°F.

a. What is the concentration of hydrocarbons in the vapor phase?

b. If the lower flammability limit for toluene in air is 1.27% and benzene is 1.4%, is the vapor phase flammable?

**8.5.7 Fuel tanks for barbecues contain propane and n-butane. At 120°F, in an essentially full tank of liquid that contains liquid and vapor in equilibrium and exhibits a pressure of 100 psia, what is the overall (vapor plus liquid) mole fraction of butane in the tank?

**8.5.8 Based on the following vapor-pressure data, construct the temperature-composition diagram at 1 atm for the system benzene-toluene, assuming ideal solution behavior.

Temperature (°C)	Vapor Pressure (mm Hg) Benzene	Toluene
80	760	300
92	1078	432
100	1344	559
110.4	1748	760

**8.5.9 Sketch a *T-x-y* diagram that shows an azeotrope, and locate and label the bubble and dew lines and the azeotrope point.

**8.5.10 Methanol has a flash point at 12°C at which temperature its vapor pressure is 62 mm Hg. What is the flash point (temperature) of a mixture of 75% methanol and 25% water? Hint: The water does not burn.

**8.5.11 You are asked to determine the maximum pressure at which steam distillation of naphtha can be carried out at 180°F (the maximum allowable temperature). Steam is injected into the liquid naphtha to vaporize it. If the distillation is carried out at 160°F, the liquid naphtha contains 7.8% (by weight) nonvolatile impurities, and the initial charge to the distillation equipment is 1000 lb of water and 5000 lb of impure naphtha, how much water will be left in the still when the last drop of naphtha is vaporized? Data: For naphtha the MW is about 107, and $p^*(180°F) = 460$ mm Hg, $p^*(160°F) = 318$ mm Hg.

***8.5.12 You are asked to remove 90% of the sulfur dioxide in a gas stream of air and sulfur dioxide that flows at the rate of 85 m^3/min and contains 3% sulfur dioxide. The sulfur dioxide is to be removed by a stream of water. The entering water contains no sulfur dioxide. The temperature is 290 K and the pressure on the process is 1 atm. Find (a) the kilograms of water per minute needed to remove the sulfur dioxide, assuming that the exit water is in equilibrium with the entering gas; and (b) the ratio of the water stream to the gas stream. The Henry's law constant for sulfur dioxide at 290 K is 43 atm/mol fraction.

***8.5.13 What are (a) the pressure in the vapor phase and (b) the composition of the vapor phase in equilibrium with a liquid mixture of 20% pentane and 80% heptane at 50°F? Assume the mixture is an ideal one at equilibrium.

***8.5.14 Two kilograms of a mixture of 50-50 benzene and toluene is at 60°C. As the total pressure on the system is reduced, at what pressure will boiling commence? What will be the composition of the first bubble of liquid?

***8.5.15 The normal boiling point of propane is −42.1°C and the normal boiling point of *n*-butane is −0.5°C.

a. Calculate the mole fraction of the propane in a liquid mixture that boils at −31.2°C and 1 atm.
b. Calculate the corresponding mole fraction of the propane in the vapor at −31.2°C.
c. Plot the temperature versus propane mole fraction for the system of propane and butane.

***8.5.16 In the system *n*-heptane–*n*-octane at 200°F, determine the partial pressure of each component in the vapor phase at liquid mole fractions of *n*-heptane of 0, 0.2, 0.4, 0.6, 0.8, and 1.0. Also calculate the total pressure above each solution.

Plot your results on a *P-x* diagram with mole fractions of C_7 increasing to the right and mole fractions of C_8 increasing to the left. The ordinate should be pressure in psia.

Read from the plotted graph the total pressure and the partial pressure of each component at a mole fraction of $C_7 = 0.47$ in the liquid.

***8.5.17 Calculate the bubble point of a liquid mixture of 80 mol % *n*-hexane and 20 mol % *n*-pentane at 200 psia.

***8.5.18 Calculate the dew point of a vapor mixture of 80 mol % *n*-hexane and 20 mol % *n*-pentane at 100 psia.

***8.5.19 A mixture of 50 mol % benzene and 50 mol % toluene is contained in a cylinder at 39.36 in. Hg absolute. Calculate the temperature range in which a two-phase system can exist.

***8.5.20 A liquid mixture of *n*-pentane and *n*-hexane containing 40 mol % *n*-pentane is fed continuously to a flash separator operating at 250°F and 80 psia. Determine (a) the quantity of vapor and liquid obtained from the separator per mole of feed, and (b) the composition of both the vapor and the liquid leaving the separator.

***8.5.21 Most combustible reactions occur in the gas phase. For any flammable material to burn, both fuel and oxidizer must be present, and a minimum concentration of the flammable gas or vapor in the gas phase must also exist. The minimum concentration at which ignition will occur is called the lower flammable limit (LFL). The liquid temperature at which the vapor concentration reaches the LFL can be found experimentally.

It is usually measured using a standard method called a "closed cup flash point" test. The "flash point" of a liquid fuel is thus the liquid temperature at which the concentration of fuel vapor in air is large enough for a flame to flash across the surface of the fuel if an ignition source is present.

The flash point and the LFL concentration are closely related through the vapor pressure of the liquid. Thus, if the flash point is known, the LFL concentration can be estimated, and if the LFL concentration is known, the flash point can be estimated.

Estimate the flash point (the temperature) of liquid *n*-decane that contains 5.0 mol % pentane. The LFL for pentane is 1.8% and that for *n*-decane is

0.8%. Assume the propane–*n*-decane mixture is an ideal liquid. Assume the ambient pressure is 100 kPa. This problem has been adapted from *Safety, Health, and Loss Prevention in Chemical Processes*, edited by J. R. Welker and C. Springer, American Institute of Chemical Engineers, New York (1990), with permission.

***8.5.22 Late in the evening of August 21, 1986, a large volume of toxic gas was released from beneath and within Lake Nyos in the Northwest Province of Cameroon. An aerosol of water mixed with toxic gases swept down the valleys to the north of Lake Nyos, leaving more than 1700 dead and dying people in its wake. The lake had a surface area of 1.48 km^2 and a depth of 200–250 m. It took 4 days to refill the lake; hence it was estimated to have lost about 200,000 tons of water during the gas emission. To the south of the lake and in the small cove immediately to the east of the spillway a wave rose to a height of about 25 m.

The conclusion of investigators studying this incident was that the waters of Lake Nyos were saturated with CO_2 of volcanic origin. Late in the evening of August 21 a pulse of volcanic gas—mainly CO_2 but containing some H_2S—was released above a volcanic vent in the northeast corner of the lake. The stream of bubbles rising to the surface brought up more bottom waters highly charged with CO_2 that gushed out, increasing the gas flow and hence the flow of water to the surface much as a warm soda bottle overflows on release of pressure. At the surface, the release of gas transformed the accompanying water into a fine mist and sent a wave of water crashing across the lake. The aerosol of water and CO_2 mixed with a trace of H_2S swept down the valleys to the north of the lake, leaving a terrible toll of injury and death in its wake.

If the solution at the bottom of the lake obeyed Henry's law, how much CO_2 was released with the 200,000 metric tons of water, and what would be the volume of the CO_2 at S.C. in cubic meters? At 25°C the Henry's law constant is 1.7×10^3 atm/mol fraction.

***8.5.23 If the pressure in the head space (gas space) in a bioreactor is 110 kPa and 25°C, and the oxygen concentration in the head space is enriched to 39.7%, what is the mole fraction of the dissolved oxygen in the liquid phase? What is the percent excess oxygen dissolved in the liquid phase compared with the saturation value that could be obtained from air alone dissolved in the liquid?

****8.5.24 One hundred moles per minute of a binary mixture of 50% A and 50% B are separated in a two-stage (serial) process. In the first stage, the liquid and vapor flow rates exiting from the stage are each 50 mol/min. The liquid stream is then passed into a second separator that operates at the same temperature as the first stage, and the respective exit streams of liquid and vapor from the second stage are each 25 mol/min. The temperature is the same for each stage, and at that temperature, the vapor pressure of A is 10 kPa and the vapor pressure of B is 100 kPa. Treat the liquids and vapors as ideal.

Calculate the compositions of all of the streams in the process, and calculate the pressure in each stage.

8.6 Multicomponent Vapor-Liquid Equilibrium

**8.6.1 Three separate waste discharge streams from a plant into a river contain the following respective chemicals in the water:

	Concentration (g/100 g water)	K
Glycerol	5.5	1.20×10^{-7}
Methyl ethyl ketone (MEK)	1.1	3.065
Phenol	2.1	0.00485

The *K*-values are from the Aspen tech process simulator at 20°C.

Estimate the concentration of the respective compounds in the gas phase above each discharge stream at 20°C. Will volatilization from the discharge stream be significant?

PART IV

ENERGY

CHAPTER 9

Energy Balances

Your objectives in studying this chapter are to be able to

1. Define or explain the following terms: energy, system, closed system, nonflow system, open system, flow system, surroundings, property, extensive property, intensive property, state, heat, work, kinetic energy, potential energy, internal energy, enthalpy, initial state, final state, state variable, cyclical process, path function, heat capacity
2. Select a system suitable for solving a problem, either closed or open, steady- or unsteady-state, and fix the system boundary
3. Convert energy in one set of units to another set
4. Quickly locate the source of property values from tables, charts, equations, and computer databases
5. Understand each term in the general energy balance
6. Simplify the general energy balance for the specifics of a particular problem
7. Apply the general energy balance to open and closed systems, and to steady-state and unsteady-state systems

Looking Forward

This chapter presents general energy balances as well as the nomenclature to support them. Finally, the general energy balance is applied to various systems that do not involve chemical reactions.

How often have you see a headline similar to

Energy Crunch Worsens

No question exists as to the increase in the long-run use of energy. Figure 9.1 shows the history and forecast of U.S. energy demand and supply to the

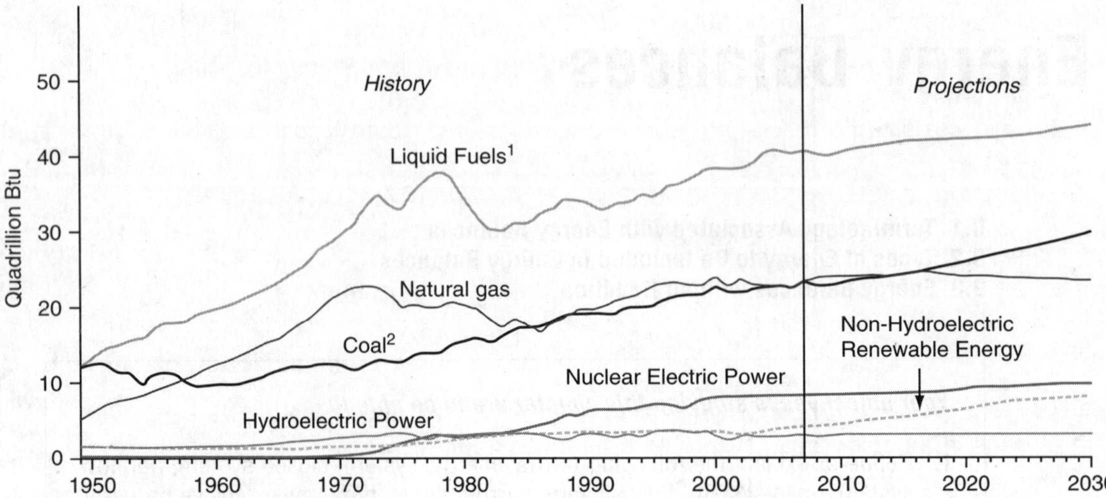

Figure 9.1 Past and predicted energy consumption by category (in quadrillion British thermal units)

[1]History: petroleum-derived fuels. Projections: petroleum-derived fuels and non-petroleum-derived fuels, such as fuel ethanol, biodiesel, and coal-based synthetic liquids.

[2]Includes net imports of coal coke.

Source: Energy Information Administration, *Annual Energy Review 2007*, DOE/EIA 0384 (2007), Washington, DC (July, 2007).

year 2020. Figure 9.2 shows the energy consumption and sources in the United States in 2007 in quadrillion British thermal units.

The answers to questions such as

- How can energy costs be reduced?
- How can "clean" energy be provided economically?
- Is thermal pollution inherently necessary?
- What is the most economical source of energy?
- What can be done with "waste" heat?

rely on valid databases and extensive analysis using the principles presented in this book.

In this chapter we discuss energy balances together with the accessory background information needed to understand and apply them correctly. Our main attention will be devoted to heat, work, enthalpy, internal energy, and carrying out energy balances in the absence of chemical reaction.

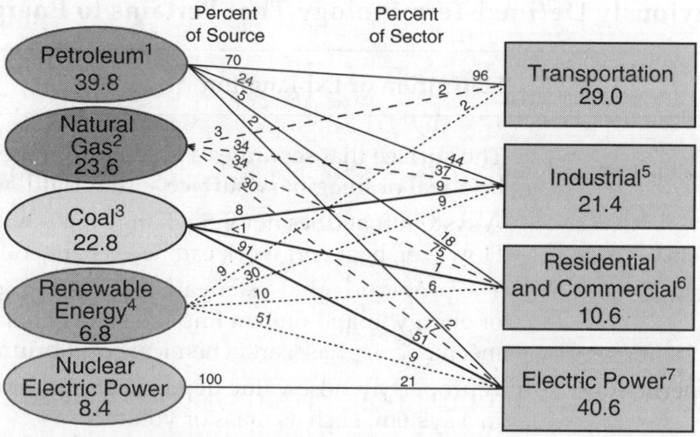

Figure 9.2 U.S. primary energy consumption by source and sector, 2007

[1]Excludes 0.6 quadrillion Btu of ethanol, which is included in "Renewable Energy."
[2]Excludes supplemental gaseous fuels.
[3]Includes 0.1 quadrillion Btu of coal coke net imports.
[4]Conventional hydroelectric power, geothermal, solar/PV, wind, and biomass.
[5]Includes industrial combined-heat-and-power (CHP) and commercial electricity-only plants.
[6]Includes commercial CHP and commercial electricity-only plants.
[7]Electricity-only and CHP plants whose primary business is to sell electricity, or electricity and heat, to the public.

Note: Sum of components may not equal 100% due to independent rounding.
Source: Energy Information Administration, *Annual Energy Review 2007*, DOE/EIA—08841 (2007), Washington, DC (2007), Tables 1.3 and 2.1b-2.1f, and 10.3.

9.1 Terminology Associated with Energy Balances

"When I use a word," Humpty Dumpty said in a rather scornful tone, "it means just what I choose it to mean, neither more nor less."
"The question is," said Alice, "whether you can make words mean so many different things."
"The question is," said Humpty Dumpty, "which is to be master—that's all."
Lewis Carroll, *Through the Looking-Glass, and What Alice Found There*

Some of the difficulty in analyzing processes involving energy balances occurs because of the failure of our language to communicate an exact meaning.

Table 9.1 Previously Defined Terminology That Pertains to Energy Balances

Term	Definition or Explanation
Boundary	The surface that separates a system from the surroundings. It may be a real or imaginary surface, either rigid or movable.
Closed system (nonflow system)	A system that does not interchange mass with the surroundings. However, heat and work can be exchanged.
Equilibrium (state)	The properties of a system are invariant in spite of flows of material or energy in and out; an implied state of balance. Types are thermal, mechanical, phase, and chemical equilibrium.
Extensive property	A property whose value depends on the amount of material present in a system, such as mass or volume.
Intensive property	A property whose value is independent of the amount of material present in a system, such as temperature or density (inverse of specific volume).
Open system (flow system)	A system that is open to interchange of mass with the surroundings. Heat and work can also be exchanged.
Phase	A part (or whole) of a system that is physically distinct and macroscopically homogeneous of fixed or variable composition, such as gas, liquid, or solid.
Property	Observable (or calculable) characteristic of a system such as pressure, temperature, volume, etc.
State	Conditions of a system (specified by the values of temperature, pressure, composition, etc.).
Steady-state	For this book, the accumulation in a system is zero. More generally, the flows in and out are constant, and the properties of the system are invariant.
Surroundings	Everything outside the system boundary.
System	The quantity of matter or region of space chosen for study enclosed by a boundary.
Unsteady-state (transient state)	The system is not in the steady state.

Many difficulties will disappear if you take care to learn the meaning of the terms listed in Tables 9.1 and 9.2. Table 9.1 lists terminology covered in previous chapters, and Table 9.2 lists new terminology that arises in connection with energy balances.

Now for some comments about the new terminology listed in Table 9.2. The terms **adiabatic**, **isothermal**, **isobaric**, and **isochoric** are useful to specify conditions that do not change in a process. Isothermal, isobaric, and isochoric are short ways of saying no temperature change, no pressure change,

Table 9.2 Additional Terminology That Pertains to Energy Balances

Term	Definition or Explanation
Adiabatic system	A system that does not exchange heat with the surroundings during a process (i.e., perfectly insulated).
Isobaric system	A system in which the pressure is constant during a process.
Isochoric system	A system in which the volume is invariant during a process.
Isothermal system	A system in which the temperature is invariant during a process.
Path variable (function)	Any variable (function) whose value depends on how the process takes place and can differ for different histories (e.g., heat and work).
State variable (point function) (state function)	Any variable (function) whose value depends only on the state of the system and not upon its previous history (such as internal energy).

and no volume change, respectively, occurs in the system; that is, the properties are invariant. Adiabatic means no heat transfer occurs between the system and its surroundings across the system boundary. Under what circumstances might a process be adiabatic?

- The system is insulated.
- Q is very small in relation to the other terms in the energy equation and may be neglected.
- The process takes place so fast that there is no time for heat to be transferred.

The concept of a **state (or point) function** or **variable** is an important concept to understand. Temperature, pressure, and all of the other intensive variables are known as state variables because between two different states, any change in value is the same no matter what path is taken between the two states. Look at Figure 9.3, which illustrates two processes, A and B. Both start at state 1 and terminate at state 2. The change in the value of a state variable is the same by both processes.

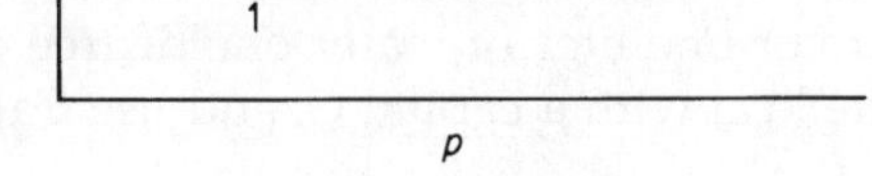

Figure 9.3 The value for the change in a state variable is the same for path A as for path B, or any other route between 1 and 2.

If two systems are in the same state, all of their state variables, such as temperature or pressure, must be identical. If the state of a system is changed, say, by heating, so that energy flows in, the values of its state variables change, and if the system is returned to its original state, say, by cooling, so that energy flows out, the values of its state variables return to their original values.

A process that proceeds first at constant pressure and then at constant temperature from state 1 to state 2 will yield exactly the same final value for a state variable as one that takes place first at constant temperature and then at constant pressure as long as the end point is the same. The concept of the state or point function is the same as that of an airplane passenger who plans to go straight to New York from Chicago but is detoured by way of Cincinnati because of bad weather. When the passenger arrives in New York, he or she is the same distance (the state variable) from Chicago whichever way the plane goes; hence, the value of the state variable depends only on the initial and final states. However, the fuel consumption of the plane may vary considerably; in analogous fashion, heat and work are two **path functions** or **variables** that are involved in an energy balance and may vary depending on the specific path chosen. If the passenger were to return to Chicago from New York, the distance from Chicago would be zero at the end of the return trip. Thus, the change in a state variable is zero for a cyclical process, which goes from state 1 to state 2 and back to state 1 again.

Let us now mention the units associated with energy. As you know, in the SI system the unit of energy is the joule (J). In the AE system we use Btu, (ft)(lbf), and (kW)(hr) among others. You can find conversion factors among the energy units listed on the inside of the front cover of this book. What about the calorie? Is the unit archaic? It seems not. Most people are concerned about calories, not British thermal units or joules. A food package gives you information about calories (look at Figure 9.4).

Dr. Lawrence Lamb gave a succinct answer to a letter to the newspaper asking if 500 calories of energy is the same as 500 kcal. He said:

> Dear Reader: I usually say calories for people like you, but I do cringe each time because it is not correct.
>
> A calorie is only enough energy to raise the temperature of one mL of water one degree centigrade.
>
> But the word calorie for your food is misused. What you think are calories are actually kilocalories (kcal). One kilocalorie (kcal) equals 1,000 real (sic thermochemical) calories, or is the amount of energy required to raise the temperature of one liter of water one degree centigrade. You are supposed to write kcal with a capital C, and use Calories instead of kilocalories.

Nutrition Facts
Serving Size 2 cookies (26 g)
Servings Per Container 8

Amount Per Serving	
Calories 110	Calories from Fat 35
	% Daily Value*
Total Fat 4g	6%
Saturated Fat 3g	15%
Cholesterol 0mg	0%
Sodium 50mg	2%
Total Carbohydrate 18g	6%
Dietary Fiber 0g	0%
Sugars 12g	
Protein 1g	

Vitamin A 0%	•	Vitamin C 0%
Calcium 0%	•	Iron 2%

*Percent Daily Values are based on 2,000 calorie diet. Your daily values may be higher or lower depending on your calorie needs:

	Calories:	2,000	2,500
Total Fat	Less than	65g	80g
Sat Fat	Less than	20g	25g
Cholesterol	Less than	300mg	300mg
Sodium	Less than	2400mg	2400mg
Total Carbohydrate		300g	375g
Dietary Fiber		25g	30g

Figure 9.4 Information about the nutritional characteristics of a cookie

Thus, if your diet consists of 2000 Calories per day, you can calculate the number of joules involved per hour:

$$\frac{2000\ \text{Calories}}{\text{day}}\left|\frac{1\ \text{kcal}}{\text{Calorie}}\right|\frac{1000\ \text{cal}}{\text{kcal}}\left|\frac{4.184\ \text{J}}{\text{cal}}\right|\frac{1\ \text{day}}{24\ \text{hr}} = 350{,}000\ \text{J/hr}$$

Your body converts the food you eat into this amount of heat or work every hour, neglecting any energy stored as fat.

Self-Assessment Test

Questions

1. What is the essential difference between the system and the surroundings? Between an open and a closed system? Between a property and a phase?
2. Can a variable be both an intensive and an extensive variable at the same time?
3. Describe the difference between a state variable and a path variable.

Problems

1. What is the value of the change in the specific volume of a gas in a closed container that is first compressed to 100 atm, then heated to increase the temperature by 20%, and finally returned to its original state?
2. If you eat food containing 1800 Calories per day, according to an advertisement you will lose weight. A handbook says that a person uses 20,000 kJ per day, given

normal waking and sleeping activities. Will the person lose weight if he or she eats food as suggested by the advertisement?

Thought Problems

1. A proposed goal to reduce air pollution from automobiles is to introduce into U.S. domestic gasoline a specified fraction of oxygenated compounds from renewable resources, one of which is ethanol grown from corn. What is your estimate of the fraction of the available U.S. cropland that would be required to replace 10% of the gasoline with alcohol in all of the annual gasoline production of about 1.2×10^{10} gal/yr? Assume 90.0 bu/acre of corn and 2.6 gal ethanol/bu.
2. Another proposal is to supply 10% of the U.S. oil usage by coal liquefaction. What is your estimate of the percentage of the coal now mined in the United States that would have to be processed in order to realize this proposal? Assume 3.26 bbl of liquid per ton of coal.

Discussion Question

Consider the following sources of energy that can be used to generate electric power:

Biomass (direct combustion)	Oil shale
Coal	Peat
Ethanol from biomass	Solar thermal
Geothermal	Solar voltaic
Hydropower	Tar sands
Methane from biomass	Vegetable oils
Natural gas	Waves
Ocean thermal	Wind
Oil	

Use reference books and the Internet to estimate the cost in dollars per kilowatt-hour of each source. Briefly discuss the potential for future usage.

9.2 Types of Energy to Be Included in Energy Balances

In this section we comment on several of the terms to be used in the energy balance—heat, work, kinetic energy, potential energy, internal energy, and enthalpy—many of which you have encountered before.

They have invented a term "energy" and the term has been enormously fruitful because it also creates a law by eliminating exceptions because it gives names to things which differ in matter but are similar in form.

H. Poincaré

Energy itself is often defined as the capacity to do work or transfer heat, a fuzzy concept. It is easier to understand specific types of energy. Two things energy is *not* are (a) some sort of invisible fluid and (b) something that can be measured directly.

9.2.1 Heat (Q)

Heat, Q, when used in the general energy balance, Equation (9.17), as a single term, is *the net amount of heat* ***transferred*** *to or from the system over a fixed time interval.* A process may involve more than one specified form of heat transfer, of course, the sum of which is Q. The rate of transfer will be designated by an overlay dot on Q thus: $\dot{Q}$, with the units of heat transfer per unit time, and the net heat transfer per unit mass would be designated by an overlay caret thus: $\hat{Q}$.

In a discussion of *heat* we enter an area in which our everyday use of the term may cause confusion, because we are going to use heat in a very restricted sense when we apply the laws governing energy changes. **Heat** (Q) is commonly defined as that part of the total energy flow across a system boundary that is caused by a temperature difference (potential) between the system and the surroundings (or between two systems). See Figure 9.5. Engineers say "heat" when meaning "heat transfer" or "heat flow." Because heat is based on the transfer of energy, heat cannot be stored. **Heat is positive when transferred to the system and negative when removed from the system.** Heat is a **path variable**.

Heat can't flow from a cooler to a hotter, you can try if you like but you'd far better notter.
Michael Flanders and Donald Swann,
"At the Drop of Another Hat," Angel Records

Keep in mind that a process in which no heat transfer occurs is an **adiabatic** process ($Q = 0$).

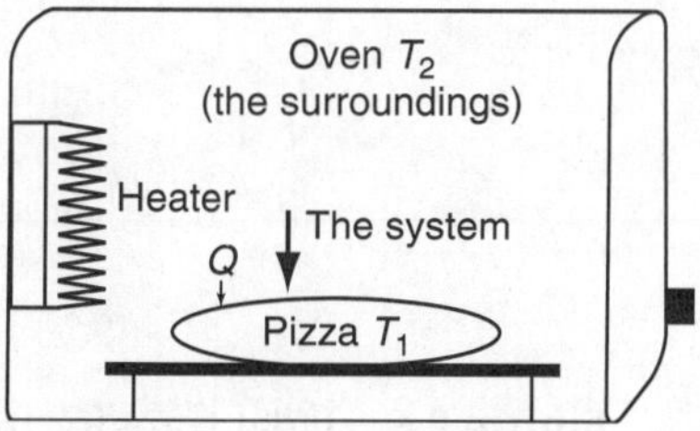

Figure 9.5 Heat transfer is energy that crosses the system boundary because of a temperature difference.

Here are some misconceptions about heat that you should avoid (saying *heat transfer* helps):

- Heat is a substance.
- Heat is proportional to temperature.
- A cold body contains no heat.
- Heating always results in an increase in temperature.
- Heat only travels upward.

Heat transfer is usually classified in three categories: conduction, convection, and radiation. To evaluate heat transfer *quantitatively*, you can apply various empirical formulas to estimate the heat transfer rate. One example of such a formula is the rate of heat transfer by convection that can be calculated from

$$\dot{Q} = U^*A(T_2 - T_1) \tag{9.1}$$

where $\dot{Q}$ is the rate of heat transfer (such as joules per second), A is the area for heat transfer (such as square meters), $(T_2 - T_1)$ is the temperature difference between the surroundings at T_2 and the system at T_1 (such as in degrees Celsius), and U^* is an empirical coefficient usually determined from experimental data for the equipment involved; it might have the units of J/(s)(m^2)(°C). For example, ignoring conduction and radiation, the convective heat transfer from a person (the system) to a room (the surroundings) can be calculated using $U^* = 7\,\text{W}/(\text{m}^2)(°\text{C})$ and the data in Figure 9.6.

$$\dot{Q} = \frac{7\text{W}}{(\text{m}^2)(°\text{C})}\left|\frac{1.6\ \text{m}^2}{}\right|\frac{(25 - 29)°\text{C}}{} = -44.8\ \text{W or } -44.8\ \text{J/s}$$

Note that $\dot{Q}$ is negative because heat is transferred from the system to the surroundings. Multiply $\dot{Q}$ by the time period in hours to get Q in watt-hours.

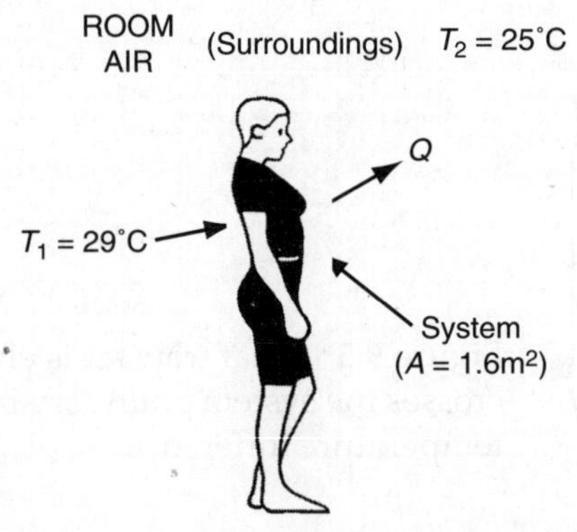

Figure 9.6 Heat transfer from a person

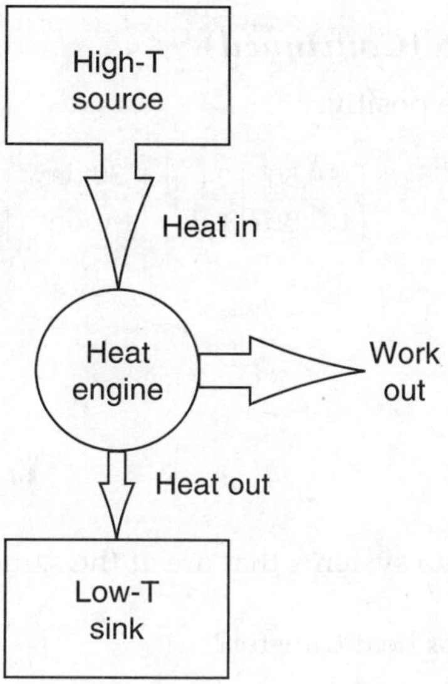

Figure 9.7 A heat engine produces work by operating between a high-temperature fluid and a low-temperature fluid.

What is the value of the rate of heat transferred into the air if the air is the system? It is +44.8 W.

A device (system) that involves a high-temperature fluid and a low-temperature fluid to do work is known as a "heat engine." Examine Figure 9.7. Examples are power plants, steam engines, heat pumps, and so on.

Example 9.1 Energy Conservation

Energy conservation is important for houses, commercial buildings, and so on. To what fraction is the heat transfer rate reduced by replacing a glass window 3 ft wide and 5 ft high with a smaller window 2 ft wide by 3 ft high? As a typical case, assume the outside temperature in the winter is 25°F and the inside temperature is 75°F. For this example assume $U^* = 5.5\ \text{Btu}/(\text{hr})(\text{ft}^2)(°\text{F})$.

If the cost of energy is $9.50/10^6 Btu, how many dollars per 30-day month are saved by changing the window size if the given temperatures are constant?

Solution

You could calculate $\dot{Q}$ for each case using Equation (9.1), but it is quicker to take a ratio:

$$\frac{\dot{Q}_2}{\dot{Q}_1} = \frac{U_2^* \, A_2 \, \Delta T_2}{U_1^* \, A_1 \, \Delta T_1} = \frac{(2)(3)}{(3)(5)} = 0.40$$

(Continues)

Example 9.1 Energy Conservation (*Continued*)

$\dot{Q}$ itself is negative but the savings will be positive.

$$\text{Savings} = -\frac{5.5\ \text{Btu}}{(\text{hr})(\text{ft}^2)(°\text{F})}\Bigg|\frac{(3)(5)\ \text{ft}^2}{}\Bigg|\frac{(25-75)°\text{F}}{}\Bigg|\frac{\$9.50}{10^6\ \text{Btu}}\Bigg|\frac{24\ \text{hr}}{1\ \text{day}}\Bigg|\frac{30\ \text{day}}{1\ \text{month}}\Bigg|\frac{0.6}{}$$

$$= \$17/\text{month}$$

Self-Assessment Test

Questions

1. There can be no heat transfer between two systems that are at the same temperature. True or false?
2. Which of the following are valid terms for heat transfer?

Heat addition	Heat generation
Heat rejection	Heat storage
Heat absorption	Electrical heating
Heat gain	Resistance heating
Heat loss	Frictional heating
Heat of reaction	Gas heating
Specific heat	Waste heat
Heat content	Body heat
Heat quality	Process heat
Heat sink	Heat source

3. Which of the following statements presents an incorrect view of heat?
   a. Los Angeles winters are mild because the ocean holds a lot of heat.
   b. Heat rises in the chimney of a fireplace.
   c. Your house won't lose much heat this winter because of the new insulation in the attic.
   d. A nuclear power plant dumps a lot of heat into the river.
   e. Close that door—don't let the heat out (in Minnesota)/or in (in Texas).

Problems

1. A calorimeter (a device to measure heat transfer) is being tested. It consists of a sealed cylindrical vessel containing water placed inside a sealed well-insulated tank containing ice water at 0°C. The water in the cylinder is heated by an electric coil so that 1000 J of energy are introduced into the water. Then the water is allowed to cool until it reaches 0°C after 15 minutes and is in thermal equilibrium with the water in the ice bath.

a. How much heat was transferred from the ice bath to the surrounding air during this test?
b. How much heat was transferred from the cylinder to the ice bath during the test?
c. If you pick two different systems composed of (1) the ice bath and (2) the cylinder, was the heat transfer to the ice bath *exactly* the same as the heat transfer from the ice bath to the cylinder?
d. Is the interaction between the surroundings and the ice bath work, heat, or both?

2. Classify the energy transfer in the following processes as work, heat, both, or neither.
   a. A gas in a cylinder is compressed by a piston, and as a result the temperature of the gas rises. The gas is the system.
   b. When an electric space heater is operating in a room, the temperature of the air goes up. The system is the room.
   c. The situation is the same as in b but the system is the space heater.
   d. The temperature of the air in a room increases because of the sunshine passing through a window.

Thought Problems

1. A piece of metal and a piece of wood are at the same temperature, yet the metal feels colder than the wood. Why?
2. The freezing point and the melting point of a substance are at the same temperature. If you put a piece of the solid into the liquid, why does it not melt? The ability to walk barefoot across a bed of red-hot embers has long been a sign of supernatural powers and recently a demonstration of confidence-raising. How can a person walk across the embers and not be injured?
3. Suppose you place a jar three-quarters full of water in a pot that rests on a stove burner. Place the jar on an upside-down saucer in the bottom of the pot. Then fill the pot with water up to the same level as the water in the jar. Heat the water in the pot so that it boils. Why will the water in the jar not boil?

Discussion Question

Worldwide interest exists in the possible global warming that occurs as a result of human activities. Prepare a report with tables that lists the sources of and mitigation options that exist for total CO_2, CH_4, CFC, and N_2O emissions in the world. For example, for CH_4 include rice cultivation, enteric fermentation, landfills, and so on. If possible, find data for the values of the annual emissions in megatons per year. For the mitigation options, list the possible additional approximate costs over current costs to implement the option per 1 ton of emitter.

9.2.2 Work (*W*)

The next type of energy we discuss is work (*W*). *Work* is a term that has wide usage in everyday life (such as "I am going to work") but has a specialized meaning in connection with energy balances. Work is a form of energy that represents a **transfer** of energy between the system and surroundings. Work cannot be stored. ***Work is a path variable. Work is positive when the surroundings perform work on the system. Work is negative when the system performs work on the surroundings.*** In some books the sign is the opposite. In this text, the symbol W refers to the net work done over a period of time, *not* the *rate* of work. The latter is $\dot{W}$, the **power**, namely, the work per unit time. The work per unit mass will be designated by $\hat{W}$.

Many types of work can take place (which we will lump together under the notation W), among which are the following:

- **Mechanical work:** Work that occurs because of a mechanical force that **moves the boundary** of a system. You might calculate W on the system (or by the system with the appropriate sign) as

$$W = \int_{\text{state 1}}^{\text{state 2}} \vec{F} \cdot d\vec{s} \tag{9.2}$$

  where $\boldsymbol{F}$ is an external force (a vector) in the direction of $\boldsymbol{s}$ (a vector) acting on the system boundary (or a system force acting at the boundary on the surroundings). However, the amount of mechanical work done by or on a system can be difficult to calculate because (a) the displacement $d\vec{s}$ may not be easy to define, and (b) the integration of $\boldsymbol{F} \cdot d\boldsymbol{s}$ does not necessarily give the amount of work actually being done on the system (or by the system) because some of the energy of W dissipates as Q into the contents of the system. For an example, look at Figure E9.2a in Example 9.2.
- **Electrical work:** Electrical work occurs when an electrical current passes through an electrical resistance in the circuit. If the system generates an electrical current (e.g., an electrical generator inside the system) and the current passes through an electrical resistance outside the system, the electrical work is negative because the electrical work is done on the surroundings. If the electrical work is done inside the system because of an applied voltage from outside the system, the electrical work is positive.
- **Shaft work:** Shaft work occurs when the system causes a shaft to turn against an external mechanical resistance. When a source of water outside the system circulates in the system and consequently causes a shaft to turn, the shaft work is positive. Look at Figure 9.8. When a shaft in the system turns a pump to pump water out of the system, the shaft work is negative. Does the shaft in Figure 9.8 rotate clockwise or counterclockwise? If water is pumped out of the system, which way does the shaft rotate?
- **Flow work:** Flow work is performed on the system when a fluid is pushed into the system by the surroundings. Look at Figure 9.9. For example,

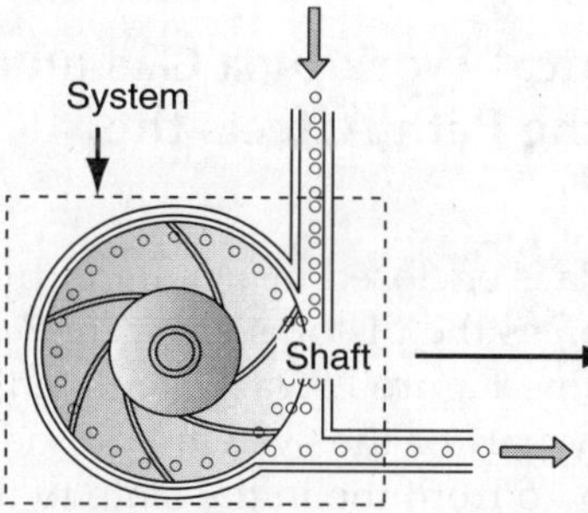

Figure 9.8 Shaft work. In the figure the force exerted on the impeller due to the water flow from the surroundings causes the shaft to rotate. Work is positive here—done by the surroundings on the system.

when a fluid enters a pipe, some work is done on the system (the water that already is in the pipe) to force the new fluid into the pipe. Similarly, when fluid exits the pipe, the system does some work on the surroundings to push the exiting fluid into the surroundings. Flow work will be described in more detail in the next subsection.

Suppose a gas in a fixed-volume container is heated so that its temperature is doubled. How much work was done on or by the gas during the process? Such a question is easy to answer: No work was done because the boundary of the system (the gas) remained fixed. Let us next look at an example in which the boundary changes.

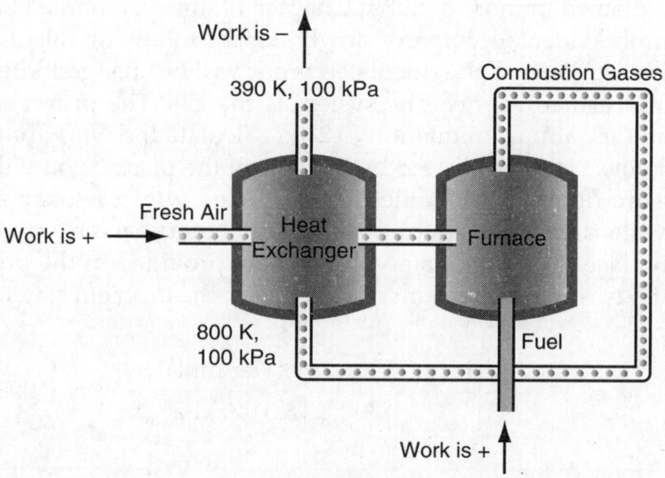

Figure 9.9 Flow work. Flow work occurs when the surroundings push an element of fluid into the system (sign is positive) or when the system pushes an element of fluid into the surroundings (sign is negative).

Example 9.2 Calculation of Mechanical Work by a Gas on a Piston Showing How the Path Affects the Value of the Work

Suppose that an ideal gas at 300 K and 200 kPa is enclosed in a cylinder by a *frictionless (ideal) piston*, and the gas slowly forces the piston so that the volume of gas expands from 0.1 to 0.2 m^3. Examine Figure E9.2a. Calculate the work done by the gas on the piston (the only part of the system boundary that moves) if two different paths are used to go from the initial state to the final state.

Path *A*: The expansion occurs at constant pressure (**isobaric**) (p = 200 kPa).

Path *B*: The expansion occurs at constant temperature (**isothermal**) (T = 300 K).

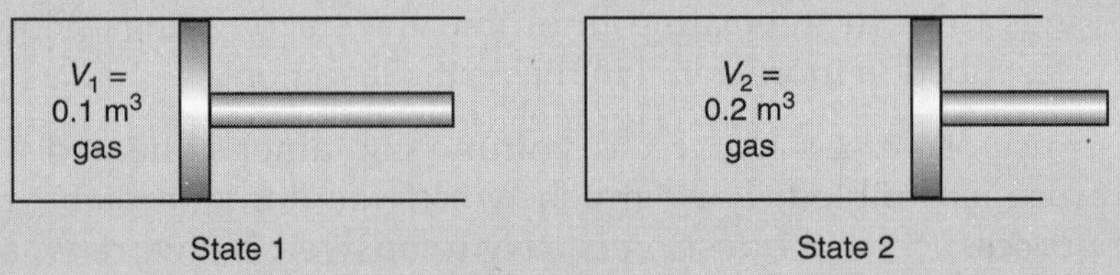

Figure E9.2a

Solution

As explained in more detail in Chapter 14, the piston must be frictionless and the process ideal (occur very slowly) for the following calculations to be valid. Otherwise some of the calculated work will be changed into a different form of unmeasured energy. The system is the gas. The piston is part of the surroundings. You use Equation (9.2) to calculate the work, but because you do not know the force exerted by the gas on the piston, you will have to use the pressure (force/area) as the driving force—which is okay since you do not know the area of the piston anyway, and because p is exerted normally on the piston face. All of the data you need are provided in the problem statement. Let the basis be the amount of gas cited in the problem statement:

$$n = \frac{200\ \text{kPa}}{} \left| \frac{0.1\ \text{m}^3}{} \right| \frac{}{300\ \text{K}} \left| \frac{(\text{kg mol})(\text{K})}{8.314\ (\text{kPa})\ (\text{m}^3)} \right. = 0.00802\ \text{kg mol}$$

Figure E9.2a illustrates the two processes: an isobaric path and an isothermal path.

The mechanical work done by the system on the piston (in moving the system boundary) *per unit area* is

$$W = -\int_{\text{state 1}}^{\text{state 2}} \left(\frac{F}{A}\right)(A\mathbf{ds}) = -\int_{V_1}^{V_2} p\, dV$$

Note that by definition, the work done by the system is *negative*. If the integral dV is positive (such as in expansion), the value of the integral will be positive and W negative (work done on the surroundings). If dV is negative, W will be positive (work done on the system).

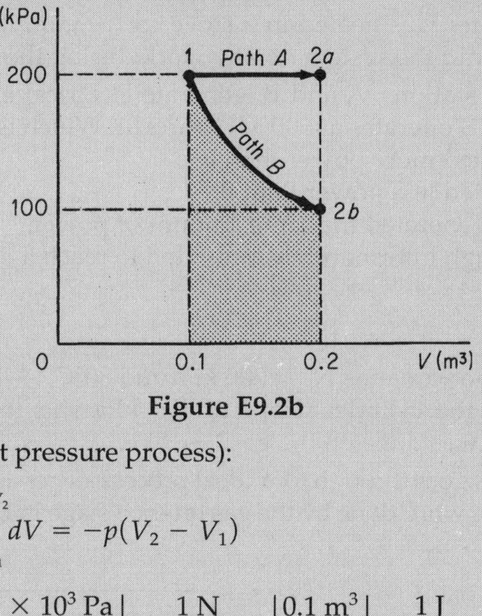

Figure E9.2b

Path A (the constant pressure process):

$$W = -p\int_{V_1}^{V_2} dV = -p(V_2 - V_1)$$

$$= -\frac{200 \times 10^3 \text{ Pa}}{} \left| \frac{1 \text{ N}}{1\ (\text{M}^2)(\text{Pa})} \right| \frac{0.1 \text{ m}^3}{} \left| \frac{1 \text{ J}}{1(\text{N})(\text{m})} \right. = -20 \text{ kJ}$$

Path B (the constant temperature process):
The gas is ideal. Then

$$W = -\int_{V_1}^{V_2} \frac{nRT}{V} dV = -nRT \ln\left(\frac{V_2}{V_1}\right)$$

$$= -\frac{0.00802 \text{ kg mol}}{} \left| \frac{8.314 \text{ kJ}}{(\text{kg mol})(\text{K})} \right| \frac{300 \text{ K}}{} \ln 2 = -13.86 \text{ kJ}$$

In Figure E9.2b the two integrals are areas under the respective curves in the p-V plot. By which path does the system do the most work?

We next discuss three types of energy that can be stored, that is, retained, because material has associated energy, namely, kinetic energy, potential energy, and internal energy.

Self-Assessment Test

Questions

1. If the energy crossing the boundary of a closed system is not heat, it must be work. True or false?
2. A pot of water was heated in a stove for 10 min. If the water was selected to be the system, did the system do any work during the 10 min?
3. Two power stations, A and B, generate electrical energy. A operates at 800 MW for 1 hr and B operates at 500 MW for 2 hr. Which is the correct statement?
   a. A generated more power than B.
   b. A generated less power than B.
   c. A and B generated the same amount of power.
   d. Not enough information is provided to reach a decision.

Problems

1. A gas cylinder contains N_2 at 200 kPa and 80°C. As a result of cooling at night the pressure in the cylinder drops to 190 kPa and the temperature to 30°C. How much work was done on the gas?
2. Nitrogen gas goes through four ideal process stages as detailed in Figure SAT9.2.2P2. Calculate the work done by the gas on each stage in British thermal units.

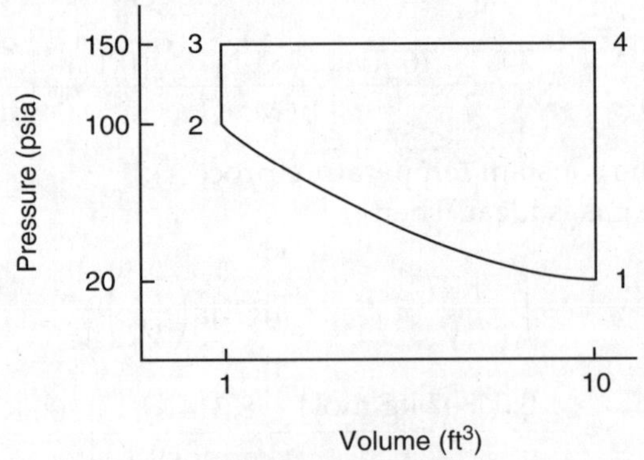

Figure SAT9.2.2P2

Thought Problems

1. In AE units power can be expressed in horsepower. Can one horse provide 1 hp?
2. How can you measure the power of an automobile engine as you drive along at a constant speed?

Discussion Question

You can find suggestions in the popular magazines that electric vehicles be covered with solar cells to reduce use of coal-generated electricity. The solar insolation outside the Earth's atmosphere is 1 kW/m^2 (or square yard, for all practical purposes), yielding 24 kWh/day. Experience has shown that an electric vehicle converted from a conventional car usually requires a minimum of about 30 kWh stored in the battery. If the surfaces of the car referred to are covered with a layer of solar cells 5 ft by 10 ft, or 50 ft^2, about 5 m^2, 120 kWh/day, will be available from the photovoltaic cells.

Carry out an evaluation of this calculation. What deficiencies does it have? Think of the entire transition from solar energy to chemical change in the battery.

9.2.3 Kinetic Energy (*KE*)

Kinetic energy (*KE*) is the energy a system, or some material, possesses because of its velocity relative to the surroundings, which are usually, *but not always*, at rest. The wind, moving automobiles, waterfalls, flowing fluids, and so on possess kinetic energy. The kinetic energy of a material refers to what is called the macroscopic kinetic energy, namely, the energy that is associated with the gross movement (velocity) of the system or material, and not the kinetic energy of the individual molecules, which belongs in the category of internal energy, U, that is discussed below.

Do you recall the equation used to calculate the kinetic energy relative to stationary surroundings? It is

$$KE = \frac{1}{2}mv^2 \tag{9.3a}$$

The kinetic energy per unit mass (the specific kinetic energy), a state variable, is

$$\widehat{KE} = \frac{1}{2}v^2 \tag{9.3b}$$

In Equation (9.3a) m refers to the center of mass of the material and v to a suitably averaged velocity of the material. The value of a *change* in the specific kinetic energy ($\Delta\widehat{KE}$) occurs in a specified time interval and depends only on the initial and final values of the mass and the velocity of the material.

Example 9.3 Calculation of the Specific Kinetic Energy for a Flowing Fluid

Water is pumped from a storage tank through a tube of 3.00 cm inner diameter at the rate of 0.001 m^3/s. See Figure E9.3. What is the specific kinetic energy of the water in the tube?

(*Continues*)

Example 9.3 Calculation of the Specific Kinetic Energy for a Flowing Fluid (*Continued*)

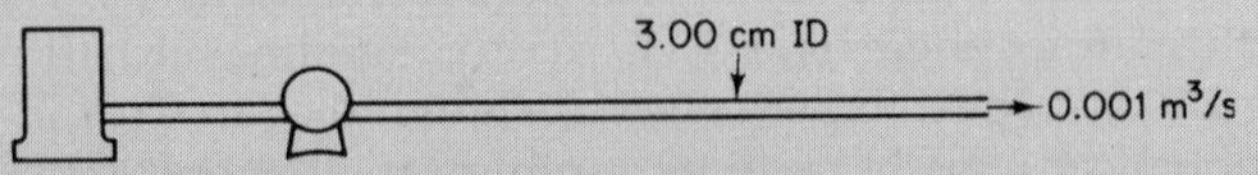

Figure E9.3

Solution

Basis: 0.001 m^3 equivalent to 1 s

Assume that

$$\rho = \frac{1000\ \text{kg}}{\text{m}^3} \quad \text{and} \quad r = \frac{1}{2}(3.00) = 1.50\ \text{cm}$$

$$v = \frac{0.001\ \text{m}^3}{\text{s}}\left|\frac{}{\pi(1.50)^2\ \text{cm}^2}\right|\left(\frac{100\ \text{cm}}{1\ \text{m}}\right)^2 = 1.415\ \text{m/s}$$

$$\widehat{KE} = \frac{1}{2}\left|\left(\frac{1.415\ \text{m}}{\text{s}}\right)^2\right|\frac{1\ (\text{N})(\text{s}^2)}{1(\text{kg})(\text{m})}\left|\frac{1\ \text{J}}{1(\text{N})(\text{m})}\right. = 1.00\ \text{J/kg}$$

Self-Assessment Test

Questions

1. Can the kinetic energy of a mass be zero if the mass has a velocity, that is, is moving?
2. Temperature is a measure of the average kinetic energy of a material. True or false?
3. The kinetic energy of an automobile going 100 mi/hr is greater than the energy stored in the battery of the automobile (300 Wh). True or false?

Problem

Calculate the kinetic energy changes in the water that occur when 10,000 lb/hr flow in a pipe that is reduced from 2 in. in diameter to 1 in. in diameter.

9.2.4 Potential Energy (*PE*)

Potential energy (*PE*) is energy a system possesses because of the force exerted on its mass by a gravitational or electromagnetic field with respect to a reference surface. When an electric car or bus goes uphill, it gains potential energy

Figure 9.10 Gain of potential energy by an electric automobile going uphill

(Figure 9.10), energy that can be recovered to some extent by regeneration—charging the batteries when the vehicle goes down the hill on the other side. You can calculate the potential energy in a gravitational field from

$$PE = mgh \tag{9.4a}$$

or the specific potential energy

$$\hat{PE} = gh \tag{9.4b}$$

where h is the distance from the reference surface, and where the overlay (^) means potential energy per unit mass. The measurement of h is made to the center of mass of a system. Thus, if a ball suspended inside a container somehow is permitted to drop from the top of the container to the bottom, and in the process it raises the thermal energy of the system slightly, we do not say work is done on the system but instead say that the potential energy of the system is reduced (slightly) because the center of mass changes slightly. The value of a *change* in the specific potential energy, $\Delta\hat{PE}$, occurs during a specified time interval and depends only on the initial and final states of the system ($\Delta\hat{PE}$ is a state variable), and not on the path followed.

Example 9.4 Calculation of Potential Energy of Water

Water is pumped from one reservoir to another 300 ft away, as shown in Figure E9.4. The water level in the second reservoir is 40 ft above the water level of the first reservoir. What is the increase in specific potential energy of the water in British thermal units per pound (mass)?

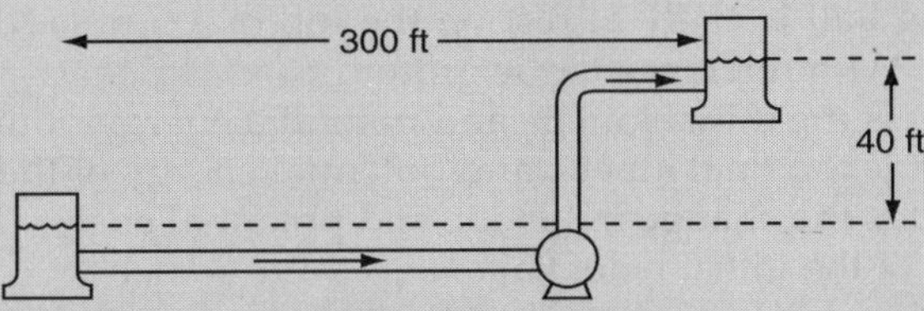

Figure E9.4

(Continues)

Example 9.4 Calculation of Potential Energy of Water (*Continued*)

Solution

Because you are asked to calculate the potential energy change of 1 lb of water and not of the whole reservoir, you can assume for this problem that the 40 ft difference in height does not change. Think of a Ping-Pong ball riding on top of the water that determines the height difference.

Let the water level in the first reservoir be the reference plane. Then $h = 40$ ft.

$$\Delta\hat{PE} = \frac{32.2 \text{ ft}}{\text{s}^2}\left|\frac{(40-0)\text{ ft}}{}\right|\frac{1\,(\text{lb}_\text{f})(\text{s}^2)}{32.2\,(\text{lb}_\text{m})(\text{ft})}\left|\frac{1 \text{ Btu}}{778.2\,(\text{ft})(\text{lb}_\text{f})}\right. = 0.0514 \text{ Btu/lb}_\text{m}$$

Self-Assessment Test

Questions

1. Indicate whether the following statements are true or false:
   a. Potential energy has no unique absolute value.
   b. Potential energy can never be negative.
   c. The attractive and repulsive forces between the molecules in a material contribute to the potential energy of a system.
2. The units of potential energy or kinetic energy in the American Engineering system are (select all of the correct expressions):
   a. $(\text{ft})(\text{lb}_\text{f})$
   b. $(\text{ft})(\text{lb}_\text{m})$
   c. $(\text{ft})(\text{lb}_\text{f})/(\text{lb}_\text{m})$
   d. $(\text{ft})(\text{lb}_\text{m})/(\text{lb}_\text{f})$
   e. $(\text{ft})(\text{lb}_\text{f})/(\text{hr})$
   f. $(\text{ft})(\text{lb}_\text{m})/(\text{hr})$

Problems

1. A 100 kg ball initially at rest on the top of a 5 m ladder is dropped and hits the ground. With reference to the ground:
   a. What is the initial kinetic and potential energy of the ball?
   b. What is the final kinetic and potential energy of the ball?
   c. What is the change in kinetic and potential energy for the process?
   d. If all of the initial potential energy were somehow converted to heat, how many calories would this amount to? How many British thermal units? How many joules?
2. A 1 kg ball 10 m above the ground is dropped and hits the ground. What is the change in *PE* of the ball?

Thought Problem

Why does a bicyclist pick up speed when going downhill even if he or she is not pedaling? Doesn't this situation violate the conservation of energy concept?

9.2.5 Internal Energy

Internal energy (U) is a *macroscopic* concept that takes into account the molecular, atomic, and subatomic energies of entities, all of which follow definite microscopic conservation rules for dynamic systems. Specific internal energy is a state variable and can be stored. Because no instruments exist with which to measure internal energy directly on a macroscopic scale, internal energy must be calculated from certain other variables that can be measured macroscopically, such as pressure, volume, temperature, and composition.

To calculate the internal energy per unit mass ($\hat{U}$) from variables that can be measured, we make use of the phase rule. For a pure component in one phase, $\hat{U}$ can be expressed in terms of just two intensive variables according to the phase rule:

$$\mathcal{F} = 2 - \mathcal{P} + \mathcal{C} = 2 - 1 + 1 = 2$$

Custom dictates the use of temperature and specific volume as the two variables. For a *single phase* and single component, we say that $\hat{U}$ is a function of only T and $\hat{V}$: $\hat{U} = \hat{U}(T, \hat{V})$.

If two components are in the phase, what is $\mathcal{F}$? $\mathcal{C} = 2$, hence $\mathcal{F} = 3$, and $\hat{U}$ would also be a function of the composition. Because $\hat{U}$ is state function, $\hat{U}$ can be differential with respect to T and V: By taking the total derivative, we find that

$$d\hat{U} = \left(\frac{\partial \hat{U}}{\partial T}\right)_{\hat{V}} dT + \left(\frac{\partial \hat{U}}{\partial \hat{V}}\right)_{T} d\hat{V} \tag{9.5}$$

By definition $(\partial \hat{U}/\partial T)_{\hat{V}}$ is the **heat capacity** (specific heat) at constant volume, given the special symbol C_v. C_v can also be defined to be the amount of heat necessary to raise the temperature of 1 kg of substance by 1 degree in a closed system and so has the SI units of J/(kg)(K), if the process is carried out at constant volume. For all practical purposes in this text the term $(\partial \hat{U}/\partial \hat{V})_T$ is so small that the second term on the right-hand side of Equation (9.5) can usually be neglected. (Note that in the steam tables, the second term on the right-hand side of Equation (9.5) cannot be neglected.) Consequently, *changes in the internal energy* over a specified time interval can usually be computed by integrating Equation (9.5) as follows:

$$\Delta\hat{U} = \hat{U}_2 - \hat{U}_1 = \int_{\hat{U}_1}^{\hat{U}_2} d\hat{U} = \int_{T_2}^{T_2} C_v dT \tag{9.6}$$

For an ideal gas $\hat{U}$ is a function of temperature only. Always keep in mind that **Equation (9.6) alone is *not* valid if a phase change occurs** during the interval.

Note that you can only calculate differences in internal energy, or calculate the internal energy relative to a reference state, *but cannot calculate absolute values* of internal energy. Look up the values of p and $\hat{V}$ for water for the reference state that has been assigned a zero value for $\hat{U}$. From the program for saturated steam, at 0°C did you get $p = 0.612$ kPa corresponding to liquid water with $\hat{V} = 0.001000$ m^3/kg? The reference internal energy cancels out when you calculate an internal energy difference as long as you use the same reference state for the variables:

$$\Delta\hat{U} = (\hat{U}_2 - \hat{U}_{\text{ref}}) - (\hat{U}_1 - \hat{U}_{\text{ref}}) = \hat{U}_2 - \hat{U}_1 \tag{9.7}$$

If you do not use the same table, chart, or equation, you can't automatically cancel the $\hat{U}_{\text{ref}}$.

What would be the value of ΔU for a constant volume system if 1 kg of water at 100 kPa was heated from 0°C to 100°C, and then cooled back to 0°C and 100 kPa? Would $\Delta\hat{U} = 0$? Yes, because it is a state variable, and the integral in Equation (9.6) would be zero because $\hat{U}_2 = \hat{U}_1$.

The internal energy of a system containing more than one component is the sum of the internal energies of each component:

$$U_{\text{tot}} = m_1\hat{U}_1 + m_2\hat{U}_2 + \cdots + m_n\hat{U}_n \tag{9.8}$$

The heat of mixing, if any (discussed in Chapter 13), is neglected in Equation (9.8).

Equations, charts, or tables for C_v are rare; hence you will usually have to calculate $\Delta\hat{U}$ by some other method than using Equation (9.6). But if you can find a relation for C_v, then getting ΔU is as simple as shown in the next example.

Example 9.5 Calculation of an Internal Energy Change Using the Heat Capacity

What is the change in internal energy when 10 kg mol of air are cooled from 60°C to 30°C in a constant volume process?

Solution

Since you don't know the value of C_v, you have to look up the value. It is 2.1×10^4 J/(kg mol)(°C) over the temperature range. Use Equation (9.6) to carry out the calculation:

$$\Delta U = 10\ \text{kg}\int_{60°\text{C}}^{30°\text{C}}\left(2.1 \times 10^4 \frac{\text{J}}{(\text{kg mol})(°\text{C})}\right)dT$$

$$= 2.1 \times 10^5(30 - 60) = -6.3 \times 10^6\ \text{J}$$

Self-Assessment Test

Questions

1. An entrance examination for graduate school asked the following two multiple-choice questions:
   a. The internal energy of a solid is equal to (1) the absolute temperature of the solid, (2) the total kinetic energy of its molecules, (3) the total potential energy of its molecules, or (4) the sum of the kinetic and potential energy of its molecules.
   b. The internal energy of an object depends on its (1) temperature only; (2) mass only; (3) phase only; (4) temperature, mass, and phase.

   Which answers would you choose?
2. If C_v is not constant over the temperature range in Equation (9.6), can you still integrate C_v to get ΔU?

Problems

1. A database lists an equation for $\hat{U}$ as

$$\hat{U} = 1.10 + 0.810T + 4.75 \times 10^{-4}\, T^2$$

   where $\hat{U}$ is in kilojoules per kilogram and T is in degrees Celsius. What are (a) the corresponding equation for Cv and (b) the reference temperature for $\hat{U}$?
2. Use the steam tables to calculate the change in internal energy between liquid water at 1000 kPa and 450 K, and steam at 3000 kPa and 800 K. How did you take into account the phase change in the water?

Thought Problem

Are there any reasons why internal energy should be used in your calculations when most problems seem to involve open systems?

Discussion Questions

1. Richard Feynman was a brilliant physicist and comedian who invested a great amount of effort in making physics accessible to students. In one of his lectures he wrote: "Let us consider a rubber band. When we stretch the rubber band, we find that its temperature ________." Fill in the missing word for the professor.
2. Would you say it is better to use a photovoltaic system of solar collectors to heat a swimming pool, or to use water heated by a thermal solar collector?

9.2.6 Enthalpy

You will find later in this chapter that the two terms $(U + pV)$ and $(\hat{U} + p\hat{V})$ do not appear in an energy balance. Instead, the variables are combined into another variable called the **enthalpy** (pronounced en-THAL-py):

$$H = U + pV \tag{9.9a}$$

where p is the pressure and V is the volume, or on a basis of a unit mass

$$\hat{H} = \hat{U} + p\hat{V} \tag{9.9b}$$

To calculate the specific enthalpy (enthalpy per unit mass), as with the specific internal energy, we use the property that the enthalpy is an exact differential. As you saw for internal energy, the state for the enthalpy for a **single phase** and single component can be completely specified by two intensive variables. We will express the enthalpy in terms of the temperature and pressure (the former a more convenient variable than the specific volume). If we let

$$\hat{H} = \hat{H}(T, p)$$

by taking the total derivative of $\hat{H}$, we can form an expression analogous to Equation (9.5):

$$d\hat{H} = \left(\frac{\partial \hat{H}}{\partial T}\right)_p dT + \left(\frac{\partial \hat{H}}{\partial p}\right)_T dp \tag{9.10}$$

By definition $(\partial \hat{H}/\partial T)_p$ is the heat capacity at constant pressure and is given the special symbol C_p. For most practical purposes $(\partial \hat{H}/\partial p)_T$ is so small at modest pressures that the second term on the right-hand side of Equation (9.10) can be neglected. Changes in the specific enthalpy can then be calculated by integration of Equation (9.10) as follows:

$$\Delta\hat{H} = \hat{H}_2 - \hat{H}_1 = \int_{H_1}^{H_2} \mathrm{d}\hat{H} = \int_{T_1}^{T_2} C_p \mathrm{d}T \tag{9.11}$$

However, in processes operating at high pressures, the second term on the right-hand side of Equation (9.10) cannot necessarily be neglected but must be evaluated from experimental data. Consult the references at the end of the chapter for details. **One property of ideal gases, but *not* of real gases, to remember is that their enthalpies and internal energies are functions of temperature only** and are not influenced by changes in pressure or specific volume. Also, the relation between C_p and C_v for an ideal gas is $C_v = C_p - R$.

Where can you get values of enthalpies or data to calculate enthalpy values? Here are some sources:

1. Heat capacity and other equations
2. Equations to estimate the enthalpy of a phase transition
3. Tables
4. Enthalpy charts
5. Computer databases

Enthalpies can also be *estimated* by generalized methods based on the theory of corresponding states or additive bond contributions, but we will not discuss these methods. Refer instead to the references at the end of the chapter for information.

As with internal energy, **enthalpy has no absolute value; only changes in enthalpy can be calculated**. Often you will use a reference set of conditions (perhaps implicitly) in computing enthalpy changes. For example, the reference conditions used in the steam tables are liquid water at 0°C (32°F) and its vapor pressure. This does not mean that the enthalpy is actually zero under these conditions but merely that the enthalpy has arbitrarily been assigned a value of zero at these conditions. In computing enthalpy changes, the reference conditions cancel out, as can be seen from the following:

Initial state of the system (1)	*Final state of the system (2)*
Specific enthalpy $= \hat{H}_1 - \hat{H}_{\text{ref}}$	Specific enthalpy $= \hat{H}_2 = \hat{H}_{\text{ref}}$

$$\text{Net specific enthalpy change} = (\hat{H}_2 - \hat{H}_{\text{ref}}) - (H_1 - \hat{H}_{\text{ref}}) = \hat{H}_2 - \hat{H}_1$$

Example 9.6 Calculation of an Enthalpy Change

Calculate the enthalpy change for the process in Example 9.5, except assume that the enthalpy change occurs in a constant pressure process.

Solution

For this example you have to look up the value of C_p at 1 atm. Assume it is a constant over the temperature range. It is 2.913×10^4 J/(kg mol)(°C) at 45°C (from the physical property tables on the CD). Use Equation (9.11) to carry out the calculation:

$$\Delta H = 10 \text{ kg mol} \int_{60°\text{C}}^{30°\text{C}} (2.9 \times 10^4) \frac{\text{J}}{(\text{kg mol})(°\text{C})} dT = 2.9 \times 10^5(30 - 60)$$

$$= -8.7 \times 10^6 \text{ J}$$

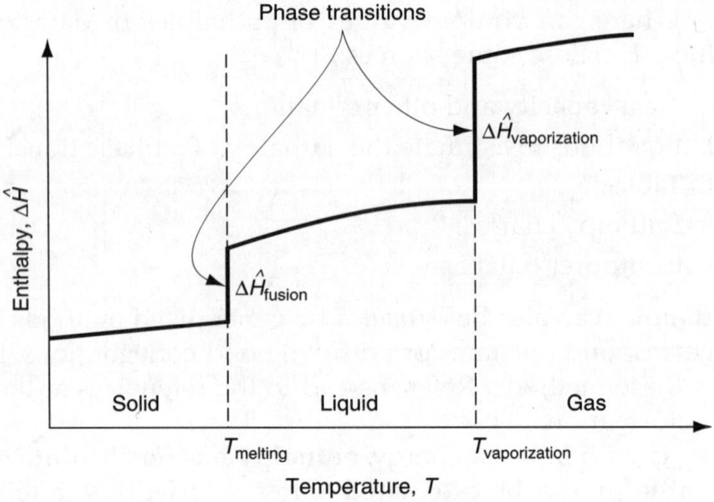

Figure 9.11 The overall enthalpy change includes the sensible heats (the enthalpy changes within a phase) plus the latent heats (the enthalpy changes of the phase transitions).

Be careful when reading a value of enthalpy from a table or chart because the symbol H usually refers to the enthalpy per unit mass (or mole), not the total enthalpy. The symbol $\hat{H}$ is not widely used except in this book, and consequently the meaning of H should be determined by looking at the units located (ideally) in a heading or footnote. A value of $H = 100$ kJ probably means a value of 100 kJ/kg from a reference value of H such as $H = 0$, or possibly another reference value.

Equation (9.11) is *not* valid if a phase change occurs. Look at Figure 9.11. The enthalpy associated with a phase change must be included to get the overall ΔH (or $\Delta\hat{H}$).

Do you recall from Chapter 8 that **phase transitions** occur from the solid to the liquid phase, and from the liquid to the gas phase, and vice versa? During these transitions very large changes in the value of the enthalpy (and internal energy) of a substance occur, changes called **latent heat** changes, because they occur without any noticeable change in temperature. Because of the relatively large enthalpy change associated with a phase transition, it is important to get accurate values of any latent heats involved when applying energy balances. For a single phase, the enthalpy varies as a function of the temperature, as illustrated in Figure 9.11. The enthalpy changes that take place within a single phase are usually called **sensible heat** changes.

The enthalpy changes for the common specific phase transitions are termed **heat of fusion** (for melting), ΔH_{fusion}, and **heat of vaporization** (for vaporization), ΔH_v. The word *heat* has been carried by custom from very old experiments in which enthalpy changes were calculated from experimental data that frequently involved heat transfer. *Enthalpy of fusion* and *vaporization* would be the proper terms, but they are not widely used. **Heat of condensation** is the negative of the heat of vaporization and the **heat of solidification** is the negative of the heat of fusion. The **heat of sublimation** is the enthalpy change from solid directly to vapor.

The overall specific enthalpy change of a pure substance, as illustrated in Figure 9.11, can be formulated by summing the sensible and latent heats (enthalpies) from the initial state to the final state.

Sensible heat of

Overall enthalpy change Solid Melting Sensible heat of liquid

$$\Delta\hat{H} = \hat{H}(T) - \hat{H}(T_{\text{ref}}) = \int_{T_{\text{ref}}}^{T_{\text{fusion}}} C_{p,\text{solid}} dT + \Delta\hat{H}_{\text{fusion at } T_{\text{fusion}}} + \int_{T_{\text{fusion}}}^{T_{\text{vaporization}}} C_{p,\text{liquid}} dT$$

Vaporization Sensible heat of vapor

$$+ \Delta\hat{H}_{\text{vaporization}} \text{ at } T_{\text{vap}} + \int_{T_{vap}}^{T} C_{p,\text{vapor}} dT \qquad (9.12)$$

The overall enthalpy of a system or stream containing more than one component is the sum of the enthalpies of each component if you ignore the heat (enthalpy change) of mixing (discussed in Chapter 13).

$$H_{\text{tot}} = m_1\hat{H}_1 + m_2\hat{H}_2 + \cdots + m_n H_n \qquad (9.13)$$

Information is information, not matter or energy.

Norbert Wiener

Self-Assessment Test

Questions

1. When you read in the steam table in two adjacent columns $\hat{U} = 4184$ J and $\hat{H} = 4184$ J, that is the water liquid or vapor?
2. Repeat question 1 for $\hat{U} = 1362$ J and $\hat{H} = 1816$ J.

Problems

1. The enthalpy change of a real gas can be calculated from

$$\hat{H}_2 - \hat{H}_1 = \int_{T_1}^{T_2} C_p dT + \int_{P_1}^{P_2} \left[\hat{V} - T\left(\frac{\partial \hat{V}}{\partial T}\right)_p \right] dp$$

What is the enthalpy change for a real gas in (a) an isothermal process, (b) an isochoric (constant volume) process, and (c) an isobaric process?

2. Will the enthalpy change be greater along the path shown by the solid line or the path shown by the dashed line in Figure SAT9.2.5P2?

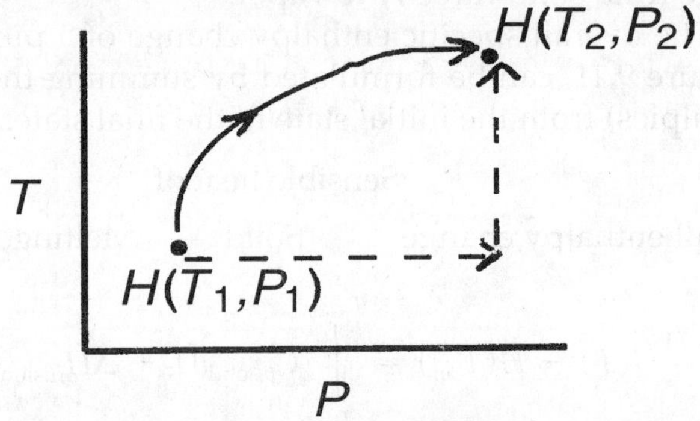

Figure SAT9.2.5P2

3. Show that for an incompressible ($d\hat{V} = 0$) liquid or solid that

$$\Delta \hat{H} = \int_{T_1}^{T_2} C_p dT + \hat{V}(p_2 - p_1)$$

Discussion Questions

1. How did the name enthalpy become attached to the sum of U plus (pV)?
2. Is the enthalpy change of a real gas undergoing a process, the enthalpy change of each of the individual molecules comprising the gas?

9.2.7 Heat Capacity

Quite a few sources exist for equations expressing C_p as a function of T, and numerous charts exist on which C_p is plotted as a function of temperature. Look at Figure 9.12.

Historically the integration of heat capacity equations has been used to calculate sensible heat (enthalpy) changes. Appendix G on the accompanying CD lists heat capacity equations for a number of common compounds, and the software on the CD contains equations for C_p for over 700 compounds. C_p, of course, is a continuous function of temperature, whereas the enthalpy change may not be. Why? Because of the phase changes that may occur.

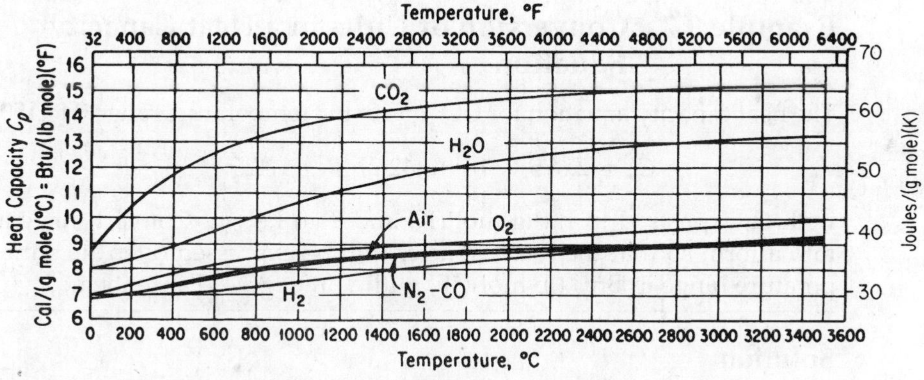

Figure 9.12 Heat capacity curves for the combustion gases at 1 atm

Consequently, it is not possible to have a heat capacity equation for a substance that will go from a low temperature up to any desired temperature. To give the heat capacity some physical meaning, you can think of C_p as representing the amount of energy required to increase the temperature of a unit mass (or mole) of a substance by 1 degree.

How are the functions such as those in Figure 9.11 determined? What you do is measure experimentally the heat capacity between the temperatures at which the phase transitions occur, and then fit the data with an equation using Polymath. If you can assume the gas is an ideal gas, the heat capacity at constant pressure is constant even though the temperature varies (examine Table 9.3).

Table 9.3 Heat Capacities of Ideal Gases

	Approximate Heat Capacity, C_p	
Type of Molecule*	**High Temperature (Translational, Rotational, and Vibrational Degrees of Freedom)**	**Room Temperature (Translational and Rotational Degrees of Freedom Only)**
Monoatomic	$\frac{5}{2}R$	$\frac{5}{2}R$
Polyatomic, linear	$\left(3n - \frac{3}{2}\right)R$	$\frac{7}{2}R$
Polyatomic, nonlinear	$(3n - 2)R$	4R

*n, number of atoms per molecule; R, gas constant.

Calories don't count when they are at room temperature.

Cathy

Example 9.7 Conversion of Units in a Heat Capacity Equation

The heat capacity equation for CO_2 gas in the temperature range 0 to 1500 K is

$$C_p = 2.675 \times 10^4 + 42.27\,T - 1.425 \times 10^{-2}\,T^2$$

with C_p expressed in J/(kg mol)(K) and T in kelvin. Convert this equation into a form so that the heat capacity will be expressed over the entire temperature range in Btu/(lb mol)(°F) with T in degrees Fahrenheit.

Solution

Changing a heat capacity equation from one set of units to another is merely a problem in the conversion of units. Each term on the right-hand side of the heat capacity equation must have the same units as the left-hand side of the equation. To avoid confusion in the conversion, you must remember to distinguish between the temperature symbol that in one usage represents temperature and in another usage represents a temperature difference. In the following conversions we shall distinguish between the temperature and the temperature difference for clarity as was done in Chapter 2.

First, multiply each side of the given equation for C_p by appropriate conversion factors to convert J/(kg mol)(Δ K) to Btu/(lb mol)(Δ°F). Multiply the left-hand side by the factor in the square brackets.

$$C_p \frac{\text{J}}{(\text{kg mol})(\Delta\,K)} \times \left[\frac{1\ \text{Btu}}{1055\ \text{J}}\bigg|\frac{1\ \Delta\,\text{K}}{1.8\ \Delta°\text{R}}\bigg|\frac{1\ \Delta°\text{R}}{1\ \Delta°\text{F}}\bigg|\frac{0.4536\ \text{kg}}{1\ \text{lb}}\right] \rightarrow C_p \frac{\text{Btu}}{(\text{lb mol})(\Delta°\text{F})}$$

and multiply the right-hand side by the same set of conversion factors.

Next, substitute the relation between the temperature in kelvin and the temperature in degrees Fahrenheit

$$T_{\text{K}} = \frac{T_{°\text{R}}}{1.8} = \frac{T_{°\text{F}} + 460}{1.8}$$

into the given equation for C_p where T appears.

Finally, carry out all the indicated mathematical operations, and consolidate quantities to get

$$C_p \frac{\text{Btu}}{(\text{lb mol})(\Delta°\text{F})} = 8.702 \times 10^{-3} + 4.66 \times 10^{-6}\,T_{°\text{F}} - 1.053 \times 10^{-9}\,T_{°\text{F}}^2$$

When you cannot find a heat capacity for a gas, you can estimate one by using one of the numerous equations that can be found in the supplementary references at the end of this chapter. Estimation of C_p for liquids is fraught with error, but some relationships exist. For aqueous solutions, you can roughly approximate C_p by using the water content only.

Self-Assessment Test

Questions

1. Indicate whether the following statements are true or false:
   a. For a real gas $\Delta\hat{H} = \int_{T_1}^{T_2} C_p dT$ is an exact expression.
   b. For liquids below $T_r = 0.75$ or a solid, $\Delta U \approx \int_{T_1}^{T_2} C_p dT$.
   c. For ideal gases near room temperature $C_p = 5/2\,R$.
   d. C_v cannot be used to calculate $\Delta\hat{H}$—you have to use C_p.
2. Is the term *specific energy* a better term to use to represent specific heat?
3. What is the heat capacity at constant pressure at room temperature of O_2 if the O_2 is assumed to be an ideal gas?

Problems

1. Determine the specific enthalpy of liquid water at 400 K and 500 kPa relative to the specific enthalpy value of liquid water at 0°C and 500 kPa using a heat capacity equation. Compare to the value obtained from the steam tables.
2. A problem indicates that the enthalpy of a compound can be predicted by an empirical equation $\hat{H}(\text{J/g}) = -30.2 + 4.25T + 0.001T^2$, where T is in kelvin. What is a relation for the heat capacity at constant pressure for the compound?
3. A heat capacity equation in cal/(g mol)(K) for ammonia gas is

$$C_p = 8.4017 + 0.70601 \times 10^{-2}\,T + 0.10567 \times 10^{-5}\,T^2 - 1.5981 \times 10^{-9}\,T^3$$

   where T is in degrees Celsius. What are the units of each of the coefficients in the equation?

Thought Problems

1. A piece of wood and a piece of metal having identical masses are removed from an oven after they reach the same temperature. Then both are placed on a block of ice. Which piece will melt more ice when the piece reaches the ice temperature?
2. Fire walkers with bare feet walk across beds of glowing coals without apparent harm. The rite is found in many parts of the world today and was practiced in classical Greece and ancient India and China, according to the *Encyclopaedia Britannica*. Explain this.

Discussion Questions

1. Dow Chemical sells Dowtherm Q, a heat transfer fluid that has an operating range of −30°F to 625°F, for $12/gal. Dowtherm Q competes with mineral oil, which costs $3/gal and operates up to 600°F. Why would a company pay so much more for Dowtherm Q?
2. In *Chemical Engineering Education* (Summer, 1994) the question was asked why soldiers in the Middle Ages poured boiling oil on attacking enemy soldiers rather than boiling water, especially when the heat capacity of oil is less than one-half that of water. What is the rationale for the use of oil?

9.2.8 Enthalpies for Phase Transitions

Where can you get values for the enthalpies of phase changes? Some common values of $\Delta\hat{H}$ for phase changes are listed in an Appendix F on the CD that accompanies this book, and over 700 are used by the physical property software on the CD. Other sources of experimental data are cited in reference books listed in the supplementary references at the end of this chapter. You can also estimate values for $\Delta\hat{H}_v$ from one of the relations such as the following three. Use of experimental values for the heat of vaporization is recommended whenever possible.

Chen's equation[1] An equation that yields values of $\Delta\tilde{H}_v$ (in kilojoules per gram mole) (the overlay tilde ($\sim$) on H designates per mole rather than per mass) to within 2% is Chen's equation:

$$\Delta\tilde{H}_v = RT_b\left(\frac{3.978\,(T_b/T_c) - 3.938 + 1.555 \ln p_c}{1.07 - (T_b/T_c)}\right) \tag{9.14}$$

where T_b is the normal boiling point of the liquid in kelvin, T_c is the critical temperature in kelvin, and p_c is the critical pressure in atmospheres.

---

[1] N. H. Chen, *Ind. Eng. Chem.*, **51**, 1494 (1959).

Riedel's equation[2]

$$\Delta\tilde{H}_v = 1.093\,R\,T_c\left[\frac{T_b\,(\ln p_c - 1)}{T_c\,(0.930 - (T_b/T_c))}\right] \tag{9.15}$$

Watson's equation[2] Watson found empirically that below the critical temperature the ratio of two heats of vaporization could be related by

$$\frac{\Delta\tilde{H}_{v_2}}{\Delta\tilde{H}_{v_1}} = \left(\frac{1 - T_{r2}}{1 - T_{r1}}\right)^{0.38} \tag{9.16}$$

where $\Delta\tilde{H}_{v_1}$ = heat of vaporization of a pure liquid at T_1
$\Delta\tilde{H}_{v_2}$ = heat of vaporization of the same liquid at T_2

Yaws[3] lists other values of the exponent for various substances.

The physical property software on the CD that accompanies this book provides the heat of vaporization at the normal boiling point (1 atm pressure). Therefore, it is recommended that when you have no data and need a heat of vaporization for an energy balance calculation, you use the heat of vaporization at the normal boiling point from the physical property software and use Equation (9.16) to correct for temperature.

Example 9.8 Comparison of an Estimate of the Heat of Vaporization with the Experimental Value

Use Chen's equation to estimate the heat of vaporization of acetone at its normal boiling point, and compare your results with the experimental value of 30.2 kJ/g mol listed in Appendix F on the CD that accompanies this book.

Solution

The basis is 1 g mol. You have to look up some data for acetone in Appendix F:

Normal boiling point:	329.2 K
T_c:	508.0 K
p_c:	47.0 atm

(Continues)

[2] K. M. Watson, *Ind. Eng. Chem.*, **23**, 360 (1931); 35, 398 (1943).

[3] C. L., Yaws, H. C. Yang, and W. A. Cawley, "Predict Enthalpy of Vaporization," *Hydrocarbon Processing*, 87–90 (June, 1990).

Example 9.8 Comparison of an Estimate of the Heat of Vaporization with the Experimental Value (*Continued*)

The next step is to calculate some of the values of the variables in the estimation equations:

$$\frac{T_b}{T_c} = \frac{329.2}{508.0} = 0.648$$

$$\ln p_c = \ln(47.0) = 3.85$$

From Equation (9.14):

$$\Delta\widetilde{H}_v = \frac{8.314 \times 10^{-3}\ \text{kJ}}{(\text{g mol})(\text{K})}\left|\frac{329.2\ \text{K}}{}\right|\frac{[(3.978)(0.648) - 3.938 + (1.555)(3.85)]}{1.07 - 0.648}$$

$$= 30.0\ \text{kJ/g mol (insignificant error)}$$

Self-Assessment Test

Questions

1. Indicate whether the following statements are true or false:
   a. The molar heat of vaporization of water is 40.7 kJ/g mol.
   b. The molar heats of vaporization you look up in a reference book or database come from experimental data.
   c. The molar heat of fusion is the amount of energy necessary to melt or freeze 1 g mol of substance at its melting point.
2. Define (a) heat of vaporization, (b) heat of condensation, and (c) heat of transition.
3. Why do engineers use the term *heat of* for the energy change that occurs in a phase transition rather than the better term *enthalpy change*?

Problems

1. Ethane (C_2H_6) has the heat of vaporization of 14.707 kJ/g mol at 184.6 K. What is the estimated heat of vaporization of ethane at 210 K?
2. At 0°C you melt 315 g of H_2O. What is the energy change corresponding to the process?
3. One hundred grams of H_2O exist in the gas phase at 395 K. How much energy will it take to condense all of the H_2O at 395 K?

Thought Problem

A tanker used in the construction of pavement was being filled with tar at a temperature of slightly less than 100°C. When half full, the tar pump failed. To clear the inlet line of solid tar, the line was blown out with steam. The driver of the tanker went to another location to fill the tanker. A few seconds after the filling commenced, hot tar erupted from the tanker manhole. Why did this occur?

Discussion Question

An advertisement in the paper said you can buy a can of instant car cooler from which you spray the product inside a car to reduce the temperature. The spray consists of a mixture of ethanol and water. The picture in the advertisement shows a thermometer "before" registering 41°C and the thermometer "after" registering 27°C. Explain to the owner of Auto Sport, an auto parts retailer, whether or not he should buy a case of the spray cans from the distributor.

9.2.9 Tables and Charts to Retrieve Enthalpy Values

The firewood lies there but every man must gather and light it himself.

The Lone Ranger

Tables listing smoothed experimental data can cover the values of physical properties well beyond the range valid for a single equation. Because the most commonly measured properties are temperature and pressure, tables of enthalpies (and internal energies) for pure compounds usually are organized in columns and rows, with T and p being the independent variables. If the intervals between table entries are close enough, linear interpolation between entries is reasonably accurate. For example, look at the following calculation for saturated steam in SI units from the steam tables on the CD in the back of this book. The units are kilojoules per kilogram. If you want to calculate the enthalpy change of saturated steam from 305 to 307 K by linear interpolation from a table graduated in 5-degree increments, as explained in Chapter 8 in connection with the steam tables, you would carry out the following computation:

$$\hat{H}_{307} = \hat{H}_{305} + \frac{\hat{H}_{310} - \hat{H}_{305}}{T_{310} - T_{305}}(T_{307} - T_{305})$$

$$= 2558.9 + \frac{2567.9 - 2558.9}{310 - 305}(307 - 305) = 2562.5$$

But using the CD, you can insert 307 K into the dialog box and directly get essentially the same value.

Usually a steam table lists values for the heat of vaporization as well as the enthalpy for liquid water and water vapor. The values of $\Delta\hat{H}_v$ are automatically included in $\hat{H}_g$. Note that the steam tables (and similar tables) include the effects of pressure changes as well as phase transition.

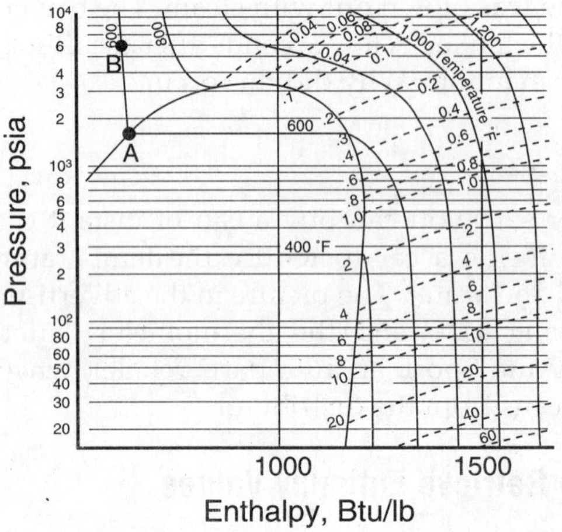

Figure 9.13 At higher pressures in which the compound is liquid, the lines of constant temperature (such as 600°F) are almost vertical, and thus the enthalpy of the saturated liquid can be substituted for the actual enthalpy if necessary.

For compressed liquids at values of pressure higher than the saturation pressure you can use the properties of the saturated liquid at the same temperature as a good approximation if needed. Look at the 600°F line in Figure 9.13, which is almost vertical. At point B you could substitute the value of $\hat{H}$ at point A.

You should look at the tables of enthalpies at 1 atm of selected gases in Appendix F, Tables F.2–F.6, on the CD that accompanies this book. Remember that enthalpy values are all relative to some reference state. What is the reference state for the gases in Table F.6? Did you decide on 273 K (0°C) and 1 atm? Remember, you can calculate enthalpy changes by subtracting the initial enthalpy from the final enthalpy for any two sets of conditions **whatever the reference state**. Thus, despite the reference state for the gas enthalpy tables being slightly different from that of the steam tables, you can choose as a reference state for a problem another temperature, such as the temperature of an entering stream, to serve as a reference for an enthalpy change.

You no doubt have heard the saying "A picture is worth a thousand words." Something similar might be said of two-dimensional charts, such as Figure 9.13, namely, that you can get an excellent idea of the values of the enthalpy and other properties of a substance in all of the regions displayed in a chart. Although the accuracy of the values read from a chart may be limited (depending on the scale and accuracy of the chart), tracing out various processes on a chart enables you to rapidly visualize and analyze what is taking place. Charts are certainly a simple and quick method for you to get data to compute approximate enthalpy changes. Figure 9.14 for *n*-butane is an example chart. A number of sources of charts are listed in the references at the end of the chapter. Appendix K on the CD that accompanies this book contains charts for toluene and carbon dioxide. The CD also has *p-H* charts

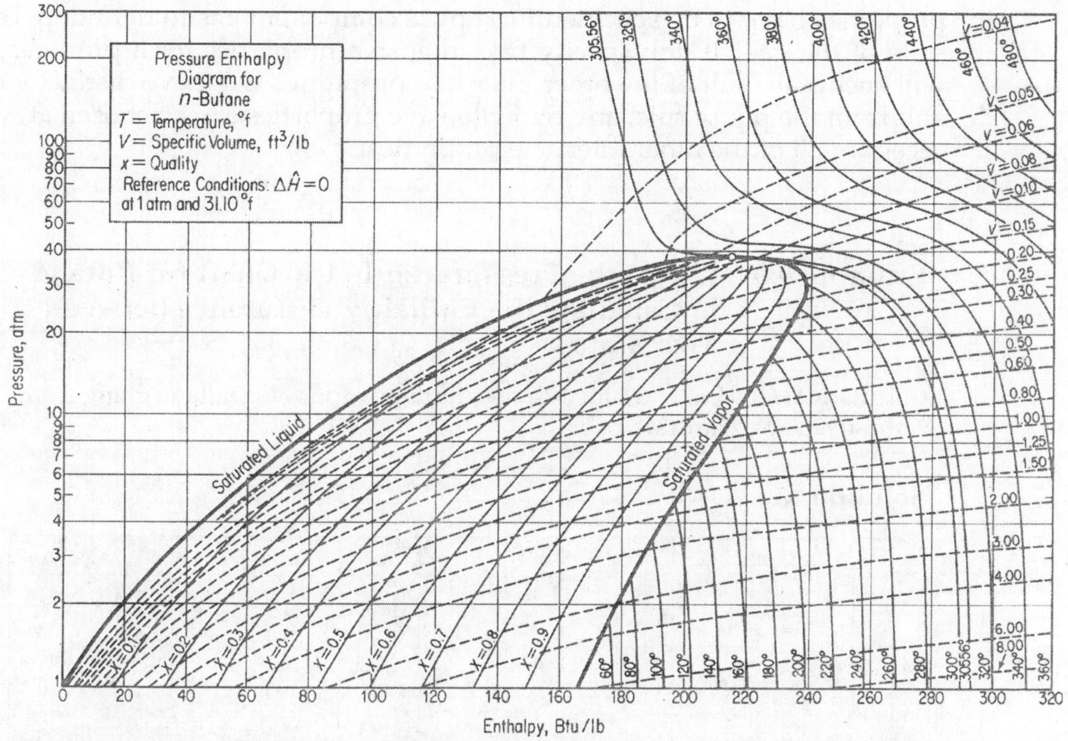

Figure 9.14 A pressure-versus-enthalpy chart for butane showing lines of constant temperature and constant specific volume

that can be expanded to read property values more accurately. A search of the Internet will turn up additional charts.

Charts are drawn with various coordinates, such as p versus $\hat{H}$, p versus $\hat{V}$, or p versus T. Since a chart has only two dimensions, the coordinates can represent only two variables. The other variables of interest have to be plotted as lines or curves of constant value across the face of the chart. Now for some remarks about the two-phase region of a compound, such as point C in Figure 9.13. You will find that in the two-phase region only values for the saturated liquid and saturated vapor are listed in a table. You have to interpolate between these saturated liquid and vapor values to get properties of vapor-liquid mixtures as explained in Chapter 8. However, if you know the quality (fraction vapor) and the temperature (or pressure), the program on the CD will return the properties of the mixture shown as parameters. Similarly, on a chart with pressure and enthalpy as the axes, lines of constant specific volume and/or temperature might be drawn as in Figure 9.14. This situation limits the number of variables that can be presented in tables. How many properties have to be specified for a pure component gas to definitely fix the state of the gas? If you specify two intensive properties for a pure gas, you will ensure that all of the other intensive properties will have distinct values, and from the phase rule, any two intensive properties can be chosen at will to ascertain all of the other intensive properties.

Example 9.9 Use of the Pressure-Enthalpy Chart for Butane to Calculate the Enthalpy Difference between Two States

Calculate $\Delta\hat{H}$, $\Delta\hat{V}$, and ΔT for 1 lb of saturated vapor of n-butane going from 2 atm to 20 atm (saturated).

Solution

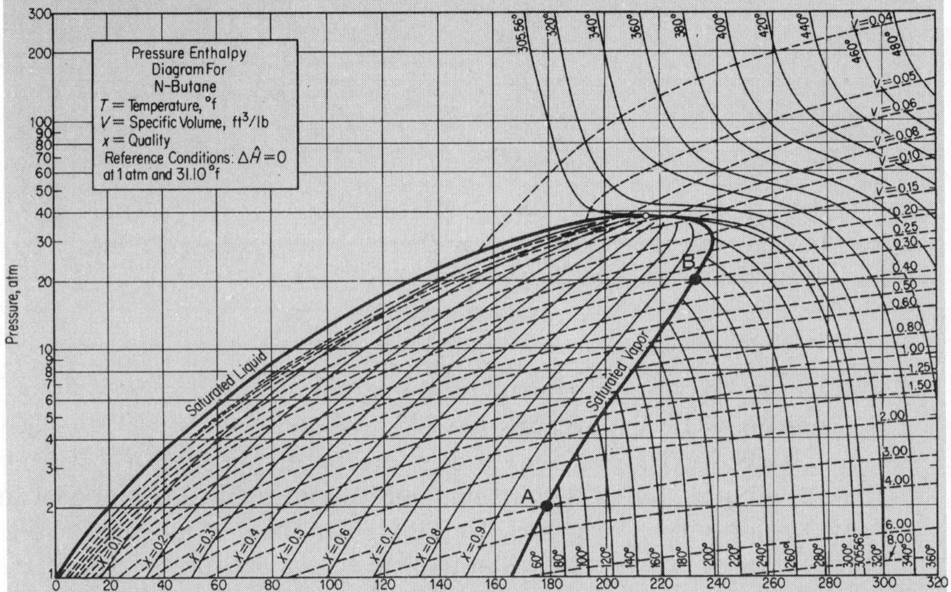

Figure E9.9

Obtain the necessary data from Figure 9.14 or Figure E9.9.

	$\hat{H}$ (Btu/lb)	$\hat{V}$ (ft^3/lb)	T (°F)
Saturated vapor at 2 atm:	179	3.00	72
Saturated vapor at 20 atm:	233	0.30	239

$$\Delta\hat{H} = 233 - 179 = 54 \text{ Btu/lb}$$

$$\Delta\hat{V} = 0.30 - 3.00 = -2.70 \text{ ft}^3/\text{lb}$$

$$\Delta T = 239 - 72 = 167°\text{F}$$

9.2.10 Computer Databases

Numbers have souls, and you can't help but get involved with them in a personal way.
Paul Auster, *The Music of Chance*

Values of the properties of thousands of pure substances and mixtures are available in databases accessible via software programs that can provide sets of values of physical properties for any given state. Thus, you avoid the need for interpolation and/or auxiliary computation, such as problems in which ΔU must be calculated from $\Delta U = \Delta H - \Delta(pV)$. The CD that accompanies this book is an example. The program is abridged so that it does not have the accuracy or cover the range of commercially available programs, but it is quite suitable for solving problems in this text. Professor Yaws has been kind enough to provide the data for most of the 750 compounds. If you want to save time in your calculations, we recommend using this software.

You can also purchase comprehensive databases and design packages that contain different ways to calculate the properties of a large number of compounds. You can get immediate access to many free information systems via the Internet. Here are several such sources of information, including tables, equations, and software (an asterisk indicates that the source is a directory of sources of data):

www.aiche.org/DIPPR*
www.gpengineeringsoft.com/pages/pdtphysprops.html*
www.indiana.edu/~cheminfo/ca_ppi.html*
www.lib.berkeley.edu/ENGI/physchemData.html
http://library.csus.edu/guides/rogenmoserd/hall/chemistryphysical.html*
http://library.njit.edu/researchhelpdesk/subjectguides/phys-prop.php*
www.molknow.com
www.nal.usda.gov/wqic/dbases.shtml
www.nist.gov/srd
www.pirika.com/chem/index.html

Some sites require that you use the software on the site. Others sell individual licenses so that you can download software to your computer.

What kind of data can you find that is pertinent to this book? Among other parameters you can locate are

Boiling point	Heat of condensation
Bubble point and dew point for binary mixtures	Heat of formation
	Heat of vaporization
Critical properties	p, $\hat{V}$, and $\hat{T}$ values for gases and liquids
Density of liquids	Vapor-liquid equilibria
Heat capacity	Vapor pressure

Some sites are quite specialized, such as those devoted to the properties of petroleum fluids, resins, solvents, and so on. Some sites consist of directories of other sites that pertain to specialized topics such as atomic and molecular data, dielectric constants, ionization energies, refraction indices, refrigerants, safety, and so on. Many sites also have helpful links to other sites.

How valid (consistent, accurate, rectified) are the data you collect from Internet sources? Are the data from a reliable source? Have the data been audited? Are they free of anomalies? What is the uncertainty? Are they easy to use? Such questions are best answered by the footnote found at one data site:

> Neither staff nor developers assume any legal responsibility for the information provided on these pages. Use at your own risk!

9.3 Energy Balances without Reaction

In Section 9.2 we discussed Q, W, U, and H but did not try to relate them in an energy balance. In this section we explain what energy balances are all about. We also show you how to formulate, simplify, and solve energy balances for variables of interest. This chapter focuses on processes without chemical reaction. Chapter 10 covers balances in which chemical reactions occur in a process.

As you know, the principle of the **conservation of energy** states that the total energy of a system *plus the surroundings* can neither be created nor destroyed.[4]

[4]Can mass be converted into energy according to $E = mc^2$? It is not correct to say that $E = mc^2$ means mass is converted into energy. The equal sign can mean that two quantities have the same value as in measurements of two masses in an experiment, or it may mean (as in general relativity) that the two variables are the same or are equivalent things. It is in the latter sense that $E = mc^2$ applies, and not in terms of converting a *rest mass* into energy. You might write $\Delta E = c^2\Delta m$. If Δm is negative, ΔE is also negative. What this means is that as the inertia decreases, ΔE decreases, and vice versa.

Die Energie der Welt ist Konstant.

Clausius

Julius Mayer (1814–1878) gave the first precise quantitative formulation of the principle of the conservation of energy. A journal refused to publish his ideas originally, and many of his contemporaries laughed at him when he explained his ideas. Ph. Von Jolly said if what Mayer proposed was true, it should be possible to heat water by shaking it! Which, of course, you can do if your experiment is performed properly.

The principle is well founded based on experimental measurements. During an interaction between a system and its surroundings, the amount of energy gained by the system must be exactly equal to the amount of energy lost by the surroundings. Rather than focus on the words the *law of the conservation of energy* in this book, we will use the words *energy balance* to avoid confusion with the colloquial use of the words *energy conservation*, that is, reduction of energy waste or increased efficiency of energy utilization. Keep in mind three important points as you read what follows:

1. We examine only systems that are homogeneous, not charged, and without surface effects, in order to make the energy balance as simple as possible.
2. The energy balance is developed and applied from the macroscopic viewpoint (overall about the system) rather than from a microscopic viewpoint (i.e., an elemental volume within the system).
3. The energy balance will be presented as a difference equation that incorporates net quantities over a time interval analogous to the mass balance in Chapter 3. Chapter 17 treats differential equations.

As America entered the First World War, in 1917, an Armenian named Garabed Giragossian petitioned Congress to investigate his miraculous and eponymous Garabed, an invention that would provide unlimited energy, "a natural force that we can utilize and have energy as we like, without toil or expense." First he secured the endorsements of the director of music in the Boston Public Schools, the president of the Board of Trustees of the Boston Public Library, and the president of a shipbuilding concern; when he began his lobbying campaign on Capitol Hill, reports about his machine appeared in the *Literary Digest* and the *St. Louis Post-Dispatch*. The House voted 234 to 14 to investigate the Garabed. The congressional investigation revealed that Giragossian had made the most elementary of errors; he had confused power and force. The Garabed was a simple flywheel set spinning manually with pulleys and kept in motion with an electric motor. Giragossian hoped to extract his "free energy" from the difference between the 10 horsepower required to stop the wheel and the 20th of 1 horsepower needed to keep it going; he failed to realize that when the wheel was stopped quickly, it spent all of the energy it had gradually stored up.

In more recent times inventors have been able to demonstrate devices that indeed put out more energy than was put in. Most such devices have been based on charging and discharging a storage battery repeatedly. Each time more electrical energy is recovered than was used to charge the battery (as long as it was not charged too rapidly). How is this outcome possible? If you do not know, read the footnote.[5]

Let us start the formulation of a general energy balance in words by reproducing the general mass balance, Equation (3.1), for a time interval t_1 to t_2:

$$\left\{\begin{array}{l}\textbf{Accumulation of}\\ \textbf{material within}\\ \textbf{the system}\\ \textbf{from } t_1 \textbf{ to } t_2\end{array}\right\} = \left\{\begin{array}{l}\textbf{Final material}\\ \textbf{in the system}\\ \textbf{at } t_2\end{array}\right\} - \left\{\begin{array}{l}\textbf{Initial material}\\ \textbf{in the system}\\ \textbf{at } t_1\end{array}\right\}$$

$$= \left\{\begin{array}{l}\textbf{Total cumulative}\\ \textbf{material flow into}\\ \textbf{the system from}\\ t_1 \textbf{ to } t_2\end{array}\right\} - \left\{\begin{array}{l}\textbf{Total cumulative}\\ \textbf{material flow}\\ \textbf{out of the system}\\ \textbf{from } t_1 \textbf{ to } t_2\end{array}\right\} \quad (3.1a)$$

$$+ \left\{\begin{array}{l}\textbf{Generation of}\\ \textbf{material within}\\ \textbf{the system}\\ \textbf{from } t_1 \textbf{ to } t_2\end{array}\right\} - \left\{\begin{array}{l}\textbf{Consumption of}\\ \textbf{material within}\\ \textbf{the system}\\ \textbf{from } t_1 \textbf{ to } t_2\end{array}\right\}$$

The last two terms, generation and consumption, apply to reactions and are discussed in the next chapter (Chapter 10); hence they will be neglected in our discussion in this section.

If you substitute the word *energy* for *material* in each term of Equation (3.1a), you will have the general energy balance!

$$\left\{\begin{array}{l}\textbf{Accumulation of}\\ \underline{\textbf{\textit{energy}}} \textbf{ within}\\ \textbf{the system}\\ \textbf{from } t_1 \textbf{ to } t_2\end{array}\right\} = \left\{\begin{array}{l}\textbf{Final } \underline{\textbf{\textit{energy}}}\\ \textbf{in the system}\\ \textbf{at } t_2\end{array}\right\} - \left\{\begin{array}{l}\textbf{Initial } \underline{\textbf{\textit{energy}}}\\ \textbf{in the system}\\ \textbf{at } t_1\end{array}\right\}$$

[5]The battery temperature goes up and can deliver more (watt) (hours) at the higher temperatures.

$$= \left\{\begin{array}{l}\textbf{Total cumulative}\\ \underline{\textbf{\textit{energy}}}\textbf{ flow into}\\ \textbf{the system from}\\ t_1 \textbf{ to } t_2\end{array}\right\} - \left\{\begin{array}{l}\textbf{Total cumulative}\\ \underline{\textbf{\textit{energy}}}\textbf{ flow}\\ \textbf{out of the system}\\ \textbf{from } t_1 \textbf{ to } t_2\end{array}\right\} + \left\{\begin{array}{l}\textbf{Generation of}\\ \underline{\textbf{\textit{energy}}}\textbf{ within}\\ \textbf{the system}\\ \textbf{from } t_1 \textbf{ to } t_2\end{array}\right\}$$

$$- \left\{\begin{array}{l}\textbf{Consumption of}\\ \underline{\textbf{\textit{energy}}}\textbf{ within}\\ \textbf{the system}\\ \textbf{from } t_1 \textbf{ to } t_2\end{array}\right\} \tag{9.17}$$

Of course, Equation (9.17) is the energy balance in words. To substitute symbols for words requires a little more explanation. Let's start with the case of a closed system and then focus on the more prevalent case of an open system. Both steady-state and unsteady-state systems will be reviewed. The last two terms on the right-hand side of Equation (9.17) will **not** be involved until Chapter 10. In this way Equation (9.17) when finally presented in symbols will not be so formidable.

9.3.1 Unsteady-State, Closed Systems

For a closed, unsteady-state system the energy balances in the symbols that we have previously defined are

$$\underset{\text{accumulation}}{\Delta E_{\text{inside}}} \equiv \Delta(U + PE + KE)_{\text{inside}} = \underset{\text{heat transfer}}{Q} + \underset{\text{work}}{W} \tag{9.18}$$

If several components are involved in the process, then U_{inside} is the sum of the mass (or moles) of each component i times the respective specific internal energy of each component i, $\hat{U}_i$. Note that $(Q + W)$ represents the total flow of energy into the closed system. We do not put a Δ representing the symbol for a change in states before Q or W because they are not state variables. Remember that Q and W are both *positive* when the net transfer is into the system, and E represents the sum of $(U + KE + PE)$ associated with *mass inside the system* itself. Be careful; in some books W is defined as positive when done by the system. Also keep in mind that each term in Equation (9.18) represents the respective *net cumulative* amount of energy over the time interval from t_1 to t_2, not the respective energy per unit time, a rate, which would be denoted by an overlay dot.

In closed systems, the values of ΔPE and ΔKE in ΔE are usually negligible or zero; hence, often you see $\Delta U = Q + W$ used as the energy balance.

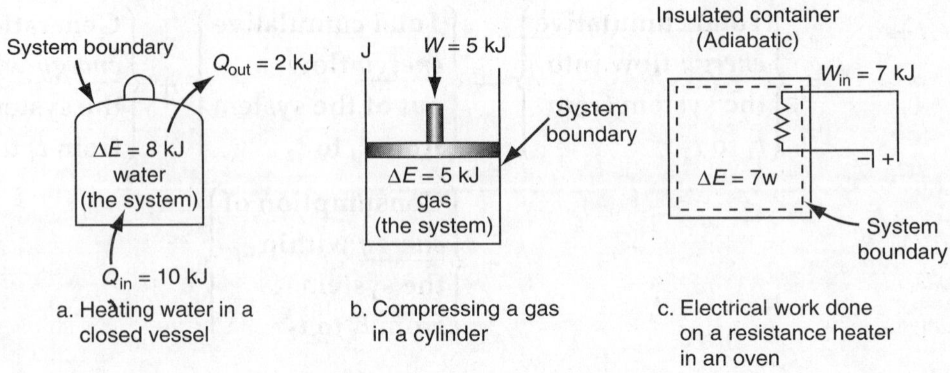

Figure 9.15 Examples of unsteady-state, closed systems that involve energy changes

If the sum of Q and W is positive, ΔE increases; if negative, ΔE decreases. Does $W_{\text{system}} = -W_{\text{surroundings}}$? Not necessarily, as you will learn subsequently. For example, in Figure 9.15c the electrical work done by the surroundings on the system degrades into the internal energy (increase of temperature) of the system, not in expanding its boundaries.

Figure 9.15 illustrates three examples of applying Equation (9.18) to simple closed, unsteady-state systems. In Figure 9.15a, 10 kJ of heat are transferred through the fixed boundary (bottom) of a vessel with 2 kJ being transferred out at the top during the same time period. Thus, $\Delta U = \Delta E$ increases by 8 kJ. In Figure 9.15b, a piston does 5 kJ of work on a gas whose internal energy increases by 5 kJ. In Figure 9.15c, the voltage difference between the system and surroundings forces a current into a system in which no heat transfer occurs because of the insulation on the system.

Equation (9.18) involves several variables whose values have to be specified or solved for: U_{inside}, $\Delta KE_{\text{inside}}$, $\Delta PE_{\text{inside}}$, Q, and W. Furthermore, the specific internal energy, $\hat{U}_{\text{inside}}$, is a function of T and $\hat{V}$, or alternatively, T and p, inside the system. The net number of variables to be considered in a degree-of-freedom analysis involving Equation (9.18) is five using U_{inside}, or six substituting T_{inside} and p_{inside} for U_{inside}. Almost always $\Delta KE_{\text{inside}}$ and $\Delta PE_{\text{inside}}$ are zero so that a practical count of variables is three or four. With only one equation, Equation (9.18), the energy balance, the number of unknowns, before specifications and material balances come into play, will be two or three.

The next example will help you to ascertain how good your grasp is of the concept of applying energy balances to closed, unsteady-state systems.

Example 9.10 Application of an Energy Balance to a Closed, Unsteady-State System without Reaction

Alkaloids are chemical compounds containing nitrogen that can be produced by plant cells. In an experiment, an insulated closed vessel 1.673 m^3 in volume was injected with a dilute water solution containing two alkaloids: ajmalicine and serpentine. The temperature of the solution was 10°C. To obtain an essentially dry residue of alkaloids, all of the water in the vessel was vaporized. Assume that the properties of water can be used in lieu of the properties of the solution. How much heat had to be transferred to the vessel if 1 kg of saturated liquid water initially at 10°C was completely vaporized to a final condition of 100°C and 1 atm? See Figure E9.10. Ignore any air present in the vessel (or assume an initial vacuum existed).

Solution

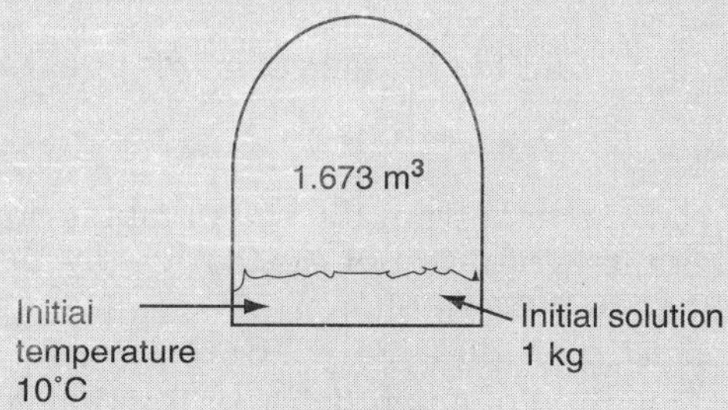

Figure E9.10

The system is the closed vessel; hence from the viewpoint of the material balance it is steady-state, but from the viewpoint of the energy balance it is unsteady-state.

Sufficient data are given in the problem statement to fix the initial state and the final state of the water. You can look up the properties of water in the steam tables or the CD in the back of this book. Note that the specific volume of steam at 100°C and 1 atm is 1.673 m^3/kg (!).

	Initial state (liquid)	Final state (gas)
p	vapor pressure	1 atm
T	10.0°C	100°C
$\hat{U}$	17.7 kJ/kg	2506.0 kJ/kg

You can look up additional properties of water such as $\hat{V}$ and $\hat{H}$, but they are not needed for the problem.

(Continues)

Example 9.10 Application of an Energy Balance to a Closed, Unsteady-State System without Reaction (*Continued*)

The system is closed, unsteady-state, so Equation (9.18) applies:

$$\Delta E = (\Delta U + \Delta PE + \Delta KE)_{\text{inside}} = Q + W$$

Because the system (the water) is at rest, $\Delta KE = 0$. Because the center of mass of the water changes so slightly on evaporation relative to Q, you can assume $\Delta PE = 0$. Calculate the value if you have any doubts. No work is involved (fixed tank boundary, no engine in the system). You can conclude, using

Basis: 1 kg of H_2O evaporated

that

$$Q = \Delta U = m\Delta\hat{U} = m(\hat{U}_2 - \hat{U}_1)$$

$$Q = \frac{1 \text{ kg } H_2O}{} \bigg| \frac{(2506.0 - 17.7) \text{ kJ}}{\text{kg}} = 2488 \text{ kJ}$$

Note that we used the data from the steam tables on the CD that accompanies this book. Some tables do not contain values of $\hat{U}$, just values of $\hat{H}$; $\hat{U}$ must then be calculated. Most often $\Delta(p\hat{V})_{\text{inside}}$ is negligible, but not always. Recall that $\hat{H}_{\text{inside}}$, $\hat{U}_{\text{inside}}$, and $(p\hat{V})_{\text{inside}}$ have an arbitrary reference state that cancels in the calculation of an enthalpy or internal energy difference. If you use two different sources of data, take care to account for differences in the reference values of each source.

How would the solution differ if the container prior to the injection contained dry air at 1 atm?

Example 9.11 Application of the Energy Balance

Ten pounds of CO_2 at room temperature (80°F) are stored in a fire extinguisher that has a volume of 4.0 ft^3. How much heat must be transferred from the extinguisher so that 40% of the CO_2 becomes liquid?

Solution

This problem involves a closed, unsteady-state system (Figure E9.11) without reaction. You can use the CO_2 chart in Appendix K on the CD that accompanies this book to get the necessary property values.

Steps 1–4

The specific volume of the CO_2 is $4.0/10 = 0.40$ ft^3/lb. The reference state for the CO_2 chart is -40°F, saturated liquid. You can locate from the CO_2 chart at $\hat{V} = 0.40$ and $T = 80$°F that CO_2 in the extinguisher is a gas at a pressure of 300 psia with the initial $\Delta\hat{H} = 160$ Btu/lb.

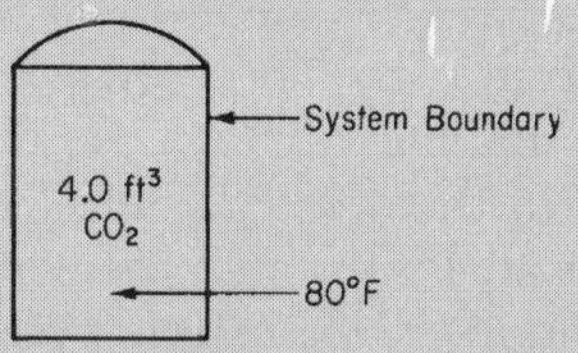

Figure E9.11

Step 5

Basis: 10 lb of CO_2

Steps 6 and 7

The material balance is easy—the mass in the system is constant at 10 lb. As to the energy balance,

$$\Delta E = Q + W$$

W is zero because the boundary of the system is fixed, no electrical work is done, and so forth; hence, with $\Delta KE = \Delta PE = 0$ inside the system:

$$Q = \Delta U = \Delta H - \Delta(pV)$$

You cannot obtain values of $\Delta\hat{U}$ from the CO_2 chart, just values of $\Delta\hat{H}$. Let us for the moment not ignore the value of $\Delta(pV)$. By following the constant volume line of 0.40 ft^3/lb to the spot where the quality is 0.6, you can locate the final state of the system, and all of the final properties can be identified or calculated. Two values from the chart are

$$\Delta\hat{H}_{\text{final}} = 81 \text{ Btu/lb} \quad \text{and} \quad p_{\text{final}} = 140 \text{ psia}$$

You can conclude that since all of the variables can be assigned values except Q, only one unknown exists, and with one equation available (the energy balance), the degrees of freedom for the problem are zero.

Steps 7–9

$$Q = \left\{(81 - 160) - \left[\frac{(140)(144)(0.40)}{778.2} - \frac{(300)(144)(0.40)}{778.2}\right]\right\}10$$

$$= -672 \text{ Btu (heat is removed)}$$

By how much would the answer of -672 Btu be in error if the term $-\Delta(pV)$ were ignored?

Example 9.12 Application of the Energy Balance to Plasma Etching

Argon gas in an insulated plasma deposition chamber with a volume of 2 L is to be heated by an electric resistance heater. Initially the gas, which can be treated as an ideal gas, is at 1.5 Pa and 300 K. The 1000-ohm heater draws current at 40 V for 5 min (i.e., 480 J of work is done on the system by the surroundings). What are the final gas temperature and pressure in the chamber? The mass of the heater is 12 g and its heat capacity is 0.35 J/(g)(K). Assume that the heat transfer through the walls of the chamber from the gas at this low pressure and in the short time period involved is negligible.

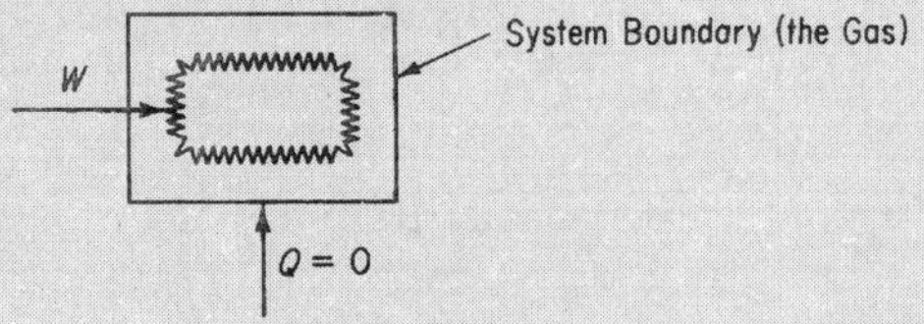

Figure E9.12

Solution

Pick the system shown in Figure E9.12. No reaction occurs. The fact that the electric coil is used to "heat" (raise the temperature of) the argon inside the system does not mean that heat transfer takes place to the selected system from the surroundings. Only work is done. The system does not exchange mass with the surroundings; hence it is steady-state with regard to mass but is unsteady-state with respect to energy.

Steps 1–4

Because of the assumption about the heat transfer from the chamber wall, $Q = 0$. W is given as +480 J (work done on the system) in 5 min.

Step 5

Basis: 5 min

Steps 6 and 7

The general energy balance (with $\Delta PE = \Delta KE = 0$ inside the system) is

$$\Delta E = \Delta U = Q + W$$

and, after assignment of known values to the respective variables, reduces to $\Delta U = 480$ J.

One way to solve the problem is to find the T and p associated with a value of ΔU equal to 480 J. A table, equation, or chart for argon would make this procedure easy. In the absence of such a data source, we will fall back on the assumption that the argon gas is an ideal gas (as is true), so that $pV = nRT$. Initially we know p, V, and T and thus can calculate the amount of the gas:

$$n = \frac{pV}{RT} = \left.\frac{1.5\ \text{Pa}}{}\right|\left.\frac{2\ \text{L}}{}\right|\left.\frac{10^{-3}\ \text{m}^3}{1\ \text{L}}\right|\left.\frac{1\ (\text{g mol})(\text{K})}{8.314\ (\text{Pa})(\text{m}^3)}\right|\frac{}{300\ \text{K}} = 1.203 \times 10^{-6}\ \text{g mol}$$

You are given the heater mass and its heat capacity of $C_v = 0.35\ \text{J}/(\text{g})(\text{K})$. The C_v of the gas can be calculated. Since $C_p = \frac{5}{2}\text{R}$ (see Table 9.3), then

$$C_v = C_p - \text{R} = \frac{5}{2}\text{R} - \text{R} = \frac{3}{2}\text{R}$$

You have to pick a reference temperature for the calculations. The most convenient reference state is 300 K. Then ΔU for the gas and the heater can be calculated assuming both the heater and the gas end up at the same temperature:

$$\text{Gas:}\quad \Delta U_g = 1.203 \times 10^{-6}\int_{300}^{T} C_v dT = 1.203 \times 10^{-6}\left(\frac{3}{2}\text{R}\right)(T - 300)$$

$$\text{Heater:}\quad \Delta U_h = 12\ \text{g}\left(\frac{0.35\ \text{J}}{(\text{g})(\text{K})}\right)(T - 300)$$

The mass balance is trivial—the mass in the chamber does not change. The unknown is T, and one equation is involved, the energy balance, so the degrees of freedom are zero.

Steps 8 and 9

Because $\Delta U = 480$ J, you can calculate T from the energy balance:

$$\Delta U = 480\ \text{J} = (12)(0.35)(T - 300) + (2.302 \times 10^{-6})\left(\frac{3}{2}\right)(8.314)(T - 300)$$

$$T = 414\ \text{K}$$

The final pressure is obtained from

$$\frac{p_2V_2}{p_1V_1} = \frac{n_2RT_2}{n_1RT_1} \quad \text{or}$$

$$p_2 = p_1\left(\frac{T_2}{T_1}\right) = 1.5\left(\frac{414}{300}\right) = 2.07\ \text{Pa}$$

Self-Assessment Test

Questions

1. Indicate whether the following statements are true or false:
   a. The law of the conservation of energy says that all of the energy changes in a system must add up to zero.
   b. The law of the conservation of energy says that no net change, or no creation or destruction of energy, is possible in a system.
   c. The law of the conservation of energy says that any change in heat that is not exactly equal to and opposite a change in work must appear as a change in the total internal energy in a system.
   d. Heat can flow into, and out of, a system from and to its surroundings. Work can be done by the system on the surroundings, and vice versa. The internal energy of a system can increase or decrease by any of these processes.
   e. For all adiabatic processes between two specified states of a closed system, the net work done is the same regardless of the nature of the closed system and the details of the process.
   f. A closed, unsteady-state system cannot be an isothermal system.
   g. Heat transfer to an unsteady-state, closed system cannot be zero.
   h. A closed system must be adiabatic.
2. Can heat be converted completely into work?

Problems

1. A closed system undergoes three successive processes: $Q_1 = +10$ kJ, $Q_2 = +30$ kJ, and $Q_3 = -5$ kJ, respectively. In the first process, $\Delta E = +20$ kJ, and in the third process, $\Delta E = -20$ kJ. What is the work in the second process, and what is the net work output of all three stages if $\Delta E = 0$ for the overall three-stage process?
2. A closed tank contains 20 lb of water. If 200 Btu are added to the water, what is the change in internal energy of the water?
3. When a batch of hot water at 140°F is suddenly well mixed with cold water at 50°F, the water that results is at 110°F. What was the ratio of the hot water to the cold water? You can use the steam tables to get the data.

Thought Problems

1. Does the law of the conservation of energy explain how energy can be extracted from uranium in a nuclear reactor?
2. Does the law of the conservation of energy state that energy can neither be created nor destroyed?
3. The current concern with "renewable" energy seems to be a paradox because of the law of the conservation of energy. Windmills extract power from the wind and convert it to electricity. Can this affect the weather in some way?

Discussion Question

The head of a small electronics firm solicited money to develop a device that uses noise from an engine, or other source, to generate electricity. He said, "By using automobile noise the device could furnish enough power to handle the lights, air conditioning, radio, and so on, everything except the power to start the car." In what term in the energy balance would the noise be accommodated? Do you think that the device would be practical?

9.3.2 Steady-State, Closed Systems

Recall that steady-state means the accumulation in the system is zero, and that closed means that no mass flow occurs across the system boundary. Only Q and W may occur during a time interval. What does Equation (9.17) reduce to?

$$\begin{aligned} \Delta KE &= 0 & \qquad \Delta U &= 0 \\ \Delta PE &= 0 & \text{hence} \quad \Delta E &= 0 \end{aligned}$$

so that

$$Q + W = 0 \tag{9.19}$$

If you set $W = -Q$, you can conclude that all of the work done on a closed, steady-state system must be transferred out as heat $(-Q)$ so that the initial and final states of the system are the same. However, ironically, the reverse is false; namely, the heat added to a *closed, steady-state system* Q *does not always equal the work done by the system* $(-W)$. If this were true, you could construct a perpetual motion machine by building a thermoelectric engine that would transfer heat from a system at high temperature, convert all of the heat to work, and convert the work back to heat at the high temperature in order to run the cycle once more. The second law of thermodynamics states the conditions that make it impossible to construct such a perpetual motion machine. You will learn about this topic when you study thermodynamics, or if you read one of the many references at the end of the chapter.

Consequently, if W_{out} is less than Q_{in}, where does the residual energy in Q go? It changes ΔE; hence the system cannot then be classified as steady-state. Figure 9.16 illustrates some examples of closed, steady-state systems with energy interchange. Contrast them with Figures 9.15a, b, and c.

In the processes shown in Figure 9.16, Q and W are as follows: (a) $W = 0$ and hence $Q = 0$; (b) $W = 5$ kJ and hence $Q = -5$ kJ; (c) $W = 7$ kJ and hence $Q = -7$ kJ.

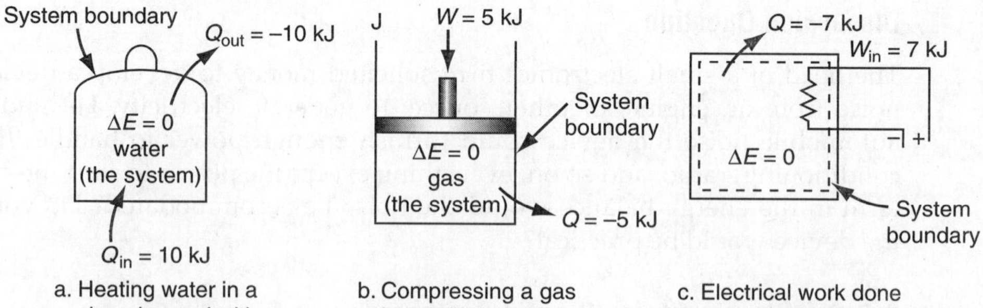

Figure 9.16 Examples of closed, steady-state systems that involve energy changes

9.3.3 General Energy Balance for Open Systems

We shall not cease from exploration, and the end of all our exploring will be to arrive where we started and to know the place for the first time.

T. S. Eliot

Now that we have discussed closed systems, it is time to focus on processes represented by open systems, which are much more common than closed systems. An open system involves mass flow in the energy balance. If mass flows in and out of a system, the mass carries energy along with it. What types of energy? Just the same types that are associated with the mass inside the system, namely, $\hat{U}$, $\hat{PE}$, and $\hat{KE}$. All you have to do, then, is include notation in the energy balance, Equation (9.17), to account for these three types of energy for each stream going in and out of the system, when applicable.

Figure 9.17 illustrates a general open system with one stream containing one component entering and leaving. We are not concerned with the internal details of the process, only with the energy transfers into and out of the system, and changes within the system as a whole. Table 9.4 lists the notation that we will use in writing the general energy balance, Equation (9.17), in symbols. The overlay caret (^) still means the energy per unit mass, and m with the subscript t_1 or t_2 denotes mass inside the system at time t_1 or t_2, respectively, whereas m with the subscript of just

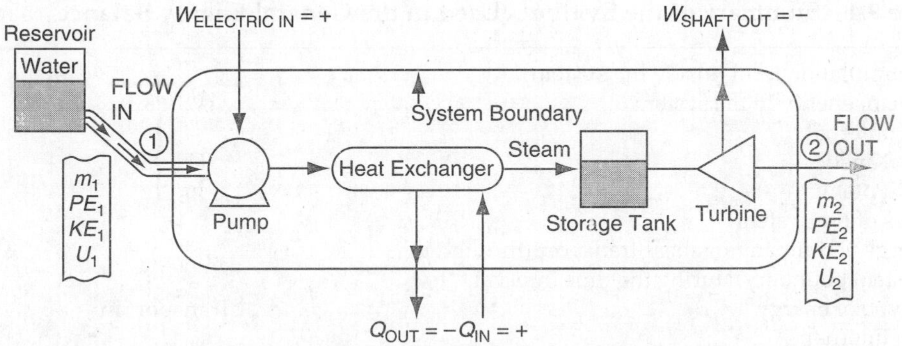

Figure 9.17 An open, unsteady-state system. The system is inside the boundary; 1 and 2 denote the points of entering and exiting mass flows, respectively. The liquid mass in the storage tank and the temperature possibly can vary inside the system.

1 or 2 denotes the cumulative flow of mass (during the interval t_1 to t_2) through the boundary at 1 or 2, respectively. Note that in Table 9.4 we have split the work term into distinct parts.

For a specified time interval, you can substitute the notation in Table 9.4 for each of the terms in the general energy balance, Equation (9.17).

Accumulation inside the system during t_1 to t_2:

$$\Delta E \equiv m_{t_2}(\hat{U} + \hat{KE} + \hat{PE})_{t_2} - m_{t_1}(U + \hat{KE} + \hat{PE})_{t_1}$$

Energy transfer in with mass during t_1 to t_2:

$$(\hat{U}_1 + \hat{KE}_1 + \hat{PE}_1)m_1$$

Energy transfer out with mass during t_1 to t_2:

$$(\hat{U}_2 + \hat{KE}_2 + \hat{PE}_2)m_2$$

Net energy transfer by heat transfer in or out during t_1 to t_2: Q
Net energy transfer by shaft, mechanical, or electrical work in or out during t_1 to t_2: W
Net energy transfer by work to introduce and remove mass during t_1 to t_2:

$$p_1\hat{V}_1 m_1 - p_2\hat{V}_2 m_2$$

Table 9.4 Summary of the Symbols Used in the General Energy Balance

Accumulation term (inside the system)		
Type of energy in the system	*At time* t_1	*At time* t_2
Internal	U_{t_1}	U_{t_2}
Kinetic	KE_{t_1} } E_{t_1}	KE_{t_2} } E_{t_2}
Potential	PE_{t_1}	PE_{t_2}
Mass of the system	m_{t_1}	m_{t_2}
Energy accompanying mass transport (through the system boundary) during the time interval t_1 to t_2		
Type of energy	Transport in	Transport out
Internal	U_1	U_2
Kinetic	KE_1	KE_2
Potential	PE_1	PE_2
Mass of the flow	m_1	m_2
Net heat exchange between the system and the surroundings during the interval t_1 to t_2	Q	
Work terms (exchange with the surroundings) during the interval t_1 to t_2		
Net shaft, mechanical, and electrical work		W { W_{shaft}, $W_{\text{mechanical}}$, $W_{\text{electrical}}$
Flow work done on the system by the surroundings to introduce material into the system	$m_1(p_1\hat{V}_1)$	
Flow work done by the system on the surroundings to remove material from the system		$-m_2(p_2\hat{V}_2)$

The quantities $p_1\hat{V}_1$ and $p_2\hat{V}_2$ probably need a little explanation. They represent the so-called "pV work," "pressure energy," "flow work," or "flow energy," that is, the work done by the surroundings to put a unit mass into the system at boundary 1 in Figure 9.17, and the work done by the system on the surroundings as a unit mass leaves the system at boundary 2, respectively. Because the pressures at the entrance and exit to the system are deemed to be constant for differential displacements of mass, the work done per unit mass by the surroundings on the system adds energy to the system at boundary point 1:

$$\hat{W}_1 = \int_0^{\hat{V}_1} p_1 d\hat{V} = p_1(\hat{V}_1 - 0) = p_1\hat{V}_1$$

where $\hat{V}$ is the volume per unit mass. Similarly, the work done by the fluid on the surroundings as the fluid leaves the system at point 2 is $\hat{W}_2 = -p_2\hat{V}_2$.

If we now introduce all of the terms listed in Table 9.4 into the energy balance, Equation (9.17), ignoring the generation and consumption terms, we get a somewhat formidable equation:

$$\Delta E = (\hat{U}_1 + \hat{KE}_1 + \hat{PE}_1)m_1 - (\hat{U}_2 + \hat{KE}_2 + \hat{PE}_2)m_2 + Q + W + p_1\hat{V}_1 m_1 - p_2\hat{V}_2 m_2 \tag{9.20}$$

where

$$\Delta E = E_{t_2} - E_{t_1} = (\hat{U} + \hat{KE} + \hat{PE})_{t_2} m_{t_2} - (\hat{U} + \hat{KE} + \hat{PE})_{t_1} m_{t_1}$$

In an open, unsteady-state system, the accumulation term (ΔE) in the energy balance can be nonzero because

1. The mass in the system changes
2. The energy per unit mass in the system changes
3. Both 1 and 2 occur

Do you want to memorize Equation (9.20)? Of course not. To simplify the notation in Equation (9.20), let us add on the right-hand side

$$p_1\hat{V}_1 m_1 \text{ to } \hat{U}_1 m_1 \quad \text{and} \quad p_2\hat{V}_2 m_2 \text{ to } \hat{U}_2 m_2$$

to get

$$\Delta E = [(\hat{U}_2 + p_2\hat{V}_2) + \hat{KE}_2 + \hat{PE}_2]m_2 - [(\hat{U}_1 + p_1\hat{V}_1) + \hat{KE}_1 + \hat{PE}_1]m_1 + Q + W \tag{9.20a}$$

Next, substitute $\hat{H}$ for the expression $(\hat{U} + p\hat{V})$ into Equation (9.20a) to get

$$\Delta E = (\hat{H}_2 + \hat{KE}_2 + \hat{PE}_2)m_2 - (H_1 + \hat{KE}_1 + \hat{PE}_1)m_1 + Q + W \tag{9.20b}$$

You can now see the origin of the variable that came to be called enthalpy.

Because Equation (9.20b) is still quite detailed, we have a suggestion. To help you memorize the energy balance in symbols, use the form found in many texts that involves the total quantities rather than specific quantities:

$$\boldsymbol{\Delta E = Q + W - \Delta(H + KE + PE)} \tag{9.21}$$

In Equation (9.21) the delta symbol (Δ) **standing for a difference** has two different applications:

1. In ΔE, Δ means final minus initial in time and applies to inside the system.
2. In $\Delta(H + PE + KE)$, Δ *means out of the system minus into the system* pertaining to fluid flows.

When you use Equation (9.21), be sure to link firmly in your mind the two uses of Δ; you should also be able to connect the terms in Equation (9.21) to the respective terms in words in Equation (9.17) to avoid any confusion. We will often use subscripts to distinguish the application of Δ to clarify a variable, such as $\Delta U_{\text{inside the system}}$.

If *more than one component exists* inside the system or in a stream flow, use Equation (9.20b) for each component. Sum the resulting terms of each component to get the total for Equation (9.21) (in Chapters 9 and 10 we ignore the heat of mixing). In applications, Equation (9.21) is almost always simplified by deleting or ignoring terms. Examine Figure 9.18 for some typical simplified equations resulting from various assumptions about a process.

If there is *more than one input and output stream* for the system, you will find it convenient to calculate the properties of each stream separately and sum the respective inputs and outputs so that Equation (9.20) becomes the **general energy balance** (without reaction occurring) for the entire system:

$$\Delta E = E_{t_2} - E_{t_1} = \sum_{\text{input streams (I) } i=1}^{M} m_i^{\text{I}}(\dot{H}_i^{\text{I}} + \hat{KE}_i^{\text{I}} + \hat{PE}_i^{\text{I}}) - \sum_{\text{output streams (O) } i=1}^{N} m_i^{\text{O}}(\hat{H}_i^{\text{O}} + \hat{KE}_i^{\text{O}} + \hat{PE}_i^{\text{O}}) + Q + W \tag{9.22}$$

Where $E_t = U_t + KE_t + PE_t$ inside the system at time t
M = number of input streams
N = number of output streams
I = input stream
O = output stream

All Equation (9.22) says is to sum the respective stream enthalpies, kinetic energies, and potential energies of each for each flow stream, and if ΔE is applicable, to sum the same respective quantities inside the system.

9.3.4 Steady-State, Open Systems

As we mentioned earlier, you will find that the preponderance of industrial processes operate under approximately continuous, open, steady-state conditions. Most processes in the refining and chemical industries are open, steady-state systems. Biological processes are more likely to be closed systems (e.g., batch systems). You will find that continuous processes are most cost-effective in producing high-volume products.

Figure 9.19 shows some examples of open, steady-state processes, and you will find information about such processes on the CD that accompanies this book.

In Figure 9.19a, a fuel is burned in a boiler to heat tubes through which water flows and becomes steam. In Figure 9.19b, a dilute liquid feed containing a solute is concentrated to a "thick liquor." Vapor from the liquid is removed overhead. To provide the necessary heat transfer, steam flows through a steam chest (heat exchanger). In Figure 9.19c, a liquid containing

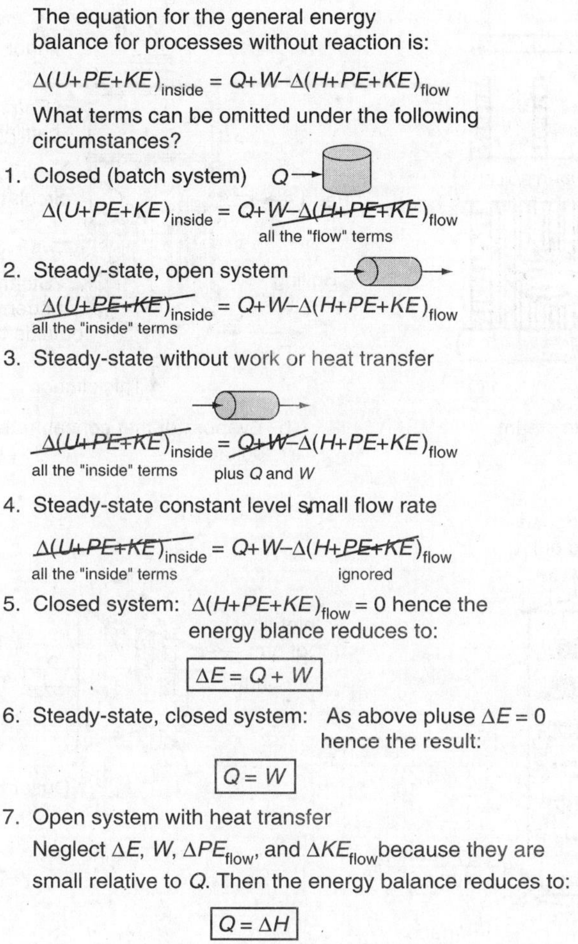

Figure 9.18 Results of various simplifications of the general energy balance

a desirable solute is passed through a column countercurrent to an immiscible solvent that favors extracting the solute from the liquid. In Figure 9.19d, gas at high pressure flows over turbine blades, causing the shaft to rotate and do work.

How do you change the general energy balance, Equation (9.21), so that it applies to steady-state, open processes? Because steady-state means that the final and initial states of the system are the same, $\Delta E = 0$. Consequently, Equation (9.21) becomes

$$Q + W = \Delta(H + PE = KE)_{\text{flows}} \tag{9.23}$$

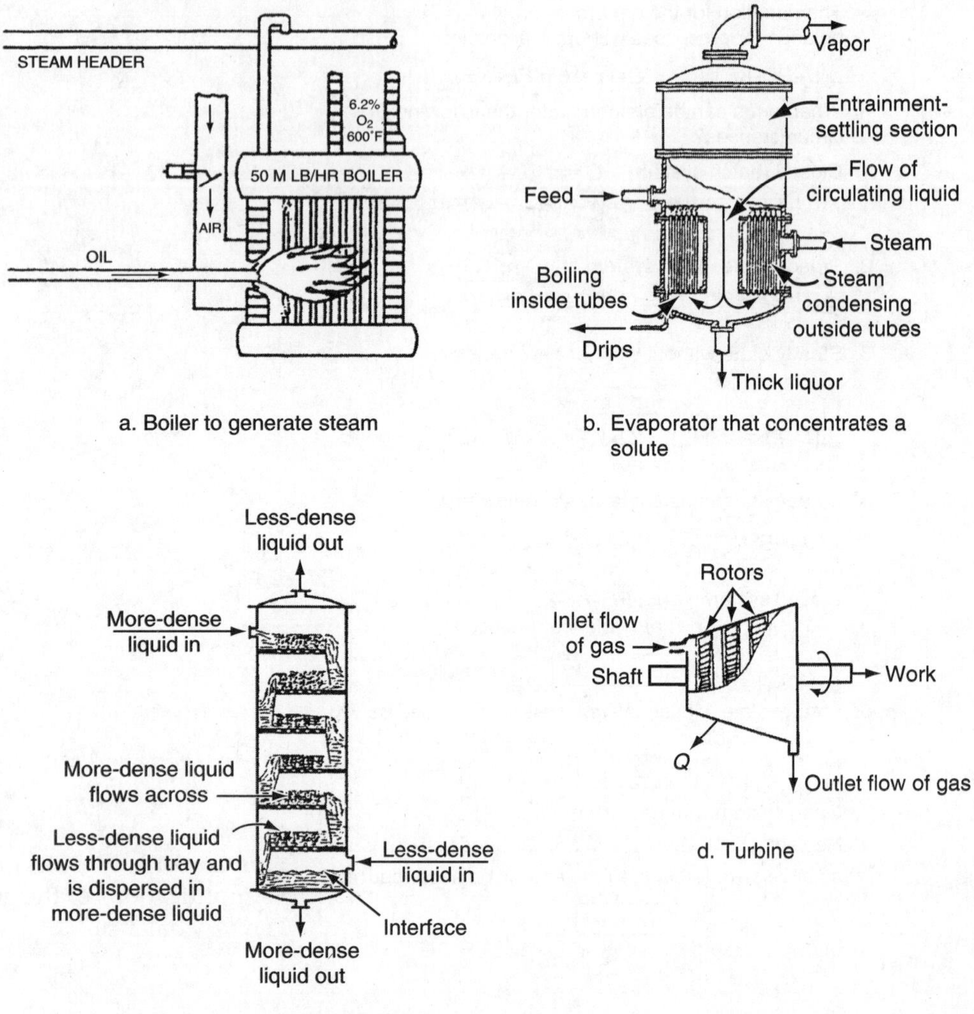

a. Boiler to generate steam

b. Evaporator that concentrates a solute

c. Plate extraction column

d. Turbine

Figure 9.19 Examples of open, steady-state processes (without reaction occurring)

When are ΔPE and ΔKE negligible? Because the energy terms in the energy balance in most open processes are dominated by Q, W, and ΔH, ΔPE and ΔKE only infrequently need to be used in Equation (9.23). For example, consider the right-hand side of Equation (9.23). An enthalpy change of 1000 J/kg is really quite small, corresponding in air to a temperature change of about 1 K. For the other terms on the right-hand side of Equation (9.23) to be equivalent to 1000 J:

1. The *PE* change would require 1 kg to go up a distance of 100 m.
2. The *KE* change would require a velocity change from 0 to 45 m/s.

As a result, the equation most commonly applied to open, steady-state processes does not include any potential and kinetic energy changes:

$$Q + W = \Delta H_{\text{flows}} \tag{9.24}$$

In Chapter 3 we outlined a strategy for solving material balance problems. The same strategy applies to solving combined material and energy balance problems. Here are two additional recommendations to employ:

1. Simplify the general energy equation, Equation (9.21), on paper by listing the assumptions you make and striking out terms that are zero or essentially zero.
2. Choose a reference state in terms of p, T, $\hat{H}$, and so on that is the same for the material inside the system and for the flows in and out.

"When a sailor doesn't know what harbor he is making for," counseled the Roman philosopher Seneca, "no wind is the right wind."

One step in the strategy for the solution of a problem, a step that becomes a bit more complex when using energy balances than when using strictly material balances, is the analysis of the degrees of freedom, because the intensive variables specifying enthalpy $\hat{H}$ (and internal energy $\hat{U}$), which are intensive variables, are themselves functions of T and p. Consequently, each stream as well as the material in the system can be characterized by its temperature and pressure. For example, you can introduce the value of $\hat{H}$ and p into the dialog box in the steam tables (really equations) on the CD in the back of the book and recover the values of T at the state that is specified by $\hat{H}$ and p. In tables, you have to interpolate. Often just the temperature of a stream is specified, and the value of the pressure is not mentioned. How can a state be specified by just one variable? In such circumstances, for a saturated liquid you could assume the value of the pressure is the vapor pressure of the liquid. If the liquid or gas is not saturated, you can fall back on assuming 1 atm when no better value is known for the pressure. For two phases you may have to assume equilibrium exists in the absence of better information to make the degrees of freedom zero in such ill-posed problems.

Example 9.13 Fluid Warmer

Figure E9.13a shows a fluid warmer that uses standard IV tubing instead of special disposable equipment. An electrically powered 250 W dry heat warmer supplies the heat transfer to the plastic tubing found in IV tube sets. The IV tubing is easily positioned in the S-shaped channel between the aluminum heating plates.

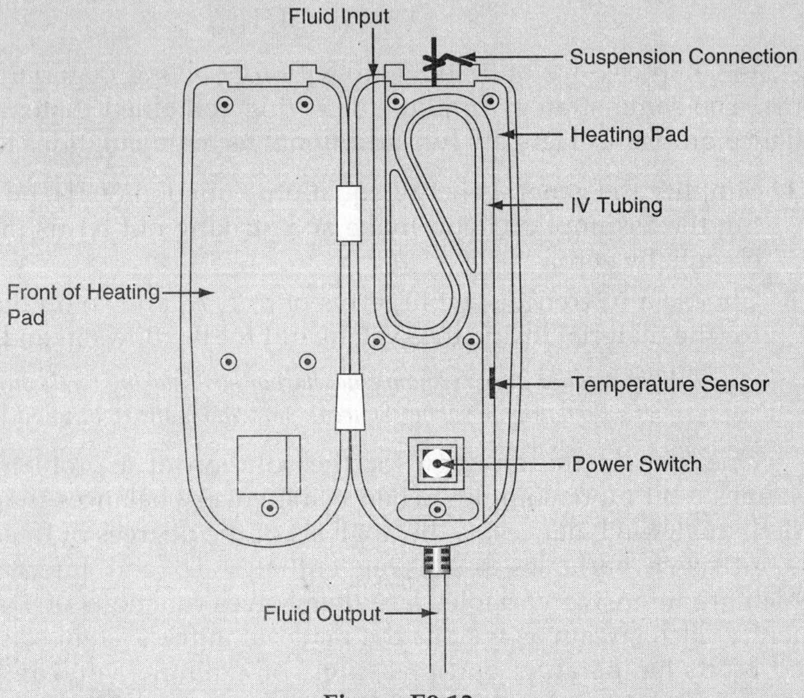

Figure E9.13a

The small size of the device permits a minimal length of tubing between the patient and the unit so that subsequent cooling of the exit fluid is negligible. The device warms up in 2 to 3 min, and then the fluid flows through the device at rates up to 12 cm^3/min.

Acidovir is an antiviral agent used for genital herpes (*Herpes simplex*) and shingles (varicella zoster—the chicken pox virus) by infusion through a vein. The solution flows through the warmer at the rate of 1.67 g/min of infusion solution. The entering fluid is at 24°C and exits at 37°C prior to infusion. The infusion solution in addition to the acidovir contains 0.45% NaCl and 2.5% glucose.

How many watts must be used by the warmer to warm the solution?

Solution

Steps 1–4

Figure E9.13b is a sketch of the process. Note that the electric work is given as power, watts (joules per second). For simplicity (and in the absence of data!), we will use the properties of sugar (glucose) in water for the fluid.

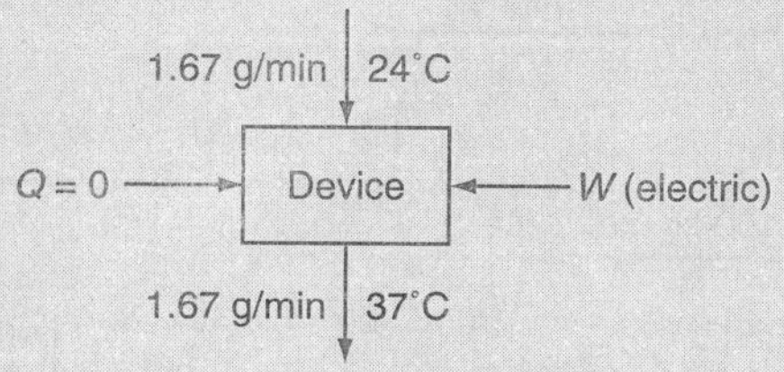

Figure E9.13b

Step 5

Basis: 1 min

Steps 6–8

What should you pick for the system? Choose the warming device. Then the system is open, steady-state (except for a short warmup interval that we will ignore), and the general energy balance can be simplified by making the following assumptions:

1. $\Delta E_{inside} = 0$ because the process is steady-state.
2. $Q = 0$ (negligible heat loss).
3. $\Delta\hat{PE} = \Delta\hat{KE} = 0$ for the fluid flowing in and out.

Thus, $Q + W = \Delta(H + PE + KE)_{flow}$ reduces to $W = \Delta H$.

$$W = \Delta H_{flow} = mC_p(T_{out} - T_{in}) = \frac{1.67\text{ g}}{\text{min}}\left|\frac{4.18\text{ J}}{(\text{g})(°\text{C})}\right|\frac{17°\text{C}}{}\left|\frac{1\text{ min}}{60\text{ s}}\right.$$

$$= 1.98\text{ J/s} \equiv 1.98\text{ W}$$

The heat capacity found on the Internet (ignoring the small amount of NaCl) is for 4% sugar solution.

The power in W is approximately 2 W.

Example 9.14 Use of Combined Material and Energy Balances

Figure E9.14 shows a hot gas stream at 500°C being cooled to 300°C by transferring heat to the liquid water that enters at 20°C and exits at 213°C. Assume that the heat exchanger is insulated. The cooling water does not mix with the gas. Calculate the value of the outlet temperature of the water.

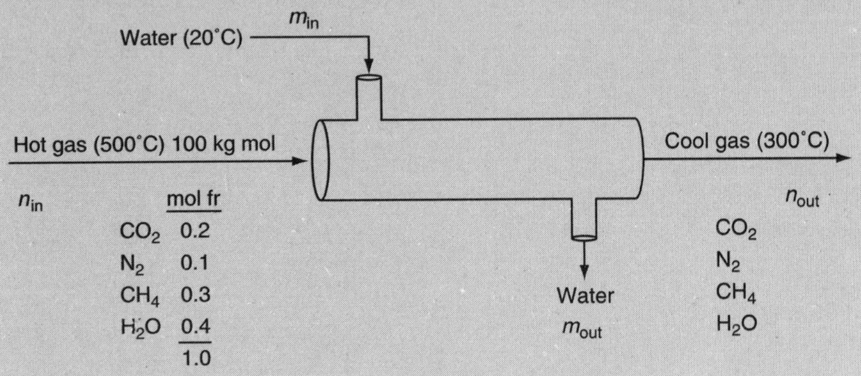

Figure E9.14

Solution

Steps 1–4

Figure E9.14 contains all of the data for the process.

Step 5

Basis: 100 kg mol of entering hot gas $\equiv$ 1 min

Steps 6–8

Pick the system as the heat exchanger. Very few unknowns exist for this system if certain easy calculations are made for the material balances:

a. Make material balances in your head to conclude that the exit gas is 100 kg mol, and that the kilogram moles of each of the components are known.

b. Make another material balance in your head to conclude that water in = water out = m_{water}.

The result is that after solving the material balances only one unknown exists, namely, m_{water}. All of the other variables can be assigned known values, but because they are obvious, we will not list them here to save space.

How can you calculate the value of m_{water}? Use the energy balance. The general energy balance

$$\Delta E = Q + W - \Delta(H + PE + KE)$$

can be greatly simplified by stating

$$\begin{aligned} \Delta E &= 0 \text{ steady-state} \\ Q &= 0 \text{ assume that the heat exchanger is insulated} \\ W &= 0 \text{ assumed (no information given)} \\ \Delta PE &= 0 \text{ assumed for all streams and inside the system} \\ \Delta KE &= 0 \text{ assumed for all streams and inside the system} \end{aligned}$$

Then $\Delta H = 0$ is the result, or in full:

$$[(n_{gas})(\hat{H}_{gas\ out}) + (m_{water})(\hat{H}_{water\ out})] - [(n_{gas})(\hat{H}_{gas\ in}) + (m_{water})(\hat{H}_{water\ in})] = 0$$

Rearrange to

$$n_{gas}\Delta\hat{H}_{gas} - m_{water}\Delta\hat{H}_{water} = 0$$

Note that the simplified equation has been written in terms of $\Delta\hat{H}$ rather than $\hat{H}$ to avoid taking into account differences in the reference states for the gas and water.

Where can you get values for $\Delta\hat{H}$ of the gas components and the water for the reduced energy balance? You can assign values to each of the components of the 100 kg mol of gas from data on the CD in the back of the book because you know the temperature and pressure (assume 1 atm) for the entrance and exit streams.

For each of the components of the gas:

$$n_{gas\ component}(\hat{H}_{gas\ component\ at\ 300°C} - \hat{H}_{gas\ component\ at\ 500°C}) \equiv n_i\Delta\hat{H}_i$$

Either integrate heat capacity equations (poor choice), or better, use the physical property software on the CD which gives $\Delta\hat{H}_i$ directly for the transition from 500°C to 300°C. You should get the following:

Component	$\Delta\hat{H}_i$(kJ/kg mol)	n_i(kg mol)	$\Delta H = n_i\Delta\hat{H}_i$(kJ)
CO_2	−9333	20	−186,660
N_2	−6215	10	−62,150
CH_4	−11,307	30	−339,210
H_2O	−7441	40	−297,640
Total		100	$-885{,}660 = \Delta H_{gas}$

The next step is easy if you use the steam tables for liquid water. Although the vapor pressure of water and the enthalpy change as the

(Continues)

Example 9.14 Use of Combined Material and Energy Balances (*Continued*)

water temperature rises, you can use the value of $\hat{H}_i$ for saturated water without much loss of accuracy.

From the values in the saturated portion of the steam tables on the CD:

$T(°C)$	$\hat{H}$(kJ/kg)
20	35.7
213	911.4

$$\Delta\hat{H} = 875.7 \text{ kJ/kg}$$

$$m_{water} = \frac{-\Delta H_{gas}}{\Delta\hat{H}_{water}} = \frac{-(-885{,}660) \text{ kJ}}{875.7 \text{ kJ/kg}} = 1011 \text{ kg}$$

$\dot{m}_{water}$1,011 kg/min (the basis was 1 min)

Example 9.15 Calculation of the Power Needed to Pump Water in an Open, Steady-State System

Water is pumped from a well (Figure E9.15) in which the water level is a constant 20 ft below the ground level. The water is discharged into a level pipe that is 5 ft above the ground at a rate of 0.50 ft^3/s. Assume that negligible heat transfer occurs from the water during its flow. Calculate the electric power required by the pump if it is 100% efficient; you can neglect friction in the pipe and the pump.

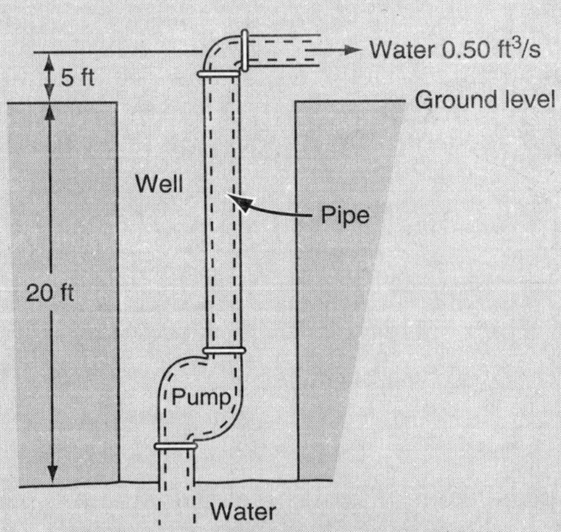

Figure E9.15

Solution

You can make the material balance for this example in your head, if you believe "what comes in must go out" is true. Let's pick as the system the pipe from the water level in the well to the place where the water that exits is at 5 ft above the ground so that the pump is included. To simplify the general energy balance, make some common assumptions:

1. The system is steady-state so $\Delta E = 0$.
2. $Q = 0$ (given assumption).
3. $\Delta KE_{\text{flow}} \cong 0$ (negligible change in KE; you can verify this statement if you have any doubts—assume a linear velocity at the top of the pipe equal to 7 ft/s, which is a typical value found in industrial applications).

What about ΔH_{flow}? Let us assume that the temperature of the water in the well is the same as the temperature of the water as it is discharged—a helpful assumption. Then $\Delta H \cong 0$ and the general energy balance reduces to

$$W = \Delta PE = mg(h_{\text{out}} - h_{\text{in}})$$

in which only one unknown exists, W. All of the other variables can be assigned known values.

Choose a basis of 1 s. The mass flow is (say, at 50°F)

$$\frac{0.50\ \text{ft}^3}{\text{s}}\left|\frac{62.4\ \text{lb}_\text{m}}{\text{ft}^3}\right. = 31.3\ \text{lb}_\text{m}\ \text{water/s}$$

$$W = PE_{\text{out}} - PE_{\text{in}} = \frac{31.3\ \text{lb}_\text{m}\ \text{H}_2\text{O}}{\text{s}}\left|\frac{32.2\ \text{ft}}{\text{s}^2}\right|\frac{25\ \text{ft}}{}\left|\frac{(\text{s}^2)(\text{lb}_\text{f})}{32.2\ (\text{ft})(\text{lb}_\text{m})}\right|\frac{1.055\ (\text{kW})(\text{s}^2)}{778.2\ (\text{lb}_\text{f})(\text{ft})}$$

$$= 1.06\ \text{kW}\ (1.42\ \text{hp})$$

In Chapter 14 (located on the accompanying CD) we discuss how to account for friction and pressure drop losses caused by pipe fittings and constrictions by using another type of balance.

Self-Assessment Test

Questions

1. Indicate whether the following statements are true or false:
   a. The shaft work done by a pump and motor located inside the system is positive.
   b. The Δ symbol in an energy balance for a steady-state system refers to the property of the material entering the system minus the property of the material leaving the system.

c. An input stream to a system does flow work on the system.
d. The input stream to a system possesses internal energy.
e. Work done by fluid flowing in a system that drives a turbine coupled with an electric generator is known as flow work.

2. What are two circumstances in which you can neglect the heat transfer term in the general energy balance?
3. What term in the general energy balance is always zero for a steady-state process?
4. What intensive variables are usually used to specify the value of the enthalpy?
5. Under what condition can (a) the *KE* term and (b) the *PE* term be ignored or deleted from the general energy balance?

Problems

1. A boiler converts liquid water to steam by letting the water flow through tubes that are heated by hot gases or another liquid. The pressure and flow rate of the water through the tubes are maintained by a regulator. For the boiler as the system, simplify the general energy balance as much as possible by deleting as many terms as feasible.
2. Calculate Q for the system shown in Figure SAT9.3.4P2.

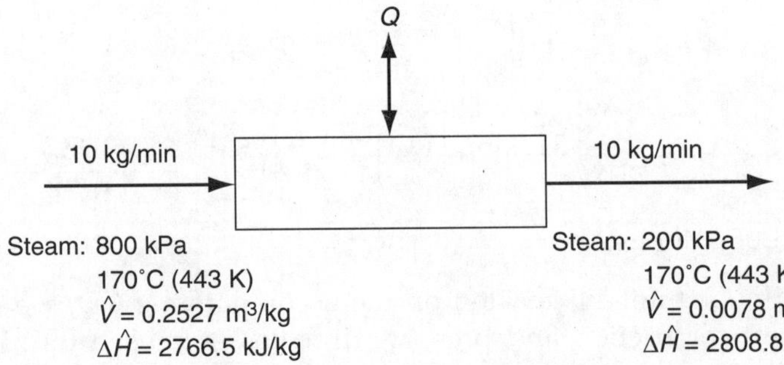

Figure SAT 9.3.4P2

3. A 3 MW steam-driven turbine operates in the steady state using 20 kg/s of steam. The inlet conditions for the steam are p = 3000 kPa and 450°C. The outlet conditions are 500 kPa, saturated vapor. The entering velocity of the steam is 250 m/s and the exit velocity is 40 m/s. What is the heat transfer in kilowatts for the turbine as the system? What fraction of the energy supplied by the steam is generated power?

Thought Problems

1. During flow through a partially opened valve the temperature of a fluid drops from 30°C to −30°C. Can the process occur adiabatically?
2. Would the temperature of air drop as it passes through a cracked valve in steady flow?
3. If an air compressor operates as a steady-state system, compare the volumetric flow rates of the input and the out streams.
4. Will the outlet temperature of the gas in an adiabatic compressor be higher or lower than the inlet temperature of the gas? Explain your answer.

Discussion Questions

1. Why are throttling valves commonly used in refrigeration and air conditioning equipment?
2. William Thompson (later Lord Kelvin) met Joule on the latter's honeymoon in the valley of Chamonix, France. Joule had a long thermometer to use in testing his theory that there should be a difference in the temperature at the top and bottom of a neighboring waterfall because of the dissipation of kinetic energy. This chance encounter cemented a warm friendship and a lifelong collaboration. Do you think Joule could have demonstrated the correctness of his theory in the indicated way?

9.3.5 Unsteady-State, Open Systems

Information is pretty thin stuff, unless it is mixed with experience.

Clarence Day

This last section treats the solution of energy balance problems for general open systems. To what kinds of processes can you apply the unsteady-state general energy balance? Look at Figure 9.20 for some examples.

In Figure 9.20a, a feed composed of salt in solution is separated by crystallization. A vacuum is created in the condenser by very cold water to remove water vapor from above the liquid solution so as to concentrate the solution. Crystals that form are periodically discharged from the bottom of the cone. In Figure 9.20b water discharges into and fills up the tank. In Figure 9.20c, two components in the liquid are separated by batch distillation. The distillation column is filled with the liquid to be separated. Steam in the heat exchanger (the reboiler) vaporizes the liquid, causing the more volatile component to accumulate in the product stream. Some of the liquid condensate produced by the condenser is withdrawn as product, and the remainder is recycled back into the column.

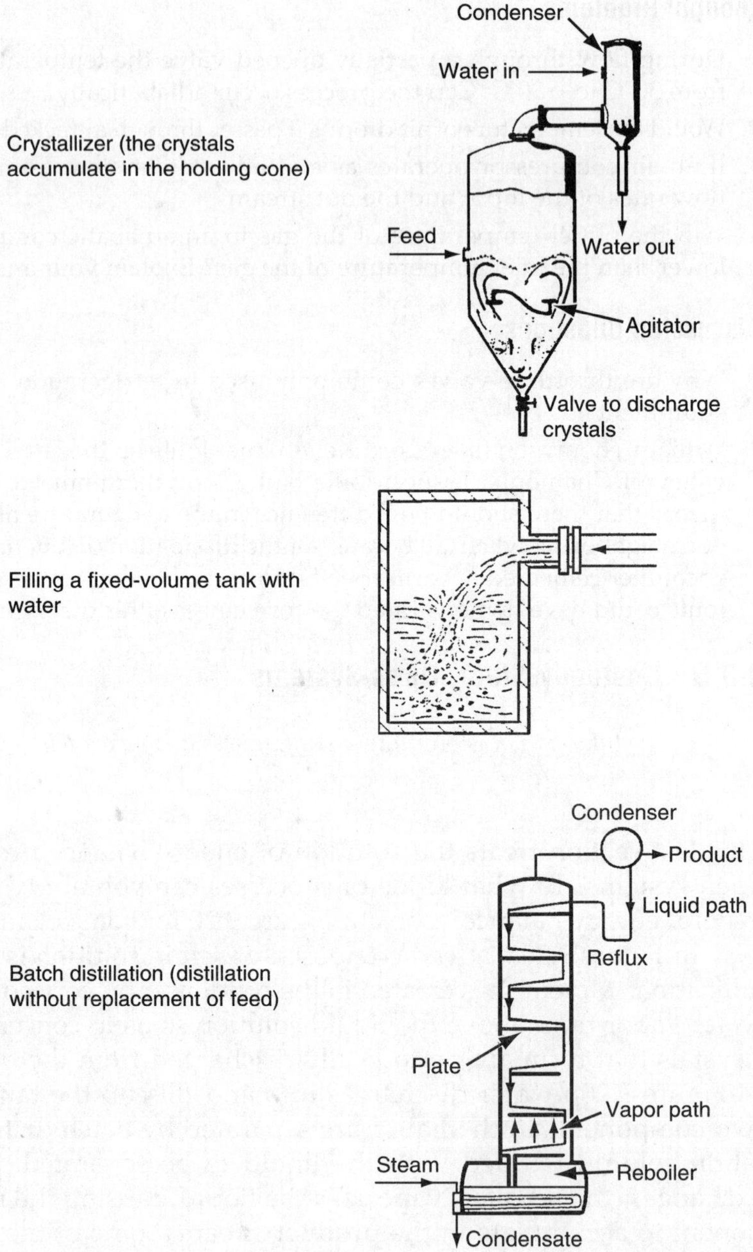

Figure 9.20 Examples of open, unsteady-state systems without reaction

Example 9.16 Unsteady-State, Open Process

For 10 min 10 lb of water at 35°F flow into a 125 ft^3 insulated vessel that initially contains 4 lb of ice at 32°F. To heat and mix the ice and water, 6 lb of steam at 250°F and 20 psia are introduced. What is the final temperature in the vessel after 10 min (assume all the material is well mixed)?

Solution

Steps 1–4

Figure E9.16 is a sketch of the process. Pick the system as the vessel. Let T_2 be the final temperature in the system.

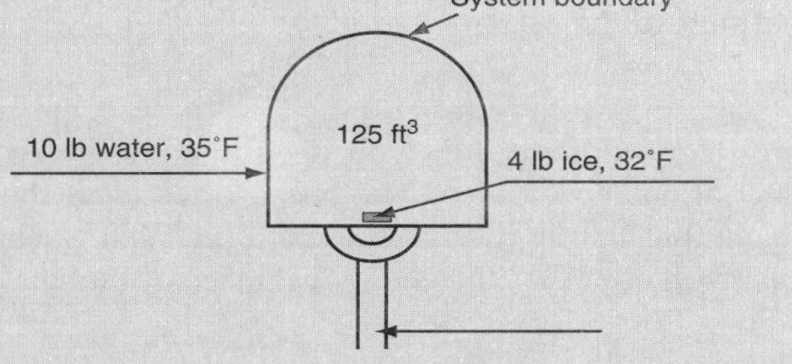

Figure E9.16

Step 5

Pick 10 min as the basis, which is equal to 20 total lb (water plus steam plus ice) in the final state of the system.

Step 6

Let us make a list of the variables to which we can assign values at this stage in the solution, and those to which we cannot (the initial unknowns).

	State 1: Initial Conditions in System (Ice)	Water Flow into System	Steam Flow into System	State 2: Known Final Conditions in System
m (lb)	4	10	0.6	20
T (°F)	32	35	250°F	
$\hat{U}$ (Btu/lb)	−143.6*	3.025	1090.26	
$\hat{H}$ (Btu/lb)	−143.6*	3.025	1167.15	
$\hat{V}$ (ft^3/lb)	Ignore	Ignore (0.016)	20.80	6.25
p (psia)	Ignore	Ignore (vapor pressure)	20	

*Heat of fusion

(Continues)

Example 9.16 Unsteady-State, Open Process (*Continued*)

The data come from the steam tables on the CD in the back of the book. One particularly important piece of information you might normally ignore is $\hat{V}_2 = \hat{V}_{\text{inside, final}} = \dfrac{125 \text{ ft}^3}{20 \text{ lb}} = 6.25 \text{ ft}^3/\text{lb}$. Remember that the values of two intensive variables must be known to fix the final conditions, and $\hat{V}_2$ can thus be one of them. What is another one?

A quick glance at the list of variables whose values have not been assigned indicates that T_2, $\hat{U}_2$, and $\hat{H}_2$ are unknowns. But the situation is not as bad as it might appear because the three unknowns are all intensive variables, they are all related to each other, and any one of them will suffice to fix the final state of the system.

Steps 7–9

To get zero degrees of freedom you need just one independent equation. Will a material balance suffice? No, because we used the overall material balance to get the 20 lb in the system at the final state. What other information can be involved? Why, the energy balance, of course!

$$\Delta E = Q + W - \Delta[m(\hat{H} + \hat{PE} = \hat{KE})]$$

To simplify the general energy balance assume:

1. All the $\hat{KE}$ and $\hat{PE}$ terms are zero inside the system and for the flow streams.
2. $Q = 0$ (insulated).
3. $W = 0$.

Then

$$\Delta U \equiv (m_2\hat{U}_2 - m_1\hat{U}_1) = -[0 - (m_{\text{water in}}\hat{H}_{\text{water in}} + m_{\text{steam in}}\hat{H}_{\text{steam in}})]$$

$$[20\,\hat{U}_2 - 4(-143.6)] = [10(3.025) + 6(1167.15)]$$

$$\hat{U}_2 = 6458.75/20 = 322.9 \text{ Btu/lb}$$

Although the values of $\hat{U}_2 = 322.9$ Btu/lb and $\hat{V}_2 = 6.25/\text{ft}^3/\text{lb}$ fix the solution, you still have to obtain the value of T_2. How can you get T_2? Not easily. If you had a way to insert the values of $\hat{U}_2$ and $\hat{V}_2$ into a computer program that would yield T_2, you would be all set. Even a chart of $\hat{V}_2$ versus $\hat{U}_2$ would be of help. Perhaps those tools will be available someday. But at the moment, the best you can do is to look at a chart that shows p, T, $\hat{V}$, and $\hat{H}$ to ascertain the value of T_2, assuming $\hat{U} \cong \hat{H}$, if the solution is in the two-phase region, which it is. The paper steam tables folded in the back of this book have such a chart (in SI units).

Alternatively, you can prepare a small region of your own chart of $\hat{V}$ versus $\hat{U}$, say, from 280 to 360 for $\hat{U}$ in Btu/lb and $\hat{V}$ from 3 to 9 ft3/lb. Use the saturated steam tables on the CD to calculate at constant pressure values of $\hat{U}$ and $\hat{V}$ surrounding the point $\hat{V} = 6.25\ \text{ft}^3/\text{lb}$ and $\hat{U} = 322.9\ \text{Btu/lb}$ by changing values of the quality of the two-phase mixture x. The lines of p and x will be straight. An answer with a reasonable number of significant figures might require a half-hour of your time. The final values of the unknowns are

$$T_2 = 197°\text{F} > 200°\text{F}$$

$$p_2 = 11.0\ \text{psia} > 11\ \text{psia}$$

$$x = 17.8\% > 18\%$$

What if the vessel contained air at atmospheric pressure when the ice was placed in the vessel? How would the solution change?

Example 9.17 Heating of a Biomass

Steam at 250°C saturated (which is used to preheat a fermentation broth) enters the steam chest. The steam is segregated from the biomass solution in the preheater and is completely condensed in the steam chest. The rate of the heat loss from the surroundings is 1.5 kJ/s. The material to be heated flows into the process vessel at 20°C and is at 45°C. If 150 kg of biomass with an average heat capacity of $C_p = 3.26\ \text{J/(g)(K)}$ flows into the process per hour, how many kilograms of steam are needed per hour?

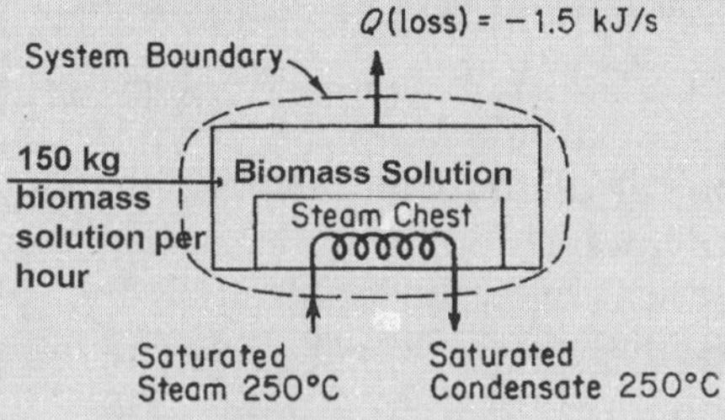

Figure E9.17

Solution

Steps 1–4

Figure E9.17 defines the system and shows the known conditions. If the system is composed of the biomass solution plus the steam chest (a heat exchanger), the process is open, unsteady-state because the temperature of the biomass increases as well as the mass in the system. Assume that the entering solution causes the biomass to be well mixed.

(Continues)

Example 9.17 Heating of a Biomass (*Continued*)

Step 5

Basis: 1 hr of operation (150 kg of biomass solution heated)

Steps 6–8

The material balance for the steam is $m_{in} = m_{out} = m$, and m is an unknown. The material balance for the biomass solution is also simple: What flows in stays in the system. Initially no biomass was in the vessel, and at the end of 1 hr 150 kg of solution existed in the vessel. As a result one or more unknowns are associated with the biomass solution.

Next, let's examine the energy balance for the system to ascertain if the balance can be used to solve for m. The energy balance is

$$\Delta E = Q + W - \Delta[(\hat{H} + \hat{KE} + \hat{PE})m] \qquad \text{(a)}$$

Let's simplify the energy balance:

1. The process is not in the steady state, so $\Delta E \neq 0$.
2. We can safely assume that $\Delta KE = 0$ and $\Delta PE = 0$ inside the system and for the flow in.
3. $W = 0$.
4. ΔKE and ΔPE of the entering and exiting material are zero.

Consequently, Equation (a) becomes

$$\Delta E = \Delta U = Q - \Delta[(\hat{H})m] \qquad \text{(b)}$$

where ΔU can be calculated from just the change in state of the biomass solution and does not include the water in the steam chest because we will assume that there was no water or steam in the steam chest at the start of the hour and none in the steam chest at the end of the hour.

$$(150\text{ kg})(\hat{U}_{\text{final}}) - (0\text{ kg})(\hat{U}_{\text{initial}}) = \frac{-1.50\text{ J}}{\text{s}}\left|\frac{3600\text{ s}}{1\text{ hr}}\right|\frac{1\text{ hr}}{} - m_{\text{steam}}(\hat{H}_2 - \hat{H}_1) \qquad \text{(c)}$$

Choose as a reference temperature either 0°C or 20°C; the p reference is 1 atm by assumption. Pick $T_{\text{reference}} = 20°\text{C}$. Assume $\Delta\hat{U} = \Delta\hat{H}$.

$$\hat{U}_{\text{final}} = \frac{3.26\text{ kJ}}{(\text{kg})(°\text{C})}\left|\frac{(45-20)°\text{C}}{}\right. = 81.5\text{ kJ/kg}$$

$\hat{U}_{\text{initial}} = 0$ (Why? 20°C is the reference temperature)

$$(\hat{H}_2 - \hat{H}_1) = \Delta\hat{H}_{\text{condensation of steam}} = -\Delta\hat{H}_{\text{vaporization}}\text{ at }250°\text{C} = -1701\text{ kJ/kg}$$

Introduce these values into Equation (c):

$$12{,}225 = -5400 - m(-1701)$$

$$m_{\text{steam}} = 10.4 \text{ kg/hr}$$

Example 9.18 Unsteady-State System (without Reaction Occurring)

A rigid, well-insulated tank is connected to two valves. One valve goes to a steam line that has steam at 1000 kPa and 600 K and the other to a vacuum pump. Both valves are initially closed. Then the valve to the vacuum pump is opened, the tank is evacuated, and the valve is closed. Next the valve to the steam line is opened so that the steam enters the evacuated tank very slowly until the pressure in the tank equals the pressure in the steam line. Calculate the final temperature of the steam in the tank. Figure E9.18 shows the process.

Solution

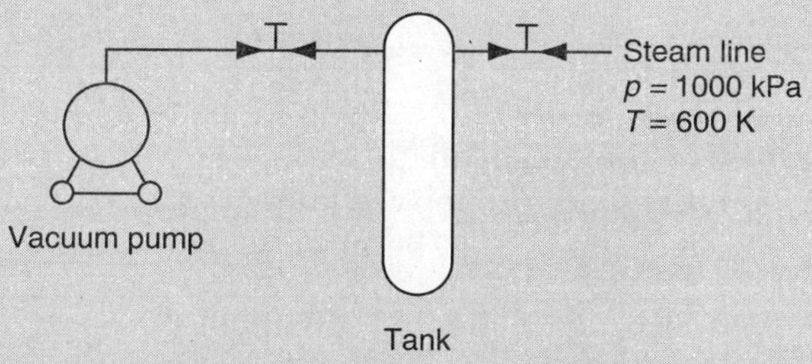

Figure E9.18

First, pick the tank as the system. If you do, the system is unsteady-state (the mass increases in the system) and open. Pick a basis of 1 kg of steam.

Next, get the data for steam at 1000 kPa and 600 K from the database on the CD in the back of this book:

$$\hat{U} = 2837.73 \text{ kJ/kg}$$

$$\hat{H} = 3109.44 \text{ kJ/kg}$$

$$\hat{V} = 0.271 \text{ m}^3\text{/kg}$$

(Continues)

Example 9.18 Unsteady-State System (without Reaction Occurring) (*Continued*)

Next, write down the general energy balance

$$E_{t_2} - E_{t_1} = Q + W - \Delta(H + KE = PE) \quad \text{(a)}$$

and begin simplifying it. You can make the following assumptions:

1. No change occurs within the system for PE and KE; hence $\Delta E = \Delta U_{\text{inside}}$.
2. No work is done on or by the system; hence $W = 0$.
3. No heat is transferred to or from the system because it is well insulated; hence $Q = 0$.
4. The ΔKE (and ΔPE) for the flow is small and assumed to be zero.
5. No stream exits the system; hence $H_{\text{out}} = 0$.
6. Initially no mass exists in the system; hence $U_{t_1,\text{ inside}} = 0$.

Consequently Equation (a) reduces to

$$U_{t_2,\text{ inside}} - 0 = -(H_{\text{out}} - H_{\text{in}})$$

and

$$\Delta U = U_{t_2} = m_{\text{in}}\hat{U}_{t_2} = H_{\text{in}} = m_{\text{in}}\hat{H}_{\text{in}} \quad \text{(b)}$$

To fix the final temperature of the steam in the tank, you have to determine two properties of the steam in the tank—any two. One value is given: $p = 1000$ kPa. What other property is known? Not T nor $\hat{V}$. But you can use Equation (b) to calculate $U_{t_2,\text{ inside}}$ because

$$\hat{U}_{t_2,\text{ inside}} = \hat{H}_{\text{in}} = 3109.44 \text{ kJ/kg} \quad \text{(c)}$$

and from interpolating in the steam tables at $p = 1000$ kPa you can find that $T = 764$ K.

You may wonder what the source of the energy is that causes the increase in temperature of the steam in the system over that in the steam line. If you look at the problem as a closed system composed of the volume of all of the steam in the steam line that eventually will be in the tank, plus the volume of the tank itself, let a hypothetical piston very slowly compress the volume of steam into the tank so that it acts as an ideal piston. The system boundary changes position during the compression so that the piston does work on the system, and thus the temperature of the steam inside the tank goes up.

Self-Assessment Test

Questions

1. Do turbines and pumps represent examples of unsteady-state systems?
2. Can a system with no moving parts be treated as an unsteady-state, open system?
3. Indicate whether the following statements are true or false:
   a. The input stream to a process possesses kinetic energy.
   b. The input stream to a process possesses potential energy.
   c. The input stream to a process possesses internal energy.
   d. The exit stream from the system does flow work.
   e. The shaft work done by a turbine that is rotated by a fluid in a system is positive.
4. What assumptions must be made to reduce the general energy balance, Equation (9.22), to $Q = \Delta H$?
5. You read in an engineering book that

$$U_2 - U_1 = Q + W_{\text{nonflow}}$$

   If you cannot find the notation list in the book, would you agree that the equation represents the energy balance for an unsteady-state, open system?

Problems

1. Simplify the general energy balance, Equation (9.22), as much as possible for each of the following circumstances (state which terms can be deleted and why):
   a. The system has no moving parts.
   b. The temperatures of the system and its surroundings are the same.
   c. The velocity of the fluid flowing into the system equals the velocity of the fluid leaving the system.
   d. The fluid exits the system with sufficient velocity so that it can shoot out 3 m.
2. Under what circumstances is $\Delta U = \Delta H$ for an open, unsteady-state system?
3. A compressor fills a tank with air at a service station. The inlet air temperature is 300 K, and the pressure is 100 kPa, the same conditions as for the initial air in the tank with a mass of 0.8 kg. After 1 kg of air is pumped into the tank, the pressure reaches 300 kPa, and the temperature is 400 K. How much heat was added to or removed from the tank during the compression? Hint: Do not forget the initial amount of air in the tank.

   Data for the air:

	$\hat{H}$(kJ/kg)	$\hat{U}$(kJ/kg)	$\hat{V}$(m^3/kg)
100 kPa and 300 K	459.85	337.75	0.8497
300 kPa and 400 K	560.51	445.61	0.3830

Thought Problems

1. The valve on a rigid insulated cylinder containing air at a high pressure is opened briefly to use some of the air. Will the temperature of the air in the cylinder change? If so, how?
2. If you open the valve on an insulated tank that is initially evacuated, and let the atmospheric air enter until atmospheric pressure is reached, will the temperature of the air in the tank be higher, the same as, or lower than the temperature of the atmosphere? Explain your answer.

Discussion Questions

1. Under what conditions can an unsteady-state flow process be solved as a unsteady-state, closed process?
2. In Kurt Vonnegut's famous novel *Cat's Cradle*, a form of ice is discovered that is more stable than normal ice, and contact with normal water eventually converts all of the normal water to a solid at room temperature with the predictable consequences for the existence of life. Can this outcome be analyzed using the general energy balance?

Looking Back

This chapter introduced a number of different forms of energy and how to formulate energy balances using them.

Glossary

Adiabatic Describes a system that does not exchange heat with the surroundings during a process.

Adiabatic process A process in which no heat transfer occurs ($Q = 0$).

Boundary Hypothetical perimeter used to define what a system is.

Closed system A system in which mass is not exchanged with the surroundings.

Conservation of energy The total energy of a system plus its surroundings is constant.

Electric work Work done on or by a system because a voltage difference forces a current to flow.

Energy The capacity to do work or transfer heat.

Energy transfer The movement of energy from one site or state to another.

Enthalpy (*H*) The sum of the variables $U + pV$.

Equilibrium state The properties of a system remain invariant under a balance of potentials.

Extensive property A property that depends on the amount of material present, such as volume or mass.

Flow system See **open system**.

Flow work Work done on a system to put a fluid element into the system, or work done by a system to push a fluid element into the surroundings.

General energy balance The change of energy inside a system is equal to the net heat and net work interchange with the surroundings plus the net energy transported by flow into and out of the system.

Heat (*Q*) Energy transfer across a system boundary that is caused by a temperature difference (potential) between the system and the surroundings.

Heat capacity Also called the specific heat. One heat capacity, C_v, is defined as the change in internal energy with respect to temperature at constant volume, and a second heat capacity, C_p, is defined as the change in enthalpy with respect to temperature at constant pressure.

Heat of condensation The negative of the heat of vaporization.

Heat of fusion The enthalpy change for the phase transition of melting.

Heat of solidification The negative of the heat of fusion.

Heat of sublimation The enthalpy change of a solid directly to vapor.

Heat of vaporization The enthalpy change for the phase transition of a liquid to a vapor.

Intensive property A property that is independent of the amount of material present, such as temperature, pressure, or any specific variable (per unit amount of material, mass, or mole).

Internal energy (*U*) Energy that represents a macroscopic account of all the molecular, atomic, and subatomic energies, all of which follow definite microscopic conservation rules for dynamic systems.

Isobaric Describes a system for which the pressure is invariant during a process.

Isochoric Describes a system for which the volume is invariant during a process.

Isothermal Describes a system in which the temperature is invariant during a process.

Kinetic energy (*KE*) Energy a system possesses because of its velocity relative to the surroundings.

Latent heat An enthalpy change that involves a phase transition.

Nonflow system See **closed system**.

Open system A system in which interchange of mass occurs with the surroundings.

Path variable (function) A variable (function) whose value depends on how the process takes place, such as heat and work.

Phase transition A change from the solid to the liquid phase, from the liquid to the gas phase, from the solid to the gas phase, or the respective reverses.

Point function See **state variable**.

Potential energy (*PE*) Energy a system possesses because of the force exerted on its mass by a gravitational or electromagnetic field with respect to a reference surface.

Power Work per unit time.

Property An observable or calculable characteristic of a system, such as temperature or enthalpy.

Reference substance plots Plots used to estimate values for physical properties based on comparison with the same properties of a substance used as a reference.

Sensible heat An enthalpy change that does not involve a phase transition.

Shaft work Work corresponding to a force acting on a shaft to turn it.

Single phase A part of a system that is physically distinct and homogeneous, such as a gas or liquid.

State Conditions of the system, such as the values of the temperature or pressure.

State variable (function) A variable (function), also called a point function, whose value depends only on the state of the system and not upon its previous history, such as internal energy.

Steady state The accumulation in a system is zero, the flows in and out are constant, and the properties of the system are invariant.

Surroundings Everything outside a system.

System The quantity of matter or region of space chosen for analysis.

Transient See **unsteady state**.

Unsteady state Not steady state (see **steady state**).

Work (*W*) Work is a form of energy that represents a transfer of energy between a system and its surroundings. Work cannot be stored.

Supplemental References

Abbott, M. M., and H. C. Van Ness. *Schaum's Outline of Thermodynamics with Chemical Applications*, 2nd ed., McGraw-Hill, New York (1989).

Blatt, F. J. *Modern Physics*, McGraw-Hill, New York (1992).

Boethling, R. S., and D. McKay. *Handbook of Property Estimation Methods for Chemicals*, Lewis Publishers (CRC Press), Boca Raton, FL (2000).

Crawly, G. M. *Energy*, Macmillan, New York (1975).

Daubert, T. E., and R. P. Danner. *Data Compilation of Properties of Pure Compounds*, Parts 1, 2, 3 and 4, Supplements 1 and 2, DIPPR Project, AIChE, New York (1985–1992).

Gallant, R. W., and C. L. Yaws. *Physical Properties of Hydrocarbons*, 4 vols., Gulf Publishing, Houston, TX (1992–1995).

Guvich, L. V., I. V. Veyts, and C. B. Alcock. *Thermodynamic Properties of Individual Substances*, Vol. 3, Parts 1 & 2, Lewis Publishers (CRC Press), Boca Raton, FL (1994).

Levi, B. G. (ed.). *Global Warming, Physics and Facts*, American Institute of Physics, New York (1992).

Pedley, J. B. *Thermochemical Data and Structures of Organic Compounds*, Vol. 1, TRC Data Distribution, Texas A&M University System, College Station, TX (1994).

Poling, B. E., J. M. Prausnitz, and J. P. O'Connell. *The Properties of Gases and Liquids*, 5th ed., McGraw-Hill, New York (2001).

Raznjevic, K. *Handbook of Thermodynamic Tables*, 2nd ed., Begell House, New York (1995).

Stamatoudis, M., and D. Garipis. "Comparison of Generalized Equations of State to Predict Gas-Phase Heat Capacity," *AIChE J.*, **38**, 302–7 (1992).

Vargaftik, N. B., Y. K. Vinogradov, and V. S. Yargin. *Handbook of Physical Properties of Liquids and Gases*, Begell House, New York (1996).

Yaws, C. L. *Handbook of Thermodynamic Diagrams*, Vols. 1, 2, 3 and 4, Gulf Publishing Co., Houston, TX (1996).

Web Sites

http://hyperphysics.phy-astr.gsu.edu/hbase/thermo/inteng.html
www.molknow.com
www.taftan.com/steam.htm
www.taftan.com/thermodynamics
http://webbook.nist.gov/chemistry

Problems

9.1 Terminology Associated with Energy Balances

***9.1.1** Convert 45.0 Btu/lb_m to the following:

a. cal/kg
b. J/kg
c. kWh/kg
d. (ft)(lbf)/lb_m

***9.1.2** Convert the following physical properties of liquid water at 0°C and 1 atm from the given SI units to the equivalent values in the listed American Engineering units:

a. Heat capacity of 4.184 J/(g)(K)	Btu/(lb)(°F)
b. Enthalpy of −41.6 J/kg	Btu/lb
c. Thermal conductivity of 0.59 (kg)(m)/(s^3)(K)	Btu/(ft)(hr)(°F)

***9.1.3** Convert the following quantities as specified:

a. A rate of heat flow of 6000 Btu/(hr)(ft^2) to cal/(s)(cm^2)
b. A heat capacity of 2.3 Btu/(lb)(°F) to cal/(g)(°C)
c. A thermal conductivity of 200 Btu/(hr)(ft)(°F) to J/(s)(cm)(°C)
d. The gas constant, 10.73 (psia)(ft^3)/(lb mol)(°R) to J/(g mol)(K)

*9.1.4 The energy from the sun incident on the surface of the Earth averages 32.0 cal/(min)(cm^2). It has been proposed to use space stations in synchronous orbits 36,000 km from Earth to collect solar energy. How large a collection surface is needed (in square meters) to obtain 108 W of electricity (equivalent to a 100 MW power plant)? Assume that 10% of the collected energy is converted to electricity. Is your answer a reasonable size?

*9.1.5 Indicate whether the following statements are true or false:

a. A simple test to determine if a property is an extensive property is to split the system in half, and if the value of the sum of the properties in each half is twice the value of the property in one half, then the property is an extensive variable.
b. The value of a state variable (point function) depends only on its state and not on how the state was reached.
c. An intensive variable is a variable whose value depends on the amount of mass present.
d. Temperature, volume, and concentration are intensive variables.
e. Any variable that is expressed as a "specific"—that is, per unit mass—quantity is deemed to be an intensive variable.

*9.1.6 Indicate whether the following statements are true or false:

a. An isobaric process is one of constant pressure.
b. An isothermal process is one of constant temperature.
c. An isometric process is one of constant volume.
d. A closed system is one in which mass does not cross the system boundary.
e. The units of energy in the SI system are joules.
f. The units of energy in the AE system can be (ft)(lb_f)/lb_m.
g. The difference between an open system and a closed system is that energy transfer can take place between the system and the surroundings in the former but not in the latter.

*9.1.7 Are the following variables intensive or extensive?

a. Partial pressure
b. Volume
c. Specific gravity
d. Potential energy
e. Relative saturation
f. Specific volume
g. Surface tension
h. Refractive index

*9.1.8 Classify the following measurable physical characteristics of a gaseous mixture of two components as an intensive property, an extensive property, both, or neither:

a. Temperature
b. Pressure
c. Composition
d. Mass

****9.1.9** A simplified equation for the heat transfer coefficient from a pipe to air is

$$h = \frac{0.026G^{0.6}}{D^{0.4}}$$

where h = heat transfer coefficient, Btu/(hr)(ft^2)(°F)
G = mass rate flow, lb$_m$/(hr)(ft^2)
D = outside diameter of the pipe, ft

If h is to be expressed in J/(min)(cm^2)(°C) and the units of G and D remain the same, what should the new constant in the equation be in place of 0.026?

****9.1.10** A problem for many people in the United States is excess body weight stored as fat. Many people have tried to capitalize on this problem with fruitless weight-loss schemes. However, since energy is conserved, an energy balance reveals only two real ways to lose weight (other than water loss): (1) reduce the caloric intake and/or (2) increase the caloric expenditure. In answering the following questions, assume that fat contains approximately 7700 kcal/kg (1 kcal is called a "dietetic calorie" in nutrition, or commonly just a "Calorie"):

a. If a normal diet containing 2400 kcal/day is reduced by 500 kcal/day, how many days does it take to lose 1 lb of fat?
b. How many miles would you have to run to lose 1 lb of fat if running at a moderate pace of 12 km/hr expends 400 kJ/km?
c. Suppose that two joggers each run 10 km/day. One runs at a pace of 5 km/hr and the other at 10 km/hr. Which will lose more weight (ignoring water loss)?

****9.1.11** Solar energy has been suggested as a source of renewable energy. If in the desert the direct radiation from the sun (say, for 320 days) is 975 W/m^2 between 10 AM and 3 PM, and the conversion efficiency to electricity is 21.0%, how many square meters are needed to collect an amount of energy equivalent to the annual U.S. energy consumption of 3×10^{20} J? Is the construction of such an area feasible?

How many tons of coal (of heating value of 10,000 Btu/lb) would be needed to provide the 3×10^{20} J if the efficiency of conversion to electricity was 70%? What fraction of the total U.S. resources of coal (estimated at 1.7×10^{12} tons) is the calculated quantity?

****9.1.12** The thermal conductivity equation for a substance is $k = a + bT$, where

k is in (Btu)/(hr)(ft)(°F)
T is temperature, °F
a, b are constants with appropriate units

Convert this relation to make it possible to introduce the temperature into the modified equation as degrees Celsius, and to have the units in which k is expressed to be (J)/(min)(cm)(K). Your answer should be the modified equation. Show all transformations.

****9.1.13** In an electric gun, suppose we want the muzzle velocity to be Mach 4, or about 4400 ft/s. This is a reasonable number because at around 3500 ft/s you begin to get hypervelocity effects, such as straws going through doors.

Now the requirement that our slug start from zero and reach 4400 ft/s in 20 ft exactly defines our environment, if we insist that the acceleration is constant, which is a good starting point.

By a simple manipulation of Newton's laws, we find that the time of launch is 9 ms. If the slug weighs a kilogram, which is much less than a tank shell but quite sufficient to wreck any tank it hits at Mach 4, then the kinetic energy of the bullet is 900,000 J. Since we have 9 ms to get this energy into the slug, we have to transfer about 100 MW into the bullet during the 9 ms launch time.

Is the energy transfer really 100 MW?

****9.1.14** How much weight can you lose by exercise, say, running?
Data: 1 kg fat = 7700 Calories = 7700 kcal = 32,000 J
Running at 5 min/km requires the expenditure of about 400 kJ/km.

****9.1.15** An overweight person decides to lose weight by exercise. Hard aerobic exercise requires 700 W. By exercising for one-quarter of an hour, can the person compensate for a big meal (4000 kJ)?

****9.1.16** A can of soda at room temperature is put into the refrigerator so that it will cool. Would you model the can of soft drink as a closed system or as an open system? Explain.

*****9.1.17** Lasers are used in many technologies, but they are fairly large devices even in CD players. A materials scientist has been working on producing a laser from a microchip. He claims that his laser can produce a burst of light with up to 10,000 W, qualifying it for use in eye surgery, satellite communications, and so on. Is it possible to have such a powerful laser in such a small package?

*****9.1.18** A letter to the editor of the local newspaper said:

> Q. I'm trying to eat a diet that's less than 30 percent fat. I bought some turkey franks advertised as "80 percent fat free."
>
> The statistical breakdown on the back of the package shows each frank has 8 grams of fat and 100 calories. If a gram of fat has nine calories, that makes the frank have about 75 percent fat. I'm so confused; is it them or is it me?

How would you answer the question?

9.2 Types of Energy to Be Included in Energy Balances

***9.2.1** Suppose that a constant force of 40.0 N is exerted to move an object for 6.00 m. What is the work accomplished (on an ideal system) expressed in the following:

a. joules
b. (ft)(lb_f)

c. calories
d. British thermal units

***9.2.2** A gas is contained in a horizontal piston-cylinder apparatus at a pressure of 350 kPa and a volume of 0.02 m^3. Determine the work done by the piston on the gas if the cylinder volume is increased to 0.15 m^3 through heating. Assume the pressure of the gas remains constant throughout the process, and that the process is ideal.

***9.2.3** A rigid tank contains air at 400 kPa and 600°C. As a result of heat transfer to the surroundings, the temperature and pressure inside the tank drop to 100°C and 200 kPa, respectively. Calculate the work done during this process.

***9.2.4** What is the potential energy in joules of a 12 kg mass 25 m above a datum plane?

***9.2.5** Which is an example of an exothermic phase change: (a) liquid to solid; (b) liquid to gas; (c) solid to liquid; (d) solid to gas?

***9.2.6** Heat is added to a substance at a constant rate and the temperature of the substance remains the same. This substance is (a) solid melting at its melting point; (b) solid below its melting point; (c) liquid above its freezing point; (d) liquid freezing at its freezing point.

****9.2.7** A horizontal frictionless piston-cylinder contains 10 lb of liquid water saturated at 320°F. Heat is now transferred to the water until one-half of the water vaporizes. If the piston moves slowly to do work against the surroundings, calculate the work done by the system (the piston-cylinder) during this process.

****9.2.8** Indicate whether the following statements are true or false:
a. The work done by a constant volume system is always zero.
b. In a cycle that starts at one state and returns to the same state, the net work done is zero.
c. Work is the interchange of energy between the system and its surroundings.
d. Work can always be calculated by the integration of $+pdV$ for a gas.
e. For a closed system, work is always zero.
f. When an ideal gas expands in two stages from state one to state two, the first stage being at constant pressure and the second at constant temperature, the work done by the gas is greater during the second stage.
g. When positive work is done on a system, its surroundings do an equal quantity of negative work, and vice versa.

****9.2.9** Often in books you read about "heat reservoirs" existing in two bodies at different temperatures. Can this concept be correct?

****9.2.10** An examination question asked: "Is heat conserved?" Sixty percent of the students said "no," but 40% said "yes." The most common explanation was (a) "Heat is a form of energy and therefore conserved." The next most common was (b) "Heat is a form of energy and therefore is not conserved." Two other common explanations were (c)" Heat is conserved. When something is cooled, it heats something else up. To get heat in the first place, though, you have to use energy. Heat is just one form of energy," and (d) "Yes, heat is transferred from a system to its surroundings and vice versa. The amount lost by one system equals the amount gained by the surroundings." Explain whether or not heat is conserved and criticize each of the four answers.

****9.2.11** Tell what is right and what is wrong with each of these concepts related to heat:

a. Heat is a substance.
b. Heat is really not energy.
c. Heat and cold are the same thing; they represent opposite ends of a continuum.
d. Heat and temperature are the same thing.
e. Heat is proportional to temperature.
f. Heat is not a measurable, quantifiable concept.
g. Heat is a medium.
h. Heat flows from one object to another or can be stored.
i. Heat is produced by burning.
j. Heat cannot be destroyed.

****9.2.12** This explanation of the operation of a refrigerator was submitted for publication in a professional news magazine:

> Recall that in the usual refrigerator, a liquid coolant with a low boiling point vaporizes while at low pressure, absorbing heat from the refrigerator's contents. This heat energy is concentrated by a compressor and the result dissipated in a condenser, with the gas converting to liquid at high pressure. As the liquid passes through an expansion nozzle back into the refrigerator chamber, the cooling cycle begins anew.

Clarify the explanation for the editor of the magazine.

****9.2.13** Indicate whether the following statements are true or false:

a. In an adiabatic process no heat transfer occurs.
b. When a gas is compressed in a cylinder, no heat transfer occurs.
c. If an insulated room contains an operating freezer, no heat transfer takes place if the room is picked as the system.
d. Heat and thermal energy are synonymous terms.
e. Heat and work are the only mechanisms by which energy can be transferred to a closed system.
f. Light is a form of heat.
g. You can measure the heat in a system by its temperature.
h. Heat is a measure of the temperature of a system.

****9.2.14** Calculate the heat transfer to the atmosphere per second from a circular pipe, 5 cm in diameter and 100 m long, carrying steam at an average temperature of 120°C if the surroundings are at 20°C. The heat transfer can be estimated from the relation

$$Q = hA\Delta T$$

where $h = 5\,\text{J}/(\text{s})(\text{m}^2)(°\text{C})$

A is the surface area of the pipe

ΔT is the temperature difference between the surface of the pipe and ambient conditions

**9.2.15 Calculate the KE of the liquid flowing in a pipe with a 5 cm inner diameter at the rate of 500 kg/min. The density of the liquid is 1.15 g/cm^3.

**9.2.16 Find the kinetic energy in (ft)(lb$_f$)/(lb$_m$) of water moving at the rate of 10 ft/s through a pipe of 2 in. inner diameter.

**9.2.17 Find the value of internal energy for water (relative to the reference state) for the states indicated:

a. Water at 0.4 MPa, 725°C
b. Water at 3.0 MPa, 0.01 m^3/kg
c. Water at 1.0 MPa, 100°C

**9.2.18 Steam is used to cool a polymer reaction. The steam in the steam chest of the apparatus is found to be at 250.5°C and 4000 kPa absolute during a routine measurement at the beginning of the day. At the end of the day the measurement showed that the temperature was 650°C and the pressure 10,000 kPa absolute. What was the internal energy change of 1 kg of steam in the chest during the day? Obtain your data from the steam tables.

**9.2.19 What is the difference between heat and internal energy?

**9.2.20 Explain why, when a liquid evaporates, the change in enthalpy is greater than the change in internal energy.

**9.2.21 One pound of liquid water is at its boiling point of 575°F. It is then heated at constant pressure to 650°F, then compressed at constant temperature to one-half of its volume (at 650°F), and finally returned to its original state of the boiling point at 575°F. Calculate ΔH and ΔU for the overall process.

**9.2.22 A gas is heated at 200 kPa from 300 K to 400 K and then cooled to 350 K. In a different process the gas is directly heated from 300 K to 350 K at 200 kPa. What difference is there in internal energy and enthalpy changes for the two processes?

**9.2.23 For the systems defined below, state whether Q, W, ΔH, and ΔU are 0, >0, or <0, and compare their relative values if not equal to 0:

a. An egg (the system) is placed into boiling water.
b. Gas (the system), initially at equilibrium with its surroundings, is compressed rapidly by a piston in an insulated nonconducting cylinder by an insulated nonconducting piston; give your answer for two cases: (1) before reaching a new equilibrium state and (2) after reaching a new equilibrium state.
c. A Dewar flask of coffee (the system) is shaken.

**9.2.24 Indicate whether the following statements are true or false:

a. In a process in which a pure substance starts at a specified temperature and pressure, goes through several temperature and pressure changes, and then returns to the initial state, $\Delta U = 0$.
b. The reference enthalpy for the steam tables ($\Delta H = 0$) is at 25°C and 1 atm.
c. Work can always be calculated as $\Delta(pV)$ for a process going from state 1 to state 2.
d. An isothermal process is one for which the temperature change is zero.
e. An adiabatic process is one for which the pressure change is zero.
f. A closed system is one for which no reaction occurs.
g. An intensive property is a property of material that increases in value as the amount of material increases.

h. Heat is the amount of energy liberated by the reaction within a process.
i. Potential energy is the energy a system has relative to a reference plane.
j. The units of the heat capacity can be (cal)/(g)(°C) or Btu/(lb)(°F), and the numerical value of the heat capacity is the same in each system of units.

****9.2.25** Indicate whether the following statements are true or false:

a. The enthalpy change of a substance can never be negative.
b. The enthalpy of steam can never be less than zero.
c. Both Q and ΔH are state (point) functions.
d. Internal energy does not have an absolute value.
e. By definition $U = H - (pV)$.
f. The work done by a gas expanding into a vacuum is zero.
g. An intensive property is a property whose value depends on the amount of material present in the system.
h. The enthalpy change for a system can be calculated by just taking the difference between the final and initial values of the respective enthalpies.
i. Internal energy has a value of zero at absolute zero.
j. A batch system and an open system are equivalent terms.
k. $\left(\frac{\partial U}{\partial p}\right)_T = 0$ for an ideal gas.
l. Enthalpy is an intensive property.
m. Internal energy is an extensive property.
n. The value of the internal energy for liquid water is about the same as the value for the enthalpy.

****9.2.26** Figure P9.2.26 shows a pure substance that is heated by a constant source of heat supply.

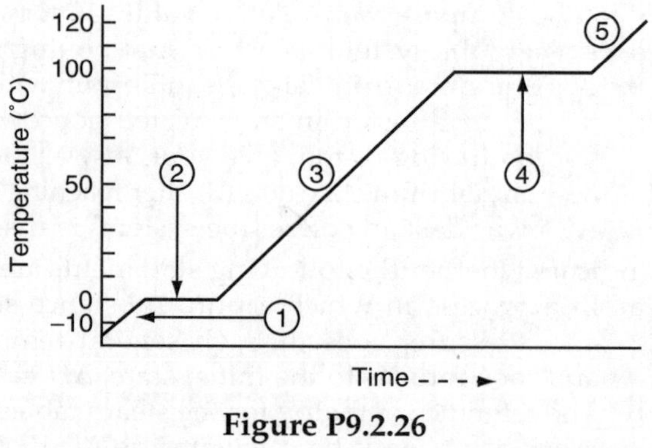

Figure P9.2.26

Use the numbers in the diagram to denote the following stages:

a. Being warmed as a solid
b. Being warmed as a liquid
c. Being warmed as a gas
d. Changing from a solid to a liquid
e. Changing from a liquid to a gas

Also, what is the boiling temperature of the substance? The freezing temperature?

**9.2.27 Estimate ΔH_v for n-heptane at its normal boiling point given $T_b = 98.43°C$, $T_c = 540.2$ K, $p_c = 27$ atm. Use Chen's equation. Calculate the percent error in this value compared to the tabulated value of 31.69 kJ/g mol.

**9.2.28 Estimate the heat capacity of gaseous isobutane at 1000 K and 200 mm Hg by using the Kothari-Doraiswamy relation

$$C_p = A + B \log_{10} T_r$$

from the following experimental data at 200 mm Hg:

C_p[J/(K)(g mol)]	97.3	149.0
Temperature(K)	300	500

The experimental value is 227.6; what is the percentage error in the estimate?

**9.2.29 Two gram moles of nitrogen are heated from 50°C to 250°C in a cylinder. What is ΔH for the process? The heat capacity equation is

$$C_p = 27.32 + 0.6226 \times 10^{-2}T - 0.0950 \times 10^{-5}\,T^2$$

where T is in kelvin and C_p is in J/(g mol)(°C).

**9.2.30 What is the enthalpy change for acetylene when heated from 37.8°C to 93.3°C?

**9.2.31 You are asked to calculate the electric power required (in kilowatt-hours) to heat all of an aluminum wire (positioned in a vacuum similar to a lightbulb filament) from 25°C to 660°C (liquid) to be used in a vapor deposition apparatus. The melting point of Al is 660°C. The wire is 2.5 mm in diameter and has a length of 5.5 cm. (The vapor deposition occurs at temperatures in the vicinity of 900°C.)

Data: For Al, $C_p = 20.0 + 0.0135T$, where T is in kelvin and C_p is in J/(g mol)(°C). The $\Delta H_{\text{fusion}} = 10{,}670$ J/(g mol)(°C) at 660°C. The density of Al is 19.35 g/cm^3.

**9.2.32 A closed vessel contains steam at 1000.0 psia in a 4-to-1 vapor-volume-to-liquid-volume ratio. What is the steam quality?

**9.2.33 What is the enthalpy change in British thermal units when 1 gal of water is heated from 60°F to 1150°F at 240 psig?

**9.2.34 What is the enthalpy change that takes place when 3 kg of water at 101.3 kPa and 300 K are vaporized to 15,000 kPa and 800 K?

**9.2.35 Equal quantities by weight of water at +50°C and of ice at −40°C are mixed together. What will be the final temperature of the mixture?

**9.2.36 A chart for carbon dioxide (see Appendix K) shows that the enthalpy of saturated CO_2 liquid is zero at −40°F. Can this be true? Explain your answer.

***9.2.37 A horizontal cylinder, closed at one end, is fitted with a movable piston. Originally the cylinder contains 1.2 ft^3 of gas at 7.3 atm pressure. If the pressure against the piston face is reduced very slowly to 1 atm, calculate the work done by the gas on the piston, assuming the following relationship to hold for the gas: $pV^{1.3}$ = constant.

***9.2.38 A windmill converts the kinetic energy of the moving air into electrical energy at an efficiency of about 30%, depending on the windmill design and speed of the wind. Estimate the power in kilowatts for a wind flowing perpendicular to a windmill with blades 15 m in diameter when the wind is blowing at 20 mi/hr at 27°C and 1 atm.

***9.2.39 Before it lands, a vehicle returning from space must convert its enormous kinetic energy to heat. To get some idea of what is involved, a vehicle returning from the moon at 25,000 mi/hr can, in converting its kinetic energy, increase the internal energy of the vehicle sufficiently to vaporize it. Obviously, a large part of the total kinetic energy must be transferred from the vehicle. How much kinetic energy does the vehicle have (in British thermal units per pound)? How much energy must be transferred by heat if the vehicle is to heat up by only 20°F/lb?

***9.2.40 The world's largest plant that obtains energy from tidal changes is at Saint-Malo, France. The plant uses both the rising and falling cycles (one period in or out is 6 hr 10 min in duration). The tidal range from low to high is 14 m, and the tidal estuary (the Rance River) is 21 km long with an area of 23 km^2. Assume that the efficiency of the plant in converting potential to electrical energy is 85%, and estimate the average power produced by the plant. (Note: Also assume that after high tide, the plant does not release water until the sea level drops 7 m, and after a low tide it does not permit water to enter the basin until the level outside the basin rises 7 m, and the level differential is maintained during discharge and charge.)

***9.2.41 Calculate the change in the internal energy of 1 mol of a monoatomic ideal gas when the temperature goes from 0°C to 50°C.

***9.2.42 You have calculated that the specific enthalpy of 1 kg mol of an ideal gas at 300 kN/m^2 and 100°C is 6.05×10^5 J/kg mol (with reference to 0°C and 100 kN/m^2). What is the specific internal energy of the gas at 300 kPa and 100°C?

***9.2.43 In a proposed molten-iron coal gasification process [*Chemical Engineering*, 17 (July, 1985)], pulverized coal of up to 3 mm size is blown into a molten iron bath, and oxygen and steam are blown in from the bottom of the vessel. Materials such as lime for settling the slag, or steam for batch cooling and hydrogen generation, can be injected at the same time. The sulfur in the coal reacts with lime to form calcium sulfide, which dissolves into the slag. The process operates at atmospheric pressure and 1400°C to 1500°C. Under these conditions, coal volatiles escape immediately and are cracked. The carbon conversion rate is said to be above 98%, and the gas is typically 65% to 70% CO, 25% to 35% H_2, and less than 2% CO_2. Sulfur content of the gas is less than 20 ppm. Assume that the product gas is 68% CO, 30% H_2, and 2% CO_2, and calculate the enthalpy change that occurs on the cooling of 1000 m^3 at 1400°C and 1 atm of gaseous product from 1400°C to 25°C and 1 atm. Use the table for the enthalpies of the combustion gases.

***9.2.44 Calculate the enthalpy change (in joules) that occurs when 1 kg of benzene vapor at 150°C and 100 kPa condenses to a solid at −20.0°C and 100 kPa.

***9.2.45 Use the steam tables to answer the following questions:

a. What is the enthalpy change needed to change 3 lb of liquid water at 32°F to steam at 1 atm and 300°F?
b. What is the enthalpy change needed to heat 3 lb of water from 60 psia and 32°F to steam at 1 atm and 300°F?
c. What is the enthalpy change needed to heat 1 lb of water at 60 psia and 40°F to steam at 300°F and 60 psia?
d. What is the enthalpy change needed to change 1 lb of a water-steam mixture of 60% quality to one of 80% quality if the mixture is at 300°F?
e. Calculate the ΔH value for an isobaric (constant pressure) change of steam from 120 psia and 500°F to saturated liquid.
f. Repeat part e for an isothermal change to saturated liquid.
g. Does an enthalpy change from saturated vapor at 450°F to 210°F and 7 psia represent an enthalpy increase or decrease? A volume increase or decrease?
h. In what state is water at 40 psia and 267.24°F? At 70 psia and 302°F? At 70 psia and 304°F?
i. A 2.5 ft^3 tank of water at 160 psia and 363.5°F has how many cubic feet of liquid water in it? Assume that you start with 1 lb of H_2O. Could the tank contain 5 lb of H_2O under these conditions?
j. What is the volume change when 2 lb of H_2O at 1000 psia and 20°F expand to 245 psia and 460°F?
k. Ten pounds of wet steam at 100 psia have an enthalpy of 9000 Btu. Find the quality of the wet steam.

***9.2.46 Use the steam tables to calculate the enthalpy change (in joules) of 2 kg mol of steam when heated from 400 K and 100 kPa to 900 K and 100 kPa. Repeat using the table in the text for the enthalpies of combustion gases. Repeat using the heat capacity for steam. Compare your answers. Which is more accurate?

***9.2.47 Use the CO_2 chart in Appendix K for the following calculations:

a. Four pounds of CO_2 are heated from saturated liquid at 20°F to 600 psia and 180°F.
   1. What is the specific volume of the CO_2 at the final state?
   2. Is the CO_2 in the final state gas, liquid, solid, or a mixture of two or three phases?
b. The 4 lb of CO_2 is then cooled at 600 psia until the specific volume is 0.07 ft^3/lb.
   1. What is the temperature of the final state?
   2. Is the CO_2 in the final state gas, liquid, solid, or a mixture of two or three phases?

***9.2.48 Calculate the enthalpy change in heating 1 g mol of CO_2 from 50°C to 100°C at 1 atm. Do this problem by three different methods:

a. Use the heat capacity equation from Appendix G.
b. Use the CO_2 chart in Appendix K.
c. Use the table of combustion gases in Appendix F.

***9.2.49 Use the chart for n-butane (Figure 9.14) to calculate the enthalpy change for 10 lb of butane going from a volume of 2.5 ft^3 at 360°F to saturated liquid at 10 atm.

***9.2.50 A propane gas tank is filled, closed, and attached to a barbecue grill. After standing for some time, what is the state of the propane inside the tank? What are the temperature and pressure inside the tank? After 80% of the propane in the tank is used, what is the pressure inside the tank after it reaches equilibrium?

****9.2.51 a. Ten pound moles of an ideal gas are originally in a tank at 100 atm and 40°F. The gas is heated to 440°F. The specific molar enthalpy of the ideal gas is given by the equation $\hat{H} = 300 + 8.00T$, where $\hat{H}$ is in British thermal units per pound mole and T is the temperature in degrees Fahrenheit.
1. Compute the volume of the container (in cubic feet).
2. Compute the final pressure of the gas (in atmospheres).
3. Compute the enthalpy change of the gas.

b. Use the equation above to develop an equation giving the molar internal energy, in joules per gram mole as a function of temperature, T, in degrees Celsius.

****9.2.52 Wet steam flows in a pipe at a pressure of 700 kPa. To check the quality, the wet steam is expanded adiabatically to a pressure of 100 kPa in a separate pipe. A thermocouple inserted into the pipe indicates that the expanded steam has a temperature of 125°C. What was the quality of the wet steam in the pipe prior to expansion?

9.3 Energy Balances without Reaction

*9.3.1 Consider the following systems:
a. Open system, steady-state
b. Open system, unsteady-state
c. Closed system, steady-state
d. Closed system, unsteady-state

For which system(s) can energy cross the system boundary?

*9.3.2 In a closed system process, 60 Btu of heat are added to the system, and the internal energy of the system increases by 220 Btu. Calculate the work of the process.

**9.3.3 Draw a picture of the following processes, draw a boundary for the system, and state for each whether heat transfer, work, a change in internal energy, a change in enthalpy, a change in potential energy, or a change in kinetic energy occurs *inside the system*. Also classify each system as open or closed, and as steady-state or unsteady-state.
a. A pump, driven by a motor, pumps water from the first to the third floor of a building at a constant rate and temperature. The system is the pump.
b. As in a, except the system is the pump and the motor.
c. A block of ice melts in the sun. The system is the block of ice.
d. A mixer mixes a polymer into a solvent. The system is the mixer.

****9.3.4** Explain specifically what the system is for each of the following processes; indicate if any energy transfer takes place by heat or work (use the symbols Q and W, respectively) or if these terms are zero.

a. A liquid inside a metal can, well insulated on the outside of the can, is shaken very rapidly in a vibrating shaker.
b. A motor and propeller are used to drive a boat.
c. Water flows through a pipe at 1.0 m/min, and the temperatures of the water and the air surrounding the pipe are the same.

****9.3.5** Draw a simple sketch of each of the following processes, and in each, label the system boundary, the system, the surroundings, and the streams of material and energy that cross the system boundary.

a. Water enters a boiler, is vaporized, and leaves as steam. The energy for vaporization is obtained by combustion of a fuel gas with air outside the boiler surface.
b. Steam enters a rotary steam turbine and turns a shaft connected to an electric generator. The steam is exhausted at a low pressure from the turbine.
c. A battery is charged by connecting it to a source of current.

****9.3.6** Is it possible to compress an ideal gas in a cylinder with a piston isothermally in an adiabatic process? Explain your answer briefly.

****9.3.7** Water is heated in a closed pot on top of a stove while being stirred by a paddle wheel. During the process, 30 kJ of heat are transferred to the water, and 5 kJ of heat are lost to the surrounding air. The work done amounts to 500 J. Determine the final energy of the system if its initial internal energy was 10 kJ.

****9.3.8** A person living in a 4 m × 5 m × 5 m room forgets to turn off a 100 W fan before leaving the room, which is at 100 kPa, 30°C. Will the room be cooler when the person comes back after 5 hr, assuming zero heat transfer? The heat capacity at constant volume for air is 30 kJ/kg mol.

****9.3.9** A vertical cylinder capped by a piston weighing 990 g contains 100 g of air at 1 atm and 25°C. Calculate the maximum possible final elevation of the piston if 100 J of work are used to raise the cylinder and its contents vertically. Assume that all of the work goes into raising the piston.

****9.3.10** Two tanks are suspended in a constant temperature bath at 200°F. The first tank contains 1 ft^3 of dry saturated steam. The other tank is evacuated. The two tanks are connected. After equilibrium is reached, the pressure in both tanks is 1 psia. Calculate (a) the work done in the process, (b) the heat transfer to the two tanks, (c) the internal energy change of the steam, and (d) the volume of the second tank.

****9.3.11** Two states, 1 and 2, are marked in Figure P9.3.11. Path A is taken from 1 to 2. Two alternative return paths from 2 to 1 are shown: B and C. Two different cycles can now be made up, each going from point 1 to point 2, and then returning to point 1. One cycle is made up from path A and path B, and the other from path A and path C. Are the following equations correct for the cycle 1 to 2 and return?

$$Q_A + Q_B = W_A + W_B \qquad\qquad Q_A + Q_C = W_A + W_C$$

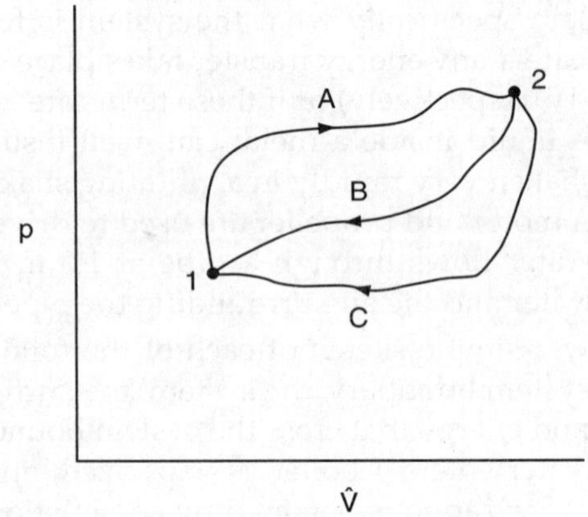

Figure P9.3.11

****9.3.12** A national mail-order firm is advertising for $699.99 ("manufacturer's suggested retail price $1,199.00") the Cold Front Portable Air Conditioner, a "freestanding portable unit" that "does not require outside venting." It is intended to be rolled from room to room by the user, who simply plugs it into an ordinary 110 V AC outlet and enjoys the cool air from its "Cold Front." It is claimed to provide 5500 Btu/hr cooling capacity for 695 W power. Is the cooling capacity correct? What is the catch to the advertisement?

****9.3.13** A large high-pressure tank contains 10 kg of steam. In spite of the insulation on the tank it loses 2050 kJ/hr to the surroundings. How many kilowatts are needed to maintain the steam at 3000 kPa and 600 K?

****9.3.14** An expensive drug is manufactured in a sealed vessel that holds 8 lb of water at 100°F. A 1/4 hp motor stirs the contents of the vessel. What is the rate of heat removal from the vessel in British thermal units per minute to maintain the temperature at 100°F?

****9.3.15** Calculate how much heat is needed to evaporate 1 kg of water in an open vessel if the water starts at 27°C. Use the steam tables. The barometer reads 760 mm Hg.

****9.3.16** Air is being compressed from 100 kPa and 255 K (where it has an enthalpy of 489 kJ/kg) to 1000 kPa and 278 K (where it has an enthalpy of 509 kJ/kg). The exit velocity of the air from the compressor is 60 m/s. What is the power required (in kilowatts) for the compressor if the load is 100 kg/hr of air?

****9.3.17** Write the simplified energy balance for the following processes:

a. A fluid flows steadily through a poorly designed coil in which it is heated from 170°F to 250°F. The pressure at the coil inlet is 120 psia, and at the coil outlet is 70 psia. The coil is of uniform cross section, and the fluid enters with a velocity of 2 ft/s.

b. A fluid is allowed to flow through a cracked (slightly opened) valve from a region where its pressure is 200 psia and 670°F to a region where its pressure is 40 psia, the whole operation being adiabatic.

List each assumption or decision by number. You do not have to solve the problems.

9.3.18 Write the appropriate simplified energy balances for the following changes; in each case the amount of material to be used as a basis of calculation is 1 lb and the initial condition is 100 psia and 370°F:

a. The substance, enclosed in a cylinder fitted with a movable frictionless piston, is allowed to expand at constant pressure until its temperature has risen to 550°F.
b. The substance, enclosed in a cylinder fitted with a movable frictionless piston, is kept at constant volume until the temperature has fallen to 250°F.
c. The substance, enclosed in a cylinder fitted with a movable frictionless piston, is compressed adiabatically until its temperature has risen to 550°F.
d. The substance, enclosed in a cylinder fitted with a movable frictionless piston, is compressed at constant temperature until the pressure has risen to 200 psia.
e. The substance is enclosed in a container that is connected to a second evacuated container of the same volume as the first, there being a closed valve between the two containers. The final condition is reached by opening the valve and allowing the pressures and temperatures to equalize adiabatically.

9.3.19 An insulated, sealed tank that is 2 ft^3 in volume holds 8 lb of water at 100°F. A 1/4 hp stirrer mixes the water for 1 hr. What is the fraction vapor at the end of the hour? Assume all of the energy from the stirrer enters the tank.

For this problem you do not have to get a numerical solution. Instead, list the following in this order:

a. State what the system you select is.
b. Specify open or closed.
c. Draw a picture.
d. Put all of the known or calculated data on the picture in the proper place.
e. Write down the energy balance (use the symbols in the text) and simplify it as much as possible. List each assumption in so doing.
f. Calculate W.
g. List the equations with data introduced that you would use to solve the problem.
h. Explain step by step how to solve the problem (but do not do so).

9.3.20 Hot reaction products (assume they have the same properties as air) at 1000°F leave a reactor. In order to prevent further reaction, the process is designed to reduce the temperature of the products to 400°F by immediately spraying liquid water into the gas stream. Refer to Figure P9.3.20.

How many pounds of water at 70°F are required per 100 lb of products leaving at 400°F?

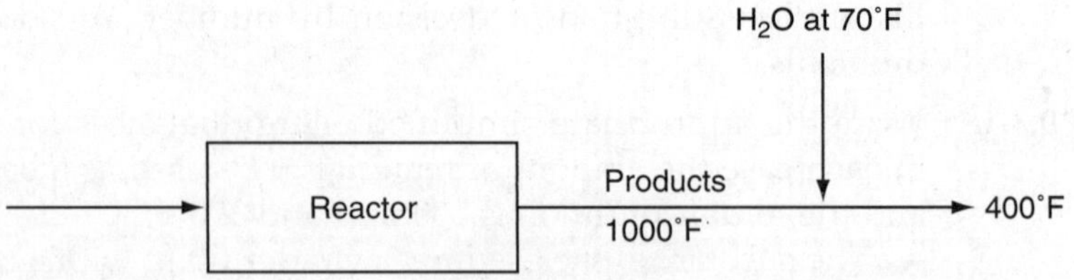

Figure P9.3.20

For this problem you do not have to get a numerical solution. Instead, list the following in this order:

a. State what the system you select is.
b. Specify open or closed.
c. Draw a picture.
d. Put all the known or calculated data on the picture in the proper places.
e. Write down the material and energy balances (use the symbols in the text) and simplify them as much as possible; list each assumption in so doing.
f. Insert the known data into the simplified equation(s) you would use to solve the problem.

****9.3.21** Write the simplified energy balances for the following changes:

a. A fluid flows steadily through a poorly designed coil in which it is heated from 70°F to 250°F. The pressure at the coil inlet is 120 psia and at the coil outlet is 70 psia. The coil is of uniform cross section, and the fluid enters with a velocity of 2 ft/s.
b. A fluid is expanded through a well-designed adiabatic nozzle from a pressure of 200 psia and a temperature of 650°F to a pressure of 40 psia and a temperature of 350°F. The fluid enters the nozzle with a velocity of 25 ft/s.
c. A turbine directly connected to an electric generator operates adiabatically. The working fluid enters the turbine at 1400 kPa absolute and 340°C. It leaves the turbine at 275 kPa absolute and at a temperature of 180°C. Entrance and exit velocities are negligible.
d. The fluid leaving the nozzle of part b is brought to rest by passing it through the blades of an adiabatic turbine rotor, and it leaves the blades at 40 psia and at 400°F.

****9.3.22** Simplify the general energy balance so as to represent the process in each of the following cases. Number each term in the general balance, and state why you retained or deleted it.

a. A bomb calorimeter is used to measure the heating value of natural gas. A measured volume of gas is pumped into the bomb. Oxygen is then added to give a total pressure of 10 atm, and the gas mixture is exploded using a hot wire. The resulting heat transfer from the bomb to the surrounding water bath is measured. The final products in the bomb are CO_2 and water.
b. Cogeneration (generation of steam for both power and heating) involves the use of gas turbines or engines as prime movers, with the exhausted

steam going to the process to be used as a heat source. A typical installation is shown in Figure P9.3.22.

c. In a mechanical refrigerator the Freon liquid is expanded through a small insulated orifice so that part of it flashes into vapor. Both the liquid and vapor exit at a lower temperature than the temperature of the liquid entering.

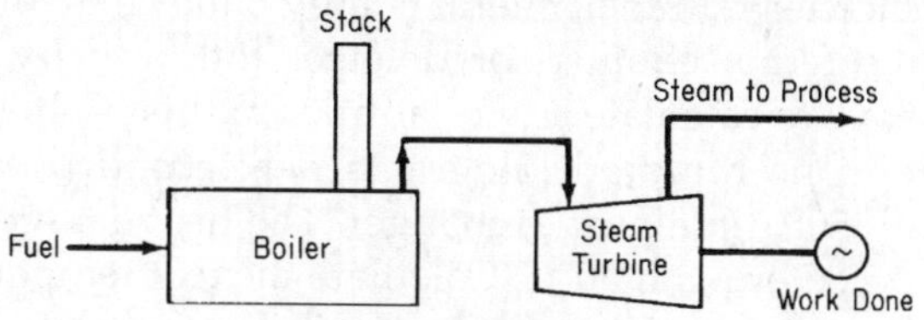

Figure P9.3.22

****9.3.23** One kilogram of gaseous CO_2 at 550 kPa and 25°C was compressed by a piston to 3500 kPa, and in so doing 4.016×10^3 J of work were done on the gas. To keep the container isothermal, the container was cooled by blowing air over fins on the outside of the container. How much heat (in joules) was removed from the system?

****9.3.24** A household freezer is placed inside an insulated sealed room. If the freezer door is left open with the freezer operating, will the temperature of the room increase or decrease? Explain your answer.

*****9.3.25** By use of the steam tables, compute the numerical values for Q, W, ΔH, and ΔU for the complete process in which 1 lb of liquid water is initially confined in a capsule at 327.8°F and 100 psia within an evacuated vessel of 4.435 ft^3 capacity; the capsule is then broken within the vessel, allowing the water to escape into the evacuated vessel; and finally the water is brought to the initial temperature (327.8°F).

*****9.3.26** Four kilograms of superheated steam at 700 kPa and 500 K are cooled in a tank to 400 K. Calculate the heat transfer involved.

*****9.3.27** A cylinder contains 1 lb of steam at 600 psia and a temperature of 500°F. It is connected to another equal-size cylinder which is evacuated. A valve between the cylinders is opened. If the steam expands into the empty cylinder, and the final temperature of the steam in both cylinders is 500°F, calculate Q, W, ΔU, and ΔH for the system composed of both cylinders.

*****9.3.28** Calculate Q, W, ΔU, and ΔH for 1 lb of liquid water which is evaporated at 212°F by (a) a nonflow process and (b) a unsteady-state flow process.

*****9.3.29** A cylinder that initially contains nitrogen at 1 atm and 25°C is connected to a high-pressure line of nitrogen at 50 atm and 25°C. When the cylinder pressure reaches 40 atm, the valve on the cylinder is closed. Assume the process is adiabatic, and that nitrogen can be treated as an ideal gas. What is the temperature in the cylinder when the valve is closed? Ideal heat capacities are listed in Table 9.3.

*****9.3.30** An insulated tank having a volume of 50 ft^3 contains saturated steam at 1 atm. It is connected to a steam line maintained at 50 psia and 291°F. Steam flows slowly into the tank until the pressure reaches 50 psia. What is the temperature in the tank at that time? Hint: Use the steam tables on the CD to make this an easy problem.

***9.3.31 Start with the general energy balance, and simplify it for each of the processes listed below to obtain an energy balance that represents the process. Label each term in the general energy balance by number, and list by their numbers the terms retained or deleted, followed by your explanation. (You do not have to calculate any quantities in this problem.)

a. One hundred kilograms per second of water are cooled from 100°C to 50°C in a heat exchanger. The heat is used to heat up 250 kg/s of benzene entering at 20°C. Calculate the exit temperature of the benzene.
b. A feed-water pump in a power generation cycle pumps water at the rate of 500 kg/min from the turbine condensers to the steam generation plant, raising the pressure of the water from 6.5 kPa to 2800 kPa. If the pump operates adiabatically with an overall mechanical efficiency of 50% (including both the pump and its drive motor), calculate the electric power requirement of the pump motor (in kilowatts). The inlet and outlet lines to the pump are of the same diameter. Neglect any rise in temperature across the pump due to friction (i.e., the pump may be considered to operate isothermally).

***9.3.32 It is necessary to evaluate the performance of an evaporator that will be used to concentrate a 5% organic solution. Assume there will be no boiling point rise.

The following information has already been obtained: U = 300 Btu/(hr)(ft^2)(°F);

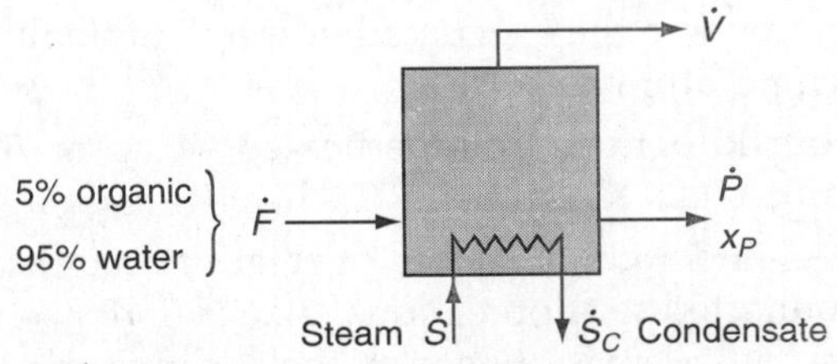

Figure P9.3.32

A = 2000 ft^2; heating steam S is available at 4.5 psia; feed enters at 140°F. Must any further measurements be made? The rate of heat transfer from the steam coils to the liquid is $\hat{Q} = UA(T_S - T_V)$.

***9.3.33 Your company produces small power plants that generate electricity by expanding waste process steam in a turbine. You are asked to study the turbine to determine if it is operating as efficiently as possible. One way to ensure good efficiency is to have the turbine operate adiabatically. Measurements show that for steam at 500°F and 250 psia:

a. The work output of the turbine is 86.5 hp.
b. The rate of steam usage is 1000 lb/hr.

c. The steam leaves the turbine at 14.7 psia and consists of 15% moisture (i.e., liquid H_2O).

Is the turbine operating adiabatically? Support your answer with calculations.

***9.3.34 In one stage of a process for the manufacture of liquid air, air as a gas at 4 atm absolute and 250 K is passed through a long, insulated 3-in.-ID pipe in which the pressure drops 3 psi because of frictional resistance to flow. Near the end of the line, the air is expanded through a valve to 2 atm absolute. State all assumptions.

a. Compute the temperature of the air just downstream of the valve.
b. If the air enters the pipe at the rate of 100 lb/hr, compute the velocity just downstream of the valve.

***9.3.35 A liquid that can be treated as water is being well mixed by a stirrer in a 1 m^3 vessel. The stirrer introduces 300 W of power into the vessel. The heat transfer from the tank to the surroundings is proportional to the temperature difference between the vessel and the surroundings (which are at 20°C). The flow rate of liquid in and out of the tank is 1 kg/min. If the temperature of the inlet liquid is 40°C, what is the temperature of the outlet liquid? The proportionality constant for the heat transfer is 100 W/°C.

***9.3.36 Air is used to heat an auditorium. The flow rate of the entering air to the heating unit is 150 m^3 per minute at 17°C and 100 kPa. The entering air passes through a heating unit that uses 15 kW for the electric coils. If the heat loss from the heating unit is 200 W, what is the temperature of the exit air?

***9.3.37 The following problem and its solution were given in a textbook: How much heat in kilojoules is required to vaporize 1.00 kg of saturated liquid water at 100°C and 101.3 kPa? The solution is

$$n = 1.00 \text{ kg}$$

$$\Delta E = Q + W + E_{\text{flow}}$$

$$\Delta U = Q + W = Q - p\Delta V$$

$$Q = \Delta H = (1 \text{ kg})(2256.9 \text{ kJ/kg}) = 2256.9 \text{ kJ}$$

Is this solution correct?

***9.3.38 A turbine that uses steam drives an electric generator. The inlet steam flows through a 10-cm-diameter pipe to the turbine at the rate of 2.5 kg/s at 600°C and 1000 kPa.

The exit steam discharges through a 25-cm-diameter pipe at 400°C and 100 kPa.

What is the expected power obtained from the turbine if it operates essentially adiabatically?

***9.3.39 A closed vessel having a volume of 100 ft^3 is filled with saturated steam at 265 psia. At some later time, the pressure has fallen to 100 psia due to heat losses from the tank. Assuming that the contents of the tank at 100 psia are in an equilibrium state, how much heat was lost from the tank?

***9.3.40 A large piston in a cylinder does 12,500 (ft)(lb_f) of work in compressing 3 ft^3 of air to 25 psia. Five pounds of water in a jacket surrounding the cylinder increased in temperature by 2.3°F during the process. What was the change in the internal energy of the air?

$$C_{p,\text{ water}} = 8.0 \frac{\text{Btu}}{(\text{lb mol})(°\text{F})}$$

***9.3.41 Carbon dioxide cylinders, initially evacuated, are being loaded with CO_2 from a pipeline in which the CO_2 is maintained at 200 psia and 40°F. As soon as the pressure in a cylinder reaches 200 psia, the cylinder is closed and disconnected from the pipeline. If the cylinder has a volume of 3 ft^3, and if the heat losses to the surroundings are small, compute (a) the final temperature of the CO_2 in a cylinder, and (b) the number of pounds of CO_2 in a cylinder.

***9.3.42 One pound of steam at 130 psia and 600°F is expanded isothermally to 75 psia in a closed system. Thereafter it is cooled at constant volume to 60 psia. Finally, it is compressed adiabatically back to its original state. For each of the three steps of the process, compute ΔU and ΔH. For each of the three steps, where possible, also calculate Q and W.

***9.3.43 In the vapor-recompression evaporator shown in Figure P9.3.43, the vapor produced on evaporation is compressed to a higher pressure and passed through the heating coil to provide the energy for evaporation. The steam entering the compressor is 98% vapor and 2% liquid, at 10 psia; the steam leaving the compressor is at 50 psia and 400°F; and 6 Btu of heat are lost from the compressor per pound of steam throughput. The condensate leaving the heating coil is at 50 psia, 200°F.

a. Compute the British thermal units of heat supplied for evaporation in the heating coil per British thermal unit of work needed for compression by the compressor.
b. If 1,000,000 Btu/hr of heat are to be transferred in the evaporator, what must be the intake capacity of the compressor in cubic feet of wet vapor per minute?

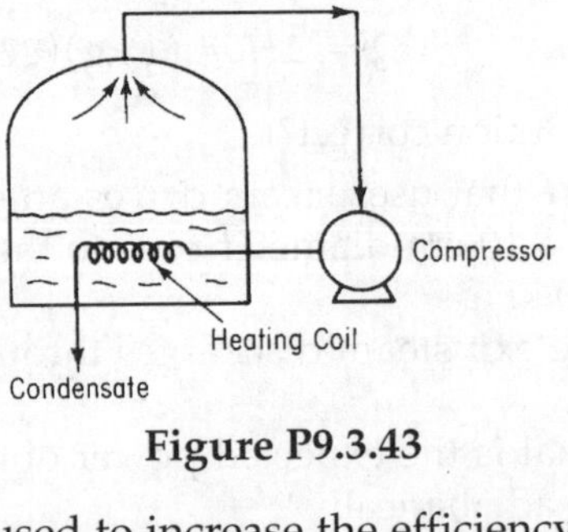

Figure P9.3.43

***9.3.44 Feed-water heaters are used to increase the efficiency of steam power plants. A particular heater is used to preheat 10 kg/s of boiler feed water from 20°C to 188°C at a pressure of 1200 kPa by mixing it with saturated steam bled from a turbine at 1200 kPa and 188°C, as shown in Figure P9.3.44. Although insulated, the heater loses heat at the rate of 50 J/g of exiting mixture. What fraction of the exit stream is steam?

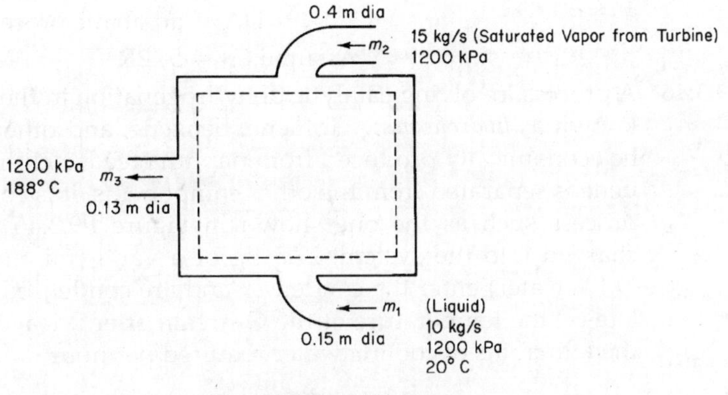

Figure P9.3.44

****9.3.45 A quantity of an ideal gas goes through the ideal cycle shown in Figure P9.3.45. Calculate:

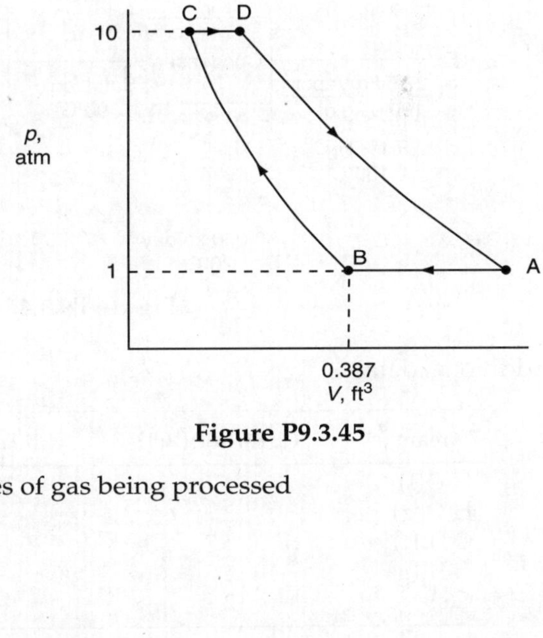

Figure P9.3.45

a. Pound moles of gas being processed
b. VD, ft³
c. W_{AB}, Btu
d. W_{BC}, Btu
e. W_{CD}, Btu
f. W_{DA}, Btu
g. W for the cycle, Btu
h. ΔH for the cycle, Btu
i. Q for the cycle, Btu

Data:

$$T_A = 170°\text{F} \; T_D = 823°\text{F}$$
$$T_B = 70°\text{F}$$
$$\text{BC} = \text{isothermal process}$$

DA = adiabatic process
Assume $C_V = 5/2R$

**** **9.3.46** A process involving catalytic dehydrogenation in the presence of hydrogen is known as *hydroforming*. Toluene, benzene, and other aromatic materials can be economically produced from naphtha feedstocks in this way. After the toluene is separated from the other components, it is condensed and cooled in a process such as the one shown in Figure P9.3.46. For every 100 kg of C charged into the system, 27.5 kg of a vapor mixture of toluene and water (9.1% water) enter the condenser and are condensed by the C stream. Calculate (a) the temperature of the C stream after it leaves the condenser, and (b) the kilograms of cooling water required per hour.

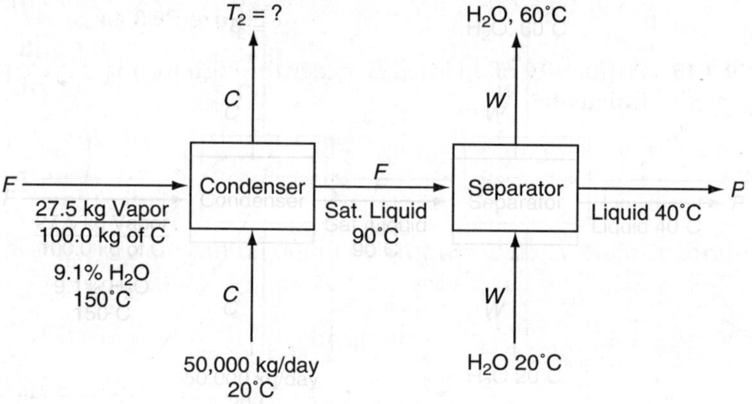

Figure P9.3.46

Additional data:

Stream	C_p[kJ/(kg)(°C)]	B.P. (°C)	ΔH_{vap}(kJ/kg)
$H_2O(l)$	4.2	100	2260
$H_2O(g)$	2.1	—	—
$C_7H_8(l)$	1.7	111	230
$C_7H_8(g)$	1.3	—	—
C(s)	2.1	—	—

*******9.3.47** A power plant operates as shown in Figure P9.3.47. Assume that the pipes, boiler, and superheater are well lagged (insulated), and that friction can be neglected. Calculate (in British thermal units per pound of steam):

a. The heat supplied to the boiler
b. The heat supplied to the superheater
c. The heat removed in the condenser
d. The work delivered by the turbine
e. The work required by the liquid pumps

Also calculate the efficiency of the entire process defined as

$$\frac{\text{Net work delivered / lb steam}}{\text{Total heat supplied / lb steam}}$$

If the water rate to the boiler is 2000 lb/hr, what horsepower does the turbine develop? Finally, what suggestion can you offer that will improve the efficiency of the power plant?

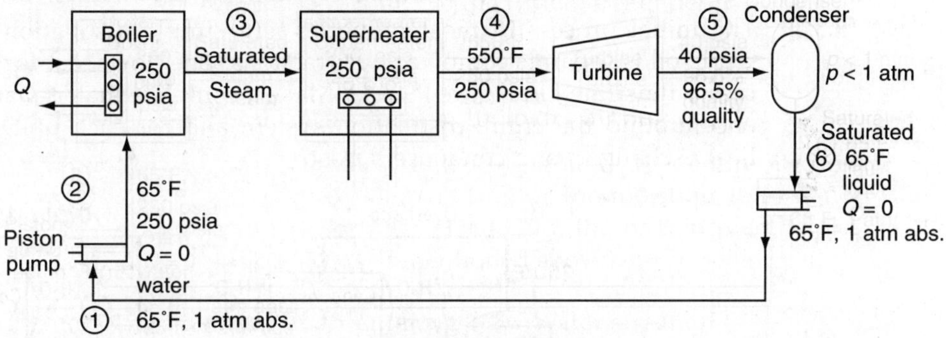

Figure P9.3.47

****9.3.48 A proposal to store Cl_2 as a liquid at atmospheric pressure was recently in the news. The operation is shown in Figure P9.3.48.

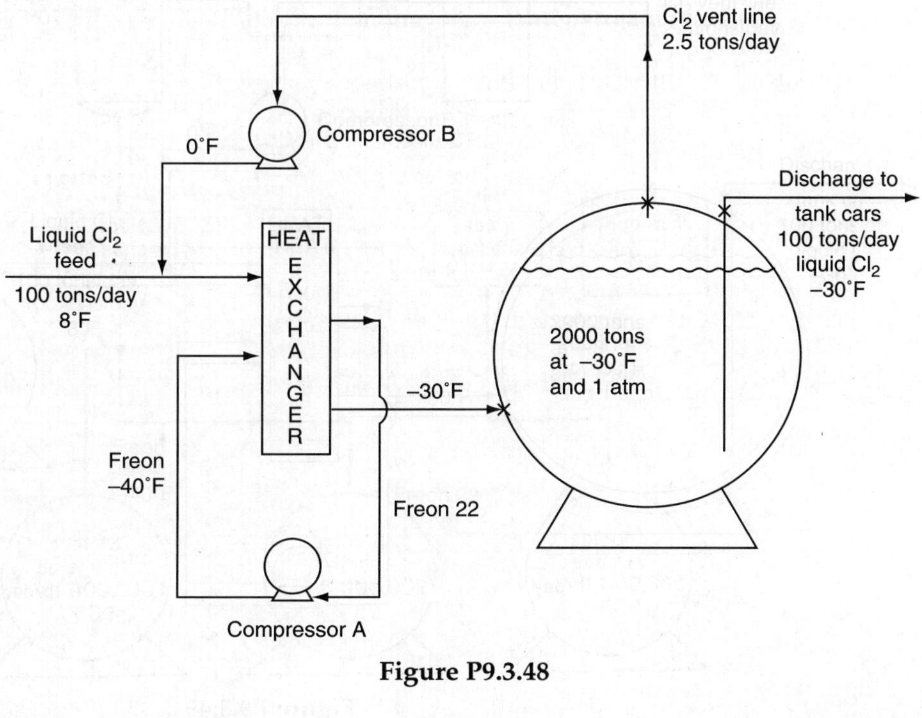

Figure P9.3.48

The normal boiling point of Cl_2 is −30°F. Vapor formed in the storage tank exits through the vent and is compressed to liquid at 0°F and returned to the feed. The vaporization rate is 2.5 tons/day when the sphere is filled to its capacity and the surrounding air temperature is 80°F.

If the compressors are driven by electric motors and are about 30% efficient, what is the horsepower input required to make this process successful? Assume lines and heat exchangers are well insulated. Use 8.1 Btu/(lb mol)(°F) for the heat capacity of liquid $Cl_2 \cdot \Delta H_{\text{vaporization}} = 123.67$ Btu/lb Cl_2.

****9.3.49 The initial process in most refineries is a simple distillation in which the crude oil is separated into various fractions. The flowsheet for one such process is illustrated in Figure P9.3.49. Make a complete material and energy balance around the entire distillation system and for each unit, including the heat exchangers and condensers. Also:

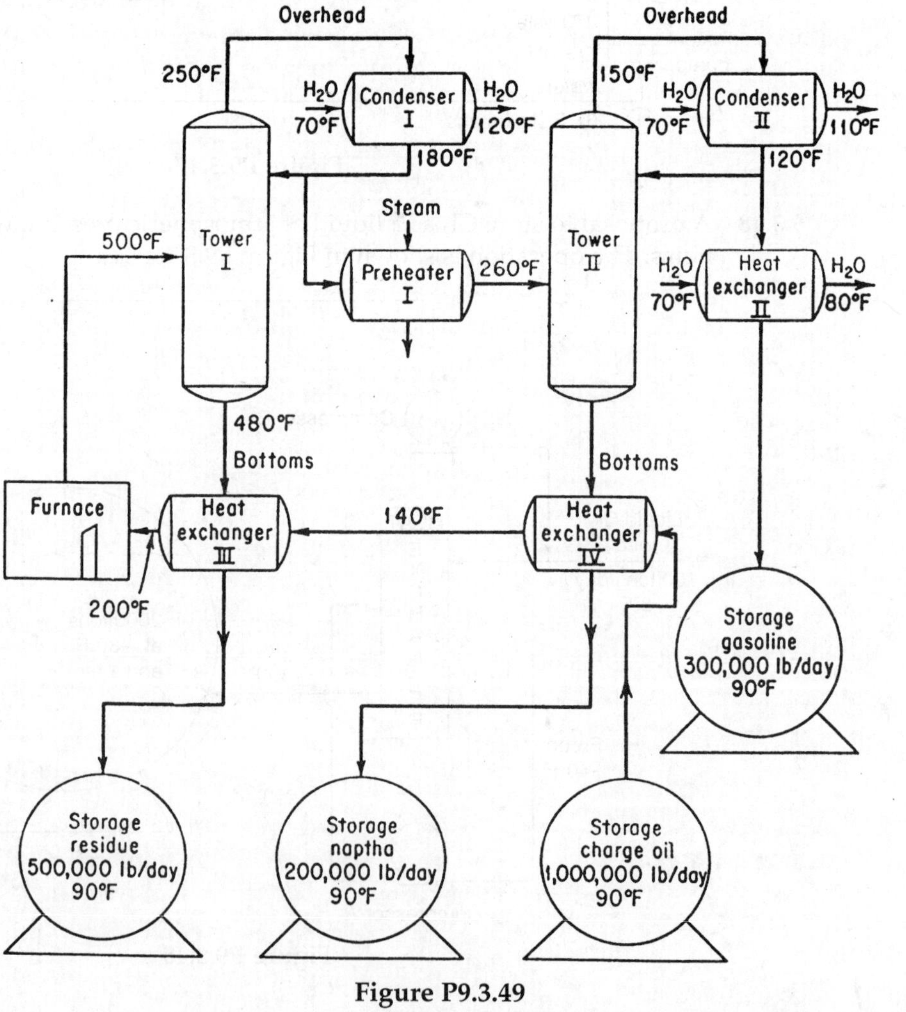

Figure P9.3.49

a. Calculate the heat load that has to be supplied by the furnace in British thermal units per hour.
b. Determine the additional heat that would have to be supplied by the furnace if the charge oil were not preheated to 200°F before it entered the furnace.

Do the calculated temperatures of the streams going into storage from the heat exchangers seem reasonable?

Additional data:

	Specific Heat of Liquid Btu/(lb)(°F)	Latent Heat of Vaporization Btu/lb	Specific Heat of Vapor Btu/(lb)(°F)	Condensation Temp. °F
Charge oil	0.53	100	0.45	480
Overhead, tower I	0.59	111	0.51	250
Bottoms, tower I	0.51	92	0.42	500
Overhead, tower II	0.63	118	0.58	150
Bottoms, tower II	0.58	107	0.53	260

The reflux ratio of tower I is 3 recycle to 1 product.
The reflux ratio of tower II is 2 recycle to 1 product.
The reflux ratio is the ratio of the mass flow rate from the condenser to the mass flow rate that leaves the top of the tower (the overhead) and enters the condenser.

****9.3.50 A boiler house flowsheet for a chemical process plant is shown in Figure P9.3.50 on page 594. The production rate of 600 psia superheated steam is 100,000 lb/hr. The return condensate flow rate is 50,000 lb/hr. Calculate:

a. The flow rate of 30 psia steam required in the deaerator (lb/hr)
b. The flow rate of makeup feed water (lb/hr)
c. The pump horsepower (hp)
d. The pump electrical consumption if the pump is 55% efficient (kW)
e. Yearly electrical cost to operate the pump (at 0.05/kWh)
f. Yearly electrical savings if the pump could be operated with a 600 psia discharge pressure
g. Heat input to the steam drum (Btu/hr)
h. Heat input to the superheater section (Btu/hr)
i. The amount of 30 psia steam lost to the atmosphere (lb/hr)
j. The amount of 30 psia condensate lost to the atmosphere (lb/hr)

****9.3.51 A distillation process has been set up to separate an ethylene-ethane mixture as shown in Figure P9.3.51 on page 595. The product stream will consist of 98% ethylene, and it is desired to recover 97% of the ethylene in the feed. The feed, 1000 lb/hr of 35% ethylene, enters the preheater as a subcooled liquid (temperature = −100°F, pressure = 250 psia). The feed experiences a 20°F temperature rise before it enters the still. The heat capacity of liquid ethane may be considered to be constant and equal to 0.65 Btu/(lb)(°F), and the heat capacity of ethylene may be considered to be constant and equal to 0.55 Btu/(lb)(°F). Heat capacities and saturation temperatures of mixtures may be determined

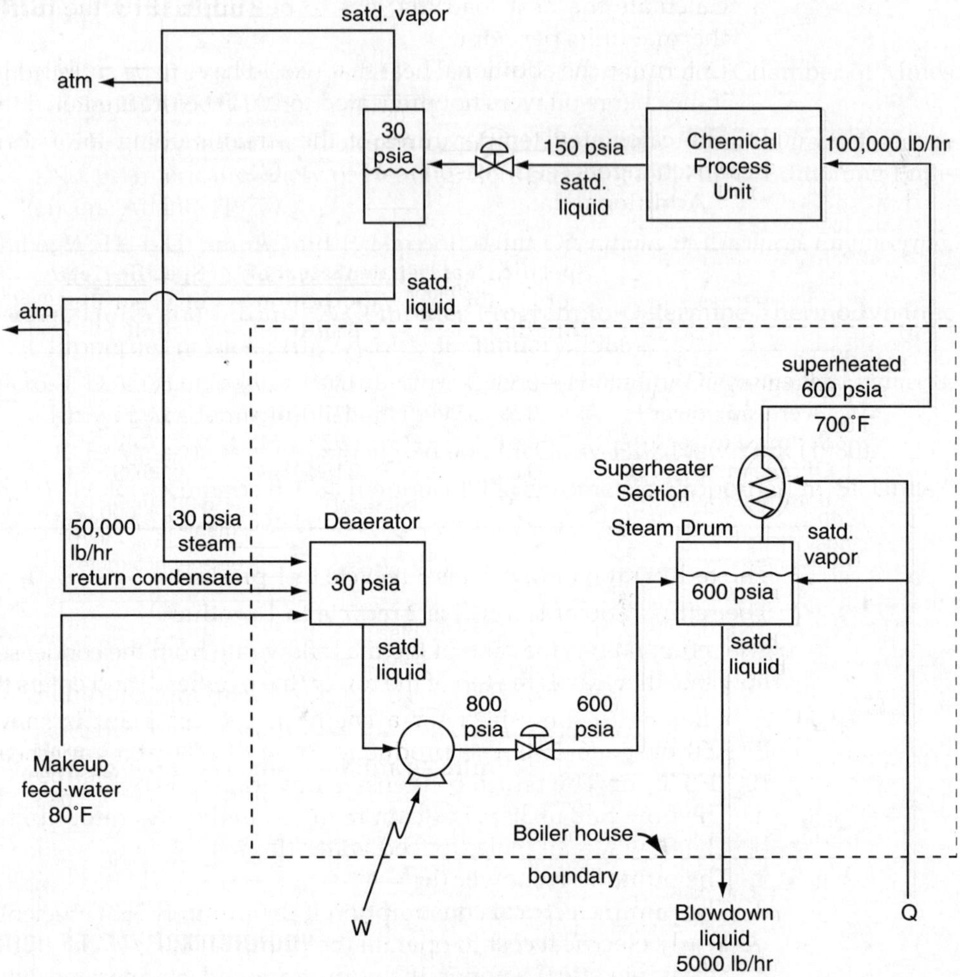

Figure P9.3.50

on a weight fraction basis. An optimum reflux ratio of 6.1 lb reflux/lb product has been previously determined and will be used. Operating pressure in the still will be 250 psia. Additional data are as follows:

Pressure = 250 psia

Component	Temp. Sat. (°F)	Heat of Vaporization (Btu/lb)
C_2H_6	10°	140
C_2H_4	−30°	135

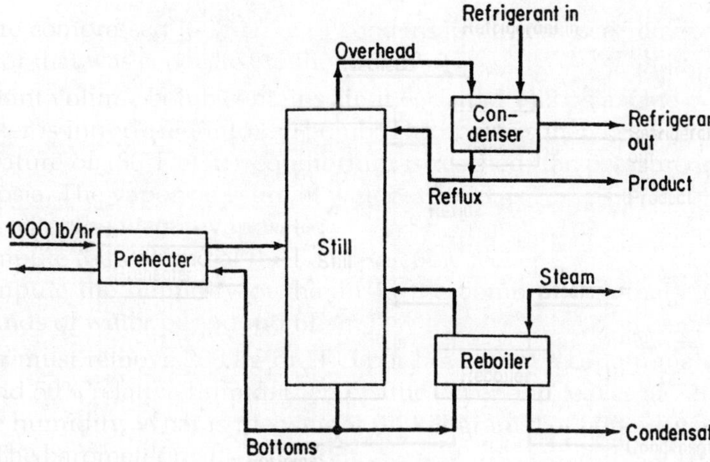

Figure P9.3.51

Determine:

a. The pounds of 30 psig steam required in the reboiler per pound of feed.
b. The gallons of refrigerant required in the condenser per hour assuming a 25°F rise in the temperature of the refrigerant. Heat capacity is approximately 1.0 Btu/(lb)(°F) and density = 50 lb/ft^3.
c. The temperature of the bottoms as it enters the preheater.

CHAPTER 10

Energy Balances: How to Account for Chemical Reaction

Your objectives in studying this chapter are to be able to

1. Explain the meaning of standard heat (enthalpy) of formation, heat (enthalpy) of reaction, and higher and lower heating values
2. List the standard conventions and reference states used for reactions associated with the standard heat of formation and the heat of combustion
3. Calculate the standard heat of reaction from tabulated standard heats of formation (or combustion) for a given reaction
4. Understand how to combine the heat of formation with sensible heat changes to solve problems involving chemical reactions
5. Solve simple material and energy balance problems involving reactions

As you probably are aware from previous comments, chemical reactions are at the heart of many industrial processes and directly affect the economics of an entire plant. Chemical reactions also are the basis of complex biological systems. We have deferred including the effects of chemical reactions in the discussion of the generation and consumption terms in energy balances to this point to avoid confusion. It's time to include them. In this chapter we explain how you can do it. We describe two closely related accounting procedures that treat the effects of energy generation or consumption due to chemical reaction. In one method, all of the energy effects are consolidated into one term in the energy balance called the heat of reaction. In the other, the reaction energy effects are merged with the enthalpies associated with each stream flowing in and out of the system. Both methods are based on a quantity known as the heat of formation, which we take up first.

10.1 The Standard Heat (Enthalpy) of Formation

The observed heat transfer that occurs to or from a closed isothermal system in which a reaction takes place represents the energy change associated with the rearrangement of the bonds holding together the atoms of the reacting molecules. For an **exothermic reaction**, heat is removed from the process in order to maintain a fixed system temperature; that is, energy is produced by the reaction to maintain isothermal conditions. The reverse is true of an **endothermic reaction**, in which heat is added to the system.

To include energy changes caused by a reaction in the energy balance, we make use of a quantity called the **standard heat** (really **enthalpy**) **of formation**, often called just the heat of formation,[1] denoted by the symbol $\Delta\hat{H}_f^\circ$. The superscript ° denotes the **standard state** (reference state) for reaction of 25°C and 1 atm, and the subscript f denotes "formation." In this chapter the overlay caret (^) will usually denote that the value is per mole.

The difficulty lies, not in the new ideas, but in escaping from the old ones, which ramify, for those brought up as most of us have been, into every corner of our minds.

John Maynard Keynes

The standard heat of formation is the name given to the special enthalpy change associated with the formation of 1 mol of a compound from its constituent elements and products in their standard state of 25°C and 1 atm. An example is the enthalpy change that occurs for the reaction of carbon and oxygen to form carbon dioxide at 25°C and 1 atm, as indicated in Figure 10.1.

$$C(s) + O_2(g) \rightarrow CO_2(g)$$

Note in the chemical reaction equation that the explicit specification of the states of the compounds is placed in parentheses.

In the reference state the enthalpies of all of the elements are assigned values of 0. Thus C and O_2 are assigned 0 values in the reaction shown in Figure 10.1. If you simplify the general energy balance, Equation (9.17), for the isothermal process in Figure 10.1 (steady-state, flow, no *KE* or *PE*, etc.), you get the standard heat of formation of CO_2, calculated from

$$-393.5 \text{ kJ/kg mol } CO_2 = Q = \Delta H = (1)\Delta\hat{H}^\circ_{f,CO_2} - (1)\Delta\hat{H}^\circ_{f,C} - (1)\Delta\hat{H}^\circ_{f,O_2}$$

$$= (1)\Delta\hat{H}^\circ_{f,CO_2} - 0 - 0 = \Delta\hat{H}^\circ_{f,CO_2}$$

[1]Historically, the name arose because the changes in enthalpy associated with chemical reactions were generally determined in a device called a calorimeter, to which heat is added or removed from the reacting system so as to keep the temperature constant.

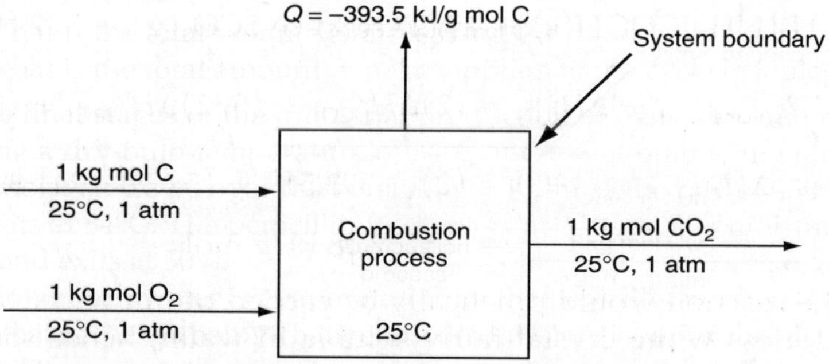

Figure 10.1 The heat transferred from a steady-state combustion process for the reaction $C(s) + O_2(g) \rightarrow CO_2(g)$ with reactants and products at 25°C and 1 atm is the standard heat of formation.

Since enthalpy $\hat{H}$ is a state variable, any state would do for the standard state of reference, but, by convention, the standard state of a substance (both reactants and products) usually is 25°C and 1 atm (absolute) pressure.[2] Fixing a reference state should cause no problem because you are always interested in calculating *enthalpy differences*, so the reference state is eliminated.

The reaction underlying $\Delta\hat{H}_f^o$ does not necessarily represent a real reaction that would proceed at constant temperature but can be a fictitious process for the formation of a compound from the elements. By defining **the heat of formation as zero in the standard state for each stable** (e.g., N_2 versus N) **element**, it is possible to design a system to express the heats of formation for all *compounds* at 25°C and 1 atm. Appendix D contains standard heats of formation for a number of molecules. On the CD in the back of this book you will find heats of formation for over 700 compounds, provided through the courtesy of Professor Yaws.[3] Remember that the values for the **standard heats of formation are negative for exothermic reactions and positive for endothermic reactions.**

For various compounds of interest to bioengineers, $\Delta\hat{H}_f^o$ may not be easy to find. Equation (5.17) shows a general substrate and cellular product that includes C, H, O, and N. To get $\Delta\hat{H}_f^o$ for $C_w H_x O_y N_z$, you have to start with a balanced equation. Use the electron balance technique shown in Chapter 5 to save time. For example, to get $\Delta\hat{H}_f^o$ for $CH_2(NH_3)COOH$ (glycine), the smallest amino acid, as shown in Chapter 5 write the balanced chemical equation

[2]The standard pressure is 1 bar (100 kPa) in some tabulations.

[3]C. L. Yaws, "Correlation for Chemical Compounds," *Chem. Engr.*, **79** (August 15, 1976).

$$CH_2(NH_3)COOH(s) + 15/2\,O_2(g) \rightarrow 2CO_2(g) + 5/2\,H_2O(g) + HNO_3(g)$$

$$\Delta H^{\circ}_{f,\text{glycine solid}} + \Delta H^{\circ}_{f,O_2} = (2)\Delta H^{\circ}_{f,CO_2} + (5/2)\Delta H^{\circ}_{f,H_2O} + (1)\Delta H^{\circ}_{f,HNO_3}$$

$$\Delta \hat{H}^{\circ}_{f,\text{glycine solid}} + 0 = (2)(-393.51) + (5/2)(-241.826) + (1)(-173.23)$$
$$= -1202.08 \text{ kJ/g mol}$$

The reaction would practically be carried out in water because glycine is an odorless white crystal fairly soluble in water; hence some solvation effects would occur.

Example 10.1 Use of Heat Transfer Measurements to Get a Heat of Formation

Suppose that you want to find the standard heat of formation of CO from experimental data. Can you prepare pure CO from the reaction of C and O_2 and measure the heat transfer? This would be far too difficult. It would be easier experimentally to find first the heat of reaction at standard conditions for the two reactions shown below for the flow process as shown in Figure E10.1 (assuming you had some pure CO to start with).

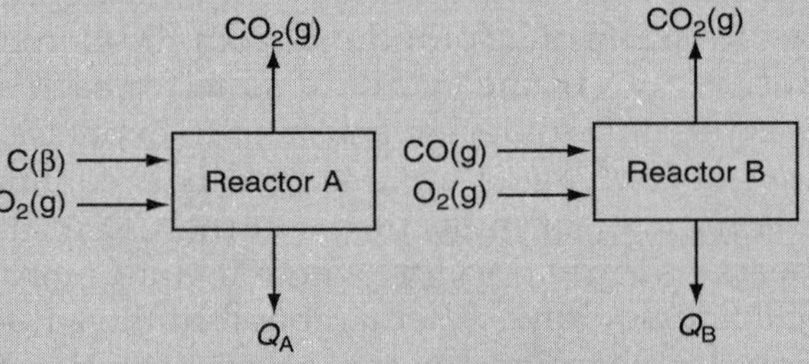

Figure E10.1 Use of two convenient reactions to determine the heat of formation for an inconvenient reaction:

$$C(s) + \frac{1}{2}O_2(g) \rightarrow CO_2(g)$$

$$C(s) + O_2(g) \rightarrow CO_2(g) \qquad Q = -393.51 \text{ kJ/g mol C} \equiv \Delta \hat{H}_A \qquad \text{(a)}$$

$$CO(g) + \frac{1}{2}O_2(g) \rightarrow CO_2(g) \qquad Q = -282.99 \text{ kJ/g mol CO} \equiv \Delta \hat{H}_B \qquad \text{(b)}$$

Basis: 1 g mol each of C and CO

According to Hess's law, you subtract Reaction (b) from Reaction (a), subtract the corresponding $\Delta\hat{H}_i$, and rearrange the compounds to form the desired chemical equation:

$$C(s) + \frac{1}{2}O_2(g) \rightarrow CO(g) \qquad \text{(c)}$$

for which the net heat of reaction per gram mole of CO is the heat of formation of CO:

$$\Delta\hat{H}^{\circ}_{f,CO} = -393.51 - (-282.99) = -110.52 \text{ kJ/g mol CO}$$

One hazard assessment of a compound is based on the potential rapid release of energy from it. A common prediction method for such release is to use the heat of formation per gram of compound as a guide. For example, what would you predict about the relative hazard of the following compounds: acetylene gas, lead azide solid, trinitroglycerine (TNT) liquid, and ammonium nitrate solid? Would you pick TNT? If you did, you would be wrong. Check the respective heats of formation and convert to a per gram basis.

Self-Assessment Test

Questions

1. If for the reaction

$$2N(g) \rightarrow N_2(g)$$

the heat transfer is $Q = -941$ kJ, how do you determine the value for the heat of formation of $N_2(g)$?

2. If the reaction for the decomposition of CO

$$CO(g) \rightarrow C(\beta) + \frac{1}{2}O_2(g)$$

takes place only at high temperature and pressure, how will the value of the standard heat of formation of CO be affected?

3. Will reversing the direction of a reaction equation reverse the sign of the heat of formation of a compound?

Problems

1. What is the standard heat of formation of HBr(g)?
2. Show that for the process in Figure 10.1 the general energy balance reduces to $Q = \Delta H$. What assumptions do you have to make?

3. Could the heat of formation be calculated from measurements taken in a batch process? If so, show the assumptions and calculations.
4. Calculate the standard heat of formation of CH_4 given the following experimental results at 25°C and 1 atm (Q is for complete reaction):

$$H_2(g) + \frac{1}{2}O_2(g) \rightarrow H_2O(1) \qquad Q = -285.84 \text{ kJ/g mol } H_2$$

$$C(\text{graphite}) + O_2(g) \rightarrow CO_2(g) \qquad Q = -393.51 \text{ kJ/g mol C}$$

$$CH_4(g) + 20_2(g) \rightarrow CO_2(g) + 2H_2O(1) \qquad Q = -890.36 \text{ kJ/g mol } CH_4$$

Compare your answer with the value found in the table of the heats of formation listed in Appendix D.

Thought Problems

1. Mercury is known to amalgamate with many metals—the recovery of gold in ancient times is well known. Normally water does not react with the aluminum in tubes in which water flows because the thin film of aluminum oxide that adheres to the aluminum surface prevents additional reaction from taking place. However, mercury contamination prevents the oxide film from functioning as a protector of the aluminum, and the aluminum readily corrodes by the reaction at room temperature:

$$2Al + 6H_2O \rightarrow 2Al(OH_3) + 3H_2$$

You might expect copper tubes to also be attacked by water with an amalgam present by the reaction

$$Cu + 2H_2O \rightarrow Cu(OH)_2 + H_2$$

but copper is not corroded. Why? Hint: The film is not protective for Cu-Hg amalgam.
2. Why are conventional automobiles undrivable using a pure methanol or ethanol fuel?

Discussion Question

Many different opinions have been expressed as to whether gasohol is a feasible fuel for motor vehicles. An important economic question is: Does 10% grain-based-alcohol-in-gasoline gasohol produce positive net energy? Examine the details of the energy inputs and outputs, including agriculture (transport, fertilizer, etc.), ethanol processing (fermentation, distilling, drying, etc.), petroleum processing and distribution, and the use of by-products (corncobs, stalks, mash, etc.). Ignore taxes and tax credits, and assume that economical processing takes place. Discuss octane ratings, heats of vaporization, flame temperatures, fuel-air ratios, volumetric fuel economy, effect of added water, and the effect on engine parts.

10.2 The Heat (Enthalpy) of Reaction

What is the difference between a method and device?
A method is a device which you use twice.
George Pólya

As we mentioned previously, one method of including the effect of chemical reaction in the energy balance makes use of the heat of reaction. The **heat of reaction** (which should be but is only rarely called the **enthalpy of reaction**) is the enthalpy change that occurs when reactants at various T and p react to form products at some T and p. The **standard heat of reaction** (ΔH°_{rxn}) is the name given to the heat of reaction when **stoichiometric quantities of reactants in the standard state** (25°C and 1 atm) **react completely** to produce products in the standard state. Do not confuse the symbol for the standard heat of reaction, ΔH°_{rxn}, with the symbol for the more general heat of reaction, ΔH_{rxn}, which applies to a process in which a reaction occurs under any conditions with any amount of reactants and products. The units of ΔH°_{rxn} are energy (kilojoules, British thermal units, etc.), and the units of the specific standard heat of reaction, $\Delta \hat{H}^{\circ}_{rxn}$, are energy for one mole of reaction based on the chemical reaction equation, so the result is energy per mole.

You can obtain the heat of reaction from experiments, of course, but it is easier to first see if you can calculate the standard heat of reaction from the known tabulated values of the heats of formation as follows: Consider a steady-state flow process with no work involved, such as the one shown in Figure 10.2 in which benzene (C_6H_6) reacts with the stoichiometric amount of H_2 to produce cyclohexane (C_6H_{12}) in the standard state:

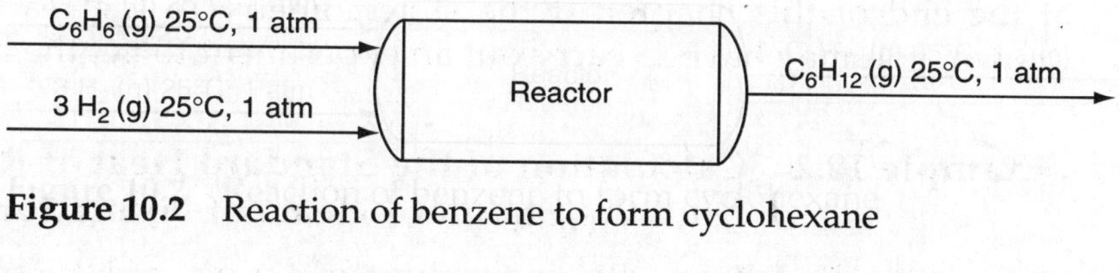

Figure 10.2 Reaction of benzene to form cyclohexane

$$C_6H_6(g) + 3\,H_2(g) \rightarrow C_6H_{12}(g)$$

The energy balance for the process reduces to $Q = \Delta H$, where ΔH is by definition the ΔH°_{rxn} for the specified chemical reaction equation.

Because we adopt for the heat of reaction the same reference conditions (0 enthalpy for the elements at 25°C and 1 atm) as used in defining the heats of formation, the values of the specific enthalpies associated with each species involved in the reaction are just the values of the respective heats of formation. For the process shown in Figure 10.2 the data are as follows:

Compound	Specific Enthalpy $\Delta\hat{H}_f^\circ$(kJ/g mol)	Number of Moles
$C_6H_6(g)$	82.927	1
$H_2(g)$	0	3
$C_6H_{12}(g)$	−123.1	1

You calculate the standard heat of reaction thus (based on 1 mol of C_6H_6 that reacts):

$$\Delta H^\circ_{\text{rxn}} = n_{C_6H_{12}}\Delta\hat{H}^\circ_{f,C_6H_{12}} - n_{C_6H_6}\Delta\hat{H}^\circ_{f,C_6H_6} - n_{H_2}\Delta\hat{H}^\circ_{f,H_2}$$

$$= (1)(-123.1) - (1)(82.927) - (3)(0) = -206.0 \text{ kJ}$$

or with H_2 as the reference:

$$\Delta H^\circ_{\text{rxn}} = -206.0 \text{ kJ}/3 \text{ g mol } H_2, \text{ hence } \Delta H^\circ_{\text{rxn}} = -68.67 \text{ kJ/g mol } H_2$$

If the reaction equation is multiplied by 2, will the heat of reaction be doubled? Yes. Will the standard heat of reaction also be doubled? No.

In general, for **complete** reaction,

$$\Delta H^\circ_{\text{rxn}} = \left(\sum_i^{\text{Products}} v_i\Delta\hat{H}^\circ_{f,i} - \sum_i^{\text{Reactants}} |v_i|\Delta\hat{H}^\circ_{f,i}\right) = \sum_i^{\text{All Species}} v_i\Delta\hat{H}^\circ_{f,i} \quad (10.1)$$

where v_i is the stoichiometric coefficient in the reaction equation. Keep in mind the sign convention for v_i. Always remember that the "heat of reaction" is actually an **enthalpy change** and not necessarily equivalent to heat transfer to or from the system. If you do not have values for the heats of formation, you can estimate them as described in some of the references listed at the end of this chapter. If the stoichiometry of the reaction is not well known, you may have to carry out an experiment to get the heat of reaction.

Example 10.2 Calculation of the Standard Heat of Reaction from the Standard Heats of Formation

Calculate $\Delta H^\circ_{\text{rxn}}$ for the following reaction of 4 g mol of NH_3 and 5 g mol of O_2:

$$4NH_3(g) + 5O_2(g) \rightarrow 4NO(g) + 6H_2O(g)$$

Solution

Basis: 4 g mol of NH_3

Tabulated data:	NH_3(g)	O_2(g)	NO(g)	H_2O(g)
$\Delta\hat{H}_f^o$ per mol at 25°C and 1 atm (kJ/g mol)	−46.191	0	+90.374	−241.826

We shall use Equation (10.1) to calculate ΔH^o_{rxn} (25°C, 1 atm) for 4 g mol of NH_3, assuming complete reaction:

$$\Delta H^o_{rxn} = [4(90.374) + 6(-241.826)] + [(-5)(0) + (-4)(-46.191)] = -904.696 \text{ kJ}$$

$$\text{Per gram mole of } NH_3,\ \Delta\hat{H}^o_{rxn} = \frac{904.646 \text{ kJ}}{4 \text{ g mol } NH_3} = -226.174 \text{ kJ/g mol } NH_3.$$

Example 10.3 Green Chemistry: Examining Alternate Processes

Green chemistry refers to the adoption of chemicals in commercial processes that reduce concern with respect to the environment. An example is the elimination of methyl isocyanate, a very toxic gas, in the production of carbaryl (1-napthalenyl methyl carbamate). In 1984 in Bhopal, India, the accidental release of methyl isocyanate in a residential area led to the death of thousands of people and the injury of many thousands more. The Bhopal process can be represented by the reaction Equations (a) and (b):

$$\underset{\text{methyl amine}}{CH_3NH_2} + \underset{\text{phosgene}}{COCl_2} \rightarrow \underset{\text{methyl isocyanate}}{C_2H_3NO} + 2\ HCl \quad \text{(a)}$$

$$\underset{\text{methyl isocynate}}{C_2H_3NO} + \underset{\text{1-napthol}}{C_{10}H_8O} \rightarrow \underset{\text{carbaryl}}{C_{12}H_{11}O_2N} \quad \text{(b)}$$

An alternate process eliminating the methyl isocyanate is represented by two other reaction equations:

$$\underset{\text{1-napthol}}{C_{10}H_8O} + \underset{\text{phosgene}}{COCl_2} \rightarrow \underset{\text{1-napthalenyl chloroformate}}{C_{11}H_7O_2Cl} \quad \text{(c)}$$

$$\underset{\text{1-napthalenyl chloroformate}}{C_{11}H_7O_2Cl} + \underset{\text{methyl amine}}{CH_3NH_2} \rightarrow \underset{\text{carbaryl}}{C_{12}H_{11}O_2N} + HCl \quad \text{(d)}$$

All of the reactions are in the gas phase. Calculate the approximate amount of heat transfer that will be required in each step of each process per gram

(Continues)

Example 10.3 Green Chemistry: Examining Alternate Processes (*Continued*)

mole of carbaryl produced overall in the process. Will there be any additional cost of heat transfer using the greener process?

Data: Note: Some of the values listed are only estimates. All of the values for $\Delta \hat{H}_f^o$ are in kilojoules per gram mole.

Component	ΔH^o_{rxn} **(kJ/g mol)**
Carbaryl	−26
Hydrogen chloride	−92.311
Methyl amine	−20.0
Methyl isocyanate	-9.0×10^4
1-Napthalenyl chloroformate	−17.9
1-Napthol	30.9
Phosgene	−221.85

Solution

Basis: 1 g mol of carbaryl

The simplest analysis is to say $\Delta H^o_{rxn} = Q$ so that the values of the standard heats of reaction will provide a rough measure of the major contributions to the energy balances for the respective processes.

Reaction (a):

$$\Delta H^o_{rxn} = [2(-92.311) + 1(-90{,}000)] - [1(-20.0) + 1(-221.85)] = -9.0 \times 10^4 \text{ kJ}$$

Reaction (b):

$$\Delta H^o_{rxn} = [1(-26)] - [1(-90{,}000) + 1(30.9)] = \underline{9.0 \times 10^4 \text{ kJ}}$$

$$\text{Total} \approx 0 \text{ kJ}$$

Reaction (c):

$$\Delta H^o_{rxn} = [1(-17.9)] - [1(30.9) + 1(-221.85)] = 173.05 \text{ kJ}$$

Reaction (d):

$$\Delta H^o_{rxn} = [1(-26) + 1(-92{,}311)] - [1(-17.9) + 1(-20.0)] = \underline{-80.41 \text{ kJ}}$$

$$\text{Total} \approx 92 \text{ kJ}$$

Both the original process and the suggested process require relatively small heat removal overall, but the Bhopal process requires considerable heat transfer on each of the two stages of the process, with Reaction (a) requiring removal and Reaction (b) requiring heating. Thus, the capital costs of the Bhopal process could be higher than those of the alternate process.

So far we have focused on the heat of reaction for complete reaction. What if the reaction is not complete? As mentioned in Chapter 5, for most processes the moles of reactants entering are not in their stoichiometric quantities, and the reaction may not go to completion, so some reactants appear in the products from the reactor. How do you calculate the heat of reaction, $\Delta H_{rxn}(25°C, 1\text{ atm})$ (*not* the standard heat of reaction—why?), in the standard state? One way is to start with Equation (10.1) and use the extent of reaction if you know it or can calculate it. For each species associated with the reaction for an unsteady-state, closed system (no flow in or out):

$$n_i^{\text{final}} = n_i^{\text{initial}} + v_i\xi$$

The equivalent for a steady-state flow system is

$$n_i^{\text{out}} = n_i^{\text{in}} + v_i\xi$$

Thus, for a steady-state flow system

$$\Delta H_{rxn}(25°C, 1\text{ atm}) = \sum_i^{\text{Flow out}} (n_i^{\text{in}} + v_i\xi)\Delta\hat{H}^{\circ}_{f,i} - \sum_i^{\text{Flow in}} (n_i^{\text{in}})\Delta\hat{H}^{\circ}_{f,i}$$

$$= \xi \sum v_i \Delta\hat{H}^{\circ}_{f,i} = \xi\ \Delta H^{\circ}_{rxn} \tag{10.2}$$

In Equation (10.2) you use the summations over all of the species associated with the reaction with $n_i = 0$ and $v_i = 0$ for any species not present as a product or reactant.

For example, for the reaction in Figure 10.2, assume the fraction conversion is 0.80 and the limiting reactant is 1 mol of C_6H_6. From Equation (5.6):

$$\xi = \frac{(-f)(n^{\text{in}}_{\text{limiting reactant}})}{v_{\text{limiting reactant}}} = \frac{-(0.80)(1)}{-1} = 0.80$$

$$\Delta H_{rxn}(25°C, 1\text{ atm}) = (0.80)[(1)(-123.1) + (-3)(0) + (-1)(82.927)]$$

$$= -164.8\text{ kJ}$$

Can you calculate the heat of reaction for a process in which the reactants enter and products exit at temperatures other than 25°C and 1 atm? The answer is yes. Recall that the enthalpy is a state (point) variable. Then you can calculate a change in enthalpy by any path that goes from the initial state to the final state.

Look at Figure 10.3. We want to calculate the ΔH_{rxn} (the enthalpy change) from state 1 to state 2. The value is the same as summing the values of all of the enthalpy changes 1 to 3, 3 to 4, and 4 to 2. The enthalpy change for the reactants and products is the combination of the sensible and latent heat (enthalpy) changes that might be taken from a table, or be calculated using a heat capacity equation for each species. The products are shown

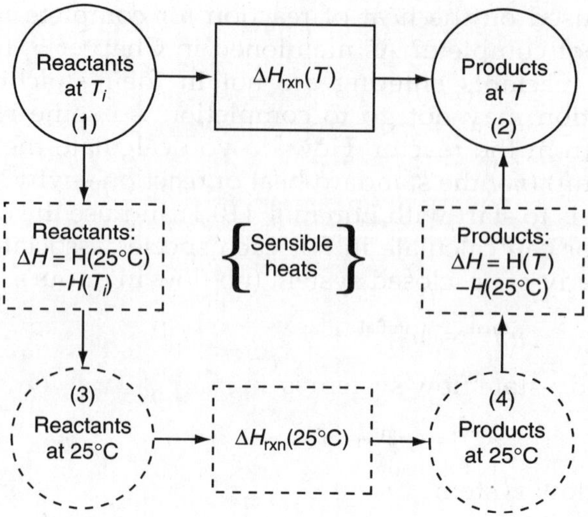

Figure 10.3 Information flow used to calculate the enthalpies constituting ΔH_{rxn} (from state 1 to state 2)

leaving the reactor at a common temperature, but each reactant might exist at a separate temperature T_i. Then the sensible heat plus any phase change for each component would be (excluding any mixing effects)

$$H_i(T_i) - H_i(25°\text{C}) = n_i \int_{25°\text{C}}^{T_i} C_{p,i} dT + n_i \Delta \hat{H}_{i,\text{phase change}} \tag{10.3}$$

If you ignore any slight pressure and mixing effects, for a steady-state system, the heat of reaction at a temperature T other than the reference temperature is

$$\Delta H_{rxn}(T) = \sum_{i}^{\text{Reactants}} n_i [\hat{H}_i(25°\text{C}) - \hat{H}_i(T)] + \Delta H_{rxn}(25°\text{C})$$
$$+ \sum_{i}^{\text{Products}} n_i [\hat{H}_i(T) - \hat{H}_i(25°\text{C})] \tag{10.4}$$

and in shorter notation

$$\Delta H_{rxn}(T) = [H(T) - H(25°\text{C})]_{\text{Products}} - [H(T) - H(25°\text{C})]_{\text{Reactants}}$$
$$+ \Delta H_{rxn}(25°\text{C}) \tag{10.4a}$$

Next, let's look at how the heat of reaction fits into the general energy balance $\Delta E = 0 = Q + W - \Delta H - \Delta PE - \Delta KE$ for a steady-state flow system.

In the energy balance the enthalpy change for the compounds involved in the reaction is

$$\Delta H = [H(25°\text{C}) - H(T^{\text{in}})]_{\text{Reactants}} + \Delta H_{\text{rxn}}(25°\text{C}) + [H(T^{\text{out}}) - H(25°\text{C})]_{\text{Products}} \quad (10.5)$$

Any compounds not involved in the reaction (such as N_2, for example, in combustion) can be included in the first and third terms on the right-hand side of the equal sign in Equation (10.5), but they make no contribution to the heat of reaction at 25°C.

The conclusion that you should reach from Equation (10.5) is that when reaction occurs in a system, just one term has to be added to the energy balance as stated previously, namely, the heat of reaction at 25°C. All of the effects of energy generation or consumption caused by reaction can be lumped into the one quantity $\Delta H_{\text{rxn}}(25°\text{C})$. What would be the analog of Equation (10.5) for an unsteady-state, batch (closed) system? Remember that $\Delta U = \Delta H - \Delta(pV)$.

You have to be a bit careful in calculating the heat of reaction at 25°C to get ΔH for an energy balance because certain conventions exist:

1. The reactants are shown on the left-hand side of the chemical equation, and the products are shown on the right; for example,

$$CH_4(g) + H_2O(l) \rightarrow CO(g) + 3H_2(g)$$

2. The conditions of phase, temperature, and pressure must be specified unless the last two are standard conditions, as presumed in the equation above, when only the phase is required. This is particularly important for a compound such as H_2O, which can exist as more than one phase under common conditions.
3. The amounts of material reacting are assumed to be the quantities shown by the stoichiometry in the chemical equation and the extent of reaction. Thus, if the reaction is

$$2\text{Fe}(s) + \frac{3}{2}O_2(g) \rightarrow Fe_2O_3(s) \quad \Delta H_{\text{rxn}} = -822.2 \text{ kJ}$$

the value of −822.2 kJ refers to 2 g mol of Fe(s) reacting and not 1 g mol. The heat of reaction for 1 g mol of Fe(s) would be −411.1 kJ/g mol of Fe.

Example 10.4 Calculation of the Heat of Reaction in a Process in Which the Reactants Enter and the Products Leave at Different Temperatures

Public concern about the increase in the carbon dioxide in the atmosphere has led to numerous proposals to sequester or eliminate the carbon dioxide. An inventor believes he has developed a new catalyst that can make the gas phase reaction

$$CO_2(g) + 4H_2(g) \rightarrow 2H_2O(g) + CH_4(g)$$

proceed with 40% conversion of the CO_2. The source of the hydrogen would be from the electrolysis of water using electricity generated from solar cells. Assume that 1.5 mol of CO_2 enter the reactor at 700°C together with 4 mol of H_2 at 100°C. Determine the heat of reaction if the exit gases leave at 1 atm and 500°C.

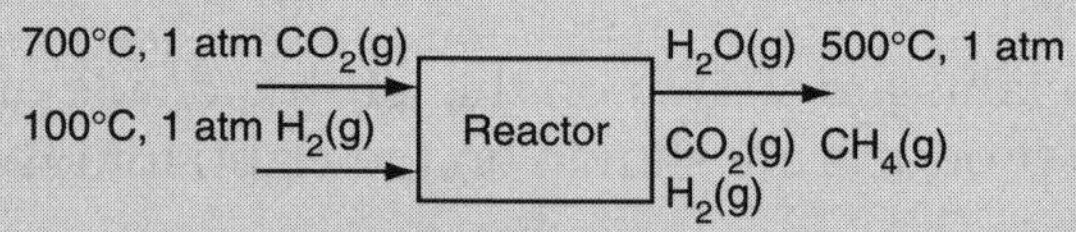

Figure E10.4

Solution

The system is steady-state and open with reaction. Assume 1 atm for the products and reactants. We will not outline each step in the solution to save space. First, complete the material balance, and then the energy balance in three phases: (1) the standard heat of reaction at 25°C, (2) then the heat of reaction at 25°C, (3) and last the sensible heats.

Basis: 1.5 g mol of $CO_2(g)$

The reference temperature is 25°C. The outcome of the material balance and some needed enthalpy data are as follows:

	g mol		Fraction	$\Delta\hat{H}^\circ_f$ (kJ/g mol)	$\Delta\hat{H}^{T^\circ C}_{25^\circ C}$ (kJ/g mol)		
Compounds	**In**	**Out**	**Reacted**	**25°C**	**100°C**	**500°C**	**700°C**
$CO_2(g)$	1.5	0.90	0.40	−393.250	20.996	30.975	
$H_2(g)$	4	1.60	0.60	0	2.123	13.826	
$H_2O(g)$	2	1.20		−241.835		17.010	
$CH_4(g)$	1	0.60		−74.848		23.126	

The first step is to calculate the standard heat of reaction at the reference temperature. The reaction is

$$CO_2(g) + 4H_2(g) \rightarrow 2H_2O(g) + CH_4(g)$$

To avoid inconsistencies, the sensible heat data in the sixth column have been taken from the CD in the back of this book instead of integrating heat capacity equations by hand. Complete reaction is specified for the standard heat of reaction. H_2 is the limiting reactant.

$$\Delta\hat{H}^\circ_{rxn} = [(1)(-74.848) + (2)(-241.835)] - [(1)(-393.250) + (4)(0)] = -165.27 \text{ kJ/g mol } CO_2$$

For 40% conversion of the CO_2 (the excess reactant), the conversion of the limiting reactant (H_2) is 0.60; hence the extent of reaction is $\xi = \dfrac{(-0.60)(4)}{-4} = 0.60$. Then

$$\Delta H_{rxn}(25^\circ C) = (0.60)(-165.27) = -99.16 \text{ kJ}$$

The next step is to calculate the enthalpy changes (sensible heats) from 298 K to the respective temperatures of the compounds entering and leaving the reactor:

Sensible Heat In

Compound	g mol	T (K)	$\Delta\hat{H}_{sensible}$ (kJ/g mol)	$\Delta H_{sensible}$ (kJ)
$CO_2(g)$	1.5	700	30.975	46.463
$H_2(g)$	4.0	100	2.123	8.492
Total				54.955

(Continues)

Example 10.4 Calculation of the Heat of Reaction in a Process in Which the Reactants Enter and the Products Leave at Different Temperatures (*Continued*)

Sensible Heat Out				
$CO_2(g)$	0.90	500	20.996	18.896
$H_2(g)$	1.60	500	13.826	28.122
$H_2O(g)$	1.20	500	17.010	20.412
$CH_4(g)$	0.60	500	23.126	13.876
Total				75.306

Water is a gas at 500 K, so a phase change is not involved. The heat of reaction is

$$\Delta H_{rxn} = 75.306 - 54.955 + (-99.161) = -78.85 \text{ kJ/g mol } CO_2$$

The standard heat of reaction for biological reactions, such as the growth of yeast with a substrate of glucose, can be calculated using the stoichiometric equation, Equation (5.17), as a guide. However, one factor that you have to keep in mind regarding such calculations is that the substrate may be in solution. You have to find ΔH_f° for the solution, not the pure solute, or solvent. For example, from F. O. Rossini et al., *Selected Values of Chemical Thermodynamic Properties* [National Bureau of Standards Circular 500, USGPU (1952)], the heat of solution of NH_3, often said to be a nitrogen source, in H_2O depends on the concentration of the NH_3 in the water.

	$\Delta \hat{H}^\circ_{\text{solution}}$ (25°C and 1 atm, kJ/g mol)	ΔH_f° (kJ/g mol)
$NH_3(g)$	0	−67.20
$NH_3(1\ H_2O)$	−7.06	−74.26
$NH_3(5\ H_2O)$	−8.03	−75.25
$NH_3(50\ H_2O)$	−8.25	−75.45
$NH_3(100\ H_2O)$	−8.28	−75.48

The value in parentheses is the total number of moles of H_2O added to 1 mol of $NH_3(g)$. Remember that the pressure of NH_3 over the solution and the pH of the solution vary with its concentration at equilibrium. The heat of formation of NH_3 in water is the sum of the heat of formation of $NH_3(g)$ plus the heat of solution.

The law is not concerned with the trifles.

Anonymous

If the $\Delta H^{\circ}_{solution}$ between different concentrations of solute is small relative to the heat of reaction, you can ignore the heat of solution, but it may be relatively large in some cases. The details of the heat of formation of compounds in solutions are presented in Chapter 13 on the CD.

Self-Assessment Test

Questions

1. Can the standard heat of reaction ever be positive?
2. Is it correct to calculate a heat of reaction for which the reaction is incomplete?
3. How does phase change, such as when water goes from a liquid to a vapor, affect the value of the heat of reaction?
4. What does it mean when the standard heat of reaction is (a) negative and (b) positive?
5. Can you choose a reference state other than 25°C and 1 atm to use in applying the energy balance?
6. What is the difference between the heat of reaction and the standard heat of reaction?

Problems

1. Calculate the standard heat of reaction for the following reaction from the heats of formation:

$$C_6H_6(g) \rightarrow 3C_2H_2(g)$$

2. Calculate the heat of reaction at 90°C for the Sachse process (in which acetylene is made by partial combustion of LPG):

$$C_3H_8(l) + 2O_2(g) \rightarrow C_2H_2(g) + CO(g) + 3H_2O(l)$$

Thought Problems

1. When lava and water are mixed, an explosion can easily occur. Of course the lava and water are not well mixed—only a short depth of the lava is contacted by the

water. Nevertheless, what is the ratio of the mass of lava equivalent to the mass of an explosive that has a ΔH°_{rxn} of about 4000 J/g? Data: C_p of lava is about 1 J/g and the temperature can be assumed to be 1100°C.

2. "Are fuel cells really green?" is the heading of an article in a chemical engineering magazine. The article says that cells fueled with H_2 only omit water. Do fuel cells really represent a green technology?

Discussion Question

Eric Cottell, a British-born inventor, does not change base metals into gold, but he does mix oil and water—and these days that may be the most welcome alchemy of all. Cottell claims that in a furnace a blend of three parts oil and one part water burns so much more cleanly and efficiently than ordinary oil that it can cut fuel consumption by at least 20% while producing almost no soot or ash. He also claims that road tests show that a car can run on 18% water and 82% gasoline, resulting in such a low output of pollutants that the engine does not need the mileage-robbing emission-control devices required on new cars.

Cottell says that when the emulsion prepared in his reactor is pumped into a furnace, the water droplets explode into superheated steam, shattering the oil droplets and exposing a maximum of the oil's surface. This provides quick, nearly complete burning.

What do you think about the claims?

10.3 Integration of Heat of Formation and Sensible Heat

You might have speculated from the tabular formation of the solution in Example 10.3 that one way to simplify the calculation procedure for the overall enthalpy change in a process in which a reaction(s) occurs is to transfer each standard heat of formation from the calculation of the standard heat of reaction and merge it with its respective sensible heat. Thus the heat of reaction does not have to be explicitly calculated. Look at Figure 10.4.

The result is that the enthalpy change term in the energy balance for a steady-state, open system can simplify to just

$$\Delta H = \sum_{\text{outputs}} n_i \Delta \hat{H}_i - \sum_{\text{inputs}} n_i \Delta \hat{H}_i \tag{10.6}$$

because the heats of formation are embedded into the H_is.

Note: All of the compounds in a stream are included in the summation "output" or "input," respectively, whether the compound reacts or not. To be consistent you can include every compound that exists in the problem in each stream, input or output, if you make its number of moles zero; for example, $n_i = 0$, when the compound is not present in a stream. Although it may seem that you have lost some essential feature in the calculation of ΔH,

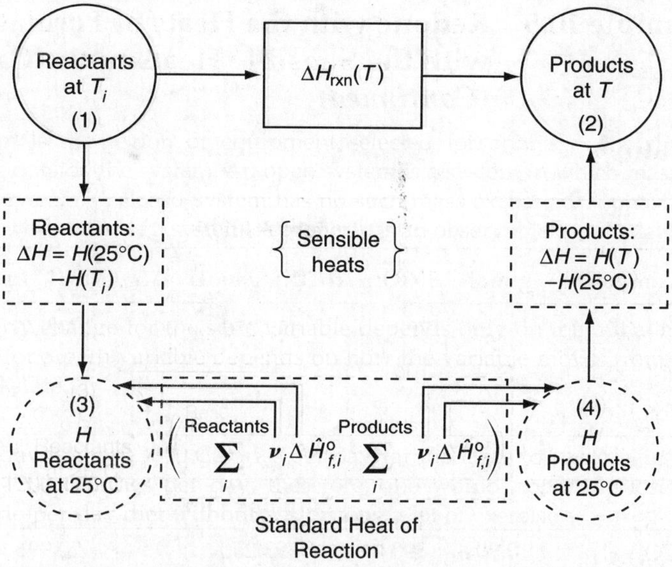

Figure 10.4 Splitting the respective heats of formation from the standard heat of reaction, and merging them with their appropriate sensible heats

Figure 10.4 shows that you have simply dispersed the heats of formation into a different sequence in the calculations so that they are no longer aggregated in a term called the "standard heat (enthalpy) of reaction." If the reaction is not complete, the material balance determines the amount of reactants appearing with the products, and the unreacted reactants become subtracted from the total reactants in the entering compounds if the calculation sequence shown in Figure 10.4 is used for the energy balance. This approach is actually easier to apply and less susceptible to errors than the heat of reaction approach, as will be demonstrated in the following example.

Example 10.5 Redone with the Heats of Formation Merged with the Sensible Heats in the Calculations

The problem and all of the data are the same as in Example 10.3. The table below lists the details of the solution.

(Continues)

Example 10.5 Redone with the Heats of Formation Merged with the Sensible Heats in the Calculations (*Continued*)

Solution

Enthalpy In

Compound	g mol	T (K)	$\Delta\hat{H}_f^o$(kJ/g mol)	$\Delta\hat{H}_{sensible}$ (kJ/g mol)	ΔH (kJ)
$CO_2(g)$	1.5	700	−393.250	30.975	−543.413
$H_2(g)$	4.0	100	0	2.123	8.492
Total					−534.546

Enthalpy Out

Compound	g mol	T (K)	$\Delta\hat{H}_f^o$(kJ/g mol)	$\Delta\hat{H}_{sensible}$ (kJ/g mol)	ΔH (kJ)
$CO_2(g)$	0.90	500	−393.250	20.996	−335.029
$H_2(g)$	1.60	500	0	13.826	28.122
$H_2O(g)$	1.20	500	−241.835	17.010	−269.790
$CH_4(g)$	0.60	500	−74.848	23.126	−31.033
Total					−613.130

$$\Delta H = -613.130 - (-534.546) = -78.209 \text{ kJ}$$

What do you have to do in solving the problem if a change occurs in a temperature of an input or output stream? A change in the exact reaction of the CO_2? Hint: What balance is made first?

Frequently Asked Questions

1. Where does the heat of reaction go if you use the tabular solution procedure of Example 10.5 in calculating ΔH? $\Delta H_{rxn}(25°C)$ still exists, but not as an agglomerated term. Its segments are decentralized and added to the respective sensible heats of the components. It can still be calculated separately, if needed, as before.
2. How does the fraction conversion or extent of reaction used to modify ΔH^o_{rxn} in Equation (10.2) become involved in calculating ΔH if the reaction is not complete? The preliminary material balance takes care of the

amount of a reactant that enters and leaves the process. When the same amount of compound is included in both the output and input—that is, no contribution to reaction occurs—the result is the only contribution to ΔH in the sensible heat and phase changes: $(n_i \Delta H^{\circ}_{f,i})_{in} - (n_i \Delta H^{\circ}_{f,i})_{out} = 0$.

3. Do you need to know the specific reaction(s) in the process to use the tabular method in Example 10.5? No. The extent of reaction is automatically taken into account in the material balance (unless it is the unknown), or from process measurements. In biological reactions, the reactions themselves and the extents of reaction may be uncertain, but the overall ΔH_{rxn} can be calculated.

Now that we have described the details of how to calculate the enthalpy change in the energy balance when reaction(s) occurs, we next give some illustrative examples. Here are some typical problems frequently posed for steady-state, open systems:

1. What is the temperature of one stream given data for the other streams?
2. How much heat has to be added to or removed from the process?
3. What is the temperature of the reaction?
4. How much material must be added to or removed from the process to give a specified value of heat transfer?

You can think of many others.

My method to overcome a difficulty is to go round it.

George Pólya

Example 10.6 Application of the General Energy Balance in a Process in Which More than One Reaction Occurs

Limestone ($CaCO_3$) is converted to calcium oxide (CaO) in a continuous vertical kiln such as that illustrated in Figure E10.6a. The energy to decompose the limestone is supplied by the combustion of natural gas (CH_4) in direct contact with the limestone using 50% excess air. The $CaCO_3$ enters the process at 25°C, and the CaO exits at 900°C. The CH_4 enters at 25°C, and the product gases exit at 500°C. Calculate the maximum number of pounds of $CaCO_3$ that can be processed per 1000 ft^3 of CH_4 measured at standard conditions (0°C and 1 atm for gases). To simplify the calculations, assume that the heat capacities of $CaCO_3$ (56.0 Btu/(lb mol)(°F)) and CaO (26.7 Btu/(lb mol)(°F)) are constant.

(Continues)

Example 10.6 Application of the General Energy Balance in a Process in Which More than One Reaction Occurs (*Continued*)

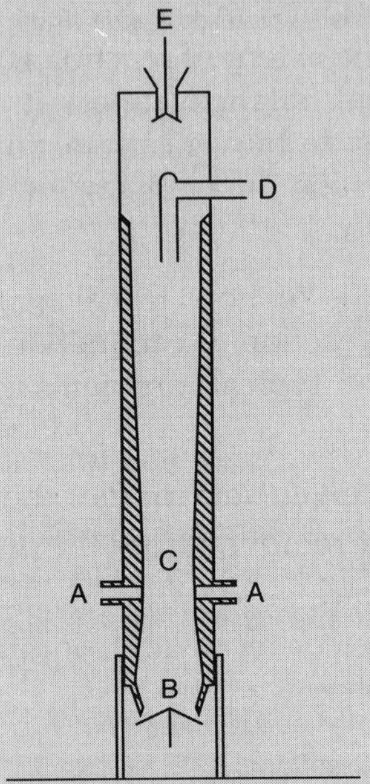

Figure E10.6a A vertical line kiln composed of a steel cylinder lined with fire brick approximately 80 ft high and 10 ft in diameter. Fuel is supplied at A, air at B, and limestone ($CaCO_3$) at E. Combustion products and CO_2 exit at D.

Solution

Steps 1–3

You have to decide whether to make the calculations using the SI or AE systems of units (or both). We will use the SI system for most of the solution for convenience, but the CD in the back of this book makes use of the AE system just as easy. We will assume that the entire process occurs at 1 atm (so that H is a function only of temperature), and that the air enters at 25°C. You also have to choose a basis.

Step 5

Let us start with a basis of 1 g mol CH_4 for convenience. Many other bases, of course, could be selected. At the end of the calculations the results can be converted to the requested units.

Steps 1–4

Figure E10.6b is a sketch of the process. Assume that the process is open and steady-state. The assumption of steady-state means that no material was in the kiln at the start or end of the process and ignores a relatively short interval for start-up and shutdown. Your first mission is to solve the material balances.

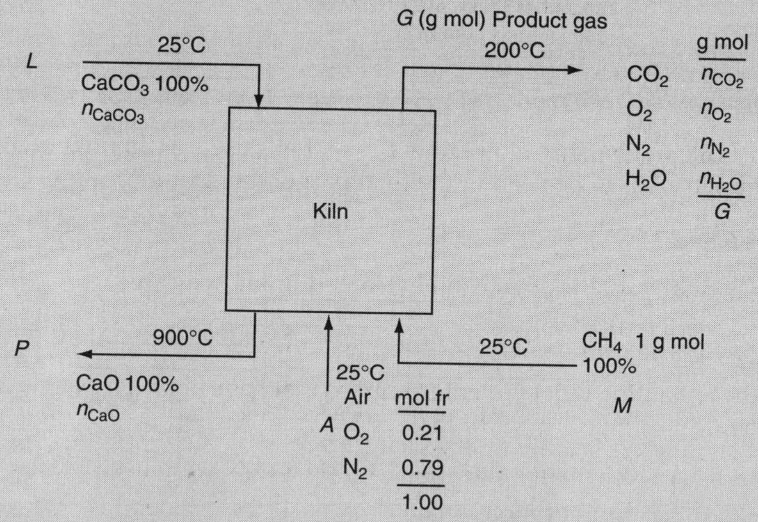

Figure E10.6b

Start by specifying the values of all of the variables you can identify or easily calculate in your head. Be sure to keep track of all of the material balances you use in your preliminary assignments so that you don't find subsequently that you have inadvertently used redundant information.

Steps 3 and 4

To get the values of the entering O_2 and N_2 you can assume that the values cited in the problem statement relate to the following reaction:

$$CH_4(g) + 2O_2(g) \rightarrow CO_2(g) + 2H_2O(g)$$

Thus, the moles of O_2 and N_2 entering are

1 mol CH_4 requires:	2 g mol O_2
50% excess:	1
Total O_2	3
Entering N_2:	$3(0.79/0.21) = 11.29$ g mol

The numbers have already been added to Figure E10.6b.

What other values can be assigned? Look at Figure E10.6b and note that the inlet amount of N_2 equals the outlet N_2 (N_2 doesn't react). Thus, $n^G_{N_2} = 11.29$ g mol. Also, all of the H from CH_4 goes into stream G; hence $n^G_{H_2O} = 2.00$ g mol. Use the implicit relation of the sum of moles in the stream to get (in gram moles)

(Continues)

Example 10.6 Application of the General Energy Balance in a Process in Which More than One Reaction Occurs (*Continued*)

$$n^{L}_{CaCO_3} = L \quad A = 14.29$$

$$n^{P}_{CaO} = P \quad M = 1.00$$

From a calcium balance, you know $L = P$. Let's use L as an unknown because that variable has to be used in the final part of the solution.

Steps 6 and 7

The residual variables that can be deemed unknowns are

$$L,\ G,\ n^{G}_{CO_2},\ n^{G}_{O_2} \quad \text{(Total 4)}$$

and the remaining independent equations that have not been used in assigning values are

Element balances: C, O (H and N were used)	2
Sum of moles: $n^{G}_{CO_2} + n^{G}_{O_2} + 11.29 + 2.00 = G$	1
Total	3

The remaining degrees of freedom are $4 - 3 = 1$.

Note that the specification of complete reaction of the $CaCO_3$ and the CH_4 would be a redundant specification because no $CaCO_3$ or CH_4 was indicated to exit the process (the variables thus are automatically assigned zero values). A specification of less than complete reaction of $CaCO_3$ or CH_4 would be inconsistent with the other information given in the problem.

If you used species balances in the solution of the material balance problem, you would start with 14 variables (L, P, A, M, G, 7 species, and 2 extents of reaction), assign values to 10 of the variables (the 2 extents of reaction are unity), the basis is $CH_4 = 1$ mol, the CH_4 reaction gives the H_2O, the $CaCO_3$ reaction gives $L = M$, and the other assignments would be as shown for the element balances, to leave 3 species balances (O_2, CO_2, $CaCO_3$) available to solve for 4 unknowns—the same result as obtained from using element balances.

Thus you need one more independent equation, the energy balance:

$$\Delta E = 0 = Q + W - \Delta(H + KE + PE)$$

Assume: $Q = 0$ (fairly well-insulated toner by fire brick—and you don't know how to calculate Q):

$$W = 0$$

$$\Delta KE = 0$$

$$\Delta PE = 0$$

Conclusion: $\Delta H = 0$.

Remember that H is a function of temperature (p has been assumed to be 1 atm), and note that the temperature is specified for each stream. Consequently, the values of $\hat{H}(T)$ can be looked up and multiplied by the respective number of moles to get $\Delta H(T)$ for each stream (the heat of mixing for a gas is essentially zero).

$$\underbrace{(L)(\Delta\hat{H}^P_{CaO}) + (n^G_{CO_2})(\Delta\hat{H}^G_{CO_2}) + (n^G_{O_2})(\Delta\hat{H}^G_{O_2}) + (11.29)(\Delta\hat{H}^G_{N_2}) + (2)(\Delta\hat{H}_{H_2O})}_{\text{Out (at 500°C and 900°C)}}$$

$$-\underbrace{-(1)(\Delta\hat{H}_{CH_4}) - (3)(\Delta\hat{H}_{O_2}) - (11.29)(\Delta\hat{H}_{N_2}) - (L)(\Delta\hat{H}_{CaCO_3})}_{\text{In (at 25°C)}} = 0$$

Select the reference state as 25°C and 1 atm.

Use a tabular format for efficiency in the calculations (an ideal application for a spreadsheet). A stream ΔH (in joules) is the sum of $(\Delta\hat{H}^\circ_f + \Delta\hat{H}_{sensible})n_i$ for the respective i compounds.

Compound	Mol	$\Delta\hat{H}^\circ_f$(kJ/g mol)	T (°C)	Sensible Heat* (kJ/g mol)	Stream ΔH (kJ)
In					
$CH_4(g)$	1	−49.963	25	0	−49.963
$O_2(g)$	3	0	25	0	0
$N_2(g)$	11.29	0	25	0	0
$CaCO_3(s)$	$n^L_{CaCO_3} = L$	−1206.9	25	0	$-1206.9L$
Out					
$CaO(s)$	$n^P_{CaO} = P = L$	−635.6	900	(0.062)(900 − 25)	$-581.35L$
$CO_2(g)$	$n^G_{CO_2}$ } $\sum = G$	−393.25	500	21.425	$-371.825\ n^G_{CO_2}$
$O_2(g)$	$n^G_{O_2}$ } $\sum = G$	0	500	15.034	15.034
$N_2(g)$	11.29 } $\sum = G$	0	500	14.241	160.781
$H_2O(g)$	2 } $\sum = G$	−241.835	500	17.010	−449.650

*Data from Appendix F, reference No. 1 (Kobe). The CD gives different values.

(*Continues*)

Example 10.6 Application of the General Energy Balance in a Process in Which More than One Reaction Occurs (*Continued*)

We have to solve the two remaining element mass balances (in moles) with the energy balance:

Balance	In		Out	
C:	$1 + L$	=	$n^{G}_{CO_2}$	(1)
O:	$3L + 2(3)$	=	$2n^{G}_{CO_2} + 2n^{G}_{O_2} + 2 + L$	(2)

$$(-581.35L - 371.825n^{G}_{CO_2} + 15.034 + 160.781 - 449.650) - (-49.963 - 1206.9L) = O \quad (3)$$

$$L = 2.56 \text{ g mol}$$

On the basis of 1 g mol of CH_4:

$$\frac{2.56 \text{ g mol } CaCO_3}{1 \text{ g mol } CH_4}\left|\frac{100.09 \text{ g } CaCO_3}{1 \text{ g mol } CaCO_3}\right. = \frac{256\text{g } CaCO_3}{1 \text{ g mol } CH_4}$$

To get the ratio asked for (assuming CH_4 is an ideal gas—a good assumption):

$$\frac{1000 \text{ ft}^3 \; CH_4}{}\left|\frac{1 \text{ lb mol } CH_4}{359.05 \text{ ft}^3}\right|\frac{256 \text{ lb } CaCO_3}{1 \text{ lb mol } CH_4} = \frac{713 \text{ lb } CaCO_3}{1000 \text{ ft}^3 \; CH_4 \text{ at S.C.}}$$

If you decided to use the exit gas from the kiln to preheat the entering air, would you increase or decrease the ratio calculated above? Would the exit gases from the entire system, if you include the heat exchange, be at a lower or higher temperature? What would the revised plan do to the sensible heats of the exit gases? Does the tabular format of calculation help you reach an answer?

A term called the **adiabatic reaction (theoretical flame or combustion) temperature** is defined as the temperature obtained from a combustion process when

1. The reaction is carried out under adiabatic conditions; that is, there is no heat interchange between the system in which the reaction occurs and the surroundings
2. No other effects occur, such as electrical effects, work, ionization, free radical formation, and so on
3. The limiting reactant reacts completely

When you calculate the adiabatic reaction temperatures for combustion reactions, you assume complete combustion occurs, but equilibrium considerations may dictate less than complete combustion in practice. For example, the adiabatic flame temperature for the combustion of CH_4 with theoretical air has been calculated to be 2010°C; allowing for incomplete combustion at equilibrium, it would be 1920°C. The actual temperature when measured is 1885°C.

The adiabatic reaction temperature tells you the temperature ceiling of a process. You can do no better, but of course the actual temperature may be less. The adiabatic reaction temperature helps you select the types of materials that must be specified for the equipment in which the reaction is taking place. Chemical combustion with air produces gases with a maximum temperature of roughly 2500 K, but the temperature can be increased to 3000 K with the use of oxygen and more exotic oxidants, and even this value can be exceeded, although handling and safety problems are severe. Applications of such hot gases lie in the preparation of new materials, micromachining, welding using laser beams, and the direct generation of electricity using ionized gases as the driving fluid.

As you have seen, the *open system, steady-state* energy balance with $Q = KE = PE = 0$ reduces to just $\Delta H = 0$. Because you do not know the flame temperature (e.g., the solution) to a problem before you start, if you use tables such as Appendix D or the physical property software on the CD in calculating the "sensible heats" of the various streams entering and leaving the reactor, the solution will involve trial and error. To find the exit temperature for which $\Delta H = 0$ in the energy balance, if tables or the CD are used as the source of the $\Delta \hat{H}$ values, this is the simplest procedure:

1. Assume a sequence of values of T selected to bracket $\Delta H = 0$ (+ and −) for the sum of the enthalpies of the outputs minus the enthalpies of the inputs.
2. Once the bracket is obtained, interpolate within the bracket to get the desired value of T when $\Delta H = 0$.

When using tables based on 0°C, don't forget to subtract the value of the enthalpy at 25°C (the usual reference temperature for the energy balance) from the value of the enthalpy at T in the tables so that effectively the 25°C reference temperature prevails.

On the other hand, instead of using tables, if you integrate the heat capacity equations from 25°C to T to obtain the sensible heats, ΔH will involve at least a final cubic or quadratic equation for ΔH to be solved for the exit temperature. Make sure you get a unique reasonable solution from the polynomial; more than one solution may exist, as well as negative or complex values.

For an *unsteady-state, closed system* with ΔKE and $\Delta PE = 0$ inside the system and $W = 0$, the energy balance reduces to

$$Q = \Delta U = U_{\text{final}} - U_{\text{initial}}$$

If you do not have values for $\hat{U}$, you have to calculate Q from

$$Q = [H(T) - H(25°C)]_{\text{final}} - [H(T) - H(25°C)]_{\text{initial}} - [(pV)_{\text{final}} - (pV)_{\text{initial}}]$$

The contribution of $\Delta(pV)$ is frequently negligible, so you can just use H in lieu of U, in which case the open and closed system calculations will give the same answer. But if $\Delta(pV)$ is significant, remember that for a constant volume process, $\Delta(pV) = V\Delta p$, and for a constant pressure but expandable closed system, $\Delta(pV) = p\Delta V$.

Example 10.7 Calculation of an Adiabatic Reaction (Flame) Temperature

Calculate the theoretical flame temperature for CO gas burned at constant pressure with 100% excess air, when the reactants enter at 100°C and 1 atm.

Solution

The solution presentation will be compressed to save space. The system is shown in Figure E10.7. We will use data from Appendix D and the CD in the back of this book. The process is a steady-state flow system. Ignore any equilibrium effects.

$$CO(g) + \frac{1}{2}O_2(g) \rightarrow CO_2(g)$$

Basis: 1 g mol of CO(g); ref. temp. 25°C and 1 atm

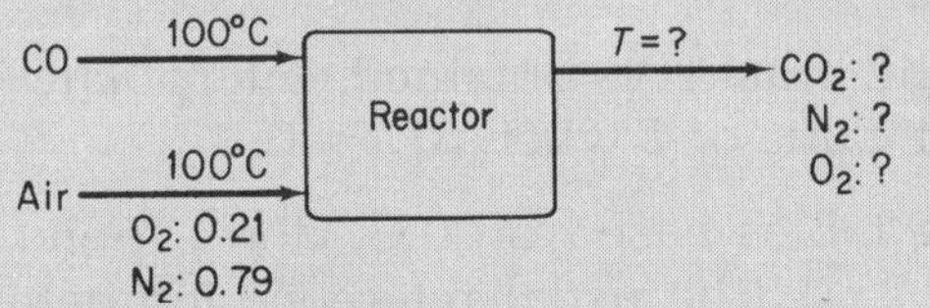

Figure E10.7

The reaction is always assumed to occur with the limiting reactant (CO) reacting completely. The excess air and the N_2 are nonreacting components, but you nevertheless have to calculate their associated sensible heats to obtain the proper total sensible heats (enthalpies) for the process streams. We will use the tabular method of bookkeeping used previously for all of the pertinent enthalpy changes for each compound entering and leaving the system. You can solve the material balances (for which the degrees of freedom are zero) independently of the energy balance. A summary of the results of the solution of the material balances is as follows:

Entering Compounds		Exit Compounds	
Component	*g mol*	*Component*	*g mol*
CO(g)	1.00	$CO_2(g)$	1.00
O_2 (req. + xs) 0.50 + 0.50	1.00	$O_2(g)$	0.50
N_2	3.76	$N_2(g)$	3.76
(Air = 4.76)			

In the first approach to the solution of the problem, the "sensible heat" (enthalpy) values have been taken from the table of the enthalpy values for the combustion gases in Appendix D. Some enthalpies could not be calculated using the physical property software on the accompanying CD as the states at some temperatures were out of range. The energy balance (with $Q = 0$) reduces to $\Delta H = 0$. Here are the data needed for the energy balance:

Component	**g mol**	T **(K)**	$\Delta\hat{H}_{\text{sens.}}$ **(J/g mol)**	$\Delta\hat{H}_f^o$ **(J/g mol)**	ΔH **(J)**
Inputs					
CO	1.00	373	(2917–728)	−110,520	−108,331
O_2	1.0	373	(2953–732)	0	2221
N_2	3.76	373	(2914–728)	0	8219
		Total			−97,891
Outputs					
Assume T = 2000 K:					
CO_2	1.00	2000	(92,466–912)	−393,510	−301,956
O_2	0.50	2000	(59,914–732)	0	29,591
N_2	3.76	2000	(56,902–728)	0	211,214
		Total			−61,151

$$\Delta H = \Delta H_{\text{outputs}} - \Delta H_{\text{inputs}} = (-61{,}151) - (-97{,}891) = 36{,}740 > 0$$

Component	**g mol**	T **(K)**	$\Delta\hat{H}_{\text{sens.}}$ **(J/g mol)**	$\Delta\hat{H}_f^o$ **(J/g mol)**	ΔH **(J)**
Assume T = 1750 K:					
CO_2	1.00	1750	(77,455–912)	−393,510	−316,977
O_2	0.50	1750	(50,555–732)	0	24,912
N_2	3.76	1750	(47,940–728)	0	177,517
		Total			−114,548

$$\Delta H = (-114{,}548) - (-97{,}891) = -16{,}657 < 0$$

(*Continues*)

Example 10.7 Calculation of an Adiabatic Reaction (Flame) Temperature (*Continued*)

Now that $\Delta H = 0$ is bracketed, we can carry out a linear interpolation to find the theoretical flame temperature (TFT):

$$\text{TFT} = 1750 + \frac{0 - (-16{,}657)}{36{,}740 - (-16{,}657)}(250) = 1750 + 78 = 1828\text{ K}\quad(1555°\text{C})$$

An alternate approach to solving this problem would be to develop explicit equations in T to be solved for TFT without trial and error. The difference from the first approach is that you would have to formulate nonlinear polynomial equations in TFT by integrating the heat capacity equations for each compound to obtain the respective sensible heats for each compound. If the heat capacity equations were cubic in T, the integrated equations in TFT would be quartic. To avoid error, you can take the already-integrated equations for each compound from the CD that accompanies this book, introduce them into the energy balance as sensible heats, merge them, and solve the resulting energy balance using an equation solver such as Polymath. You can get a rough preliminary solution by first truncating the quartic equations to quadratic equations and solving the latter.

Example 10.8 Production of Citric Acid by a Fungus

Citric acid ($C_6H_8O_7$) is a well-known compound that occurs in living cells of both plants and animals. The citric acid cycle is a series of chemical reactions occurring in living cells that is essential for the oxidation of glucose, the primary source of energy for the cells. The detailed reaction scheme is far too complicated to show here, but from a macroscopic (overall) viewpoint, for the commercial production of citric acid in a batch (closed) process, three different phases occur for which the stoichiometries are slightly different:

Early idiophase (occurs between 80 and 120 hr), initial reaction:

$$\begin{aligned}&1\text{ g mol glucose} + 1.50\text{ g mol }O_2(g) \rightarrow 3.81\text{ g mol biomass}\\&\quad + 0.62\text{ g mol citric acid} + 0.76\text{ g mol }CO_2(g) + 0.37\text{ g mol polyols}\end{aligned}$$

Medium idiophase (occurs between 120 and 180 hr), additional glucose consumed:

$$\begin{aligned}&1\text{ g mol glucose} + 2.40\text{ g mol }O_2(g) \rightarrow 1.54\text{ g biomass} + 0.74\text{ g mol citric acid}\\&\quad + 1.33\text{ g mol }CO_2(g) + 0.05\text{ g mol polyols}\end{aligned}$$

Late idiophase (occurs between 180 and 220 hr), additional glucose consumed:

$$1 \text{ g mol glucose} + 3.91 \text{ g mol } O_2(g) + 0.42 \text{ g mol polyols} \rightarrow$$
$$0.86 \text{ g mol citric acid} + 2.41 \text{ g mol } CO_2$$

In an aerobic (in the presence of air) batch process, a 30% glucose solution at 25°C is introduced into a fermentation vessel. Citric acid is to be produced by using the fungus *Aspergillus niger*. Stoichiometric sterile air is mixed with the culture solution by a 100 hp aerator. Only 60% overall of the glucose supplied is eventually converted to citric acid. The early phase is run at 32°C, the middle phase at 35°C, and the late phase at 25°C.

Based on the given data, hcw much net heat has to be added or removed from the fermenter during the production of a batch of 10,000 kg of citric acid?

Solution

The detailed steps of the solution will be merged. Let's omit the inclusion of the respective heats of solution of glucose and citric acid as having minor impact but a complicating role. Because you are interested only in values at the final state and the initial state, the details of the intermediate stage can be ignored for this specific problem (but not for designing the equipment for the dynamic process in which both heating and cooling may be needed). Let citric acid be denoted by CA and the biomass by BM. Given the composition of the biomass (roughly), you can make a material balance for the overall process in which the overall net reaction of the three steps is

$$3 \text{ glucose} + 7.81\ O_2 \rightarrow 5.35 \text{ BM} + 2.22 \text{ CA} + 4.50\ CO_2$$

CA material balance:

Basis: 10,000 kg of CA produced

$$\frac{10{,}000 \text{ kg CA}}{} \left| \frac{1 \text{ kg mol CA}}{192.12 \text{ kg CA}} \right. = 52.05 \text{ kg mol CA produced}$$

$$\frac{52.05 \text{ kg mol CA}}{} \left| \frac{3 \text{ kg mol glucose}}{2.22 \text{ kg mol CA}} \right| \frac{1.00 \text{ kg mol glucose at start}}{0.60 \text{ kg mol glucose consumed}}$$

$$= 117.23 \text{ kg mol glucose}$$

$$\frac{117.23 \text{ kg mol glucose}}{} \left| \frac{180.16 \text{ kg glucose}}{1 \text{ kg mol glucose}} \right| \frac{1.00 \text{ kg soln}}{0.30 \text{ kg glucose}}$$

$$= 70{,}400 \text{ kg of 30\% solution needed at the beginning of the reaction}$$

Note that the water serves as a culture medium (sort of a constant background) and does not enter the overall stoichiometry. To save space we show only a summary of the material balances.

(*Continues*)

Example 10.8 Production of Citric Acid by a Fungus (*Continued*)

Component	Initial (kg mol)	Final (kg mol)
Glucose (70,400)(0.3)/180.16 =	117.23	46.92
BM (52.05)(5.35/2.22) =		125.44
CA		52.05
O_2 (117.23)(7.8/3) =	305.03	
CO_2 (117.23)(4.5/3)(0.60) =		105.59

We will assume that the air (O_2 and N_2) drawn into the system and the CO_2 and N_2 leaving the system are deemed to be part of the initial state of the system and the final state, respectively, to maintain the presumption of a closed system. The assumption about no flow is not correct, of course, because gas flows in at temperatures not specified and flows out at various temperatures. However, if you were to calculate the difference in the energy associated with the gas flows in an open system, you would find it to be quite insignificant relative to the work done on the system and the heat of reaction.

Energy balance:

For the closed system:

$$\Delta U = Q + W$$

because the changes in *KE* and *PE* inside the system are zero.

The work done is

$$W = \frac{100\text{ hp}}{} \left| \frac{745.7\text{ J}}{1\text{ (hp)(s)}} \right| \frac{220\text{ hr}}{} \left| \frac{3600\text{ s}}{1\text{ hr}} \right| \frac{1\text{ kJ}}{1000\text{ J}} = 5.906 \times 10^7\text{ kJ}$$

Because we do not have a value for U for this system, we will have to assume that

$$\Delta U = \Delta H - \Delta(pV) \cong \Delta H \text{ because } \Delta(pV) \text{ is negligible}$$

Then:

The next step is to calculate the enthalpy change. The available data for $\Delta\hat{H}$ are:

MW		ΔH_f^o(kJ/g mol)
d, $\acute{\alpha}$ glucose ($C_6H_{12}O_6$)	180.16	−1266
Citric acid ($C_6H_8O_7$)	192.12	−1544.8
Dry cells (biomass)	28.6	−91.4

The reference temperature will be 25°C. The initial state is 25°C and the final state is also 25°C, so the sensible heats are zero. We will omit including the nitrogen in the energy balance because the nitrogen in equals the nitrogen out and the temperature in and out is 25°C.

Component	kg mol	$\Delta\hat{H}_f^\circ$ (kJ/g mol)	ΔH (kJ)
In			
Glucose	117.32	−1266	$-148{,}530 \times 10^3$
O_2	305.03	0	0
Total			$-148{,}530 \times 10^3$
Out			
Glucose	46.93	−1266	$-59{,}410 \times 10^3$
BM	125.44	−91.4	$-11{,}470 \times 10^3$
CA	52.05	−1544.8	$-80{,}410 \times 10^3$
CO_2	105.59	−393.51	$-41{,}550 \times 10^3$
Total			$-192{,}840 \times 10^3$

$$\Delta H = [(-192{,}840) - (-148{,}530)] \times 10^3 = -44{,}310 \times 10^3 \text{ kJ}$$

$$Q = -4.43 \times 10^7 - 5.91 \times 10^7 = -1.03 \times 10^8 \text{ kJ (heat removed)}$$

The reward of a thing well done, is to have done it.

Example 10.9 Application of the Energy Balance to a Process Composed of Multiple Units

Figure E10.9a shows a process in which CO is burned with 80% of the theoretical air in Reactor 1. The combustion gases are used to generate steam, and also to transfer heat to the reactants in Reactor 2. A portion of the combustion gases that are used to heat the reactants in Reactor 2 are recycled. SO_2 is oxidized in Reactor 2. You are asked to calculate the pound moles of CO burned per hour in Reactor 1. Note: The gases involved in the SO_2 oxidation do not come in direct contact with the combustion gas used to heat the SO_2 reactants and products.

Data for Reactor 2 pertaining to the SO_2 oxidation are

Reactants	Mol fr.	T (°F)
SO_2	0.667	77
O_2	0.333	77
	1.00	

(*Continues*)

Example 10.9 Application of the Energy Balance to a Process Composed of Multiple Units (*Continued*)

Products		
SO_3	0.586	1000
SO_2	0.276	1000
O_2	0.138	1000
	1.000	

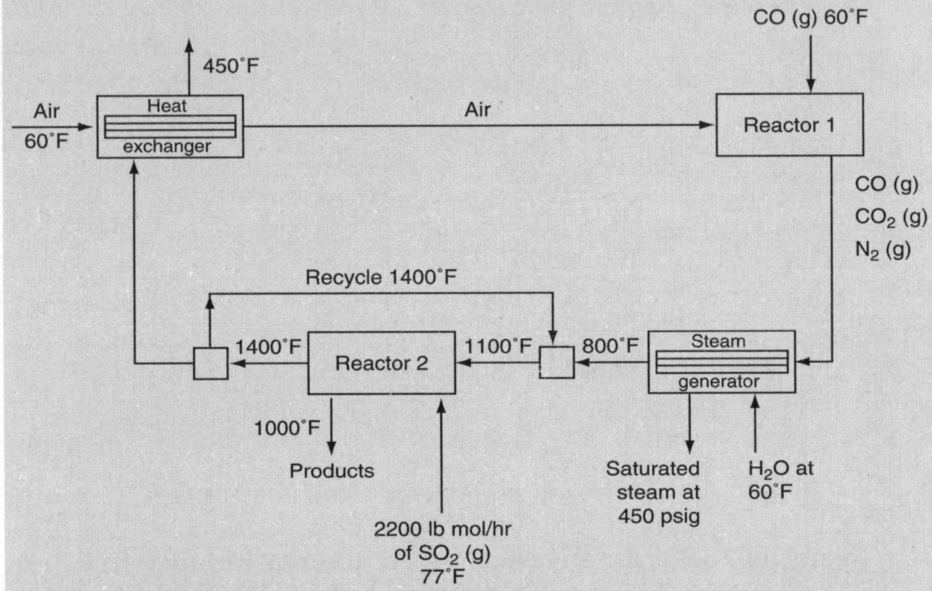

Figure E10.9a

Solution

A number of key decisions must be made before starting any calculations to solve this problem.

1. What units should you use? Although all of the data in the diagram are in the AE system, the convenient enthalpy data are in the SI system; hence we will use the latter. The temperature conversions are

T (°F)	T (K)
77	298
800	700
1000	811
1400	1033

2. What system should you choose to start with in the calculations? Selection of the entire process as the system will lead to about 15 unknowns and require 15 independent equations. It makes sense to look for a system to begin with that involves fewer unknowns and equations. If you examine some of the possible subsystems in the process, the selection of Reactor 2 excluding the local recycle stream proves to be a reasonable choice. It includes SO_2 directly, and the input and output temperatures are known, but we must first determine the composition of the exit stream from Reactor 1.
3. What basis should be used? The given amount of material is the 2200 lb mol/hr of SO_2. A more convenient basis would be 1 g mol of CO entering so that we can first determine the composition of the combustion gas stream leaving Reactor 1. Both bases are suitable—we will use the latter, and at the end of the calculations we can convert to a pound mole of CO per hour based on 2200 lb mol/hr of SO_2.

In addition to these three decisions, we have to make some assumptions:

1. The process and its components are continuous, steady-state flow systems; hence $E = 0$ throughout.
2. The pressure everywhere is 1 atm.
3. No heat loss occurs ($Q = 0$ in any energy balances).
4. $\Delta KE = \Delta PE = W = 0$ in any energy balances.

Step 5

Basis: 1 g mol of CO entering Reactor 1

Step 3

On the selected basis you have to calculate the O_2, N_2, CO, and CO_2 in the combustion gases from Reactor 1 that pass through the entire process. No O_2 exists in the combustion products because excess air was not used. Thus, the combustion gases are:

Basis: 1 g mol CO(g) entering Reactor 1

Reaction: $CO + 1/2\,O_2 \rightarrow CO_2$

Entering O_2: $1\,(0.5)(0.8) = 0.4$ g mol

Entering N_2: $0.4\,(79/21) = 1.5$ g mol

Summary of the output of Reactor 1

Component	g mol	Mol fr.
CO	$1(1 - 0.8) = 0.2$	0.32
CO_2	$1(0.8) = 0.8$	0.08
N_2	$= 1.5$	0.60
Total	2.5	1.00

(Continues)

Example 10.9 Application of the Energy Balance to a Process Composed of Multiple Units (*Continued*)

Let's next look at Reactor 2. A figure helps in the analysis—examine Figure E10.9b. The system selected excludes the recycle stream.

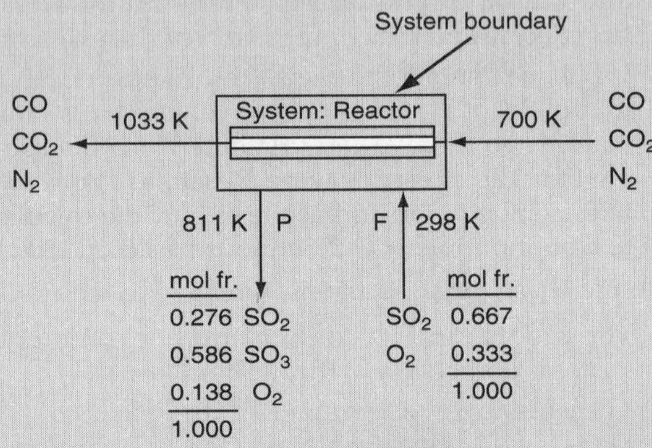

Figure E10.9b

Steps 6 and 7

The degree-of-freedom analysis for Reactor 2 is as follows: We can assign values to the entering and exiting gases in the CO, CO_2, N_2 stream from previous calculations. We know the entering and exit temperatures and pressures. We also know the values of the mole fractions of the input and output streams F and P, hence can calculate the values of $n^F_{SO_2}, n^F_{O_2}, n^P_{SO_2}, n^P_{SO_3}, n^P_{O_2}$ from the specified mole fractions, for example, $n^F_{SO_2} = 0.667\,F$. The temperatures in and out are specified.

The *unknowns* for the system in Figure E10.9b are $n^F_{SO_2}, n^F_{O_2}, n^P_{SO_2}, n^P_{SO_3}, n^P_{O_2}, \xi, F$, and P, a total of 8. How many *independent equations* are there? A typical analysis for the material balances would be as follows:

a. SO_2, SO_3, and O_2 species balances: 3
b. Sum of component moles to get P and F (redundant): 0
c. Specifications ($n^F_{SO_2}, n^F_{O_2}, n^P_{SO_2}, n^P_{SO_3}, n^P_{O_2}$): 5

But wait! The three species balances are

$$SO_2: \quad P(0.276) - F(0.667) = -1(\xi) \tag{a}$$

$$SO_3: \quad P(0.586) - F(0) = 1(\xi) \tag{b}$$

$$O_2: \quad P(0.138) - F(0.333) = -0.5(\xi) \tag{c}$$

Only two are independent. Try to solve Equations (a)–(c), and you will see that this statement is true. Thus, one more equation is needed. What equation can you use? What is the topic of this chapter? Use an energy balance!

If you carry out the degree-of-freedom analysis using element balances, seven unknowns exist (ξ is not involved), and the equations comprise two element balances plus five specifications.

Steps 8 and 9

To make an energy balance you have to get information about the heats of formation and the sensible heats. The data used below have been taken from the CD in the back of this book. With the assumptions made, the energy balance reduces to $\Delta H = 0$.

Data and Calculations for the Energy Balance

Comp	g mol	T (K)	$\Delta\hat{H}_f^\circ$(kJ/g mol)	$\Delta\hat{H}_{\text{sensible}}$(kJ/g mol)	ΔH (kJ)
Out					
CO(g)	0.2	1033	−109.054	35.332	−14.744
CO_2(g)	0.8	1033	−393.250	35.178	−286.458
N_2(g)	1.5	1033	0	22.540	33.810
SO_2(g)	$n^P_{SO_2}$	811	−296.855	20.845	$-276.010\ n^P_{SO_2}$
SO_3(g)	$n^P_{SO_3}$	811	−395.263	34.302	$-360.961\ n^P_{SO_3}$
O_2(g)	$n^P_{O_2}$	811	0	16.313	$16.313\ n^P_{O_2}$
In					
CO_2(g)	0.8	700	−393.250	17.753	−300.398
N_2(g)	1.5	700	0	11.981	17.972
SO_2(g)	$n^F_{SO_2}$	298	−296.855	0	$-296.855\ n^F_{SO_2}$
O_2(g)	$n^F_{O_2}$	298	0	0	0

The energy balance is

$$-276.010n^P_{SO_2} - 360.961n^P_{SO_3} + 296.855n^F_{SO_2} + 16.313n^P_{O_2} = -33.41n^F_{SO_2} \qquad \text{(d)}$$

Step 9

Substitute for the variables in Equation (d) the following to get Equation (d) in terms of F and P:

$$n^P_{SO_2} = P(0.276) \qquad n^P_{O_2} = P(0.138)$$

$$n^P_{SO_3} = P(0.586) \qquad n^F_{SO_2} = F(0.667)$$

Solve Equation (d) in terms of F and P together with Equations (a) and (b) (using Polymath) to get (the n_i are in gram moles):

(Continues)

Example 10.9 Application of the Energy Balance to a Process Composed of Multiple Units (*Continued*)

$n^P_{SO_2} = 0.312$ $n^F_{SO_2} = 0.974$
$n^P_{SO_3} = 0.662$ $n^F_{O_2} = 0.486$
$n^P_{O_2} = 0.156$
$P = 1.33$ $\xi = 0.66$
$F = 1.46$

Step 10

Check the solution using the redundant oxygen balance:

$$\text{O:}\quad 2n^P_{SO_2} + 3n^P_{SO_3} + 2n^P_{O_2} - 2n^F_{O_2} - 2n^F_{SO_2} = 0$$

$$2(0.312) + 3(0.662) + 2(0.156) - 2(0.486) - 2(0.974) = 0$$

Finally, we have to calculate the pound moles of CO per hour that flow into Reactor 1. Because no loss of combustion gases occurs up to and through Reactor 2, you know (on the basis of 1 g mol of CO entering Reactor 1) that

$$\frac{1 \text{ lb mol CO}}{0.974 \text{ lb mol } SO_2}\bigg|\frac{2200 \text{ lb mol } SO_2}{\text{hr}} = 2259\,\frac{\text{lb mol CO}}{\text{hr}}$$

Self-Assessment Test

Questions

1. Explain why omitting the term for the heat of reaction at S.C. in the energy balance does not prevent you from obtaining the proper value of ΔH to use in the energy balance.
2. What are some of the advantages and disadvantages of calculating the enthalpy change of a compound in the process for use in the energy balance by the method described in Section 10.3 versus that in Section 10.2?

Problems

1. What is the heat transfer to or from a reactor in which methane reacts completely with oxygen to form carbon dioxide gas and water vapor? Base your calculations on a feed of 1 g mol of CH_4(g) at 400 K plus 2 g mol of O_2(g) at 25°C. The exit gases leave at 1000 K. Use the method of this section.
2. Repeat problem 1, but assume that the fraction conversion is only 60%.
3. Calculate the heat added to or removed from a reactor in which stoichiometric amounts of CO(g) and H_2(g) at 400°C react to form CH_3OH(g) (methanol) at 400°C.

Thought Problem

A review of additives to gasoline to give blends that improve its octane rating shows that oxygenated compounds necessarily contain lower energy (British thermal units per gallon). Methanol was the lowest, having a heat of reaction approximately one-half that of gasoline. Methanol costs \$2.80/gal, whereas unleaded premium gas costs \$3.60/gal. Even though methanol has only half the energy content of gasoline, it costs about 75% as much as gasoline. Evaluate the merits of methanol as a replacement for gasoline.

Discussion Question

An automobile owner sued his insurance company because his (lead-acid) battery exploded, damaging the hood and motor. The adjuster from the insurance company, after inspecting the remains of the battery, stated that his company was not liable to pay the claim because the battery grossly overheated.

Can a battery overheat? How? If overheating took place, why would the battery explode? What would be the most likely mechanism of the explosion?

10.4 The Heat (Enthalpy) of Combustion

An older method of calculating enthalpy changes when chemical reactions occur is via **standard heats (enthalpies) of combustion**, $\Delta\hat{H}^{\circ}_{c}$, which have a different set of reference conditions from the standard heats of formation. The conventions used with the standard heats of combustion are as follows:

1. The compound is oxidized with oxygen or some other substance to the products $CO_2(g)$, $H_2O(l)$, $HCl(aq)$, and so on.
2. The reference conditions are still 25°C and 1 atm.
3. Zero values of $\Delta\hat{H}^{\circ}_{c}$, are assigned to certain of the oxidation products, for example, $CO_2(g)$, $H_2O(l)$, $HCl(aq)$, and to $O_2(g)$ itself.
4. If other oxidizing substances are present, such as S, N_2 or Cl_2, it is necessary to make sure that states of the products are carefully specified and are identical to (or can be transformed into) the final conditions that determine the standard state.
5. Stoichiometric quantities react completely.

The major difference between the standard heats of formation and the standard heats of combustion is item 3 in the list above. Certain products have zero values for the heat of combustion while certain reactants have zero values for reactants.

The heat of combustion has been proposed as one of the criteria to determine the ranking of the incinerability of hazardous waste. The rationale is that if a compound has a higher heat of combustion, it can release more energy than other compounds during combustion and would be easier to incinerate. Heats of combustion are frequently reported for bio reactions rather than heats of reaction.

How to convert a value of the standard heat of combustion of a compound to the corresponding value for the standard heat of formation is shown in the next example.

Example 10.10 Conversion of a Standard Heat of Combustion to the Corresponding Standard Heat of Formation

The heat of combustion of liquid (>17°C) lactic acid ($C_3H_6O_3$) varies somewhat in the literature. We will use −1361 kJ/g mol [*Electronic J. Biotech.*, 7, No. 2 (2004)]. What is the corresponding standard heat of formation?

Solution

ΔH_f° is determined by the chemical equation

$$3\,C(\beta) + 3\,H_2(g) + 1.5\,O_2(g) \rightarrow C_3H_6O_3(l) \qquad \text{(a)}$$

The procedure is to start with the chemical equation that gives ΔH_c° (Equation (b) below) and add or subtract other chemical equations whose standard heat(s) of combustion is known so that the final result is the desired Equation (a).

Equation	$\Delta\hat{H}_c^\circ$(kJ/g mol)	
$C_3H_6O_3(l) + 3\,O_2(g) \rightarrow 3\,CO_2(g) + 3\,H_2O(l)$	−1361	(b)
$C(\beta) + O_2(g) \rightarrow CO_2(g)$	−393.51	(c)
$H_2(g) + \frac{1}{2}O_2(g) \rightarrow H_2O(l)$	−285.84	(d)

Equation (d) includes a phase change for water (see Figure 10.5).

The following—(b) + 3(c) + 3(d)—gives Equation (a), or

$$1361 + 3\,(-393.51) + 3\,(-285.84) = \Delta\hat{H}_f^\circ = -677 \text{ kJ/g mol}$$

Based on this modus operandi, can you convert a standard heat of formation to a standard heat of combustion?

$$
\begin{array}{l}
\Delta H^{\circ}_{\text{vap}} \text{ at } 25^{\circ}\text{C} \\
\text{and 1 atm} \\
= 44{,}000 \text{ J/g mol}
\end{array}
\left\{
\begin{array}{ll}
\text{H}_2\text{O(l) } 25^{\circ}\text{C, 1 atm} & \text{H}_1 = 0 \text{ J/g mol} \\
\quad \downarrow \quad \Delta\text{H}_1 \cong 0 \text{ J/g mol} & \\
\text{H}_2\text{O(l) } 25^{\circ}\text{C, vapor pressure at } 25^{\circ}\text{C} & \text{H}_2 = 0 \text{ J/g mol} \\
\quad \downarrow \quad \Delta\text{H}_2 = \Delta\text{H}_{\text{vap}} \text{ at the vapor pressure of water} & \\
\qquad\qquad (\text{p} = 3.17 \text{ kPa}) = 44{,}004 \text{ J/g mol} & \\
\text{H}_2\text{O(g) } 25^{\circ}\text{C, vapor pressure at } 25^{\circ}\text{C} & \text{H}_3 = 44{,}004 \text{ J/g mol} \\
\quad \downarrow \quad \Delta\text{H}_3 = -4 \text{ J/g mol} & \\
\text{H}_2\text{O(g) } 25^{\circ}\text{C, 1 atm} & \text{H}_4 = 44{,}000 \text{ J/g mol}
\end{array}
\right.
$$

Figure 10.5 The enthalpy change that occurs when H_2O(l) goes from 25°C and 1 atm to H_2O(g) at 25°C and 1 atm. Look at the steam tables on the sheet in the back of the book if the steps are not clear.

The precise heat of vaporization for a compound such as water at 1 atm and 25°C can be calculated as shown in Figure 10.5, but the value of heat of vaporization of a compound at 25°C and the vapor pressure of the compound will suffice for most engineering calculations. Note that the heat of combustion of H_2O(l) is zero but for H_2O(g) is −44.00 kJ/g mol—you have to subtract the heat of vaporization of water at 25°C and 1 atm from the value for H_2O(l). (Look at Figure 10.5 as to the source of the $\Delta\hat{H}_{\text{vap}} = 44.00$ kJ/g mol.)

For a fuel such as coal or oil, the negative of the standard heat of combustion is known as the heating value of the fuel. Both a lower (net) heating value (LHV) and a higher (gross) heating value (HHV) exist, depending upon whether the water in the combustion products is in the form of a vapor (for the LHV) or a liquid (for the HHV). Examine Figure 10.6.

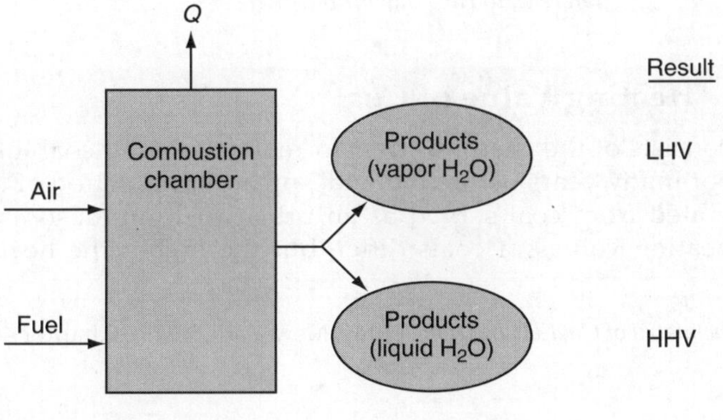

Figure 10.6 The classification of LHV or HHV for a fuel depends on the state of the water exiting from the system.

$$\text{HHV} = \text{LHV} + (n_{\text{H}_2\text{O(g)}} \text{ in product})(\Delta\hat{H}_{\text{vap}} \text{ at } 25°\text{C and 1 atm})$$

You can calculate a standard heat of reaction using the heats of combustion by an equation analogous to Equation (10.1). For **complete combustion**:

$$\Delta H^{\circ}_{\text{rxn}}(25°\text{C}) = -\left(\sum^{\text{Products}} n_i \Delta\hat{H}^{\circ}_{\text{c},i} - \sum^{\text{Reactants}} n_i \Delta\hat{H}^{\circ}_{\text{c},i}\right) \tag{10.7}$$

Note: The minus sign in front of the summation expression occurs because the choice of reference states is zero for the right-hand products of the standard reaction. Refer to Appendix D for values of $\Delta\hat{H}^{\circ}_{\text{c}}$.

As an example of the calculation of $\Delta H^{\circ}_{\text{rxn}}(25°\text{C})$ from heat of combustion data, we will calculate $\Delta H^{\circ}_{\text{rxn}}(25°\text{C})$ for the reaction (the data are taken from Appendix D in kilojoules per gram mole).

$$\text{CO(g)} + \text{H}_2\text{O(g)} \rightarrow \text{CO}_2\text{(g)} + \text{H}_2\text{(g)}$$

$$\begin{aligned}\Delta H^{\circ}_{\text{rxn}}(25°\text{C}) &= -\{[(1)(0) + (1)(-285.84)] \\ &\quad - [(1)(-282.99) + (1)(-44.00)]\} = -41.15 \text{ kJ}\end{aligned}$$

Calculation of the heat of reaction for fuels and compounds that have a complicated analysis requires you to use empirical formulas to estimate $\Delta\hat{H}^{\circ}_{\text{rxn}}$. You can estimate the heating value of a coal within about 3% from the Dulong formula[4]:

Higher heating value (HHV) in Btu per pound

$$= 14{,}544\text{ C} + 62{,}028\left(\text{H} - \frac{\text{O}}{8}\right) + 4050\text{ S}$$

where C, H, S, and O are the respective *weight* fractions in the fuel.

$$\text{net Btu/lb coal} = \text{gross Btu/lb coal} - (91.23)(\%\text{ total H by weight})$$

The HHV of fuel oils in British thermal units per pound can be approximated by

$$\text{HHV} = 17{,}887 + 57.5°\text{API} - 102.2\ (\%\text{S}).$$

There's no fuel like an old fuel.

Example 10.11 Heating Value of Coal

Coal gasification consists of the chemical transformation of solid coal into a combustible gas. For many years before the widespread introduction of natural gas, gas generated from coal served as an urban fuel (and also as an illuminant). The heating values of coals differ, but the higher the heating

[4]H. H. Lowry (ed.), *Chemistry of Coal Utilization*, Wiley, New York (1945), Chapter 4.

value, the higher the energy value of the gas produced. The analysis of the following coal has a reported heating value of 29,770 kJ/kg as received. Assume that this is the gross heating value at 1 atm and 25°C obtained in an open system.[5] Use the Dulong formula to check the validity of the reported value.

Component	Percent
C	71.0
H_2	5.6
N_2	1.6
Net S	2.7
Ash	6.1
O_2	13.0
Total	100.0

Solution

$$\text{HHV} = 14{,}544(0.71) + 62{,}028\left[(0.056) - \frac{0.130}{8}\right] + 4050(0.027) = 12{,}901 \text{ Btu/lb}$$

Note that 0.056 lb of H_2 is still 0.056 lb of H, and 0.130 lb of O_2 is 0.130 lb of O, because the percent values are the mass of the element no matter how many atoms are joined together as a molecule.

$$\frac{12{,}901 \text{ Btu}}{\text{lb}}\left|\frac{1 \text{ lb}}{0.454 \text{ kg}}\right|\frac{1.055 \text{ kJ}}{1 \text{ Btu}} = 29{,}980 \text{ kJ/kg}$$

The two values are quite close.

Example 10.12 Selection of a Fuel to Reduce SO_2 Emissions

SO_2 emissions from power plants are the primary source of acid rain. SO_2 in the atmosphere is absorbed by tiny water droplets that eventually end up as rain, thus increasing the acidity of lakes and rivers. Consider the two fuels listed in the table below. Determine which fuel would be preferred to provide 106 Btu of thermal energy from combustion while minimizing the

(Continues)

[5]If the reported value was obtained in a closed system as might be the case, the value reported might be $\Delta\hat{U}$, not $\Delta\hat{H}$.

Example 10.12 Selection of a Fuel to Reduce SO_2 Emissions (*Continued*)

SO_2 emissions. SO_2 removal equipment for flue gas can be installed to reduce the SO_2 discharge, but at additional cost, bringing into play another important factor in choosing a fuel.

	No. 6 Fuel Oil	No. 2 Fuel Oil
Density (lb/ft^3)	60.2	58.7
Lower heating value (Btu/gal)	155,000	120,000
Carbon (wt %)	87.2	87.3
Hydrogen (wt %)	10.5	12.6
Sulfur (wt %)	0.72	0.62
Ash (wt %)	0.04	<0.01

Solution

Basis: 10^6 Btu from combustion

For No. 6 heating oil:

$$\frac{10^6 \text{ Btu}}{} \bigg| \frac{1 \text{ gal}}{155{,}000 \text{ Btu}} \bigg| \frac{60.2 \text{ lb fuel}}{1 \text{ gal}} \bigg| \frac{0.0072 \text{ lb S}}{1 \text{ lb fuel}} = 2.80 \text{ lb S}$$

For No. 2 heating oil:

$$\frac{10^6 \text{ Btu}}{} \bigg| \frac{1 \text{ gal}}{120{,}000 \text{ Btu}} \bigg| \frac{58.7 \text{ lb fuel}}{1 \text{ gal}} \bigg| \frac{0.0062 \text{ lb S}}{1 \text{ lb fuel}} = 3.03 \text{ lb S}$$

The No. 6 heating oil should be selected because its combustion will generate less SO_2 emissions, even though it has a higher weight percent S.

Self-Assessment Test

Questions

1. Explain why in calculating ΔH using heats of formation you subtract the ΔH of the inputs from the ΔH of the outputs, whereas in using heat of combustion you subtract the ΔH of the outputs from that of the inputs.
2. Can the HHV ever be the same as the LHV?
3. Can the HHV ever be lower than the LHV?

4. Do you have to use heats of combustion to calculate an HHV or an LHV?
5. Is it true for the reaction $H_2(g) + 1/2\, O_2(g) \rightarrow H_2O(g)$ at 25°C and 1 atm, which has a cited heat of reaction of −241.83 kJ, that if you write the reaction as $2H_2(g) + O_2(g) \rightarrow 2H_2O(g)$, the calculated heat of reaction will be 2(−241.83) kJ?
6. What is the difference in HHV and LHV when CO is burned with O_2 at 25°C and 1 atm?

Problems

1. A synthetic gas analyzes 6.1% CO_2, 0.8% C_2H_4, 0.1% O_2, 26.4% CO, 30.2% H_2, 3.8% CH_4, and 32.6% N_2. What is the heating value of the gas in British thermal units per cubic foot measured at 60°F, saturated when the barometer reads 30.0 in. Hg?
2. Calculate the standard heat of reaction using heat of combustion data for the reaction

$$CO(g) + 3H_2(g) \rightarrow CH_4(g) + H_2O(l)$$

Thought Problems

1. How does the presence of a diluent or excess reactant alter the heat effects associated with an exothermic reaction?
2. What is the effect of nonstoichiometric quantities of reactants on the standard heat of reaction?

Discussion Question

As an interest in synthetic fuel projects develops, allegations about the health and environmental risks associated with these projects increase. From process descriptions it is not always clear why certain projects gain rapid acceptance whereas others do not.

For example, coal gasification and liquefaction processes are cited as contributing to a "greenhouse effect" by releasing more CO_2 to the atmosphere than oil or natural gas. On the other hand, the use of alcohol as a fuel is considered to be an acceptable CO_2 release.

A recent issue of *Energy Resources & Technology* describes plans for a 60,000 bbl/day ethanol plant for fuel production. The plant will consume 86,600 Btu of energy to produce 1 gal of ethyl alcohol, plus a cattle feed by-product having "an energy equivalent of 14,500 Btu." Since combustion of 1 gal of alcohol yields 76,500 Btu, the total energy produced on burning the alcohol is 4400 Btu greater than that consumed in its production.

Calculations show that this overall process yields 7500 Btu of energy per mole of CO_2 produced. Efficient coal conversion processes produce 140,000 Btu for each mole of CO_2 produced. This is almost 19 times the energy produced from alcohol at equivalent CO_2 production.

Is fuel production from alcohol attractive?

Looking Back

This chapter introduced the concept of the heat of formation and showed how it can be used to solve energy balances with reaction using two techniques: (1) the heat of reaction approach or (2) an approach that combines the heat of formation with sensible heat changes. Finally, the heat of combustion, which contains the same information as the heat of formation but on a different basis, was presented and used to solve energy balance problems with reaction.

Glossary

Endothermic reaction A reaction for which heat must be added to the system to maintain isothermal conditions.

Exothermic reaction A reaction for which heat must be removed from the system to maintain isothermal conditions.

Heating value The negative of the standard heat of combustion for a fuel such as coal or oil.

Heat of reaction Enthalpy change that is associated with a reaction.

Higher (gross) heating value (HHV) Value of the negative heat of combustion when the product water is a liquid.

Lower (net) heating value (LHV) Value of the negative of the heat of combustion when the product water is a vapor.

Reference state For enthalpy it is the state at which the enthalpy is zero.

Sensible heat The quantity $(\hat{H}_{T,P} - \hat{H}^{\circ}_{\text{ref}})$ that excludes any phase changes.

Standard heat of combustion Enthalpy change for the oxidation of 1 mol of a compound at 25°C and 1 atm. By definition the water and carbon dioxide on the right-hand side of the reaction equation are assigned zero values.

Standard heat of formation Enthalpy change for the formation of 1 mol of a compound from its constituent elements at 25°C and 1 atm.

Standard heat of reaction Heat of reaction for components in the standard state (25°C and 1 atm) when stoichiometric quantities of reactants react completely to give products at 25°C and 1 atm.

Standard state For the heat of reaction, 25°C and 1 atm.

Supplemental References

Benedek, P., and F. Olti. *Computer Aided Chemical Thermodynamics of Gases and Liquids—Theory, Model and Programs*, Wiley-Interscience, New York (1985).

Benson, S. W., et al. "Additivity Rules for the Estimation of Thermophysical Properties," *Chem Rev.*, **69**, 279 (1969) [cited in Reid, R. C., et al. *The Properties of Gases and Liquids*, 4th ed., McGraw-Hill, New York (1987)].

Danner, R. P., and T. E. Daubert. "Manual for Predicting Chemical Process Design Data," *Documentation Manual*, Chapter 11—"Combustion," American Institute of Chemical Engineers, New York (1987).

Daubert, T. E., et al. *Physical and Thermodynamic Properties of Pure Chemicals: Data Compilation*, American Institute of Chemical Engineers, New York, published by Taylor and Francis, Bristol, PA, periodically.

Garvin, J. "Calculate Heats of Combustion for Organics," *Chem. Eng. Progress*, 43–45 (May, 1998).

Kraushaar, J. J. *Energy and Problems of a Technical Society*, John Wiley, New York (1988).

Poling, B. E., J. M. Prausnitz, and J. P. O'Connell. *The Properties of Gases and Liquids*, 5th ed., McGraw-Hill, New York (2001).

Rosenberg, P. *Alternative Energy Handbook*, Association of Energy Engineers, Lilbum, GA (1988).

Seaton, W. H., and B. K. Harrison. "A New General Method for Estimation of Heats of Combustion for Hazard Evaluation," *Journal of Loss Prevention*, **3**, 311–20 (1990).

Vaillencourt, R. *Simple Solutions to Energy Calculations*, Association of Energy Engineers, Lilbum, GA (1988).

Web Sites

http://webbook.nist.gov/

http://en.wikipedia.org/wiki/Standard_enthalpy_change_of_formation_(data_table)

Problems

10.1 The Standard Heat (Enthalpy) of Formation

***10.1.1** Which of the following heats of formation would indicate an endothermic reaction?

a. −32.5 kJ
b. 32.5 kJ
c. −82 kJ
d. 82 kJ
e. Both a and c
f. Both b and d

***10.1.2** Which of the following changes of phase is exothermic?

a. Gas to liquid
b. Solid to liquid
c. Solid to gas
d. Liquid to gas

*10.1.3 Exothermic reactions are usually self-sustaining because of which of the following explanations?

a. Exothermic reactions usually require low activation energies.
b. Exothermic reactions usually require high activation energies.
c. The energy released is sufficient to maintain the reaction.
d. The products contain more potential energy than the reactants.

**10.1.4 How would you determine the heat of formation of gaseous fluorine at 25°C and 1 atm?

**10.1.5 Determine the standard heat of formation for FeO(s) given the following values for the heats of reaction at 25°C and 1 atm for the following reactions:

$$2Fe(s) + \tfrac{3}{2}O_2(g) \longrightarrow Fe_2O_3(s): -822{,}200\ J$$
$$2FeO(s) + \tfrac{1}{2}O_2(g) \longrightarrow Fe_2O_3(s): -284{,}100\ J$$

***10.1.6 The following enthalpy changes are known for reactions at 25°C and 1 atm:

No.				$\Delta H°$(kJ/g mol)
1	$C_3H_6(g) + H_2(g)$	→	$C_3H_8(g)$	−123.8
2	$C_3H_8(g) + 5O_2(g)$	→	$3CO_2(g) + 4H_2O(l)$	−2220.0
3	$H_2(g) + \frac{1}{2}O_2(g)$	→	$H_2O(l)$	−285.8
4	$H_2O(l)$	→	$H_2O(g)$	43.9
5	C(diamond) + $O_2(g)$	→	$CO_2(g)$	−395.4
6	C(graphite) + $O_2(g)$	→	$CO_2(g)$	−393.5

Calculate the heat of formation of propylene (C_3H_6).

***10.1.7 In a fluidized bed gasification system you are asked to find out the heat of formation of a solid sludge of the composition (formula) C_5H_2 from the following data:

			ΔH(kJ/g mol)
$C(s) + \frac{1}{2}O_2(g)$	→	$CO(g)$	−110.4 kJ/g mol C
$C(s) + O_2(g)$	→	$CO_2(g)$	−394.1 kJ/g mol C
$H_2(g) + \frac{1}{2}O_2(g)$	→	$H_2O(g)$	−241.826 kJ/g mol H_2
$CO(g) + \frac{1}{2}O_2(g)$	→	$CO_2(g)$	−283.7 kJ/g mol CO
$H_2O(g) + CO(g)$	→	$H_2(g) + CO_2(g)$	−38.4 kJ/g mol H_2O
$C_5H_2(s) + 5\frac{1}{2}O_2(g)$	→	$5CO_2(g) + H_2O(l)$	−2110.5 kJ/g mol C_5H_2
ΔH vaporization H_2O at 25°C			+43.911 kJ/g mol H_2O_2

***10.1.8 Look up the heat of formation of (a) liquid ammonia, (b) formaldehyde gas, (c) acetaldehyde liquid. If you can't find the value, get data to calculate it.

10.2 The Heat (Enthalpy) of Reaction

***10.2.1** Calculate the standard (25°C and 1 atm) heat of reaction per gram mole of the first reactant on the left-hand side of the reaction equation for the following reactions:

a. $NH_3(g) + HCl(g) \rightarrow NH_4Cl(s)$
b. $CH_4(g) + 2O_2(g) \rightarrow CO_2(g) + 2H_2O(l)$
c. $C_6H_{12}(g) \rightarrow C_6H_6(l) + 3H_2(g)$
   cyclohexane → benzene

***10.2.2** Calculate the standard heat of reaction for the following reactions:

a. $CO_2(g) + H_2(g) \rightarrow CO(g) + H_2O(l)$
b. $CaO(s) + 2MgO(s) + 4H_2O(l) \rightarrow 2Ca(OH)_2(s) + 2Mg(OH)_2(s)$
c. $Na_2SO_4(s) + C(s) \rightarrow Na_2SO_3(s) + CO(g)$
d. $NaCl(s) + H_2SO_4(l) \rightarrow NaHSO_4(s) + HCl(g)$
e. $NaCl(s) + 2SO_2(g) + 2H_2O(l) + O_2(g) \rightarrow 2Na_2SO_4(s) + 4HCl(g)$
f. $SO_2(g) + \frac{1}{2}O_2(g) + H_2O(l) \rightarrow H_2SO_4(l)$
g. $N_2(g) + O_2(g) \rightarrow 2NO(g)$
h. $Na_2CO_3(s) + 2Na_2S(s) + 4SO_2(g) \rightarrow 3Na_2S_2O_3(s) + CO_2(g)$
i. $CS_2(l) + Cl_2(g) \rightarrow S_2Cl_2(l) + CCl_2(l)$
j. $C_2H_4(g) + HCl(g) \rightarrow CH_3CH_2Cl(g)$
   ethylene → ethyl chloride
k. $CH_3OH(g) + \frac{1}{2}O_2(g) \rightarrow H_2CO(g) + H_2O(g)$
   methyl alcohol → formaldehyde
l. $C_2H_2(g) + H_2O(l) \rightarrow CH_3CHO(l)$
   acetylene → acetaldehyde
m. $n\text{-}C_4H_{10}(g) \rightarrow C_2H_4(g) + C_2H_6(g)$
   butane → ethylene + ethance

****10.2.3** J. D. Park et al. [*JACS* **72**, 331–3 (1950)] determined the heat of hydrobromination of propene (propylene) and cyclopropane. For hydrobromination (addition of HBr) of propene to 2-bromopropane, they found that $\Delta H = -84{,}441$ J/g mol. The heat of hydrogenation of propene to propane is $\Delta H = -126{,}000$ J/g mol. N.B.S. Circ. 500 gives the heat of formation of HBr(g) as $-36{,}233$ J/g mol when the bromine is liquid, and the heat of vaporization of bromine as 30,710 J/g mol. Calculate the heat of bromination of propane using gaseous bromine to form 2-bromopropane using these data.

****10.2.4** In the reaction

$$4FeS_2(s) + 11O_2(g) \rightarrow 2Fe_2O_3(s) + 8SO_2(g)$$

the conversion of $FeS_2(s)$ to $Fe_2O_3(s)$ is only 80% complete. If the standard heat of reaction for the reaction is calculated to be -567.4 kJ/g mol $FeS_2(s)$, what value of ΔH°_{rxn} will you use per kilogram of FeS_2 burned in an energy balance?

10.2.5 M. Beck et al. [*Can J. Ch.E.*, **64, 553 (1986)] described the use of immobilized enzymes (E) in a bioreactor to convert glucose (G) to fructose (F).

$$G + E \leftrightarrow EG \leftrightarrow E + F$$

At equilibrium the overall reaction can be considered to be $G + E \leftrightarrow E + F$.

Glucose		*Fructose*
CHO		CH_2OH
|		|
CHOH		C = O
|		|
CHOH	$\Rightarrow$	CHOH
|		|
CHOH		CHOH
|		|
CHOH		CHOH
|		|
CH_2OH		CH_2OH
$\Delta\hat{H}_f^\circ$: 0.990×10^9 J/g mol		$\Delta\hat{H}_f^\circ$: 1.040×10^9 J/g mol

The fraction conversion is a function of the flow rate through the reactor and the size of the reactor, but for a flow rate of 3×10^{-3} m/s and a bed height of 0.44 m, the fraction conversion on a pass through the reactor was 0.48. Calculate the heat of reaction at 25°C per mole of G converted.

**10.2.6 A consulting laboratory is called upon to determine the heat of reaction at 25°C of a natural gas in which the combustible is entirely methane. They do not have a Sargent flow calorimeter but do have a Parr bomb calorimeter. They pump a measured volume of the natural gas into the Parr bomb, add oxygen to give a total pressure of 1000 kPa, and explode the gas-O_2 mixture with a hot wire. From the data they calculate that the gas has a heating value of 39.97 kJ/m^3. Should they report this value as the heat of reaction? Explain. What value should they report?

**10.2.7 Is it possible to calculate the heat of reaction by using property tables that are prepared using different reference states? How?

**10.2.8 A fat is a glycerol molecule bonded to a combination of fatty acids or hydrocarbon chains. Usually, the glycerol bonds to three fatty acids, forming a triglyceride. A fat that has no double bonds between the carbon atoms in the fatty acid is said to be saturated.

How are fats treated by the human digestive system? They are first enzymatically broken down into smaller units, fatty acids and glycerol. This is called digestion and occurs in the intestine or cellularly by lysosomes. Next, enzymes remove two carbons at a time from the carboxyl end of the chain, a process that produces molecules of acetyl CoA, NADH, and $FADH_2$. The acetyl CoA is a high-energy molecule that is then treated by the citric acid cycle which oxidizes it to CO_2 and H_2O.

Calculate the heat of reaction when tristearin is used in the body.

Data:	$\Delta\hat{H}_f^\circ$ (kJ/gmol)
Stearic acid (s) ($C_{18}H_{36}O_2$), a fatty acid	−964.3
Glycerol (1) ($C_3H_8O_3$)	−159.16
Tristearin(s) ($C_{63}H_{132}O_{15}$), a triglyceride	−3.820

**10.2.9 Answer the following questions briefly (in no more than three sentences):

a. Does the addition of an inert diluent to the reactants entering an exothermic process increase, decrease, or make no change in the heat transfer to or from the process?

b. If the reaction in a process is incomplete, what is the effect on the value of the standard heat of reaction? Does it go up, go down, or remain the same?

c. Consider the reaction $H_2(g) + \frac{1}{2}O_2(g) \rightarrow H_2O(g)$. Is the heat of reaction with the reactants entering and the products leaving at 500 K higher, lower, or the same as the standard heat of reaction?

**10.2.10 Compute the heat of reaction at 600 K for the following reaction:

$$S(1) + O_2(g) \rightarrow SO_2(g)$$

**10.2.11 Calculate the heat of reaction at 500°C for the decomposition of propane:

$$C_3H_8 \rightarrow C_2H_2 + CH_2 + H_2$$

**10.2.12 Calculate the heat of reaction of the following reactions at the stated temperature:

a. $\underset{\text{methyl alcohol}}{CH_3OH(g)} + \frac{1}{2}O_2(g) \xrightarrow{200°C} \underset{\text{formaldehyde}}{H_2CO(g)} + H_2O(g)$

b. $SO_2(g) + \frac{1}{2}O_2(g) \xrightarrow{300°} SO_3(g)$

**10.2.13 In a new process for the recovery of tin from low-grade ores, it is desired to oxidize stannous oxide, SnO, to stannic oxide, SnO_2, which is then soluble in a caustic solution. What is the heat of reaction at 90°C and 1 atm for the reaction $SnO + \frac{1}{2}O_2 \rightarrow SnO_2$? Data are as follows:

	$\Delta\hat{H}_f^\circ$ (kJ/gmol)	C_p [J/(g mol)(K)], T in K
SnO	−283.3	$39.33 + 15.15 \times 10^{-3}\,T$
SnO_2	−577.8	$73.89 + 10.04 \times 10^{-3}\,T - \dfrac{2.16 \times 10^4}{T^2}$

**10.2.14 One hundred gram moles of CO at 300°C are burned with 100 g mol of O_2 which is at 100°C; the exit gases leave at 400°C. What is the heat transfer to or from the system in kilojoules?

**10.2.15 The composition of a strain of yeast cells is determined to be $C_{3.92}H_{6.5}O_{1.94}$, and the heat of combustion was found to be -1518 kJ/g mol of yeast. Calculate the heat of formation of 100 g of the yeast.

***10.2.16 Physicians measure the metabolic rate of conversion of foodstuffs in the body by using tables that list the liters of O_2 consumed per gram of foodstuff. For a simple case, suppose that glucose reacts

$$C_6H_{12}O_6(\text{glucose}) + 6O_2(g) \rightarrow 6H_2O(l) + 6CO_2(g)$$

How many liters of O_2 would be measured for the reaction of 1 g of glucose (alone) if the conversion were 90% complete in your body? How many kilojoules per gram of glucose would be produced in the body?

Data: $\Delta\hat{H}_f^\circ$ of glucose is -1260 kJ/g mol of glucose. Ignore the fact that your body is at 37°C and assume it is at 25°C.

***10.2.17 Calculate the heat of reaction in the standard state for 1 mol of $C_3H_8(l)$ for the following reaction from the given data:

$$C_3H_8(l) + 5O_2(g) \rightarrow 3CO_2(g) + 4H_2O(l)$$

Compound	$-\Delta\hat{H}_f^\circ$ (kcal/g mol)	Vaporization at 25°C (kcal/g mol)
$C_3H_8(g)$	24.820	3.823
$CO_2(g)$	94.052	1.263
$H_2O(g)$	57.798	10.519

***10.2.18 Hydrogen is used in many industrial processes, such as the production of ammonia for fertilizer. Hydrogen also has been considered to have a potential as an energy source because its combustion yields a clean product and it is easily stored in the form of a metal hydride. Thermochemical cycles (a series of reactions resulting in a recycle of some of the reactants) can be used in the production of hydrogen from an abundant natural compound—water. One process involving a series of five steps is outlined in Figure P10.2.18. State assumptions about the states of the compounds.

$$3Fe_2O_3 + 18HCL \xrightarrow{120°C} 6FeCl_3 + 9H_2O(g) \quad (1)$$

$$6FeCl_3 \xrightarrow{420°C} 6FeCl_2 + 3Cl_2 \quad (2)$$

$$6FeCl_2 + 8H_2O(g) \xrightarrow{650°C} 2Fe_3O_4 + 12HCl + 2H_2 \quad (3)$$

$$2Fe_3O_4 + \tfrac{1}{2}O_2 \xrightarrow{350°C} 3Fe_2O_3 \quad (4)$$

$$3H_2O(g) + 3Cl_2 \xrightarrow{800°C} 6HCl + \tfrac{3}{2}O_2 \quad (5)$$

a. Calculate the standard heat of reaction for each step.
b. What is the overall reaction? What is its standard heat of reaction?

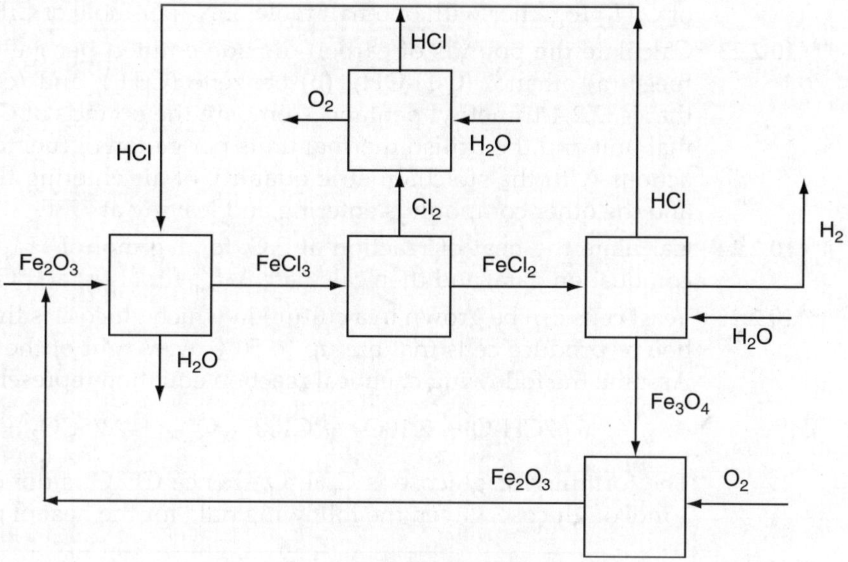

Figure P10.2.18

*****10.2.19** About 30% of crude oil is processed eventually into automobile gasoline. As petroleum prices rise and resources dwindle, alternatives must be found. However, automobile engines can be tuned so that they will run on simple alcohols. Methanol and ethanol can be derived from coal or plant biomass, respectively. While the alcohols produce fewer pollutants than gasoline, they also reduce the travel radius of a tank of fuel. What percent increase in the size of a fuel tank is needed to give an equivalent travel radius if gasoline is replaced by alcohol? Base your calculations on 40 kJ/g of gasoline having a specific gravity of 0.84, and with the product water as a gas. Make the calculations for (a) methanol; (b) ethanol.

*****10.2.20** Calculate the heat transfer for the following reaction if it takes place at constant volume and a constant temperature of 25°C:

$$C_2H_4(g) + 2H_2(g) \rightarrow 2CH_4(g)$$

*****10.2.21** If 1 lb mol of Cu and 1 lb mol of H_2SO_4 (100%) react together completely in a bomb calorimeter, how many British thermal units are absorbed (or evolved)? Assume that the products are $H_2(g)$ and $CuSO_4$. The initial and final temperatures in the bomb are 25°C.

*****10.2.22** Calculate the standard heat of reaction for the conversion of cyclohexane to benzene:

$$C_6H_{12}(g) \rightarrow C_6H_6(g) + 3H_2(g)$$

If the reactor for the conversion operates at 70% conversion of C_6H_{12}, what is the heat removed from or added to the reactor if (a) the exit gases leave at 25°C and (b) the exit gases leave at 300°C and the entering materials consist of C_6H_{12} together with one-half mole of N_2 per mole of C_6H_{12}, both at 25°C?

***10.2.23 Calculate the pounds of carbon dioxide emitted per gallon of fuel for three fuels: (a) ethanol (C_2H_5OH), (b) benzene (C_6H_6), and (c) isooctane (C_8H_{18}), that is, 2,2,4-trimethyl pentane. Compare the pounds of CO_2 per British thermal unit and the British thermal units per gallon of fuel for the respective reactions with the stoichiometric quantity of air entering the process at 100°C and the other compounds entering and leaving at 77°F.

***10.2.24 Calculate the heat of reaction at S.C. for 1 g mol of H_2(g) using the heat of combustion data, and then calculate ΔH_{rxn} for H_2(g) at 0°C.

***10.2.25 Yeast cells can be grown in a culture in which glucose is the sole source of carbon to produce cells that are up to 50% by weight of the glucose consumed. Assume the following chemical reaction equation represents the process:

$$6.67CH_2O + 2.10O_2 \rightarrow C_{3.92}H_{6.5}O_{1.94} + 2.75CO_2 + 3.42H_2O(l)$$

The formula for glucose is $C_6H_{12}O_6$; hence CH_2O is directly proportional to 1 mol of glucose. Given the following data for the heat of combustion:

	$\Delta \hat{H}_c^\circ$(kJ/g mol)	MW
Dry cells ($C_{3.92}H_{6.5}O_{1.94}$)	−1517	84.58
Glucose (CH_2O)	−2817	30.02

calculate the standard heat of reaction per 100 g of dry cells.

10.3 Integration of Heat of Formation and Sensible Heat

***10.3.1 A synthesis gas at 500°C that analyzes 6.4% CO_2, 0.2% O_2, 40.0% CO, 50.8% H_2, and the balance N_2 is burned with 40% dry excess air which is at 25°C. The composition of the flue gas which is at 720°C is 13.0% CO_2, 14.3% H_2O, 67.6% N_2, and 5.1% O_2. What was the heat transfer to or from the combustion process?

***10.3.2 Dry coke composed of 4% inert solids (ash), 90% carbon, and 6% hydrogen at 40°C is burned in a furnace with dry air at 40°C. The solid refuse at 200°C that leaves the furnace contains 10% carbon and 90% inert, and no hydrogen. The ash does not react. The Orsat analysis of the flue gas which is at 1100°C gives 13.9% CO_2, 0.8% CO, 4.3% O_2, and 81.0% N_2. Calculate the heat transfer to or from the process. Use a constant C_p for the inert of 8.5 J/g.

****10.3.3 An eight-room house requires 200,000 Btu per day to maintain the interior at 68°F. How much calcium chloride hexahydrate must be used to store the energy collected by a solar collector for one day of heating? The storage process involves heating the $CaCl_2 \cdot 6H_2O$ from 68°F to 86°F and then converting the hexahydrate to dehydrate and gaseous water:

$$CaCl_2 \cdot 6H_2O(s) \rightarrow CaCl_2 \cdot 2H_2O(s) + 4H_2O(g)$$

The water from the dehydration is evaporated during the process. Use the following data:

	$\Delta\hat{H}_f^\circ$(kJ/g mol)	C_p (J/(g)(°C)
$CaCl_2 \cdot 6H_2O(s)$	−2607.89	1.34
$CaCl_2 \cdot 2H_2O(s)$	−1402.90	0.97

10.4 The Heat (Enthalpy) of Combustion

**10.4.1 Estimate the higher heating value (HHV) and lower heating value (LHV) of the following fuels in British thermal units per pound:

a. Coal with the analysis C (80%), H (0.3%), O (0.5%), S (0.6%), and ash (18.60%)

b. Fuel oil that is 30°API and contains 12.05% H and 0.5% S

**10.4.2 Is the higher heating value of a fuel ever equal to the lower heating value? Explain.

**10.4.3 Find the higher (gross) heating value of $H_2(g)$ at 0°C.

**10.4.4 The chemist for a gas company finds a gas analyzes 9.2% CO_2, 0.4% C_2H_4, 20.9% CO, 15.6% H_2, 1.9% CH_4, and 52.0% N_2. What should the chemist report as the gross heating value of the gas?

***10.4.5 What is the higher heating value of 1 m^3 of *n*-propylbenzene measured at 25°C and 1 atm and with a relative humidity of 40%?

***10.4.6 An off-gas from a crude oil topping plant has the following composition:

Component	Vol. %
Methane	88
Ethane	6
Propane	4
Butane	2

a. Calculate the higher heating value on the following bases: (1) Btu per pound, (2) Btu per lb mole, and (3) Btu per cubic foot of off-gas measured at 60°F and 760 mm Hg.

b. Calculate the lower heating value on the same three bases indicated in part a.

**10.4.7 The label on a 43 g High Energy ("Start the day with High Energy") bar states that the bar contains 10 g of fat, 28 g of carbohydrate, and 4 g of protein.

The label also says that the bar has 200 calories per serving (the serving size is 1 bar). Does this information agree with the information about the contents of the bar?

Data:

Component	$\Delta\hat{H}_c^\circ$(kJ/g)
Carbohydrate	17.1
Fat	39.5
Protein	14

**10.4.8 Under what circumstances would the heat of formation and the heat of combustion have the same value?

***10.4.9 One of the ways to destroy chlorinated hydrocarbons in waste streams is to add a fuel (here toluene waste) and burn the mixture. In a test of the combustion apparatus, 1200 lb of a liquid mixture composed of 0.0487% hexachloroethane (C_2Cl_6), 0.0503% tetrachloroethane (C_2Cl_4), 0.2952% chlorobenzene (C_6H_5Cl), and the balance toluene was burned completely with air. What was the HHV of the mixture calculated using data for the heats of combustion in British thermal units per pound? Compare the resulting value with the one obtained from the Dulong formula. The observed value was 15,200 Btu/lb. What is one major problem with the incineration process described?

**10.4.10 Gasohol is a mixture of ethanol and gasoline used to increase the oxygen content of fuels and thus reduce pollutants from automobile exhaust. What is the heat of reaction for 1 kg of the mixture calculated using heat of combustion data for a fuel composed of 10% ethanol and the rest octane? How much is the heat of reaction reduced by adding the 10% ethanol to the octane?

CHAPTER 11

Humidity (Psychrometric) Charts and Their Use

Your objectives in studying this chapter are to be able to

1. Define and understand humidity, dry-bulb temperature, wet-bulb temperature, humidity chart, moist volume, and adiabatic cooling line
2. Use the humidity chart to determine the properties of moist air
3. Calculate enthalpy changes and solve heating and cooling problems involving moist air

Looking Ahead

As you know, air is a mixture of various gases of constant composition plus water vapor in amounts that vary from time to time and place to place. What we will do in this chapter, after first presenting some new terminology, is discuss how to apply simultaneous material and energy balances to solve problems involving air and water, such as humidification, air conditioning, water cooling, and the like.

11.1 Terminology

Following are new terms that relate to humidification and drying.

a. **Relative saturation (relative humidity)** is defined as

$$\mathcal{RS} = \frac{p_{\text{vapor}}}{p^*} = \text{relative saturation} \tag{11.1}$$

where p_{vapor} = partial pressure of the vapor in the gas mixture.

p^* = partial pressure of the vapor in the gas mixture if the gas is saturated at the given temperature of the mixture (i.e., the vapor pressure of the vapor component)

Then, for brevity, if the subscript 1 denotes the vapor:

$$\mathcal{RS} = \frac{p_1}{p^*_1} = \frac{p_1/p_{\text{tot}}}{p^*_1/p_{\text{tot}}} = \frac{V_1/V_{\text{tot}}}{V_{\text{satd}}/V_{\text{tot}}} = \frac{n_t}{n_{\text{satd}}} = \frac{\text{mass}_1}{\text{mass}_{\text{satd}}} \tag{11.2}$$

You can see that relative saturation, in effect, represents the fractional approach to the total saturation. If you listen to the radio or TV and hear the announcer say that the temperature is 25°C (77°F) and the relative humidity is 60%, he or she implies that

$$\frac{p_{H_2O}}{p^*_{H_2O}}(100) = \%\mathcal{RH} = 60$$

with both the p_{H_2O} and the $p^*_{H_2O}$ being measured at 25°C. Zero percent relative saturation means no vapor in the gas. What does 100% relative saturation mean? It means that the partial pressure of the vapor in the gas is the same as the vapor pressure of the substance.

b. **Humidity** $\mathcal{H}$ (specific humidity) is the mass (in pounds or kilograms) of water vapor per unit mass (in pounds or kilograms) of bone-dry air (some texts use moles of water vapor per mole of dry air as the humidity):

$$\mathcal{H} = \frac{m_{H_2O}}{m_{\text{dry air}}} = \frac{18p_{H_2O}}{29(p_{\text{total}} - p_{H_2O})} = \frac{18n_{H_2O}}{29(n_{\text{total}} - n_{H_2O})} \tag{11.3}$$

c. **Dry-bulb temperature** (T_{DB}) is the ordinary temperature you always have been using for a gas in degrees Fahrenheit or Celsius (or Rankine or kelvin).

d. **Wet-bulb temperature** (T_{WB}) is something new. Suppose that you put a wick, or porous cotton cloth, on the mercury bulb of a thermometer and wet the wick. Next you either (1) whirl the thermometer in the air as in Figure 11.1 (this apparatus is called a sling psychrometer when a wet-bulb and a dry-bulb thermometer are mounted together), or (2) set up a fan to blow rapidly on the bulb at a high linear velocity. What happens to the temperature recorded by the wet-bulb thermometer?

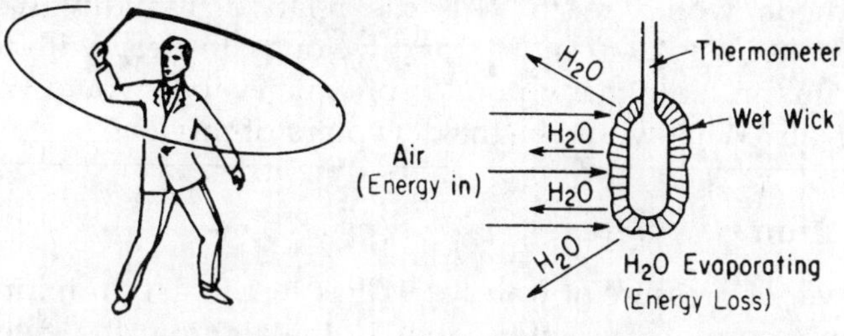

Figure 11.1 Wet-bulb temperature obtained with a sling psychrometer

As the water from the wick evaporates, the wick cools down and continues to cool until the steady-state rate of energy transferred to the wick by the air blowing on it equals the steady-state rate of loss of energy caused by the water evaporating from the wick. We say that the temperature of the bulb when the water in the wet wick is at equilibrium with the water vapor in the air is the wet-bulb temperature. (Of course, if water continues to evaporate, it eventually will disappear, and the wick temperature will rise to the dry-bulb temperature.) The equilibrium temperature for this process will lie on the 100% relative humidity curve (saturated-air curve). Look at Figure 11.2.

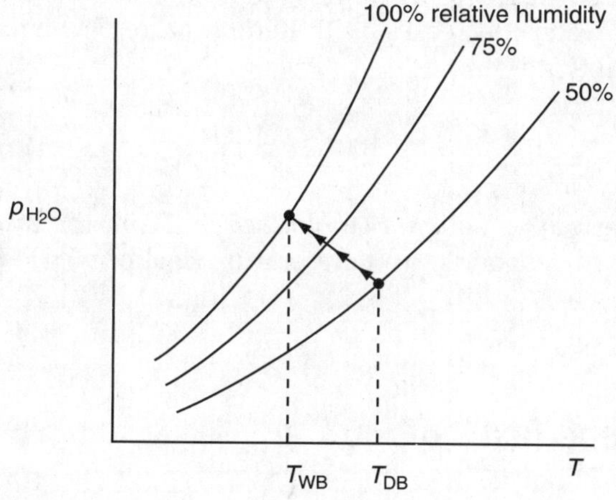

Figure 11.2 Evaporative cooling of the wick initially at T_{DB} causes a wicked thermometer at equilibrium to reach the wet-bulb temperature, T_{WB}.

Example 11.1 Application of Relative Humidity to Calculate the Dew Point

The weather report on the radio this morning was that the temperature this afternoon would reach 94°F, the relative humidity would be 43%, the barometer was 29.67 in. Hg, partly cloudy to clear, with the wind from SSE at 8 mi/hr. How many pounds of water vapor would be in 1 mi^3 of afternoon air? What would be the dew point of this air?

Solution

The vapor pressure of water at 94°F is 1.61 in. Hg. You can calculate the partial pressure of the water vapor in the air from the given percent relative humidity:

$$p_{H_2O} = (1.61 \text{ in. Hg})(0.43) = 0.692 \text{ in. Hg}$$

Basis: 1 mi^3 of water vapor at 94°F and 0.692 in. Hg

$$\frac{1 \text{ mi}^3}{} \left| \left(\frac{5280 \text{ ft}}{1 \text{ mi}}\right)^3 \right| \frac{492°\text{R}}{555°\text{R}} \left| \frac{0.692 \text{ in. Hg}}{29.92 \text{ in. Hg}} \right| \frac{1 \text{ lb mol}}{359 \text{ ft}^3} \left| \frac{18 \text{ lb H}_2\text{O}}{1 \text{ lb mol}} \right.$$

$$= 1.52 \times 10^8 \text{ lb H}_2\text{O}$$

Now the dew point is the temperature at which the water vapor in the air will first condense on cooling at *constant pressure and composition*. As the gas is cooled, you can see from Equation (11.1) that the relative humidity increases since the partial pressure of the water vapor is constant while the vapor pressure of water decreases with temperature. When the relative humidity reaches 100%

$$100 \frac{p_{H_2O}}{p^*_{H_2O}} = 100\% \quad \text{or} \quad p_{H_2O} = p^*_{H_2O} = 0.692 \text{ in. Hg}$$

the water vapor will start to condense. From the steam tables you can see that this corresponds to a temperature of about 68°F–69°F.

Self-Assessment Test

Questions

1. On a copy of the p-H chart for water plot the locus of (a) where the dry-bulb temperatures can be located; (b) where the wet-bulb temperatures can be located.
2. Would the dew point temperatures be the locus of the wet-bulb temperatures?

3. Can psychrometric charts exist for mixtures other than water-air?
4. What is the difference between the wet- and dry-bulb temperatures?
5. Can the wet-bulb temperature ever be higher than the dry-bulb temperature?

Problems

1. Apply the Gibbs phase rule to an air–water vapor mixture. How many degrees of freedom exist (for the intensive variables)? If the pressure on the mixture is fixed, how many degrees of freedom result?
2. Prepare a chart in which the vertical axis is the humidity (Equation (11.1) in kilograms of H_2O per kilogram of dry air) and the horizontal axis is the temperature (in degrees Celsius). On the chart plot the curve of 100% relative humidity for water.

11.2 The Humidity (Psychrometric) Chart

The **humidity chart**, more formally known as the **psychrometric chart**, relates the various parameters involved in making combined material and energy balances for moist air. In the Carrier chart, the type of chart we will discuss, the vertical axis (usually placed on the right-hand side) is the (specific) humidity, and the horizontal axis is the dry-bulb temperature. Examine Figure 11.3. On this chart we want to construct several other lines and curves featuring different parameters.

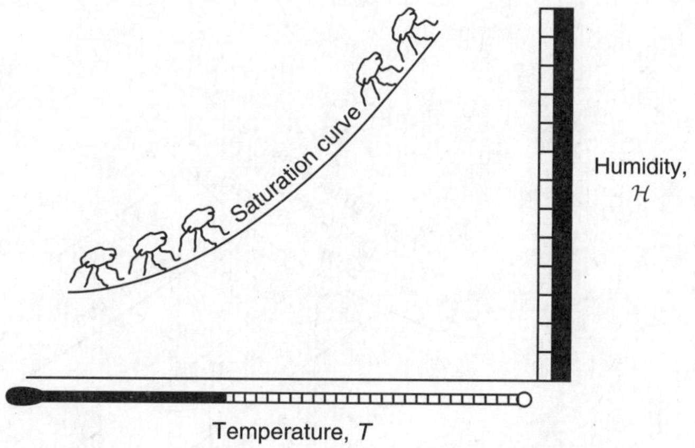

Figure 11.3 Major coordinates of the humidity chart

We will start with two of the new concepts and relations you read about in Section 11.1.

11.2.1 Wet-Bulb Line (Equation)

The equation for the **wet-bulb** lines is based on a number of assumptions, a detailed discussion of which is beyond the scope of this book. Nevertheless, as mentioned in Section 11.1, the idea of the wet-bulb temperature is based on the equilibrium between the *rates* of energy transfer to the bulb and the evaporation of water. The fundamental idea is that a large amount of air is brought into contact with a little bit of water, and that presumably the evaporation of the water leaves the temperature and humidity of the air unchanged. Only the temperature of the water changes. The equation for the wet-bulb line evolves from an energy balance that equates the heat transfer to the water to the heat of vaporization of the water. Figure 11.4 shows several such lines on the psychrometric chart for air–water. The plot of $\mathcal{H}_{WB}$ versus T_{WB}, the wet-bulb line, is approximately straight and has a negative slope.

What use is the wet-bulb line (equation)? You require two pieces of information to fix a state (point A in the humidity chart). One piece of information can be T_{DB}, $\mathcal{H}_{DB}$, $\mathcal{RS}$, and so on. The other can be T_{WB}. Where do you locate the value of T_{WB} on the humidity chart? The Carrier chart lists values of T_{WB} along the saturation curve (such as point B). You can also find the same values by projecting vertically upward from a temperature (point C) on the horizontal axis to the saturation curve (point B). If the process is a wet-bulb temperature process, all of the possible states of the process fall on the wet-bulb line.

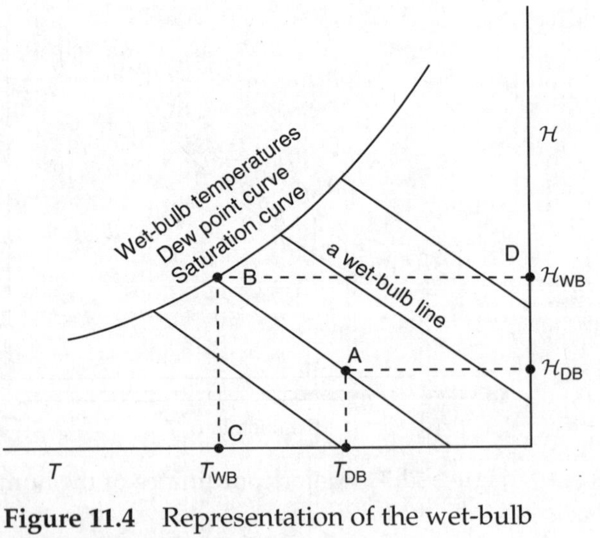

Figure 11.4 Representation of the wet-bulb process on an $\mathcal{H}$ -T chart

For example, for a known wet-bulb process, given an initial T_{DB} (along the horizontal axis) and $\mathcal{H}_{DB}$ (along the vertical axis), the combination fixes a point on the graph (A). You can follow (up to the left) the wet-bulb process (line) to the saturation curve (B). Then project vertically downward onto the temperature axis to get the related T_{WB} (C), and project horizontally to the right to the related $\mathcal{H}_{WB}$ (D). All points along the wet-bulb line are fixed by just one additional piece of information.

11.2.2 Adiabatic Cooling Line (Equation)

Another type of process of some importance occurs when **adiabatic cooling or humidification** takes place between air and water that is recycled, as illustrated in Figure 11.5. In this process the air is both cooled and humidified (its water content rises) while a little bit of the recirculated water is evaporated. Thus, makeup water is added. At **equilibrium**, in the steady state, the temperature of the exit air is the same as the temperature of the water, and the exit air is saturated at this temperature. By making an overall energy balance around the process (with $Q = 0$), you can obtain the equation for the adiabatic cooling of the air.

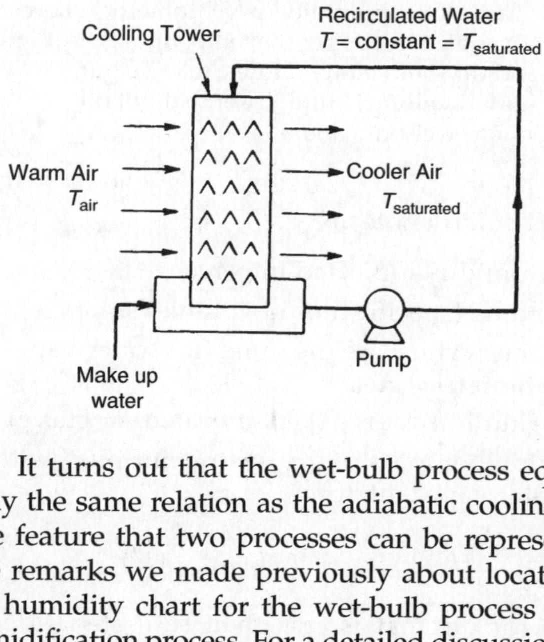

Figure 11.5 Adiabatic humidification with recycle of water

It turns out that the wet-bulb process equation, *for water only*, is essentially the same relation as the adiabatic cooling equation. Thus, you have the nice feature that two processes can be represented by the same set of lines. The remarks we made previously about locating parameters of moist air on the humidity chart for the wet-bulb process apply equally to the adiabatic humidification process. For a detailed discussion of the uniqueness of this coincidence, consult any of the references at the end of the chapter. For most other substances besides water, the respective equations will have different slopes.

In addition to the wet-bulb line being identical to the adiabatic cooling line, let's look at some of the other details found on a humidity chart in which the vertical axis is $\mathcal{H}$ and the horizontal axis is T. Look at Figure 11.6.

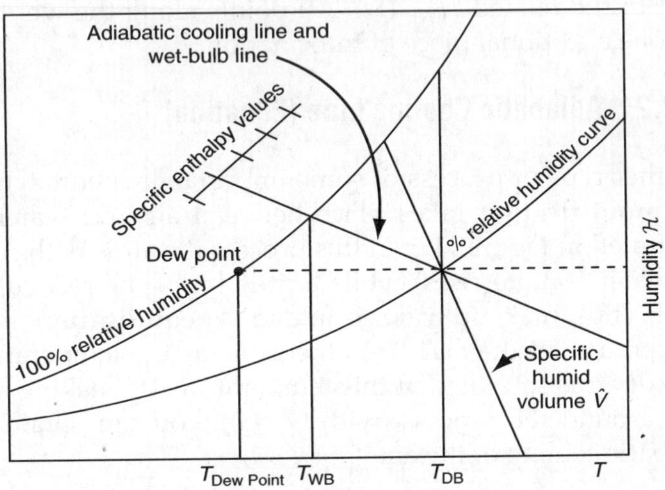

Figure 11.6 A skeleton humidity (psychrometric) chart showing typical relationships of the temperatures, dew point, wet- and dry-bulb temperatures, relative humidity, specific humid volume, humidity enthalpy, and adiabatic cooling/wet-bulb line

Lines and curves to particularly note are

1. Constant relative humidity indicated in percent
2. Constant moist volume (specific humid volume)
3. Adiabatic cooling lines which are the same (for water vapor only) as the wet-bulb or **psychrometric lines**
4. The 100% relative humidity curve (i.e., saturated-air curve)
5. The specific enthalpy per mass of dry air (not air plus water vapor) for a saturated air–water vapor mixture:

$$\Delta\hat{H} = \Delta\hat{H}_{\text{air}} + \Delta\hat{H}_{\text{H}_2\text{O vapor}}(\mathcal{H}) \tag{11.4}$$

Enthalpy adjustments for air that is less than saturated (identified by a minus sign) are shown on the chart itself by a series of curves.

Figures 11.7a and 11.7b are reproductions of the Carrier psychrometric charts.

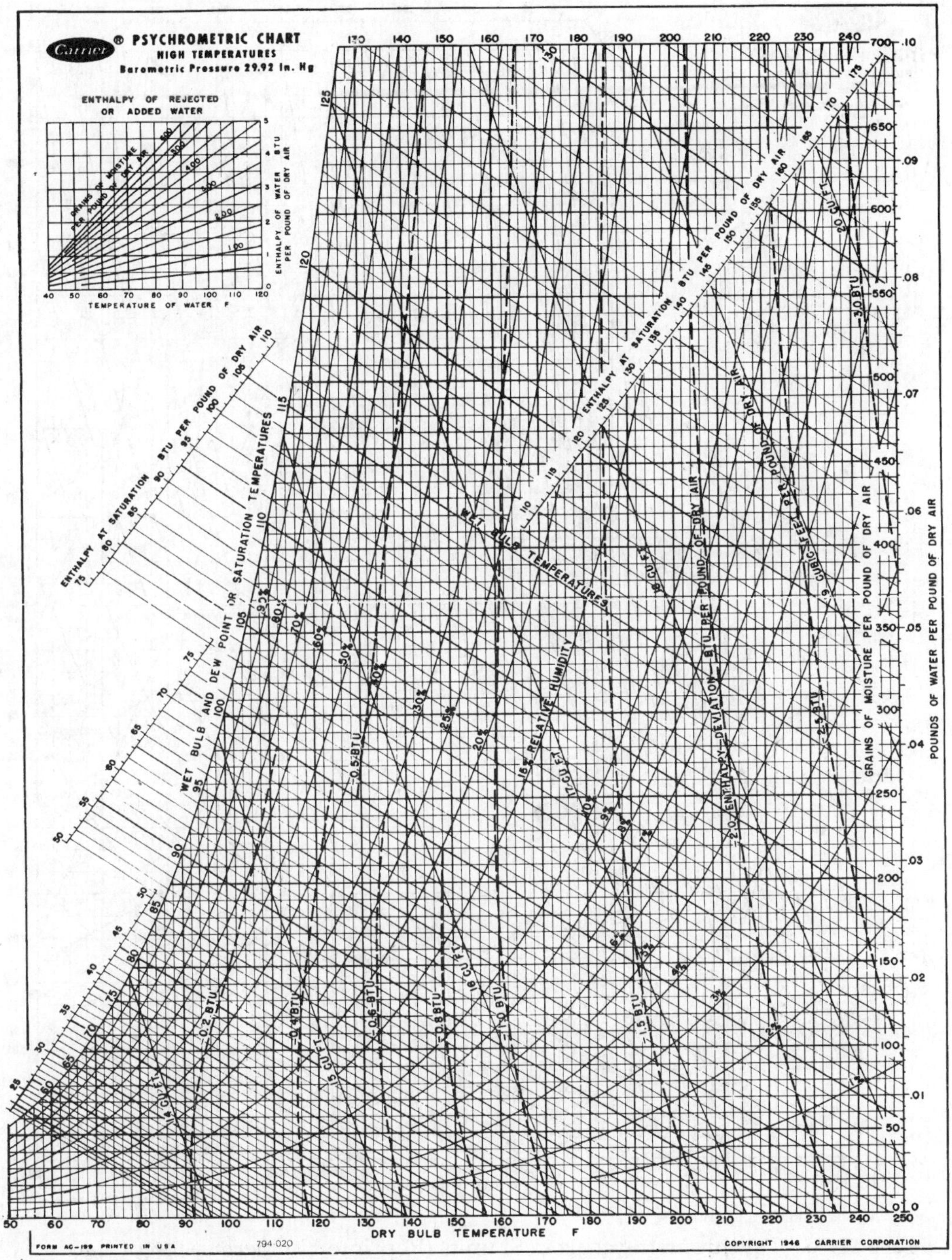

Figure 11.7a Psychrometric chart in American Engineering units (Reprinted by permission of Carrier Corporation)

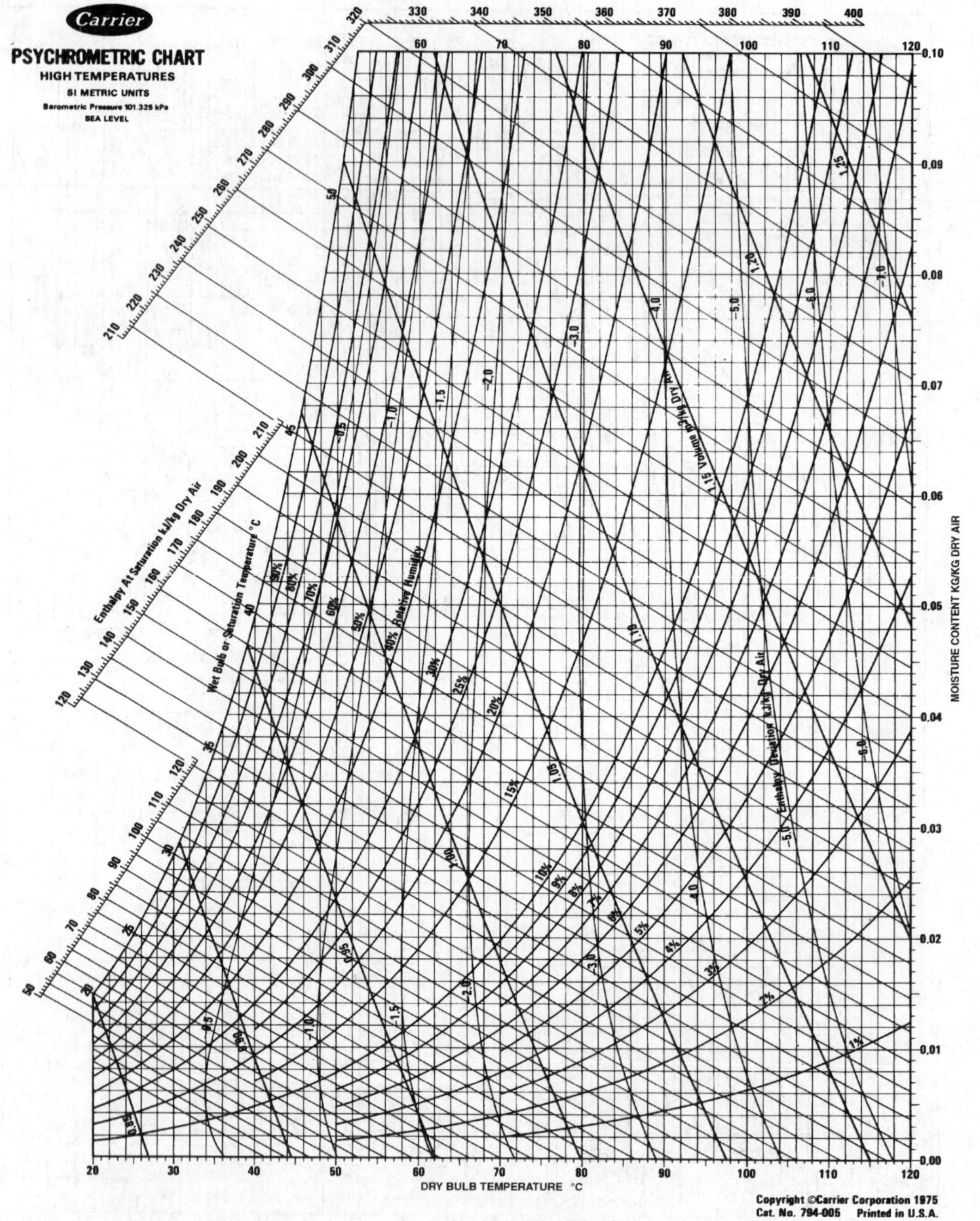

Figure 11.7b Psychrometric chart in SI units (Reprinted by permission of Carrier Corporation)

If you analyze the degrees of freedom for the intensive variables via the phase rule, you find

$$\mathcal{F} = 2 - \mathcal{P} + \mathcal{C} = 2 - 1 + 2 = 3$$

Is this a paradox in view of our previous remarks stating that only two parameters are required to fix a point on the humidity chart? No.

A presumption for the chart is that the *pressure is fixed* at 1 atm. Consequently, $\mathcal{F} = 2$, and specification of any two values of the various variables shown on the chart fix a specific point. Values for all of the other variables are correspondingly fixed.

The adiabatic cooling lines are lines of almost constant enthalpy for the entering air–water mixture, and you can use them as such without much error (1% or 2%). Follow the line on the chart up to the left where the values for the enthalpies of saturated air appear. If you want to correct a saturated enthalpy value for the deviation that exists for a less than saturated air–water vapor mixture, you can employ the enthalpy deviation lines that appear on the chart, which can be used as illustrated in the following examples.

Any process that is not a wet-bulb process or an adiabatic process with recirculated water can be analyzed using the usual material and energy balances, taking the basic data for the calculations from the humidity charts (or equations). If there is any increase or decrease in the moisture content of the air in a psychrometric process, the small enthalpy effect of the moisture added to the air or lost by the air may be included in the energy balance for the process to make it more exact, as illustrated in the examples.

Tables of the properties shown in the humidity charts exist if you need better accuracy than can be obtained via the charts, and computer programs that you can find on the Internet can be used to retrieve psychrometric data. Although we have been discussing humidity charts exclusively, charts can be prepared for mixtures of any two substances in the vapor phase, such as CCl_4 and air or acetone and nitrogen.

Now let's look at an example of getting data from the humidity chart. On the CD in the back of this book you will find an animated example of using the humidity chart to get data.

Example 11.2 Determining Properties of Moist Air from the Humidity Chart

List all of the properties you can find on the humidity chart in American Engineering units for moist air at a dry-bulb temperature of 90°F and a wet-bulb temperature of 70°F. It is reasonable to assume that someone could measure the dry-bulb temperature using a typical mercury-in-glass thermometer and the wet-bulb temperature using a sling psychrometer.

(*Continues*)

Example 11.2 Determining Properties of Moist Air from the Humidity Chart (*Continued*)

Solution

A diagram will help explain the various properties that can be obtained from the humidity chart. See Figure E11.2.

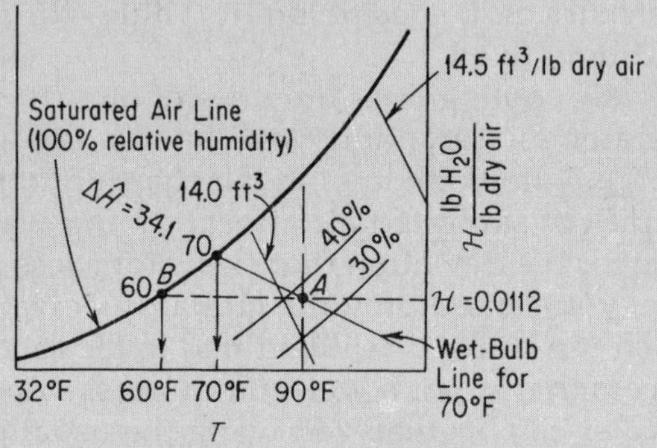

Figure E11.2

You can find the location of point A for 90°F DB (dry-bulb) and 70°F WB (wet-bulb) by following a vertical line from T_{DB} = 90°F until it crosses the wet-bulb line for 70°F. You can locate the end of a wet-bulb line by searching along the 100% humidity curve (saturation curve) until the label for the saturation temperature of 70°F is reached. Alternatively, you can proceed up a vertical line at 70°F until it intersects with the 100% humidity line.

Next, from the wet-bulb temperature of 70°F follow the adiabatic cooling line (which is the same as the wet-bulb temperature line on the humidity chart) down to the right until it intersects the 90°F DB line. Now that point A has been fixed, you can read the other properties of the moist air from the chart.

a. **Dew point.** When the air at A is cooled at constant pressure (and in effect at *constant humidity*), it eventually reaches a temperature at which the moisture begins to condense. This process is represented by a horizontal line, a constant humidity line, on the humidity chart, and the dew point is located at B, or about 60°F.

b. **Relative humidity.** By interpolating between the 40% $\mathcal{RH}$ and 30% $\mathcal{RH}$ lines, you can find that point A is at about 37% $\mathcal{RH}$.

c. **Humidity ($\mathcal{H}$).** You can read the humidity from the right-hand ordinate as 0.0112 lb H_2O/lb of dry air.

d. **Humid volume.** By interpolation again between the 14.0 ft^3/lb and the 14.5 ft^3/lb lines, you can find the humid volume to be 14.1 ft^3/lb of dry air.

e. **Enthalpy.** The enthalpy value for saturated air with a wet-bulb temperature of 70°F is $\Delta\hat{H} = 34.1$ Btu/lb of dry air (a more accurate value can be obtained from psychrometric tables if needed). The enthalpy deviation shown by the dashed curves in Figure 11.7a (not shown in Figure E11.1) for less than saturated air is about -0.2 Btu/lb of dry air; consequently, the actual enthalpy of air at 37% $\mathcal{RH}$ is $34.0 - 0.2 = 33.8$ Btu/lb of dry air.

Frequently Asked Questions

1. For what range of temperatures can a humidity chart be applied? Air conditioning applications range from about -10°C to 50°C. Outside this range of temperatures (and for other pressures), applications may require amended charts.
2. Can air and water vapor be treated as ideal gases in that range? Yes, with an error less than 0.2%.
3. Will the enthalpy of the water vapor be a function of the temperature only? Yes, since it is assumed to be an ideal gas.
4. How can you achieve greater accuracy than the values read from the humidity chart? Use the equations listed in Sections 11.1 and 11.2, or use a computer program.

Self-Assessment Test

Questions

1. What properties of an air–water vapor mixture are displayed on a humidity chart?
2. Explain what happens to both the air and water vapor when an adiabatic cooling process occurs.
3. Wet towels are to be dried in a laundry by blowing hot air over them. Can the Carrier humidity chart be used to help design the dryer?
4. Can a humidity chart use values per mole rather than values per mass?

Problems

1. Air at a dry-bulb temperature of 200°F has a humidity of 0.20 mol H_2O/mol of dry air.
   a. What is its dew point?
   b. If the air is cooled to 150°F, what is its dew point?

2. Air at a dry-bulb temperature of 71°C has a wet-bulb temperature of 60°C.
   a. What is its percentage relative humidity?
   b. If this air is passed through a washer-cooler, what would be the lowest temperature to which the air could be cooled without using refrigerated water?
3. Estimate for air at 70°C dry-bulb temperature, 1 atm, and 15% relative humidity:
   a. kg H_2O/kg of dry air
   b. m^3/kg of dry air
   c. dew point (in °C)
4. Calculate the following properties of moist air at 1 atm and compare with values read from the humidity chart:
   a. The humidity of saturated air at 120°F
   b. The enthalpy of air in part a per pound of dry air
   c. The volume per pound of dry air of the air in part a
   d. The humidity of air at 160°F with a wet-bulb temperature of 120°F

Thought Problem

The use of home humidifiers has been promoted in advertisements as a means of providing more comfort in houses with the thermostat turned down. "Humidification makes life more comfortable and prolongs the life of furniture." Many advertisers describe a humidifier as an energy-saving device because it allows lower temperatures (4°F or 5°F lower) with comfort.

Is this true?

Discussion Questions

1. In cold weather, water vapor exhausted from cooling towers condenses and fog is formed as a plume. What are one or two *economically practical* methods of preventing such cooling tower fog?
2. In winter in the North, ice that forms at the intake of a rotating air compressor can break off and be ingested by the compressor with resulting damage to the blades and internal parts. What can you recommend doing to prevent such accidents?

11.3 Applications of the Humidity Chart

Quite a few industrial processes exist for which you can involve the properties found on a humidity chart, including

1. Drying: Dry air enters and moist air leaves the process.
2. Humidification: Liquid water is vaporized into moist air.
3. Combustion: Moist air enters a process and additional water is added to the moist air from the combustion products.
4. Air conditioning: Moist air is cooled.
5. Condensation: Moist air is cooled to the saturation temperature.

Example 11.3 Heating at Constant Humidity in a Home Furnace

Moist air at 38°C and 49% $\mathcal{RH}$ is heated in your home furnace to 86°C. How much heat has to be added per cubic meter of initial moist air, and what is the final dew point of the air?

Solution

As shown in Figure E11.3, the process has an input state at point A at T_{DB} = 38°C and 49% $\mathcal{RH}$, and an output state at point B which is located at the intersection of a horizontal line of constant humidity with the vertical line up from T_{BD} = 86°C. The dew point is unchanged in this process because the humidity is unchanged and is located at C at 24.8°C.

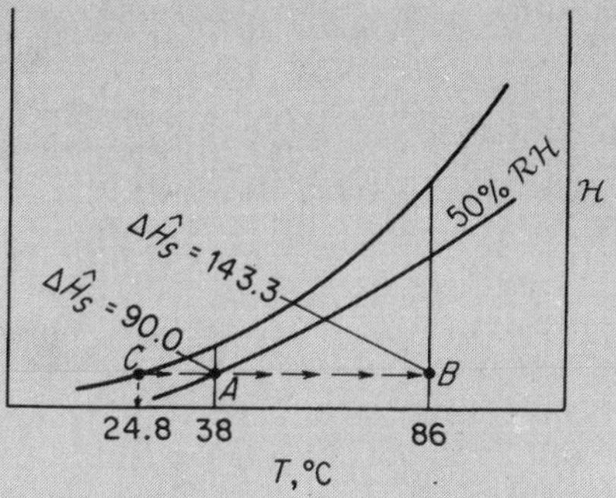

Figure E11.3

The enthalpy values are as follows (all in kilojoules per kilogram of dry air):

Point	$\Delta\hat{H}_{satd}$	δH	$\Delta\hat{H}_{actual}$
A	90.0	−0.5	89.5
B	143.3	−3.3	140.0

The value for the reduction of $\Delta\hat{H}_{satd}$ via δH is obtained from the enthalpy deviation lines (not shown in Figure E11.2) whose values are printed about a quarter of the way down from the top of the chart. Also, at A the volume of the moist air is 0.91 m^3/kg of dry air. Consequently, the heat added is (the energy balance reduces to $\hat{Q} = \Delta\hat{H}$) 140.0 − 89.5 = 50.5 kJ/kg of dry air.

$$\frac{50.5\text{ kJ}}{\text{kg dry air}}\bigg|\frac{1\text{ kg dry air}}{0.91\text{ m}^3} = 55.5\text{ kJ/m}^3\text{ initial moist air}$$

Example 11.4 Cooling and Humidification Using a Water Spray

One way of adding moisture to air is by passing it through water sprays or air washers. See Figure E11.4a. Normally, the water used is recirculated rather than wasted. Then, in the steady state, the water is at its adiabatic saturation temperature, which is the same as the wet-bulb temperature. The air passing through the washer is cooled, and if the contact time between the air and the water is long enough, the air will be at the wet-bulb temperature also.

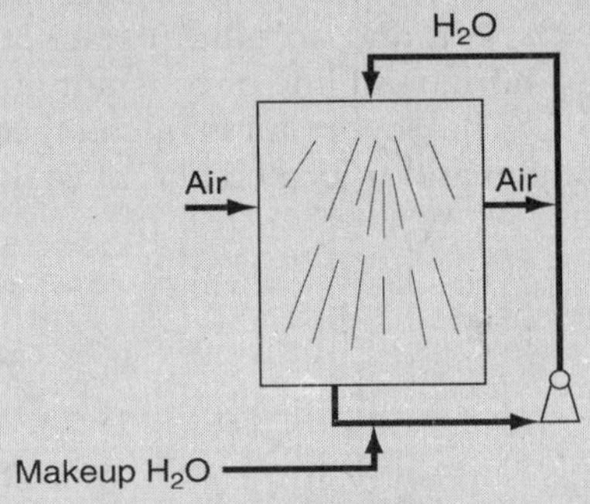

Figure E11.4a

However, we shall assume that the washer is small enough so that the air does not reach the wet-bulb temperature; instead, the following conditions prevail:

	T_{DB}(°C)	T_{WB}(°C)
Entering air:	40	22
Exit air:	27	

Find the moisture added in kilograms per kilogram of dry air going through the humidifier.

Solution

The whole process is assumed to be adiabatic. As shown in Figure E11.4b, the process inlet conditions are at *A*. The outlet state is at B, which occurs at the intersection of the adiabatic cooling line (the same as the wet-bulb line) with the vertical line at T_{DB} = 27°C. The wet-bulb temperature remains constant at 22°C. Humidity values (from Figure 11.7b) are

$$\mathcal{H}_B = 0.0145 \qquad \mathcal{H}_A = 0.0093 \qquad \text{where } \mathcal{H}\left(\frac{\text{kg H}_2\text{O}}{\text{kg dry air}}\right)$$

$$\text{Difference: } 0.0052 \frac{\text{kg H}_2\text{O}}{\text{kg dry air}} \text{ added}$$

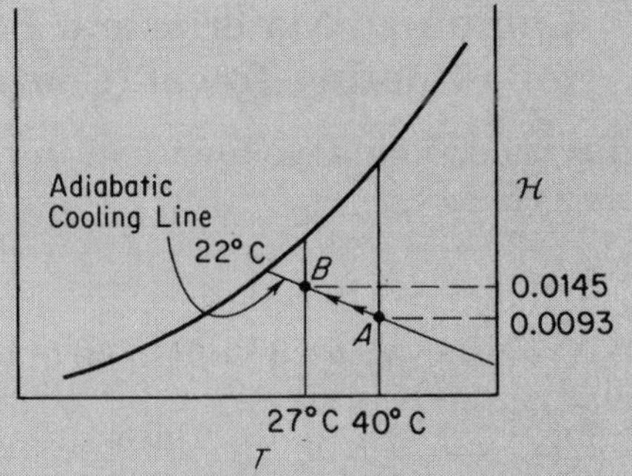

Figure E11.4b

Example 11.5 Combined Material and Energy Balances for a Cooling Tower

You have been requested to redesign a water cooling tower that has a blower with a capacity of 8.30×10^6 ft^3/hr of moist air. The moist air enters at 80°F and a wet-bulb temperature of 65°F. The exit air is to leave at 95°F and 90°F wet-bulb. How much water can be cooled in pounds per hour if the water to be cooled enters the tower at 120°F, leaves the tower at 90°F, and is not recycled?

Solution

Figure E11.5 shows the process and the corresponding states on the humidity chart.

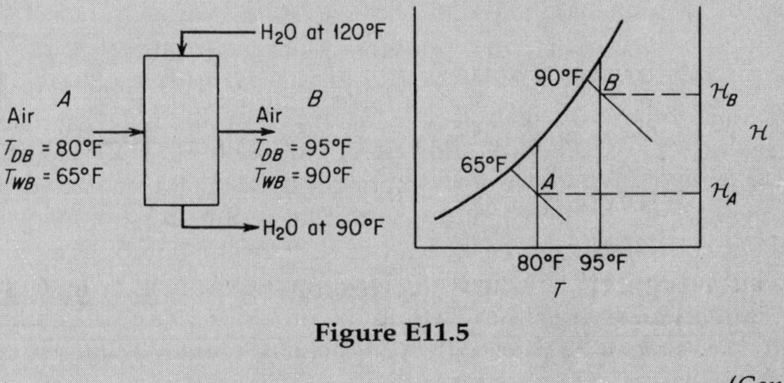

Figure E11.5

(Continues)

Example 11.5 Combined Material and Energy Balances for a Cooling Tower (*Continued*)

Enthalpy, humidity, and humid volume data for the air taken from the humidity chart are as follows:

	A	B
$\mathcal{H}\left(\frac{\text{lb H}_2\text{O}}{\text{lb dry air}}\right)$	0.0098	0.0297
$\mathcal{H}\left(\frac{\text{grains H}_2\text{O}}{\text{lb dry air}}\right)$	69	208
$\Delta\hat{H}\left(\frac{\text{Btu}}{\text{lb dry air}}\right)$	$30.05 - 0.12 = 29.93$	$55.93 - 0.10 = 55.83$
$\hat{V}\left(\frac{\text{ft}^3}{\text{lb dry air}}\right)$	13.82	14.65

The cooling water exit rate can be obtained from an energy balance around the process.

$$\text{Basis: } 8.30 \times 10^6 \text{ ft}^3/\text{hr of moist air} \equiv 1 \text{ hr}$$

The entering air is

$$\frac{8.30 \times 10^6 \text{ ft}^3}{} \left| \frac{1 \text{ lb dry air}}{13.82 \text{ ft}^3} \right. = 6.01 \times 10^5 \text{ lb dry air}$$

The relative enthalpy of the entering water stream per pound is (the reference temperature for the water stream is 32°F and 1 atm)

$$\Delta\hat{H} = C_{p_{\text{H}_2\text{O}}}\Delta T = (120 - 32)(1) = 88 \text{ Btu/lb H}_2\text{O}$$

and that of the exit stream is $90 - 32 = 58$ Btu/lb H_2O. (The value from the steam tables at 120°F for liquid water of 87.92 Btu/lb H_2O is slightly different since it represents water at its vapor pressure (1.69 psia) based on reference conditions of 32°F and liquid water at its vapor pressure.) Any other reference datum could be used instead of 32°F for the liquid water. For example, if you chose 90°F, one water stream would not have to be taken into account because its relative enthalpy would be zero. In any case, the enthalpies of the reference state will cancel when you calculate enthalpy differences.

The transfer of water to the air is

$$0.0297 - 0.0098 = 0.0199 \text{ lb } H_2O/\text{lb dry air}$$

a. Material balance for the liquid water stream: Let W = pounds of H_2O entering the tower in the water stream per pound of dry air. Then

$$W - 0.0199 = \text{lb } H_2O \text{ leaving tower in the water stream per lb dry air}$$

b. Material balance for the dry air:

$$6.01 \times 10^5 \text{ lb in} = 6.01 \times 10^5 \text{ lb dry air out}$$

c. Energy balance (enthalpy balance) around the entire process (although the reference temperature for the moist air (0°F) is not the same as that for liquid water (32°F), the reference enthalpies cancel in the calculations as mentioned above):

Air and water in air entering **Water stream**

$$\frac{29.93 \text{ Btu}}{1 \text{ lb dry air}}\left|\frac{6.01 \times 10^5 \text{ lb dry air}}{}\right. + \frac{88 \text{ Btu}}{1 \text{ lb } H_2O}\left|\frac{W \text{ lb } H_2O}{1 \text{ lb dry air}}\right|\frac{6.01 \times 10^5 \text{ lb dry air}}{}$$

Air and water in air leaving **Water stream leaving**

$$= \frac{55.83 \text{ Btu}}{\text{lb dry air}}\left|\frac{6.01 \times 10^5 \text{ lb dry air}}{}\right. + \frac{58 \text{ Btu}}{\text{lb } H_2O}\left|\frac{(W - 0.0199)1 \text{ lb } H_2O}{\text{lb dry air}}\right|\frac{6.01 \times 10^5 \text{ lb dry air}}{}$$

Simplifying this expression:

$$29.93 + 88W = 55.83 + 58(W - 0.0199)$$

$$W = 0.825 \text{ lb } H_2O/\text{lb dry air}$$

$$W - 0.0199 = 0.805 \text{ lb } H_2O/\text{lb dry air}$$

The total water leaving the tower is

$$\frac{0.805 \text{ lb } H_2O}{\text{lb dry air}}\left|\frac{6.01 \times 10^5 \text{ lb dry air}}{\text{hr}}\right. = 4.83 \times 10^5 \text{ lb/hr}$$

Self-Assessment Test

Questions

1. In combustion calculations with air in previous chapters you usually neglected the moisture in the air. Is this assumption reasonable?
2. Is the use of the humidity charts restricted to adiabatic cooling or wet-bulb processes?

Problem

A process that takes moisture out of the air by passing the air through water sprays sounds peculiar but is perfectly practical as long as the water temperature is below the dew point of the air. Equipment such as that shown in Figure SAT11.3P1 would do the trick. If the entering air has a dew point of 70°F and is at 40% $\mathcal{RH}$, how much heat has to be removed by the cooler, and how much water vapor is removed, if the exit air is at 56°F with a dew point of 54°F?

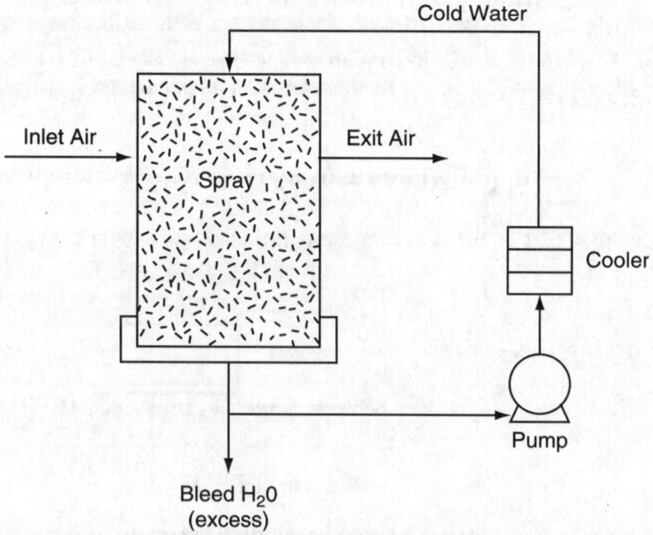

Figure SAT11.3P1

Thought Problems

1. Cooling systems known as "swamp coolers" are a low-cost, environmentally friendly technology based on evaporative cooling—the natural evaporation of water, the same process that cools a sweating human body. In evaporative air conditioning's simplest form, hot air is drawn through a water spray or through wet

porous pads. The exit air becomes cooler and more humid as it takes up water. The only moving parts requiring electrical power are a fan and a small water pump that moves the water to the top of the system.

Would a swamp cooler be an effective cooler in a swamp?

2. In adiabatic operation of a cooling tower, makeup water has to be added to compensate for the water evaporated. As a result, a number of operating problems occur that would not occur without the vaporization of the water. What are some that you think might occur?

Discussion Question

In a small office building or a house, you have a choice between using central air conditioning and window air conditioners. List some of the factors that govern the use of one type of system versus the other; for example, under what circumstances are window units preferred to central air conditioning?

Looking Back

In this chapter we described the structure of and information obtained from humidity charts and showed how they can be used in material and energy balance calculations.

Glossary

Dry-bulb temperature The usual temperature of a gas or liquid.

Humid heat The heat capacity of an air–water vapor mixture per mass of bone-dry air.

Humidity chart See **psychrometric chart.**

Humid volume The volume of air including the water vapor per mass of bone-dry air.

Psychrometric chart A chart showing the humidity versus temperature along with all of the other properties of moist air.

Psychrometric line See **wet-bulb line.**

Wet-bulb line The representation on the humidity chart of the energy balance in which the heat transfer to water from the air is assumed to equal the enthalpy of vaporization of liquid water.

Wet-bulb temperature The temperature reach at equilibrium for the vaporization of a small amount of water into a large amount of air.

Supplementary References

Barenbrug, A. W. T. *Psychrometry and Psychrometric Charts*, 3d ed., Chamber of Mines of South Africa (1991).

Bullock, C. E. "Psychrometric Tables," in *ASHRAE Handbook Product Directory*, paper No. 6, American Society of Heating, Refrigeration, and Air Conditioning Engineers, Atlanta (1977).

McCabe, W. L., J. C. Smith, and P. Harriott. *Unit Operations of Chemical Engineering*, McGraw-Hill, New York (2001).

McMillan, H. K., and J. Kim. "A Computer Program to Determine Thermodynamic Properties of Moist Air," *Access*, **36** (January, 1986).

Shallcross, D. C. *Handbook of Psychrometric Charts—Humidity Diagrams for Engineers*. Kluwer Academic Publishers (1997).

Treybal, R. E. *Mass Transfer Operations*, 3d ed., McGraw-Hill, New York (1980).

Wilhelm, L. R. "Numerical Calculation of Psychrometric Properties in SI Units," *Trans. ASAE*, **19**, 318 (1976).

Web Sites

www.chempute.com/psychro.htm
www.handsdownsoftware.com
www.usatoday.com/weather/whumcalc.htm

Problems

Section 11.1 Terminology

***11.1.1** If a gas at 60.0°C and 101.6 kPa absolute has a molal humidity of 0.030, determine (a) the relative humidity and (b) the dew point of the gas (in degrees Celsius).

****11.1.2** What is the relative humidity of 28.0 m^3 of wet air at 27.0°C that is found to contain 0.636 kg of water vapor?

****11.1.3** Air at 80°F and 1 atm has a dew point of 40°F. What is the relative humidity of this air? If the air is compressed to 2 atm and 58°F, what is the relative humidity of the resulting air?

****11.1.4** If a gas at 140°F and 30 in. Hg absolute has a molal humidity of 0.03 mole of H_2O per mole of dry air, calculate (a) the relative humidity (percent) and (b) the dew point of the gas (degrees Fahrenheit).

****11.1.5** A wet gas at 30°C and 100.0 kPa with a relative humidity of 75.0% was compressed to 275 kPa, then cooled to 20°C. How many cubic meters of the original gas were compressed if 0.341 kg of condensate (water) was removed from the separator that was connected to the cooler?

***11.1.6 A constant volume bomb contains air at 66°F and 21.2 psia. One pound of liquid water is introduced into the bomb. The bomb is then heated to a constant temperature of 180°F. After equilibrium is reached, the pressure in the bomb is 33.0 psia. The vapor pressure of water at 180°F is 7.51 psia.

a. Did all of the water evaporate?
b. Compute the volume of the bomb in cubic feet.
c. Compute the humidity of the air in the bomb at the final conditions in pounds of water per pound of air.

***11.1.7 A dryer must remove 200 kg of H_2O per hour from a certain material. Air at 22°C and 50% relative humidity enters the dryer and leaves at 72°C and 80% relative humidity. What is the weight (in kilograms) of bone-dry air used per hour? The barometer reads 103.0 kPa.

***11.1.8 Thermal pollution is the introduction of waste heat into the environment in such a way as to adversely affect environmental quality. Most thermal pollution results from the discharge of cooling water into the surroundings. It has been suggested that power plants use cooling towers and recycle water rather than dump water into streams and rivers. In a proposed cooling tower, air enters and passes through baffles over which warm water from the heat exchanger falls. The air enters at a temperature of 80°F and leaves at a temperature of 70°F. The partial pressure of the water vapor in the air entering is 5 mm Hg and the partial pressure of the water vapor in the air leaving the tower is 18 mm Hg. The total pressure is 740 mm Hg. Calculate:

a. The relative humidity of the air–water vapor mixture entering and of the mixture leaving the tower
b. The percentage composition by volume of the moist air entering and of that leaving
c. The percentage composition by weight of the moist air entering and of that leaving
d. The percent humidity of the moist air entering and leaving
e. The pounds of water vapor per 1000 ft^3 of mixture both entering and leaving
f. The pounds of water vapor per 1000 ft^3 of vapor-free air both entering and leaving
g. The weight of water evaporated if 800,000 ft^3 of air (at 740 mm and 80°F) enters the cooling tower per day

Section 11.2 The Humidity (Psychrometric) Chart

*11.2.1 Autumn air in the deserts of the southwestern United States during the day is typically moderately hot and dry. If a dry-bulb temperature of 27°C and a wet-bulb temperature of 17°C are measured for the air at noon:

a. What is the dew point?
b. What is the percent relative humidity?
c. What is the humidity?

***11.2.2** Under what conditions are the dry-bulb, wet-bulb, and dew point temperatures equal?

***11.2.3** Explain how you locate the dew point on a humidity chart for a given state of moist air.

***11.2.4** Under what conditions can the dry-bulb and wet-bulb temperatures be equal?

***11.2.5** Under what conditions are the adiabatic saturation temperature and the wet-bulb temperature the same?

***11.2.6** Moist air has a humidity of 0.020 kg of H_2O/kg of air. The humid volume is 0.90 m^3/kg of air.

a. What is the dew point?
b. What is the percent relative humidity?

***11.2.7** Use the humidity chart to estimate the kilograms of water vapor per kilogram of dry air when the dry-bulb temperature is 30°C and the relative humidity is 65%.

***11.2.8** Urea is produced in cells as a product of protein metabolism. From the cells it flows through the circulatory system, is extracted in the kidneys, and is excreted in the urine. In an experiment the urea was separated from the urine using ethyl alcohol and dried in a stream of carbon dioxide. The gas analysis at 40°C and 100 kPa was 10% alcohol by volume and the rest carbon dioxide. Determine (a) the grams of alcohol per gram of CO_2, and (b) the percent relative saturation.

***11.2.9** What is the difference between the constant enthalpy and constant wet-bulb temperature lines on the humidity chart?

***11.2.10** What does a home air conditioning unit do besides cool the air in the house?

***11.2.11** Moist air at 100 kPa, a dry-bulb temperature of 90°C, and a wet-bulb temperature of 46°C is enclosed in a rigid container. The container and its contents are cooled to 43°C.

a. What is the molar humidity of the cooled moist air?
b. What is the final total pressure in atmospheres in the container?
c. What is the dew point in degrees Celsius of the cooled moist air?

***11.2.12** Humid air at 1 atm has a dry-bulb temperature of 180°F and a wet-bulb temperature of 120°F. The air is then cooled at 1 atm to a dry-bulb temperature of 115°F. Calculate the enthalpy change per pound of dry air.

***11.2.13** What is the lowest temperature that air can attain in an evaporative cooler if it enters at 1 atm, 29°C, and 40% relative humidity?

****11.2.14** The air supply for a dryer has a dry-bulb temperature of 32°C and a wet-bulb temperature of 25.5°C. It is heated to 90°C by coils and blown into the dryer. In the dryer, it cools along an adiabatic cooling line as it picks up moisture from the dehydrating material and leaves the dryer fully saturated.

a. What is the dew point of the initial air?
b. What is its humidity?

c. What is its percent relative humidity?
d. How much heat is needed to heat 100 m^3 of initial air to 90°C?
e. How much water will be evaporated per 100 m^3 of air entering the dryer?
f. At what temperature does the air leave the dryer?

Section 11.3 Applications of the Humidity Chart

**11.3.1 A rotary dryer operating at atmospheric pressure dries 10 tons/day of wet grain at 70°F, from a moisture content of 10% to 1% moisture. The airflow is countercurrent to the flow of grain, enters at 225°F dry-bulb and 110°F wet-bulb temperature, and leaves at 125°F dry-bulb. See Figure P11.3.1. Determine (a) the humidity of the entering and leaving air if the latter is saturated; (b) the water removal in pounds per hour; (c) the daily product output in pounds per day; (d) the heat input to the dryer.

Assume that there is no heat loss from the dryer, that the grain is discharged at 110°F, and that its specific heat is 0.18.

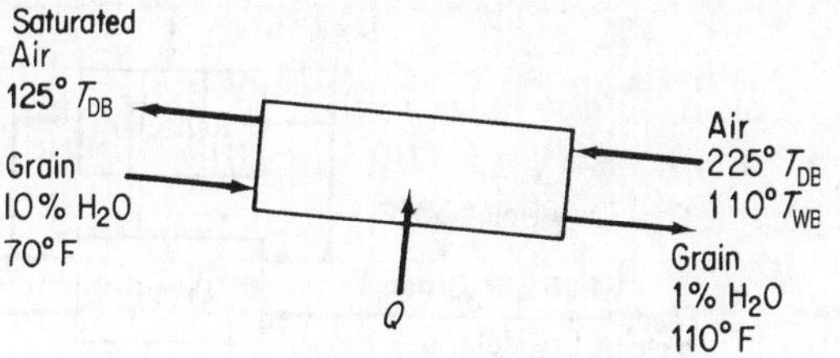

Figure P11.3.1

**11.3.2 A stream of warm air with a dry-bulb temperature of 40°C and a wet-bulb temperature of 32°C is mixed adiabatically with a stream of saturated cool air at 18°C. The dry air mass flow rates of the warm and cool airstreams are 8 and 6 kg/s, respectively. Assuming a total pressure of 1 atm, determine (a) the temperature, (b) the specific humidity, and (c) the relative humidity of the mixture.

**11.3.3 Temperatures (in degrees Fahrenheit) taken around a forced-draft cooling tower are as follows:

	In	Out
Air	85	90
Water	102	89

The wet-bulb temperature of the entering air is 77°F. Assuming that the air leaving the tower is saturated, calculate (a) the humidity of the entering air, (b) the pounds of dry air through the tower per pound of water into the tower, and (c) the percentage of water vaporized in passing through the tower.

**11.3.4 A person uses energy simply by breathing. For example, suppose that you breathe in and out at the rate of 5.0 L/min, and that air with a relative humidity of 30% is inhaled at 25°C. You exhale at 37°C, saturated. What must be the heat transfer to the lungs from the blood system in kilojoules per hour to maintain these conditions? The barometer reads 97 kPa.

***11.3.5 A dryer produces 180 kg/hr of a product containing 8% water from a feed stream that contains 1.25 g of water per gram of dry material. The air enters the dryer at 100°C dry-bulb and a wet-bulb temperature of 38°C; the exit air leaves at 53°C dry-bulb and 60% relative humidity. Part of the exit air is mixed with the fresh air supplied at 21°C, 52% relative humidity, as shown in Figure P11.3.5. Calculate the air and heat supplied to the heater, neglecting any heat lost by radiation, used in heating the conveyor trays, and so forth. The specific heat of the product is 0.18.

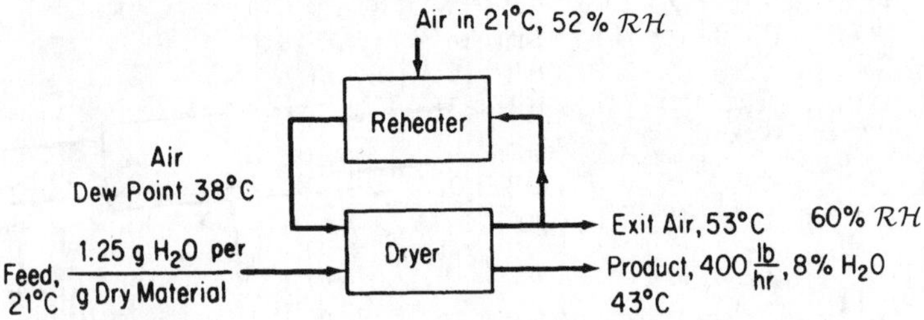

Figure P11.3.5

***11.3.6 Air, dry-bulb 38°C, wet-bulb 27°C, is scrubbed with water to remove dust. The water is maintained at 24°C. Assume that the time of contact is sufficient to reach complete equilibrium between air and water. The air is then heated to 93°C by passing it over steam coils. It is then used in an adiabatic rotary dryer from which it issues at 49°C. It may be assumed that the material to be dried enters and leaves at 46°C. The material loses 0.05 kg H_2O per kilogram of product. The total product is 1000 kg/hr.

a. What is the humidity (1) of the initial air? (2) after the water sprays? (3) after reheating? (4) leaving the dryer?
b. What is the percent humidity at each of the points in part a?
c. What is the total weight of dry air used per hour?

d. What is the total volume of air leaving the dryer?
e. What is the total amount of heat supplied to the cycle in joules per hour?

***11.3.7 In the final stages of the industrial production of penicillin, air enters a dryer having a dry-bulb temperature of 34°C and a wet-bulb temperature of 17°C. The moist air flows over the penicillin at 1 atm at the rate of 4500 m^3/hr. The air exits at 34°C. The penicillin feed enters at 34°C with a moisture content of 80% and exits at 50%.

a. How much water is evaporated from the penicillin per hour?
b. What is the enthalpy change in kilojoules per hour of the air from inlet to exit?
c. How many kilograms of the entering penicillin are dried per hour?

Assume that the properties of the wet penicillin are those of the water content of the mixture.

***11.3.8 A plant has a waste product that is too wet to burn for disposal. To reduce the water content from 63.4% to 22.7%, it is passed through a rotary kiln dryer. Prior to flowing through the kiln, the air is first preheated in a heater by steam coils. To conserve energy, part of the exit gases from the kiln are recirculated and mixed with the heated airstream as they enter the kiln. An engineering intern stood by the entrance to the air heater and measured a dry-bulb temperature of 80°F and a wet-bulb temperature of 54°F. She then moved to the exit of the kiln and found that the dry-bulb temperature was 120°F and the wet-bulb temperature was 94°F. She next drew a sample of the moist air entering the kiln itself and determined that the humidity was 0.0075 lb water per lb of dry air. Finally, she looked a the weather data on the TV and saw that the barometer reading was 29.92 in. Hg.

Calculate (a) the percentage of the air leaving the kiln that is recirculated; (b) the pounds of inlet air per ton of dried waste; (c) the cubic feet of moist air leaving the kiln per ton of dried waste.

****11.3.9 Air is used for cooling purposes in a plant process. The scheme shown in Figure P11.3.9 has been proposed as a way to cool the air and continuously reuse it. In the process, the air is heated and its water content remains constant. It is estimated that the air leaving the process will be at 234°F regardless of the entering temperature or the throughput rate. The air must remove 425,000 Btu/hr from the process. Water for the spray tower is available at 100°F. This tower operates adiabatically with the air leaving saturated. Determine the following: (a) the wet-bulb temperature and percent relative humidity of the air leaving the process; (b) the temperature and dew point of the air leaving the cooler; (c) the psychrometric chart for this cycle; (d) the air circulation rate needed in pounds bone-dry air (BDA) per hour.

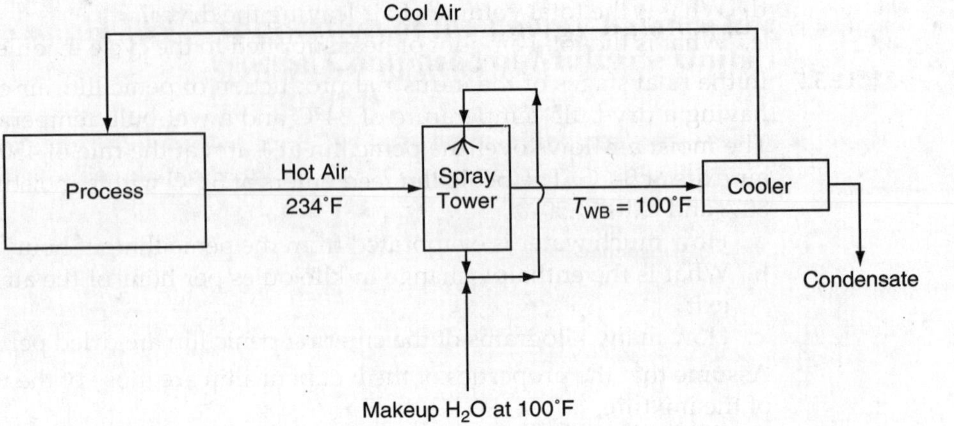

Figure P11.3.9

PART V

SUPPLEMENTARY MATERIAL

See the CD for pages 683–826, Chapters 12 through 17

APPENDIXES

Chapters 12 through 17, pages 683 through 826, are found on the CD

APPENDIX A

Answers to Supplemental Questions and Problems

Chapter 2

Section 2.1

Q1. (c)
Q2. (a)
Q3. Derived
P1. (a), (d), and (e) are correct.

Section 2.2

Q1. A conversion factor in the American Engineering system of units
Q2. Yes
Q3. lb_f is force and lb_m is mass, and the units are different.
Q4. In SI the magnitudes of many of the units are scaled on the basis of 10, but not in the AE system. Consequently, the units are often ignored in making conversions in SI.
Q5. In the AE system of units, 1 lb_f; yes; no
Q6. 1000 kg
P1. 1, dimensionless
P2. 5.38 hr
P3. (a) 5.96 kg/m; (b) 16.0 kg/(m^3)(s)
P4. 1.06×10^{-3} lb_m/(ft)(s)
P5. 4.16×10^{-3} m^3

Section 2.3

Q1. All additive terms on the right-hand side of an equation must have the same dimensions as those on the left-hand side.
Q2. All the units cancel out.
Q3. Yes

Q4. (a) Divide by the radius or diameter; (b) divide by (volume of the tank divided by a volumetric flow rate).

P1. c is dimensionless.

P2. A has the same units as k; B has the same units as T.

Section 2.4

Q1. Retain extra digits in the calculations.

Q2. No

P1. (a) in 1.0 – 2; (b) in 0.353 – 3; (c) in 1,000.0 – 4; (d) in 23 – 2; (e) in 1,000 – 1; (f) in 1,000.0 – 5

P2. (a) 5760—3 sign. fig.; (b) 2.22—3 sign. fig.

P3. 1380 gal/hr (3 sign. fig. in 87.0)

P4. 10,000 yen has at least 5 sign. fig. even though no decimal is shown. The given conversion rate has at least 3 sign. fig. (in reality more). The answer should be at least $78.0.

P5. 74.8

Section 2.5

Q1. Check to ensure that your answer seems reasonable according to your understanding of the physical system.

Q2. Errors: Keying in the wrong number, confusing the proper decimal point of a number, reading an intermediate number incorrectly, transposing two numbers, and not being careful about the units are among the most common errors. Correcting: Carefully repeat your calculations using your calculator or computer, perhaps in a different sequence.

Q3. It reduces a wide variety of errors.

P1. Assume π is equal to 3, and after canceling terms you get 5/6 or 0.84 as an approximate solution. The actual solution is 0.872. Did you make the proper unit conversions?

Section 2.6

Q1. (a) T; (b) T; (c) T

Q2. 60.05

P1. (a) 7010 g; (b) 2.05 g mol; (c) 7010 lb; (d) 2.05 lb mol

P2. 0.123 kg mol NaCl/kg mol H_2O

P3. 1.177 lb mol

P4. 9

P5. (b) SO_2

P6. O_2 0.62; SO_2 0.19; SO_3 0.19

P7. Either (a) or (d) which have the same molar ratios

Section 2.7

Q1. See text.

Q2. For convenience or to simplify the calculations

P1. Problem 2.6.3, 100 lb; problem 2.6.7, 100 kg mol; problem 2.6.8, 100 kg mol

Section 2.8

Q1. (a) T; (b) T; (c) T

Q2. 13.6 g/cm^3

Q3. The statement means that the density at 10°C of liquid HCN is 1.2675 times the density of water at 4°C.

Q4. (a) F—the units differ; (b) T; (c) T; (d) T

P1. 0.5 m^3/kg

P2. Measure the mass of the water (it should be about 500 g) and add it to 50 g sugar. Measure the volume of the solution and divide the total mass by the volume.

P3. 0.79389 g/cm^3 (assuming the density of water is also at 60°F)

P4. 8.11 ft^3

P5. 871 kg HNO_3/m^3 solution

Section 2.9

Q1. For gases, but not for liquids and solids

Q2. No

Q3. (f) 0.001

Q4. Yes

Q5. Mass percentages

Q6. They are percentages of a total and are dimensionless.

P1. 12,000 mg/L

P2. Report 1400 FC/100 mL.

P3. 0.00183 ppm or 1.83 ppb

Section 2.10

Q1. (a) 0°C and 100°C; (b) 32°F and 212°F

Q2. Δ°F(1.8) = Δ°C; therefore, divide change in Δ°F by 1.8 to get change in Δ°C

Q3. Yes; yes

P1.

°C	°F	K	°R
−40.0	−40.0	233	420
25.0	77.0	298	537
425	796	698	1256
−234	−390	38.8	69.8

P2. $C_p = 93.05 + 0.198\ T_F$

Section 2.11

Q1. Both pots will hold the same amount of coffee if the diameters of the pots and the levels of the spouts are the same.

Q2. All are true.

Q3. Barometric pressure − vacuum pressure = absolute pressure

Q4. No

P1. (a) 15.5; (b) 106.6; (c) 1.052; (d) 35.6

P2. (a) Gauge pressure; (b) barometric pressure; (c) absolute pressure

P3. In the absence of a barometric pressure, assume 101.3 kPa (1 atm). The absolute pressure is 61.3 kPa.

P4. The Hg is static. (a) 3.21 kPa; (b) 3.47 kPa.

Section 2.12

P1. 132 min

P2. 0.654 kg mol/hr

Chapter 3

Section 3.1

Q1. The conservation of mass focuses on the invariance of material in an all-encompassing system, whereas a material balance focuses on ensuring that the flows in and out of a more limited system and the material in the system are equated.

Q2. On a very short timescale when all of the valves are closed, the system behaves as a closed system. On a longer timescale it is an open system.

Q3. See text.

Q4. Yes; depletion

Q5. Two conditions: (a) no reaction and a closed system; (b) no reaction and flow of a component in and out are equal.

Q6. A transient process is an unsteady-state process. See question 5.

Q7. Yes

Q8. Yes

P1. A system that you pick will be somewhat arbitrary, as will be the time interval for analysis, but (a) and (c) can be closed systems (ignoring evaporation) and (b), open.

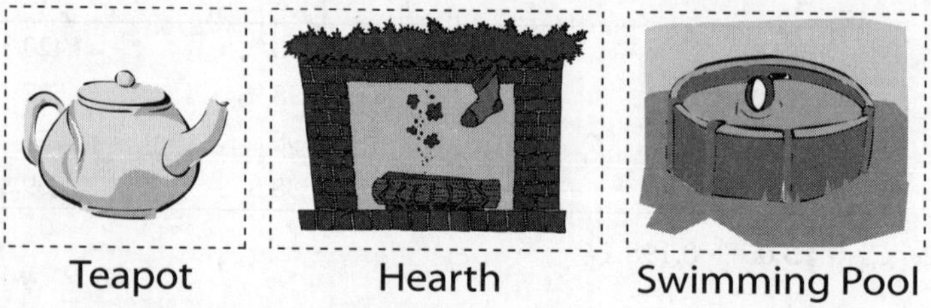

P2. (a) Closed when no material is added or removed; (b) open (flow) when it is being flushed; (c) open (flow) while the car is running; (d) closed during the fermentation step

P3. (a) The beer inside the bottle; (b) everything outside the bottle; (c) closed

P4. 1.49×10^{-3} mg/L

P5. No, the amount left is not a conservation variable.

Section 3.2

Q1. A solution means a (possibly unique) set of values for the unknowns in a problem that satisfies the equations formulated in the problem.

Q2. (a) one; (b) three; (c) three

Q3. Linear equations are independent if the vectors formed from the row coefficient in the equation are independent (see Appendix M, Section M.1). For nonlinear equations, no simple definition exists.

Q4. Delete nonpertinent equations, or find additional variables not included in the analysis.

Q5. The sum of the mass or mole fraction in a stream or inside a system is unity.

Q6. Obtain more equations or specifications, or delete variables of negligible importance.

Q7. Seven (wives) + 7 × 7 (sacks) + 7 × 7 × 7 (cats) + 7 × 7 × 7 × 7 (kits)

P1. (a) Two; (b) two of these three: acetic acid, water, total; (c) two; (d) feed of the 10% solution (say, F) and mass fraction ω of the acetic acid in P; (e) 13% acetic acid and 87% water

P2. Not for a unique solution because only two of the equations are independent.

P3. $F, D, P, \omega_{D_2}, \omega_{P_1}$

P4. Three unknowns exist. Because only two independent material balances can be written for the problem, one value of F, D, or P must be specified to obtain a solution. Note that specifying values of ω_{D_2} or ω_{P_1} will not help.

Chapter 4

Section 4.1

Q1. (a) T; (b) F; (c) T; (d) F

Q2. When they are not independent

P1. 33.3 kg

P2. 179 kg/hr

P3. Salt: 0.00619; oil: 0.99391

Chapter 5

Section 5.1

Q1. When writing a chemical equation, you can use the stoichiometric ratios of two species, but 1 mol MnO_4^- is not equal to 5 mol Fe^{2+} and their ratio is not equal to a dimensionless 1. That line in the solution is wrong, but the calculation is correct.

P1. (a) $C_9H_{18} + \frac{27}{2}O_2 \rightarrow 9CO_2 + 9H_2O$; (b) $4FeS_2 + 11O_2 \rightarrow$ (c) $2Fe_2O_3 + 8SO_2$

P2. 3.08

P3. No

Section 5.2

Q1. A specific chemical reaction and changes in the moles of a specific species

Q2. Based on complete reaction of each species in the feed, the species with the smallest extent of reaction is the limiting reactant.

P1. (a) C_2H_6 is the limiting reactant for the first reaction while hydrogen is the limiting reactant for the second reaction; (b) none for the first reaction while propane is the excess reactant for the second reaction; (c) fraction conversion = 0.184; (d) 63% for reaction (a) and 29% for reaction (b); (e) 1.22; 0.45; (f) based on reactant in the feed: 0.45, based on reactant consumed: 0.84, based on theory: 0.50; (g) Reaction (a) is 33 mol reacting and Reaction (b) is 13.5 mol reacting, both based on 100 mol product.

Section 5.3

Q1. (a) T; (b) T if you are thinking of C as an element but F if you are thinking of C as a compound because the exit product contains CO_2, not C; (c) F (the reaction may be reversible and the compound changes in going through the process, or several reactions may occur with the effect of no net change in the compound)

Q2. See text.

Q3. The value of the extent of reaction or conversion, or data to calculate one of the two values and the stoichiometric coefficient for the compound.

Q4. Total mass: yes, yes
Total moles: yes, no
Mass of a pure compound: yes, no
Moles of a pure compound: yes, no
Mass of an atomic species: yes, yes
Moles of an atomic species: yes, yes

Q5. See Equation (5.10).

Q6. Because it will indicate whether a reaction scheme is composed of independent reactions.

P1. 886 lb

P2. C_6H_{12} = 192 mol/hr; H_2O = 478 mol/hr; CO = 322 mol/hr;
H_2 = 640 mol/hr; C_6H_{14} = 4.5 mol/hr

P3. 186 kg

Section 5.4

Q1. No. If the units of each term are mass, dividing each term by the molecular weight makes the units moles, and the reverse. For element balances it make no difference whether the balance is done in mass or moles.

Q2. The degrees of freedom should be the same if the problem is properly formulated.

Q3. (a) Get the rank of the coefficient matrix; (b) put the equations in an equation solver; or (c) just examine them for possible redundancy.

Q4. No

P1. Two

P2. Three

P3. Three

P4. Four

P5. See the answers for P1, P2, and P3 from Section 5.3.

Section 5.5

Q1. Orsat (dry basis) does not include the water vapor in the analysis.

Q2. SO_2 is not included in the analysis.

Q3. See text.

Q4. Yes

Q5. More

Q6. (a) T; (b) T; (c) F; (d) F

P1. 4.5%

P2. 11.0% CO_2; 7.5% O_2; 81.5% N_2

P3. 1

P4. (a) 252; (b) 1.001; (c) 2.65; (d) 34.4%

Chapter 6

Section 6.2

Q1. Yes

Q2. Yes, even if the equipment is continuous

Q3. It does not have to, but usually it is useful to do so.

Q4. No, if the degrees of freedom are calculated correctly

P1. Assume that the compositions in the figure are mass fractions. Then:

	lb	Mass Fraction
Toluene	396	0.644
Benzene	19.68	0.032
Xylene	200	0.325

P2. 959 kg air/100 kg S burned

	Converter	Burner
SO_2	0.5%	9.4%
SO_3	9.4	—
O_2	7.4	11.6
N_2	82.7	79.0

P3. (a) 1.20; (b) 2232 lb; (c) 21.5%

Section 6.3

Q1. A recycle stream is used to improve the performance and economics of a process.
Q2. Under unsteady-state conditions, particularly during start-up and shutdown, or during changes in the steady-state operating conditions
Q3. Essentially all systems; see text for some examples.
Q4. Yes
Q5. The stoichiometric ratio
Q6. All true
P1. 2(X_2 and X_4)
P2. \$2250
P3. (a) 547.4 lb/hr; (b) 452.6 lb/hr; (c) 0.568
P4. (a) C_4H_6 = 37.5 mol/hr and F = 50.5 mol/hr; (b) 0.65
P5. (a) 960 kg/hr (b) 887 kg/hr

Section 6.4

Q1. Bypassing is a stream that skips some intermediate processing unit (or system) and joins the process downstream.
Q2. (a) T; (b) F; (c) F
Q3. Not necessarily. A purge is considered a stream that removes a small amount of a component to prevent it from building up in the system. A waste stream usually contains a large amount of a component.
P1. (a) 890 recycled and 3.2 purged; (b) 9.2% conversion (errors can be caused by loss of significant figures)
P2. (a) 1.49 mol/hr; (b) Cl_2: 0.658; C_2H_4: 0.338; $C_2H_4Cl_2$: 0.0033

Chapter 7

Section 7.1

Q1. T: absolute temperature in degrees; p: absolute pressure in mass/(length)(time)^2; V: (length)^3/mole; n: mole; R: (mass)(length)^2/(time)^2 (mole)(degree)
Q2. See text.
Q3. Using the ideal gas law, $\rho = Mn/V = Mp/RT$
Q4. Yes
Q5. Lower
Q6. Yes
P1. 1883 ft^3
P2. 2.98 kg
P3. 1.32
P4. 28.3 m^3/hr
P5. (a) N_2: 0.276 psia; CH_4: 10.9 psia; C_2H_6: 2.62 psia; (b) same as mol%
P6. (a) 2733 ft^3/hr; (b) 5040 ft^3/hr; (c) 22,457 ft^3/hr; (d) 31,110 ft^3/hr

Section 7.2

Q1. Each equation is explicit in p or T but implicit in V.
Q2. At low pressure and high temperature
Q3. b is in $m^3/kg\ mol$; a is $(K)^{0.5}(m)^6(kPa)/(kg\ mol)^2$
P1. KO: $z = 1 + B'p + \ldots$ Holborn: $z = 1 + B/\hat{V} + \ldots$
P2. 314 K
P3. (a) 50.7 atm; (b) 34.0 atm

Section 7.3

Q1. $\hat{V}_{ci} = RT_c/p_c$; it can be used as a different parameter on the compressibility charts when a value of p or T is not known.
Q2. All are true except (f).
Q3. It means that z is a function of T_r and p_r.
Q4. 1
P1. (a) Yes; (b) yes; (c) no ($z = 0.98$)
P2. 1.65 kg
P3. 14.9 atm

Chapter 8

Section 8.2

Q1. The point representing ambient temperature and pressure falls below the liquid region, where only solid and vapor exist in equilibrium.
Q2. Intensive: any of p, T, c, ρ, etc.; extensive: any of V, m, n, etc.
Q3. All true
Q4. 1 phase: 2; two phases: 1; 3 phases: 0. Note that the number of variables that can be adjusted is equal to the number of degrees of freedom for the system.
P1. (a) 2; (b) 2; (c) 4
P2. If equilibrium is maintained, the pressure is the vapor pressure at 610 mm Hg.
P3. See Figures 8.2 and 8.3.

Section 8.3

Q1. The vapor pressure increases.
Q2. Yes
Q3. Using a Cox chart
Q4. Yes, if you use another substance as the reference substance

P1. Ice at its vapor pressure changes to liquid water at 32°F (0°C) and 0.0886 psia (0.6113 kPa). Then liquid and vapor exist (at equilibrium) as the temperature goes up to 250°F.

P2. The experimental value is 219.9 mm Hg; the calculated value is 220.9 mm Hg.

P3. 80.1°C

P4. Look at Figure 8.9.

Section 8.4.1

Q1. At equilibrium the partial pressure of the condensable component in the gas equals the vapor pressure of the condensable component.

Q2. (a) Both gas; (b) some liquid water with the residuum gas; (c) both gas; (d) some liquid water with the residuum gas

Q3. Dilute it with inert gas

Q4. At saturation

P1. 0.0373

P2. Water vapor is 0.024 and air is 0.976.

P3. 7.85 L at S.C.

Section 8.4.2

Q1. No, the dew point is a temperature, not a pressure.

Q2. Reduce the temperature, raise the pressure, reduce the volume.

Q3. Yes; see the answer to the previous question.

P1. −15°C; benzene

P2. 2.59 lb in the oxygen cylinder and 5.18 lb in the hydrogen cylinder

Section 8.4.3

Q1. No

Q2. (a) No; (b) yes; (c) no

P1. 2.55 ft^3 at 10 atm pressure

P2. 452.3 lb/hr

P3. A slight increase of 0.25%

P4. 531 m^3

Section 8.5

Q1. Henry's law for a gas dissolved in a solvent; Raoult's law for a solvent in an ideal solution

Q2. Butane

Q3. Yes, because the solution acts as an ideal solution

P1. 135°C

Chapter 9

Section 9.1

Q1. The system is the region or equipment selected for analysis. The surroundings are everything outside the system. An open system is a system in which mass enters and/or leaves the system. A closed system has no such mass exchange. A phase is a physically distinct part or whole of a system. A property is an observable or calculable characteristic of a system.

Q2. No

Q3. The property change for the state variable depends only on the initial and final states. The value for a path variable depends on how the variable moves from the initial state to the final state.

P1. No change

P2. 20,000 kJ/day is about 4800 Calories per day, far too large to be realistic. Most diets call for about 1200 Calories per day; therefore, one would not likely lose weight on an 1800-Calorie-per-day diet without performing a lot of exercise.

Section 9.2.1

Q1. At equilibrium, true.

Q2. Any of the terms that explicitly or implicitly denote transfer of energy are OK; those that denote generation or storage are definitely wrong; those that give a mechanism of heat transfer do not denote heat but just the mechanism for transfer. Words such as *heat of reaction, specific heat, heat quality, heat sink* or *source*, and *body heat* are wrong.

Q3. (a), (b), (d), (e)

P1. (a) Zero J (insulated); (b) 1000 J; (c) no, work was done on the cylinder but heat was transferred to the ice bath; (d) work only because of insulation.

P2. (a) Work; (b) work; (c) both; (d) heat

Section 9.2.2

Q1. True

Q2. Yes, in vaporization—pushing against the atmosphere

Q3. (a)

P1. If the cylinder is the typical cylinder you see with a valve that is closed at the top, no work is done on the gas (ignoring minute contraction of the cylinder itself).

P2. The work is 1 to 2: −0.678 Btu; 2 to 3: 0 since the volume is constant; 3 to 4: 0.771 Btu; 4 to 1: 0 since the volume is constant.

Section 9.2.3

Q1. No

Q2. True

Q3. False for cars less than 2350 lb; true for cars with more mass

P1. 8.63 ft-lb_m/s

Section 9.2.4

Q1. (a) T; (b) F; (c) F

Q2. (a); note that (c) is the *specific PE*.

P1. (a) $KE(0) = 0$, $PE(0) = 4900$ J; (b) KE (final) $= 0$, PE (final) $= 0$; (c) $\Delta KE = 0$, $\Delta PE = -4900$ J; (d) 1171 cal, 4.64 Btu, 4900 J

P2. −98.0 J

Section 9.2.5

Q1. Remember that only the internal energy *change* can be calculated, not the absolute value of the internal energy. Consequently, (a) none of the answers is really adequate to include all of the possible properties of the molecules; (b) none, but 4 would be acceptable if the pressure were included.

Q2. Yes

P1. (a) $C_v = 0.810 + 9.50 \times 10^{-4}\,T$; (b) $T \sim -1.36$°C

P2. (a) 2756 kJ/kg; (b) the steam tables automatically incorporate any phase changes into the listed data.

Section 9.2.6

Q1. Liquid

Q2. Vapor

a. $$\hat{H}_2 - \hat{H}_1 = \int_{p_2}^{p_2}\left[\hat{U} - T\left(\frac{\partial \hat{V}}{\partial T}\right)_p\right]dp$$

b. $$\hat{H}_2 - \hat{H}_1 = \int_{T_1}^{T_2} C_p\,dT + \int_{p_1}^{p_2} \hat{V}\,dp$$

c. $$\hat{H}_2 - \hat{H}_1 = \int_{T_1}^{T_2} C_p\,dT$$

P2. Neither because the enthalpy change is the same by either path

Section 9.2.7

Q1. All false except (b), which is true

Q2. Probably it would technically be correct but it is rarely used.

Q3. 7/2 R

P1. If the liquid water has a constant heat capacity of 4.184 J/(g)(°C), 531 J/(g)/(C); from the steam tables, 532 J/g

P2. $C_p = 4.25 + 0.002T$

P3. (a) cal/(mol)(Δ K); (b) cal/(g mol)(Δ K)(K); (c) cal/(g mol)(Δ K)$(K)^2$; (d) cal/(g mol)(Δ K)$(K)^3$

Section 9.2.8

Q1. (a) F (it is a function of T); (b) T; (c) F (it is for melting)
Q2. See text.
Q3. Tradition
P1. By using the Watson equation, 13.445 kJ/g mol (the experimental value is 13.527 kJ/g mol)
P2. 105 kJ
P3. −219.7 kJ

Section 9.3.1

Q1. All false except (e), which is true
Q2. No
P1. Work in second process: −30 kJ; net work of three processes: −35 kJ
P2. 200 Btu
P3. Assume $Q = 0$. Mass hot water/mass cold water = 2.

Section 9.3.4

Q1. (a) If done on the surroundings, F; (b) F (it is out minus in); (c) T; (d) T; (e) F
Q2. (1) Insulated system; (2) no temperature difference between system and surroundings; (3) it is negligible relative to the other terms in the equation.
Q3. ΔE
Q4. Temperature and pressure (and composition)
Q5. (a) KE: no velocity of fluid flow, $KE_{in} = KE_{out}$, KE is negligible relative to other terms; (b) PE: no height difference exists for streams above the reference plane, $PE_{in} = PE_{out}$, PE is negligible relative to other terms.
P1. Result is $Q = \Delta H$
P2. 423 kJ
P3. Heat transfer: −9649 kW; fraction generated power: 23.7%

Section 9.3.5

Q1. No, usually they operate in the steady state, but not always.
Q2. Yes
Q3. (a) T; (b) possibly T, depending on the location above a reference plane; (c) T; (d) T; (e) F
Q4. All of the other terms in the equation are eliminated by being zero, equal, or negligible.
Q5. No, no flow term exists in the equation.
P1. (a) No term dropped; (b) $Q = 0$; (c) $\Delta KE = 0$; (d) no term dropped
P2. $\Delta KE = \Delta PE = 0$ inside the system, $Q = W = 0$. And $\Delta KE = \Delta PE = 0$ for the flow in and out of the system.
P3. $Q = (445.6)(1.8) - 337.75(0.8) - 560.51(1.0) = -28.6$ kJ

Chapter 10

Section 10.1

Q1. It is zero by definition.

Q2. It won't because by definition the standard heat of formation is at 25°C and 1 atm.

Q3. No

P1. 36.4 kJ/g mol HBr

P2. Assume an open, steady-state system with negligible change in *KE* and *PE*, and $W = 0$.

P3. Yes. $\Delta E = \Delta U = \Delta H - \Delta(pV)$. If $W = 0$, $\Delta KE = 0$, $\Delta PE = 0$ inside the system, and V is constant, all at 25°C and 1 atm.

P4. −74.83 kJ/g mol CH_4 compared with −74.84 in Appendix D

Section 10.2

Q1. Yes

Q2. No

Q3. It should not change the value if the chemical reaction is written with the proper phases and temperatures and pressures denoted for each species.

Q4. (a) Exothermic reaction; (b) endothermic reaction

Q5. Yes, but it may not prove to be convenient.

Q6. The heat of reaction is the enthalpy change at some temperature and pressure for a given number of moles; the standard heat of reaction is the enthalpy change at S.C. for complete reaction according to the reaction equation per unit mole of a compound.

P1. 597.32 kJ/g mol Benzene

P2. −70.94 kJ/g mol

Section 10.3

Q1. In using the heat of reaction in the energy balance, the heats of formation are lumped together. When the heats of formation are merged with the enthalpies of the sensible heats and phase changes, the components of the heat of reaction are split up and no longer appear as an amalgamated term.

Q2. Advantages: easier to use for problems in which (1) multiple reactions occur, (2) the reaction equation(s) is unknown, (3) standard software is used. Disadvantages: (1) heats of formation of a compound are unknown and cannot be estimated, (2) only experimental data is available for the reaction enthalpy.

P1. −720.82 kJ (removed)

P2. −416.68 kJ (removed)

P3. −107.84 kJ/g mol CH_3OH

Section 10.4

Q1. The reference for the heats of combustion is zero for the right-hand side of the chemical equation whereas the reference for the heats of formation is zero for the left-hand side of the chemical equation.

Q2. Yes, if none of the compounds in the chemical equation can undergo a phase transition

Q3. No

Q4. No

Q5. If you pick the same basis for the heat of reaction, namely, 1 g mol of H_2, it makes no difference. However, if you want the heat of reaction for the chemical equation as written, the calculated heat of reaction will be two times the first calculation if the original chemical equation is multiplied by two.

Q6. None (no water is involved)

P1. 234 Btu/ft^3 at 60°F and 30.0 in. Hg

P2. 250.2 kJ/g mol CO

Chapter 11

Section 11.1

Q1. See text.

Q2. Yes

Q3. Yes

Q4. The wet-bulb temperature is the temperature of the evaporating water that is in equilibrium with the ambient air that has a temperature called the dry-bulb temperature.

Q5. No

P1. Degrees of freedom: 3; at fixed pressure: 2

P2. See Figure 11.3.

Section 11.2

Q1. See text.

Q2. See text.

Q3. Yes

Q4. Yes

P1. (a) 133°F; (b) the same

P2. (a) 61%; (b) 60°C

P3. (a) 0.03 kg H_2O/kg dry air; (b) 1.02 m^3/kg dry air; (c) 31.5°C

P4. (a) $\mathcal{H}$ = 0.0808 lb H_2O/lb dry air; (b) H = 118.9 Btu/lb dry air; (c) V = 16.7 ft^3/lb dry air; (d) $\mathcal{H}$ = 0.0710 lb H_2O/lb dry air

Section 11.3

Q1. Usually it is OK, but in a flue gas, for example, you want to make sure that condensation in the exit duct does not occur, so entering moisture may be important to include in the calculations.

Q2. No

P1. Heat: 18 Btu/lb dry air; water vapor: 7×10^{-3} lb H_2O/lb dry air

APPENDIX B

Atomic Weights and Numbers

Table B.1 Relative Atomic Weights, 1965 (Based on the Atomic Mass of $^{12}C = 12$)
The values for atomic weights given in the table apply to elements as they exist in nature, without artificial alteration of their isotopic composition, and, further, to natural mixtures that do not include isotopes of radiogenic origin.

Name	Symbol	Atomic Number	Atomic Weight	Name	Symbol	Atomic Number	Atomic Weight
Actinium	Ac	89	—	Mercury	Hg	80	200.59
Aluminum	Al	13	26.9815	Molybdenum	Mo	42	95.94
Americium	Am	95	—	Neodymium	Nd	60	144.24
Antimony	Sb	51	121.75	Neon	Ne	10	20.183
Argon	Ar	18	39.948	Neptunium	Np	93	—
Arsenic	As	33	74.9216	Nickel	Ni	28	58.71
Astatine	At	85	—	Niobium	Nb	41	92.906
Barium	Ba	56	137.34	Nitrogen	N	7	14.0067
Berkelium	Bk	97	—	Nobelium	No	102	—
Beryllium	Be	4	9.0122	Osmium	Os	75	190.2
Bismuth	Bi	83	208.980	Oxygen	O	8	15.9994
Boron	B	5	10.811	Palladium	Pd	46	106.4
Bromine	Br	35	79.904	Phosphorus	P	15	30.9738
Cadmium	Cd	48	112.40	Platinum	Pt	78	195.09
Cesium	Cs	55	132.905	Plutonium	Pu	94	—
Calcium	Ca	20	40.08	Polonium	Po	84	
Californium	Cf	98	—	Potassium	K	19	39.102
Carbon	C	6	12.01115	Praseodym	Pr	59	140.907
Cerium	Ce	58	140.12	Promethium	Pm	61	—
Chlorine	Cl	17	35.453[b]	Protactinium	Pa	91	—
Chromium	Cr	24	51.996[b]	Radium	Ra	88	—
Cobalt	Co	27	58.9332	Radon	Rn	86	—
Copper	Cu	29	63.546[b]	Rhenium	Re	75	186.2
Curium	Cm	96	—	Rhodium	Rh	45	102.905
Dysprosium	Dy	66	162.50	Rubidium	Rb	37	84.57
Einsteinium	Es	99	—	Ruthenium	Ru	44	101.07
Erbium	Er	68	167.26	Samarium	Sm	62	150.35
Europium	Eu	63	151.96	Scandium	Sc	21	44.956
Fermium	Fm	100	—	Selenium	Se	34	78.96
Flourine	F	9	18.9984	Silicon	Si	14	28.086
Francium	Fr	87	—	Silver	Ag	47	107.868
Gadolinium	Gd	64	157.25	Sodium	Na	11	22.9898
Gallium	Ga	31	69.72	Strontium	Sr	38	87.62
Germanium	Ge	32	72.59	Sulfur	S	16	32.064
Gold	Au	79	196.967	Tantalum	Ta	73	180.948
Hafnium	Hf	72	178.49	Technetium	Tc	43	—
Helium	He	2	4.0026	Tellurium	Te	52	127.60
Holmium	Ho	67	164.930	Terbium	Tb	65	158.924
Hydrogen	H	1	1.00797	Thallium	Tl	81	204.37
Indium	In	49	114.82	Thorium	Th	90	232.038
Iodine	I	53	126.9044	Thulium	Tm	59	168.934
Iridium	Ir	77	192.2	Tin	Sn	50	118.69
Iron	Fe	26	55.847	Titanium	Ti	22	47.90
Krypton	Kr	36	83.80	Tungsten	W	74	183.85
Lanthanum	La	57	138.91	Uranium	U	92	238.03
Lawrencium	Lr	103	—	Vanadium	V	23	50.942
Lead	Pb	82	207.19	Xenon	Xe	54	131.30
Lithium	Li	3	6.939	Ytterbium	Yb	70	173.04
Lutetium	Lu	71	174.97	Yttrium	Y	39	88.905
Magnesium	Mg	12	24.312	Zinc	Zn	30	65.37
Manganese	Mn	25	54.9380	Zirconium	Zr	40	91.22
Mendelevium	Md	101	—				

SOURCE: *Comptes Rendus*, 23rd IUPAC Conference, 1965, Butterworth's, London, 1965, pp. 177–178.

APPENDIX C

Tables of the Pitzer Z^0 and Z^1 Factors

Table C.1 Values of z^0 (from Lee, B. I., and M. G. Kessler, *AIChEJ*, 21, 510–518 (1975)).

	p_r														
T_r	0.010	0.050	0.100	0.200	0.400	0.600	0.800	1.000	1.200	1.500	2.000	3.000	5.000	7.000	10.000
0.30	0.0029	0.0145	0.0290	0.0579	0.1158	0.1737	0.2315	0.2892	0.3470	0.4335	0.5775	0.8648	1.4366	2.0048	2.8507
0.35	0.0026	0.0130	0.0261	0.0522	0.1043	0.1564	0.2084	0.2604	0.3123	0.3901	0.5195	0.7775	1.2902	1.7987	2.5539
0.40	0.0024	0.0119	0.0239	0.0477	0.0953	0.1429	0.1904	0.2379	0.2853	0.3563	0.4744	0.7095	1.1758	1.6373	2.3211
0.45	0.0022	0.0110	0.0221	0.0442	0.0882	0.1322	0.1762	0.2200	0.2638	0.3294	0.4384	0.6551	1.0841	1.5077	2.1338
0.50	0.0021	0.0103	0.0207	0.0413	0.0825	0.1236	0.1647	0.2056	0.2465	0.3077	0.4092	0.6110	1.0094	1.4017	1.9801
0.55	0.9804	0.0098	0.0195	0.0390	0.0778	0.1166	0.1553	0.1939	0.2323	0.2899	0.3853	0.5747	0.9475	1.3137	1.8520
0.60	0.9849	0.0093	0.0186	0.0371	0.0741	0.1109	0.1476	0.1842	0.2207	0.2753	0.3657	0.5446	0.8959	1.2398	1.7440
0.65	0.9881	0.9377	0.0178	0.0356	0.0710	0.1063	0.1415	0.1765	0.2113	0.2634	0.3495	0.5197	0.8526	1.773	1.6519
0.70	0.9904	0.9504	0.8958	0.0344	0.0687	0.1027	0.1366	0.1703	0.2038	0.2538	0.3364	0.4991	0.8161	1.1241	1.5729
0.75	0.9922	0.9598	0.9165	0.0336	0.0670	0.1001	0.1330	0.1656	0.1981	0.2464	0.3260	0.4823	0.7854	1.0787	1.5047
0.80	0.9935	0.9669	0.9319	0.8539	0.0661	0.0985	0.1307	0.1626	0.1942	0.2411	0.3182	0.4690	0.7598	1.0400	1.4456
0.85	0.9946	0.9725	0.9436	0.8810	0.0661	0.0983	0.1301	0.1614	0.1924	0.2382	0.3132	0.4591	0.7388	1.0071	1.3943
0.90	0.9954	0.9768	0.9528	0.9015	0.7800	0.1006	0.1321	0.1630	0.1935	0.2383	0.3114	0.4527	0.7220	0.9793	1.3496
0.93	0.9959	0.9790	0.9573	0.9115	0.8059	0.6635	0.1359	0.1664	0.1963	0.2405	0.3122	0.4507	0.7138	0.9648	1.3257
0.95	0.9961	0.9803	0.9600	0.9174	0.8206	0.6967	0.1410	0.1705	0.1998	0.2432	0.3138	0.4501	0.7092	0.9561	1.3108
0.97	0.9963	0.9815	0.9625	0.9227	0.8338	0.7240	0.5580	0.1779	0.2055	0.2474	0.3164	0.4504	0.7052	0.9480	1.2968
0.98	0.9965	0.9821	0.9637	0.9253	0.8398	0.7360	0.5887	0.1844	0.2097	0.2503	0.3182	0.4508	0.7035	0.9442	1.2901
0.99	0.9966	0.9826	0.9648	0.9277	0.8455	0.7471	0.6138	0.1959	0.2154	0.2538	0.3204	0.4514	0.7018	0.9406	1.2835
1.00	0.9967	0.9832	0.9659	0.9300	0.8509	0.7574	0.6353	0.2901	0.2237	0.2583	0.3229	0.4522	0.7004	0.9372	1.2772
1.01	0.9968	0.9837	0.9669	0.9322	0.8561	0.7671	0.6542	0.4648	0.2370	0.2640	0.3260	0.4533	0.6991	0.9339	1.2710
1.02	0.9969	0.9842	0.9679	0.9343	0.8610	0.7761	0.6710	0.5146	0.2629	0.2715	0.3297	0.4547	0.6980	0.9307	1.2650
1.05	0.9971	0.9855	0.9707	0.9401	0.8743	0.8002	0.7130	0.6026	0.4437	0.3131	0.3452	0.4604	0.6956	0.9222	1.2481
1.10	0.9975	0.9874	0.9747	0.9485	0.8930	0.8323	0.7649	0.6880	0.5984	0.4580	0.3953	0.4770	0.6950	0.9110	1.2232
1.15	0.9978	0.9891	0.9780	0.9554	0.9081	0.8576	0.8032	0.7443	0.6803	0.5798	0.4760	0.5042	0.6987	0.9033	1.2021
1.20	0.9981	0.9904	0.9808	0.9611	0.9205	0.8779	0.8330	0.7858	0.7363	0.6605	0.5605	0.5425	0.7069	0.8990	1.1844

1.30	0.9985	0.9926	0.9852	0.9702	0.9396	0.9083	0.8764	0.8438	0.8111	0.7624	0.6908	0.6344	0.7358	0.8998	1.1580
1.40	0.9988	0.9942	0.9884	0.9768	0.9534	0.9298	0.9062	0.8827	0.8595	0.8256	0.7753	0.7202	0.7761	0.9112	1.1419
1.50	0.9991	0.9954	0.9909	0.9818	0.9636	0.9456	0.9278	0.9103	0.8933	0.8689	0.8328	0.7887	0.8200	0.9297	1.1339
1.60	0.9993	0.9964	0.9928	0.9856	0.9714	0.9575	0.9439	0.9308	0.9180	0.9000	0.8738	0.8410	0.8617	0.9518	1.1320
1.70	0.9994	0.9971	0.9943	0.9886	0.9775	0.9667	0.9563	0.9463	0.9367	0.9234	0.9043	0.8809	0.8984	0.9745	1.1343
1.80	0.9995	0.9977	0.9955	0.9910	0.9823	0.9739	0.9659	0.9583	0.9511	0.9413	0.9275	0.9118	0.9297	0.9961	1.1391
1.90	0.9996	0.9982	0.9964	0.9929	0.9861	0.9796	0.9735	0.9678	0.9624	0.9552	0.9456	0.9359	0.9557	1.0157	1.1452
2.00	0.9997	0.9986	0.9972	0.9944	0.9892	0.9842	0.9796	0.9754	0.9715	0.9664	0.9599	0.9550	0.9772	1.0328	1.1516
2.20	0.9998	0.9992	0.9983	0.9967	0.9937	0.9910	0.9886	0.9865	0.9847	0.9826	0.9806	0.9827	1.0094	1.0600	1.1635
2.40	0.9999	0.9996	0.9991	0.9983	0.9969	0.9957	0.9948	0.9941	0.9936	0.9935	0.9945	1.0011	1.0313	1.0793	1.1728
2.60	1.0000	0.9998	0.9997	0.9994	0.9991	0.9990	0.9990	0.9993	0.9998	1.0010	1.0040	1.0137	1.0463	1.0926	1.1792
2.80	1.0000	1.0000	1.0001	1.0002	1.0007	1.0013	1.0021	1.0031	1.0042	1.0063	1.0106	1.0223	1.0565	1.1016	1.1830
3.00	1.0000	1.0002	1.0004	1.0008	1.0018	1.0030	1.0043	1.0057	1.0074	1.0101	1.0153	1.0284	1.0635	1.1075	1.1848
3.50	1.0001	1.0004	1.0008	1.0017	1.0035	1.0055	1.0075	1.0097	1.0120	1.0156	1.0221	1.0368	1.0723	1.1138	1.1834
4.00	1.0001	1.0005	1.0010	1.0021	1.0043	1.0066	1.0090	1.0115	1.0140	1.0179	1.0249	1.0401	1.0747	1.1136	1.1773

The shaded region contains values for the liquid phase.

Table C.2 Values of z^1 (from Lee, B. I. and M. G. Kessler, *AIChEJ*, 21, 510–518 (1975)).

	p_r														
T_r	0.10	0.50	0.100	0.200	0.400	0.600	0.800	1.000	1.200	1.500	2.000	3.000	5.000	7.000	10.000
0.30	−0.0008	−0.0040	−0.0081	−0.0161	−0.0323	−0.0484	−0.0645	−0.0806	−0.0966	−0.1207	−0.1608	−0.2407	−0.3996	−0.5572	−0.7915
0.35	−0.0009	−0.0046	−0.0093	−0.0185	−0.0370	−0.0554	−0.0738	−0.0921	−0.1105	−0.1379	−0.1834	−0.2738	−0.4523	−0.6279	−0.8863
0.40	−0.0010	−0.0048	−0.0095	−0.0190	−0.0380	−0.0570	−0.0758	−0.0946	−0.1134	−0.1414	−0.1879	−0.2799	−0.4603	−0.6365	−0.8936
0.45	−0.0009	−0.0047	−0.0094	−0.0187	−0.0374	−0.0560	−0.0745	−0.0929	−0.1113	−0.1387	−0.1840	−0.2734	−0.4475	−0.6162	−0.8606
0.50	−0.0009	−0.0045	−0.0090	−0.0181	−0.0360	−0.0539	−0.0716	−0.0893	−0.1069	−0.1330	−0.1762	−0.2611	−0.4253	−0.5831	−0.8099
0.55	−0.0314	−0.0043	−0.0086	−0.0172	−0.0343	−0.0513	−0.0682	−0.0849	−0.1015	−0.1263	−0.1669	−0.2465	−0.3991	−0.5446	−0.7521
0.60	−0.0205	−0.0041	−0.0082	−0.0164	−0.0326	−0.0487	−0.0646	−0.0803	−0.0960	−0.1192	−0.1572	−0.2312	−0.3718	−0.5047	−0.6928
0.65	−0.0137	−0.0772	−0.0078	−0.0156	−0.0309	−0.0461	−0.0611	−0.0759	−0.0906	−0.1122	−0.1476	−0.2160	−0.3447	−0.4653	−0.6346
0.70	−0.0093	−0.0507	−0.1161	−0.0148	−0.0294	−0.0438	−0.0579	−0.0718	−0.0855	−0.1057	−0.1385	−0.2013	−0.3184	−0.4270	−0.5785
0.75	−0.0064	−0.0339	−0.0744	−0.0143	−0.0282	−0.0417	−0.0550	−0.0681	−0.0808	−0.0996	−0.1298	−0.1872	−0.2929	−0.3901	−0.5250
0.80	−0.0044	−0.0228	−0.0487	−0.1160	−0.0272	−0.0401	−0.0526	−0.0648	−0.0767	−0.0940	−0.1217	−0.1736	−0.2682	−0.3545	−0.4740
0.85	−0.0029	−0.0152	−0.0319	−0.0715	−0.0268	−0.0391	−0.0509	−0.0622	−0.0731	−0.0888	−0.1138	−0.1602	−0.2439	−0.3201	−0.4254
0.90	−0.0019	−0.0099	−0.0205	−0.0442	−0.1118	−0.0396	−0.0503	−0.0604	−0.0701	−0.0840	−0.1059	−0.1463	−0.2195	−0.2862	−0.3788
0.93	−0.0015	−0.0075	−0.0154	−0.0326	−0.0763	−0.1662	−0.0514	−0.0602	−0.0687	−0.0810	−0.1007	−0.1374	−0.2045	−0.2661	−0.3516
0.95	−0.0012	−0.0062	−0.0126	−0.0262	−0.0589	−0.1110	−0.0540	−0.0607	−0.0678	−0.0788	−0.0967	−0.1310	−0.1943	−0.2526	−0.3339
0.97	−0.0010	−0.0050	−0.0101	−0.0208	−0.0450	−0.0770	−0.1647	−0.0623	−0.0669	−0.0759	−0.0921	−0.1240	−0.1837	−0.2391	−0.3163
0.98	−0.0009	−0.0044	−0.0090	−0.0184	−0.0390	−0.0641	−0.1100	−0.0641	−0.0661	−0.0740	−0.0893	−0.1202	−0.1783	−0.2322	−0.3075
0.99	−0.0008	−0.0039	−0.0079	−0.0161	−0.0335	−0.0531	−0.0796	−0.0680	−0.0646	−0.0715	−0.0861	−0.1162	−0.1728	−0.2254	−0.2989
1.00	−0.0007	−0.0034	−0.0069	−0.0140	−0.0285	−0.0435	−0.0588	−0.0879	−0.0609	−0.0678	−0.0824	−0.1118	−0.1672	−0.2185	−0.2902
1.01	−0.0006	−0.0030	−0.0060	−0.0120	−0.0240	−0.0351	−0.0429	−0.0223	−0.0473	−0.0621	−0.0778	−0.1072	−0.1615	−0.2116	−0.2816
1.02	−0.0005	−0.0026	−0.0051	−0.0102	−0.0198	−0.0277	−0.0303	−0.0062	0.0227	−0.0524	−0.0722	−0.1021	−0.1556	−0.2047	−0.2731
1.05	−0.0003	−0.0015	−0.0029	−0.0054	−0.0092	−0.0097	−0.0032	0.0220	0.1059	0.0451	−0.0432	−0.0838	−0.1370	−0.1835	−0.2476
1.10	−0.0000	0.0000	0.0001	0.0007	0.0038	0.0106	0.0236	0.0476	0.0897	0.1630	0.0698	−0.0373	−0.1021	−0.1469	−0.2056
1.15	0.0002	0.0011	0.0023	0.0052	0.0127	0.0237	0.0396	0.0625	0.0943	0.1548	0.1667	0.0332	−0.0611	−0.1084	−0.1642
1.20	0.0004	0.0019	0.0039	0.0084	0.0190	0.0326	0.0499	0.0719	0.0991	0.1477	0.1990	0.1095	−0.0141	−0.0678	−0.1231

1.30	0.0006	0.0030	0.0061	0.0125	0.0267	0.0429	0.0612	0.0819	0.1048	0.1420	0.1991	0.2079	0.0875	0.0176	−0.0423
1.40	0.0007	0.0036	0.0072	0.0147	0.0306	0.0477	0.0661	0.0857	0.1063	0.1383	0.1894	0.2397	0.1737	0.1008	0.0350
1.50	0.0008	0.0039	0.0078	0.0158	0.0323	0.0497	0.0677	0.0864	0.1055	0.1345	0.1806	0.2433	0.2309	0.1717	0.1058
1.60	0.0008	0.0040	0.0080	0.0162	0.0330	0.0501	0.0677	0.0855	0.1035	0.1303	0.1729	0.2381	0.2631	0.2255	0.1673
1.70	0.0008	0.0040	0.0081	0.0163	0.0329	0.0497	0.0667	0.0838	0.1008	0.1259	0.1658	0.2305	0.2788	0.2628	0.2179
1.80	0.0008	0.0040	0.0081	0.0162	0.0325	0.0488	0.0652	0.0816	0.0978	0.1216	0.1593	0.2224	0.2846	0.2871	0.2576
1.90	0.0008	0.0040	0.0079	0.0159	0.0318	0.0477	0.0635	0.0792	0.0947	0.1173	0.1532	0.2144	0.2848	0.3017	0.2876
2.00	0.0008	0.0039	0.0078	0.0155	0.0310	0.0464	0.0617	0.0767	0.0916	0.1133	0.1476	0.2069	0.2819	0.3097	0.3096
2.20	0.0007	0.0037	0.0074	0.0147	0.0293	0.0437	0.0579	0.0719	0.0857	0.1057	0.1374	0.1932	0.2720	0.3135	0.3355
2.40	0.0007	0.0035	0.0070	0.0139	0.0276	0.0411	0.0544	0.0675	0.0803	0.0989	0.1285	0.1812	0.2602	0.3089	0.3459
2.60	0.0007	0.0033	0.0066	0.0131	0.0260	0.0387	0.0512	0.0634	0.0754	0.0929	0.1207	0.1706	0.2484	0.3009	0.3475
2.80	0.0006	0.0031	0.0062	0.0124	0.0245	0.0365	0.0483	0.0598	0.0711	0.0876	0.1138	0.1613	0.2372	0.2915	0.3443
3.00	0.0006	0.0029	0.0059	0.0117	0.0232	0.0345	0.0456	0.0565	0.0672	0.0828	0.1076	0.1529	0.2268	0.2817	0.3385
3.50	0.0005	0.0026	0.0052	0.0103	0.0204	0.0303	0.0401	0.0497	0.0591	0.0728	0.0949	0.1356	0.2042	0.2584	0.3194
4.00	0.0005	0.0023	0.0046	0.0091	0.0182	0.0270	0.0357	0.0443	0.0527	0.0651	0.0849	0.1219	0.1857	0.2379	0.2994

The shaded region contains values for the liquid phase.

APPENDIX D

Heats of Formation and Combustion

Table D.1 Heats of Formation and Heats of Combustion of Compounds at 25°C*†
Standard states of products for $\Delta\hat{H}_c^\circ$ are CO_2(g), H_2O(l), N_2(g), SO_2(g), and HCl(aq). To convert to Btu/lb mol, multiply by 430.6.

Compound	Formula	Mol. wt.	State	$\Delta\hat{H}_f^\circ$ (kJ/g mol)	$\Delta\hat{H}_c^\circ$ (kJ/g mol)
Acetic acid	CH_3COOH	60.05	l	−486.2	−871.69
			g		−919.73
Acetaldehyde	CH_3CHO	40.052	g	−166.4	−1192.36
Acetone	C_3H_6O	58.08	aq, 200	−410.03	
			g	−216.69	−1821.38
Acetylene	C_2H_2	26.04	g	226.75	−1299.61
Ammonia	NH_3	17.032	l	−67.20	
			g	−46.191	−382.58
Ammonium carbonate	$(NH_4)_2CO_3$	96.09	c		
			aq	−941.86	
Ammonium chloride	NH_4Cl	53.50	c	−315.4	
Ammonium hydroxide	NH_4OH	35.05	aq	−366.5	
Ammonium nitrate	NH_4NO_3	80.05	c	−366.1	
			aq	−339.4	
Ammonium sulfate	$(NH_4)SO_4$	132.15	c	−1179.3	
			aq	−1173.1	
Benzaldehyde	C_6H_5CHO	106.12	l	−88.83	
			g	−40.0	
Benzene	C_6H_6	78.11	l	48.66	−3267.6
			g	82.927	−3301.5
Boron oxide	B_2O_3	69.64	c	−1263	
			l	−1245.2	
Bromine	Br_2	159.832	l	0	
			g	30.7	

Compound	Formula	Mol. wt.	State	$\Delta\hat{H}^\circ_f$ (kJ/g mol)	$\Delta\hat{H}^\circ_c$ (kJ/g mol)
n-Butane	C_4H_{10}	58.12	l	−147.6	−2855.6
			g	−124.73	−2878.52
Isobutane	C_4H_{10}	58.12	l	−158.5	−2849.0
			g	−134.5	−2868.8
1-Butene	C_4H_8	56.104	g	1.172	−2718.58
Calcium arsenate	$Ca_3(AsO_4)_2$	398.06	c	−3330.5	
Calcium carbide	CaC_2	64.10	c	−62.7	
Calcium carbonate	$CaCO_3$	100.09	c	−1206.9	
Calcium chloride	$CaCl_2$	110.99	c	−794.9	
Calcium cyanamide	$CaCN_2$	80.11	c	−352	
Calcium hydroxide	$Ca(OH)_2$	74.10	c	−986.56	
Calcium oxide	CaO	56.08	c	−635.6	
Calcium phosphate	$Ca_3(PO_4)_2$	310.19	c	−4137.6	
Calcium silicate	$CaSiO_3$	116.17	c	−1584	
Calcium sulfate	$CaSO_4$	136.15	c	−1432.7	
			aq	−1450.5	
Calcium sulfate (gypsum)	$CaSO_4 \cdot 2H_2O$	172.18	c	−2021.1	
Carbon	C	12.01	c Graphite (β)	0	−393.51
Carbon dioxide	CO_2	44.01	g	−393.51	
			l	−412.92	
Carbon disulfide	CS_2	76.14	l	87.86	−1075.2
			g	115.3	−1102.6
Carbon monoxide	CO	28.01	g	−110.52	−282.99
Carbon tetrachloride	CCl_4	153.838	l	−139.5	−352.2
			g	−106.69	−384.9
Chloroethane	C_2H_5Cl	64.52	g	−105.0	−1421.1
			l	−41.20	−5215.44
Cumene (isopropylbenzene)	$C_6H_5CH(CH_3)_2$	120.19	g	3.93	−5260.59
			c	−769.86	
Cupric sulfate	$CuSO_4$	159.61	aq	−843.12	
			c	−751.4	
Cyclohexane	C_6H_{12}	84.16	g	−123.1	−3953.0
Cyclopentane	C_5H_{10}	70.130	l	−105.8	−3290.9
			g	−77.23	−3319.5
Ethane	C_2H_6	30.07	g	−84.667	−1559.9
Ethyl acetate	$CH_3CO_2C_2H_5$	88.10	l	−442.92	−2274.48
Ethyl alcohol	C_2H_5OH	46.068	l	−277.63	−1366.91
			g	−235.31	−1409.25
Ethyl benzene	$C_6H_5 \cdot C_2H_5$	106.16	l	−12.46	−4564.87
			g	29.79	−4607.13
Ethyl chloride	C_2H_5Cl	64.52	g	−105	
Ethylene	C_2H_4	28.052	g	52.283	−1410.99
Ethylene chloride	C_2H_3Cl	62.50	g	31.38	−1271.5
3-Ethyl hexane	C_8H_{18}	114.22	l	−250.5	−5470.12
			g	−210.9	−5509.78

(Continues)

Table D.1 Heats of Formation and Heats of Combustion of Compounds at 25°C*† (*Continued*)

Compound	Formula	Mol. wt.	State	$\Delta\hat{H}^\circ_f$ (kJ/g mol)	$\Delta\hat{H}^\circ_c$ (kJ/g mol)
Ferric chloride	$FeCl_3$		c	−403.34	
Ferric oxide	Fe_2O_3	159.70	c	−822.156	
Ferric sulfide	FeS_2	*see* Iron sulfide	*see* Iron sulfide		
Ferrosoferric oxide	Fe_3O_4	231.55	c	−1116.7	
Ferrous chloride	$FeCl_2$		c	−342.67	−303.76
Ferrous oxide	FeO	71.85	c	−267	
Ferrous sulfide	FeS	87.92	c	−95.06	
Formaldehyde	H_2CO	30.026	g	−115.89	−563.46
n-Heptane	C_7H_{16}	100.20	l	−224.4	−4816.91
			g	−187.8	−4853.48
n-Hexane	C_6H_{14}	86.17	l	−198.8	−4163.1
			g	−167.2	−4194.753
Hydrogen	H_2	2.016	g	0	−285.84
Hydrogen bromide	HBr	80.924	g	−36.23	
Hydrogen chloride	HCl	36.465	g	−92.311	
Hydrogen cyanide	HCN	27.026	g	130.54	
Hydrogen sulfide	H_2S	34.082	g	−20.15	−562.589
Iron sulfide	FeS_2	119.98	c	−177.9	
Lead oxide	PbO	223.21	c	−219.2	
Magnesium chloride	$MgCl_2$	95.23	c	−641.83	
Magnesium hydroxide	$Mg(OH)_2$	58.34	c	−924.66	
Magnesium oxide	MgO	40.32	c	−601.83	
Methane	CH_4	16.041	g	−74.84	−890.4
Methyl alcohol	CH_3OH	32.042	l	−238.64	−726.55
			g	−201.25	−763.96
Methyl chloride	CH_3Cl	50.49	g	−81.923	−766.63†
Methyl cyclohexane	C_7H_{14}	98.182	l	−190.2	−4565.29
			g	−154.8	−4600.68
Methyl cyclopentane	C_6H_{12}	84.156	l	−138.4	−3937.7
			g	−106.7	−3969.4
Nitric acid	HNO_3	63.02	l	−173.23	
			aq	−206.57	
Nitric oxide	NO	30.01	g	90.374	
Nitrogen dioxide	NO_2	46.01	g	33.85	
Nitrous oxide	N_2O	44.02	g	81.55	
n-Pentane	C_5H_{12}	72.15	l	−173.1	−3509.5
			g	−146.4	−3536.15
Phosphoric acid	H_3PO_4	98.00	c	−1281	
			aq ($1H_2O$)	−1278	
Phosphorus	P_4	123.90	c	0	
Phosphorus pentoxide	P_2O_5	141.95	c	−1506	
Propane	C_3H_8	44.09	l	−119.84	−2204.0
			g	−103.85	−2220.0

Compound	Formula	Mol. wt.	State	$\Delta\hat{H}^\circ_f$ (kJ/g mol)	$\Delta\hat{H}^\circ_c$ (kJ/g mol)
Propene	C_3H_6	42.078	g	20.41	−2058.47
n-Propyl alcohol	C_3H_8O	60.09	g	−255	−2068.6
n-Propylbenzene	$C_6H_5 \cdot CH_2 \cdot C_2H_5$	120.19	l	−38.40	−5218.2
			g	7.824	−5264.5
Silicon dioxide	SiO_2	60.09	c	−851.0	
Sodium bicarbonate	$NaHCO_3$	84.01	c	−945.6	
Sodium bisulfate	$NaHSO_4$	120.07	c	−1126	
Sodium carbonate	Na_2CO_3	105.99	c	−1130	
Sodium chloride	$NaCl$	58.45	c	−411.00	
Sodium cyanide	$NaCN$	49.01	c	−89.79	
Sodium nitrate	$NaNO_3$	85.00	c	−466.68	
Sodium nitrite	$NaNO_2$	69.00	c	−359	
Sodium sulfate	Na_2SO_4	142.05	c	−1384.5	
Sodium sulfide	Na_2S	78.05	c	−373	
Sodium sulfite	Na_2SO_3	126.05	c	−1090	
Sodium thiosulfate	$Na_2S_2O_3$	158.11	c	−1117	
Sulfur	S	32.07	c (rhombic)	0	
			c (monoclinic)	0.297	
Sulfur chloride	S_2Cl_2	135.05	l	−60.3	
Sulfur dioxide	SO_2	64.066	g	−296.90	
Sulfur trioxide	SO_3	80.066	g	−395.18	
Sulfuric acid	H_2SO_4	98.08	l	−811.32	
			aq	−907.51	
Toluene	$C_6H_5CH_3$	92.13	l	11.99	−3909.9
			g	50.000	−3947.9
Water	H_2O	18.016	l	−285.840	
			g	−241.826	
m-Xylene	$C_6H_4(CH_3)_2$	106.16	l	−25.42	−4551.86
			g	17.24	−4594.53
o-Xylene	$C_6H_4(CH_3)_2$	106.16	l	−24.44	−4552.86
			g	19.00	−4596.29
p-Xylene	$C_6H_4(CH_3)_2$	106.16	l	−24.43	−4552.86
			g	17.95	−4595.25
Zinc sulfate	$ZnSO_4$	161.45	c	−978.55	
			aq	−1059.93	

*Sources of data are given at the beginning of Appendix D, References 1, 4, and 5.

†Standard state HCl(g).

APPENDIX E

Answers to Selected Problems

Chapter 2

2.1.1 (b) 4; (c) 1; (g) 1

2.2.1 (a) 4.17×10^9 m^3; (b) 449 gal/min

2.2.3 (a) 88 ft/sec; (b) 3.52×104 kg/m^2; (c) 4.79 nm/sec^2

2.2.6 2 h

2.2.9 1.49×10^4 kJ/(day)(m^2)(°C/cm)

2.2.12 The object has a mass of 21.3 kg, and the force to support this mass corresponds to the weight.

2.3.3 No. The g should be g_c.

2.3.6 $u' = 2.57k(\tau'/e')$

2.3.7 0.943 has no associated units.

2.4.1 Two

2.4.4 569.8 cm^2

2.6.3 $C_{12}O_{11}H_{22}$

2.7.1 (a) 100 or 1 g mol (use SI units); (d) use 1 or 100 mol (SI or AE)

2.7.3 134.2 lb Cl or 1 day (10.7×10^6 gal water)

2.8.2 152 ft^3

2.9.1 Na: 0.22; Cl: 0.33; O: 0.45

2.9.3 CO_2: 0.56; N_2: 0.44 (mole fraction)

2.9.12 Mass fraction $H_2S = 2.7 \times 10^{-4}$

2.10.1 474°R, 264 K, 14°F

2.10.4 (a) 50°F; (c) 241.3 K

2.11.1 (a) 15,000 kg; (b) 1.47 kPa (0.21 psi)

2.11.7 Neither (the pressure varies continuously)

2.11.9 A (18.4 mm Hg difference)

Chapter 3

3.1.1 System boundary includes both the pumps and soil at the end of the pipes.

3.1.5 (a) Closed; (b) open; (c) open; (d) closed

3.1.8 (a) Open, unsteady state; (b) open, steady state; (c) open, steady state

3.1.9 (a): 4, 6; (b) 1, 5; (c) 1, 5 (excluding vaporization) or 4 (with vaporization), 6

3.1.11 1200 kg
3.1.15 No (in = 2.5×10^6, out = 2.278×10^6)
3.1.19 The balance on NaCl gives 4690 in approx. = 4696 out, hence closure is good.
3.2.2 Three components plus 1 total; only 3 are independent
3.2.4 (a) No, and thus no solution; (b) they are independent but there are two solutions.
3.2.6 (a) No
3.2.9 Examine the C_3H_8 row. No concentration reaches 50%.
3.2.12 The degrees of freedom are equal to 5. You can make any set of measurements that results in independent equations (assuming equal accuracy).
3.2.14 1.40

Chapter 4

4.1.1 Yes
4.1.4 Water = 5 mL, urea = 141 mg; urea: 0.0934 mg/L
4.1.7 26.5 lb
4.1.10 A = 0.600, B = 0.350, C = 0.05; there are an infinite number of choices.
4.1.13 8.33×10^4 lb/hr
4.1.16 51.8 g $Na_2B_4O_7$/100 g H_2O
4.1.19 (a) 23.34 lb; (b) 66.5%
4.1.22 35 lb
4.1.25 Fraction recovery: 0.64; amount: 655 g/min

Chapter 5

5.1.1 (a) 7.33 g $BaCl_2$; (e) 3.04 g Na_2SO_4; (h) 8.21 lb $BaSO_4$
5.1.3 (a) $a_1 = 3, a_2 = 4, a_3 = 28, a_4 = 28, a_5 = 6, a_6 = 9$
5.1.8 $BaI_2 \cdot 2H_2O$
5.1.10 H_2SO_4: 2.08 lb; Cl_2: 0.445 lb; seawater: 15,400 lb; $C_2H_4Br_2$: 1.176 lb
5.1.12 12.4%
5.1.15 Very little, about 1.5%
5.2.4 Extent of reaction = 0.557; 1.341 g mol
5.2.7 (a) 145%; (b) 70.7%; (c) 0.205
5.2.9 0.619
5.2.13 (a) 33% excess C; (b) 86.0% Fe_2O_3; (c) 938 lb CO
5.2.16 (a) CO is limiting; (b) H_2O is excess; (c) 0.514; (f) 0.60 mol CO_2
5.3.1 Mole fraction: 0.25; extent of reaction: 0.5
5.3.3 O and H balances not exact but close
5.3.6 20%
5.3.10 NH_4OH = 232 g; $Cu(NH_3)_4Cl_2$ = 336.9 g

5.5.1 CO_2: 13%; H_2O: 14.3%; N_2: 67.6%; O_2: 5.1%
5.5.3 Excess air: 17%; composition: CH_4: 89.8%, N_2: 10.2%
5.5.5 22%
5.5.7 (a) 137.8 mol; (b) in percent, values are CO_2: 14.3; O_2: 3.7; N_2: 72.4; H_2O: 9.6.
5.5.10 Yes. 0.236 mol air leaked/mol exhaust gas
5.5.13 Percent conversion: 92.2%; percent excess air: 10.8%

Chapter 6

6.1.1 2
6.1.5 Variables: In total 8 stream flows plus 19 mass fractions; unknowns: 6 stream flows plus 10 mass fractions; independent material balances: 8. Note the independent specifications of sum of mass fractions is equal to 8.
6.2.2 (a) W = 500 kg/hr, D = 500 kg/hr; (b) A = 281 kg/hr, B = 717 kg/hr, C = 219 kg/hr
6.2.7 Stream S^1: Bz = 0.97, Xy = 0.03; Stream F^1: Bz = 0.082, Xy = 0.018; Stream S^2: Bz = 0.082, Xy = 0.18
6.2.9 241 kg
6.2.12 (a) 91.2%; (b) 0.44 lb Cl_2/lb product; (c) 1.22 lb H_2O
6.3.1 (a) 1; (c) 0
6.3.4 The calculated values are not correct.
6.3.7 7670 kg/hr
6.3.10 Recycle/H_2 = 1.25
6.3.11 (a) 111 kg (0.703 kg mol); (b) 760 kg
6.3.19 (a) 50 lb/hr; (b) 764 lb/hr
6.3.22 Recycle = 111 kg mol
6.4.2 By inspection you can see that the flow of Cl_2 and H_2 into the separator is going the wrong way.

Chapter 7

7.1.1 0.944 lb H_2O
7.1.4 Mass specific volume = 13.56 ft^3/lb; molal specific volume = 392.8 ft^3/lb mol
7.1.7 221 psig
7.1.8 41.4 m
7.1.12 (c) 10.73; (d) 8.314
7.1.15 1.48
7.1.22 51.9 psig
7.1.25 2.99 ft diameter
7.1.27 Volume: 0.047 ft^3 S.C.; time: 1.1×10^{-7} hr
7.1.31 10.6%

7.1.34 (a) 5.28%; (b) 30.3 lb/lb mol; (c) 1.045; (d) 0.190; (e) 7.54

7.1.35 Partial pressures: in mm Hg, CO_2 148; O_2 444; N_2 148; at 40°C: yes; CO_2 158; O_2 474; N_2 158

7.1.37 366 kPa

7.1.40 8.46×10^6 m^3/min

7.1.43 Flow: 1270 m^3/day; removed: 575 mol C_4H_{10}/day

7.1.51 0.284

7.1.54 (a) 3980 ft^3/min; (b) 20,500 ft^3/min

7.2.2 (e) (assuming you have a handbook nearby)

7.2.9 2460 psia

7.2.12 266 kg

7.2.15 (a) VdW: 17.8 lb; (b) compressibility factor: 23.4 lb

7.2.17 (a) VdW: 10.6 g mol; (b) RK: 10.6 g mol

7.3.1 (a) 141 atm; (b) 144 atm

7.3.4 1189°R (729°F)

7.3.7 0.608 lb

7.3.10 0.0127 m^3/kg

7.3.12 102 cm^3 (C_2H_5Cl is a *liquid* at the stated conditions)

7.3.15 (a) 4.157×10^{-4} lb mol/hr; (b) 0.98 hr; (c) 1.75×10^{-5} lb mol/ft^3; (d) the CO alarms would sound a warning of a leak.

7.4.1 2.88 ft^3 (at 180°F and 2415 psia)

7.4.5 Select 1A because it is the cheapest. Use 30 cylinders.

Chapter 8

8.2.1 (a) (1); (b) (2); (c) (1)

8.2.3 2

8.2.7 2

8.2.10 (e)

8.2.12 (a) $P = 4$ (max); (b) $C = 3$

8.3.1 All true

8.3.5 (a) $p^* = 70.51$ mm Hg versus 70.55 from the CD.

8.3.8 No. A typo occurred (equate the equations to get $T = 98.8$ °C).

8.3.11 (a) 0.835 m^3/kg; (d) 1.296 ft^3/lb

8.3.15 145 ft/s

8.3.18 Mass liquid = 0.198 lb; mass vapor = 1.826; vol. liquid = 0.00328 ft^3; vol. vapor = 9.997 ft^3; quality = 0.908 (from steam tables)

8.3.19 $a = -1199.26$; $b = 532.27$; $c = -79.03$; $d = 3.9638$

8.3.23 Approx. 3.3 atm

8.4.2 (a) 104.8 kPa; (b) 0.0349

8.4.4 13.95 kg

8.4.7 9.62 lb H_2O
8.4.11 At 1.4%: −11.5°C (−15°C from CD); for 8%: 15.4°C
8.4.16 1.93×10^{-2} m^3/min (at 100 kPa and 20°C)
8.4.17 5.4×10^{-4} psia (the Hg will not condense)
8.4.20 22 kg
8.4.22 104°F (40°C)
8.4.26 3940 ft^3 (at 800 mm Hg and 200°F)
8.5.3 Yes (use plot or curve fitting program)
8.5.6 (a) Mole fract., benzene: 0.0316, toluene: 0.0126; (b) yes
8.5.13 (a) 80.95 mm Hg (10.8 kPa); (b) pentane: 0.90, heptane: 0.10
8.5.15 (a) 0.56; (b) 0.88
8.5.18 $T = 413$ K
8.5.21 $T = 242$ K

Chapter 9

9.1.1 (a) 2.5×10^4 cal/kg; (b) 1.048×10^5 J/kg; (c) 0.0291 kWh/kg; (d) 35,000 (ft)(lb_f)/(lb_m)
9.1.8 (a), (b), and (c) are intensive; (d) is extensive.
9.1.9 8.86×10^{-4} $(G^{0.6})(J)/(D^{0.4})(min)(cm^2)(°C)$
9.1.13 Yes
9.1.15 One-quarter hour of exercise uses 630 kJ, an insufficient time
9.1.17 Yes. In a short interval, energy/time will be large.
9.2.2 −45.5 kJ
9.2.4 2940 J
9.2.7 -3.16×10^5 (lb_f)(ft)
9.2.10 Heat is not conserved; it is transferred. The transfer from one system can equal the transfer to another, but this is not conservation of heat.
9.2.13 (a) F, (b) F, (c) T, (d) F, (e) T, (f) T but debatable, (g) F, (h) F
9.2.16 1.55 (ft)(lb_f)
9.2.21 ΔH and ΔU are zero because they are state variables.
9.2.25 (a) F, (b) F, (c) F, (d) T
9.2.26 (a) 1, (b) 3, (c) 5, (d) 2, (e) 4. Boiling is at 4 by extending the scale to the left. Freezing is at 1 to the left by extending the scale.
9.2.27 $\Delta H_v = 58{,}070$ J/g mol
9.2.42 624 J
9.3.1 All of them
9.3.4 (a) System: can plus liquid so $Q = 0$, W not equal to 0; (b) system: motor so $Q = 0$ and W not equal to 0 but may get hot; (c) system: pipe plus water so $Q = 0$ and W not equal to 0.
9.3.8 Final $T = 45°C > 30°C$

9.3.10 (a) $W = 0$; (b) and (c) $\Delta U = 0.105$ Btu $= Q$; (d) 10.7 ft^3

9.3.13 $W = 0.57$ kW

9.3.16 0.61 kW

9.3.21 (a) (1) ignore PE, KE $\Delta PE = \Delta PE = 0$; (2) No reaction; (3) $W = 0$; (4) $\Delta E = 0$ (steady state), hence $\Delta H = Q$
(b) (1) ignore PE, $\Delta PE = 0$; (2) No reaction; (3) $W = 0$; (4) $\Delta E = 0$ (steady state), hence $\Delta H + \Delta KE = Q$

9.3.23 −27.86 kJ/kg CO_2 (removed)

9.3.25 $Q = 807.3$ Btu/lb, $W = 0$, $\Delta H = 889$ Btu/lb, $\Delta U = 807$ Btu/lb

9.3.28 (a) $Q = 970$ Btu, $W = -72.8$ Btu, $\Delta U = 897$ Btu, $\Delta H = 970.0$ Btu;
(b) $Q = 970$ Btu, $W = 0$, $\Delta U = -180$ Btu, $\Delta H = 1150.3$ Btu

9.3.33 $Q = -4.1 \times 10^4$ Btu, hence not adiabatic

9.3.36 $T = 21.9$°C

9.3.42 All values are Btu/lb. Step 1: $\Delta U = 2.5$, $\Delta H = 3.5$, $Q = ?$, $W = ?$; step 2: $\Delta U = -73.8$, $\Delta H = -93.4$, $Q = -73.8$, $W = 0$; step 3: $\Delta U = 71.3$, $\Delta H = 93.4$, $Q = 0$, $W = 71.3$

Chapter 10

10.1.1 (b), (d), (f)

10.1.5 20.1 kJ/g mol

10.2.3 −10,030 J/g mol C_3H_8

10.2.4 −4728 kJ/kg FeS_2

10.2.5 5×10^7 J/g mol G converted

10.2.11 263 kJ/g mol

10.2.19 (a) 113% larger; (b) 59% larger

10.2.21 15,980 Btu/lb mol

10.2.22 (a) 1.44×10^5 J/g mol C_6H_{12}; (b) 1.68×10^5 J/g mol C_6H_{12}

10.3.1 −16,740 kJ (leaving)

10.4.3 −286.26 kJ

10.4.5 HHV $= 2.11 \times 10^5$ kJ/m^3

Chapter 11

11.1.1 (a) 15.7%; (b) 24.8°C

11.1.4 (a) 23%; (b) 87.8°F

11.1.6 (a) Yes; (b) 53 ft^3; (c) 0.17

11.1.7 941 kg

11.2.1 (a) $\cong$ 10°C; (b) 38%; (c) 0.79 kg H_2O/kg air

11.2.4 At 100% relative humidity (saturated air)

11.2.6 (a) 27°C, (b) 57%

11.2.8 (a) 0.116 g alcohol/g CO_2; (b) 55.9%

See the CD for pages 861–928, Appendixes F through N

INDEX

In this index the page numbers in *italic* refer to pages in Chapters 12 through 17, and Appendixes F through N, which are located on the CD that accompanies this book. Pages in the problem Workbook are indexed separately, and that index will be found in the Workbook itself as well as proceeding this index..

D

N

S

PROBLEMS WORKBOOK INDEX

CD-ROM Warranty

Prentice Hall warrants the enclosed CD-ROM to be free of defects in materials and faulty workmanship under normal use for a period of ninety days after purchase (when purchased new). If a defect is discovered in the CD-ROM during this warranty period, a replacement CD-ROM can be obtained at no charge by sending the defective CD-ROM, postage prepaid, with proof of purchase to:

Disc Exchange
Prentice Hall
Pearson Technology Group
75 Arlington Street, Suite 300
Boston, MA 02116
Email: disc.exchange@pearson.com

Prentice Hall makes no warranty or representation, either expressed or implied, with respect to this software, its quality, performance, merchantability, or fitness for a particular purpose. In no event will Prentice Hall, its distributors, or dealers be liable for direct, indirect, special, incidental, or consequential damages arising out of the use or inability to use the software. The exclusion of implied warranties is not permitted in some states. Therefore, the above exclusion may not apply to you. This warranty provides you with specific legal rights. There may be other rights that you may have that vary from state to state. The contents of this CD-ROM are intended for personal use only.

More information and updates are available at:
informit.com/ph